CUTTHROAT: NATIVE TROUT OF THE WEST

Cutthroat: Native Trout of the West

SECOND EDITION, REVISED AND UPDATED

Patrick Trotter

ILLUSTRATED BY JOSEPH TOMELLERI

UNIVERSITY OF CALIFORNIA PRESS

BERKELEY · LOS ANGELES · LONDON

Published by University of California Press
2120 Berkeley Way
Berkeley, CA 94704-1012

Produced by
Scott & Nix, Inc.
150 West 28th Street, Suite 1103
New York, NY 10001-6103

Library of Congress Cataloging-in-Publication Data

Trotter, Pat.
    Cutthroat: native trout of the West / Patrick Trotter.—2nd ed., rev. and updated.
        p.   cm.
    Includes bibliographical references and index.
    ISBN 978-0-520-25458-9 (cloth : alk. paper)   1. Cutthroat trout—West (U.S.)   2. Trout fishing—West (U.S.)   I. Title.
    QL638.S2T76 2008
    597.5'7—dc22   2008003982

The paper of this book meets the guidelines for the permanence and durability
of the Committee on Production Guidelines for Book Longevity of the Council on Library Resources.

10   9   8   7   6   5   4   3   2   1

15   14   13   12   11   10   09   08

Printed in China

Contents

Preface   *vii*

CHAPTER 1
Cutthroat: Native Trout of the West   1

CHAPTER 2
Evolution and Prehistoric Distribution   27

CHAPTER 3
Coastal Cutthroat Trout   59

CHAPTER 4
Westslope Cutthroat Trout   103

CHAPTER 5
Lahontan Cutthroat Trout   145

CHAPTER 6
Humboldt Cutthroat Trout   191

CHAPTER 7
Willow/Whitehorse Cutthroat Trout   215

CHAPTER 8
Paiute Cutthroat Trout   231

CHAPTER 9
Yellowstone Cutthroat Trout   245

CHAPTER 10
Finespotted Snake River Cutthroat Trout   295

CHAPTER 11
Bonneville Cutthroat Trout   321

CHAPTER 12
Colorado River Cutthroat Trout   359

CHAPTER 13
Greenback Cutthroat Trout   389

CHAPTER 14
Rio Grande Cutthroat Trout   421

CHAPTER 15
Extinct Subspecies   449

Epilogue   465

Bibliography   471

Conversion Table   535

Common and Scientific Names
of North American Fishes
(Other Than Cutthroat Trout)
Mentioned in the Text   537

Index   539

Colophon   548

Preface

In 1987, after a four-year search for a publisher, the good people at Colorado Associated University Press (now University Press of Colorado) brought out the first edition of *Cutthroat: Native Trout of the West*. It was a handsome book in its gray dust jacket with a Robert Friedli painting of a westslope cutthroat trout on the cover and additional paintings by Friedli along with maps and illustrations by artist Dan Berglund on the inside. I had written the book with anglers in mind, even though it was not a "how-to, where-to" manual on fishing for cutthroat trout. Rather, it was the story of the creatures themselves as I had come to know them: the various forms of the species that occur; how they evolved; where and how they are distributed; their life histories, habits, and habitats; what has happened to them in historical times; and what problems and prospects they face in the future. Being an angler myself, I believe that the more we know about our quarry the more appreciative and protective we become, not only of the trout themselves, but also of the environments in which they live. So yes, I hoped that other anglers would

share that belief and find the book of particular interest. But I also hoped that environmentalists, ecologists, other nature lovers, students—indeed, anyone who holds a love and concern for nature and the outdoors—would find something of value in its pages as well.

If staying power of a book is any measure, those hopes of mine were fulfilled. Although royalties did taper off over time, they continued to come in for several years longer than a book of this sort normally endures. Also, those who bought it evidently thought highly enough of it to keep it on their shelves. I never saw a used one in a book store, and it wasn't until 2004 that copies started turning up on the Internet—at prices four times or more what they originally sold for.

Early-on in the book's history another readership developed that, while unexpected, was every bit as gratifying because I count myself among their numbers as well. While it wasn't aimed directly at professional biologists or scientists, *Cutthroat: Native Trout of the West* also attracted their attention. It received excellent reviews from their professional societies and publications, and was even carried on the American Fisheries Society book list for a time. I also learned that it was widely cited as a reference (and still is for that matter) in dissertations, professional reports, and papers in scientific journals.

But its use, as a reference work, especially after twenty years' time, began to trouble me. So much has transpired in the last twenty years—new techniques applied to old questions, new information and insights gained into life histories, habits, and habitat relationships, new attitudes and respect for the trout, new policies affecting how they are managed, not to mention a major change in scientific classification that moved the cutthroat trout along with all other western North American trouts from the genus *Salmo* to the genus *Oncorhynchus*—all these things added up to a book that was sadly out of date.

This is a new, thoroughly revised and updated edition. It is still written primarily for anglers, naturalists, and others who like cutthroat trout and the places they are found, and who want to learn more about this remarkable, beautiful, and fragile western native. But much new material from the biological and ecological sciences is included here, material that is usually published in journals and books intended for specialists. I've done what I can to make this material understandable to lay persons, including taking liberties with certain scientific conventions. For example, scientists and their

publications universally employ the metric system for weights, measures, temperatures, and the like. I chose to stick with English units, which are still the most familiar to readers in the U.S., although occasionally I did include both in the text when I thought it would improve understanding. If you do prefer metrics, or if your own writing or studies require them, there's a conversion table on page 535. Also, I've kept the use of the Latin scientific names for fishes to a minimum although again, for reference, there is a table of common and scientific names on page 537.

For those of you who may use this second edition as a reference work, I've done a couple of other things that I hope will increase its value for that purpose. I have: 1) scrupulously annotated the text (footnotes are aligned along the right-hand margin of each page), and 2) greatly expanded the bibliography. Regarding the footnotes, you can read this book all the way through and get the full story of the cutthroat trout without once referring to a footnote. But I hope you will want to. Personally, I get annoyed with an author who piques my curiosity about a topic to the point that I want to dig deeper, but doesn't provide me with the way points for doing so. That's what my footnotes are for: to provide additional or supplemental information, fuller explanations of particular topics, and, above all, a direct route to all my sources so you can read them for yourselves if you choose.

In the footnotes, I refer to books, reports, magazine articles, and scientific papers by author and year of publication, e.g., Behnke (2002). If the same author(s) appear on more than one publication in the same year, they are designated thus: Hilderbrand and Kershner (2000a, 2000b). Complete citations, designated this same way and arranged alphabetically, will be found in the bibliography.

About the bibliography, this is greatly expanded from the first edition. This is a direct reflection of the tremendous amount of new knowledge that has been gained since 1983 when the original manuscript was completed. I haven't made an actual count, but my guess is that close to 75 percent of the references cited in this new bibliography were published after 1983. You'll need access to a good library to find them all for yourselves. I'm fortunate in living within easy bus rides of a major university library and an excellent city public library, but any facility with decent interlibrary loan service should be able to do the job for you.

A few words now about the way the book is organized. The first Chapter introduces you to the cutthroat trout, how to recognize one when you encounter it, and where its various forms are distributed in western North America. It also provides some information that sets the stage for things you will read about in later Chapters. The second Chapter describes the evolution of the cutthroat trout and its subspecies, and then discusses how and to some extent when those subspecies came to occupy their respective historical ranges. Then follow Chapters on each extant cutthroat subspecies, twelve in all, and a final Chapter that deals with two subspecies that did exist historically but are now extinct. A short epilogue wraps up the package.

Regarding the subspecies Chapters, each of these (save for the final Chapter) begins with a header giving the formal description complete with an accurate illustration of the subspecies by the noted artist, Joseph Tomelleri. Following this are an introductory section that sometimes contains personal anecdotes or observations about the subspecies, a section describing its historical range, a section covering its life history and ecology, and one on its status and future prospects. A couple of reviewers, I think from the scientific community, have criticized my inclusion of life history information in each of the subspecies Chapters as repetitive (one even called it annoyingly repetitive), suggesting instead that it all be consolidated into one summary Chapter up front in the book. I appreciate the comments, but I have two very good reasons for doing it the way that I did. First, the different cutthroat trout subspecies (and sometimes even populations within a subspecies) have each adapted in their own ways to the unique landscapes and conditions in which they exist, and I wanted to highlight that to the extent that I could in each subspecies account. Second, reader feedback I received on the first edition indicated that the majority of readers appreciated finding everything they wanted to know about an individual subspecies right there in one chapter, so they did not have to search back through the book. So my way of organizing the subspecies Chapters stands. I beg your forbearance if you come upon material that you think you have read before.

Every book has to have a cutoff point, a date beyond which no new material can be included. That date for this edition appears below where I signed off on this preface. I hope this book, like the first edition, has another 20-year run. But it too will drift out of date. That's the nature of the biological and ecological sciences: to continually provide new insights into how things work in the natural realm. It will be up to you to keep abreast of these new discoveries and interpret for yourselves what they may mean for the cutthroat trout.

At this point I want to pay tribute to all the scientists, management biologists, conservationists, and volunteers who work so tirelessly for the restoration and protection of our native trout species and their habitats. A great many of you helped me directly in one way or another in the preparation of this book. I thank you for that, and I hope you will find this new edition worthy of your contributions.

—Patrick Trotter
Seattle, Washington
March 5, 2007

NOTE ADDED IN PROOF

Back in April, 2007, shortly after the cutoff point for adding new material to this book, Dr. Robert Behnke and I agreed to write a short research paper in which we examined all evidence for setting apart the Humboldt cutthroat trout (listed in Chapter 6 of this book as an unnamed subspecies) as a separate, named subspecies. In our paper, we wrote out a formal description of this subspecies and named it *Oncorhynchus clarkii humboldtensis* in accord with international rules of scientific nomenclature. Owing to the length of the peer review process for most scientific journals and to the usually long queue of accepted papers awaiting publication, I fully expected this book to be out long before our paper could be published. But I was wrong. Our paper sailed through the peer review process, and I just received word that it is scheduled to be published in *Western North American Naturalist* in April, 2008, before the publication of this book. The publication of our *humboldtensis* paper will affect how you interpret what you read in Chapter 6, and parts of Chapters 5 and 7 as well.

—Patrick Trotter
Seattle, Washington
March 28, 2008

CUTTHROAT: NATIVE TROUT OF THE WEST

CHAPTER 1

Cutthroat: Native Trout of the West

"These trout are from 16 to 23 inches in length, precisely resemble our mountain or speckled trout in form and the position of their fins, but the specks on these are of a deep black instead of the red or gold of those common in the U' States. These are furnished with long teeth on the pallet and tongue and have generally a small dash of red on each side behind the front ventral fins; the flesh is of a pale yellowish red, or when in good order, of a rose red."

—MERIWETHER LEWIS,
Journals of the Lewis and Clark Expedition,
JUNE 13, 1805

ENCOUNTERS

A blustery spring morning a very long time ago. A stockily built man making his way along the shore of a vast body of water pauses at the top of a rise. He looks out over a primitive, timeless scene. Forests of pine come down nearly to the water's edge. In places across the water are blue outlines of distant land. The man watches as wind-whipped patches of water change color, slowly like a kaleidoscope: deep blue, green, turquoise, and back to blue again as clouds scudded across the sun.

Several thousand years later, the explorer John C. Fremont will come upon a shrunken remnant of this lake. The country 'round about will then be desert. A crusty deposit will cover everything and weird tufa formations will stud the shoreline. Fremont will remark on one of those formations jutting several hundred feet out of the water in "a pretty exact outline of the great pyramid of Cheops," thus giving the lake its modern name.[1]

But of course the man who stands there now has no comprehension of all this. He belongs to a culture that pre-dates even the American Indians, who will call his people simply "the ancient ones," or "the ones who came before."

The man walks on, then pauses again as he spies a huge fish gliding by in the shallows. Presently there comes another, not quite so large. The man hurries on now, eager to join the others in manning the nets. The fish will be entering the river to spawn. It is time for the people to secure the season's food supply.[2]

Many centuries later and several hundred miles to the northwest, the coast of what will one day be called Oregon lies under a late summer sun. The deep forests are tinder dry. The small coastal river is reduced to a thin flow.

A boy from the village at the mouth of the river, arrested by a splash he had heard near the tail of the pool, stands quietly among the bordering alders. This is tidewater, and as the boy waits, a leaf which had been floating motionless for a time in the middle of the pool begins to move almost imperceptibly back upstream.

The tidal flow gains momentum, carrying the leaf ever more quickly back the way it had come. A silvery fish rolls under the alders just below where the youth stood. It is followed by another. And now the lad sees others, visible in the thin shafts of sunlight that slants into the water. He hesitates only a moment more. Spirits racing, he rushes back to the village to spread the news. The messengers had returned—the harbingers, foretelling the coming of the salmon. Preparations must be made. Ceremonies of thanks must be performed. The prophets had come back to the river.[3]

Late April, 1541. Out of the Rio Grande valley they came, the meager sunlight glinting off burnished hel-

1. Fremont, brevet Captain John C. 1845. Report of the exploring expedition to the Rocky Mountains in the year 1842 and to Oregon and North California in the years 1843–44. U.S. Senate, 28th Congress, 2nd Session, Executive Document 174. Washington, D.C., Gales and Seton Printers [reprinted 2002 (without maps and some sketches) by The Narrative Press, Santa Barbara, California].

2. The evidence upon which this account is based comes from numerous geological and archaeological studies. Description of the climate that existed during and just after the last glacial period can be found in: Hansen (1947), Blackwelder (1948), Weier et al. (1974), Allen (1979), and Wright (1983). Although the people of the last glacial period were known to have been hunters of mammoth, mastodon, giant bison, and other game animals, those living in shelters around the shores of ancient lakes also caught fish in nets. For accounts of the lifestyles of these peoples, see Heizer (1951), Orr (1956), LaRivers (1962), Wheeler (1974), Marshack (1975), Wright (1983), Aikens (1984) and Touhy (1990).

3. There are dozens of books and unpublished manuscripts about the lifestyle and customs of the Native Americans of the Pacific Northwest

mets and metal breastplates. It was an imposing sight. The army of Francisco Vasquez de Coronado was on the move, heading eastward, climbing slowly through the scrub oak and pinion toward a low defile known today as Glorieta Pass.

The year before, Coronado and his conquistadors had come up out of Mexico seeking the riches of the fabled "Seven Cities of Cibola." They had found the seven cities, but no riches. Cibola proved to be nothing more than a series of poor Pueblo Indian villages in the Zuni Valley near the present border of New Mexico and Arizona. The conquistadors had subjugated these villages, then crossed the Great Divide to winter in the pueblos of the Rio Grande valley.

Now with spring upon the land, and lured by tales of treasure still further to the east, the army was marching again. The trail they followed was already ancient. Glorieta Pass had long been used by Indian traders, hunting parties and raiders; and it will continue to be used, becoming famous in the years to come as the Santa Fe Trail, one of the major gateways to the Mexican and later American southwest.

Coronado and his men dropped down the east side of the pass into a broad valley where the waters of the Pecos River flowed southward out of the Sangre de Cristo Mountains to join the Rio Grande. Turning north up the valley, they soon came to the maize, squash and melon fields of the Pueblo city of Cicuye. Nearby were the pueblos themselves, two of them, each one four stories high. They were built so that each level was set back from the one beneath, with ladders leading to the top of each level. The inhabitants gained entry through holes in the roofs.

Pedro de Castaneda de Najera, a member of Coronado's army, later wrote a history of the expedition. In it he described Cicuye as a city about 500 warriors strong, and much feared throughout that land. About the surrounding area Castaneda wrote:

> "Cicuye is located in a small valley between snowy mountain ranges and mountains covered with big pines. There is a little stream which abounds in excellent trout and otters. Big bears and fine falcons multiply in this region."

Of course every schoolchild knows that Coronado never found his treasure. He was returned to Spain in disgrace. But the "excellent trout" referred to so briefly in Castaneda's history is the first written record of trout in the New World. [4]

Winter, 1776. Earlier this year, on the east coast of North America in a town called Philadelphia, 13 rebellious colonies had declared their independence from the British Crown, and symbolized by the ringing of a liberty bell on July 4th, had embarked on a long, bitter struggle to establish a new nation. Twenty-five days later and about 2,500 miles to the southwest, Francis-

coast. Among the references I consulted for this made-up story are Parish (1931), Clark (1953), Newman (1959), Sauter and Johnson (1974), Beckham (1977), Minor (1983), and Aikens (1984).

4. A history of the Coronado expedition was written by one of its members, Pedro de Castaneda de Najera, and published in Seville, Spain on October 26, 1596. The text of this history is contained in Hammond and Rey (1940) and is also cited in Winship (1964).

Cutthroat country: Kelly Creek, Idaho.
PHOTO BY AUTHOR

can friar Francisco Atanasio Dominguez and his junior, Fray Silvestre Velez de Escalante, embarked on a long and arduous journey of their own. Leading a party of ten men including themselves, with Fray Escalante serving as scribe, they set out from Santa Fe to explore the country and open an overland route to the newly established missions in California.

The first leg of the journey, northward to the Colorado River, was over known trails. Spanish explorers, including some of their own men, had been this far before. But from there they were on their own. Fortunately, several Indians who knew the country joined the expedition along the way and helped guide the small party across these unknown tracts. They continued north until they struck the White River, then bent their course to the west past Musket Shot Springs (a campsite that got its name because the two springs that supplied their water were "a musket shot apart") to another waterway they called the San Buenaventura (present-day Green River). From the Green River they continued west, skirting the southern foothills of the Uinta Mountains to yet another stream. Fray Escalante wrote:

"Then we went down to a medium-sized river in which good trout breed in abundance, two of which Joaquin the Laguna killed with arrows and caught, and each one must have weighed more than two pounds.... The guide told us that in it for some time there had dwelt a portion of Lagunas who depended on the said river's fishing for their more regular sustenance..."

But there was more to come. Bearing west from there and then a bit south, the party came to another river the Indians called "Timpanogus," which led them through the Wasatch Range to a broad valley with a large bluewater lake. The journey this far had taken months. It was winter now, and the party was running low on resolve. Faced with the cold and the harsh, seemingly interminable desert country to the west, they decided to abandon their quest for the California missions and return to Santa Fe by striking south. But before doing so, they tarried awhile at the lake of the Timpanogus, which Fray Escalante described as follows:

"The Lake of the Timpanogitzes has great quantities of various kinds of food fish, geese, beaver and other amphibious animals which we had no opportunity to see. Round about it are a great number of these Indians who live on the abundant supply of fish in the lake. For this reason the Yutas Sabnaganas call them 'fish eaters'."

The "medium-sized river" in the first passage bears the modern name Trout Creek, but the site where the Franciscans struck it now lies under the northeast bay of Utah's Strawberry Reservoir. As for the "fish-eater Indians," Mormon settlers would later follow their example and seine trout out of the lake of the Timpanogus by the ton. Still later, visitors would describe taking trout up to "a yard in length" from that lake and its companion river, which we now call Utah Lake and the Provo River, respectively. [5]

5. For a recent, authoritative translation of the Dominguez-Escalante journal, see Chavez and Warner (1995). An earlier translation can be found in Bolton (1951), and a general account of the expedition was given in Lavender (1968). Escalante's description of Utah Lake was also recited by Tanner (1936) and Robinson (1950) in their studies of the native trout of the Bonneville basin.

Cutthroat country: Buffalo Ford on the Yellowstone River is popular with anglers as well as buffalo. PHOTO BY AUTHOR

June, 1805. The small band of explorers, under commission from President Thomas Jefferson to explore the Louisiana Purchase to the headwaters of the Missouri and beyond, has come to a major fork in the river. They had been told of a Great Falls on the river they should be following, but not about any major fork. Was the cataract they sought on the north fork or the south fork? The party was at an important crossroad. Either fork could be the Missouri. Which one should they follow?

After reconnoitering, the two men who shared command, captains Meriwether Lewis and William Clark, decide to push on up the south fork. But before the party can break camp, another crisis occurs. The young Shoshone Indian girl, Sacagawea, who has already helped the party in so many ways, has become gravely ill. The captains talk it over. They agree that while Clark holds the main party in camp and tries to nurse Sacagawea back to health, Lewis will go on ahead with a few men and try to locate the Great Falls.

Lewis isn't feeling so well himself on the morning of departure. Nevertheless, he sets out in the company of four men: George Drouillard, Joseph Field, George Gibson, and Silas Goodrich. The march that day is pure agony for Lewis, and that night he decides to dose himself with a thick, bitter decoction of boiled choke-cherry. About a pint of the vile stuff and his insides are turning over. But he senses the pain is easing, so taking the view that if a little is good a lot must be better, he downs a couple more pints. His journal does not record how much time he spent on the latrine that night but the next morning he awakes a new man. It is the morning of June 13th. Today's march will be a pleasure!

The little group hears the great cataract long before they see it. A thrill goes through them all. They push on, eager for the first view. At last, about noontime, they come upon the site—not just one falls, as Lewis will shortly ascertain, but five in all in the next 12 miles upstream. It will take the assembled party 11 days to complete the portage.

But meanwhile, that first afternoon when the work of setting up camp was finished, young Silas Goodrich, who was very fond of fishing and slipped away whenever he could, did so again to try his luck in the pool at the base of the first cataract. Goodrich had already come up with two new species of "white fish" that Lewis and Clark had recorded in their journals, but when he returned that evening he had something new: a half-dozen very fine trout.

The trout were of a kind never seen before, at least by white Americans. Lewis would describe them in his journal, and about the feast enjoyed by the little band that night he would write:

Cutthroat country: Mary's River in the Jarbidge Wilderness, Nevada.

"The fare was really sumptuous this evening; buffaloe's hump, tongues and marrowbones. Fine trout parched meal pepper and salt, and a good appetite the last is not considered the least of the luxuries."

As you can see, Lewis didn't worry much about punctuation. [6]

Late April, 1834. A party of men bearing trade goods for the fur trappers in the mountains sets out from Independence, Missouri, under the leadership of Nathaniel Wyeth. Accompanying them are two naturalists, John Kirk Townsend and Thomas Nuttall. Many of the bird specimens these two will collect on this trip will be painted by John James Audobon for his classic, *Birds of America.*

On June 30th, the party reaches the Green River in present-day Wyoming, where Townsend would write in his journal:

"The river, here, contains a number of large trout, some grayling, and a small narrow-mouthed white fish, resembling a herring. They are all frequently taken with the hook, and, the trout particularly, afford excellent sport to the lovers of angling. Old Isaac Walton would be in his glory here, and the precautionary measures which he so strongly recommends in approaching a trout stream, he would not need to practice, as the fish is not shy, and bites quickly and eagerly at a grasshopper or minnow."

The fishing got better and better. On July 4th, on a branch of the Bear River, in the midst of an Independence Day celebration by the party members, Townsend wrote:

"In this little stream, the trout are more abundant than we have yet seen them. One of our sober men took, this afternoon, upward of thirty pounds. These fish would probably average fifteen or sixteen inches in length, and weigh three-quarters of a pound; occasionally, however, a much larger one is seen."

On July 10th, they met up with the Bonneville party, resting after the fatigue of a long march. Captain Benjamin Bonneville, exploring to the Rockies and beyond, had led his men through this country two years before and was now returning to civilization. They too had encountered the trout. Washington Irving, in writing of Bonneville's adventures, described the bright little lakes and fountainhead streams of the Wind River Range:

"So transparent were these waters that the trout with which they abounded could be seen gliding about as if in the air."

But even the early explorers could have their problems with the fishing. On July 12, 1834, the Wyeth party reached the Snake River watershed where Townsend observed:

"The stream contains an abundance of excellent trout. Some of these are enormous, and very fine eating. They bite eagerly at a grasshopper or minnow, but the largest fish are shy, and the sportsman requires to be carefully concealed in order to take them."

Even in those early times, when the sight of an angler was rare indeed, the larger trout learned quickly. [7]

Summer, 1882. "What boy of fifteen ever had his dream of adventure and sport fulfilled better than I

6. My source for all things pertaining to the Lewis and Clark expedition is the multi-volume set edited by Gary E. Moulton and published by the University of Nebraska Press (see Moulton 1986–1997). This set contains the full text of the journals and all other associated writings of members of the expedition. The passage containing Lewis's description of the trout caught by Silas Goodrich at the Great Falls of the Missouri is in Volume 4 at pages 286 and 287.

7. Townsend's account of these experiences can be found in *Narrative of a Journey across the Rocky Mountains to the Columbia River* (Townsend 1839 [1905, 1978]). Bonneville's expedition was chronicled by Washington Irving in the book, *Adventures of Captain Bonneville* (Irving 1837 [1868, 1954, 2001]).

when my father asked me to accompany him on a trip out West...to visit Yellowstone Park?" asked Edward R. Hewitt, one of the "grand old men" of American fly-fishing. The Yellowstone region had been set aside as a National Park ten years earlier, but was still remote. The Northern Pacific railroad extended no further west than Billings, Montana, some 180 miles short of the park. Even so, by 1882 the park was receiving about a thousand visitors a year.

Young Hewitt and his father were in a party of ten that included U.S. Senator Thomas Francis Bayard and other notables of the period. For the overland journey from Billings, the army provided an escort of 30 cavalrymen, two supply wagons, and a buckboard for the senator. Camp was made each evening along the Yellowstone River. The river literally teemed with trout running from two to four-and-one-half pounds. Young Hewitt, ever eager to try his chances at the fishing, drew the pleasant chore of feeding the camp.

The party entered the park itself at Mammoth Hot Springs. The streams and lakes the party came to were full of trout, and young Hewitt had no trouble at all keeping the camp supplied.

Upon leaving the Park on the western side, the party camped on a small creek called Boulder Creek to rest their horses. There they met a large detachment of cavalry. The colonel in charge asked young Hewitt if he would get his men some trout. Accompanied by two soldiers and three pack mules with sacks to carry the fish, he headed off upstream.

"The stream was just alive with trout, which seemed to run from three to four pounds apiece," he wrote later. The soldiers cleaned the fish and put them in the sacks as fast as Hewitt could catch them. About three o'clock in the afternoon they said the sacks were full. "I was not sorry to quit as I was really tired out," Hewitt related. "For once I had caught all the fish I could take out in one day. There must have been between four hundred and fifty and five hundred pounds of cleaned trout, but the soldiers polished them off in two meals." [8]

Mid-August, 1946. A hip-booted boy, 11 years old, having ventured off from his family's picnic site on a shady bench of land overlooking the river, stands at the edge of a pool a short distance downstream. Above and directly behind him, the Spirit Lake Highway makes a sharp, treacherous turn and crosses the river on a high, narrow, rickety old bridge. Log truck drivers have to gear down quickly and pay close attention here. It's all too easy to plunge over to a fiery crash on the river boulders some 40 feet below.

All this looms in the back of the boy's mind and he winces with apprehension each time a vehicle clatters across overhead. But his uncle has told him this pool is a

8. "There can be little doubt that on this trip I did the first fly-fishing ever done in Yellowstone Park, and probably in the river," Hewitt wrote years later in his classic book, *A Trout and Salmon Fisherman for Seventy-five Years* (Hewitt 1948 [1966]). Hewitt was wrong about this, and about several other assertions in his narrative. But he may very well have been among the more successful fly-fishermen. Historian Paul Schullery, in his recent book, *Cowboy Trout: Fly Fishing As If It Matters*, (Montana Historical Society Press, 2006), assigns credit for being the first to use flies for the trout in both the Yellowstone River and the park to Warren Gillette, a member of the Washburn-Langford-Doane expedition that made the first official exploring trip to the park area in 1870, but it's not clear if Gillette actually caught any fish on his flies. General William Strong, a member of a small military excursion, had also sampled the fly-fishing six years prior to Hewitt's visit, but also to no avail. General Strong had to resort to grasshoppers for bait and buckshot for sinkers in order to take any trout. For an account of Strong's excursion, see Strong (1968).

good place to find harvest trout at this time of year, and it's a good place to practice his fly casting, so he grimly sticks it out.

He's making tough work of it, though. He has only been at this fly-fishing business a short time now, and the heavy tubular steel fly rod is a cumbersome thing. About the only good parts of his outfit at all are the line, a double-tapered variety purchased at his uncle's suggestion to balance the old rod, and the size 6 Royal Coachman bucktail fly affixed to the end of his leader. The boy tied that fly himself, and it's a good job, too. He does have a talent for fly-tying, at least.

As long as he has the pool to himself, the boy doesn't mind the frustration of his awkward efforts. He concentrates on the motions of casting, as his uncle showed him; and when the fly is in the water, he tries to make it twitch and dart as it swings down and across below him. But the boy has a shy streak and when another angler, a grown man, wades into the pool across from him, he feels self-conscious and clumsy—especially when the man executes a series of effortless false casts, then lays out a perfect line.

The boy's back cast catches in a bush. The next one snags in the back of his shirt. The man across from him is grinning broadly. The boy's ears begin to burn and his cheeks flush. Flustered and confused, he wants only to reel in and beat a hasty retreat. But his fly is out in

the water now and beginning to swing across. Red-faced, he twitches it along.

There's a sudden, vicious boil and the line comes taut as a heavy fish turns away with the fly! Taken completely by surprise, the boy stands momentarily frozen as the rod tip arcs too far, too quickly toward the water and the leader strains too close to the breaking point. But somehow, miraculously, the tackle stays together and the fish stays hooked. It fights a deep, rugged battle. Recovered now but still excited, the boy half plays, half horses it onto the gravel.

Completely forgotten are the clumsy attempts at casting. Forgotten, too, is the man across the river. The boy is down on his knees beside the fish. There it lies, his fly in its mouth, silvery sides thickly spotted, glistening, a faint wash of yellow on its lower sides, indistinct reddish slashes under its jaws. There it lies, 18 inches of beautiful, beautiful trout....

Each of these incidents describes an encounter with the cutthroat trout, truly the original "native" trout of western North America. The first two stories are made up of course, but the evidence suggests that they could have happened that way. The other encounters are real and are based on historical fact—even the last one. I know, because the 11-year-old boy in that story was me.

In its own way, the story of the cutthroat trout is the story of man's movement into the western part of

Cutthroat country: Smith's Fork of the Bear River, Wyoming. PHOTO BY AUTHOR

North America. The cutthroat's ancestors were contemporaries of the first Stone Age people to venture over the land bridge from Asia. The fish provided food for the Paleo-Indians. It figured into ceremonies of some of the coastal tribes. When the conquistadors came, the trout were there. They sustained the fur trappers and emigrants. They were there to be exploited by the miners and hangers-on who populated the western boom-towns. And they were there to provide pleasure for countless anglers who made their way into the western back country in pursuit of their sport. It was easy for some of these early sportsmen to get carried away by the cutthroat as this little verse by Canadian angler/author Francis C. Whitehouse will testify:

"A cutthroat! Nay, a gallant fellow he,
As proud a spotted beauty
As ever you'd wish to see."[9]

WHAT IS A CUTTHROAT TROUT?

To an angler, a simple answer to this question might be an easy-to-catch but increasingly rare native trout, characterized by two streaks of red pigment under its jaws (Fig. 1-1), that resides in streams and lakes of the Rocky Mountains, the Great Basin west, and the Pacific Northwest coast. To a naturalist or scientist, the simple answer would be a fish belonging to the order Salmoniformes, family Salmonidae, subfamily Salmoninae,

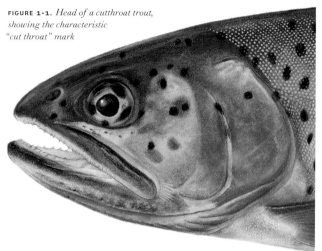

FIGURE **1-1**. *Head of a cutthroat trout, showing the characteristic "cut throat" mark*

genus *Oncorhynchus*, species *clarkii*. And, depending on where you find it, it would also belong to one of 12 extant subspecies listed later in Table 1-1.

The scientists' system of classifying and naming members of the animal kingdom was invented by Swedish naturalist Carolus Linnaeus, who published it in *Systema Naturae*, his great catalog of the natural world, back in 1758. Linnaeus was attempting to arrange all living things into a hierarchical scheme (a *phylogeny*) that would reflect degrees of relatedness and an evolutionary progression. That he succeeded is itself reflected in the fact that his basic scheme is still in use today. [10] If we trace the cutthroat trout's lineage

9. Whitehouse, Francis C. 1946. *Sport Fishes of Western Canada, and Some Others.* Vancouver, British Columbia. Published by the author.

10. To specify a particular group in the Linnaean system, such as the cutthroat trout, we customarily truncate to just the binomial genus and species names, *Oncorhynchus clarkii*, or, if a subspecies name is tacked on, the trinomial *Oncorhynchus clarkii clarkii* or *O. c. lewisi* for example. It's OK to abbreviate the genus and species names when using trinomials once it is clear, as in this case, that we are referring to *Oncorhynchus clarkii*. Note that in this system all the names down through genus are capitalized, but species and subspecies names are not. Also, genus, species, and subspecies names are always written in italics, or underlined if italics aren't available.

back through the Linnaean system, we find that at the family level it shares a relationship to the smelts of the family Osmeridae. At the subfamily level, the cutthroat's relatives include the whitefishes and ciscoes (subfamily Coregoninae) and the grayling (subfamily Thymallinae) At the genus level, the cutthroat's relatives include: *Coregonus* and *Prosopium* (31 species in the Northern Hemisphere), *Hucho* (the huchen, the taimen, and two sister species of Eurasia), *Brachymystax* (one species in Siberia), *Salmo* (the Atlantic salmon and brown trout), *Salvelinus* (the Arctic char, bull trout, Dolly Varden, brook trout and lake trout), and *Stenodus* (the inconnu, also known as the sheefish). Species-level relatives, i.e., those within its genus, include the seven (or maybe only six) species of Pacific salmon (pink, chum, coho, sockeye, chinook, masu, and, although not everybody considers this one separate from the masu, the amago salmon), the rainbow trout group (which includes the redband and other interior forms of rainbow trout, the coastal rainbow, the rainbow trout of Kamchatka, and California's State fish, the golden trout), the Gila and Apache trouts, the Mexican golden trout, and several additional rainbow-like trouts of Mexico whose classification is not yet certain. [11]

For the purpose of this book, I include the 14 subspecies of cutthroat trout recognized by Dr. Robert J. Behnke, Professor Emeritus at Colorado State University. Dr. Behnke's name will be mentioned often in

TABLE 1-1. *Behnke's Fourteen Subspecies of Cutthroat Trout*

COMMON NAME	SCIENTIFIC NAME	ENDANGERED SPECIES ACT STATUS
Coastal cutthroat	*Oncorhynchus clarkii clarkii*	Petitioned but not listed
Westslope cutthroat	*O. c. lewisi*	Petitioned but not listed
Lahontan cutthroat	*O. c. henshawi*	Threatened
Humboldt cutthroat	Unnamed subspecies*	Threatened*
Willow/Whitehorse cutthroat	Unnamed subspecies*	Threatened*
Paiute cutthroat	*O. c. seleneris*	Threatened
Alvord Basin cutthroat	*O. c. alvordensis*	Extinct
Yellowstone cutthroat	*O. c. bouvieri*	Petitioned but not listed
Finespotted Snake R. cutthroat	*O. c. behnkei*	Included in the Yellowstone petition
Bonneville cutthroat	*O. c. utah*	Petitioned, status pending
Colorado River cutthroat	*O. c. pleuriticus*	Petitioned but not listed
Greenback cutthroat	*O. c. stomias*	Threatened
Rio Grande cutthroat	*O. c. virginalis*	Petitioned but not listed
Yellowfin cutthroat	*O. c. macdonaldi*	Extinct

* The U.S. Fish and Wildlife Service and collaborating agencies include these as Lahontan cutthroat trout rather than separate subspecies. As such, they have threatened status under the U.S. Endangered Species Act.

11. For a more complete phylogeny of North American trouts and salmon, see Behnke (1992, 2002). Be warned, however, that there has been considerable debate over the years about the correctness of the phylogeny of the family Salmonidae. One of the most recent papers on the subject that I am aware of was published in 2004 by Bernard Crespi and Michael Fulton of Simon Fraser University in British Columbia. It's a difficult paper for lay persons to wade through, but it does give interesting insights into ancestral relationships within this family, and plenty of references to others' work. See Crespi and Fulton (2004) in the bibliography for the complete citation.

these pages because I consider him to be the foremost living authority on the systematics of North America's native trouts, and he's probably done more work with cutthroat trout than anybody.[12] I have listed the 14 cutthroat subspecies by common and trinomial scientific name in Table 1-1. And, for easy reference later, I have also included their status relative to the U.S. Endangered Species Act.

The cutthroat trout was first named and described for science by Sir John Richardson of the University of Edinburgh in 1836, from two specimens he had received from Meredith Gairdner, a young physician serving at Fort Vancouver, the far-off Hudson's Bay Company post on the lower Columbia River. Richardson knew about the Lewis and Clark expedition and of the leaders' descriptions of the fishes they had encountered. Thus, in 1836, Richardson named the new trout *Salmo clarkii* "…as a tribute to the memory of Captain Clarke, who noticed it in the narrative of the Expedition to the Pacific, of which he and Captain Lewis had a joint command, as a dark variety of salmon-trout…."[13]

Other formal names and descriptions for members of the cutthroat species were published subsequently. As a matter of fact, between 1792 (when *Salmo mykiss*, the rainbow trout of Kamchatka was described) and 1972 (when species status was given to *Salmo apache*, the most recent of the western North American trouts

12. Robert J. Behnke, Ph.D. is a professor emeritus of fisheries and conservation in the Department of Fishery and Wildlife Biology at Colorado State University, Fort Collins. A native of Stamford, Connecticut, he served in the U.S. Army in Korea, then returned to study fishes on the G.I. Bill at the University of Connecticut. He compiled an outstanding record, and upon completing his degree, received a call from Paul Needham of the University of California, who wanted a high-caliber assistant for a major study of western trout. The idea, basically, was to fly fish their way from California to Alaska and back, collecting trout as they went. Out of this work came his Master's thesis in 1960 on the taxonomy of the cutthroat trout of the Great Basin, and plenty of material for his two later monographs on the native trouts of western North America. For his Ph.D. dissertation in 1966, he expanded his scope and wrote on the salmonid fishes of the world. He has been an exchange scientist at the Soviet Academy of Sciences, a research biologist for the U.S. Fish and Wildlife Service, and an assistant unit leader of the Cooperative Fisheries Research Unit at Colorado State. He became a professor of fisheries and conservation at that institution in 1974. Since 1983, he has written the

"About Trout" series that appears in Trout, the journal of coldwater fishery conservation published by Trout Unlimited.

13. Sir John Richardson published his description of *Salmo clarkii* in his comprehensive treatise, "Fauna Boreali–Americana" (Richardson 1836 [1978]). William Clark's description of the dark salmon-trout, which Richardson alluded to in naming *Salmo clarkii*, is found at page 413, Volume 6 of the Moulton edition of the Lewis and Clark journals. The history of Meredith Gairdner, the young physician who figured in this story, can be found in Harvey (1945). Gairdner was recognized for his contribution by having his name applied to specimens of steelhead, which he had also sent to Richardson; thus *Salmo gairdnerii* (which later, when the rainbow trout name was changed to *Oncorhynchus mykiss*, became a subspecies name for one of the interior forms of rainbow trout and steelhead, *O. m. gairdnerii*). The double-i ending used by Richardson in the species names *clarkii* and *gairdnerii* was quite common in older literature, but had pretty much given way to the single-i in most contemporary literature. However, since it is the original, the double-i is the ending required by international rules of

nomenclature, and professional societies and journal editors have recently become more forceful in requiring compliance. For more information on nomenclatural procedure and taxonomic writing, see International Commission on Zoological Nomenclature (1999), Winston (1999), and Nelson et al. (2004). For more on the history of the discovery of the cutthroat trout, see Trotter and Bisson (1988).

to be so endowed), something like 50 "species" of trout were formally described and named from western North America. Most of these were based simply on variations in color and spotting pattern observed by the early collectors as they moved from stream to stream or from one watershed to the next.

George Suckley, a U.S. Army surgeon/naturalist who collected many western trout specimens himself while participating in the Pacific railroad survey, made the first attempt at a taxonomic arrangement of these "species." Although Suckley's manuscript was completed in 1861, it wasn't published until 1874, following his death. Suckley was the first to propose the genus name *Oncorhynchus* to separate Pacific salmon from rainbow and cutthroat trout. In 1853, he also fly-fished the same water below that first Great Falls cataract where Silas Goodrich had fished 48 years earlier, and named the trout he caught there *Salar lewisi* (now *Oncorhynchus clarkii lewisi*) in honor of Meriwether Lewis's description.

After Suckley, from about 1880 to 1930, the work of the noted ichthyologist and educator David Starr Jordan and his coworkers dominated the field. Trouble was, Jordan changed his mind often, first classifying all cutthroat trout as *Salmo purpuratus* then later including them in *Salmo mykiss* before finally adopting Richardson's name *Salmo clarkii* in 1898. Later still,

toward the end of his period of influence, Jordan took to recognizing every described form as a new species, culminating in 1930 with a checklist that still contained 32 "species" of western trout, many of which were actually subspecies of cutthroat trout. J. Otterbein Snyder published a classification of western trout in 1940 that listed nine subspecies in the cutthroat series. That was followed by Robert Rush Miller's classification in 1950 that bumped the number of cutthroat subspecies up to 12. Most recently, Dr. Behnke's work, appearing in published form in 1979, 1981, 1988, 1992, and 2002, has raised the number of recognized cutthroat subspecies to 14 (12 still with us but some just barely, and two now extinct). This list, shown in Table 1-1, continues to be scrutinized and tweaked as new methods give researchers ever sharper insights into evolutionary relationships, but no fundamental changes have resulted. [14]

Two major changes in trout classification did occur in 1989, just two years after the first edition of this book appeared. One was the transfer of all western North American trouts from the genus *Salmo* to the genus *Oncorhynchus*, thus undoing Suckley's proposal and rendering necessary a name change for cutthroat trout from *Salmo clarkii* to *Oncorhynchus clarkii*. The other was the change of the rainbow trout's name, both genus and species, from *S. gairdneri* to *O. mykiss*. [15] The rainbow

14. For those interested in more complete details of this taxonomic history, citations for the original publications are given in the bibliography. See Suckley (1874), Jordan and Evermann (1898), Jordan et al. (1930), Snyder (1940), Miller (1950), and Behnke (1979 [1981], 1988a, 1992, 2002).

15. Smith and Stearley (1989) summarized these changes and the evidence supporting them. These changes were quickly adopted by the American Fisheries Society Committee on Names of Fishes and most other scientists around the world. Only the Russians objected and continue to hold out. In 1963, V.D. Vladykov proposed *Parasalmo* as a subgenus name to set apart the cutthroat and rainbow trouts (and subsequently, all the other western North American trouts) and emphasize their distinctions from other species of the genus *Salmo* such as Atlantic salmon *S. salar* and brown trout *S. trutta* (Vladykov 1963). The Russians use *Oncorhynchus* for the Pacific salmon, but continue to use *Parasalmo* for the rainbow, cutthroat, and all other western trouts. For more on the Russian perspective on the genus issue, see Reshetnikov et al. (1997) and Mednikov et al. (1999).

species change was made because the rules of scientific nomenclature give priority to the first name used for the taxon. Thus *mykiss*, first used in 1792, has priority over *gairdneri*, not employed until 1836. This is also why the species name *clarkii*, first employed in 1836, is used today for all cutthroat trout instead of one of the myriad of other names (such as *purpuratus*) that appeared later.

But what about the common name, cutthroat trout? How did that come about? Although it was 1805 when Meriwether Lewis first recorded the twin slashes of reddish color in the grooves beneath the lower jaw that gave rise to this name, the name itself did not come into general use until the late 1880s. One of the earliest names for the fish was salmon-trout. The journals of Lewis and Clark contain many references to salmon-trout, but they apparently intended it to mean steelhead, the sea-going form of rainbow trout, which they first encountered in the Lemhi River country of Idaho. Nowhere in the journals did they link the name salmon-trout with the fish they encountered at the Great Falls of the Missouri. Even so, perhaps because Sir John Richardson made the name association, it wasn't long before "salmon-trout" was clearly being attached to the cutthroat. John C. Fremont used this name for the big Lahontan cutthroat trout the Indians brought to his camp at Pyramid Lake, Nevada, recognizing them from the red slashes as the same type of

fish he had known while quartered at Fort Vancouver, where sea-run cutthroat were abundant. Other early common names for the cutthroat trout were red-throat trout, Clark's trout, Rocky Mountain trout, black-spotted trout, speckled trout, brook trout, native trout, or just plain "native." Snobbish fishing writers, wanting to lend a scholarly air to their writings, borrowed from David Starr Jordan and used the names *Salmo mykiss* or *Salmo purpuratus*, or whatever Jordan was calling them at the time.

The name "cutthroat" was first used by Charles Hallock, a well known outdoor writer of the late 1800s. The name appeared in an article Hallock wrote for the October 4, 1884 issue of "The American Angler," about a fishing trip his party had taken from the Crow Indian Agency at Stillwater, Montana. The party trekked up Rosebud Creek to an unnamed lake at the entrance to Rosebud Canyon, where, Hallock reported, the fishing was excellent for trout weighing a uniform two pounds. He described the fish as follows:

> "It resembles the iridea of Colorado in respect to the metallic black markings scattered like lustrous grains of coarse black powder over its shoulders and body; but it lacked the rainbow lateral stripe. Its distinctive feature, however, was a slash of intense carmine across each gill cover, as large as my little finger. It was most striking. For lack of a better description we called them cutthroat trout."

Cutthroat country: South Fork Carnera Creek, Colorado.
PHOTO BY AUTHOR

Fishery scientists abhorred the name "cutthroat." But the angling public liked it and it caught on quickly despite their objections, and today is recognized officially by such organizations as the American Fisheries Society Committee for Common and Scientific Names.

HOW TO TELL IT'S A CUTTHROAT, AND IF SO, WHICH SUBSPECIES

OK, those red slashes may have given the fish its name, but they aren't always reliable guides to identifying a trout as a cutthroat. If you happen to be in the Columbia River basin east of the Cascade Mountains, or in other areas of the west inhabited by the interior forms of rainbow trout known as redbands, you can run into problems. I have caught specimens of interior rainbow trout from Oregon's Deschutes River that, in addition to the intense band of magenta-red along their sides that give them the local name "redsides," also had very prominent red cutthroat slashes. Rainbow trout of the Kuskokwim River system of Alaska may also have red pigment under the jaws. In addition, hybrids between cutthroat trout and introduced rainbow trout may also have cutthroat markings. On the other hand the slashes may be so faint on some specimens of true cutthroat, especially sea-run fish taken in salt water, that they go completely unnoticed.

However, that red pigmentation does seem to have a function in trout society. A trout social unit has been described as "a hierarchy with infrequent subordinate revolt." What this means is that there is a pecking order of sorts. The most dominant fish in a piece of water assumes the best lie, the second most dominant fish occupies the second best lie, and so on down the line. But this dominance is put to the test from time to time. The term used by scientists for the ritual displays and occasional nipping, chasing, and direct fighting that the trout use in challenging and threatening one another within this hierarchy is *agonistic behavior*.

Agonistic behavior patterns are pretty much the same for all salmonids. What's more, the different species of trout, salmon and charr seem to understand one another's threat displays. It is in these threat displays that a trout's pigmentation seems to come into play. There are two principal types, the frontal display and the lateral display.

In the frontal threat display (Fig. 1-2), the threatening fish approaches its opponent with its back arched, dorsal fin flattened, all other fins fully extended, gill covers flared out, mouth opened, and the bottom of the mouth pushed downward. As it fixes its opponent with its eyes, its head appears to be enlarged, much as a dog or cat would look with its hair standing on end. *The opponent gets a full head-on view of the red cutthroat slashes.*

In the lateral threat display (Fig. 1-3), the fish turns side-on to its opponent, assuming a parallel position,

FIGURE 1-2. *Frontal threat display*

16. For another easy overview of agonistic behavior, read Chapter 7 of Bill Willers' book, *Trout Biology: a Natural History of Trout and Salmon.* (Willers 1991). To dig deeper, Willers recommends Kalleberg (1958) and Jenkins (1969) as classic studies that should be read by anybody interested in trout. Interactions of cutthroat trout with one another and with other species are covered in papers by Newman (1960), Griffith (1972), Schutz and Northcote (1972), Glova and Mason (1977), Glova (1978, 1986, 1987), Jacobs (1981), Nilsson and Northcote (1981), Tripp and McCart (1983) Mitchell (1988), Mesa (1991), Sabo (1995), and Sabo and Pauley (1997).

FIGURE 1-3.
Lateral threat display

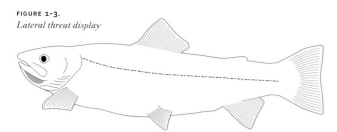

both fish facing upstream. It stretches out its body and assumes a rigid pose with the line of the back straight, the mouth and gill region dilated, and all fins, including the dorsal, fully extended. Sometimes the fish may actually appear to quiver. In this type of threat display the cutthroat slashes are visible, but less so than in the frontal threat display. On the other hand, the rose-colored band of a rainbow trout, or the reddish colors some subspecies of cutthroat trout develop on the gill plates and lower sides, would be fully visible in the lateral threat display.

Some scientists who have studied this behavior feel that the frontal threat display carries more aggressive intent than the lateral display. In many species, the frontal threat display often culminates in an attack, with the threatening fish suddenly darting in to nip or chase its opponent. Lateral threat displays, on the other hand, just peter out more often than not. This has led a few observers to suggest that the lateral display may even be an act of submission to a more dominant fish in some instances, much as a subordinate wolf or wild dog might lie down, throat exposed, at the feet of the leader of the pack.

Cutthroat trout appear to favor the frontal threat display. In aquarium studies where cutthroat and rainbow trout were put in the same tank, only the cutthroats used the frontal display whereas the rainbows always used the lateral display. This is evidence for my assertions about the role of coloration in agonistic behavior, yet it's curious in another regard, namely that in study after study of mixed populations, rainbow trout have always proven the more aggressive species. Rainbow trout dominate similar size cutthroat trout almost totally, even though they use the less aggressive threat display. Only when the cutthroat is a markedly larger fish than the rainbow does the cutthroat ever dominate. Even cutthroats that have dominated other cutthroats are stressed to death, often in less than a night, when paired with a rainbow trout of similar size. [16]

In these kinds of studies, observers invariably report that the cutthroat trout appear to be the calmer of the two species. The rainbow trout are described as "bold," "brash," and "aggressive" whereas the cutthroat trout draw terms such as "slow," and "deliberate," not only in their interactions with other fish, but also in their feeding behavior. Rainbow trout, when observed taking food at or near the surface in these study situations, are usually described as making a bold, noisy, leaping approach. Cutthroats, on the other hand, are described as approaching cautiously, quietly, and allowing only their noses to penetrate the surface.

But we were talking about how to identify cutthroat trout. Another way that's often advocated is to reach back to the base of the fish's tongue with a finger or pencil eraser and feel for little teeth. The scientists call these *basibranchial teeth* (Fig 1-4). You may also see them referred to as *hyoid teeth* in older literature. Fishery biologists often count basibranchial teeth, bringing them out for easy viewing by staining them with alizarin. At any rate, cutthroat trout have them, other western trout species do not—or so it has often been asserted. Typically the first inkling a biologist gets of introgression of a cutthroat population by rainbow trout is when there is a reduction or complete loss of basibranchial teeth in the population. But again, you can be led astray in redband country since some races

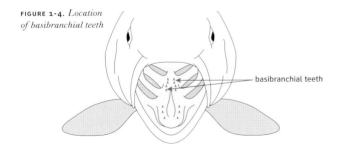

FIGURE 1-4. *Location of basibranchial teeth*

basibranchial teeth

of redband trout may also possess a few basibranchial teeth. Some Apache trout and Gila trout have vestigial basibranchial teeth as well. [17]

Still another morphological trait mentioned in some fish-identification keys is the length of the maxillary (Fig. 1-5). In cutthroat trout, the maxillary extends well beyond the posterior margin of the eye, giving the fish a long-jawed look. I've found that this isn't too bad a field guide for distinguishing cutthroat trout from coastal rainbows, provided that the specimens I'm working with are less than, say, 20 inches in length. In smaller specimens of coastal rainbow (and perhaps in some of the other members of the rainbow trout group as well), the maxillary typically extends no further back than the posterior margin of the eye. But larger-size rainbow trout can, and often do, develop longer maxillaries.

The cutthroat trout has a propensity to develop a bewildering array of colors, which differ from watershed to watershed. As I noted above, this led many of the early collectors to describe a myriad of western trout "species." Among the cutthroat subspecies, coastal cutthroats run to silvery, coppery, or brassy tones, often with a lemon- to rose-colored wash on the body and pectoral, ventral, and anal fins that can be bright orange or even reddish in some stream-resident populations. Sea-run populations are more silvery, but they darken in fresh water and the males may take on spectacular

17. Trout taxonomists have reason to be interested in basibranchial teeth. These scientists are always on the lookout for unique evolutionary changes, i.e., changes in character state that occur in one phylogenetic branch but not others, because these can be used to mark points of divergence and trace evolutionary progression. Characters that best express this evolutionary progression are those where the change in character state is (1) irreversible, i.e., once the change occurs, it does not "evolve back" to its previous state, and (2) not subject to independent evolution in more than one line. Behnke uses the example of the loss of basibranchial teeth as a change in a character state that meets the first test, but perhaps not the second. Basibranchial teeth occur as vestiges in some redband trout and in some Apache trout and Gila trout as well. Thus one can infer that all redband, Apache, and Gila trout had a common ancestor with basibranchial teeth. One could also infer that cutthroat trout, which retain basibranchial teeth, are more primitive than redband trout in which basibranchial teeth occur only as vestiges, and redbands are in turn more primitive than coastal rainbow trout, which have lost all vestiges of basibranchial teeth. Meristic characters, defined as morphological

pinkish-orange colors on the gills and bellies when spawning time draws near. Westslope cutthroats typically develop yellow, orange, and red colors, as do the Colorado River and greenback subspecies of Colorado, which are by far the most brilliantly colored and beautiful of all the cutthroat subspecies. In contrast, the Yellowstone cutthroat and all of the Great Basin subspecies tend to be dull in coloration. In general, when it comes to identifying a fish as a cutthroat or sorting one into a subspecies, it is wise to remember the old ichthyologist's admonition: "Ninium ne crede colori" (place little credence in color)!

Spotting patterns too can be quite variable in cutthroat trout. Patterns can range from the profusion of irregular small to medium spots all over the body of the coastal cutthroat to the few very large, rounded spots of the greenback cutthroat. The finespotted Snake River cutthroat is so named because it exhibits a "heavy-sprinkling-of-pepper" effect, while the Yellowstone cutthroat tends to have moderately large, rounded spots concentrated more heavily on the caudal peduncle. Then there's the rare little Paiute cutthroat, which has parr marks but usually no spots on its body at all.

Fortunately, fishery scientists have other methods, not only to tell cutthroat trout from other trouts of the genus *Oncorhynchus*, but also to distinguish the

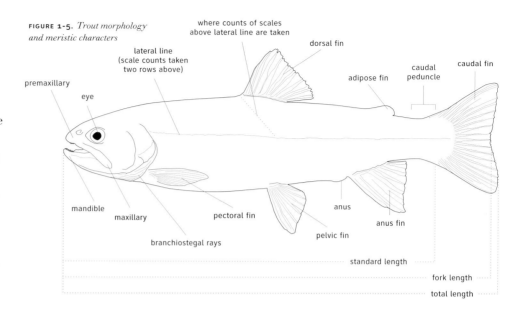

FIGURE 1-5. *Trout morphology and meristic characters*

various subspecies of cutthroat trout. For one, they can carefully measure certain *meristic characters*. Meristic characters are those physical features that have countable elements in a series, the number of elements being mainly determined by the organism's genetic programming for embryo development. These vary within and among species, and so are useful for describing and classifying fishes. For many years, analysis of meristic characters was the workhorse method for taxonomic studies of fishes. Dr. Behnke and his students, along with many of their predecessors, used the method

characters having countable elements in a series, such as scales, vertebrae, and fin rays, also change irreversibly as a line evolves, but again they do not meet the second test. Meristic characters are subject to independent evolution of similar character values in separate lines. Even so, scientists have employed them extensively in trout taxonomy. When there is another basis for inferring which line is the more primitive and which the more recent, low numbers of meristic elements are found to be associated with the primitive state and higher

extensively to slot the western trouts into the species and subspecies we recognize today.

The use of meristic characters has been criticized because several environmental factors, such as temperature, light, and dissolved oxygen concentration, can affect the process of embryo development and so influence the number of countable elements actually found in specimens. This has been verified by a host of laboratory studies. Thus, the critics say, the outcome isn't always based on genetics, as it should be to be a reliable classification tool. Dr. Behnke and his students acknowledge this, but counter with the argument that meristic characters are not as readily modified in nature as may be indicated by laboratory studies, where conditions are often pushed beyond what the embryos would experience in the wild. They point out that most trout spawn and the embryos develop under comparable environmental regimes throughout the range of a species or subspecies, so would receive essentially the same environmental cues. Therefore, when consistent character differences are found between disjunct groups of trout (especially when, as Behnke can boast, thousands of specimens have been examined), it can be assumed that these differences validly reflect genetic factors. Behnke et al. have gone so far as to compare introduced populations of golden trout, Lahontan cutthroat trout, and Yellowstone cutthroat trout reared

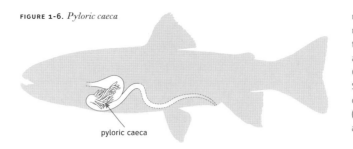

FIGURE 1-6. *Pyloric caeca*

pyloric caeca

numbers with the more advanced or recent state. For further information on this subject, consult Mayr and Ashlock (1991), de Queiroz and Gauthier (1992), and Winston (1999). See also the discussions of meristic characters in Schreck and Behnke (1971), Hickman and Behnke (1979), and Behnke (1988a, 1992, 2002).

in many different settings with their parent stocks to demonstrate the relative stability of meristic counts over a wide range of natural environments.

Meristic characters that Behnke and students have found to be most useful for classifying the various species and subspecies of *Oncorhynchus* are: number of scales in the lateral series (usually the count is made two rows up from the lateral line), number of scales above the lateral line (counted diagonally up to the front of the dorsal fin), number of vertebrae (usually counted on an x-ray of the specimen), pyloric caeca count (Fig. 1-6; pyloric caeca are slender, fingerlike pouches emanating from the first short segment of the intestine), and number of rays in the pelvic fin. Other meristic characters that have been used include the number of branchiostegal rays, number of gill rakers, and numbers of rays in the various other fins. Computer programs do the job of crunching all the data and displaying the results. Readouts are usually in the form of a Hubbs and Hubbs diagram, named after the investigators who first proposed displaying the data in this way. Figure 1-7 is a Hubbs and Hubbs diagram showing actual measurements of one meristic character, in this case scales in the lateral series, for rainbow trout, six authentic cutthroat trout subspecies, and eleven "unknown" cutthroat populations taken from creeks and lakes in Colorado. In this particular study the investigators wanted to know which if any of the eleven Colorado populations could be classified as *O. c. pleuriticus*. Which would you select?[18]

I mentioned earlier that hybridization between cutthroat trout and rainbow trout is often first detected by a reduction in the basibranchial teeth. As hybridization spreads and intensifies, scale counts decrease and the numbers of pyloric caeca and vertebrae increase. A trained eye can sometimes pick up changes in coloration and spotting pattern as well, but as indicated above, colors and spotting are tricky. I have seen even trained eyes have difficulty telling a true rainbow from a true cutthroat on the basis of color and spotting alone.

Increasingly, over about the last three or four decades, scientists have moved beyond meristic characters to methods based on biochemistry and the relatively new field of molecular biology. These methods make it possible to examine the genetic composition of species and subspecies more directly,

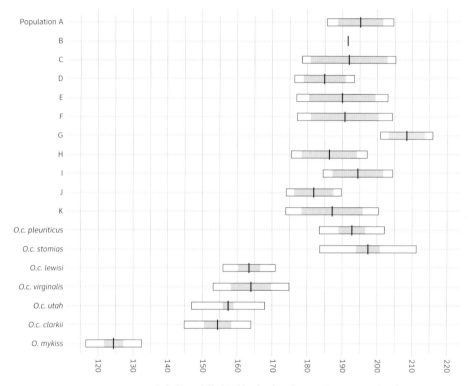

FIGURE 1-7. *Scales in lateral-line series (Hubbs and Hubbs diagram). This example presents measurement of scales in the lateral series for rainbow trout, six documented cutthroat subspecies, and eleven cutthroat populations from creeks and lakes in Colorado. For each entry, the base line represents the range of values for all specimens in the sample, the vertical center line marks the mean value, and the shaded bar indicates the 95 percent confidence limits of the mean.*

18. Carl L. and Clark Hubbs developed the graphical method of displaying large amounts of meristic character data shown in Fig. 1-7. Their original paper is Hubbs and Hubbs (1953). The example diagram used for Fig. 1-7 was taken from a report issued by the Colorado Division of Wildlife (see Lytle et al. 1982). The investigators who wrote that report gave the nod to populations G and I as likely being pure *O. c. pleuriticus*.

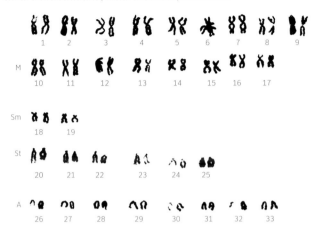

and thus distinguish among them. To grasp how these methods work and what we have learned from them, let's start with some definitions and concepts. [19]

The number of chromosomes possessed by an organism, along with the form or morphology of its chromosomes, is referred to as its *karyotype* (Fig. 1-8). In terms of chromosome number, the different subspecies of cutthroat trout have one of three karyotypes: 68 chromosomes, 66 chromosomes, or 64 chromosomes. Chromosome number is usually expressed as *diploid number*, e.g., 2N =68, to signify that chromosomes occur in a cell in sets of two (34 sets of two for the 2N =68 karyotype, 33 sets of two for the 2N =66 karyotype, and

32 sets of two for the 2N =64 karyotype). When mating and reproduction took place, one-half of these chromosomes (one chromosome from each set) came from the organism's mother and the other half (the other chromosome of each set) came from its father.

Each organism carries its complement of chromosomes tucked away in the nuclei of its cells, i.e., each cell nucleus carries a complete complement of chromosome pairs (except for egg and sperm cells which have only the half-complements contributed by the mother and father respectively). Packed into each chromosome is a single, extremely long, two-stranded molecule of DNA (short for 2-deoxyribonucleic acid). Each of the two strands of the DNA molecule is composed of repeating units of a simple sugar, 2-deoxyribose, joined linearly through sugar-phosphate linkages. The two strands spiral around one another in the familiar double helix arrangement shown in Fig. 1-9, and are linked together by bonded pairs of organic bases. These so-called *base pairs* consist of four organic bases: adenine (labeled A), thymine (labeled T), guanine (labeled G) and cytosine (labeled C), with A's always bonded to

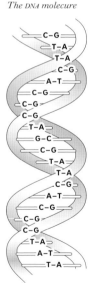

FIGURE 1-9.
The DNA molecure

Of course, they had the benefit of comparisons of the other meristic characters to help them narrow their selection. By the way, the convention is to make scale counts, gill raker counts, and counts of rays on paired fins on the left side of the fish (Behnke 1992). There is nothing wrong with counting both sides, and for certain kinds of studies it is done as a matter of routine, with the numbers reported as a set separated by a hyphen, left side first, e.g., 12–13. (Strauss and Bond 1990).

19. A ton of material is condensed in this section. To learn more about the chromosomes and karyotypes of fishes including cutthroat trout, begin with Chapter 6 of Schreck and Moyle (1990) then move on to Gold (1977), Gold et al. (1977, 1980), Loudenslager and Thorgaard (1979), and Hartley (1987). For allozyme electrophoresis, read Chapter 2 of Ryman and Utter (1987) and Chapter 5 of Schreck and Moyle (1990), then Loudenslager and Kitchin (1979), Loudenslager and Gall (1980), Martin et al. (1985), Leary et al. (1987), and Allendorf and Leary (1988). Ferris and Berg's Chapter 11 and Gyllensten and Wilson's Chapter 12 in Ryman and Utter (1987) are good ones on mtDNA. Other books I have relied on for these and especially the newer DNA methods include Hoelzel and

T's and G's always bonded to C's as shown in the figure. The lengths of DNA molecules and of fragments of DNA are almost invariably expressed in terms of base pairs (bp), e.g., "A major histocompatibility complex class II gene recently isolated from coastal cutthroat trout is between 809 and 826 bp long."

Here and there along the length of the DNA molecule the base pairs are organized into sequences that code for the synthesis of proteins. These sequences are called *genes*, and the location of a gene on its chromosome is called its *locus* (plural *loci*). The full complement of chromosomes in a cell nucleus may contain somewhere in the order of 50,000 genes.

Genes can have alternate forms. These are called *alleles*. As we have seen, diploid organisms carry two copies of each gene in their cell nuclei, one copy on each of the two chromosomes where the locus occurs. If both copies are the same allele, the individual is said to be *homozygous* for that locus. If the two copies are different alleles, the individual is *heterozygous* for that locus.

Even though individuals can have no more than two alleles for a given locus, in a population there can be many alleles for that locus. When there is more than one allele for a particular locus in a population, the population is said to be *polymorphic* at that locus. If only one allele occurs for that locus throughout the population, then the population is said to be monomor-phic or *fixed* at that locus. The proportion of each allele at a given locus in the population is called its *allele frequency* or *allelic frequency*, and is an important measure in population genetic studies. The allele frequency of a fixed allele is of course 1.00 or 100 percent.

An individual's genetic composition, i.e., the specific list of alleles it possesses at each locus, is called its *genotype*. The individual's array of observable characteristics is called its *phenotype*. The genotype, in conjunction with the environment in which the individual occurs, determines the phenotype. We noted this above in our discussion of meristic characters, where we learned that environmental factors can affect the number of countable elements actually observed. We also noted Dr. Behnke's argument that even though they are phenotypes, certain meristic characters reliably reflect the genotypes that produced them. Still, scientists would benefit from a method to study genes more directly, and here is one such method.

Since genes code for the synthesis of proteins and each allele at a given locus will code for a slightly different form of that protein, one can isolate the proteins and study them as surrogates for the alleles themselves. Two kinds of proteins are synthesized from genes. Some genes code for *structural proteins*, such as those in muscle tissue or those governing features associated with appearance or morphology. Other genes code for

Dover (1991), which is a primer on the subject, Ayala (1982) and Li and Graur (1991), which are textbooks complete with homework questions and problems, and Hillis and Moritz (1990), which is a comprehensive treatment. Garte (1994) has a couple of Chapters on the use of molecular genetic techniques to address conservation questions and manage natural fish populations. Two additional books that focus on fishes and explain more about how the techniques outlined in the text are applied are Powers (1991) and Hallerman (2002).

enzymes, which catalyze steps in the various biochemical processes essential for life. It is the enzymes that are most readily isolated and therefore most often studied for trout taxonomy, using a process called *protein electrophoresis* or more specifically, *allozyme electrophoresis*. Protein electrophoresis is a lab method of separating proteins for identification. *Allozymes* are the alternate forms of an enzyme that are synthesized from the different alleles at that enzyme's locus. Thus, allozyme electrophoresis separates the allelic proteins associated with a particular locus so that their individual allelic frequencies can be determined.

If two subspecies never or only rarely share electrophoretically detectable alleles at a locus, that locus is said to be *diagnostic*, because it enables one subspecies to be distinguished from the other directly, using electrophoresis. You may also see the term *marker* applied to diagnostic loci and their associated alleles. Even when subspecies do share their electrophoretically detectable alleles at all loci, sometimes the frequency of one or more alleles may be characteristic for a particular subspecies, which still allows it to be distinguished from others by allozyme electrophoresis.

The western trouts including cutthroat trout may just possibly have received more intensive study using allozyme electrophoresis than any other group of fishes. A prodigious amount of work is also being done on these fishes using even newer DNA methods.[20] In fact, the meaning of the term *locus* has been broadened to now encompass, in addition to genes, the locations of other types of marker sequences in the DNA molecule that also vary within and among species, subspecies, and even populations. For example, lying in between genes on the DNA molecule are extensive stretches, often many hundreds or thousands of base pairs long, which do not code for anything and do not appear to serve any other particular purpose. Yet arrays of base pairs have been found in these regions that are likewise diagnostic and have already proved valuable for trout taxonomy. Also, located out in front of the genes and also at the end of genes are arrangements of base pairs that signal the protein–synthesizing mechanism of the cell that this is the place to start its work and then, at the end, to stop and disengage. DNA markers that may be useful for trout taxonomy have recently been found in each of these regions. In addition, within some genes are arrangements of base pairs, called *introns*, that interrupt the coding sequence. DNA markers of potential value for trout taxonomy have recently been found in introns as well.

The term *locus* is now being used to signify the location of any of these new types of markers, as well as the location of a particular gene on the chromosome. Likewise, the meaning of the term *allele* has been broadened to include variants that occur at these other

20. Another big advantage of the newer DNA methods is that, owing to the small amounts of tissue required, they can be employed non-lethally. Karyotyping, meristic character analysis, and allozyme electrophoresis each utilize either preserved fish, large amounts of tissue, or types of tissue that cannot be obtained without sacrificing specimens, often 50 to 100 or more per population depending upon the objectives of the study. When dealing with increasingly rare and endangered subjects as we are with most subspecies of cutthroat trout, sacrificing fish for science can no longer be justified.

DNA marker loci. A lab procedure called PCR (which stands for *polymerase chain reaction*) is used to build up a sufficient quantity of these DNA marker fragments for analysis, and then, akin to what is done for proteins, electrophoresis is used to separate the fragments for identification and determination of their allelic frequencies. Practitioners of these molecular biology techniques have coined acronyms and given names to the various classes of DNA markers, such as *minisatellites, microsatellites, VNTRs, RAPDs, SINEs, PINEs,* and *SNPs.* This is a rapidly moving field, so by the time you read this there may be even more markers and acronyms.

Still another procedure that can be used on nuclear DNA is to digest it into fragments using different members of a class of enzymes called *restriction enzymes,* which cut the DNA molecule at specific locations. These so-called *restriction fragments,* which can also be separated by electrophoresis, often exhibit length polymorphisms called (you guessed it) *restriction fragment length polymorphisms* or *RFLPs* for short, which can also be useful in trout taxonomy studies.

To this point, we have been talking about genes and DNA markers that occur on chromosomes. Although 99 percent of an organism's DNA is indeed found in its chromosomes, the remaining 1 percent that is not can also be very important in trout taxonomy. This small amount of DNA is found in *mitochondria,* which are

tiny organelles that float around in a cell's cytoplasm and serve as the energy-producing factories of the cell.

Unlike the very long, linear DNA molecules found in chromosomes, the double-stranded molecules of *mitochondrial DNA* (*mtDNA*) are short (only 15,000 to 18,000 bp or so in length) and are joined end to end in a circular shape. Mitochondrial DNA molecules each contain only 37 genes, but unlike nuclear DNA molecules of which there are only two copies of each per cell, there are many copies of the mtDNA molecules per cell. Like nuclear DNA, mtDNA can be digested into fragments using different restriction enzymes to generate RFLPs, and some of the other molecular techniques outlined above can also be used. But what makes mtDNA of particular interest to scientists is that it is inherited from the mother only, thus providing a way to track maternal lineages.

So scientists now have a range of sophisticated methods at their disposal for directly examining genetic relationships and sorting specimens not only into subspecies, but into even finer subdivisions, at the level of individual stocks, for example. For all that, the evidence produced to date for cutthroat trout is rather anticlimactic. At the subspecies level of organization, which is our interest here, these newer methods agree reasonably well with the phylogeny Behnke already proposed based mainly on the older method of meristic

Cutthroat country: A beaver pond in the heart of the coastal cutthroat range. PHOTO BY AUTHOR

character analysis. There may be a couple of places where recent results cloud Dr. Behnke's picture, but these can be deferred to later Chapters. Now it is time for an initial look at where the cutthroat trout subspecies live—or rather, where they did live 160 to 200 years ago or so, when the westward migration of people of largely European descent began to hit its peak.

HISTORICAL DISTRIBUTION

When the original range of cutthroat trout is displayed on a map, as I have done in Fig. 1-10, it leaves little doubt that this species was the most widely distributed of all the western trouts. The range extends along the north Pacific coast from Gore Point near Prince William Sound in Alaska to the Eel River in northern California, and inland to the Rockies and beyond. Within this overall distribution, the ranges of the individual cutthroat subspecies segregate reasonably well, i.e., their respective ranges are delineated by the geographic boundaries of river basins. Although many have native distributions that are contiguous, there is little if any overlap of one with another.

First is the coastal cutthroat, *O. c. clarkii*, whose native range coincides closely with the great Pacific coast coniferous forest belt that extends roughly from Gore Point, Alaska to the Eel River, California, and inland to the Cascade crest.

FIGURE 1-10. *Historical distribution of cutthroat trout (all subspecies) in Western North America*

Next comes *O. c. lewisi*, the so-called westslope cutthroat. The historical range of this subspecies was greatest of all the subspecies, extending roughly 1,000 miles from north to south and about the same distance from west to east. It covers east-slope Cascade drainages (Yakima, Wenatchee, Entiat, Chelan, Methow, Sanpoil, Pend Oreille, Salmo, and Kootenay) of the upper Columbia River basin including the Clearwater and Salmon River systems of Idaho and the John Day River system of Oregon, and straddles the Continental Divide to include headwaters of the South Saskatchewan and the Jefferson, Madison, and upper Judith rivers of the Missouri River basin.

The native territory of the Yellowstone cutthroat *O. c. bouvieri* also straddled the Continental Divide, extending up the Snake River and its tributary watersheds from Shoshone Falls in Idaho to the headwaters, then down the Yellowstone River system to (and including) the Tongue River watershed. The finespotted Snake River cutthroat *O. c. behnkei* occurs in the upper Snake River and its tributaries from Palisades Reservoir on the Wyoming-Idaho border upstream to Jackson Lake.

The Great Basin area is the historical range of six of the subspecies listed in Table 1-1. Three of these subspecies, the Lahontan cutthroat *O. c. henshawi*, the Paiute cutthroat *O. c. seleniris*, and the Humboldt cutthroat (unnamed) are associated with the ancient Lahontan basin, which occupies roughly the northwest corner of Nevada. The now extinct Alvord cutthroat *O. c. alvordensis* and the Willow/Whitehorse cutthroat (unnamed) are associated with two separate but contiguous basins with "arms" up into southeastern Oregon. The sixth subspecies, the Bonneville cutthroat *O. c. utah*, lives on the fringes of the ancient Bonneville basin, of which the Great Salt Lake is a remnant.

The Rocky Mountain region of Wyoming, Colorado, and New Mexico is the native range of four subspecies. The Colorado River cutthroat *O. c. pleuriticus* is found in the upper Colorado and Green River drainages. These drainages are on the west side of the Continental Divide. The east side of the Divide, in the headwaters of the South Platte and Arkansas Rivers, is the native range of *O. c. stomias*, the greenback cutthroat. The extinct yellowfin cutthroat *O. c. macdonaldi* was once found in the upper Arkansas drainage as well, in the Twin Lakes near Leadville. The upper Rio Grande drainage, also east of the Continental Divide, is considered the native range of *O. c. virginalis*, the Rio Grande cutthroat, although this subspecies may have extended south into Texas and perhaps even into the upper Rio Conchos (a Rio Grande tributary) in Mexico.

Which state can boast of having the most native cutthroat subspecies? That would be Wyoming, with six: *O. c. bouvieri* in the Yellowstone and Snake River drain-

ages, *O. c. lewisi* in the northwest corner of Yellowstone
Park, the finespotted Snake River cutthroat around
Jackson Hole, *O. c. pleuriticus* in the upper Green River
drainage, *O. c. stomias* in Wyoming tributaries of the
South Platte, and *O. c. utah* in the Bear River drainage
(*O. c. stomias* is now extinct in Wyoming, and so too
is *O. c. lewisi*).

    The story of how cutthroat trout came to be where
they were found historically is a fascinating one, and
that is the subject of the next Chapter. Let me just close
this one by saying that not all of the subspecies have
fared well since European man migrated west. Two
subspecies, *O. c. macdonaldi* of Colorado's Twin Lakes
and *O. c. alvordensis*, the subspecies native to the Alvord
Basin of southeast Oregon and northern Nevada, have
gone extinct in that time. Several others are teeter-
ing on the brink and have been listed as endangered
or threatened under the U.S. Endangered Species Act
at various times since people have become concerned
about such matters. Still others have been pushed back
alarmingly, to the point where they occupy only a small
percentage of their historical range. The cutthroat may
be the most sensitive of all the trouts to habitat deterioration and especially to the introduction of non-native
species of trout. Dr. Behnke has likened it to the canary
in the mine: in the face of environmental disturbance,
it is usually the first species to go.

CHAPTER 2

Evolution and Prehistoric Distribution

"There's nothing constant in the universe.
 All ebb and flow, and every shape that's born bears
 in its womb the seeds of change."

—OVID,
Metamorphosis

THE CONCEPT OF GEOLOGICAL TIME

Just as life scientists have the Linnaean system to classify the biological world in a hierarchical order, so earth scientists have a hierarchical system for categorizing the various intervals in Earth's geological history. At the most encompassing level is the Era. For example, the Cenozoic Era, beginning about 65 million years ago and extending to the present, is popularly known as the Age of Mammals. Within eras are Periods. The Triassic, Jurassic, and Cretaceous Periods comprised the Mesozoic Era, popularly known as the Age of Dinosaurs, that ended 65 million years ago with the advent of the Age of Mammals. And within periods are Epochs. We now live in the Holocene Epoch of the Quaternary Period of the Cenozoic Era. For future reference, this system is diagrammed in Fig. 2-1. [1]

As you can see, the earth scientists' system deals in time spans of millions, tens of millions, and hundreds of millions of years. Let's just take a moment and try to comprehend how long a million years is. It's tough, because we have no frame of reference for such a vast passage of time. Yet a million years represents only one 46-hundredth of Earth's history—a mere fraction of a second in one of our hours. Time so vast, and yet so short in terms of the Ages: this is the concept of geological time.

Perhaps we can grasp it better this way. Suppose

FIGURE 2-1. *Intervals of geological time*

ERA	PERIOD	EPOCH	TIME (in millions of years)
Cenozoic	Quaternary	Holocene	
			0.1
		Pleistocene	
			2
	Tertiary	Pliocene	
			5
		Miocene	
			24
		Oligocene	
			34
		Eocene	
			53–54
		Paleocene	
			65
Mesozoic	Cretaceous		
			145
	Jurassic		
			210
	Triassic		
			245
Paleozoic	Permian		
			280
	Carboniferous	Pennsylvanian	
		Mississippian	
			380
	Devonian		
			410
	Silurian		
			430–440
	Ordovician		
			500
	Cambrian		
			570
	Ediacarian		
			700
Precambrian	Proterozoic		
			4600

1. Have a care in using Fig. 2-1. The way the chart had to be constructed, with relative time intervals drawn with equal spacing, could mislead one into thinking they covered equal spans of real time. They did not. The actual time column is what counts here; it shows, for example, that the Miocene Epoch actually spanned 19 million years whereas the Pliocene Epoch spanned only 3 million years. Speaking of Epochs, Paul J. Crutzen of the Scripps Institution of Oceanography, University of California at San Diego, suggested not long ago that a new geological Epoch ought to be defined to denote human dominance of the global environment (see Crutzen 2000). He believes this new Anthropocene Epoch began in the late 1700s, possibly coincident with the advent of James Watt's steam engine in 1784 and the beginning of the so-called Industrial Revolution. Analysis of air trapped in polar ice cores points to that approximate date as the beginning of growing global concentrations of carbon dioxide and methane. Thomas Fleischner of Prescott College, Arizona, suggested a much earlier event, the beginning of agriculture, as the point where humans' relationship to the natural world changed fundamentally (see Fleischner 2003). But neither

we speed up the passage of time so that a whole year flashes by in just one second. Consider all the things you have done in the last year and imagine them passing in one second. Now consider how long ago some major events in world history occurred in this accelerated time scale. The average human lifespan is a little over one minute long. The American Civil War was fought not quite two and a half minutes ago. The pilgrims landed at Plymouth Rock six minutes ago. Columbus sighted the New World just over eight minutes ago. The prophet Muhammad was born 24 minutes ago and Jesus of Nazareth 37 minutes ago. To go back one million years on this scale, you would have to count the seconds steadily for eleven and one-half days—no time off for eating or sleeping—while to get as far back as 100 million years you'd be counting the seconds steadily for three full years. Three full years compared to our 70-second life span: that's the concept of geological time.

One hundred million years. In Earth's history, that's time enough for the continents to shape and reshape, for countless cycles of vulcanism, greenhouse episodes, droughts, ice advances, and flooding to occur, and for great mountain ranges to form, erode away completely, and form again. It is time enough for nature to experiment with different life forms, rejecting some and redistributing others in accordance with changing climates and conditions, nudging them along the arduous path of evolution that led to the plants, the animals, and the fishes that share our planet today.

ORIGIN OF *ONCORHYNCHUS* AND EARLY SALMON AND TROUT SPECIES

One hundred million years ago dinosaurs roamed Earth. At the beginning of the Age of Dinosaurs 245 million years ago, the continents were joined into a single, vast land mass known as Pangaea. Over time, Pangaea separated into two super-continents, each encircling the globe: Laurasia in the northern hemisphere and Gondwana (or Gondwanaland) in the southern hemisphere. By the dawn of the Cretaceous Period 145 million years ago, Gondwana had itself separated into land masses that would become South America, Africa, Australia, and Antarctica, but Laurasia remained linked although it, too, "strained at the seams" from time to time, showing signs of its own eventual separation.

During much of the Cretaceous Period, shallow seas encroached upon these land masses, leaving some large portions cut off from one another and others surrounded by water and isolated as islands. That part of Laurasia that would become North America existed as two such "island" land masses divided south to north by one of those broad, shallow seas, and cut off from Europe and Asia by other shallow encroachments of

Crutzen nor Fleischner was the first to propose a new interval in the geological time chart to denote the advent of human influence. That distinction belongs to the Italian geologist Antonio Stoppani, who, in 1873, proposed that an Anthropozoic Era be defined to mark the "Age of Mankind." How much of an impact are humans having? For one analysis of the human footprint on the planet, see Sanderson et al. (2002). Another study, published by Mathis Wackernagel et al. (2002) in the Proceedings of the National Academy of Sciences (U.S.A.), concluded that humans began consuming beyond Earth's natural regenerating capacity around 1980, and we are now exceeding that capacity by about 20 percent. The most recent assessment, and the most comprehensive assembled to date, is the United Nations-sponsored Millennium Ecosystem Assessment, completed in 2005. You can access this report on the Internet at *www.millenniumassessment.org.*

water (Fig. 2-2a). The eastern land mass was rather calm geologically. The Appalachian Mountains had formed earlier and were already eroding and assuming their present shape. Across the shallow sea, the western land mass was more active. About 70 million years ago, the first of the great uplifts that eventually formed the Rocky Mountains began there, and the inland sea slowly drained away. By 65 million years ago, that sea was largely gone, with only a small embayment extending up to about where Houston, Texas is today. The two "islands" of proto-North America were now one again, and were once again joined to Asia and to Europe (Fig. 2-2b). [2]

Fishes of the Order Salmoniformes existed in Earth's oceans, and possibly also in its freshwater systems, during at least part of the Age of Dinosaurs. Scientists who study these matters speculate that sometime between 100 and 50 million years ago, an event occurred in a species of one of the northern hemisphere members of the order that resulted in the origin of the family Salmonidae. Possibly this was a tetraploid event (a doubling of chromosome numbers), since all species in the family Salmonidae have about twice the amount of DNA in their chromosomes as do species in other families of the order Salmoniformes. Other evolutionary divergences followed in short order, resulting in three subfamilies: Coregoninae (the whitefishes),

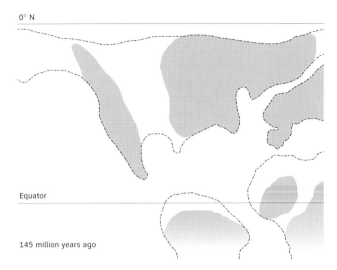

0° N

Equator

145 million years ago

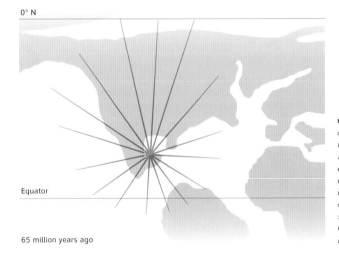

0° N

Equator

65 million years ago

2. This brief description and the maps shown in Figures 2-5a and b were derived from the books, *The Fossil Record and Evolution* (Scientific American, 1982), *Historical Geology* (Foster, 1991), and *The Eternal Frontier: an Ecological History of North America and its People* (Flannery, 2001).

FIGURES 2-2. *North American continent as it appeared* (TOP) *during the Cretaceous Period approximately 145 million years ago and* (BOTTOM) *at the Cretaceous-Tertiary boundary 65 million years ago. The dashed black lines in the top map are areas of continents subsided beneath shallow seas. The area devastated by the Chicxulub impact is shown in the bottom map.*

Thymallinae (the graylings), and Salmoninae (the trouts, salmon, and charrs). [3]

Based on the fact that surviving members of these subfamilies are coldwater fishes, it is likely that these divergences occurred in the northernmost latitudes of the northern hemisphere. Earth was in greenhouse mode during the Age of Dinosaurs. There was no polar ice, and even at 85 degrees north latitude the climate was mild, the mean annual air temperature being about what the Pacific Northwest states experience today.

But whether prior to, during, or after these subfamily divergences, the family Salmonidae had to survive some cataclysmic moments along the way. The first of these occurred 65 million years ago (slightly less than two years ago on our speeded-up time scale), when something happened to this planet that brought the Age of Dinosaurs to an abrupt and jarring end. In not much more than a snap of the fingers in geological time, the great dinosaurs vanished from the planet along with their cousins, the pterosaurs, plesiosaurs and mosasaurs. The entire lineage comprising the ammonites also went suddenly extinct. Other lineages were hard-hit as well. For example, birds, marsupials, plankton, and plants each suffered as much as 75 percent extinction at the family level and somewhere around 75 to 80 percent extinction at the species level.

The bony fishes fared better. Only 10 percent of bony fish families went extinct during the cataclysm.

When I wrote this Chapter for the first edition back in the early 1980s, it was not clear what had happened. We know better now, although the details remain fuzzy. The evidence points to a catastrophic impact with an extraterrestrial object. Fallout from the debris cloud that circled the globe following that impact dusted the surface with a layer of material enriched in the element iridium, much more abundant in extraterrestrial objects than in Earth's crust, and this, in 65 million years' time, would provide the telltale clue to the killer-object's identity. [4]

It now appears that Earth suffered not one but a series of extraterrestrial impacts plus at least one immense outburst of volcanism at the end of the Age of Dinosaurs. The sequence begins 65.3 million years ago when an asteroid, a great chunk of rock measuring more than 6 miles in diameter, struck the planet where the fishing village of Chicxulub Puerto now stands on the northern tip of Mexico's Yucatan Peninsula (Fig. 2-2). Only then, Ground Zero lay just offshore of the land mass under about 30 feet of water, right at the entrance to what remained of that shallow sea that had earlier divided the two North American "islands." At about this same time in what is now western India, huge volcanic vents opened up to unleash havoc of

3. Another thing that can be done with the molecular genetic techniques discussed in Chapter 1 is to use DNA sequence differences as a molecular clock to estimate how long ago divergences occurred. This approach is fraught with assumptions and sources of error, so must be used with great caution. Even so, it has been applied to the question of when the family Salmonidae arose (and by inference, the approximate age of divergence of the subfamilies Coregoninae, Thymallinae, and Salmoninae), to the timing of the split between *Oncorhynchus* and *Salmo* which is discussed later in this Chapter, and to the timing and sequence of divergence of the various species of the genus *Oncorhynchus*. Some papers to read to learn more about this subject are: Lim et al. (1975), Lim and Bailey (1977), Allendorf and Thorgaard (1984), Shedlock et al. (1992), Devlin (1993), Murata et al. (1993), McKay et al. (1996), Bermingham et al. (1997), and Arbogast et al. (2002). An excellent short discussion of the origin and speciation of Pacific salmon and trout can be found in McPhail (1997). Other authoritative papers on this subject are Stearley (1992), Stearley and Smith (1993), and Montgomery (2000).

4. The father-son team of Luis and Walter Alvarez and their co-workers at the University of California Berkeley

their own, pouring incomprehensibly massive amounts of gases and dust into the atmosphere and burying the land in basaltic lava to form the giant plateau-like steps known as Deccan Traps.

For several years, scientists leaned toward these two events as having ended the Age of Dinosaurs, with most of the focus on the Chicxulub impact. The Chicxulub object certainly packed enough punch to produce a major cataclysm. It came in from a little east of south at an angle of 20 to 30 degrees from the horizontal and a speed possibly exceeding 56,000 mph. It is estimated to have released an amount of energy equivalent to 100 million megatons of TNT upon impact. Approximately 200,000 cubic kilometers of Earth's crust—that's more than 7 *trillion* cubic yards of material!—was instantly vaporized, melted, or ejected from the crater. A super-hot cloud of vapor and debris spread out from the blast zone, directing itself predominately to the northwest, enveloping most if not all of the North American continent. It moved at incredible speed, more than 13,000 mph, annihilating and blowing away everything in its path and essentially carbonizing the surface of North America. Divots of material rained down, particularly along the southern fringes of the continent. Within a few hours, immense tsunamis rolled onto the land. Great wildfires following the blast may have spread around the globe. Among the longer-range effects, immense clouds of gas, smoke, and debris enveloped the planet, blotting out the sun, chilling the air, and setting off episodes of acid rain—an "impact winter" scenario on a mega-scale.

But then, in 2004, Dr. Gerta Keller of Princeton University and several colleagues demonstrated, by precise dating of layers of core taken from within the crater, that the Chicxulub impact occurred 300,000 years too early to be the actual "Crater of Doom." The Chicxulub impact and Deccan Trap volcanism probably did spell the beginning of the end for Earth's Cretaceous flora and fauna, but it took yet another impact, at 65.0 million years ago, to complete the mass extinction. So, the search goes on for the new Ground Zero.[5]

Earth's environment gradually healed itself of the effects produced by the events that ended the Age of Dinosaurs. Over the next 10 million years, known as the Paleocene Epoch of geological time, plants came back and small mammals occupied and thrived on the land masses. Mammals had existed during the Age of Dinosaurs, but played an underdog role. According to vertebrate paleontologist Tim Flannery, none had been able to evolve into anything larger than a domestic cat. But small, burrowing forms of mammals survived the blast even on the North American land mass, and now, along with creatures of aquatic environments (amphibians, turtles and terrapins, and crocodiles), they filled

first proposed that the impact of an extraterrestrial object caused the catastrophic mass extinction based on their discovery of the iridium-rich layer at the Cretaceous/Tertiary boundary. Their original paper is cited in Alvarez et al. (1980). Luis Alvarez died in 1988, but his son, Walter, published a history and explanation of their work in a 1997 book titled *T Rex and the Crater of Doom* (Alvarez 1997). For information about the extent of extinction of the various taxa, I relied on Jablonski (1997), Courtillot (1999), and Ravilious (2002).

5. Papers focusing on the Chicxulub impact and details of the effects it produced on North America and the planet include Tschudy et al. (1984), Wolbach et al. (1985), Wolfe (1987, 1991), Crutzen (1987), Bourgeois et al. (1988), Melosh et al. (1990), Sheehan et al. (1991), Sigurdsson et al. (1992), D'Hondt et al. (1994), Kyte (1998), Flannery (2001), and Vajda et al. (2001). Evidence that the Chicxulub impact predates the Cretaceous/Tertiary mass extinction is in Keller et al. (2004).

the vacant niches. However, the mammals were archaic forms that have no direct descendants today. Modern forms of mammals, ancestors of the kinds we do recognize today, came into being during this epoch in an Asian enclave, but were blocked from migrating to North America, evidently by cool climate barriers that developed in the far-northern latitudes where the land bridge then lay.

Then, 55 million years ago at the close of the Paleocene Epoch, a second calamity occurred in the form of a sudden and extraordinary burst of global warming. The driver for this transition may have been a colossal "methane burp" on the sea floor that released as much as 15 trillion tons of ice and methane, the latter being oxidized to CO_2 once it reached the atmosphere.[6] The planet got so hot that seawater temperatures reached the 82-degree Fahrenheit (28 degrees Celsius) range in the North Sea, and 57 degrees Fahrenheit (14 degrees Celsius) at the poles. Deep-ocean waters of the South Atlantic reached 54 degrees Fahrenheit (12 degrees Celsius). Either the "methane burp" itself or the heating that followed is thought to have produced another massive oceanic extinction, this time among the deep-sea plankton. On the positive side, the warming is credited with allowing an array of modern mammals to cross polar lands and achieve dominance in North America.

With regard to the family Salmonidae, at least a

6. Buried in seabed sediments on continental slopes are enormous quantities of methane hydrate, an ice-like solid composed of methane and water. This is quite stable at the high pressures and low temperatures that normally prevail in the seabed environment. However, if deep-sea temperatures rise above a certain threshold, this hydrate can be released in a cascade of bursts, each one causing more warming that triggers further releases. For whatever reason, perhaps a violent outburst of volcanic activity on the sea bed at the end of the Paleocene Epoch (a rift in the sea floor between Greenland and Norway was occurring at about that time), deep-ocean and high-latitude ocean surface temperatures suddenly shot up by 6 to 14 degrees Fahrenheit (4 to 8 degrees Celsius). This event, called the Late Paleocene Thermal Maximum (LPTM) by James Zachos and colleagues of the University of California Santa Cruz (see Zachos et al. 1993), caused the dissociation and rapid release of virtually all of the seabed's stored methane. Evidence for this "methane burp" comes from studies reported by Dickens (1999), Katz et al. (1999), Norris and Rohl (1999), and Thomas (2002). Transitions produced in the plant and animal life of North America are described in Janis (1993), Flannery (2001), and Bowen et al. (2002). A couple of recent reports (Kennett et al. 2000 and Hinrichs et al. 2003) implicate smaller-scale seabed methane releases in the warming periods that occurred between major glacier advances during the Pleistocene ice age.

few members of the family obviously survived these catastrophes, because about a third of the way into the Eocene Epoch, after Earth had cooled down a bit from its unprecedented hot spell, the first member of the family to leave a fossil record of its existence appeared. This was *Eosalmo driftwoodensis*, whose fossils were first discovered in 45- to 50-million-year-old lakebed sediments around Driftwood Creek, Skeena River drainage, near Smithers, B.C. Additional *Eosalmo driftwoodensis* discoveries followed quickly, in other 45- to 50-million-year-old lakebed sites ranging southward to the town of Republic in north-central Washington.

Eosalmo driftwoodensis (Fig. 2-3) was a stout-bodied, blunt-headed fish that dwelled in freshwater lakes where it attained sizes ranging from 6 to about 16 inches in length. Like modern trout, it may have spawned in tributary streams and its young probably reared in those streams before dropping down to the lakes to feed and grow. *Eosalmo driftwoodensis* did have some trout-like bony features while other features were more like those of the subfamily Thymallinae. Overall, however, its bony structures were primitive, so modern scientists consider it to be an archaic member of the family Salmonidae. [7]

After *Eosalmo driftwoodensis*, there's a gap of about 25 million years before a member of the family Salmonidae appears again in the fossil record. We have to fast-forward through the remainder of the Eocene Epoch, all of the Oligocene Epoch, and about the first

FIGURE 2-3. *Fossil of* Eosalmo driftwoodensis *found in 45- to 50-million-year-old lakebed sediments at Driftwood Creek near Smithers, British Columbia. This is specimen UALVP 12326.*
PHOTO COURTESY OF DR. MARK V.H. WILSON, DEPT. OF BIOLOGICAL SCIENCES, UNIVERSITY OF ALBERTA

quarter of the Miocene Epoch before the record picks up again about 20 million years ago. [8] Much of significance happened during this interval. For one thing, the land bridge across Greenland that joined North America to the European portion of Laurasia was severed. This happened 45 million years ago, about the time of the youngest age assigned to *Eosalmo driftwoodensis*. Earth continued to cool throughout the first half of the interval, including one particularly abrupt and severe, albeit short-lived, cooling spike about 34 million years ago

7. *Eosalmo driftwoodensis* fossils and the lake environments in which this archaic trout lived are described in a series of reports by Mark V.H. Wilson of the University of Alberta (see Wilson 1977, 1980, 1996; and Wilson and Williams 1982). *Eosalmo*'s place in the evolution of modern trout and salmon is discussed in the Wilson papers, and also in papers by Behnke (1990), Stearley (1992), and Stearley and Smith (1993).

8. These statements may not be strictly true. In 1986, a report came out of the old USSR describing at least one Eocene-age fossil (perhaps another species of *Eosalmo*) and still other salmonid fossils purportedly dating to the Oligocene Epoch, all found in Asia. This report does not appear to have received much attention from scientists, perhaps because it is a difficult one to find. I located only one reference to it in the literature, but was not successful in tracking down the report itself. I can tell you from the citation that I did see that it's a long report and may be written in Russian, although the title was given in English. For those who would like to try finding it on your own, see Sytchevskaya (1986) for the complete citation.

that may have ended the Eocene Epoch and ushered in the Oligocene. Earth scientists think this was triggered by a major shift in southern ocean currents and worldwide climate pattern brought on by movement of the continent of Australia, although the planet also suffered a three-punch combination of asteroid impacts at about the same time. [9] Over the interval, mountain ranges continued to build in western North America, seasonality developed, and ice sheets formed in Antarctica.

Global warming struck again in the late Oligocene, about 26 million years ago, and, although the peak itself was short-lived, the first half of the Miocene Epoch that followed was generally warm. However, unlike previous global warming spells, which were tropical and humid, this one was dry. Much of North America became a wooded savanna. Indeed, paleontologist Tim Flannery likened it to Africa's Serengeti. Mid-Miocene time saw the opening of great fissures along what would now be the Oregon-Idaho and Washington-Idaho borders. About 13 to 16 million years ago, immense quantities of basaltic lava pumped out across the land in the greatest outpouring of such lava ever to occur in North America. Some of these flows even reached the ocean. Eventually, more than 50,000 square miles of land were covered to depths up to a mile, creating what we now call the Columbia Plateau and forcing the ancient Columbia River northward into the course the modern river now follows to the sea.

Another phenomenon that appeared at about the same time and in the same general vicinity as the fissures that generated these lava flows was a hotspot in Earth's mantle beneath the continental plate. The land above the hotspot bulged upward, forming a broad, high plateau marked by geothermal activity and the occasional period of violent volcanic eruption. This bulge has appeared to migrate northeastward over time (at just under two-tenths of an inch per year), although it is really the continental plate sliding slowly to the southwest over the hotspot's fixed position that accounts for this migration. Caldera and other evidence of volcanism left behind from its eruptive periods serve to pinpoint the bulge's track across the landscape, from its supposed origin on the Oregon-Nevada border around 16 million years ago (the McDermott volcanic field) to the Owyhee-Humboldt volcanic field in the region where the borders of Oregon, Idaho, and Nevada meet (active from 13.9 to 12.8 million years ago), and so on through the Bruneau-Jarbidge (11.5 to 10 million years ago), the Twin Falls (10 to 8.6 million years ago), the Picabo (10.3 million years ago), and the Heise (6.5 to 4.3 million years ago) volcanic fields to, finally, its present location beneath the Yellowstone Park region, where it started lifting up the land about 2 million years ago. The last major eruption occurred there about 600,000 years ago.

Throughout much if not all of this interval, the high plateau above the hotspot's position also marked the

9. Impact craters dating to about 35.5 million years ago have been found at Chesapeake Bay, off the New Jersey coast, and at Popigai, Siberia. These three objects may not have hit all at once, but on the geological time scale they were pretty close, maybe within about 100,000 years of one another. The Chesapeake Bay object is thought to have been the first to strike, and left a crater 53 miles in diameter. The much smaller object that landed off the New Jersey coast left a crater only 9 miles in diameter. The Popigai object is thought to have struck last, and left a crater 62 miles in diameter. So far, these impacts have not aroused anywhere near the level of scientific interest as the Chicxulub impact. To learn what is known about them, see Poag et al. (1994) and Poag (1995) for the Chesapeake Bay and New Jersey coast impacts, and Bottomley et al. (1997) for the Popigai impact.

location of the continental divide on the landscape, just as the Yellowstone Plateau defines its location today. Thus, as the bulge has moved inexorably northeastward, so too has the continental divide. And after the bulge had passed by, the land behind it subsided, forming the broad, flat plain we know today as the Snake River Plain.

Both of these changes—the eastward progression of the continental divide and subsidence of the land in the hotspot's wake—had profound effects on the region's drainage patterns. Drainages around the bulge itself would have radiated away from its highlands in all directions. But the north- and south-flowing streams left behind with its passage reversed their direction once the land subsided, flowing back into the plain to contribute their waters initially to a lake that formed in a closed basin northwest of the Bruneau-Jarbidge volcanic field and then to a westward drainage pattern (a proto-Snake River drainage) whose headwaters extended ever northeastward with the migrating bulge. Drainage reversal did lag the progression of the hotspot track by a few million years however; the Owyhee River and its headwaters in north-central Nevada didn't become a northward draining system until about 7 million years ago, for example, and the Wood River and Salmon Falls Creek captures didn't occur until 4 million and 3 million years ago respectively, both in Pliocene time. The streams of southeastern Idaho, including the South Fork of the Snake, a stream that some believe flowed off to the south and then to the east during Miocene time, were not captured until about 2.5 million years ago, well after the hotspot had passed beyond the Heise volcanic field.[10] Climate-wise, global cooling commenced again over the last half of the Miocene Epoch. The Antarctic ice sheet became a permanent feature during this time, and ephemeral ice sheets also developed in the Arctic region.[11]

It's too bad there is such a wide gap in the fossil record for Salmonidae, because the evidence we do have indicates that the roughly 25 million years between *Eosalmo driftwoodensis* and the end of the Miocene was a vital interval in the family history. For example, that molecular clock methodology I told you about in footnote 3 of this Chapter points to a time about 20 million years ago (early Miocene) for the emergence of *Salmo* and then the sister genera *Oncorhynchus* and *Salvelinus*. Fish paleontologist Ralph Stearley opined that these splits probably occurred in Arctic waters. Cold water conditions that developed in the Arctic region during the Miocene would have provided the isolating mechanism that enabled the genus *Salmo* to emerge in western Europe and eastern North America and the genus *Oncorhynchus* to emerge in western Asia, Beringia, and western North America.

Other genera had emerged within the subfamily Salmoninae as well by this 20-million-year juncture. In fact, it was a member of one of these, the genus *Hucho*,

10. There is much more on the track of the Yellowstone hotspot in the following references: Rodgers et al. (1990), Pierce and Morgan (1992), Fritz and Sears (1993), Smith and Braille (1994), Smith and Siegel (2000), and Beranek et al. (2006). Some of these references also have good discussions of drainage reversals and stream captures in the hotspot's wake, but for more on this you should also consult Sadler and Link (1996), Link and Fanning (1999), Pierce and Morgan (1999), and Link et al. (2002).

11. For overviews of Earth's climates and their implications for global ecology over geological time, see Flannery (2001) and Zachos et al. (2001).

that left the first salmonid fossil to be found in North America after *Eosalmo driftwoodensis*—a single well-preserved specimen measuring 26.3 inches (668 mm) in total length found in 20-million-year-old lakebed sediments near Clarkia in the St. Maries River valley of the northern Idaho panhandle. [12] The significance of this find is that *Hucho* has advanced bony structures. It is not an archaic form, like *Eosalmo driftwoodensis*, but one of the modern members of the subfamily Salmoninae along with *Salvelinus*, *Oncorhynchus*, and *Salmo*. It's a good sign that these other genera had emerged by then or were in the process of emerging as well.

As we come forward in time and approach the end of the Miocene Epoch, the fossil record from Idaho, Oregon, and the western Great Basin area becomes richer and richer in members of the subfamily Salmoninae. *Salvelinus* appears at about the 7 million year point in lake sediments of the Chalk Hills Formation, the site of that ancient lake, known as Lake Idaho, that had formed in an apparently closed basin where the Snake River Plain meets the border of southwestern Idaho and eastern Oregon. But by then the genus *Oncorhynchus* was also prominent in the region. The original fish fauna of Lake Idaho had been warmwater species, reflecting the warm conditions that prevailed when the lake was formed, but cooling conditions later in the Miocene enabled the salmonid fishes to migrate

down from the north. Those same Chalk Hills sediments where *Salvelinus* was found also yielded fossils of *Oncorhynchus salax*, whose characters were much like today's sockeye salmon. Another find from other deposits of that period was *O. keta*, our present-day chum salmon. There is even a late-Miocene record of *Oncorhynchus* from marine sediments of what is now southern California. Other Miocene-age members of the subfamily Salmoninae, all now classified as *Oncorhynchus* although they were given other names when first discovered and described, include (using their original names) *Smilodonichthys rastrosus*, *Rhabdofario lacustris*, and *Salmo cyniclope*. [13]

Some of these extinct fishes were ocean migrants that, if they were alive today, we would have no qualms about fishing for as Pacific salmon—which brings us to as good a place as any to consider how, why, and when the sea-run life-history might have evolved. First some definitions:

- Diadromous fishes migrate between the sea and fresh water.
- Potamodromous fishes migrate wholly within fresh water.

Diadromous behavior can be broken down into these additional categories:

- Anadromous fishes spend much of their lives at sea but migrate to fresh water to spawn. Chinook, coho, sockeye, chum, pink, and masu salmon are anadromous. So are Atlantic salmon. The steelhead is an anadromous life-history form of rainbow trout.

12. The Clarkia fossil beds received worldwide publicity in the 1985–1990 period, not so much for fossil fishes as for the splendid array of plant fossils that were unearthed there. From a couple of these, one a particularly well-preserved *Magnolia* leaf and the other a specimen of bald cypress, research groups claimed to have extracted and sequenced large pieces of ancient plant genes. At 20 million years of age, these ranked as the oldest pieces of DNA to be recovered up to that time. These claims were not without controversy, however, as other scientists challenged their validity. To read more about these claims, as well as some of the pitfalls of working with ancient specimens, see Giannasi (1990), Golenberg et al. (1990), Golenberg (1991), Sidow et al. (1991), Soltis et al. (1992), and Logan et al. (1993). To learn more about the fossil *Hucho* and the other fossil fishes found at the Clarkia site, see Smiley and Rember (1985) and Smith and Miller (1985). Presently, *Hucho* is found only in Eurasia and has a southern distribution, suggesting a tolerance for warmer waters than the other members of the subfamily Salmoninae can put up with. This is consistent with what we know about the Clarkia site 20 million years ago, which one of the botanists working there likened to present-day southern Appalachia.

13. Reports of the fossil fishes of the subfamily Salmoninae are scattered about in the scientific literature of several fields,

- Catadromous fishes are the opposite, spending much of their lives in fresh water but migrating to sea to spawn. The American eel is catadromous.
- Amphidromous fishes migrate between salt and fresh water, not just for spawning, but also during other phases of their life history. Some populations of coastal cutthroat trout (Chapter 3) exhibit amphidromous behavior.

Available information favors the hypothesis that the original ancestor of the family Salmonidae—prior to divergence of the three subfamilies—occupied freshwater habitat. The fact that all living species of Salmonidae reproduce in freshwater argues for this view. However, the present distribution of the Salmonidae shows a clear poleward increase in the proportion of species that are diadromous. Even species that are variably diadromous exhibit a poleward increase in the incidence of sea-run populations. In 1988, Mart Gross of the University of Toronto and his colleagues called attention to a differential in productivity of the world's oceans, noting that oceans are typically more productive than freshwaters in temperate and polar latitudes, while the reverse is true in tropical latitudes. Cooling and the development of seasonality in the mid-Cenozoic Era undoubtedly brought on a poleward differential in ocean vs. freshwater productivity much as we see today. Gross and his colleagues proposed that diadromy evolved to track this differential in productivity. Diadromy has proba-bly evolved many times as global climates have shifted over geological time. [14]

Other fossil salmonids of late Miocene time were freshwater-resident forms. Take *Rhabdofario* (now *Oncorhynchus*) *lacustris*, for example, whose fossils have been found in the Snake River Plain and north-western Nevada. Dr. Gerald R. Smith, Museum of Zoology and Museum of Paleontology, University of Michigan, is one of the foremost authorities in the field of fossil fishes and biogeography. Dr. Smith and his colleagues have written that these freshwater forms show a composite of cutthroat and rainbow trout bony char-acteristics, which may denote either the ancestor prior to the rainbow-cutthroat divergence or the ancestral redband trout. [15] Either way, if *Rhabdofario lacustris* were alive today, we would not hesitate to entice it with our dry flies, nymphs, or streamers.

The Miocene Epoch ended approximately 5 million years ago with accelerated cooling coupled with the onset of an extreme, global-scale drought, attributed mostly to the rapid growth of the south-polar ice cap (ice sheets formed in the north polar region as well), which sucked water from Earth's oceans and land masses and may have contributed to a phenomenon known as the Messinian salinity crisis, wherein the entire Mediter-ranean Basin dried up, altering global ocean salinity conditions and intensifying global cooling and drought

including paleontology, ichthyol-ogy, geology, and geography. Here are some references that provide summaries of the various fossil finds from which you can trace specific leads: Smith (1981, 1992), Bon-nichsen and Breckenridge (1982), Cavender (1986), Hocutt and Wiley (1986), Stearley (1992), Stearley and Smith (1993), and Smith et al. (2002). For that report of the southern California marine fossils, see Barnes et al. (1985).

14. The definitions I posted in this paragraph are from Myers (1949), Gross (1987), and McDowall (1987). For additional thoughts on the evolution of diadromy, see Gross et al. (1988), McDowall (1988), and Stearley (1992).

15. A key reference to Dr. Smith's work is a paper he produced with five colleagues in 2002 titled "Biogeog-raphy and Timing of Evolutionary Events among Great Basin Fishes." See Smith et al. (2002) for the full citation. Other, even earlier papers that also pointed out how similar these late Miocene trouts were in bony structure to modern cutthroat and redband trout include Smith (1981) and Taylor and Smith (1981). The late Miocene produced still another interesting trout, *Salmo* (now *Oncorhynchus*) *esmeralda*,

conditions.[16] One casualty of this drought was that Miocene-age Lake Idaho where the early *Salvelinus* as well as *Oncorhynchus salax* had swum. Going extinct as well were many of the Salmonine fish species mentioned in the paragraphs above, although some, such as *Rhabdofario lacustris*, did persist into the Pliocene Epoch. Most species of large mammals that inhabited North America at the beginning of the Pliocene Epoch also went extinct, but their places were taken by new species migrating across Beringia in another faunal interchange. Joining them were other mammal species coming up from South America, which by this point in time had joined to North America via the Isthmus of Panama. Other major geological changes that occurred in North America during the Pliocene Epoch included the uplift of the Colorado Plateau and the rise of the modern Cascade Range of mountains.

Rains came again to North America over about the last million or so years of the Pliocene Epoch, adding to the snows and ice at higher elevations and northern latitudes, and no doubt aiding in the refilling of Lake Idaho in that old lakebed at the western edge of the Snake River Plain. The history of this particular stand of Lake Idaho may have extended to the beginning of the Pleistocene Epoch. I have grown increasingly certain that Lake Idaho's existence at this juncture of geological time—and perhaps of even greater importance, the associated story of the ancient Snake River's course to the sea at the time—plays an important role in the story of the emergence and early distribution of the cutthroat trout.

EMERGENCE AND EXODUS OF THE CUTTHROAT TROUT

When I completed the first edition of this book back in the early 1980s, scientists were offering only one explanation for the origin and radiation of our contemporary native western trouts (i.e., those that were in place when people of largely European extraction first ventured into western North America). Based on observations that go back more than 100 years, this explanation focused on an emergence of cutthroat trout (and rainbow trout as well) in late Pliocene or early Pleistocene time in waters draining to the Pacific Ocean, and their inland radiation by way of the Columbia River and its main tributary, the Snake. This was still the established explanation a decade later when Dr. Behnke, whom we met in Chapter 1, published his monograph, *Native Trout of Western North America* in 1992, and a decade more after that, in 2002, when his book, *Trout and Salmon of North America* appeared. Dr. Behnke's monograph and book refined and elaborated on this long-accepted explanation. But that very same year, 2002, saw the publication of a paper by

described in 1966 by Ira LaRivers of the Biological Society of Nevada, from a fossil fragment found in lakebed deposits of the Esmeralda Formation, western Nevada (see LaRivers 1966). What made this specimen so remarkable was its size. From the length of its fossilized mandible, which was the only piece found, LaRivers estimated a standard length for the intact fish of 36 inches. If you add in the length of the tail, that trout was well over 40 inches in total length! LaRivers originally assigned a Pliocene age to the lakebed sediments where the fossil was discovered, but they have since been dated at 5 to 17 million years old, which makes *O. esmeralda* a Miocene-age trout.

16. The Messinian Salinity Crisis began about 6 million years ago and persisted until about 5 million years ago, at the end of the Miocene Epoch. You can read more about it in Hsü et al. (1973), Adams et al. (1977), Prothero (1998), and Krijgsman et al. (1999).

Dr. Smith and his associates, "Biogeography and Timing of Evolutionary Events among Great Basin Fishes," that offered up a new explanation, one that placed the emergence of cutthroat and rainbow trout and their divergence from one another much farther back in time, to a time roughly 8 million years ago in the late Miocene, and a radiation outward of the cutthroat trouts from an interior source in what is now the Bonneville Basin. Let's examine both explanations.

Although opinions are not unanimous, some scientists believe that the earliest of the contemporary western trouts to appear was *O. chrysogaster*, the Mexican golden trout. This beautiful little trout is presently found only in high-elevation headwater tributaries of the Rio Fuerte, Rio Sinaloa, and Rio Culiacán drainages of Mexico, which flow off heights in excess of 10,900 ft and enter the Sea of Cortez (aka Gulf of California) between 25 degrees and 28 degrees north latitude. These scientists also believe that the divergence of the cutthroat lineage and then the rainbow trouts followed closely thereafter. On the other hand, Dr. Behnke has written that *O. chrysogaster* has its closest relationship to the rainbow trout and represents the most primitive form of the rainbow line. Dr. Behnke, in keeping with the established explanation, believes that even as the rainbow trout lineage was emerging in those southern waters, the cutthroat trout lineage

was also emerging, or perhaps had done so already, in waters farther north, in the vicinity of the Columbia River at around 40 to 50 degrees north latitude. One thing is certain though: climate conditions at the time these divergences occurred were cold enough that trout could indeed exist in Sea of Cortez drainages, and even in waters further south. The fossil of another trout, *Salmo* (now *Oncorhynchus*) *australis*, was found in Mexico's Lake Chapala basin near 20 degrees north latitude. [17]

With regard to the early evolutionary divergence and inland dispersal of the cutthroat trouts, I think it was David Starr Jordan who, in 1894, first suggested that these events were associated with the Columbia River and its largest tributary, the Snake. [18] The Columbia River and its tributaries drain an area of more than a quarter of a million square miles, encompassing parts of Washington, Oregon, Idaho, Montana, Wyoming, Nevada, Utah, and British Columbia, and is second only to the Mississippi River in the volume of water it carries. What's important to the story of the cutthroat trout, as Jordan saw it, is that this river network occupies the only northwest valley system leading to the interior part of the continent—a "Magnificent Gateway," as author John Eliot Allen put it, into the heart of western North America.

Jordan's view received strong support in the 1960s

17. Dr. Behnke's view of the origin and inland radiation of cutthroat trout can be found in his monograph, *Native Trout of Western North America* (Behnke 1992) and also in his latest book, *Trout and Salmon of North America* (Behnke 2002). The discovery of *Salmo* (now *Oncorhynchus*) *australis* is described in Cavender and Miller (1982).

18. See Jordan (1894). Although Jordan believed the Columbia and Snake rivers were key to the spread of the cutthroat trouts inland in North America, he also held the view that the species itself had originated in Asia and migrated to this continent.

and 1970s from work done on the evolution of chromosomes in salmonids, and the cutthroat trouts in particular. To understand how, we need to know a bit more about chromosomes.

One of the functions of the DNA within an organism's cells is to replicate itself when it's time for cell division, or *mitosis* as the biologists call it, to take place. When this process begins, each chromosome organizes itself into a rod-like structure with a constriction, called a *centromere*, located somewhere along its length. If the centromere is located at or very close to one end of the chromosome structure with the rod-like part (called the chromosome "arm") trailing out behind, then we have an *acrocentric* chromosome. Acrocentric chromosomes have one chromosome arm. If the centromere is located away from the end or in the middle of the structure with arms extending out in either direction, we have a *metacentric* chromosome. Metacentric chromosomes have two chromosome arms. So far, so good? I have illustrated these structures in Fig. 2-4. Now then, as the DNA in the chromosome commences to replicate, a sister arm begins to form and gradually separate from the original arm until the two remain attached only at the centromere. This is the so-called *metaphase* stage of cell division. At this stage the chromosome structures have grown large enough to be visible under high magnification in a light micro-

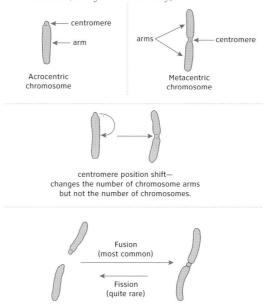

FIGURE 2-4. *Chromosome types and rearrangements during chromosome evolution (colors for illustration only)*

← centromere
← arm

Acrocentric chromosome

arms ← → centromere

Metacentric chromosome

centromere position shift—
changes the number of chromosome arms
but not the number of chromosomes.

Fusion (most common)

Fission (quite rare)

19. The karyotype photo I showed you back in Chapter 1 displays that trout's chromosomes at metaphase. As an exercise, can you count the number of acrocentric and meta-centric chromosomes?

scope, and they can be isolated for study. When thus viewed under a microscope, the acrocentric chromosomes appear as little V-shaped or inverted V-shaped structures while the metacentric chromosomes appear to be little Xes. [19] When the centromere DNA has also completed replicating, the sister chromosomes separate and, as the cell division process proceeds to completion, comprise the new cell's complement of chromosomes.

But, like the organisms themselves, chromosomes

can also evolve, leading to changes in chromosome number in the karyotype and also to changes in the number of chromosome arms. The position of the centromere can sometimes shift within a chromosome, converting what had been an acrocentric chromosome into a metacentric chromosome, or vice versa. When this happens, the number of chromosomes does not change, but the number of chromosome arms either increases or decreases depending on which way the change occurred. Another, much more common type of chromosome change results in a change in the number of chromosomes without changing the number of chromosome arms. This is the *Robertsonian rearrangement*, named for the American insect geneticist, W.R.B. Robertson, who first described this type of transformation in grasshoppers back in 1916.[20] There are two types of Robertsonian rearrangement, one type very common in vertebrate animals, such as mammals and bony fishes, but the other type quite rare. By far and away the most common type is the Robertsonian *fusion*, in which two acrocentric chromosomes fuse at their centromeres to form a single metacentric chromosome. Robertsonian *fission*, in which a metacentric chromosome splits at the centromere into two acrocentric chromosomes is also possible, but appears to have occurred only rarely in the evolution of such vertebrate animals as mammals and bony fishes. I have also dia-

grammed these changes for you in Fig. 2-4.

So what's all this got to do with the origin and dispersal of cutthroat trout? It all goes back to that tetraploid event I mentioned earlier in this Chapter wherein the number of chromosomes in the family Salmonidae was doubled. Because $2N = 48$ acrocentric or mostly acrocentric chromosomes are so common a karyotype in the fishes, particularly among the older taxa, scientists have presumed that this was also the primitive karyotype in the Salmonidae, and the tetraploid event doubled that number to $4N = 96$ acrocentric or mostly acrocentric chromosomes. Members of the family Salmonidae have been reverting to the diploid state ever since, albeit at different rates in the different genera and species. Although a few changes of the type that increases the number of chromosome arms must have also occurred (for example, in *Oncorhynchus* the most common arm number now is 104), most of this reversion has taken place via Robertsonian fusions that have reduced the number of chromosomes over time. Thus, in a series, a larger number of chromosomes is deemed to be the more primitive condition and a smaller number the more recent condition. In the cutthroat trout series, the karyotypes are $2N = 68$, $2N = 66$, and $2N = 64$ chromosomes, all with 104 chromosome arms, most likely arrived at by a nice, tidy series of Robertsonian fusions. Since the presumed most primi-

20. Robertson's original work is cited in Robertson (1916). Other key references on chromosome evolution in fishes and other vertebrates, and particularly in the western North American trouts, include Simon and Dollar (1963), Fredga (1977), Gold et al. (1977), Loudenslager and Thorgaard (1979), Hartley (1987), and Ferguson and Allendorf (1992).

"The Magnificent Gateway."
The Columbia River Gorge photographed from the Oregon side. Was this the portal to the ancient waterway that led to the inland radiation of the cutthroat trout?
PHOTO BY AUTHOR

tive karyotype of $2N = 68$ is that of the coastal cutthroat trout and the derived $2N = 66$ and $2N = 64$ karyotypes belong to the interior subspecies, the chromosome evidence supports Jordan's view of the inland dispersal of cutthroat trout from Pacific coastal waters. [21]

Dr. Behnke's take on this is that an ancestral cutthroat trout with $2N = 68$ chromosomes is the primitive form, represented today by the coastal cutthroat trout. A change to a $2N = 66$ form occurred in the interior Columbia River basin and persisted in the northern part of that basin to become the westslope cutthroat trout. In the southern part of the basin, probably in the Snake River system, the chromosome number changed to $2N = 64$. This 64-chromosome ancestor soon separated, with one branch gaining access to the Lahontan basin and evolving into the Lahontan cutthroat trout and four closely related subspecies, and the other branch remaining in but also radiating out from the upper Snake River to eventually become the Yellowstone cutthroat and six closely related subspecies.

As to when these evolutionary divergences might have occurred, in 1987 Fred Allendorf and his colleagues at the Department of Zoology, University of Montana, used allozyme electrophoresis to estimate the amount of genetic divergence among seven subspecies of cutthroat trout and rainbow trout. They discovered that there is as much genetic distance between the modern coastal cutthroat, westslope cutthroat, Lahontan cutthroat, and a group consisting of the modern Yellowstone, finespotted, Colorado River, and greenback cutthroats (all members of this group clustered essentially as one) as there is between any of these four and modern rainbow trout. In other words, the genetic distance separating these four major cutthroat lines is of the same magnitude as that which typically separates full species. Molecular clock estimates based on the allozyme divergence data of Allendorf et al. gave a range of 1 to 2 million years ago, or anywhere from the beginning of the Pleistocene Epoch to mid-Pleistocene time, for the rainbow-cutthroat divergence, with the divergence of the four cutthroat clades quickly following. More recent estimates based on sequence divergences in nuclear and mitochondrial DNA made by scientists at Simon Fraser University in British Columbia and at Brigham Young University in Utah give older dates for the rainbow-cutthroat divergence ranging from 3.5 to 8.3 million years, which would push this back into Pliocene or even Miocene time. [22]

As a believer in the inland radiation of the cutthroat trouts from an origin in Pacific coastal waters, I had come to favor a late Pliocene time estimate (say somewhere around 3.5 million years ago) for the rainbow-cutthroat divergence, and for the major divergences of the cutthroat lineage. I had a couple

21. In addition to the references I cited in footnote 20 above, you can dig deeper into the tetraploid event and subsequent karyotype evolution of the family Salmonidae in Ohno et al. (1969), Ohno (1970a, 1970b, 1974), Gold et al. (1980), Sola et al. (1981), and Allendorf and Thorgaard (1984).

22. The genetic distance results of Allendorf and his colleagues were published in two papers; see Leary et al. (1987) and Allendorf and Leary (1988). Papers by McKay et al. (1996) and Smith et al. (2002) have the information on the nuclear and mitochondrial DNA results. For more on the relationship between genetic distance and evolutionary time, see the literature on molecular clocks which I referred you to in footnote 3 of this Chapter. Evolutionary geneticist Masatoshi Nei has also written about this subject; see Nei (1987).

23. Evidence that the Snake River was not a Columbia River tributary prior to the close of the Pliocene but reached the Pacific Ocean by an independent route is presented and discussed in a string of papers by Lupher and Warren (1942), Hubbs and Miller (1948), Wheeler and Cook (1954), Cook and Larrison (1954), Taylor (1960, 1985), Miller (1965), Taylor and Smith (1981), Repenning et al. (1995), Link and Fanning (1999), Dowling et al. (2002), Link et al. (2002), Wood and Clemens (2002), Hershler and Liu (2004), and Beranek et al. (2006).

FIGURE 2-5. *Pliocene drainage patterns of Snake River before (hypothesized) and after capture by the Columbia River*

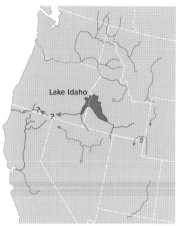  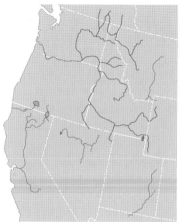

A. *Outlet via Owyhee arm to Klamath or Sacramento drainage*

B. *Route via Humboldt to Klamath or Sacramento drainage*

C. *After capture*

of reasons for this. First, dates any older than late Pliocene would mean that these divergences occurred prior to or even during the period of extreme drought that prevailed at the opening of the Pliocene, which the fossil record indicates was a period of extinction for many of the then-existing western North American *Oncorhynchus* species. I doubt if you would have seen much divergence and inland radiation of aquatic animals then. Second, a late Pliocene date fits nicely with geological and zoogeographical evidence for the connectedness—or lack thereof—of the Columbia and Snake River systems of that period.[23] This evidence indicates that up until just before the close of the Pliocene, the Snake River was not connected to

the Columbia at all, but discharged to the southwest, either across southeastern Oregon as an outlet from the Pliocene stage of Lake Idaho as illustrated in Fig. 2-5a, or south into the Humboldt River system and across northern Nevada to what is now the Modoc County area of California as illustrated in Fig. 2-5b. The recent evidence favors the latter course. From there the route to the sea could have connected with either the Klamath River or the Sacramento River system, but again, the recent evidence favors the latter. The same evidence that points to the Humboldt-Sacramento course also suggests that after the Yellowstone hotspot track had migrated beyond the Picabo and Heisse volcanic fields, sometime between 4.7 and 4.1 million years ago, the river shifted back into the now-subsided Snake River Plain and into Lake Idaho. That pattern lasted until about 2.5 million years ago, when Lake Idaho either spilled northward or a sill was eroded northward through Hell's Canyon and the entire drainage was captured by the Columbia River (Fig. 2-5c). These dates open the possibility that the 2N = 66 cutthroat of the Columbia River system and the 2N = 64 forms of the Snake River and Lahontan basins arose from two separate divergences in (at the time) different drainages.

This scenario also provides a straightforward explanation for what had heretofore been a puzzle. The Salmon and Clearwater rivers are major tributaries of the Snake River and thus, in Dr. Behnke's view of the inland radiation, the 2N = 64 Yellowstone cutthroat should be their native cutthroat. Yet the upper reaches of both basins have the 2N = 66 westslope cutthroat as their indigenous subspecies. The usual explanation offered for this is that Yellowstone cutthroats *were* originally native but were displaced by later-invading rainbow trout; then the systems were re-invaded by westslope cutthroat via headwater transfers from east of the Bitterroot Range. To me, it's tidier to think of the original cutthroat divergences occurring prior to the capture of the Snake, at a time when the Clearwater and Salmon river drainages were upper Columbia tributaries. In this scenario the 2N = 66 ancestor, being the only one present, could access both drainages directly,

no headwater transfers required. This theory was given a boost recently by the discovery of apparently ancient westslope cutthroat mitochondrial DNA in steelhead native to the Tucannon River, southeastern Washington, another tributary of the lower Snake River that lies between the John Day drainage of Oregon (a Columbia River tributary) and the Clearwater River drainage of Idaho, where westslope cutthroat populations still exist today. The scientists who made this discovery, from Washington State University and the Nez Perce Tribe, postulate that a westslope cutthroat population once inhabited the Tucannon River, but was extirpated by introgression with later-invading rainbow trout. [24]

But before we get too far with this, let's consider the alternative explanation of the biogeography of the cutthroat lineage put forth by Dr. Smith and his colleagues. In addition to a close examination of the fossil salmonids from the Great Basin and adjacent Snake River Plain, their evidence includes molecular clock estimates derived from mtDNA sequence divergence data that in turn was inferred from RFLP analysis (to refresh yourselves about what RFLPs are, go back to Chapter 1). [25] Their results point to a much earlier date, about 8 million years ago, for the divergence of the rainbow and cutthroat lines, which would make this a late-Miocene event roughly coincidental with the appearance of *Salvelinus* and *Oncorhynchus salax* in the Miocene stage of Lake Idaho. They took no posi-

24. Reported in Brown et al. (2004). Another puzzler of cutthroat trout distribution occurs in this same general region. Waha Lake, in the lower Clearwater drainage near Lewiston, Idaho, was once occupied by cutthroat trout described as large-spotted trout. Up until recently, it was generally accepted that these trout were Yellowstone cutthroats. If so, and these trout were indeed relics of the original fish fauna of the Clearwater drainage, it argues for the alternative scenario of divergence of the 2N = 64 cutthroat following the Snake-Columbia connection, and for the more complicated explanation of the presence of westslope cutthroat trout in the Clearwater and Salmon river drainages that Dr. Behnke proposed. However, as I will explain in Chapter 9, other elements of the description of these Waha Lake trout have lately raised misgivings about their initial identification and opened the possibility that they might have been westslope cutthroats instead.

25. The key reference for this work is Smith et al. (2002). For their molecular clock estimates, they used a sequence divergence rate of 0.5 percent per million years for the mtDNA of salmonids. See their paper for the full details of their methodology.

tion as to where this cutthroat-rainbow divergence might have taken place, but they did conclude that cutthroat trout have been in the Great Basin far longer than Dr. Behnke has proposed. Their molecular clock data indicate that interbasin differentiation of the cutthroat trouts may go back to the early Pliocene, 4 to 4.5 million years ago, perhaps, and furthermore, may have proceeded from the Bonneville basin outward rather than from the Columbia River inward.

If I'm interpreting their data correctly, the ancestral cutthroat trout gave rise to no existing subspecies until about 4 million years ago when a divergence occurred that led to the cutthroat populations of the Bonneville Basin (excluding populations associated with the Bear River) in one line, and a sister line from which all the other existing cutthroat subspecies would emerge. About 3 million years ago, a pair of divergences occurred in this sister line, one line leading to the Bear River, Snake River, and Yellowstone populations and the other leading to the populations of the Colorado River system. Another divergence from this sister line occurred at 2.8 million years ago that led to the coastal cutthroat subspecies, and at 1.6 million years ago, yet another divergence led to the westslope cutthroat subspecies. The remaining members of the sister line differentiated into the subspecies of the Lahontan Basin of Nevada and California and the Coyote Basin

of southeastern Oregon. By this interpretation, the subspecies of the Lahontan/Coyote basins have been separated from the Bonneville Basin line by about 4 million years, but are a sister group to the westslope cutthroat subspecies from which they diverged only about 1.6 million years ago. The coastal cutthroat represents an intermediate clade in this interpretation, emerging from the sister lineage after the Snake River-plus-Colorado River groups had diverged from the line, but before the separation of the westslope from the Lahontan/Coyote basin group. On a finer scale of differentiation, the molecular clock evidence indicates that the Bear River populations separated from the Snake River and Yellowstone populations about 700,000 years ago, and the Snake and Yellowstone populations about 200,000 years ago. 200,000 years ago is also the divergence time indicated by the data for the differentiation of the Lahontan and Coyote basin subspecies.

When I compare these molecular clock divergence times with the drainage basin histories outlined in the paragraphs above, they match up well enough for radiation outward from the Bonneville and Lahontan basins to be a plausible interpretation for the dispersal and interbasin differentiation of the cutthroat subspecies. But then again, these drainage basin histories also are consistent with inland dispersal from the Columbia River system. More troubling for this new interpreta-

Shoshone Falls on the Snake River, Idaho. The falls are dry in this photo because water was being held in upstream reservoirs for irrigation. This and other major barrier falls isolated ancestral cutthroat from new infusions of fish from below. PHOTO BY AUTHOR

tion is the chromosome evidence. To derive the $2N = 66$ westslope and $2N = 68$ coastal cutthroat karyotypes from a Great Basin lineage where all extant subspecies have $2N = 64$ chromosomes would require a series of Robertsonian fissions to have occurred, which, as I explained earlier, are rare events in the evolution of salmonid chromosomes and therefore are less probable than the far more common Robertsonian fusions in the cutthroat series. Before completely buying into this new interpretation, I would like to see more discussion of this point among scientists to reconcile what the chromosomes and the mtDNA are trying to tell us. Also, in light of the intense drought conditions that are supposed to have prevailed at the end of the Miocene after about 6 million years, a period where the fossil record seems to indicate extinction of coldwater fishes, I think more discussion is in order on the likelihood that an ancestral line of cutthroat trout would have survived in the Great Basin to initiate the radiation indicated by the mtDNA results.

But I digress. Regardless of which interpretation you choose to favor for the dispersal and interbasin differentiation of the cutthroat trout, the fit of the forms we find today into their respective niches in western North America must certainly have been influenced by events of the Pleistocene Epoch of geological time.

The Pleistocene Epoch was a period in which the

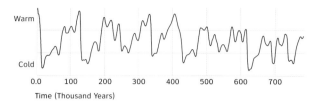

FIGURE 2-6. *Pleistocene climates recorded by heavy oxygen ($\delta^{18}O$) data. Warm periods had enriched ^{18}O levels and cold periods had depleted ^{18}O levels.*

global climate cycled between cold conditions that saw the advance of great continental ice sheets and alpine glaciers, and warmer intervals that saw the ice retreat. Earth experienced numerous cycles of glaciation in the last 2 million years, with as many as nine of these cycles occurring within the last 800,000 years (Fig. 2-6).[26] The immense accumulations of ice during glacial advances drew sea levels down by as much as 400 feet. Yet paradoxically, these were pluvial periods in the interior of western North America with heavy winter precipitation and cool, cloudy summers. Winter snow lines were anywhere from 3,000 to 3,300 feet lower than snow lines are today, and life zones were pushed correspondingly farther south. In the Great Basin and what would now be the southwestern United States, woodlands occurred in places that today support only desert scrub. Valleys filled with lakes, many of them quite large. Some of these lakes drained catastrophically as ice dams broke or the rims of basins gave way, releas-

26. The amount of ^{18}O, the heavy isotope of oxygen, varies in the air and in precipitation depending on surface air temperature, making anything that will preserve an oxygen isotope record, such as the deep-sea sediment record shown in the figure, an excellent surrogate for past surface temperatures. A continuous climate record from mid-Pleistocene time to the present is now available from ^{18}O data collected from deep-sea sediment cores (Imbrie et al. 1984), from Antarctic ice cores (Petit et al. 1999), and from cores from a calcite vein in Nevada's Great Basin (Winograd et al. 1992). The nine major glacial advances shown in this record are five more than signs left behind in the rocks had revealed to earlier geologists. Old names for major glacial advances, e.g., the Nebraskan (corresponding to the cold spike at about 750,000 years in the figure), the Kansan (corresponding to the 450,000-year spike), the Illinoian (corresponding to the 150,000-year spike) and the Wisconsin (corresponding to the most recent spike) while not exactly passé, are, well, archaic, and have given way in many scientific publications to numbered stages of the oxygen isotope record. Scientist who study the details of these ^{18}O climate records point out that the ice advances have always

ing their waters in some of the most devastating floods the world has ever known. Others existed for thousands of years, desiccating only gradually over time. Rivers flooded and changed course, often altering entire drainage patterns. Flooding and runoff were exacerbated at the onset of glacial retreat, but during the interglacial periods themselves, the climate and the forests were probably not greatly different from what they are today. One pair of authors has even written that we live in Pleistocene scenery, minus the glaciers. [27]

While the stream flows were heavy and lake levels were high there were many interconnections between drainages. Some streams were undoubtedly captured, becoming segments of other drainages. Others divided their flow into separate drainages. And some headwater creeks migrated back and forth over the alluvial cones that formed across low divides, now emptying into the drainage on one side, now emptying into the drainage on the other. It was these watery highways that enabled ancestral cutthroat trout to radiate into the heart of western North America to distribute into the areas which comprise the species' historical range. It's possible that at least some of this radiation may have been completed in early to mid-Pleistocene time. The fossil evidence indicates that cutthroat trout were present as far inland and as far south as the Rio Grande drainage by 740,000 years ago. [28]

ended abruptly, with the ice sheets melting away very quickly into warm interglacial periods, which then gradually cool back down as the ice slowly builds for the next advance. The cycle of major ice sheet advances has been about one every 100,000 years over the period of record with lesser cycles of about 41,000 and 21,000 years embedded within the major cycle. What is responsible for these cycles? The scientists say it's changes in summer *insolation*, which is defined as the amount of solar radiation received at the top of the atmosphere as measured at a particular latitude (65 degrees north latitude is a common standard). This varies as Earth's orbit around the sun transitions from more-circular to more-elliptical and back again (the 100,000-year cycles) and with the tilt of Earth's axis (the 41,000 year cycles) and its wobble as the planet spins around it (the 21,000 year cycles). To read all about this, see Hays et al. (1976) and Broecker and Denton (1990). The ice cores mentioned above also capture records of greenhouse gas levels in the atmosphere. When these records were examined, it was discovered that atmospheric CO_2 levels have tracked the surface temperature record almost exactly, peaking (at 280–300 ppm) at or near the peak warming points of

interglacials and falling to their lowest levels (about 180 ppm) as the glaciers reached their maximum advance. For the last 400,000 years, which is as far back as the ice core records of CO_2 levels go, the swing in atmospheric CO_2 concentration has been about 100 ppm from low to high for each glacial-interglacial cycle—up until the present one, that is. At the beginning of the present interglacial, atmospheric CO_2 level did climb about 100 ppm, as it should have if it were continuing on the track shown in the record. But then, in the late 1700s, it began a steep, unprecedented rise that has carried it another 100 ppm higher (to its present level of 379.1 ppm) with no sign of leveling off. To me, this is the most compelling evidence for the reality of human-caused global warming. Referring back to footnote 1 of this Chapter, it is this sharp rise in atmospheric CO_2 levels that prompted Paul Crutzen to propose that we have entered a new Epoch of geological time, one he calls the Anthropocene Epoch to denote human dominance of the global environment.

27. Much has been written about conditions in North America during the Pleistocene Epoch. Sources I used for the material summarized in this paragraph included Macklin and Cary (1965), Wright and Frey (1965), McPhail and Lindsey (1970), Wright (1983), Richmond and Fullerton (1985), Hocutt and Wiley (1986), Broecker and Denton (1990), Grayson (1993), Whitlock and Bartlein (1997), Rohling et al. (1998), and Flannery (2001). These will lead you to many additional sources.

28. Karel L. Rogers of Adams State College, Alamosa, Colorado, and a group of co-workers from other institutions involved in studies of ancient climates of the San Luis Valley, Rio Grande drainage, Colorado, recovered fossil fishes identified as cutthroat trout from 740,000-year-old strata. If you match this date against Fig. 2-6, you see it was near the end of a major glacial advance. At the time, according to Rogers and co-workers, the San Luis Valley floor supported a montane forest and deep, permanent aquatic habitats. For details, see Rogers et al. (1985, 1992).

The 2N =68 cutthroat dispersed along the Pacific northwest coast but gave rise to only one subspecies, that being the coastal cutthroat trout, which will be the subject of the next Chapter. However, dispersal northward beyond, say, the Chehalis River in western Washington would have had to occur since the end of the last glacial period, since most all of the spawning and rearing streams north of that point were covered by Cordilleran ice save for three possible ice-free refugia, one that may have existed in the Brooks Peninsula area on the northwestern coast of Vancouver Island, a second (which may have been contiguous with the Brooks Peninsula refugium) on an area of exposed continental shelf between the northern tip of Vancouver Island and the mouth of the Bella Coola River, and a third in the Queen Charlotte Islands and adjacent area of exposed continental shelf in Hecate Strait.[29] The southern limit of this subspecies' range also may have been displaced northward in late-Pleistocene time by the rainbow trout lineage, which was expanding its range northward during this period.[30]

The 2N =66 cutthroat likewise did not beget any surviving subspecies other than the westslope cutthroat trout (Chapter 4). And likewise as well, the dispersal of this subspecies into its historical range may reflect conditions imposed by the most recent glacial and post-glacial periods. If you will flip forward a few pages to

29. There is evidence from several studies that parts of the British Columbia coast and areas of its continental shelf were exposed and remained ice-free during the last glacial maximum and may have provided refugia for fish and other fauna—and, according to one reference, even for human habitation. For information about the Brooks Peninsula refugium, see Pojar (1980), Ogilvie (1989), and Haggerty (1997). Clague et al. (1982), Luternauer (1989), Josenhans et al. (1995), Clague and James (2002), and Hetherington et al. (2004) discuss exposure of British Columbia continental shelf areas during glaciation. And more on the Queen Charlotte Islands-Hecate Strait refugium can be found in Fladmark (1979), Warner et al. (1982), Warner (1984), Barrie et al. (1993), Josenhans et al. (1993), O'Reilly et al. (1993), Deagle et al. (1996), and Byun et al. (1997).

30. Dr. Behnke's view of the radiation of the rainbow trout lineage is best articulated in his 1992 monograph, *Native Trout of Western North America* (Behnke 1992), in which he lays out a northerly progression from an origin in cold waters around 20 to 28 degrees north latitude. In this view, the most primitive forms of the lineage occur in Sea of Cortez drainages and in waters associated with the lower Colorado River basin. These include the Mexican golden trout (*O. chrysogaster*), several other rainbow-like trouts found in headwaters of northwestern Mexico, and the Gila and Apache trouts (*O. gilae gilae* and *O. gilae apache*, respectively) of Arizona and New Mexico. The next most primitive members occur in the Sacramento-San Joaquin basin and are represented by the golden trout of California and the McLeod River redband trout. Members of the lineage probably did not spread into the Columbia River basin and adjacent waters until late-Pleistocene time, say 60,000 to 32,000 years ago, between the formation of upriver barrier falls in the Columbia and Snake river systems and the final advance of the Pleistocene glaciers. In 1965, Harold Malde of the U.S. Geological Survey published an estimate of anywhere from 60,000 to 30,000 years ago for the formation of Shoshone Falls on the Snake River (see Malde 1965), which is the limit of interior distribution of the rainbow lineage. The interior redband subspecies of eastern Oregon, eastern Washington, and the Kamloops region of British Columbia probably originated during this interval. *O. mykiss irideus*, the coastal form of rainbow trout, may be no more than 11,000 to 12,000 years old, originating after the last disappearance of the Beringian land bridge.

Chapter 4, the map in Fig. 4-1 shows the relationships between the maximum advance of the ice sheet and the rivers up which this subspecies had to migrate to reach the limits of its present distribution. When conditions permitted, the 2N = 66 cutthroat occupied the major Columbia River tributary basins on the west side of the Continental Divide, which include the Spokane, Kootenay, Pend Oreille, and Clark Fork systems. It also crossed the Continental Divide to occupy upper reaches of the Missouri and South Saskatchewan systems.

Did the large glacial lakes that existed during this period in the upper Columbia basin, e.g., Glacial Lake Missoula, aid in the spread of cutthroat trout into these drainages? Glacial Lake Missoula has often been mentioned as a refugium for the 2N = 66 cutthroat trout, from which it colonized or recolonized ice-bound stream systems when the great glacier decayed. But the latest evidence suggests that this may not have been the case after all.

Glacial Lake Missoula was formed a little more than 15,000 years ago when a lobe of the Cordilleran glacier, then advancing robustly to near its maximum extent, moved down the Purcell Valley of northern Idaho and crossed the valley of the Clark Fork, damming that river near the town of Sandpoint. The impounded water quickly filled an area of about 2,900 square miles in the Clark Fork and Bitterroot valleys

of western Montana. It also filled the lower Flathead River valley up to the foot of another lobe of the glacier, which had advanced down that valley. At its maximum filling, the lake's volume was an estimated 530 *cubic miles* of water. It was about 2,000 feet deep at the dam, about 950 feet deep in the Missoula Valley, and its highest shoreline stood at an elevation of 4,150 feet above today's sea level.

According to the reference books, Glacial Lake Missoula existed from a bit more than 15,000 to 13,000 years ago, or a span of about 2,000 years. But this figure is misleading, in the sense that there wasn't really a lake there for all of that period. That's because glacial lakes have an interesting property: they can float their ice dams. The specific gravity of ice is about 10 percent less than that of water (which is why ice cubes and icebergs float); so, if the lake level rises higher than 90 percent of the height of the dam, the dam will float and then break apart catastrophically. That's what happened with Glacial Lake Missoula—not just once but at least three dozen times according to David Alt, Professor of Geology at the University of Montana, Missoula, and Richard B. Waitt, Jr. of the U.S. Geological Survey.[31] Each breach released a devastating torrent of stored water. However, within a few decades the still-advancing glacier formed a new dam and a new lake filled until it too was deep enough to float its

31. The existence of Glacial Lake Missoula was established by geologist J.T. Pardee early in the 20th century; see Pardee (1910, 1942). The results of the catastrophic floods that occurred when its ice dams gave way were first recognized by another geologist, J. Harlen Bretz, in the early 1920s. His initial papers describing his findings ignited a huge controversy in geology, since they flew in the face of then-existing dogma that geological features were the result of long, slow processes of change, not rapid, cataclysmic events. Bretz published many papers full of evidence for his position over the years; but I have cited only two in the bibliography. Bretz (1923, 1969) and also Allen and Burns (1986) will give you good summaries of both the evidence and the controversy. Geologist David Alt has studied Glacial Lake Missoula since the 1960s. His book, *Glacial Lake Missoula and its Humongous Floods* (Alt 2001), is a comprehensive, not-too-technical account of all that is currently known, including the number of high stands and floods (see also Chambers 1971, Waitt 1980, 1984, 1985, and Clague et al. 2003), and will point you to dozens of additional references. Did other Glacial Lake Missoulas exist during earlier periods of glacial advance? And did they too produce episodes

dam. This occurred through at least 36 cycles, with the lake refilling each time to a lower level than the time before, until the glacier went into full retreat and could form no more dams. Studies of the succeeding layers of lake and river sediments by Alt and his student, R.L. Chambers, indicate a span of about 1,000 years for 36 fillings of the lake, or an average of about 27 years of actual lake existence per cycle (the actual range of values they reported was 58 to 59 years). The rest of the time the site was not flooded. Had Glacial Lake Missoula ever overflowed through a low pass in any of its surrounding drainage divides, it could not have floated its ice dam and would have instead stabilized at the level of the low pass and endured until the glacier finally melted.

So Glacial Lake Missoula actually filled its valley system 36 or more times, with each high stand lower and less extensive than its predecessor, and each lake relatively short-lived. Could any of these short-lived lakes have harbored fish? Dr. Alt seems to think not. He points out that no fish fossils, not even scales, have been found in lake sediments. Also, the milky suspension of rock flour that must have permeated the water especially in summer would have made poor habitat for most kinds of fish native to the region today. And, since the lake had no outlets over the rim of the valley system it occupied, it could not itself have served to lift

cutthroat trout across divides into adjacent drainages.

If Glacial Lake Missoula itself was not a refugium, then what was? Perhaps it was the remainder of the upper Clark Fork and its tributary stream network upstream (south) of the glacier and the glacial lake. Radiation of the 2N = 66 cutthroat most probably occurred from this network during the abrupt (and wet) final retreat of Cordilleran ice. The trout ascended the newly accessible Flathead River to Maria's Pass, from which they could cross the Continental Divide to tributaries of the upper Missouri River. Movement further north up the Flathead brought them to a cross-over into the upper South Saskatchewan River. Isolated headwater populations of westslope cutthroat trout, found in the Fraser River drainage of south-central British Columbia and originally classified as mountain cutthroat trout *Salmo clarkii alpestris* by Canadian scientists, probably accessed that region during ice retreat as well, when the Thompson River basin of the present Fraser River system temporarily flowed to the Columbia River. [32]

That Maria's Pass connection between the Columbia and Missouri River drainages would still exist today if it weren't for a manmade obstruction. Summit Lake, located on the Continental Divide just south of Glacier National Park, was tributary to both the Flathead River (Columbia drainage) via Bear Creek, and Maria's

of flooding? Yes to both questions. P.C. Patton and his colleagues (see Patton et al. 1978) as well as McDonald and Busacca (1988) found evidence of earlier floods that swept through the channeled scablands of eastern Washington, but weren't able to pin down the dates. More recently, Bjornstad et al. (2001) and Pluhar et al. (2006) have documented a long history of floods dating to mid-Pleistocene time based on their studies of flood-scoured unconformities in Washington's Palouse region and a well-preserved record of flood deposits at the Cold Creek bar in the Pasco Basin, also in Washington.

32. The mountain cutthroat trout *Salmo clarkii alpestris* was described in publications by J.R. Dymond of the University of Toronto; see Dymond (1931, 1932). Descriptions of interior British Columbia drainage patterns during the last ice retreat are found in Mathews (1944) and McPhail and Lindsey (1970).

River (Missouri drainage) via Summit Creek. In the days before the Interstate freeways, motorists towing trailers considered Maria's Pass, on U.S. Highway 2 just south of Glacier National Park, to be the easiest of the passes across the Continental Divide. Fish too would have found it relatively easy to swim up to Summit Lake. However, the Bear Creek outlet was dammed in the early 1890s by the Great Northern Railroad, thus closing this passageway.

The early $2N = 64$ cutthroat isolated in the Lahontan Basin gave rise to five subspecies, the Lahontan cutthroat itself (Chapter 5), two thus far unnamed subspecies indigenous to the Humboldt River system of Nevada (Chapter 6), and the Coyote Basin of southeastern Oregon (Chapter 7) that federal and state agencies have recently been lumping together with the Lahontan subspecies, the Paiute cutthroat (Chapter 8), and the now extinct Alvord basin cutthroat (discussed in Chapter 15). The radiation of this particular group of cutthroat subspecies is straightforward. The drainages in which they are found either remain connected with the Lahontan Basin or are located in basins that are contiguous with the Lahontan Basin and shared connectivity during pluvial periods. For example, Dr. Marith Reheis of the U.S. Geological Survey and co-workers have completed studies showing that Lake Lahontan stood at its highest level around 600,000 to 700,000 years ago, and at that level spilled over into the Alvord Basin. Also shown in her map of these studies is a spillover from the Alvord into the Coyote Basin, to which Whitehorse and Willow Creeks, home of the Willow/Whitehorse cutthroat trout, are tributary. Recently, Dr. Jennifer Nielsen and her colleague, George Sage, also with the U.S. Geological Survey, examined the population genetic structure of this group of fishes using microsatellite DNA markers and affirmed that at least three members of this group, the Lahontan, Humboldt River, and Paiute cutthroat forms, are separated by subspecies-level magnitudes of genetic distance. [33]

If you adhere to Dr. Behnke's interpretation of the inland dispersal and interbasin differentiation of the cutthroat trouts, the early $2N = 64$ cutthroat of the upper Snake River region gave rise to the seven remaining subspecies which include the Yellowstone cutthroat (Chapter 9), the finespotted Snake River cutthroat (Chapter 10), the Bonneville cutthroat (Chapter 11), the Colorado River cutthroat (Chapter 12), the greenback cutthroat (Chapter 13), and the Rio Grande cutthroat (Chapter 14). If you choose to go with the interpretation of Dr. Smith and his associates, the Bonneville cutthroat (excluding the Bear River and probably also the Bear Lake populations) came first, and a sister line that led to the Lahontan Basin subspecies also gave rise to all the others—the Bear River, Snake River, and Yellowstone subspecies in a diver-

33. The work of Reheis and co-workers on the high stand, age, and connections of Pluvial Lake Lahontan has been published in three papers which are listed in the bibliography under Reheis and Morrison (1997), Reheis (1999), and Reheis et al. (2002). A paper by Nielsen and Sage (2002) describes the population genetic structure of the Lahontan group of cutthroat subspecies.

gence that occurred around 3 million years ago, and the Colorado River, Greenback, Rio Grande subspecies in another divergence that also occurred about 3 million years ago. Either way, however, the dispersal of this particular group into the areas comprising their respective historical ranges is much trickier to explain (it is also not clear just what relationship existed between these subspecies and the now-extinct yellow-fin cutthroat [Chapter 15] that occupied Twin Lakes, Colorado, along with greenback cutthroats as recently as the turn of the 20th century).

As an adherent (for now) of Dr. Behnke's interpretation of the evidence, one way I have tried to get a handle on this is to consider the lay of the land in the western United States. If one could get an astronaut's view of the region, so that the landscape could be seen in relief, it is possible to pick out the relatively low, easy routes through the tortuous mountain ranges that might have connected water courses in wetter climes. Paleolithic human groups and their descendants found these routes and followed them. Fur trappers scouted them, and later they were used by emigrant wagon trains in the great western exodus. Even today, railroads and highways follow these routes.

Perhaps the best known of these natural highways is the Oregon Trail. Who does not recall from grade school history classes how the wagon trains pulled out of Independence, Missouri and followed the Platte River across the seemingly endless Great Plains? After crossing the Continental Divide at low, sweeping South Pass, Wyoming, and perhaps a short swing south for a rest stop at Fort Bridger, the wagons would head north along the Bear River to Soda Springs, then pick up the Portneuf River and follow it out to the Snake, emerging at Fort Hall near present-day Pocatello, Idaho. In later years a cutoff led from Lander, Wyoming, over the Salt River Range at Commissary Ridge, then north up the Star Valley to the present town of Freedom, then west again to Fort Hall. From there it was down the Snake to the Columbia and on to the lush green valleys of Oregon.

Let's backtrack a portion of this route from the perspective of our ancestral $2N = 64$ upper Snake River cutthroat. Our starting point is an enclave that includes the Snake River upstream from Shoshone Falls, especially the reach in the vicinity of Fort Hall, and also takes in the Portneuf and Bear river systems. This general region, inscribed as sort of a wavy T in Fig. 2-7, makes sense to me as a point of origin for this form of $2N = 64$ cutthroat trout for three reasons. First, although recent evidence suggests the Bear River may have been diverted on one or more occasions earlier, prior to 34,000 years ago the Bear River followed the Portneuf channel and was a Snake River tributary.[34] Snake River trout would not have been excluded from any of its waterways. Second, I have always been

34. Diversion of the Bear River from its former connection with the Portneuf and Snake rivers has often been proposed as the route by which the ancestral $2N = 64$ large-spotted cutthroat trout gained access into the Bonneville basin. Recent evidence indicates the Bear River may have been diverted into the Thatcher Basin more than once, the earliest occurrence traced to date being about 140,000 years ago. It is not clear from available evidence that Lake Thatcher spilled over into the Bonneville Basin at that time, but it may have done about 100,000 years ago when lava flows blocked a northern outlet. More lava flows, occurring somewhere between 34,000 and 50,000 years ago, again broke the Bear-Portneuf connection and diverted the Bear River south into the Thatcher Basin for a final time. Lake Thatcher overflowed and drained into Pluvial Lake Bonneville once again, thus rerouting the Bear River into its present course into the Bonneville Basin. Then, about 15,000 years ago, swollen by the diversion of the Bear River, Lake Bonneville breached its northern rim at a place called Red Rock Pass, sending a wall of water back down the Portneuf channel into the Snake. More information on the history of the Bear River diversion, and of Pluvial Lake Bonneville and its possible connections with other drainages, can be found in Bright (1967),

impressed with historical accounts that testify to the abundance of trout in this particular region. Even the fur trappers, who were hard-core red meat eaters, fished for trout in this enclave. [35] Third, being south of the continental ice and away from alpine glaciers, this region was ideally situated to be a fountainhead and springboard for Pleistocene dispersal, which is consistent even with the molecular clock evidence of Dr. Smith and his associates for when this dispersal occurred. It provided a perfect portal of entry into both the Bonneville Basin (or out of, as the case may be) and the Green River/Colorado River system, and later, whenever glacial ice retreated from Snake River headwaters and the Yellowstone region, across the Continental Divide into the Yellowstone River drainage.

Other possible waterways that the large-spotted ancestral cutthroat might have followed into the Green River drainage exist in this same general area. [36] Wallace R. Hansen of the U.S. Geological Survey postulated an ancient diversion of the upper Bear River into Muddy Creek, a tributary of Black's Fork of the Green River. The Grays and Hoback rivers, both Snake River tributaries, also head up in this country and their headwaters too are in close, rather easy proximity to headwater tributaries of the Green River. Even today cutthroat trout occupy these headwaters, and it isn't hard to imagine how the waterways could have intermingled in wetter climates. The Grays River divide

Scott et al. (1983), Taylor (1985), Jarrett and Malde (1987), Bouchard et al. (1998), Oviatt et al. (1987, 1999), and Mock et al. (2004).

35. The enclave encompassing the Bear River, Portneuf River, and the upper Snake and its tributaries in the vicinity of Fort Hall was singled out by many narrators of the westward migration for its abundance of trout. Robert Stuart and his small band of lost and desperate fur traders averted starvation here in 1812 by fishing for trout in Bear River tributaries and in places along the Snake. The naturalist, John Kirk Townsend, traveling with Nathaniel Wyeth's party of fur traders in 1834, wrote of a stop on Bear River over the 4th of July:

> "In this little stream, the trout are more abundant than we have yet seen them. One of our *sober* men took, this afternoon, upward of thirty pounds. These fish would average fifteen or sixteen inches in length, and weigh three-quarters of a pound; occasionally, however, a much larger one is seen."

Major Osborne Cross, quartermaster of an army regiment marching to Fort Vancouver in 1849, wrote of being plied with trout by Indian boys during a Bear River rest stop:

"The young boys, who were not over seven years old, brought us great quantities of fish for a few trifling presents. These consisted of the brook and salmon trout which are found very abundantly in this river, as well as in all the mountain streams between here and the Columbia River. They were extremely fine and the first I have seen since coming into the mountains. Having little to do while remaining here we resorted to fishing and were very successful, keeping our mess very abundantly supplied with the finest kind."

Two weeks later, now on the Snake River near its Raft River and Goose Creek tributaries, Major Cross wrote:

> "From Bear River to this place every stream abounds in fish of the finest kind. The speckled as well as the salmon-trout can be caught in great quantities. Everyone who could find time resorted to his hook-and-line, and we fared sumptuously...."

William Henry Jackson, the photographer, described a sport-fishing trip he took to Bear River in 1869 with men from a survey party:

> "With a group of men we went over to Bear River to catch some of the fine trout that

abound in the stream.... These days...were filled with interesting experiences and good fishing, and we came back with two horses packed with trout."

These and many other accounts of trout abundance in the Bear-Portneuf-upper Snake enclave can be found in Townsend (1839 [1978]), Johnson and Winter (1846), Watson (1851 [1985]), Stansbury (1852 [1988]), Wilson (1919 [1991]), Russell (1923), Jackson (1929), Rollins (1935), Goodhart (1940), Settle (1940 [1989]), and Spaulding (1953). Trout were also plentiful across the divide from Bear River in Black's Fork of the Green River near Fort Bridger. Old Jim Bridger himself wrote about this, as did many of the immigrants who stopped there on their way to Oregon, California, or the Salt Lake valley. Some of these accounts can be found in Palmer (1847), Bryant, (1848 [2000]), Suckley (1874), Pringle (1905 [1993]), Cowley (1916 [1964]), Gowan and Campbell (1975), and Moon (1982).

36. Dr. Hansen's work on possible interbasin transfer points for fish migration can be found in his 1985 publication, "Drainage Development of the Green River Basin in Southwestern Wyoming and its Bearing on Fish Biogeography, Neotectonics, and Paleoclimates" in the journal, The Mountain Geologist. Check the bibliography under Hansen (1985) for the complete citation.

Postulated original enclave

Possible routes of dispersal

FIGURE 2-7. *Ancestral 2N=64 cutthroat trout of upper Snake-Portneuf-Bear River*

is up on the eastern flanks of the Salt River Range. But here, in an open grassy meadow separated by only a very gentle rise, the very headwaters of the Grays lie only about a quarter-mile from LaBarge Creek, a trout-filled tributary of the Green River. This is an easy crossing as evidenced by the fact that the Lander cutoff of the Oregon Trail passed through here. The divide between the Hoback and the Green River is easy as well, save for a short, fairly steep climb on the Hoback side just short of the divide itself, which is called The Rim, elevation 7,921 feet. The Green River side of The Rim is one of those long, high valleys where the descent is so gradual you hardly realize it's there.

Other possible avenues by which the large spotted cutthroat might have entered the Green/Colorado system were from the Bonneville basin via headwater transfers in the Wasatch Range. Rivers such as the Heber and Provo head up in topography where headwater stream transfers could have easily taken place to headwater streams of the Colorado system. But here the fish would have had to await the melt-off of ice from the divides, because this region was subjected to alpine glaciation during major glacial maxima.

However it happened, the large-spotted cutthroat did invade the Green/Colorado drainage. From here, the route followed the Colorado River back upstream to the confluence of the Eagle River, a major tributary, then eastward up the Eagle to one or another tributary creek that leads to Fremont Pass, elevation 11,115 feet, or Tennessee Pass, elevation 10,424 feet. Fremont Pass, the site of the massive open-pit Climax molybdenum mine, is pretty high, but the fish made the climb. A couple of miles below the pass is Clinton Reservoir. In a high, glacier-scoured basin at the head of this reservoir is tiny Clinton Creek, the home of a population of the rare Colorado River cutthroat. Standing at Fremont Pass facing west, one can look at the Eagle River drainage, then look over the left shoulder at the headwaters of the Arkansas River.

Tennessee Pass, a few miles further south, is perhaps a more logical crossover point. It is lower for one thing, and the climb up from the Eagle River is more gradual. On the Arkansas River side of the pass is Tennessee Park, a high, glacier-rounded valley that descends ever so gently to the Arkansas River.

About halfway down the Arkansas valley, a gap in the mountains at Trout Creek Pass leads east to South Park and the headwaters of the South Platte River. And at the lower end of the valley, where the Arkansas River turns east to avoid the Sangre de Cristo Range, another gap at Poncha Pass leads into the San Luis Valley and the upper Rio Grande.

Thus you have a glimpse of at least one route by which the large spotted 2N = 64 cutthroat might have dispersed into the Bonneville Basin and the Green River/Colorado River system, and from the latter into the South Platte and Arkansas, and thence to the Rio Grande. These trout also crossed the Continental Divide from the headwaters of the Snake into the Yellowstone River system. During glacial periods, the entire Yellowstone plateau area was covered by alpine glacial ice, so this crossover could not have occurred until ice-free conditions prevailed. The likely crossover point is a fascinating spot located several miles south of Yellowstone Lake, in fact, south of the boundary of Yellowstone National Park, called Two Ocean Pass.

In mountain man Osborne Russell's day, circa 1836, Two Ocean Pass was one of the routes used by the fur trappers to get in and out of Jackson Hole. Russell described it in his journal as:

> "...a smooth prairie about two miles long and half a mile wide lying east and west surrounded by pines. On the south side about midway of the prairie stands a high snowy peak from whence issues a stream of water which after entering the plain it divides equally one half running west and the other east thus bidding adieu to each other, one bound for the Pacific and the other for the Atlantic Ocean."

B.W. Evermann, a fish biologist, visited the area in 1891 and described Two Ocean Pass pretty much as Russell had, with the "dividing spring" lying out in the meadow. Evermann reported actually seeing trout make the transit from Pacific Creek to Atlantic Creek, so it was still an active cross-over point at the turn of the 20th century. But today the little creeks issuing into the meadow flow differently. The "dividing spring" is no longer located out on the meadow floor, but back in the timber a short distance up on the north slope. North Two Ocean Creek comes down off the plateau, and just about where the gradient starts to flatten, it divides into two almost equal flows. One branch becomes Pacific Creek and drains into the Snake River. The other branch becomes Atlantic Creek and flows to the Yellowstone. The stream Osborne Russell described as coming down from the south side is still there, but now it's a tributary of Atlantic Creek.

I traveled across Two Ocean Pass with a party on horseback in early July, 1980. With the aid of Polaroid glasses I saw trout in the meadow section of Atlantic Creek within a half-mile or less of "Parting of the Waters," and fishermen from other parties were taking trout from all along this reach. Try as we might, however, we saw no trout in Pacific Creek.

This does not mean they aren't there. In 1967, Dr. Behnke collected several cutthroat trout from Pacific Creek near the divide and also from North Two Ocean Creek. These were resident, non-migratory populations, whereas the trout Dr. Behnke and my party saw in Atlantic Creek were fish on a spawning run from Yellowstone Lake. The two populations apparently do not interbreed. The waterway is still open for crossover, but it is unlikely that trout use it today. [37]

37. Osborne Russell's description of Two Ocean Pass (as well as his other adventures in the Jackson Hole and Yellowstone regions) can be read in his book, *Journal of a Trapper, or, Nine Years in the Rocky Mountains, 1834–1843* (Russell 1986). B.W. Evermann's description is in the Bulletin of the U.S. Fish Commission for 1891–92; see Evermann (1893). Dr. Behnke's observations and conclusions about the lack of interbreeding among the trout in Two Ocean Pass can be found in his monograph; see Behnke (1992). Also, I have a letter dated February 13, 1980, from Clarence Murdock, then Buffalo District Ranger, Bridger–Teton National Forest, that states: "Just to let you know what we have observed at Two Ocean Pass. The Yellowstone cutthroat are spawning in Atlantic Creek in late June and early July and you can see them very near Parting of the Waters. However, it hasn't been very evident that the cutthroat in Pacific Creek are near Parting of the Waters at this time."

Coastal Cutthroat Trout, *Oncorhynchus clarkii clarkii*, sea-run form

CHAPTER 3

Coastal Cutthroat Trout

Oncorhynchus clarkii clarkii: Chromosomes, $2N = 68$. Scales in lateral series typically 140–180 (there are some coarse-scaled resident strains with 120–140 scales in the lateral series). Scales above lateral line 30–40. Gill rakers 15–21, typically 17 or 18, short and blunt in structure. Pyloric caeca 24–55, mean value 40. Vertebrae 59–64, typically 61 or 62. Densely spotted with small to medium size, irregularly shaped spots that are more or less evenly distributed over the sides of the body and onto the head, and quite often also onto the ventral surface and anal fin. Coloration of sea-run fish is silvery, sometimes with a faint yellowish or brassy wash.

Resident freshwater fish are darker, with brassy or coppery olive backs shading through gold on the sides to grayish white on the bellies, often with a faint pinkish band along the lateral line. Smaller fish may exhibit grayish-violet parr marks. The dorsal fins of freshwater residents may have a white to cream-colored tip and the lower fins may be yellow to red-orange, often with white tips on the pelvic and anal fins. More prominent rose-colored tints may appear on the gill plates, sides, and ventral region as spawning time draws near. Sea-run fish darken and take on these colors after a time back in fresh water.

Coastal Cutthroat Trout,
Oncorhynchus clarkii clarkii, stream-resident form

COASTAL CUTTHROAT COUNTRY

"Cath-la-pootl!" A strange sounding word to a European ear, this was an Indian name given by the whites
to a river that heads up in the snow fields of Mt. Adams,
then drains the high meadows and rugged foothills of
the Cascade Range in southwest Washington before
joining the mighty Columbia River near the present-
day town of Woodland. We now call it the Lewis River.
But when the Hudson Bay Company held sway over
the territory, operating out of Fort Vancouver about 25
miles upstream from the mouth of Cath-la-pootl, the
Indian name was in common use. [1]

    Cath-la-pootl flows out of the mountains just south

of Mt. St. Helens, once a thing of beauty and serenity,
its perfect snow-capped cone surrounded by green forests and sparkling lakes and streams. It's still a ruggedly
beautiful landscape although the mountain itself sits
broken now, its top blown away in the major eruption
of May, 1980. But a new dome has been building inside
the crater, drawing countless visitors to the Forest
Service viewpoints, and the land around it is well on
its way to recovery, much to the marvel of the cadre of
scientists who have been observing the process.

    The Indians have a story about this country that
explains the origin of the Cascade peaks and hints at
their volcanic nature. It is said that in ancient times

1. Actually, "Cath-la-pootl" (or "Quath-
lapotle" as it was spelled in their
journals) was the name Lewis and
Clark ascribed to a large village
of native people that they found
situated near the mouth of the river.
The site is located on the Ridgefield
National Wildlife Refuge near the
town of Ridgefield, Washington, and
has been excavated to study the life
ways of the Chinookian people who
lived there (and I should add here
that the archaeologists, the U.S.
Fish and Wildlife Service, and the
modern Chinook Nation all prefer
"Cathlapotle" as the spelling of the
village name). A replica longhouse
has also been erected nearby to
commemorate the site. But getting
back to the river itself, according to
Lewis and Clark, the Indians referred
to that as "Chawahnahiooks." But
that name seems to have vanished,
as the people of Cath-la-pootl
(or Cathlapotle if you prefer)
themselves vanished, swept away
by an epidemic of "intermittent
fever" (thought now to have been
malaria) that ravaged the territory in
the early 1830s. Having no natural
resistance, over 90 percent of the
native inhabitants of the lower
Columbia region succumbed to the
disease. The name, Cath-la-pootl,
having become attached to the river,
endured into the 1860s, although
by then it was also being called

Sahalee Tyee, the Great Spirit, built a great stone bridge across the Columbia River so that the people on the north side could mingle and trade with the people on the south. A wrinkled old crone, Loo-wit by name, was made guardian of the bridge. The old woman served faithfully and well. But she had a young daughter who grew ever more beautiful, and it wasn't long before the girl attracted two most powerful and jealous suitors: Patoe, the solidly built chief of the north, and tall, proud Wyeast, chief of the south. The daughter secretly favored Patoe, but she egged them both on with her flirtatious ways. The chiefs set to quarreling over the maiden. They hurled fire and hot rocks at one another, burning villages and forests, driving away the game, and darkening the skies with their anger. Loo-wit tried her best to keep the peace, but to no avail.

Angered at the devastation the two chiefs were bringing upon their people, Sahalee Tyee smashed the Bridge of the Gods into the river and froze the two rivals in perpetual sleep right where they stood. You can see them to this day. Patoe is Mt. Adams, on the Washington side of the river. Over in Oregon stands Wyeast in the form of Mt. Hood. Loo-wit's daughter was frozen too, as Sleeping Beauty Mountain, which lies in Washington not far from Mt. Adams. His anger mollified, Sahalee Tyee decided to reward old Loo-wit for her faithful service. He transformed her into a thing

of great beauty. Loo-wit thus became Mt. St. Helens, the most beautiful and symmetrical of all the snow-covered Cascade peaks.

But Loo-wit has not always slumbered peacefully. Scientists have documented many eruptive periods over the millennia. During the historical period, prior to the major eruption of May 18, 1980, Mt. St. Helens was active off and on for some 26 years between 1831 and 1857. Pioneer journals are full of accounts of "immense and beautiful scrolls of steam," "huge columns of black smoke," "days memorable for the shower of ash," "snow melted off the mountain" and "streams of lava." [2]

Meredith Gairdner, who was serving at Fort Vancouver at the time, described one of the mountain's eruptions in 1835. He wrote:

> "There was no earthquake or preliminary noise here. The first thing which excited my notice was a dense haze for two or three days, accompanied with a fall of minute flocculi of ashes which, on clearing off, disclosed the mountain destitute of its cover of everlasting snow, and furrowed deeply by what (through the glass) appeared to be lava streams."

A contemporary of Charles Darwin, Gairdner was vitally interested in the natural history of the Pacific Northwest—so much so, the record shows, that he bitterly resented the time his medical duties took away from collecting. Even so, he did collect or arrange for the collection of many specimens which he sent back to

Lewis's River or simply Lewis River, its modern name. By the way, the Lewis for whom the river is named is not Meriwether Lewis of the Lewis and Clark expedition as is often supposed, but rather Adolphus Lee Lewis (or Le Lewes as some historians insist his name was pronounced and spelled), a retiree from the Hudson Bay Company who took up a land claim near the mouth of the river in 1854. For more on the early history and pre-history of Lewis River country, see McClellan (1853), Overmeyer (1941), Jones (1972), Dryer (1980), Parsons (1983), Boyd and Hajda (1987), Silverstein (1990), and Urrutia (1998).

2. Accounts of Mt. St. Helens eruptions and the early settlers' reactions to them can be found in Jillson (1917), Holmes (1955), and Folsom (1970).

his mentors for scientific study, including the specimen of rainbow trout originally named *Salmo gairdneri* in his honor by Dr. John Richardson.

Among the fish that Gairdner collected were two specimens of trout from the river called Cath-la-pootl. Dr. Richardson received Gairdner's specimens in Edinburgh and recognized them as the same type of trout described earlier in the Lewis and Clark journals. Richardson wrote out the first scientific description of the species and assigned the name *Salmo clarkii* in honor of Clark.

Of course, we already know that neither the coastal cutthroat nor the Lewis and Clark specimens (see westslope cutthroat, Chapter 4) were the first of the cutthroat line to be seen by Europeans on the North American continent. That honor goes to the Rio Grande subspecies (Chapter 14), which was first noted by the conquistadors of Francisco Coronado's party in 1541. But the coastal cutthroat does have the distinction of being the first member of the species to be described for science. And the river called Cath-la-pootl, now the Lewis River, is the type locality for the species.

HISTORICAL RANGE

Coastal cutthroat trout occupy about a 2,000-mile strip of the Pacific Northwest coast extending from the lower Eel River drainage of California to Gore Point on the Kenai Peninsula of Alaska. Inland, the distribution rarely extends more than about 100 miles from the coast, which coincides in general with the Cascade crest in Oregon and Washington and the Coast Range crest in British Columbia. The farthest inland the range extends is about 175 miles from the coast in the Skeena and Stikine river drainages of British Columbia. In the Columbia River system, sea-run fish were once reported to have migrated inland as far as the mouth of the Klickitat River, about 140 straight-line miles from the coast and a bit east of the Cascade crest. A report in the literature dating back to 1935 that coastal cutthroat reached as far up the Columbia system as Spokane, where it overlapped the range of the westslope cutthroat, was based on misidentified specimens of the latter subspecies.

I have mapped the historical distribution of coastal cutthroat trout in Fig. 3-1. The overlapped format you see in the figure resulted from an accidental but I think interesting and significant discovery. One day, shortly after the first edition of this book appeared, I was searching in a back issue of the journal *Science*, for a paper I needed for a work assignment. As I riffled through the pages, a map flipped past that at first glance looked like the coastal cutthroat range map my publisher had used in the book. Wondering if I had somehow missed a key reference to coastal cutthroat, I quickly turned back and was surprised to find instead a paper on forest ecology. What its map showed was the extent of the great Pacific coast temperate rain forest. Still, it appeared to match closely the historical range of the coastal cutthroat trout. I determined to compare the two maps.

The map my publisher used in the first edition of this book had been redrawn from the distribution map in Dr. Behnke's first monograph, published in 1979, the same year as the *Science* paper. I asked a friend in my company's cartography department if he could overlay the two maps at as close to the same scale as possible. He was skeptical, but even he was surprised at the goodness of the resulting fit. You are looking at a new rendition of that original pairing in Fig. 3-1. As you can see, only in the Skeena and Stikine river drainages and a minor bulge inland through the Columbia River Gorge do coastal cutthroat extend beyond the mapped rain forest influence, and only on the Kenai Peninsula around 60 degrees north latitude does the rain forest extend beyond the cutthroats. Perhaps not so obvious at the scale shown, coastal cutthroat trout are also absent from nearly all streams of Morseby Island, the southern major island of the Queen Charlotte Island group.[3]

Remarkable, I thought. I showed this overlay to several colleagues who specialize in aquatic and landscape ecology, and they thought it remarkable as well. None could offer specific reasons why the correspondence of

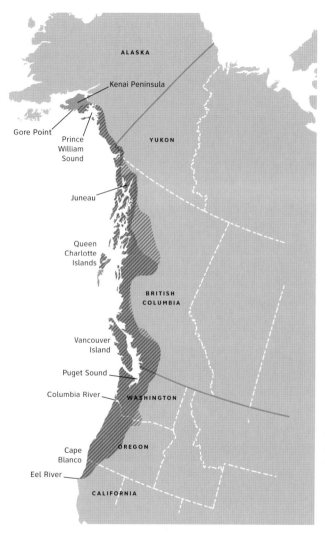

3. The original maps used for the overlay shown in Fig. 3-1 may be found in Behnke (1979) for the coastal cutthroat distribution and Waring and Franklin (1979) for the Pacific coast temperate rainforest. These maps show the divergence of ranges in the Skeena-Stikine drainages and the Gore Point vicinity, and to this I added information about the distribution in the Queen Charlotte Islands from Northcote et al. (1989), and the inland extension of the coastal cutthroat range beyond the Cascade crest in the Columbia River system from Bryant (1949), Blakley et al. (2000), and Connolly et al. (2002). The upstream distribution of coastal cutthroat trout in the Columbia River extends only to the Klickitat River, although Schultz (1935) mistakenly listed it as occurring as far upstream as Spokane. In the southern part of the range, some authorities include the entire length of the coastal redwood distribution, which extends from southern Oregon to Big Sur in California, as part of the Pacific coast rainforest (it is also often referred to as the *fog-drip*

FIGURE 3-1. *Historical distribution of Coastal Cutthroat Trout and overlap with Pacific coast coniferous rainforest*

■ Rain Forest

▨ Coastal Cutthroat Trout

▨ Rain Forest and Cutthroat Overlap

ranges should be so close, but all agreed it was too good a fit to be a coincidence. No other salmonid species or subspecies has a historical range that so closely matches a particular defined ecoregion. We concluded that the coastal cutthroat trout must truly be a fish of the Pacific coast coniferous rain forest.

The native habitat of the coastal cutthroat is a land of heavily timbered hills. In the northern part of the range, these hills drop steeply into fjord-like sounds and inlets. Coming south, forests of cedar, hemlock, spruce, and Douglas-fir give way to the redwoods of northern California as the average rainfall diminishes and the climate warms, but fog-drip precipitation becomes appreciable. Creeks and rivers in this region either run directly into saltwater or are tributary to other streams that do. Fertile river valleys cut through the ranges. A broad lowland bisects the country north to south from Puget Sound to about the middle of the State of Oregon, separating Washington's Olympic Mountains and Willapa Hills, and Oregon's Coast Range from the higher Cascades. Lakes are found here and there, often in clusters, in both the lowlands and the hills. Along the central Oregon coast there is an area of sand dunes, just behind which sit other lakes with outlets to the sea.

The coastal cutthroat is present throughout this range, although it shares the region with the coastal rainbow trout, which is also native. As we shall see in later Chapters, hybridization with introduced rainbow trout is a major reason for the decline in genetically pure populations of most of the interior cutthroat subspecies. But here in the coastal belt the two species live compatibly. Although there is recent evidence that hybridization does indeed take place and may be more common than previously thought, it does not appear to have threatened the integrity of either species. The reason it is not more widespread probably has something to do with the preference shown by the cutthroats for the smaller tributary streams, and, among resident cutthroat populations, the upper reaches of a watershed. Resident rainbow trout and their sea-run relatives, the steelhead, spawn at about the same time as the coastal cutthroats, but the rainbow/steelhead like somewhat larger streams, more robust waters and, where resident populations of the two species occupy the same stream system, the downstream reaches. Thus, in a natural setting, the two native species maintain an ecological separation that also contributes to reproductive isolation. [4]

LIFE HISTORY AND ECOLOGY

Individuals and populations of coastal cutthroat trout display one of four basic life history strategies:

- The amphidromous, or sea-run, life history, wherein fish migrate from fresh to salt water to feed and grow, then back again after only a short period to overwinter and/or to spawn. In some areas, amphidromous cutthroats make springtime forays into fresh water to feed on downstream migrating juveniles of other salmon species.

zone because fog, rather than rain, provides most of the precipitation in that region). If we were to do this as well, then the coastal rainforest would extend much farther south than the coastal cutthroat distribution, which ends at about the Eel River. Regarding the range-overlap issue, there is a field of scientific inquiry known as *coevolutionary biology* that studies how animals have accommodated to plants and vice versa. The coastal cutthroat trout's close conformance to the Pacific coast coniferous forest zone may be a grand-scale example of this. But what specific interactions bring about this close association—interactions that evidently do not play out in the same way between this particular ecosystem and other *Oncorhynchus* species and subspecies—are not known.

4. Campton and Utter (1985), Hawkins (1997), Baker (2001), and Young et al. (2001) have documented the occurrence of hybrids between coastal cutthroat trout and steelhead. Hawkins' study found that most hybrids produced in nature were from male coastal cutthroat sneaks on larger steelhead females. In the laboratory, she found that survival rates of hybrids beyond the embryo stage were the same as the maternal parents. However, she also noted differences in egg size and development rate between coastal cut-

- A lacustrine, or lake-associated, life history, wherein the fish feed and grow in lakes, then migrate to small inlet or outlet streams to spawn.

- A fluvial, or riverine (potamodromous), life history, wherein the fish feed and grow in main river reaches, then migrate to small tributaries of these rivers where spawning and early rearing occur.

- A stream-resident life history, wherein the trout complete all phases of their life cycle within the streams of their birth, often in a short stream reach and often in the headwaters of the drainage.

The sophisticated genetic analysis methods available to scientists today show only slight genetic differences governing these life history strategies, if they detect any at all. Even so, there is compelling evidence that each has a hereditary basis. What is often confusing is that populations exhibiting several of these life histories can and do exist in the same bodies of water. In 1995, in the town of Reedsport on the Oregon coast, I participated in a conference on the biology, management, and future conservation of sea-run cutthroat trout. [5] Several speakers at that conference touched on the diversity of expression of life history strategies among individuals and populations of coastal cutthroat trout. They were, I believe, mainly trying to emphasize the point that individuals and populations with different life history strategies often co-occur in the same waters. But they may have left the impression that intrinsic within the coastal cutthroat subspecies is the propensity for wholesale switching among life history strategies, i.e., for individuals in a stream-resident

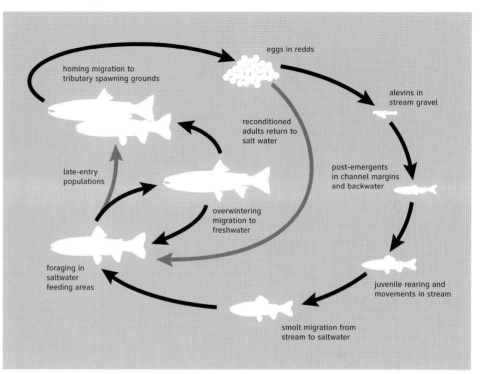

FIGURE 3-2. *Life cycle highlights, sea-run cutthroat trout*

Labels: homing migration to tributary spawning grounds; eggs in redds; alevins in stream gravel; reconditioned adults return to salt water; post-emergents in channel margins and backwater; late-entry populations; overwintering migration to freshwater; juvenile rearing and movements in stream; foraging in saltwater feeding areas; smolt migration from stream to saltwater

throat and steelhead, which resulted in mismatches in development of embryos, and this, she felt, would set the stage for selection against hybrids in natural settings. Even though hybrids do occur in nature, this may be one reason why they are not more widespread. A paper by Hawkins and Foote (1998) gives additional details on the early survival and development of the two parent species and reciprocal hybrids. A paper by Hawkins and Quinn (1996) talks about swimming performance and associated morphology.

5. The proceedings of this conference were published by the Oregon Chapter American Fisheries Society in 1997; see Hall et al. (1997). In the fall of 2005, a follow-up conference,

population to volitionally become sea-run or vice versa, or for resident populations to produce large numbers of sea-run progeny or vice versa. This kind of switching can and does take place among individuals in other salmonid species (it has been best studied in Arctic charr and brown trout), but not, I argue, to such an extent that "diversity of life history expression" is itself a life history strategy. [6]

The Sea-Run Cutthroat Trout

Figure 3-2 illustrates the basic life-cycle template of the amphidromous or sea-run cutthroat trout. There is diversity among populations even in the expression of this life history strategy. Let me walk you through this cycle, explaining as we go along, beginning with the spawning ritual in the upper left-hand corner. [7]

Although coastal cutthroats, along with all the other western North American trouts, are considered to be spring spawners, winter-spring would be a more apt description of the spawning timing of sea-run cutthroat populations. Winter is generally mild in the river valleys and foothills along the Pacific Northwest coast, and spring-like conditions come early to the region. Up in small tributaries with gentle gradients, pairs of dark trout, mostly 13 to 14 inches in length but a few running to 16 to 18 inches or more, gather in foot-deep riffles over pea- to walnut-size gravels

"The 2005 Coastal Cutthroat Trout Symposium: Status, Management, Biology, and Conservation," was held in Port Townsend, Washington, with new information being presented in the areas of population genetics, habitat use, movements, and life history strategies. As of February, 2007, the proceedings of this second symposium had not yet been published, but the program along with abstracts of many of the papers can be viewed on the U.S. Fish and Wildlife Service Columbia River Fisheries Program website at *www.fws.gov/ columbiariver/*. Navigate to Coastal Cutthroat Trout Symposium, then to Symposium Program Packet. Powerpoint presentations from the symposium are also available on that site.

6. Are life history strategies hereditary? If they were not, all hatchery programs for steelhead, Atlantic salmon, sea-run cutthroat trout, sea-trout (diadromous brown trout) and other migratory forms would long since have failed. To be sure, all of these programs produce fish that do not migrate but remain as "residuals" in the streams where they are released, but the percentage is typically small. Also, in wild populations of European brown trout and Canadian Atlantic salmon, genetic analysis has indeed detected differentiation between diadromous and non-diadromous populations, which is the expected result if these populations are maintaining reproductive isolation even in sympatry (see Allendorf et al. 1976, Jonsson 1982, Skaala and Naevdal 1989, Birt et al. 1990, Näslund 1993, and Pettersson et al. 2001). So what about the "life history switching" that I said has been observed in wild populations? These may be responses to environmental cues. In his paper on the evolution of diadromy, cited earlier (see Gross 1987), Mart R. Gross of the University of Toronto proposed that for a migratory life history pattern to exist, the gain in fitness for the individual or population (defined as the product of lifetime reproductive success and survivability) *minus* the cost of moving must be greater than staying in only one habitat. Several studies have shown that if food availability or other productivity factors are lowered in a system, the proportion of fish migrating from that system will increase (see Nordeng 1983, Jonsson and Jonsson 1993, Näslund et al. 1993, and Bohlin et al. 1996). In the food-limited situations, body size or growth may be limited, and migration to areas of higher food supply could allow growth to continue (see Forseth et al. 1994, 1999).

7. Other detailed descriptions of sea-run cutthroat life history and ecology can be found in Giger (1972), Jones (1972–1976), Johnston and Mercer (1976), Johnston (1982), Michael (1989), Pauley et al. (1989), Trotter (1989, 1997), Wenburg (1998), Wydoski (2003), and Slaney and Roberts (2005). For less detailed overviews, consult McPhail and Lindsey (1970), Scott and Crossman (1973), Wydoski and Whitney (1979 [2003]), Morrow (1980), Behnke (1992, 2002), Mecklenberg et al. (2002), and Moyle (2002). For lighter treatments still, see Smith (1984 [1994]), Trotter (1991), and Johnson (2004).

to prepare for the spawning ritual. Many slightly smaller sea-runs, fish of 10 to 11 inches in length, may have accompanied these larger fish and they show interest in the proceedings, but are not pairing up for spawning themselves. In total, these trout are part of a movement into fresh water that began as early as mid-summer of the previous year in some parts of the range, peaked about September or October, and extended on through November, December, and even into January in other parts of the range. In some streams, fish might not enter from salt water until as late as March. Depending upon the locale and whether the fish are early or late arrivals, actual spawning may take place anywhere from about mid-December on. There is a south-to-north cline in spawning time; peak spawning generally occurs in February in the southern to middle portions of the range, but not until April or May in southeastern Alaska.

Let's look in on one pair of trout holding near the lower end of a small riffle as the female, whose fork length approaches 14 inches, prepares to excavate a pocket for her eggs in the streambed gravel. The female does all the actual digging while the male, smaller than the female by about an inch, hovers nervously nearby, having already fought off several other males for her favors. We see her turn on her side and press her tail against the gravel. She flips it vigorously about a

half-dozen times, and a small cloud of silt and pebbles swirls away downstream. After a brief rest, she repeats the process. This continues for two or three hours until she has fashioned a pit about four to six inches deep and just a little shorter than her length.

From time to time toward the end of her digging, the female cutthroat has settled into the depression as if to test the fit. She settles in again, but this time something's different. Some indefinable signal must have passed because now the male cutthroat moves to join her. He bumps her side gently, persistently, as she presses her anal fin into the gravel. Then both fish quiver, their mouths gape, and eggs and milt are extruded into the depression.

The number of eggs produced will vary with the size and age of the fish and with the locality, but a trout the size of the female we are watching will typically carry from 700 to 1,200 eggs. She has discharged only about one-third of them into her first depression, and now she moves upstream a bit where she begins to sweep out a new depression. The gravel from this second excavation simultaneously covers up the fertilized eggs in the first pocket. Occasionally, rather than spawning again in this new depression, the female may select another spot entirely and start digging anew. And once in awhile, one of the fish will abandon its partner completely after the first deposition of eggs. Either fish

Before and after—Effects of poor logging practices. Coastal cutthroat trout use small tributaries such as this one for spawning and for nurseries. Poor logging and forest road building practices can so devastate this habitat that trout will not be able to use it for several years.

PHOTOS BY AUTHOR

may then complete its spawning with another or several other members of the opposite sex. But regardless of how they choose to do it, the ritual continues until all the eggs are safely covered in mounds of gravel, called redds. The whole process could extend out over a couple of days.

What will happen to the pair of fish we were watching after they have completed spawning? The rigors of spawning are great, but sea-run cutthroat trout withstand these stresses rather well. They recover their condition quickly, and chances are good that they will survive to spawn again. I have depicted this with the red arrow back to salt water in Fig. 3-2. A survival study carried out on a coastal Oregon population some years ago indicated that in the absence of any kind of harvest fishery, 39 percent of the initial spawners survived to spawn a second time, 17 percent survived to spawn a third time, and 12 percent came back for a fourth spawning run. [8] In the face of a harvest fishery, survival percentages are much lower, of course, but natural survival is generally quite high. Return of spawned-out fish (kelts) to salt water peaks around the end of March or the first of April in Washington and Oregon, and maybe three to four weeks later in Alaska. This is just about a month earlier than the peak downstream migration of salmon, steelhead, and sea-run cutthroat smolts in the coastal rivers of these locales. [9]

8. The survival figures used here were reported by Sumner (1953) for an unfished sea-run population in Sand Creek, Oregon. Johnston and Mercer (1976) reported that 41 percent of adult spawners (averaged over several Washington populations) recovered from spawning and returned to salt water, but they did not report on what percentage made it back to spawn again. The effect of a harvest fishery can be seen in results from another coastal Oregon study reported by Giger (1972). At the time of Giger's study, Oregon's Alsea River sea-run population supported a popular recreational fishery. In that heavily fished population, only 14 percent of initial spawners returned to spawn a second time.

9. Johnston (1982) suggested that selection may have favored this timing because it places the post-spawning adult sea-runs in position to intercept outmigrating juvenile salmonids in the estuaries, especially in those systems producing abundant chum and pink salmon fry. Sumner (1972) told of sea-run adults captured in early spring along the lower Kilchis River, Oregon (then one of that state's most important spawning streams for chum salmon), with stomachs bulging with juvenile chums. The fishing writer Roderick Haig-Brown (1947) described springtime forays of sea-run cutthroat trout into the lower reaches of the Campbell River to feed on juvenile pink salmon. At one time the juvenile pink salmon migration also supported popular springtime sport fisheries for sea-runs in north Puget Sound rivers, and also in British Columbia's Bella Coola River (Hume 1998). Jauquet (2002) has also documented selective feeding by sea-run cutthroat trout on outmigrating juvenile chum salmon at the mouths of several small tributary streams in south Puget Sound.

It will be six or seven weeks before the eggs our trout deposited hatch out. The infant trout will be only about one-half inch long when they hatch, and will not yet be able to feed or move about. For the next couple of weeks they will remain in the gravel and draw their sustenance from an attached yolk sac. During this stage, they are called alevins (pronounced 'a-lə-vəns). By the time their yolk sacs are depleted, they will have grown enough to work their way out of the gravel and seek their own food. They will then be called fry. Swimup—that is, emergence of the fry from the gravel—may occur anywhere from March through June, depending upon the locale. [10]

The newly emerged fry are still little more than one inch long, and are not yet able to handle strong currents. They spend the first couple of weeks or so of their free-swimming lives in the quieter nooks and crannies along the channel-margins and stream edges. They are also discovering that they live in a predatory, competitive world. They are preyed upon by larger salmonids, by sculpins, and even by aquatic insects.

Sculpins have often been singled out as efficient predators of emerging salmonids. They patrol the streambed, hunting alertly for anything that moves. In his book *Fisherman's Fall*, Roderick Haig-Brown described how sculpins kept in an aquarium reacted when they detected movement. They would perch on

stiffened ventral fins, he wrote, cocking their heads downward or sideways to follow the movement. They might adjust their position two or three times as they zeroed in on their prey, then they would pounce, quickly and violently. Sculpins abound on most every stream bottom, so you can imagine how effective they might be against defenseless alevins working their way out of the spawning gravel. But cutthroat trout and sculpins evolved together in these coldwater stream habitats without the one overwhelming the other. Even though sculpins may take their share of the newly emerged cutthroat fry, their impact is probably no more severe than any other predator.

Aside from outright predation, the little cutthroats face some stiff competition just for living space. After a few weeks of feeding and growing in the channel margin habitats, the young-of-the-year cutthroats move out into deeper pool and pocket habitats where they must find their places in the hierarchy of existence in the summer nursery stream. Perhaps their most assertive rivals at this period of their lives are fry of the coho salmon. Despite their larger body size, adult coho will use the same kinds of small tributaries for spawning that adult sea-runs select. Also, juvenile coho rear in streams for a year (two years in northern climes) before going to sea, so the young of the two species frequently exist in sympatry. But coho fry have a competitive edge

10. Biologists often express development time to hatching and swimup in terms of degree-days, aka thermal units or temperature units, determined by multiplying the cumulative number of degrees that the water temperature is above a threshold of zero degrees Celsius by the number of 24-hour days above that threshold. When expressed in this way, eggs of sea-run cutthroat trout have required anywhere from 362 to 500 degree-days to hatch and another 100 to 350 degree-days for development to the swimup stage (Merriman 1935, Lavier 1963, Turman 1972). More detailed discussions of the relationship between water temperature and development can be found in Elliott (1978), Elliott et al. (1987), Crisp (1981, 1988), and Beacham and Murray (1990).

because they emerge earlier and are somewhat larger, and also because their body shape at this size is optimal for superior swimming performance and maneuvering in the slow waters of pools, whereas the body shape of cutthroat fry confers less of an advantage. The juvenile coho also seem by nature to be more aggressive little fish. The coho fry commandeer the choicest pool habitats, forcing the little cutthroat fry to reside primarily in the riffles, where they remain until high winter flows reduce the aggressive tendencies of the coho. Similarly, where young-of-the-year steelhead are also part of the stream fish community, they tend to dominate young-of-the-year cutthroat trout in riffles. This is again most likely because of a body morphology more favorable than that of the young-of-the-year cutthroats for swimming ability in riffles, plus again, a higher level of innate aggressiveness in agonistic encounters. These kinds of exclusionary interactions may set natural limits on the size of coastal cutthroat populations in streams where either cutthroats and coho or cutthroats, coho, and steelhead co-occur. Ratios of juvenile coho to juvenile cutthroats may be as high as 4 to 1, 10 to 1, or even 100 to 1 in undisturbed coastal streams. [11]

When winter comes and aggressive tendencies subside, juvenile cutthroats that have survived their first summer of rearing and have not abandoned the stream reach for better rearing opportunities elsewhere, enter

11. In 1988, Peter A. Bisson and two of his colleagues published a fascinating paper on the body forms of juvenile coho, cutthroat, and steelhead and how they relate to channel hydraulics and habitat use in streams (see Bisson et al. 1988). They found that young-of-the-year coho have relatively deep, laterally compressed (i.e., flattened) bodies with large median and paired fins, a body form ideal for making quick turns in slow water, which facilitates feeding, territorial defense, and escape from predators in slow pools. Young-of-the-year steelhead bodies are more cylindrical with short median fins but long paired fins, a body form more suited to sustained swimming and position-holding in swift currents. Young-of-the-year cutthroats are intermediate between these two body forms, leaving them at a disadvantage to coho in slow pools and to steelhead in riffles. The observations of Tripp and McCart (1983) and Trotter et al. (1993) along with the extensive research of Glova (1984, 1986, 1987) will give you fuller details on agonistic interactions and dominance of juvenile coho over coastal cutthroats. Hartman and Gill (1968) were among the first to study the distribution of cutthroat trout and steelhead in streams. Mitchell (1988) has also done research on interactions and dominance of juvenile steelhead and cutthroats.

the pools with the juvenile coho and steelhead, but they seek shelter amid boulder and cobble interstices, within log jams, and under overhanging banks.

Despite being forced to reside in less than optimal stream habitat during the summer, the little cutthroats grow rather rapidly. By September they will average 2 to 3 inches in length. They are also becoming handsome little fish: heavily spotted, olive-colored backs shading through gold on the sides to silvery on the bellies, grayish-violet parr marks along their sides, and orange fins. At this stage they are hard to distinguish from coho fry, but a close observer will notice that the latter have edges of white lined with black on their anal and dorsal fins, which the cutthroats do not possess.

Two years of freshwater rearing is the norm in many populations, and in other populations that can extend to three years or even longer before the juvenile sea-runs migrate to salt water for the first time. The longest recorded stay in fresh water before the initial seaward migration was a fish that remained in Oregon's Alsea River for six years. Hatcheries can and commonly do rear sea-run cutthroats to smolt size [12] in one year, but this is not the norm for populations reared in the wild. During their time in fresh water, juvenile sea-runs may move upstream or down within the stream to take advantage of seasonal resources or, as they grow, to find more suitable habitat. British

Columbia scientist T.G. Northcote calls these *trophic* and *refuge* migrations, respectively. [13]

Movement down to the sea begins in some streams as early as March and peaks in mid-May in Washington and Oregon. Further north, migration to salt water is later, with the peak in Alaska occurring from late May to early June. The size of the fish entering salt water for the first time ranges from about 6 inches in fork length for a 2-year-old migrant to 8 to 10 inches in fork length for first-time migrants of age 3 and older. The movement of out-migrants in one Alaskan stream studied over several years occurred during the hours of darkness on moderate stream flows. Migration stopped during extremes of high or low water. [14]

Sea-run cutthroats grow at a rate of about one inch per month while in the salt chuck. But they do not embark on long-distance oceanic migrations like the salmon and steelhead. Along the Oregon and Washington coasts they frequent a zone that ranges from 6 to about 30 miles out, well beyond the surf zone but often still within the influence of the Columbia River plume, which is extensive along these coasts in the summer months. The farthest off these shores a sea-run cutthroat has been captured is 41 miles, and that fish, too, was still in the river plume. In contrast, in those parts of the range where the saltwater areas are sheltered, e.g., Puget Sound, Hood Canal, the Strait of Georgia, and the

12. A smolt is a juvenile trout or salmon that is undergoing or has just completed a process of physiological change that will enable it to survive in salt water. The process usually takes place in the spring of the year and is typically accompanied by a change in body color to a silvery hue. The fish may assume a more slender body form as well. In sea-run cutthroat, the process also appears to be size-related in that there is a minimum size that needs to be achieved before high survival in sea water is ensured. For those wanting to learn more about this process, a good starting point would be three review papers written by Folmar and Dickhoff (1980), Wedemeyer et al. (1980), and Schreck (1981), and a book Chapter written by Hoar (1988). For studies specific to smoltification in sea-run cutthroat trout, see Giger (1972), Tipping (1986), and Yeoh et al. (1991).

13. See Northcote (1992, 1997a). Giger (1972) and Hartman and Brown (1987) also published observations on the movements of sea-run cutthroat trout within the freshwater environment.

14. There appears to be a relationship between the size (and age) of sea-run cutthroat smolts and the type of saltwater environment they

island archipelago of southeastern Alaska, the sea-runs favor waters that are close inshore indeed, and exhibit a reluctance to cross even narrow bodies of open water. In the Puget Sound region, F.M. Utter and five colleagues discovered genetic differences between sea-run cutthroat populations separated by bodies of deep albeit narrow salt water, indicating that such areas can indeed act as barriers to gene flow among these populations. [15]

Some of these sheltered saltwater areas have long been places where saltwater angling for sea-runs is most popular. The fish appear to follow a nomadic life style, traveling about mostly in pods of, say, five to fifteen fish, but occasionally in quite large schools. They forage along the gravel beaches, around oyster beds, and in the patches of eel grass, moving in and out with the tides, looking for scuds, sand fleas, shrimp, insects that fall upon the water, or the occasional stickleback, sculpin, sand lance, herring, or other small baitfish. Anglers take advantage of this inshore foraging by searching the gravel beaches and eel grass pockets, or by casting off the mouths of creeks and even the little beach trickles. Anglers prospect for the fish by wading and casting to these places if beach access is permitted, or by rowing along within casting distance if it is not. One old angler's adage on how and where to position yourself for proper fishing is to row in until you can just see the bottom beneath your boat or float tube, then *cast inshore toward the beach*! [16]

15. Details of the saltwater distribution and ecology of sea-run cutthroat trout can be found in papers by Loch and Miller (1988), Pearcy et al. (1990), and Pearcy (1997). Other sources of information on saltwater diets and feeding ecology of sea-run cutthroats in various parts of their range include Clemens and Wilby (1946), Armstrong (1971), Bax et al. (1980), Fresh et al. (1981), and Jauquet (2002). The population genetics studies of Utter and his colleagues that demonstrate how deep areas of salt water can be barriers to gene flow among some sea-run cutthroat populations are reported in Utter et al. (1980).

will be entering. For example, in the protected waters of Puget Sound, the Strait of Georgia, and the Columbia River estuary, smolts are predominately age 2 and a little over 6 inches (about 160 mm) in fork length (see Johnston 1979 and Michael 1989). But in streams along the coasts of Oregon and Washington that are exposed to pounding surf, smolts are a year or more older and 8 to 10 inches (203 to 254 mm) in fork length (see Sumner 1962, Giger 1972, and Johnston 1982). Jim Johnston of the old Washington State Game Department (now Washington Department of Fish and Wildlife) suggested that the physical and biological characteristics of the marine environment encountered by sea-run cutthroat trout has selected for differences in smolt age and size among populations (see Johnston 1982). The studies of Alaska's coastal cutthroat trout that are discussed in the last part of this paragraph are reported in a series of papers by Jones (1972–1976).

16. My favorite book on angling for sea-runs in salt water is Steve Raymond's *The Estuary Flyfisher* (Portland, Oregon, Frank Amato Publications, 1996). Les Johnson's book, *Fly-Fishing for Coastal Cutthroat Trout* (Portland, Oregon, Frank Amato Publications, 2004), is also good on the how-to aspects. The late Enos Bradner's book, *Northwest Angling* (Portland, Oregon, Binfords and Mort, 1969), has a Chapter on the sea-run cutthroat as a sport fish, and a little paperback published in British Columbia a few years ago by Karl Bruhn and others entitled *A Cutthroat Collection* (Vancouver, British Columbia, Special Interest Publications, 1984) also provides how-to information.

Not only do sea-run cutthroats stay close inshore or in coastal waters during their time in the salt, seldom do they range very far from the mouths of their parent streams. There is a record of one fish that originated from an Oregon hatchery that did turn up 83 miles away from the stream in which it was planted, but less than half that distance is the normal range.

Sea-run cutthroats return from saltwater in the late summer, fall, or winter of the very year they enter it; in other words, only two to eight or nine months after first migrating to the salt. Depending upon the month of return, these fish either overwinter or spend what remains of the winter in fresh water. But not all of them do so for the purpose of spawning. That's the reason I drew the little black arrow from the overwintering box back to the saltwater feeding box in Fig. 3-2—to depict that in some populations, a sizable proportion of the cutthroats returning from the salt are sexually immature fish that entered the salt for the first time just that spring, and will not be capable of spawning during this initial return to fresh water. These fish will have to go back to the salt for a second season of feeding and growth. They will not ripen and spawn for the first time until their second return to fresh water.[17] This has an important management implication, which I shall discuss later in this Chapter.

The overwintering behavior of the non-spawners may account for reports of high rates of straying in sea-run cutthroat trout. In a study conducted at Petersburg Creek, Alaska, back in the 1970s, fish tagged as smolts on their initial migration to salt water were recaptured the following fall in 13 different streams. Many of these fish were recaptured in Petersburg Creek the second fall, but not the first fall after release. When it comes time to spawn, however, if genetic studies carried out in Washington streams can be believed, homing to their natal streams is very precise.[18]

One interesting variation on this overwintering theme is exhibited by many sea-run cutthroat populations in the island archipelago and adjacent mainland area of southeastern Alaska. Many of these populations overwinter in lakes rather than streams, but not always lakes in systems where they will eventually spawn. In the spring of the year, prior to spawning, these fish exit the overwintering lakes and migrate back down the outlet streams into salt water. Once back in the salt, they disperse along the shoreline in both directions until they come to their natal systems. They travel in the nearshore, intertidal zone at a rate of about a mile or two per day. About one-third of these traveling fish dip into one or more non-natal tributaries enroute, but quickly retreat to the salt to resume their journeys. Some fish travel as far as 32 miles to reach the tributaries in which they will finally spawn.

17. The proportion of sexually immature fish in a run varies with the locality, from a low of 5 percent in one study carried out in coastal Oregon (Sumner 1953) to fully 50 percent of the run as reported from Alaska (the series of reports by Jones 1972–1976), and is keyed to the age of the fish when they undergo smoltification and enter the salt for the first time. Studies made by State of Washington biologists several years ago (Johnston 1982, Fuss 1982) revealed that sea-run cutthroat females rarely spawn for the first time before they reach age 4 (in some populations on Washington's Olympic Peninsula and other populations in Alaska, first-time spawning did not occur until age 5). Age-2 smolts return to fresh water for the first time as age-2 sub-adults going on age 3, too young to spawn during that first return. Age-3 smolts return as age-3 sub-adults that are beginning to mature sexually, and these fish likely will spawn during their first return as they reach age 4.

18. The Petersburg Creek study referred to here is reported in Jones (1975). The genetic studies indicating precise homing were carried out by Campton and Utter (1987) and Wenberg and Bentzen (2001). These findings reinforce earlier results

Another interesting feature is that once in their spawning tributaries, some of these fish ascend to the tiniest headwater reaches, trickles really, where the stream width is often no greater than half a foot, to engage in the spawning act. [19]

If a sea-run cutthroat survives its first return to fresh water, whether for spawning or not, it will return annually, about the same time each year, for the rest of its life. As already indicated, immigration takes place from mid-summer through winter, with the peak in the fall in many places. Fly-fishermen in Washington and Oregon often say that the first late-summer emergence of flying termites, when these insects come boiling out of the rotting streamside stumps, is the signal to start limbering up the fly rod for the "harvest trout" season. The time of entry is fairly consistent from year to year for specific streams, but it varies widely between streams, and it seems to be a bit earlier on the average in the northern part of the range.

In Puget Sound, Hood Canal, and the sheltered waters of the Strait of Georgia in southern British Columbia, sea-run cutthroat populations of the larger rivers that empty into these areas return according to this late summer-fall entry timing. This has come to be referred to as *early-entry* timing. But the sea-run populations of most of the small creeks discharging directly to the salt in these same areas return distinctly later in the year. December and into March, with the peak in January or February, characterizes the so-called *late-entry* timing of these populations. It is hard to label this behavior as overwintering in fresh water as the box in Fig. 3-2 implies. But even the non-spawners of these small systems move back into their freshwater confines if only briefly, hence the red arrow drawn in the figure. Anglers who fish the salt in these regions find sea-runs every month of the year, it is true, but their catch per unit of fishing effort is always lowest during that month or so when most of the late entry fish have returned to their little streams.

What delineates the "larger" rivers with their early-entry timing from the "smaller" streams where the late-entry timing prevails? Former State of Washington biologist Jim Johnston, who studied the migratory behavior of Puget Sound and Hood Canal sea-run populations, found that a water flow criterion would work. Streams with summer base flows of 20 cubic feet per second (cfs) or less, and discharging directly into salt water, invariably host runs with the late-entry timing, he reported, whereas streams with summer base flows greater than 70 cfs host early-entry populations. Curiously though, he also found that there are some streams along Hood Canal with summer base flows between 30 and 70 cfs that receive both early-entry and late-entry runs with a measurable lull between the two.

reported from Oregon by Bulkley (1966) and Giger (1972), which also indicated precise homing.

19. Cheryl Seifert of the Alaska Department of Fish and Game Division of Sport Fish spent two years tracking these sea-run populations using radio telemetry. Her report on the findings of this study is in the proceedings of the Reedsport, Oregon conference I referred to earlier in this Chapter. See Jones and Seifert (1997).

Johnston's paper, presented at a conference in 1981 but not widely available until the proceedings were published a year later, highlighted another interesting finding. In tests to determine if wild stocks of early-entry sea-runs could be held year-round in saltwater net pens following introduction to these pens as smolts, 98 percent of the fish died after six months. But late-entry sea-runs held in the same pens suffered only 50 percent mortality after six months.[20] So the difference in entry timing may have something to do with how long seawater tolerance endures in the two stocks. This also suggests that, unlike their cousins, the steelhead, whose ability to function in seawater extends out through one or more years, sea-run cutthroat trout may require periods of return to fresh water in order to regenerate saltwater tolerance. As far as I know, this is an aspect of sea-run cutthroat physiology that has never been investigated.

Now for a word on the longevity and maximum size of sea-run cutthroat trout. Data compiled from throughout their range indicate that fish lucky enough to survive the pressures of predation and harvest angling to die of old age will probably max out at age 7 or 8, and will attain a fork length of about 20 inches. Fish in this age bracket will have spawned probably three or four times. The two oldest sea-run cutthroat trout on record were both taken at Sand Creek on the Oregon coast, one at a State-operated migrant trap in the late 1940s and the other by an angler a few years later. The trapped fish was 10 years old, but despite its advanced age, was only 17 inches in fork length. The angler-caught fish was either 13 or 14 years old. It was captured 11 years after having been fin-clipped as a smolt at the Sand Creek trap. Assuming a smolt age of 3 years, most common in Sand Creek, the fish was 14 years old when captured. Assuming a smolt age of 2 years, less common in Sand Creek but still possible, would age the fish at 13 years. It was 20 ¼ inches in fork length, and its weight (dressed) was 3 ¼ lbs. The largest fish in terms of weight that appears in any record is the Washington State hook-and-line record sea-run from Carr Inlet in Puget Sound. That fish weighed an even 6 lbs. Neither its age nor its length were recorded.

When the sea-runs come back into fresh water following their few months' sojourn in the salt, they are bright, silvery fish with the cutthroat slashes under their jaws so faint as to be nearly invisible. But they darken as the time in fresh water lengthens, and especially the males take on brilliant amber and pinkish orange colors that easily rival the beauty of the autumn foliage. I grew up fishing for sea-runs in the rivers and creeks of southwest Washington, where the early-entry run timing prevailed. I always cast my flies to the slower, quieter pools or places where the water

20. See Johnston (1982). Further information about the saltwater net pen experiments can be found in reports by Mercer and Johnston (1979) and Mercer (1980).

deepened and slowed—often places that other fisher-men would pass by, places with a studding of boulders on the bottom perhaps, or some kind of obstruction in the water, or a log jam or overhanging brush and trees along the far bank. If the fish were there and in a tak-ing mood, they could always be depended upon to put up a solid battle on light fly tackle.

Lacustrine or Lake-Associated Coastal Cutthroats

The well-watered Pacific coastal ecoregion is endowed with thousands of lakes. A catalog compiled by the State of Washington lists 3,813 lakes and reservoirs west of the Cascade crest in that state alone. Lakes of every size can be found, from truly imposing bodies of water (such as Lakes Washington and Sammamish near Seattle; Whatcom near Bellingham; Crescent, Ozette, and Quinault on the Olympic Peninsula; Har-rison, Cowichan, Sproat, and Great Central in British Columbia; and Siltcoos, Takanich, and Ten-Mile on the Oregon coast) to the smallest of bog lakes and beaver ponds. Most of the natural lakes that hosted fish populations at all had coastal cutthroat trout as the dominant native species, although in some lakes the cutthroat trout cohabited with rainbow trout and (in British Columbia especially) Dolly Varden charr.[21] In addition, many man-made lakes and reservoirs were formed when major coastal rivers or their tributaries were dammed for hydroelectric power, water storage, or flood control. Remnants of their sea-run or riverine cutthroat populations adapted to a landlocked, lake-dwelling lifestyle.

Most of the literature on lacustrine populations of coastal cutthroat trout focuses on the period actually spent in the lake.[22] However, we do know that these populations can spawn efficiently in the small inlet tributaries, and sometimes also in lake outlet streams. In lakes having more than one spawning tributary, homing fidelity is maintained by the individual stocks of the lake.

Some scientists draw a distinction between inlet and outlet spawning populations, referring to those that spawn in lake inlets as *lacustrine-adfluvial* populations and to those that spawn in lake outlets as *allacustrine* populations. Names aside, setting them apart is appro-priate because each has a hereditary basis. I am not aware of any studies demonstrating this specifically for lacustrine coastal cutthroat trout, but the hereditary basis for inlet vs. outlet orientation has been shown for Yellowstone Lake populations of the Yellowstone cutthroat subspecies (Chapter 9) and for lacustrine populations of rainbow trout in two British Columbia Lakes. In each of those studies, young produced by outlet spawners oriented and moved in an upstream direction, which they would need to do to reach the

21. Native coastal cutthroats can still be found in some of these lakes, but it seems that every year you have to get farther away from civilization to find them. Closer in, particularly in areas where urban and suburban develop-ment has encroached, most lakes have been converted to put-and-take trout fisheries supported by stocked rainbow or brown trout, or to warm-water fisheries based on introduced species such as bass, perch, crappie, sunfishes, and brown bullhead.

22. Details of the life history and ecology of lake-dwelling populations from across the coastal cutthroat range are published in Armstrong (1971), Frenette and Bryant (1993), and Bry-ant et al. (1996) for Alaska (the Alaska Department of Fish and Game Divi-sion of Sport Fish has also issued a long series of reports with additional notes on lake-dwelling populations); Andrusak and Northcote (1971), Schutz and Northcote (1972), Narver (1975), and Nilsson and Northcote (1981) for British Columbia; and Pierce (1984), Beauchamp et al. (1992), Nowak (2000), Nowak and Quinn (2002), Meyer and Fradkin (2002), and Nowak et al. (2004) for Washington. Shepherd (1974) has published on the activities of coastal cutthroat trout in small bog lakes.

lake; and young produced by inlet spawners oriented and moved downstream, which *they* would need to do to reach the lake. In his 1992 monograph, Dr. Behnke cited a negative example as additional evidence. He referred to the Master's thesis studies of Leo Lentsch, completed in 1985, of a trout population that had been introduced into Colorado's Emerald Lake less than 100 years before (about 25 trout generations) from an inlet spawning stock. Although some adults had begun to use the lake outlet for spawning as well as the inlet, a large proportion of the young produced in the outlet moved downstream, over a large falls, and were lost to the population. Evidently, even 25 generations had not been enough for natural selection to reverse the inlet-spawning programming of the juveniles for downstream movement.

Most studies agree that lacustrine coastal cutthroat juveniles spend considerable time rearing in their natal streams before their initial migration to the lake, recruiting to the lake for the first time at ages ranging from 1 to 4 years and fork lengths ranging from 5 to 7 or 8 inches. Stream growth may be a little slower in northern climes. At Margaret Lake, Alaska, U.S. Forest Service biologists reported that the size range for initial recruits was 2.8 to 5.5 inches in fork length but the age range of these fish was the same, from 1 to 4 years.

Once in the lake, feeding and behavior of the small cutthroats appear to depend on whether they are the only salmonid predator present, or if they share the lake with another salmonid, such as rainbow trout or Dolly Varden charr, in numbers great enough for encounters to be common. Coastal cutthroats living as the sole salmonid species in a lake make broad use of its habitat and range from shallow littoral (inshore) to open limnetic (offshore) areas (except in small bog lakes where the fish may adopt fixed focal points of activity that appear associated more with cover—under floating mats of *Sphagnum*, for example—than with food availability). Midwater food items (zooplankton, *Chironomus* pupae, and the larvae of *Chaoborus* species) comprise the bulk of the diet, but surface food (terrestrial insects and floating or emerging aquatic insects) and bottom-dwelling prey are eaten as well.[23] But in lakes where rainbow trout or Dolly Varden also occur and encounters could be commonplace, the behavior patterns of the cutthroat trout change and feeding zones are partitioned. Under these conditions, the little cutthroats tend to congregate in the inshore areas, leaving the offshore areas to the competing species. They also abandon the midwater prey items to the rainbow trout and take most of their food off the bottom or the surface—more off the surface than the bottom when Dolly Varden charr are their competitors.[24]

When lake-dwelling coastal cutthroats reach 9 to

23. Dr. Behnke has pointed out that many lake-dwelling populations of coastal cutthroat trout have developed morphological specialization for lacustrine life. For example, the numerous gill rakers in cutthroat trout from Lake Crescent, Washington, enable efficient feeding on zooplankton. In nearby Lake Sutherland, the trout exhibit high numbers of basibranchial teeth. What basibranchial teeth contribute to the quality of lacustrine life for these fishes is not known at present, but it is a trait they share with lake populations of Lahontan cutthroat trout (Chapter 5) and also Yellowstone cutthroat trout (Chapter 9).

24. Partitioning of the available habitat in this way is called *interactive segregation*, and is covered in more detail by Nilsson (1967) as it applies to fishes. It also occurs among other taxa. For example, one of the papers considered to be a classic on the subject describes interactive segregation among ants of the genus *Myrmica* (see Brian 1956).

12 inches in fork length, they become more piscivo-rous in their selection of food items, although some researchers have reported that only a modest degree of piscivory develops under allopatric conditions. They also spend more of the year in the limnetic zone of the lake. In lakes large and deep enough to stratify during the summer months, the larger cutthroat trout may often be found just beneath the thermocline in a zone of water temperature ranging around 45 to 50 degrees Fahrenheit. Three-spine sticklebacks are a preferred prey of these larger cutthroats despite their sharp spines. Sculpins are also eaten, as are kokanee and juvenile sockeye salmon if and when they are present. Yellow perch, a warmwater, spiny-ray species widely introduced into Pacific Northwest lakes and waterways in the late 1890s and early 1900s, appear quite com-monly in the diets of the larger cutthroats in lakes where the two species co-occur; for example, in Lake Washington near Seattle.

Lake Washington is also noted for a landlocked population of longfin smelt, normally an anadromous species, that developed in the lake in the early 1960s and quickly became a preferred prey of the lake's cutthroat trout population. Researchers from the University of Washington, who have tracked the move-ments and feeding of the lake's cutthroats, say that seasonal changes in the distribution of the cutthroats

follow closely the movements of the longfin smelt, with the large cutthroats moving into the shallows primarily during the month of May when the smelt are in the shallows and out to deeper water when the smelt move offshore. Being a dyed-in-the-wool fly-fish-erman, I am reminded here of the fabled springtime fly-fishing in lakes of the northeastern U.S. where anglers troll or cast streamer flies for the landlocked salmon and large brook trout that mass in the shallow water to feed on rainbow smelt. I'm not aware that local fly-fishermen have caught onto this possibility in Lake Washington. A few anglers target the lake's large cutthroat trout using trolling hardware, but I haven't heard of many fly-fishers out there.

In the larger lakes with adequate food supplies, coastal cutthroat trout can grow to impressive sizes. Lake Crescent, a large, deep lake on the northern boundary of Olympic National Park in Washington, has long been known as a producer of large trout for anglers with perseverance.[25] These trout attain their large size on a diet consisting primarily of kokanee after first subsisting on zooplankton and the occasional terrestrial and aquatic insect until they reach a length of about 9 to 12 inches. Although a 7- or 8-pound cutthroat is considered to be a very large fish by Lake Crescent standards, in July of 1961 a local angler named Billy Welsh took a 12-pounder on a slow, deep-trolled

25. Lake Crescent gained fame for its extraordinarily large trout in the late 1800s and early 1900s. There were two types. One, known locally as the "blueblack" or "Beardslee trout," reached weights of 15 to 16 pounds. The other, called "speckled trout," seldom topped 7 to 8 pounds. David Starr Jordan examined specimens of these trout in 1896. He first named the blueblack *Salmo gairdneri beardsleei*, both to acknowledge that the fish was a form of rainbow trout, and to recognize Admiral Lester Beardslee, who had sent him the specimens. Later Jordan changed his mind and elevated this fish to full species status with the name *Salmo beardsleei*. Thinking that the speckled trout was also derived from rainbow trout, Jordan named it *Salmo gairdneri crescentis*, then later gave it full species status as well with the name *Salmo crescentis*. Jordan was right the first time about the Beardslee trout. It is indeed a form of rainbow trout. But he erred with regard to *crescentis*, which is a local race of coastal cutthroat trout. In Jordan's defense, distinguishing between Beardslee rainbows and *crescentis* cutthroats was never very easy. Only when the fish were ready to spawn did they take on true rainbow or cutthroat colorations. The rest of the time they were both, in Jordan's words,

silver plug. That fish, which measured 32 inches in total length, stands as the Washington State hook-and-line record for the coastal cutthroat trout. However, as impressive as that fish was, it was outdone weight-wise in the summer of 2002 by a cutthroat trout from Lake Washington that weighed 14.9 pounds (it was weighed on an unofficial scale, then dressed, so didn't qualify for state record status). That fish, a male, measured 29 ¼ inches in length and 20 ¼ inches in girth. It was 6 years old according to an analysis of its scales, and had spawned twice in its lifetime, at ages 4 and 5 (it had evidently skipped spawning at the onset of age 6, the year it was captured). According to the angler who caught it, the behemoth cutthroat had only two food items in its stomach, both 9-inch yellow perch.

Hydropower and water storage reservoirs can also harbor populations of coastal cutthroat trout that land-lock and adopt a lake-dwelling life style. Remember Cath-la-pootl, aka Lewis River, the type locality for the cutthroat trout? Three hydroelectric dams, Merwin, Yale, and Swift, completed in 1931, 1954 and 1958, respectively, now impound reservoirs along the north fork of that river. Although a trap-and-haul operation was conducted at Merwin, the lowermost dam, for the purpose of moving salmon to upstream release sites, that operation was closed down when the other two dams were completed. The three reservoirs were stocked for years with hatchery-reared rainbow trout. Kokanee were introduced and became self-sustaining in Merwin and Yale reservoirs. More recently, large-mouth bass, bluegills, brown bullheads, and even tiger muskies have been stocked into Merwin Reservoir. Bull trout, native to the Lewis River system, are also present in Yale and Swift reservoirs. Sea-run cutthroats return to the lower river, augmented until recently by releases of hatchery-reared smolts, but the status of the wild stock is now uncertain.

I was made aware of the cutthroat populations in these reservoirs by Suzy Graves, a field biologist for the old Washington State Game Department. Suzy's job in the years when she worked on the reservoirs was to quantify the composition of their fish populations and determine which species mix provided the optimum recreational fishery. It was during this study that she discovered native cutthroat populations thriving in Merwin and Yale reservoirs, with many of the fish reaching the 17- to 23-inch size class. She also uncovered a bit of local history that convinced her that these cutthroats were derived from the original Lewis River sea-run stock.

The years following World War I were halcyon years for sea-run cutthroat angling in this part of the Pacific Northwest. Anglers would ride the "fishermen's special" trains out of Portland, then hike to the hold-

"...a deep, steel blue in color with fine specks, and without red at the throat." The only way a fisherman could tell the difference was by the heavier spotting of the *crescentis*. Lake Crescent still grows its fish large, and the locals still call them Beardslee trout and *crescentis*. And in its crystal-clear waters the fish still take on those deep blue and silver colors that make it so hard to tell them apart. But the history of Lake Crescent fishery management has been a jaded one. Over the years since the 1920s, the lake has been a dumping ground for about every strain of hatchery rainbow and non-native cutthroat you could imagine. Despite this, the original strains somehow survived, and are now the focus of what one would hope is more enlightened management. For detailed accounts of Lake Crescent, its fishes, and fishery management, see Jordan (1896), Jordan and Evermann (1902), Scheffer (1935), Garlick (1948), Pierce (1984), Steelquist (1984), Baker (1998, 2000), and Meyer and Fradkin (2002).

ing waters from the end of the rails to fish for "harvest trout" ("blueback" was another local name) in the streams of north-coastal Oregon and southwest Washington. Among the anglers who fished the Lewis River in the late summer and fall of those pre-Merwin years were a few who headed for a point of land at the confluence of Siouxon (pronounced "Soo' sohn") Creek, a mile or so upstream from the location where Yale Dam would eventually be built. The water deepened and slowed here and there was an eddy pool, conditions that attracted migrating sea-runs, and this made it a choice spot for "in-the-know" anglers.

But Suzy learned that a few anglers had continued to visit that spot in the late summer and fall to fish for large, migrating cutthroat trout even after Merwin Dam was completed. These fish couldn't have been true sea-runs, of course, because the dam blocked those migrations and sea-runs were not captured in the Merwin salmon trap. But the run timing of the migrating fish was the same, and the anglers persisted in calling them "harvest trout"—evidence enough for Suzy that they were a landlocked stock. This location was flooded when Yale Reservoir was filled, and its large migratory cutthroats might have been forgotten. But then Suzy went to work on her project and found that not only did cutthroat trout still exist in Merwin and Yale reservoirs, they still made late summer and fall migrations into certain tributaries, presumably to overwinter until spawning. What's more, large cutthroats also ran up out of Yale Reservoir into the old Lewis River channel below Swift Dam on that same late summer and fall schedule. That convinced her that landlocked cutthroats, derived from the original sea-run stock, persisted in Merwin and Yale reservoirs.

The life history of these reservoir trout is strikingly similar to their sea-run brethren. Adult fish move into the feeder creeks and the old river channel between Yale and Swift in the fall, and they remain there until they spawn, which might take place anytime from February through April. After emergence, the new generation of cutthroats spends two to three years in the creeks, then moves down to the reservoirs. Most of them return to spawn for the first time in their fourth year of life, then every year thereafter, with occasional skips. [26]

Like Suzy's landlocked reservoir populations, most of the fish in true lake-evolved populations spawn for the first time at age 4, with perhaps a few reaching sexual maturity a year earlier. However, unlike the landlocks, lake-evolved populations return to spawning tributaries later in the year. Late December, January, and February see most of these fish entering the tributaries, akin to the late-entry timing described earlier for the small-stream sea-runs of the Puget

26. Suzy Graves' work on the Lewis River reservoirs was written up as an agency report in 1982 (see Graves 1982). To flesh out this story of the reservoir cutthroats, I relied on that report and the notes I took during several interviews and two field trips with Suzy, plus copies of other notes and field book pages she would send me from time to time. In the early 1950s, during construction of Yale Dam, archaeologists discovered and excavated a Native American site on a terrace behind the Siouxon Creek sea-run cutthroat fishing hole. It apparently had been abandoned prior to 1800, but evidence was found that indicated its occupants had fished this water as well. That story can be read in two reports authored by Alan Bryan of the University of Washington (see Bryan 1955, 1992). The archaeological site as well as the old sea-run cutthroat fishing hole are now covered by about 100 feet of water.

Sound region. These fish then spawn anytime from mid-January to late May. Further north, as in southeastern Alaska, spawning runs and actual spawning may extend even later into the spring.

Fluvial or River-Migratory Coastal Cutthroats

Many of the principal rivers (perhaps even all of them) within the coastal cutthroat range have populations of cutthroats that do not migrate to salt water but instead spend all of their adult lives, except for spawning, in the main rivers. These are river-migratory (i.e., potamodromous) trout that utilize small tributary streams as nursery areas and the mainstems of large rivers and their principal tributaries for feeding and growth. In some river systems, such as the Willamette in Oregon and the Snoqualmie in Washington, it is easy to show that these populations are not sea-runs because they occur above historically impassible falls. But even rivers open to the sea have their fluvial populations. Tori Tomasson, while a graduate student at Oregon State University in 1978, documented the presence of such a population in Oregon's Rogue River. And I have myself captured many coastal cutthroats exhibiting typical freshwater coloration from the Cowlitz River in Washington at times when silvery sea-run fish of about the same general size were also in the river. Those I captured were so distinctly different from the sea-runs

both in color and in their choice of holding water that there is no doubt in my mind I was tapping a fluvial stock, despite a lack of concrete evidence.

There are many life history similarities between potamodromous cutthroats and sea-run fish. The potamodromous populations occurring in the Willamette Valley of Oregon have probably been studied the most and could probably be considered typical of river-migratory populations throughout the coastal cutthroat range. [27] These fish make regular spawning runs, migrating from the main streams into the smallest tributaries anytime from November to March or April, depending on the tributary. Actual spawning takes place between January and June. Homing fidelity to their natal tributaries is high. After spawning, the adult cutthroats drop back to the main river.

Juveniles of fluvial coastal cutthroat populations remain in their natal tributaries until they reach age 1, 2, or 3. Tributary life for these juveniles is pretty much the same as already described for sea-run and lacustrine juveniles, with the early weeks of their swimup existence being spent in quiet channel-margin habitats, then trophic movements into more productive pool and pocket habitats as they grow larger. Fish that overwinter in the natal tributaries may move downstream a short distance to deeper pools and pockets that provide good refuge habitat.

27. Details about the life history and ecology of potamodromous coastal cutthroat populations are found in Dimick and Merryfield (1945), Nicholas (1978a, 1978b), and Moring et al. (1986), all of whom studied populations of the Willamette River system in Oregon; Tomasson (1978), who conducted research on the Rogue River, Oregon population; and Pfeifer (1985), who worked with fluvial populations in the upper Snoqualmie River, Washington. For a fuller understanding of potamodromy in salmonids, you should also read the excellent review paper on the subject by Northcote (1997).

Even when they are finally ready to move out of the natal tributaries, most predominately at ages 2 and 3, downstream migration into mainstem reaches may be a protracted process that can stretch from June through November. But once in the main stream, the fish grow rapidly. Willamette River fish living upstream from the city of Corvallis were found to feed predominantly on bottom fish, with suckers, pikeminnows and dace comprising the bulk of their diet. Lengths of 16 to 22 or 23 inches are not uncommon among fluvial coastal cutthroats, and fish up to 4 and 5 pounds have been reported from the North Fork Snoqualmie River in Washington.

Fluvial coastal cutthroats typically spend a couple of years feeding and growing in mainstem reaches before returning to the tributaries to spawn for the first time. Spawning migrations in the Willamette River system can span distances of up to 30 miles. The life span of these fish appears to be about the same as for sea-run cutthroat populations, with a maximum age of 6 years reported for the Willamette River system.

Fluvial coastal cutthroats often find themselves sharing the river system with resident or anadromous rainbow trout. When this occurs, the two species partition the habitat. Rainbows are trout of the faster water and heavier riffles. River cutthroats will more often be found close in to steep banks (or under them if they overhang), under logs and overhanging bushes, within root wads, or in the slower, darker waters of deep-flowing pools. Rainbow trout will take up stations above a rock or a riffle; river cutthroats will usually lie a bit downstream of such obstructions. Rainbow trout will generally be more numerically dominant in the lower reaches of a shared river while the cutthroats will usually be more abundant higher up. There may be a hybrid zone where the two species overlap. However, like their sea-run brethren, river cutthroats avoid hybridization by spawning somewhat earlier than the rainbows and in smaller tributaries in different locations in the watershed.

The Stream-Resident Life History Strategy

Like the brook trout of the northeastern U.S. and Canada, the coastal cutthroat is the "native" trout of the upper creeks and beaver ponds of western Oregon, Washington, British Columbia, and southeastern Alaska. [28] The small tributary creeks, characterized by summer flows of 1 to 5 cubic feet per second (cfs), wind down out of coastal hills or off the west slope of the Cascade Range. Sometimes they stair-step their way beneath canopies of fir, cedar, and hemlock, their banks hemmed in by vine maple, alder, huckleberry, and salmonberry. Sometimes they plunge through rocky gorges lined with devil's club. Sometimes, too, they glide through waist-high fern meadows. Beavers

28. Stream-resident populations of coastal cutthroat trout have been and continue to be studied extensively, especially in regard to the effects of forest practices in headwater stream reaches. For reviews of the literature on stream-resident life history and ecology, I refer you first to a couple of my own papers (see Trotter 1989, 2000), then suggest you also look up and read Wyatt (1959), Aho (1977), June (1981), Fuss (1982), House (1985), Wilzbach (1985), Moore and Gregory (1988), Heggenes et al. (1991a, 1991b), Harvey (1998), Harvey et al. (1999), Rosenfeld et al. (2002), Boss and Richardson (2002), Romero et al. (2005), and Gresswell and Hendricks (2007). Specific to logging impacts, the Alsea Watershed Study in coastal Oregon, reported on by Moring and Lantz (1975), Hall et al. (1987), and Connolly and Hall (1999), and the Carnation Creek Study in British Columbia, reported on by Hartman and Scrivener (1990), were both long-term multidisciplinary studies of the effects of logging activities on small headwater stream ecosystems. Additional studies of logging impacts on coastal cutthroat trout populations include Lestelle (1978), Osborn (1981), Murphy and Hall (1981), Bisson and Sedell (1984), Fausch and Northcote (1992), Reeves et al. (1993), Connolly (1996), Connolly and Hall (1999), Young et al. (1999), Latterell (2001), Latterell et al. (2003), and De Groot et al. (2007).

often occupy the hollows and flat places, wherever they can get sufficient pool depth to support their lodges. These streams generally feed into larger rivers or their principal tributaries that support runs of salmon, steelhead, and sea-run cutthroat if they are free of migration barriers that would block these returns. As pointed out in the last section, populations of fluvial cutthroat trout may also be present in these systems, and may migrate into tributaries that are accessible for spawning and early rearing. But often the upper creeks are separated from the waters they feed by some sort of barrier: falls too high for fish to leap, perhaps, or a long reach of steep cascade, or a high-gradient reach of smooth bedrock over which the water flows swiftly in a thin sheet. Or these upper creeks may themselves be segmented into one or more reaches by barriers that block upstream migration from segments below.

For many years, stream ecologists believed that adults of stream-resident trout species led a sedentary existence, spending most if not all of their lives within a short span of stream, maybe 20 to 200 yards at most. A paper by Shelby Gerking in 1959 formalized this theory into what became known as the *restricted movement paradigm*, or RMP. Many published studies support the RMP, not only for stream-resident cutthroat trout, but for stream-resident populations of other species as well. However, in the early to mid 1990s,

research in the midwest and Rocky Mountain regions revealed surprisingly long-distance trophic and refuge movements among stream-resident populations of brook, brown, and interior cutthroat trout, thus challenging the universal applicability of the RMP. By now, quite a robust body of literature has developed around this subject, and some scientists are striving for a middle ground. For example, Dr. Behnke pointed out in 1997 that if all survival needs can be met without long-distance movements, the risk to a trout is less if it stays put. But a trout in an unproductive stream with great seasonal flow fluctuations and marginal habitat faces greater risk by staying put than by moving in search of better conditions.

The concept of *habitat complementation* may come into play here. Habitat complementation refers to the proximity of different but non-substitutable habitats required by a species to complete its life cycle. When habitat complementation is low, critical habitats are relatively far apart from one another and individuals must migrate long distances to find them. When habitat complementation is high, the critical habitats are in close proximity and movement distances are minimal.[29]

Some researchers group stream-resident trout populations into two distinct kinds: (1) *unrestricted resident populations* in which the fish reside in tributaries having no barriers and therefore no physical restrictions

29. The following papers cover all aspects of this discussion from the origin of the restricted movement paradigm to habitat complementation: Stefanich (1952), Miller (1957), Gerking (1959), Wyatt (1959), Bachman (1984), Nakano et al. (1990), Heggenes et al. (1991), Dunning et al. (1992), Riley et al. (1992), Schlosser (1995), Dingle (1996), Gowan et al. (1994), Gowan and Fausch (1996), Young (1996, 1998), Behnke (1997), Rodriguez (2002), and Novick (2006). One of the patterns that often appears in studies of trout movements is that larger fish tend to move farther to satisfy their quest for suitable habitats than smaller fish. This has been documented in brown trout populations (Clapp et al. 1990, Meyer et al. 1992, Young 1994) and in the Bonneville cutthroat population of the Thomas Fork of Bear River, Wyoming (Schrank and Rahel 2004; Chapter 11).

on seasonal movement or ranging behavior for feed-ing, growth, and overwintering; and (2) *reproductively isolated populations* in which all life history phases are restricted to reaches above natural barriers and fish movements occur only within these restricted reaches. Fish in unrestricted resident populations can be quite mobile, with seasonal movements of individual fish extending up to many miles, usually in a downstream direction, in their search for suitably productive feeding stations or refuge habitat for overwintering. This some-times blurs the distinction between these populations and fluvial populations in the same drainage basin, but usually the resident trout stop short of migrating all the way to the larger river mainstems for feeding and growth, even though it is physically possible for them to do so. On the other hand, the trophic and refuge movements of fish in reproductively isolated popula-tions are necessarily restricted to whatever reach sizes exist above the barriers that isolate them. Even in unre-stricted streams, the closer the fish live to the upper extent of fish distribution, the more restricted their year-round range seems to be. There is always some movement from the patches of spawning gravel to the pools and pockets where the feeding stations and refuge habitats are found, but the home territories of these reproductively isolated fishes are generally quite small.

The trout living in these above-barrier reaches may possess altogether different genotypes from those in populations downstream, a result of what British Columbia scientists T.G. Northcote and G.F. Hart-man call *knife-edge selection*. This arises because any fish passing downstream over the lip of the barrier is irrevocably lost from the upstream population, setting up a stringent one-way barrier to gene flow. Therefore, in order for populations to persist in these above-barrier reaches, selection must be strong for traits that would oppose migration. Research on above-barrier popula-tions of coastal cutthroat and rainbow trout by Dr. Northcote and his colleagues indeed shows that these populations possess greater swimming stamina than below-barrier populations, and they also exhibit a strong upstream movement response to current whereas below-barrier populations show either a weaker upstream response or a downstream response. Research on brown trout populations in the British Isles and Scandinavian countries has yielded similar findings. Other characteristics typically exhibited in above-barrier and uppermost trout populations include small body size, early maturation, short life span, retention of juvenile fin and body form, and retention of juvenile coloration (i.e., retention of parr marks) throughout maturity. These have been called *juvenilization* char-acters, and the suggestion has been made that they are responses to long isolation or insularization. [30, 31]

30. Many investigators have now reported distinct genetic differences between above-barrier and below-barrier trout populations (Northcote et al. 1970, Jonsson 1982, Parkinson et al. 1984, Ferguson 1989, Cur-rens et al. 1990, Hindar et al. 1991, Marshall et al. 1992, S. Phelps, Washington Department of Fish and Wildlife, personal communica-tion 1994, Griswold 1996, Latterell 2002, Guy (2004), and Wofford et al. 2005), and for at least one of these differences the adaptive significance has been worked out. Northcote et al. (1970) first reported that rainbow trout from above and below a waterfall on Kokanee Creek, B.C. exhibited differences in both meristic characters and lactate dehydrogenase (*Ldh*) genotype. The above-falls *Ldh* isozyme was subse-quently found to be more efficient in lactic acid conversion than the below-falls isozyme, thereby confer-ring greater swimming stamina to the above-falls trout (Tsuyuki and Williscroft 1973, 1977). In labora-tory studies, juvenile rainbow trout homozygous for the above-falls form of *Ldh* responded to water current by moving upstream against the flow, whereas trout homozy-gous for the below-falls *Ldh* form tended to move downstream with the flow (Northcote and Kelso 1981). Greater swimming stamina

The trout that populate the small upper tributaries seldom grow very large. Six or 7 inches is about as large as one could expect. A 10-inch trout would be a real trophy. Nor do they live to a very old age. A typical population age distribution might be 95 percent of the fish three years old or younger, 4 percent age 4, and 1 percent or less age 5. They make up for this low longevity by maturing early, with the majority being ready to spawn for the first time at age 2 and the rest certainly at age 3. The number of eggs produced per female is low in these populations, running typically in the 50 to 250 range (versus 500 to 1,700 eggs per female for the sea-run, fluvial, and lacustrine life histories). But the eggs are relatively large compared to the body size of the females. Thus, if fecundity is expressed as number of eggs per gram of female body weight, these stream-resident populations may rank closely with the larger females of the other life history forms.[32] Spawning occurs in the spring, with movement to the spawning gravels commencing when water temperatures reach about 41 degrees Fahrenheit. Depending on the location, the altitude, and the timing and amount of snow runoff, this could be anywhere from late March to the middle of June, or perhaps even later in the northern portions of the range.

Small-stream cutthroats are predominately drift feeders, i.e., they subsist on food items brought to them

and strong upstream movement response to current are just the sorts of traits that would be selected for under knife-edge selection, and would confer advantage to trout living in above-falls stream reaches (Northcote and Hartman 1988). In a study reported in 2001, researchers from Oregon State University found that even though coastal cutthroat trout living in a segment of Camp Creek (Oregon Coast Range) located above a barrier to upstream migration were themselves quite migratory within the basin upstream from the barrier, only about 1 percent of the trout ever went over the falls and were lost to the population (see Hendricks and Gresswell 2001 for details).

31. Population densities of resident trout living in the highest headwater stream reaches, whether restricted by impassable barriers or not, are often 5 to 100 times lower than population densities farther downstream where habitats are more favorable and the stream is more productive. In these headwater populations, population density and survival may not be governed by density-dependent factors as they are in the downstream populations. In a classic series of studies of the quantitative ecology of brown trout, J.M. Elliott (see Elliott 1989, 1994)

found that a low-density population of brown trout isolated above a waterfall in a small stream in England's Lake District exhibited little movement, losses were due entirely to mortality rather than migration, and the number of survivors at different stages of the life cycle increased linearly with parent stock size. In contrast, in the more favorable environs downstream where population densities were higher, trophic and refuge movements were more extensive, losses were due to migration as well as mortality, and the number of survivors at different stages of the life cycle initially increased and then declined as parent stock size increased. Citing a hypothesis originally proposed by Haldane (see Haldane 1956), Elliott concluded that selection favors those genotypes adapted to density-dependent factors in favorable habitats that allow for high population densities, but only those genotypes most adapted to density-*independent* factors are selected in less favorable habitats that allow for only low population densities.

32. The large egg size relative to the body size of headwater stream-resident coastal cutthroat females is based on my own observations and those of a colleague who has worked with them for a number

of years (Brian Fransen, Weyerhaeuser Company, Federal Way, Washington, personal communication 2000). Other biologists (Blackett 1973, Watson 1993) have published confirming observations for other salmonid species. Blackett (1973) studied a resident population and two nearby anadromous populations of Dolly Varden charr in southeastern Alaska. In the resident population, fecundity averaged 66 eggs per female compared with 1,888 eggs per female for each of the two anadromous populations. But the average egg diameter of 3.6 millimeters for resident females fell well within the range for the anadromous females, as did the number of eggs per gram of female body weight. Larger egg size generally confers a survival advantage to alevins in other salmonid populations (Bams 1969, Fowler 1972, Wallace and Aasjord 1984, Beacham and Murray 1990). If this holds true for headwater cutthroat populations as well, it means that headwater females pack a great deal of alevin-survival potential into the few eggs they do produce. Elliott (1989) considered early maturation and the opportunity to spawn more than once in their lifetimes despite their short life spans to be selective advantages for trout living in headwater and above-barrier stream

by the currents. But they do keep an eye on the surface for emerging or egg-laying aquatic insects or for hapless terrestrial insects that might fall into the stream. The trout may take up positions near the tailout of a pool if there is a structure or trench to soften the gathering flow and provide sufficient depth against overhead predators, or somewhere up alongside the main current tongue (called the *thalweg* by stream hydrologists), or in the deeper water beneath the main current tongue where erosional forces have excavated the stream bed. Sometimes, in late summer when stream flows are lowest and waters warmest, I have found the trout under and around the white tongues of current that spill into the heads of the pools, with the larger fish generally just below the surface surge. In streams with active beaver ponds, some of the largest cutthroats I've seen, aside from those in the ponds themselves, have been holding in the stream below the dam, but lying well up under its spillover.

When something looking like food comes along, these little cutthroats go for it quickly and opportunistically. They become masters at hit-and-run feeding. A fisherman working the pools and pockets of one of these little streams will experience quick, darting strikes, but there may be only one opportunity to connect because the fish are so very skittish and easily spooked. Stealth is needed for this kind of angling—in

spades! Many are the times I have drifted black ant flies through a pool with only the tip of my rod and a foot or two of leader poking out from behind a streamside rock, or stretched out flat on an overhanging log to drop my fly in a likely pocket.

Curiously enough, however, fishing in the upper creeks can often be best when the sun is on the water. Perhaps it's because the warmth of the sunlight filtering through the overhead canopy stirs up more insect activity that then becomes part of the drift. Or maybe the light provides more visibility to fish holding near or under the frothy current tongues. Whatever the reason, my fishing diary is clear on this point. Unlike trout inhabiting more open waters further downstream, where bright sunlight falling upon the water usually spells the end of the fishing, this is when small stream cutthroats often respond best. [33]

STATUS AND FUTURE PROSPECTS

In terms of geographical area, the coastal cutthroat trout is the most broadly distributed of any of the cutthroat subspecies. It may also be the most abundant numerically when all of its life history forms are taken into account. As of 1989, when I was invited to speak at the Wild Trout IV Symposium sponsored by the U.S. Fish and Wildlife Service and others at Mammoth Hot Springs, none of its life history forms were viewed as

environments. Large egg size relative to female body size (and the attendant high potential for alevin survival) would be another such advantage.

33. Other prominent anglers have also remarked on the propensity for small-stream trout to bite better in the sunshine. For easy reads on the how-to aspects of fly-fishing small streams, I recommend Black (1988), Gierach (1989), and Hughes (2002).

being threatened or even of special concern, at least not by the region's state and federal management agencies. But when one talked to the region's anglers, a bleaker picture emerged. They talked of depressed returns of the sea-run life history form to rivers all up and down the coastal range. As an example, Washington's Stillaguamish River, with an annual run size of more than 29,000 fish in some years during the 1960s, had always had a reputation among anglers as a top producer of sea-run cutthroats. But in 1978, an angler had to fish an average of 7 hours to catch even one sea-run.

Now, to put that number in perspective, you have to understand that fishing for sea-run cutthroats in a river is not like fishing over a resident trout population. Sea-run fish are moving through to spawning tributaries or to overwintering refuge habitats. Small groups of them may linger in a resting pool for awhile, but soon they move on. One morning you will fish a pool and find nothing. The next morning it will be alive with trout. You might enjoy a day or maybe two of fast action and return expecting more, only to find the pool deserted again. You have to do a lot of prospecting when you fish the sea-run cutthroat runs, and you have to expect to take a few skunks during a season. But even when these normally fishless periods are factored in, the average angler should expect a stream with a healthy sea-run return to produce about a fish

for every 1.5 hours of effort, not the 7 hours per fish expended on the 1978 Stillaguamish. [34]

To its credit, the State of Washington did respond to these complaints by initiating a status review of its sea-run cutthroat populations that was published in 1980. But for the majority of entries, stock status was listed as "unknown." For those, the prevailing attitude was "no data, no problem." Recommendations to address the few situations where stocks *were* acknowledged to be in decline, such as the Stillaguamish, included a proposal for hatchery releases to increase run sizes (never implemented) plus modest reductions in harvest to be achieved by tinkering with the fishing regulations—not too much reduction in harvest, mind, because that might put off license buyers, a major source of revenue for the Washington State Game Department of the 1980s. The Stillaguamish sea-run population did eventually rebound, but not until years later, after more drastic steps were taken to protect returning spawners. This makes for a good case study, which I will revisit later.

Priority for wild sea-run cutthroat trout ranked a distant third in Washington State, as it also did in Oregon. After all, there were populations of the larger and more glamorous salmon and steelhead to be managed. British Columbia did have a cutthroat-friendlier Salmonid Enhancement Program (SEP for short) under way by the late 1970s, primarily in urban areas on Van-

34. The Puget Sound Task Force of the Pacific Northwest River Basins Commission, a federally mandated entity, published run size estimates of 29,200 sea-run cutthroat trout per year for the Stillaguamish River for the period 1962 through 1966, citing data provided them by the Washington State Game Department (see Puget Sound Task Force 1970). The Washington State Game Department published its 1978 Stillaguamish River creel census in 1979 (see Johnston 1979). Neither report contained information on angler catch rates prior to the decline, so I used data from my own fishing experience to estimate a value for angler expectation. For many years, my "home stream" was the Toutle River, a tributary of the Cowlitz River in southwest Washington. I fished it often for sea-runs in the morning or evening hours from the first of August through the end of October each year, and I kept detailed records of each trip. The seasons of 1962 through 1976 saw good sea-run returns to the Toutle. Some years were better than others to be sure, but overall this was a period of healthy runs. My seasonal catch rates over this period ranged from 0.34 fish per hour of effort to 1.5 fish per hour of effort, with a grand overall average for the 15 seasons of 0.65 fish per hour of effort, or 1.5

couver Island and the lower mainland. But priorities weren't much different there. Although some of the projects around Victoria included habitat restoration for wild stocks, most of the emphasis was on harvest regulation and hatchery programs. Fishery managers throughout the region appeared content with their status quo policies and practices that revolved around hatchery programs even for salmon and steelhead, and seemed largely inattentive to declines in wild stocks.

But all that was about to change, at least on the U.S. side of the border. Throughout the 1980s, the winds of this change were ruffling through the feathers of a bird, a rarely seen inhabitant of the old-growth forests of the Pacific Northwest known as the northern spotted owl. The battle to preserve and protect this bird and its old-growth forest habitat reached epic proportions, involving a multi-billion dollar industry, high-stakes politics, and equally high-stakes court cases; and it didn't stop even when, in 1990, the northern spotted owl was finally listed as threatened under the U.S. Endangered Species Act. [35]

Some of the environmental groups who had brought the petition and fought the fight to list the spotted owl were also taking an independent look at available data on the status and plight of anadromous salmon and trout populations, and they did not like what they were seeing. Even the prestigious American

hours of fishing for each sea-run brought to hand. Stillaguamish River anglers could have expected to do at least that well during their river's halcyon years. To complete the Toutle River story, the bottom dropped out of my fishing in 1977. That year, my catch per hour of effort fell to 0.22, which translated to about four-and-a-half hours of fishing for each sea-run brought to hand. But rather than increase again the following year as it had always done, my catch rate declined even further. In 1979, I caught less than 0.1 sea-run per hour expended. And then of course, in the spring of 1980, Mt. St. Helens erupted, obliterating all of the river's fisheries and bringing my Toutle River diary to a close.

35. Although published a year prior to the owl listing and written with an environmental leaning, Chapter 10, "The Billion-Dollar Bird," in Keith Ervin's book, *Fragile Majesty: The Battle for North America's Last Great Forest* (Seattle, The Mountaineers, 1989), does present a factual account of how political maneuvering effectively trumped good science throughout much of the spotted owl battle. One can see many of the same tactics in play today as the agencies grapple with current Endangered Species Act listings. By the way, for everything you

wanted to know about the Endangered Species Act but were afraid to ask, the Stanford University Law School's Environmental Law Society has written a user-friendly guide to the Act that explains its history, its structure and provisions, and how it is supposed to work. You can get a paperback copy for $19.95; see Stanford Environmental Law Society (2001). Briefly, an *endangered species* is defined as any species (or subspecies or, for vertebrates, distinct population segment or evolutionarily significant unit) that is in danger of extinction throughout all or a significant portion of its range. A *threatened species* is any species (or etc.) that is likely to become an endangered species within the foreseeable future (the Endangered Species Act does not define "foreseeable future"). The Act provides that listing of a species (or etc.) as endangered or threatened be based on an assessment of threats from any one or more of these five factors: 1) the present or threatened destruction, modification, or curtailment of its habitat or range; 2) overutilization for commercial, recreational, scientific, or educational purposes; 3) disease or predation; 4) the inadequacy of existing regulatory mechanisms; and 5) other natural or manmade factors affecting its continued existence.

The Act also specifies that recovery plans be developed and implemented for listed species, but does not specify what is meant by "recovery." Since 1990, the U.S. Fish and Wildlife Service has used the following broad definition of recovery: "the process by which the decline of a threatened or endangered species is arrested or reversed, and threats to its survival are neutralized, so that its long-term survival in nature can be ensured" (see U.S. Fish and Wildlife Service 1990). A more recent book, *The Endangered Species Act at Thirty: Renewing the Conservation Promise* (Washington, D.C., Island Press, 2005), examines, among other things, the Endangered Species Act's track record since its inception. This is the first of two volumes (the second is promised for 2006), and will cost you $35.00 in paperback, $70.00 in hard cover. See Goble et al. (2005) for the complete citation. For you readers in Canada, your country's Species At Risk Act (SARA for short) is too new (enacted in 2002 but not actually in effect until 2004) to have developed a track record. But Canada has had a government-mandated but independent and scientific Committee on the Status of Endangered Wildlife in Canada (COSEWIC for short) that has met each year since 1978 to identify and assess the risk

Fisheries Society weighed in on the subject, eventually issuing three reports on the health of Pacific salmon, steelhead, and sea-run cutthroat stocks.[36] The first of these, titled "Pacific Salmon at the Crossroads: Stocks at Risk from California, Oregon, Idaho and Washington," appeared in 1991 and hit the region like the smack of the proverbial two-by-four between the eyes. Fishery managers, land management agencies, and development officials at all levels of government were jolted out of any complacency they might have had regarding the true condition of the stocks under their purview. Suddenly the realization was there—they had a real problem, and the scramble was on to get ahead of the curve.

The "Pacific Salmon at the Crossroads" report organized its assessment by fish stock or group of stocks as articulated by the late William E. Ricker in 1972, where the term *stock* describes "the fish spawning in a particular lake or stream (or portion of it) at a particular season, which fish to a substantial degree do not interbreed with any group spawning in a different place, or in the same place at a different season." Organized this way, the report recognized 15 groups of sea-run cutthroat stocks in California, Oregon, and Washington: two already extinct; two at high risk of extinction; three at moderate risk of extinction; and the remaining eight of special concern (in a subsequent report listing only healthy stocks, issued in 1996, not a single stock of sea-run cutthroat trout was identified as healthy). Within just two years of the Wild Trout IV Symposium where all was thought to be well in the coastal cutthroat world, we find all of the sea-run stocks in California, Oregon, and Washington identified as depressed, many at moderate to high risk of extinction and two already extinct!

Petitions for protection of anadromous salmon and trout stocks under the Endangered Species Act followed in rapid order, including one, filed in 1993, to list the Umpqua River, Oregon sea-run cutthroat stock as threatened or endangered. The National Marine Fisheries Service (NMFS, a branch of the National Oceanic and Atmospheric Administration now known as NOAA Fisheries), whose Endangered Species Act jurisdiction includes anadromous salmonids, completed its formal status review of this stock in 1994 and agreed with the petitioners that it was at high risk of extinction. In 1996, the Umpqua River sea-run stock was listed as endangered.

There was, however, a small proviso. The Endangered Species Act provides for listing named species and subspecies, and it also allows listing of *distinct population segments* of species and subspecies. But the Act is silent on how to define these distinct population segments. NMFS opted not to use the stock concept as the Ameri-

of extinction of species or *designatable units* of species (a designatable unit is Canada's way of dealing with population segments below the taxonomic species level, much as the U.S. uses *distinct population segments* or *evolutionarily significant units*). When, on the recommendation of the Minister of Environment, the Canadian Cabinet moves a species or designatable unit from the COSEWIC list to the SARA legal list, Canada's government is then obligated to begin measures for the protection and recovery of that species or designatable unit. For more on COSEWIC and SARA, see Green (2004), Irvine et al. (2005), and these websites: www.cosewic.gc.ca/, and www.sararegistry.gc.ca/.

36. The three American Fisheries Society reports examined stocks at risk in California, Oregon, Idaho, and Washington (Nehlsen et al. 1991); British Columbia and Yukon (Slaney et al. 1996); and southeastern Alaska (Baker et al. 1996). A separate report, also published in 1996, surveyed healthy stocks in Oregon, Idaho, Washington, and California (Huntington et al. 1996).

can Fisheries Society had done, but rather, adopted an *evolutionarily significant unit* (ESU) definition that includes not only the reproductive isolation aspect, but also the evolutionary legacy of the segment under consideration. [37] NMFS concluded that the Umpqua River sea-run cutthroat stock belongs to a broader ESU whose boundaries and overall status they could not then determine. They went ahead with the Umpqua River listing, but proposed to revisit that decision and determine the status of the broader ESU at a later time.

Meanwhile, as I noted, state and federal agencies scrambled to complete their own stock inventories and status reviews. Several major reports were issued that were instrumental in setting new policy for watershed management in the region to protect fish populations. [38] For sea-run cutthroat, these all came together in 1995 at that Reedsport, Oregon conference I mentioned early in this Chapter, where status reviews from Oregon, Washington, California, British Columbia, and Alaska were presented. The findings of these reviews are summarized below:

- *California coastal drainages, Eel River northward to Oregon:* Sea-run population likely extinct in Eel River. All other sea-run stocks depressed from historical levels of the 1940s and 1950s but apparently have been stable at present low levels since the mid-1980s.
- *Oregon coastal drainages, lower Willamette tributaries, and lower Columbia tributaries:* Coast sea-run populations in steady

37. NOAA Fisheries published its ESU definition in 1991 (see Waples 1991). An ESU is a population or group of populations that 1) is substantially reproductively isolated from conspecific populations or groups, and 2) represents an important component of the evolutionary legacy of the biological species (the U.S. Fish and Wildlife Service uses essentially the same definition but employs the term *distinct population segment*, or DPS, rather than ESU). The term *evolutionary legacy* is used in the sense of inheritance, i.e., as a result of past evolutionary events, the population or group comprising the ESU has inherited a definable packet of the total genetic variability of the species (other ESUS will have inherited other definable packets) and this packet is the reservoir for future evolution of the ESU. When the ESU definition was published, it triggered debate among evolutionary biologists. The papers compiled in American Fisheries Society Symposium 17 titled *Evolution and the Aquatic Ecosystem* (Nielsen 1995) will give you a feel for the different points of view. For references to the petition to list the Umpqua River sea-run cutthroat stock and the associated NOAA Fisheries status review, see Oregon Natural Resources Council et al. (1993) and Johnson et al. (1994).

38. Among the many inventories that were produced, the following addressed coastal cutthroat trout and sea-run stocks. For California: Higgins et al. (1992), Moyle et al. (1995), and Gerstung (1997). For Oregon: Nickelson et al. (1992), Kostow (1995), and Hooton (1997). For Washington: DeShazo (1980) and Leider (1997). For British Columbia: Slaney et al. (1996, 1997). For Alaska: Baker et al. (1996) and Schmidt (1997). Appendix C of the report, "Forest Ecosystem Management: an Ecological, Economic and Social Assessment" (FEMAT 1993) addressed at-risk stocks within the range of the northern spotted owl, but relied on data available at the time from state and federal agencies. The FEMAT report set the stage for the so-called Northwest Forest Plan (also called the President's Forest Plan), signed by President Clinton in 1994, that established guidelines for managing late-successional and old-growth forests within the range of the northern spotted owl. A key component of this plan, intended to address issues surrounding listed and unlisted but depleted fish stocks, was an Aquatic Conservation Strategy (acronym PACFISH) that has governed activities in watersheds in the Pacific Northwest for the last decade, and also influenced watershed management on other federal lands throughout the cutthroat trout's range. The Bush administration has been acting to reverse many of the constraints imposed by this plan.

decline since the mid-1980s with present angler catches only 1 to 10 percent of historical levels in some streams. Lower Willamette and lower Columbia populations seriously depressed. Sandy and Hood River stocks of lower Columbia system may be extinct. Few if any sea-run cutthroat return to the smaller streams tributary to the lower Willamette in the greater Portland area.

- *Washington lower Columbia tributaries, coastal drainages, and Puget Sound drainages:* Two lower Columbia stocks known to be extinct, all others seriously depressed. Coastal populations a mix of healthy and depressed stocks. Puget Sound stocks also a mix of healthy and depressed, with the healthiest stocks from Snohomish River north. As of the year 2000, the status of 80 percent of Washington's coastal cutthroat stocks remained "unknown."

- *British Columbia coastal drainages north to Alaska, including Vancouver and Queen Charlotte islands:* Of 612 total sea-run stocks identified, no information is available for 492 stocks occurring outside the Georgia Strait/lower Fraser River area. Of the 120 stocks for which data are available (most occurring within the Georgia Strait/lower Fraser River area) 15 stocks are already extinct, 16 are at high risk, 5 are at moderate risk, and 30 are of special concern.

- *Alaska coastal drainages from southeastern Alaska to Prince William Sound:* Limited data indicate southeastern sea-run stocks may be increasing. Some Prince William Sound stocks were depressed by the Exxon Valdez oil spill of 1989.

Although new status reviews were presented at the conference in Port Townsend in 2005, it quickly became clear that little had changed in the 10 years since Reedsport regarding population status and trends, nor was there anything of consequence to report on the restoration and conservation fronts, particularly for sea-run

populations. So the summary presented above pretty much remains the best coast-wide picture of the status and health of the sea-run cutthroat life history form. [39]

The proceedings of the Reedsport conference were not published until 1997. The ink had hardly dried before a second petition was filed with NMFS to list *all* stocks of sea-run cutthroat in California, Oregon, and Washington as threatened or endangered, calling special attention to the lower Willamette and lower Columbia tributaries. The petitioners argued that the plight of sea-run populations in those two areas was even more serious than that of the already endangered Umpqua River stock. NMFS did not complete its formal status review in response to this second petition until 1999, but again the agency agreed with the petitioners, at least with regard to the lower Willamette and lower Columbia sea-run stocks. NMFS lumped these stocks into a single Southwest Washington/Columbia River ESU, and proposed to list that ESU as threatened. [40]

The 1999 NMFS status review reached four additional important conclusions. First, it concluded that there are only six ESUs of coastal cutthroat trout in Washington, Oregon, and California (down from the 15 sea-run stock groups identified in the "Pacific Salmon at the Crossroads" report). These are shown on the map in Fig. 3-3. [41] Second, and of greatest significance, each ESU was deemed to include not only any sea-run

39. New status reviews from Washington, Oregon, British Columbia, and Alaska were presented at the 2005 conference in Port Townsend (the California speaker was a no-show). Three of these, from Washington, Oregon, and British Columbia, respectively, were available in report form when I wrote this in October, 2006; see Blakley et al. (2000), Oregon Department of Fish and Wildlife (2005), and Costello and Rubidge (2005). The Oregon report was only available as a public draft, with the promise that a final version would be posted on the Department website at www.dfw.state.or.us/fish/. You can view Powerpoint presentations of each of these reviews, including Alaska's, at the U.S. Fish and Wildlife Service Columbia River Fisheries Program website at www.fws.gov/columbiariver/.

40. For this particular petition and associated NOAA Fisheries status review, see Oregon Natural Resources Council et al. (1997) and Johnson et al. (1999).

41. This map has been updated to show not only the distinct population segments originally identified by NOAA Fisheries for Washington, Oregon, and California (see Johnson et al 1999), but also the new designatable units proposed for British Columbia in Canada's COSEWIC status report (see Costello and Rubidge 2005).

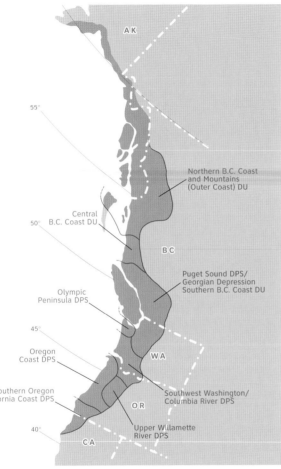

stocks within its boundaries, but *all non-diadromous populations as well*. In other words, all populations existing within a defined ESU boundary share the same evolutionary legacy regardless of life history form. In reaching this conclusion, the biological review team declined to recognize hereditary differences between migratory and resident populations as significant, or differences in genotype between above-barrier and uppermost headwater populations and populations below barriers. Third, it concluded that the Umpqua River sea-run stock and all other Umpqua River coastal cutthroat populations belong to the broader Oregon Coast ESU. Finally, although the Umpqua River sea-run stock remained at high risk of extinction (and still remains so) and all other sea-run stocks in the Oregon Coast ESU were depressed as well (and still are), available data indicated that non-migratory populations in this ESU (and most other coastal cutthroat ESUs for that matter) are both ubiquitous and abundant. Therefore, the ESU is neither threatened nor endangered and the Umpqua River sea-run stock was removed from the endangered species list.

The inclusion of both resident and sea-run populations in the same ESU had one additional ramification. Following the 1999 review, NMFS and the U.S. Fish and Wildlife Service agreed to a "change of venue" for coastal cutthroat trout, assigning all Endangered

Species Act responsibility for the coastal cutthroat subspecies in all of its life-history forms to the latter agency. The most recent action taken by the U.S. Fish and Wildlife Service under this new mandate was to choose not to list the Southwest Washington/Columbia River ESU (or DPS as it is now called) under the Endangered Species Act.[42]

Despite the apparent ubiquity and abundance of resident populations of coastal cutthroat trout, the fact remains that sea-run populations are depressed—some seriously so—in most DPSs. In the Oregon Coast DPS, no correlation has been found between the abundance of stream-resident populations and sea-run populations sharing the same stream. Instead, resident stocks appear to be healthy and stable even while sea-run stocks in the same river systems have declined, some to as low as 1 to 10 percent of historical levels.

Five major reasons are typically cited for these declines. The first four are largely anthropogenic and have come to be referred to as "The Four Hs:" 1) habitat loss or alteration that is adverse to the well-being of the stock; 2) harvest, specifically overharvest in recreational fisheries; 3) hatchery impacts; and 4) hydropower and water development. In addition, the significant decline in near-ocean productivity that occurred beginning in the mid-1970s and extended to about 2005 when the first signs of an improve-

ment were detected, has also been flagged as a likely contributor to the decline of coastal Oregon sea-run populations, and likely coastal Washington populations as well.

Habitat Loss or Alteration

For sea-run cutthroat trout, this refers to freshwater habitats vital for spawning and juvenile rearing, to estuary, delta, and nearshore habitats important for saltwater feeding and growth, and to mainstem passage and flow problems faced by migrating fish.

Sea-run cutthroat trout spawn in the first- and second-order streams and rear in the first-, second-, and higher-order streams within the anadromous zones of Pacific Northwest watersheds. Logging and forest road building have often been blamed for reduction in habitat quality and for outright destruction of habitat in these important freshwater reaches. Here are some reasons why.

Forest trees in upland areas are good at intercepting the water that falls on the catchment during rain showers and storms. The trees transpire much of this water back into the atmosphere (conifers are especially good at this). The rest usually infiltrates into the ground and is metered gradually into the stream as groundwater, providing reasonably dependable flows the year round. Groundwater reflects the temperature of the ground,

42. Williams, S., signatory. 2002a. Endangered and threatened wildlife and plants; withdrawal of proposed rule to list Southwestern Washington/Columbia River distinct population segment of the coastal cutthroat trout as threatened. Federal Register 67, no. 129: 44933–44961.

so contributes to cool and more even temperatures the year round as well. Removing the trees makes more water available to reach the stream (i.e., that not transpired), and more of this moves as overland flow, carrying with it increased loads of fine sediment that fill pools and spawning gravels. Overland flow tends to come in a rush after each storm, so stream flows become flashier, more scouring, and less dependable. Over time, trout populations react to these changes, usually by decreasing in abundance.

The small streams used by coastal cutthroat trout are sensitive to changes in riparian vegetation as well. Trees and other vegetation growing in the riparian zone help to maintain healthy stream habitat even after uplands have been logged. They do this by trapping fine sediment that would otherwise flush into the stream, by stabilizing streambanks and channels, by maintaining undercut banks, and by modulating water temperatures through shading. Streamside vegetation also provides structural components to the stream channel in the form of large woody debris (fallen trees or snags, large pieces of tree trunk or large branches, root wads, and the like, collectively referred to as LWD). LWD influences channel morphology. It provides the structural elements for pool formation and the creation of habitat variety within the stream. A loss of LWD results in fewer pools and less varied instream habitat types, and a transition to simplified riffle habitat. That in turn has been related to declines in the number of age-1 and older coastal cutthroat trout in affected streams.

Some studies have shown that coastal cutthroat abundance can actually increase following logging of deeply shaded streams, presumably because primary productivity of these streams is boosted by more exposure to sunlight. But other studies, notably the Alsea Watershed and Carnation Creek studies referred to in footnote 28 of this Chapter, showed that cutthroat populations declined considerably after logging and remained depressed more than 5 decades later. [43]

For sea-run cutthroat trout, habitat impacts on important stream reaches due to dairy farming and other agricultural practices may be every bit as important as forest practices. This is an area that has scarcely been touched in the fishery science literature.

Some of the most productive habitat for sea-run cutthroat spawning and rearing has been lost to culverts that block fish migration. An inventory taken in Washington in 1995 concluded that more than 3,000 miles of spawning and rearing habitat is inaccessible owing to an estimated 2,400 blocking culverts in that state alone. In a specific example from the Southern Oregon/California Coast DPS, highway construction for U.S. 101 in the 1960s created passage barriers that extirpated sea-run

43. The literature on impacts of forest practices on stream habitat and salmonid abundance has grown to be quite voluminous. Footnote 28 of this Chapter contains a good but pared-down list of references. For another extensive treatment (that is itself dated) see Meehan (1991), particularly Chapters 6, 8, and 14.

cutthroat trout populations in several small tributaries between the Pistol and Chetko rivers.

The loss of estuary and delta habitats of importance to sea-run cutthroat trout is particularly well-documented in northern California. The Eel River delta and estuary have been severely degraded by water diversions, marsh and tideland reclamation, stream channelization, removal of riparian vegetation, and heavy siltation associated with agricultural activities. The same is true for the Humboldt Bay system, Mad River estuary, and the once very productive Redwood Creek estuary. Estuary development may not be as well cataloged for Oregon and Washington, but has happened apace. In Washington, the last remaining relatively unaltered estuary on Puget Sound is the Nisqually Delta, now set aside as a National Wildlife Area. All others have been developed extensively for industrial or agricultural uses.

Harvest

Overharvest in recreational fisheries has been a serious problem in places. In the past, both Oregon and Washington allowed liberal harvest of trout 6 inches in total length and larger in recreational fisheries. In coastal Oregon, sea-run cutthroat trout were harvested at three life-history stages. First, a spring fishery harvested reconditioning sea-run adults as they returned to saltwater. Second, harvest of sea-run cutthroat juveniles was allowed if they reached legal-size during the normal summer fishing season, which they usually did at age 2. Third, beginning in July and extending into the fall months, a major harvest fishery took sea-run adults and sub-adults returning to streams after completing their season in saltwater. In Washington, the situation was much the same in coastal streams. But in Puget Sound streams, it was even more difficult for the sea-runs.

The Stillaguamish River, which flows into Puget Sound north of Everett, Washington, provides a good case study. This stream had once been one of the premier sea-run cutthroat fisheries in the State. But, you'll recall, in 1978 an angler had to fish many hours to catch even one sea-run. A basic life-history study showed why. Stillaguamish River sea-run juveniles migrate to salt water for the first time at age 2, and they typically return to the river that fall still at age 2 and about 11 inches in fork length. The females among these first-return fish are too young and too small to spawn during that first return, and must go back to the salt for another season of growth before approaching sexual maturity. They do not spawn for the first time until their second return to the river. Now, in the Puget Sound region, angling for sea-runs in the sheltered salt waters of the Sound is a popular sport that can extend for eight or nine months. State biologists realized that Stillaguamish River sea-runs had to run the gauntlet of four fisheries, two in the salt and two in the stream, before the females could spawn for the first time. Very few females had been surviving that gauntlet to do so. Washington biologists estimated that harvest might need to be reduced by as much as 80 percent in order to reverse the decline of the Stillaguamish River cutthroat population.

Thus, liberal harvest in sport fisheries can seriously reduce the escapement of naturally spawning sea-run cutthroat trout and has surely driven populations down. Both Oregon and Washington have taken steps to reduce harvest of sea-run cutthroat stocks. In 1997, Oregon imposed catch-and-release angling regulations for all trout angling in coastal Oregon streams with sea-run cutthroat populations, but in 2001 changed that to allow a harvest of two fish per day over 8 inches in length on the south coast (Salmon River south to the California line) and in 2004 proposed (but thankfully did not implement) the same harvest regulation for north coast streams. The State of Washington started earlier, but tip-toed its way toward effective reduction of sea-run cutthroat harvest. In Washington, based on its Stillaguamish River studies, harvest of sea-runs was restricted to two fish over 12 inches in total length in most waters in an effort to give at least 70 percent of the fish a chance to spawn at least once. However, it was subsequently found that only 40 percent of first-time female spawners were in fact being protected under this regulation, which wasn't enough, so the State moved to two fish over 14 inches as a harvest limit. The Stillaguamish River sea-run population rebounded quickly under this restriction, but little or no improvement was noted elsewhere. Washington has since required that all wild cutthroats be released in all but a few State waters.

Hatchery Impacts

Hatchery stocking programs pose several kinds of threats to naturally produced sea-run cutthroat trout. The first is intrinsic in sea-run cutthroat augmentation programs where hatchery reared sea-runs are stocked to provide fish for recreational harvest, which has been the major objective of such programs in the Pacific Northwest. Aside from any genetic risks and loss of fitness that might result from the interbreeding of released hatchery-reared fish with natural fish, a significant problem with sea-run cutthroat augmentation programs is that harvest levels for these fisheries have typically been set quite liberal, so as not to "waste" the hatchery component. Natural stocks have been harvested to excess as a result. Over time, returns to augmented streams have shifted over predominately or totally to hatchery-reared fish. And even for these fisheries, returns of the hatchery-reared fish have not always held up over time.

In the Pacific Northwest, releases of hatchery-reared sea-run cutthroat trout have occurred primarily in Oregon coast streams, lower Columbia River tributaries, the Grays Harbor, Hood Canal and Puget Sound areas of Washington, and the lower Fraser River and Strait of Georgia areas of British Columbia. There have never been any major hatchery programs to enhance or supplement sea-run cutthroat populations in northern California, although several small cooperative projects have been undertaken from time to time. Likewise, hatchery releases of sea-run cutthroat trout have occurred only intermittently in Alaska, with no artificial propagations projects operating there as of the 1999 NMFS status review.

In British Columbia, under the SEP program, local stocks have been propagated to supplement native populations since 1979. By 1999, the program had released more than 3.1 million fish from 17 hatcheries in the lower Fraser River and Strait of Georgia areas.

In Oregon, artificial propagation of sea-run cutthroat trout began in the late 1950s, and for decades relied on a broodstock derived from Alsea River stock, with occasional use of stocks from other Oregon coast streams. A hatchery stock from Big Creek, a lower Columbia River tributary, was the primary stock used for releases into Oregon's lower Columbia River tributaries. As it became clear that hatchery fish were replacing natural populations, not augmenting them, and as concern grew for the well-being of the natural populations, Oregon scaled back its sea-run cutthroat hatchery programs. Planting of hatchery-reared cutthroats in lower Columbia tributaries on the Oregon side of the river was discontinued in 1994. Since 1997, no hatchery-reared cutthroats have been planted in coastal streams containing sea-run cutthroats.

The State of Washington began hatchery production of sea-run cutthroat trout in 1958 at its Beaver Creek Hatchery on the Elochoman River, a lower Columbia tributary. Fish from this hatchery stock were released into lower Columbia and Willapa Bay tributaries, and were also used in stock transfers to Grays Harbor, Puget Sound, and Hood Canal tributaries, but without much success in augmenting any of these recreational fisheries. A stock was developed later for Grays Harbor tributaries that did maintain angler harvest levels in that area, but experiments with hatchery production of local Puget Sound and Hood Canal stocks were unsuccessful. No plants of hatchery sea-run cutthroat trout have been made in north Puget Sound streams since 1985, and none have been planted in south Puget Sound or Hood Canal streams since 1994. Only in the lower Columbia area has Washington continued to release hatchery-reared sea-runs. These come from its Cowlitz Trout Hatchery, a facility built to mitigate for the hydroelectric dams that eliminated natural production of sea-runs from significant portions of the Cowlitz River basin.

Hatchery programs for coho salmon and steelhead can also pose problems for wild sea-run cutthroat stocks. Wherever sea-run cutthroat trout occur in sympatry with coho and steelhead, even in the absence of hatchery programs, there is apportionment of living space such that coho and steelhead densities are often much higher than sea-run cutthroat densities in the stream. As I discussed earlier in this Chapter, this process may set natural limits on sea-run cutthroat population size in streams where these species occur sympatrically. Where coho salmon and steelhead enhancement and supplementation programs also operate, especially the coho programs, the practice is often to stock large numbers of hatchery juveniles without regard to the numbers of wild juveniles already present, or to the potentially negative impacts of these releases on the process of apportioning living space among all the species present. There is ample evidence, unfortunately most of it unpublished and circumstantial, that sea-run cutthroat populations can all but disappear under these circumstances. [44]

Hydropower and Water Development

Dams have not often been thought of as posing the problems for sea-run cutthroat trout that they do for other stocks of Pacific salmon, but dams have indeed brought about the demise of some sea-run cutthroat populations and have landlocked others. Examples include the Sandy and Clackamas rivers in Oregon and the upper Cowlitz River and Lewis River in Washington. Historically, the Sandy River had a modest-sized run of sea-run cutthroat trout, most of which may have originated from the Bull Run River, a major tributary that also serves as the City of Portland's municipal water supply. This run was eliminated back near the turn of the 20th century when the City began constructing its municipal water project. On Oregon's Clackamas River, the sea-run cutthroat migration once extended upstream past the City of Estacada, but Cazadero Dam blocked all upstream passage from 1917 to 1939, and no sea-runs have been seen in that reach ever since.

In Washington, I've already told you about the landlocked cutthroats of the Lewis River reservoirs. Sea-run cutthroats also ascended the Cowlitz River as far upstream as the town of Packwood. The Tilton and Cispus rivers, major tributaries of the Cowlitz, were destination fisheries, with the Tilton in particular being "famed in song and fable" among sea-run

44. Observations that have been published include Tripp and McCart (1983), House and Boehne (1986), and those referred to in Trotter et al. (1993) and Johnson et al. (1999).

cutthroat anglers. But construction of Mayfield and Mossyrock dams on the Cowlitz River and subsequent closure of the Mayfield fishway blocked the upper river to all anadromous fishes, cutting off the sea-run cutthroat runs to the Tilton and to all other upstream tributaries. Migratory cutthroat trout still move between the upper Cowlitz and the reservoir behind Mossyrock Dam, and I'm told that in a recent experiment one fish out of a group from this population released downstream from the dams showed evidence of a successful seaward migration and return. So genes for migration to saltwater may not yet be totally extinct in the upper basin should it ever become a priority of management to restore the full sea-run life history. However, this does not appear to be on anybody's "to do" list at present.

Near-Ocean Conditions

The productivity of the near-shore ocean waters along the Oregon, Washington, and northern California coasts correlates with the degree of wind-driven upwelling that occurs in spring and summer. Upwelling brings cold, nutrient-rich water up to the surface. Plankton biomass increases when this happens, and so in turn does salmonid production. Good upwelling conditions prevail when the climate is cool and wet—what we natives of the Pacific Northwest think of as "normal." But upwelling shuts off when warmer and drier conditions prevail and warm water spreads north, as it does when El Niño events occur. Indices that track El Niños tell us that they have been coming much more frequently and in greater intensities than average since 1976. At the Reedsport conference in 1995, Dr. William Pearcy of Oregon State University speculated that these unfavorable near-shore ocean conditions may have resulted in poor saltwater growth and survival of sea-runs, and may be one cause of their low abundance along the Oregon, Washington and northern California coasts.

But it's not just sea-run cutthroat populations that have been in low abundance along the Oregon, Washington, and northern California coasts since 1976. Coho and chinook salmon and steelhead runs have also been seriously depressed, at least in part because of poor ocean conditions. Here may be a working example of the old ecological axiom that everything is connected to everything else. A major source of nutrients in Pacific Northwest streams is marine-derived carbon, nitrogen, and phosphorus released from the bodies of returning salmon when they spawn and die. Direct consumption of salmon carcasses and eggs, and consumption of other items in the aquatic food web that are themselves enriched by these marine-derived nutrients, may have been especially important in the otherwise nutrient-poor tributaries where sea-run cutthroats spawn and rear. Declines in the number of Pacific salmon and the nutrients they release into stream ecosystems, brought about by the same poor ocean conditions that have limited sea-run cutthroat growth and survival in the salt, may have also limited their production in fresh water. [45]

So what of the future for sea-run cutthroat trout?

In the U.S., without the hammer provided by an Endangered Species Act listing for a coastal cutthroat DPS, responsibility for protection and recovery of sea-run populations shifts back to federal directives already in place, such as the President's Forest Plan, and to recovery plans written by state and local governments in response to listings of other salmon and steelhead ESUs. State and local recovery plans assume even greater importance when one realizes that two-thirds or more of the freshwater habitat of sea-run cutthroat trout in Washington, Oregon, and northern California is on nonfederal lands. Two concerns, voiced mostly by the environmental community and by fishing industry and angler groups, comprise the flip-side of this picture: 1) that state and local recovery plans are highly politicized and underfunded, and lack teeth as a result; and 2) that sea-run cutthroat trout may once again fall off the radar screen of state and local agencies.

For better or worse, the States of California, Oregon, and Washington *have* developed recovery plans for their salmon populations and watersheds. [46] These

45. Two indices commonly used to track El Niño events are the Southern Oscillation Index (Bakun 1973, 1975, 1990), which gives measures of strength and timing, and the Pacific Northwest Index (Ebbesmeyer and Strickland 1995) that tracks warm-dry vs. cool-wet periods along the coast. However, there is much more to the story. Ocean conditions in a particular region are governed by complex processes and cycles that may have their origins in other regions altogether, and may take decades to play out. The weekly journal *Science News* carried a short explanation for lay persons in 1997 (see Monastersky 1997), and I'm sure there are others out there that I have missed. Accounts given in the scientific literature require deeper study. One of the most recent of these is by Chavez et al. (2003). Others who have written learned papers on the subject include Pearcy (1984, 1997), Hayward (1997), Mantua et al. (1997), Beamish et al. (1999), and Finney et al. (2002). These will lead you to many more. Regarding the decreased levels of marine-derived nutrients being brought back to streams as a result of declines in Pacific salmon returns, in 2001 an international conference was held in Eugene, Oregon, to address how to restore these lost nutrients and revitalize depressed freshwater stream productivity. All of the pertinent papers from that conference were published recently in "Nutrients in Salmonid Ecosystems: Sustaining Production and Biodiversity" (American Fisheries Society Symposium 34, Stockner 2003). Two other recent papers worth reading are Naiman et al. (2002) and Wipfli et al. (2003).

46. For those wishing to peruse the California, Oregon, and Washington recovery plans, your best bet is to access them on the Internet. I located an undated version of the California plan, "California Coastal Salmon and Watersheds Program," at *http://resources.ca.gov/coastal_salmon_plan.html.* Three websites provide access and information about the "Oregon Plan for Salmon and Watersheds." These are: *http://governor.oregon.gov/Gov/exec.orders.shtml#1995%20-%202002, www.oregon-plan.org,* and *www.wrd.state.or.us.* Washington's plan is written up in a document titled "Extinction is Not an Option: The Statewide Strategy to Recover Salmon" (GSRO 1999). It can be downloaded at *www.governor.wa.gov/gsro/publications.htm.* For coastal cutthroat trout in Canada, a COSEWIC status report has been written (see Costello and Rubidge 2005), and a check of the COSEWIC website at *www.cosewic.gc.ca/* revealed that the Committee will meet in May, 2007 for a formal species assessment. The Oregon Plan has received at least one serious critique (see Spain 1997), so a new Oregon initiative may have been published by the time you read this. When the title of the Washington plan was announced back in 1999, another cynic is said to have quipped, "Yeah, extinction is not an option, it's the preferred alternative!"

plans run heavily to the Habitat component of The Four Hs listed above. Forest practice rules governing federal, state, and private lands have been beefed up to provide better buffers for streams, more adequate riparian protection in reaches utilized by salmonids during the critical juvenile life-history stages, and generally more sensitive logging of uplands. The State of Washington's forest practices rules are regarded as being the strongest while Oregon's are largely voluntary and compliance is viewed as occurring only reluctantly if at all. Even so, to the extent that these habitat strategies benefit any of the Pacific salmon species, they will benefit coastal cutthroat trout as well.

Harvest and Hatcheries, although perhaps not articulated as strongly as Habitat in these state recovery plans, have been dealt with by the respective fishery management agencies. Harvest has been restricted in Oregon and Washington through the use of catch-and-release angling regulations (although, as noted above, Oregon is softening its restrictions in coastal streams), and releases of hatchery-produced sea-run cutthroat trout have been eliminated or severely curtailed. Even the wholesale releases of hatchery-produced coho juveniles have been scaled back in important sea-run cutthroat streams. Only British Columbia's hatchery programs in the lower Fraser and Strait of Georgia areas continue apace.

Finally, fishery and oceanography scientists are now telling us that the interdecadal cycle that has brought poor conditions to the ocean waters where Washington, Oregon, and northern California salmon and steelhead stocks roam may have turned the corner at last. Perhaps so. The 2002 and 2003 seasons saw huge rebounds in the number of returning coho and chinook salmon to coastal streams in Oregon and Washington, and the 2004 and 2005 seasons were decent ones as well. So something positive does seem to be happening out there, and renewed infusions of marine-derived nutrients into spawning and rearing tributaries is beginning to occur. Will improvements also be seen in near-ocean conditions that favor the growth and survival of sea-runs in the salt? Resting pool counts of sea-run cutthroat adults in Oregon's north coastal streams had been showing an increase in the years from 2001 to 2004, but a running data set of smolt trap counts kept by the Oregon Department of Fish and Wildlife was not showing any corresponding increase in sea-run cutthroat smolt production in these streams over that same period. [47] And in September of 2006, NOAA's Climate Prediction Center issued a new warning of El Niño conditions developing in the eastern Pacific. So it may be too early to tell.

47. This information is contained in a report written in 2004 titled "Biology, Status and Management of Coastal Cutthroat Trout on the North Oregon Coast (Neskowin Creek-Necanicum River)." It was prepared by the Oregon Department of Fish and Wildlife North Coast Watershed District, Tillamook, Oregon, and was sent to me by that District's Fish Biologist.

Westslope Cutthroat Trout, *Oncorhynchus clarkii lewisi*, spawning male

CHAPTER 4

Westslope Cutthroat Trout

Oncorhynchus clarkii lewisi: Chromosomes, $2N = 66$. Scales in lateral series 150 to 200 or more, mean values generally 165–180 (there is quite a variation between watersheds; fish from the Clearwater and Salmon river drainages have the highest counts among populations within the core distribution, but populations of the outlier "mountain cutthroat trout" have the highest counts of all at 195–210). Scales above lateral line 30–40.

Vertebrae 59–63, typically 60 or 61. Pyloric caeca 25–50, mean values 30–40. Gill rakers 17–21, usually 18 or 19. In most populations, spots are small and irregularly shaped, and occur in a characteristic pattern, the body being free, or nearly free, of spots within an arc drawn to extend above the lateral line from the anal fin to the pectoral fin. However, individuals in some eastern Washington populations exhibit large spots and spotting patterns that stray from this norm. This subspecies can develop bright yellow, orange, and red colors, particularly among males during spawning season. Pink to red colors on the lower sides and belly, especially on males, is not uncommon.

Westslope Cutthroat Trout,
Oncorhynchus clarkii lewisi, stream-resident form,
John Day River drainage

A CURIOUS ENCOUNTER

Silas Goodrich of the Lewis and Clark expedition became the first person from the United States to catch a cutthroat trout on hook-and-line when he slipped away to fish the pools and pockets at the Great Falls of the Missouri River the afternoon of June 13, 1805. The fish he caught that day were what we now call westslope cutthroat trout, even though he found them on the east side of the Continental Divide. My own first encounter with this subspecies occurred one hundred forty-six years and nine days later, on June 22, 1969. But the place I found them was far to the west of the Great Falls, and, in fact, well outside their native range.

The day was cool, overcast, and blustery, fairly typical conditions for that time of year in the Pacific Northwest high country. My brother and I had taken a long weekend to fish several of the lakes in the district south of Mt. Adams in southwest Washington. Fishing in the other lakes had ranged from poor to only so-so. This lake was the last stop on our itinerary and it wasn't starting off any better. No fish were rising and we were plagued by the gusting wind. We tried a variety of wet and dry flies, but to no avail.

As we fished, a hatch of gray mayflies began to come off the water and trout started rising all over the lake. But they weren't rising to our offerings. The problem seemed to be that the gusty breeze scudded the real insects across the surface or blew them helter-skelter into the air, but our artificials behaved as if tethered to the surface, as indeed they were by our lines and the snubbed-back leaders we were using to combat the wind.

Then I remembered an old *Field & Stream* article I had read in which A.J. McClane, the magazine's fishing editor, talked about "bouncing spiders"—sparsely tied dry flies with stiff, oversized hackles fashioned on small, light hooks. I had carried flies of this type ever since reading that article and my kit included some gray ones about the right color for the hatching mayflies. We lengthened our leaders with long, fine tippets, tied on the spiders, and went back to work. Rather than casting, we just let the breeze catch our fine leaders and carry the wispy spiders where it would. They jounced and bounced on the surface, as much in the air as on the water, still not quite like the natural insects but apparently close enough. The fish showed *much* more enthusiasm and we had fast action as long as the hatch lasted. It turned out to be the finest afternoon's fishing of our trip.

The fish we caught were cutthroat trout, but not like any we had ever seen before firsthand. We were totally familiar with the coastal cutthroat, of course, having grown up fishing for them in the rivers, lakes, and beaver ponds of southwest Washington. But these fish were markedly different. Their backs were olive, their fins orange, their sides and bellies fawn-colored. There was a faint red wash along the lower sides from the ventral area forward, with red also on the gill covers. The spots were small, but rather than covering the entire body as they do on the coastal cutthroat, these were mostly concentrated on the posterior. Also, you could have drawn a compass arc from the vent to the pectoral fins, below which there were no spots at all.

I recalled hearing that the Washington State Game Department had stocked several waters with a type of cutthroat trout referred to as "Montana black-spotted trout." Upon arriving home, I sent off an inquiry, and the response confirmed that we had indeed been fishing a stocked lake. But the letter also revealed that our fish had not come from Montana at all. Instead, they were from a strain called the Twin Lakes cutthroat found right here in the State of Washington!

There are many sets of Twin Lakes, but the set that produced our trout lies in a narrow, glacier-carved basin in the central Cascade Mountains, just east of the Cascade crest. The outlet drains precipitously out of the northwest end of the basin into the Napeequa River, which flows out of the spectacular Glacier Peak Wilderness to join the White River, a major tributary of the Wenatchee River drainage. The Wenatchee, of course, empties into the Columbia River near the city of Wenatchee, Washington. At the other end of the basin, the southeast end, only a low divide separates the Twin Lakes from the waters of Big Meadow Creek, which drains off gradually to the southeast to join the Chiwawa River, another tributary of the Wenatchee.

The State of Washington has used the Twin Lakes as broodstock lakes for its Twin Lakes strain of cutthroat trout since 1915. You can visit the spawning station by hiking up a trail above the Napeequa River to a point where it intersects the outlet creek. From there you climb steeply uphill to the first and smallest of the lakes, a total distance of 3.4 miles. The trail skirts around this first lake on the north side, through heavy underbrush that affords only occasional glimpses of the water, and soon leads to the egg-taking station at the outflow of the second and larger lake.

You can also reach the Twin Lakes from the other end of the basin by driving the forest road up Big Meadow Creek to the base of that low divide. From there, it's an easy half-mile hike over the hill to the southeast end of the larger lake. It's well worth the trip to see the lake from this end, just for the view. There before you are the spectacular snow fields of the White Mountains, their images mirrored in the waters of the pretty alpine lake. When I last visited the Twin Lakes, on a warm, clear July 4th morning, trout were dim-pling all over the surface. But, being a broodstock lake, no fishing is allowed.

The Twin Lakes cutthroat has been described as a classic model of what the westslope cutthroat trout should look like. Its genetics have been tested using all of the modern methods and indeed it does key out as pure *Oncorhynchus clarkii lewisi*. Yet the Twin Lakes are located far to the west of what has always been thought of as the subspecies' historical range. The question is, how did the Twin Lakes cutthroat trout get there? There are several stories. The Indians say they were always present. Other reports say they were stocked from eyed eggs taken at the Stehekin River, a tributary of Lake Chelan, just a couple of watersheds north of the Wenatchee River system. [1] But that drainage too is much further west than previously-drawn boundaries of westslope cutthroat country. So again, what's the story?

HISTORICAL RANGE

What has normally been thought of as westslope cutthroat country is the upper Columbia basin of northeastern Washington, Idaho, Montana, and southeastern British Columbia up to the Continental Divide (this has sometimes been called the westslope or intermountain region) and the upper Missouri and upper South Saskatchewan river basins east of the Continental Divide

1. Bruce Crawford, who later rose to the position of Assistant Director for Fisheries Management in the Washington Department of Fish and Wildlife (formerly Washington State Game Department), wrote up a history of the Department's hatchery trout broodstocks (see Crawford 1979). Washington maintains two westslope cutthroat broodstocks: the Twin Lakes cutthroat, which is stocked in Cascade Mountain waters on both sides of the Cascade crest, although not as commonly as it once was, and the Kings Lake cutthroat, derived originally from a Priest Lake, Idaho, population that is stocked in waters of northeastern Washington. In addition to Crawford's report, I also have in my files a 5-page report on the Twin Lakes station titled "Management of a Cutthroat Trout Egg Taking Station," written in 1949 by J.M. Johansen. Johansen's report relates the Indians' assertion that cutthroat trout were native to the lakes and the Indians were well acquainted with the fishing there. Another argument for the indigenous origin of the Twin Lakes population is based on its biology. It is an outlet-spawning stock (*allacustrine*), whereas Lake Chelan populations spawned in inlet streams (*lacustrine-adfluvial*). Harking back to Chapter 3, these behaviors have a hereditary basis,

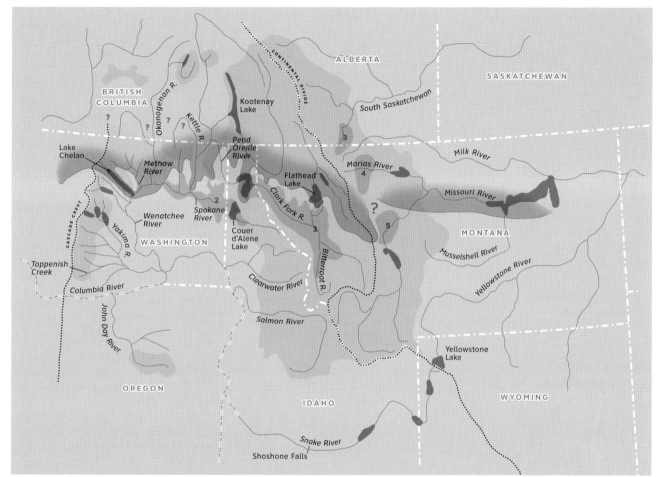

as has been shown for Yellowstone Lake populations of the Yellowstone cutthroat subspecies (Raleigh and Chapman 1971) and for lacustrine populations of rainbow trout in two British Columbia Lakes (Kelso et al. 1981). If the Twin Lakes stock was truly introduced from Lake Chelan, it would have had only 11 or 12 years at most—sometime between 1903 and 1904, when egg-taking commenced at Lake Chelan, and 1915, when egg-taking started at the Twin Lakes station—to make the adaptive switch from the inlet-spawning to the outlet-spawning life history.

Legend:

- ■ Existing Streams, Lakes, and Reservoirs
- **1** Glacial Lakes
- ■ Front of Cordilleran and Laurenticle Ice Sheets
- **?** Approximate Juncture of Cordilleran and Laurenticle Ice Sheets
- **?** Unknown Extent of Historical Range
- ■ Approximate Boundary of Historical Range

FIGURE 4-1. *Historical distribution of Westslope Cutthrout Trout, shown in relation to maximum extent of last Pleistocene ice sheet advance and location of glacial lakes* (1) *Missoula,* (2) *Columbia,* (3) *St. Mary's,* (4) *Cut Bank, and* (5) *Great Falls.*

in Montana and Alberta (Fig. 4-1). [2] Westslope cutthroat trout were observed in the past throughout the Kootenay River drainage of British Columbia (spelled Kootenai on the U.S. side of the border), the Pend Oreille River (known as the Clark Fork upstream of Lake Pend Oreille), and the Flathead River drainages of Washington, Idaho, and Montana; also in the Spokane River upstream of Spokane Falls in Washington and Idaho; and in the Coeur d'Alene, Clearwater, and Salmon river drainages of Idaho. East of the Continental Divide, westslope cutthroat trout occurred throughout the upper Missouri River basin downstream to the vicinity of Fort Benton, Montana in the mainstem, and in headwater tributaries of the Milk, Marias, Judith, and Musselshell rivers, which are tributaries of the Missouri entering downstream from Fort Benton. In the South Saskatchewan system, westslope cutthroats are or were found in headwater tributaries in Montana and Alberta, from the St. Mary's River north to the Bow River. [3]

As a historical note, for many years the subspecies name *lewisi* was associated with Yellowstone cutthroat trout (Chapter 9) as well, even though biologists who worked with these trouts had long recognized that the two forms were different. This association goes back to the days of David Starr Jordan. At the turn of the 20th century, Jordan's publications were considered the final word on trout classification and distribution.

2. The information shown on this map was distilled from many sources, including Cooper (1870), Gilbert and Evermann (1894), Vick (1913), Dymond (1931, 1932), Alden (1932), Horberg (1954), Hanzel (1959), Hewkin (1960), Lemke et al. (1965), Richmond et al. (1965), Roscoe (1974), Porter et al. (1983), Waitt and Thorson (1983), Allen and Burns (1986), Atwater (1986, 1987), Carrara et al. (1986), Nelson and Paetz (1992), Behnke (1992, 2002), Trotter et al. (1999, 2000, 2002), Alt (2001), and Shepard et al. (2003, 2005). A recently compiled Multi-State Assessment Report authored by Shepard et al. (2003) differs somewhat from Fig. 4-1 in its delineation of historical range. Shepard et al. (2003) excluded the Milk River headwaters and upper Musselshell drainage from their delineation, and also restricted their historical range in Washington to just the Pend Oreille, Methow, and Chelan drainages. I am uncertain whether the Cordilleran ice sheet actually spanned the Continental Divide in Montana, as I've shown in Fig. 4-1, or whether the Continental Divide was engulfed in alpine glaciers that coalesced with the Cordilleran ice sheet. However, somewhere east of the Continental Divide was a cleft where this ice sheet butted against the even more massive Laurentide

ice sheet (I've noted my uncertainty about its exact location with a question mark on the map). When opened, this cleft was one route humans may have used to populate the heart of the continent. Often referred to in both scientific and popular writing as "the ice-free corridor," this route later became known as The Old North Trail or The Great North Trail. It was sacred to Native Americans and served them also as an avenue for trading and raiding along the eastern front of the Rocky Mountains. To learn more about the Old North Trail, see McClintock (1910 [1999]), Ewers (1958), Cushman (1966), and Stark (1997).

3. Roscoe (1974) has a good summary and bibliography of early observations and collections of westslope cutthroat trout throughout the core historical range. Among these early references are two reports of trout that may have been westslope cutthroats occurring in the Snake River drainage of southern Idaho. In Ferdinand Hayden's Geological Survey report of 1872 (see Hayden 1872), mention was made of a trout with small spots found in Medicine Lodge Creek, tributary to the Big Lost River. Barton Evermann also reported that small-spotted cutthroat trout were quite abundant in the headwaters of the Big Wood River during surveys

conducted in 1894 and 1895 (see Evermann 1876). But both of these systems are in the Snake River drainage, now considered outside the westslope cutthroat historical range. Streams of the Lost River drainage flow out of the mountains south of the Salmon River divide and sink into the ground before they reach the Snake River, so fish in their headwaters are completely isolated from the mainstem Snake. Behnke (1992) pointed out that other fishes native to these isolated streams are present in the Salmon River drainage but not in the upper Snake River. His interpretation is that Pleistocene volcanic activity eliminated all fish life from the Lost River streams and buried their connections with the upper Snake River. Subsequently, headwater stream transfers from the Salmon River drainage established the present fish community. If this is so, then westslope cutthroat trout could very well have been present in the Lost River and Big Wood systems when Hayden and Evermann conducted their respective surveys. However, confounding this interpretation are recent reports by Idaho Department of Fish and Game biologists of the occurrence of Yellowstone cutthroat populations (Chapter 9) in Medicine Lodge Creek and two additional tributaries of the Lost River system,

In 1902, he and Barton Evermann published a book that recognized the upper Missouri River as the type locality for *lewisi*, but, in the erroneous belief that upper Missouri trout had spread upstream from the Yellowstone River, they also applied that name to the large-spotted native cutthroat trout of the Yellowstone River drainage. Since Jordan and Evermann said both of these forms were *lewisi*, then *lewisi* they would be, despite the obvious differences, for the next seventy-plus years. The common name, Yellowstone cutthroat, also came to be used for both forms, as in "*Salmo clarkii lewisi*, the Yellowstone cutthroat trout." It wasn't until the 1970s that the taxonomic confusion this created was finally cleared away. In 1973, Dr. Behnke made the case for giving the Yellowstone cutthroat its own subspecies name, which he chose to be *bouvieri*. By 1979, *bouvieri* had become widely-enough accepted that it was being used in scientific publications. [4]

In addition to the core distribution on either side of the Continental Divide, Fig. 4-1 also shows several discontinuous areas of westslope cutthroat distribution lying to the west of the core area. In the three disjunct areas in British Columbia, two of which are in the upper Columbia River drainage and the third and westernmost in the Fraser River drainage, all known populations occur in small streams isolated above waterfalls. In 1931, J.R. Dymond named these as a new subspecies, *Salmo clarkii alpestris*, the mountain cutthroat trout. However, more recent researchers, such as Dr. Behnke, have determined that they are isolated populations of westslope cutthroat trout.

In eastern Oregon, a disjunct area of historical westslope cutthroat distribution centers on headwater tributaries of the John Day River, a Columbia River tributary that drains a portion of Oregon's Blue Mountains. The uppermost headwaters of a number of creeks tributary to the mainstem John Day River are inhabited by native westslope cutthroat populations. Westslope cutthroat populations also occur in North Fork John Day tributaries but were introduced. Wild populations of interior rainbow trout also reside in these streams, but in reaches downstream from the cutthroats. Bull trout, too, occur in some of these waters. The John Day system is also used by anadromous salmonids and supports steelhead spawning and rearing, but, again, in reaches downstream from the cutthroats.

The fact that cutthroat populations exist in the John Day system has been known for quite some time. Records show that trout from headwater populations of mainstem tributaries were used for transplants into North Fork John Day tributaries as early as the late 1950s or early 1960s (Twin Lakes cutthroats from Washington were also released into Olive Lake in the North Fork drainage sometime in the 1970s). I

Beaver and Camas creeks. These are considered *core populations* of the Yellowstone cutthroat subspecies, i.e., populations that are native and still genetically pure, and are written into new management plans to restore Yellowstone cutthroat trout to at least 20 miles of historical stream habitat in these closed basins (see Northwest Power and Conservation Council 2004).

4. The book referred to in this paragraph is *American Food and Game Fishes* (Jordan and Evermann 1902). Two of their earlier publications (Jordan and Evermann 1896, 1898) may have also assigned the subspecies name *lewisi* to both the upper Missouri and Yellowstone River forms of cutthroat trout. A Rare and Endangered Species Report on westslope cutthroat trout, written by Dr. Behnke in 1973, records the first use of the subspecies name *bouvieri* for what he called the "true Yellowstone" cutthroat (see Behnke 1973). By 1979, that name was well on its way to acceptance in scientific nomenclature. Confirmatory evidence that the upper Missouri and Yellowstone drainage cutthroats are indeed separate subspecies is found in the studies of Loudenslager and Thorgaard (1979) on karyotype differences, and Loudenslager and Gall (1980) on genetic

was first told about these populations in 1977 by R.E. "Prof" Dimick, who was then Professor Emeritus in the Department of Fish and Wildlife at Oregon State University (he, no doubt influenced by the nomenclature of Jordan and Evermann, called them Yellowstone cutthroats). But it wasn't until 1981 that specimens were finally sent to an expert for examination. That expert was Dr. Behnke, who found them to be virtually pure *O. c. lewisi*. By then, of course, the Yellowstone cutthroat had been formally recognized as a separate subspecies and given its own scientific name.

In Washington, the historical distribution of westslope cutthroat trout extends across the northern tier of the state west to the Cascade crest (and perhaps north as well into British Columbia), then southward along the east side of the Cascade crest to include the upper reaches of the Methow, Chelan, Entiat, Wenatchee, and Yakima river drainages. The southernmost populations of pure westslope cutthroat trout found in Washington to date occur in the high country of the Yakama Indian Reservation in upper tributaries of the South Fork of Toppenish Creek, a Yakima River tributary.

Interruptions in the distribution of westslope cutthroat trout in Washington occur along the Okanogan and Columbia River corridors, which were conduits for later-invading rainbow trout. Interior Columbia River rainbow trout replaced westslope cutthroats in most

areas where they came into contact, including tributaries along these corridors, relegating the cutthroat populations to isolated headwater reaches. In addition to these areas in Washington, sympatric occurrence of native interior rainbow and native westslope cutthroat trout is known only in the Salmon and Clearwater drainages of Idaho and the upper John Day drainage of Oregon. In all of these places, the two native species have developed ecological distinctions that favor reproductive isolation. One way they do this is by partitioning the streams where they co-occur, the native cutthroat trout occupying the upper headwater reaches and the native rainbow trout the lower stream reaches.

Waterfalls or other hydraulic barriers sometimes serve as the isolating mechanism preventing incursion and replacement of westslope cutthroat populations by rainbow trout. Oftentimes, these barriers were formed by alpine glaciers that advanced down the mainstem valleys during the Pleistocene, leaving upper tributaries to flow in hanging valleys, isolated from the mainstem by long, precipitous drops that fish could not negotiate from below. But in streams lacking barriers, habitat partitioning and persistence of westslope cutthroat populations may be based on differences in the thermal tolerance of the two species. In the late 1980s and early 1990s, James Mullan of the U.S. Fish and Wildlife Service and several of his colleagues made extensive stream

differences revealed by allozyme electrophoresis. Verification that westslope cutthroat populations on the east and west sides of the Continental Divide belong to the same subspecies is provided in the work of Zimmerman (1965) on meristic characters, Loudenslager and Thorgaard (1979) on karyotypes, and Loudenslager and Gall (1980a) and Phelps and Allendorf (1982) on allozyme electrophoresis.

Kelly Creek (right) confluence with North Fork Clearwater River in Idaho. In the ea[rly] 1970s, these two streams figured in importe[d] experiments that demonstrated the efficac[y] of special regulations in improving the lot of wild native westslopes while still provid[ing] plenty of angling recreation. PHOTO BY AUTHOR

temperature measurements and fish observations, mostly in the Methow River drainage but also in the Entiat and Wenatchee river drainages of Washington. They observed that in barrier-free streams containing both rainbow and westslope cutthroat trout, the westslope cutthroats persisted without incursion only in the higher reaches that accumulated fewer than 1,600 thermal units on an annual basis. Reaches that could accumulate more thermal units than that in a year were invariably populated only with rainbow trout. [5]

Mapping the historical range of westslope cutthroat trout in Washington as broadly as I have shown is perhaps the most controversial aspect of Fig. 4-1. Even though cutthroat trout were acknowledged as native to Lake Chelan and two of its tributaries, the Stehekin River and Railroad Creek, as early as 1903 by no less an authority than the State Superintendent of Hatcheries, full recognition that the westslope cutthroat is a broader-based Washington native has come only reluctantly. An authoritative book on the inland fishes of Washington, published in 1979, stated that the coastal cutthroat trout is the only native subspecies. It recognized the presence in the state of an "intermountain" or "Cascade" form of cutthroat trout, but stated that all known populations were introduced. The Washington Department of Fish and Wildlife reinforced that view, at least in part, as recently as 1998 in a submittal to a U.S. Fish and

5. The findings of Mullan et al. were reported in 1992 in a U.S. Fish and Wildlife Service document titled "Production and Habitat of Salmonids in Mid-Columbia River Tributary Streams" (Mullan et al. 1992). The thermal unit definition used by Mullan et al. is the same as in footnote 10 of Chapter 2; but Mullan et al. toted them up over a full year. Other scientists who have studied longitudinal partitioning of stream habitat by introduced and native salmonids have cited gradient and temperature as major factors governing partitioning. But Mullan et al. concluded that rainbow trout can exclude westslope cutthroat trout (bull trout and introduced brook trout, too) from stream reaches that warm to more than 1,600 thermal units annually, regardless of gradient. A recent report by Sloat et al. (2001) of westslope cutthroat distribution in the Madison River basin noted that westslope cutthroats were associated with habitats where average and maximum daily stream temperatures remained below 54 and 61 degrees Fahrenheit, respectively, in summer. Also, stream temperatures were significantly colder at sites occupied by westslope cutthroats than at sites occupied by nonnative salmonids. For a sampling of other studies on this subject, see Vincent and Miller (1965), Fausch (1989), De Staso and Rahel (1994), and Taniguchi et al. (1998).

Wildlife Service team reviewing the status of westslope cutthroat for possible listing under the Endangered Species Act. The state's submittal asserted that:

> "Native cutthroats outside of Lake Chelan, the Methow and Pend Oreille river basins cannot be documented with the information at hand. Seemingly native populations in the most remote, rugged areas invariably were stocked there long ago, as stocking records attest."

Actually, the stocking records supplied by the state to support that assertion did *not* always so attest. Many known but remote populations were unaccounted for in those stocking records, and other such populations, also not accounted for in the records, have been discovered and cataloged since. But that was the official view on the matter as of 1998. [6]

Some of the earliest reports of what may have been native westslope cutthroat trout in waters on the east slope of the Cascade crest came from 1850s-era explorers seeking routes across that crest for the railroads. Lakes Keechelus, Kachess, and Cle Elum are three glacier-formed lakes located in the upper Yakima River drainage just east of Snoqualmie and Yakima passes, in country that George B. McClellan explored briefly during the Pacific railroad survey of 1853. McClellan was impressed by the fact that:

> "In all the lakes...are found salmon-trout; canoes are carried up the river (the Yakima) to these lakes, and the best fisheries are either on or near the lakes."

On the 9th of September, 1853, McClellan even tried his hand at fly-fishing for the salmon-trout in Lake Kachess, "but the wretches would not rise to the fly."

Sixteen years later, pioneer Yakima Valley stockman A.J. Splawn had better luck on Hyas Lake, a trout-fishing lake at the head of the Cle Elum River. In the summer of 1869, Splawn was taken to Hyas Lake by an elderly Indian acquaintance, using an old Indian trail. He wrote:

> "Making our camp in a beautiful mountain meadow, we proceeded to catch the mountain trout. No sooner would our hooks touch the water than hundreds would rush to grab the bait. We remained here for three days and for once, I had fish enough."

The problem with these narratives is that terms such as salmon-trout and mountain trout were used quite loosely. McClellan was familiar with sea-run cutthroat trout from his time at Vancouver Barracks on the Columbia River. His diaries indicated that he had fished for them there and referred to them as salmon-trout. One could presume he would use the same term for any red-throated trout he encountered elsewhere, as his predecessor, John C. Fremont, had done in 1844 at Pyramid Lake. But still, these narratives leave no real way to pin down the identity of the reported trout.

By 1904, however, people were becoming more specific in their descriptions. A correspondent identified only as "Chelano" sent this item to *Pacific Sportsman*,

6. The official who identified cutthroat trout as being native to Lake Chelan and Stehekin River, and to Railroad Creek, another Lake Chelan tributary, was Washington State Superintendent of Hatcheries John M. Crawford. Crawford visited the area several times and authorized egg-taking and a hatchery at the Stehekin River in 1903. He also wrote about these native fishes in an article for *Forest and Stream* magazine in 1912. The magazine misspelled Crawford's name, so the reference is listed in the bibliography under Cranford (1912). The first edition of *Inland Fishes of Washington* (Wydoski and Whitney 1979) is the book that listed coastal cutthroat trout as the state's only native subspecies, but that oversight was corrected in a new edition that came out in 2003. The Washington Department of Fish and Wildlife's submittal to the U.S. Fish and Wildlife Service on the status of westslope cutthroat trout in Washington is in Crawford (1998).

an early outdoor magazine, on the fishing in Lake Chelan and one of its tributaries:[7]

> "Fishing in Railroad Creek some six miles from the mouth is excellent and twelve or fifteen miles up late in August or early in September is perfect. The little brook trout (cut-throats) are the sweetest and most delicate of all, and for a lazy day of rest and sport, afford much pleasure."

In the years since that report, Lake Chelan has been managed for recreational fisheries based on any number of introduced and exotic gamefish species, including kokanee, rainbow trout, and chinook salmon. The high quality cutthroat fishery that existed at the turn of the 20th century has long since disappeared. However, in the summer of 1982, Washington State Game Department biologist Larry Brown began a search of headwater tributaries of the Stehekin River that turned up specimens of cutthroat trout subsequently identified by Dr. Behnke as pure *O. c. lewisi.* Their location in remote, isolated tributaries and the absence of stocking records make it likely, in Behnke's view, that these are relict populations of the original native fish fauna. Following that came the work of Mullan et al., discussed above, mostly in headwater tributaries of the Methow drainage but also a bit in the upper Entiat and upper Wenatchee drainages, that revealed even more such populations. The U.S. Fish and Wildlife Service and most recently the U.S. For-est Service, the Northwest Power Planning Council (now Northwest Power and Conservation Council), and Bonneville Power Administration made additional surveys in each of these drainages and extended down the east slope of the Cascade crest into upper tributaries of the Yakima River drainage. Still more genetically pure populations were discovered in this work that have not been accounted for in stocking records despite diligent searching and with no way the fish could have swum there from somewhere else.

I had the privilege of participating in the Northwest Power Planning Council/Bonneville Power Administration cataloging project during each of its three field seasons of work. I don't mean to convey the impression that unstocked, native, genetically pure westslope cutthroat populations are ubiquitous throughout the range in Washington State; that's definitely not the case. But those populations that have been found in recent field work are well distributed across the region I mapped in Fig. 4-1. As noted above, the southernmost pure populations so far found in Washington occur in headwater tributaries of the South Fork of Toppenish Creek in the Yakima River drainage.[8]

One final note regarding westslope cutthroat distribution in Washington: field workers have encountered a surprisingly large number of individuals with spots that appear larger than normal and in patterns that are

7. The historical accounts cited here are from McClellan (1853), Splawn (1917), and "Chelano" (1904). Other early accounts that gave trout fishing nearly as much ink as they did the true purpose of the expeditions include Charles Wilson's diary of the 1858–1862 survey of the boundary between Canada and the U.S. (see Stanley 1970) and Lt. Henry Pierce's military exploration from Fort Colville to Puget Sound via Lake Chelan in 1882 (see Pierce 1883).

8. Biologist Larry Brown's collection work in the upper Stehekin drainage is reported in Brown (1984), and the work of Mullan and his colleagues in the upper Methow and elsewhere is in Mullan et al. (1992). Results of the U.S. Fish and Wildlife Service collections from mid-Columbia basin tributaries are in reports by Proebstel and Noble (1994), Proebstel et al. (1996), Ringel (1997), and Proebstel (1998). Results of the Northwest Power Planning Council/Bonneville Power Administration cataloging project are written up in three reports by Trotter et al. (1999, 2000, 2002). Recent additional collections and genetic analysis of populations from Lake Chelan tributaries were sponsored by the National Park Service. This work has not been published, but was presented at the 2002 Annual General Meeting of the

not typical of the subspecies, especially (but not exclusively) among Yakima drainage populations. Alerted to this by earlier collectors, the field team working on the Northwest Power Planning Council/Bonneville Power Administration cataloging project made careful records of spotting phenotype when we surveyed Yakima drainage headwaters. We recorded five consistently-recurring spotting phenotypes, which are sketched in Fig. 4-2. Atypical spotting is often cited as a sign of introgression. However, none of the atypically-spotted trout examined by us or by earlier investigators using modern genetic techniques showed any evidence of introgression, nor was introgression detected in the populations from which they came. All were pure *O. c. lewisi*.

Several self-reproducing populations of westslope cutthroat trout have been established outside of the subspecies' historical range. I have already mentioned those present in the North Fork John Day River drainage of Oregon. All but one of these were established by translocating trout from tributary streams of the main John Day drainage, which is part of their historical distribution. The population in Olive Lake and its outlet stream, also in the North Fork John Day drainage, resulted from stocking of Twin Lakes strain cutthroat trout from Washington.

In Washington, self-reproducing populations have become established in several drainages west of the Cascade crest, within the range of the coastal cutthroat subspecies. I have personal knowledge of six of these populations and there are undoubtedly others that await discovery. Each appears to have become established following release of Twin Lakes cutthroat trout.

For many years, the Washington Department of Fish and Wildlife has stocked high lakes on both sides of the Cascade crest with either Twin Lakes cutthroat or brook trout to provide angling opportunities, although in recent years they have favored the use of the Mt. Whitney strain of rainbow trout for this purpose. According to department stocking policy, lakes are picked where the fish cannot gain access to inlet streams to reproduce, and they are isolated from downstream reaches by waterfalls or long, steep-gradient reaches where presumably they will not escape. Escapes have occurred nevertheless, and self-reproducing populations have become established in downstream reaches in at least a few cases. [9] Also, for reasons that are not recorded, the department saw fit to stock Twin Lakes cutthroat trout into several westside streams in years past, and again, in a few cases, self-reproducing populations became established. I am personally familiar with stocked stream populations (no lakes anywhere in these sub-basins) in upper Jackman Creek of the Skagit River drainage and Silver and Pinto creeks of the Cowlitz River drainage. McCoy Creek, also in the Cowlitz River

Western Division American Fisheries Society, Spokane, Washington, by C. Ostberg and J. Rodriguez. By the way, sharp readers will notice that I spelled the name of the Yakima River with an *i*, but earlier I spelled the Yakama Indian Nation with an *a*. That's in keeping with modern usage. The Native Americans call themselves the Yakama, but the English spelling, Yakima, still applies to the river.

9. To me, one of the more interesting populations that apparently was established by escapees from a stocked high lake occurs in a City of Seattle water supply reservoir on the South Fork Tolt River. Crater Lake, at the head of a very steep tributary of the South Fork Tolt upstream from the reservoir, was stocked many times over the years with Twin Lakes cutthroat trout. In 1998, snorkelers conducting a late spring stream survey of the upper South Fork observed many large trout (16 to 19 inches in length) moving up out of the reservoir on a spawning run. These were identified as westslope cutthroat trout, and most probably originated from Crater Lake escapees that took up residence in the reservoir. The water supply reservoir is not open to the angling public, so the trout are not disturbed there and have evidently found conditions much to their liking.

drainage, was stocked with Twin Lakes cutthroat trout the same year as Silver and Pinto creeks (that year was 1960), but I found only coastal cutthroat trout in the stocked reach of McCoy Creek on a recent visit.

LIFE HISTORY AND ECOLOGY

Populations of westslope cutthroat trout exhibit three life history strategies that are similar to the corresponding forms of the coastal cutthroat trout (Chapter 3): [10]

- A lacustrine, or lake-associated, life history, wherein the fish feed and grow in lakes, then migrate to tributary streams to spawn.
- A fluvial, or riverine (potamodromous), life history, wherein the fish feed and grow in main river reaches, then migrate to small tributaries of these rivers where spawning and early rearing occur.
- A stream-resident life history, wherein the trout complete all phases of their life cycle within the streams of their birth, often in a short stream reach and often in the headwaters of the drainage.

There is, of course, no amphidromous or sea-run form of westslope cutthroat trout.

The Lacustrine Life-History Strategy

When the last great glacier retreated from westslope cutthroat country, it left behind several large lakes as reminders of the region's Ice Age history: Lake Chelan in Washington, Priest, Coeur d'Alene, and Pend Oreille lakes in Idaho, and Flathead Lake in Montana. The westslope cutthroat is native to all of these lakes, as well as to many of the myriad smaller lakes that are

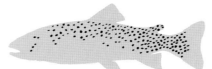

A. *Classic fine-spotted*

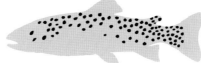

B. *Classic large-spotted*

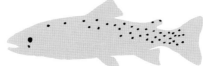

C. *Minimal fine-spotted*

D. *Minimal large-spotted*

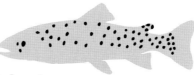

E. *Leopard-spot*

FIGURE 4-2. *Spotting phenotypes observed in Yakima basin Westslope Cutthroat Trout (dorsal and caudal spotting not shown).*

10. Overall, summaries of westslope cutthroat life history and ecology covering each of the three principal strategies can be found in Roscoe (1974), Likness and Graham (1988), and Behnke (1992, 2002). Reports and papers dealing with lacustrine life history (although occasionally an author would refer to these as *adfluvial*) include Block (1955), Bjornn (1957, 1961), Johnson (1961, 1963), Averett and MacPhee (1971), Likness (1978), and Shepard et al. (1984). A series of agency reports produced by Huston (1969–1973, 1984) looked at westslope cutthroat trout in reservoirs. The same reference string cited for lacustrine populations also provides details about fluvial life history and ecology, since many of these were basin-wide studies that examined populations of both life history strategies. Other studies of fluvial populations are Bjornn and Mallet (1964), Brown and Mackay (1995a, 1995b), Dunnigan (1997), Schmetterling (2000, 2001, 2002), and Fraley and Shepard (2005). Stream-resident life history and ecology are the subjects of papers by Miller (1957), Magee et al. (1996), Downs et al. (1997), Jakober et al. (1998, 2000), Brown (1999), and Hilderbrand (2003).

scattered throughout the range. Many high alpine lakes remained fishless after glacier retreat if waterfalls or other barriers were formed that prevented trout from colonizing them. For example, in the Flathead River drainage of Glacier National Park, westslope cutthroat trout attained dominance only in low- to mid-elevation waters. Forty-nine-acre Cerulean Lake, which sits at an altitude of 4,659 ft in the North Fork Flathead drainage, is the highest lake in the Park known to have an indigenous population of westslope cutthroat trout.

In addition to the natural lakes, many man-made reservoirs were formed when major rivers or their tributaries were dammed for hydroelectric power, water storage, or flood control. In some cases, fluvial populations of westslope cutthroat trout have adopted a lake-dwelling lifestyle to accommodate to this change in environment.

Most of what we know about the lacustrine life history strategy of westslope cutthroat trout comes from studies of populations of the large glacier-relict lakes such as Priest, Coeur d'Alene, and Flathead Lake. The Flathead Lake and river system has been especially well studied. Populations of these systems are invariably referred to as *lacustrine-adfluvial*, because their spawning migrations are directed toward inlet streams. Although I'm sure there must be others, I am personally aware of only one *allacustrine*, or outlet-spawning,

stock of westslope cutthroat trout, and that is the State of Washington's Twin Lakes broodstock. Another stock that may have spawned in the outlet stream of its lake in the far-distant past, but is now unique in another way, occurs in Picklejar Lake number 2, one of a string of four Picklejar Lakes in the Bow River basin of Alberta. This lake has no surface water inlet and its outlet creek was buried by a rock slide before any historical record of the lake was made. The westslope cutthroat population of Picklejar Lake number 2 has maintained itself through all these years by spawning on a shoal in the lake. [11]

Spawning runs from the large lakes may travel great distances (one stock of fish from Flathead Lake makes a spawning run of 132 miles), progressing upstream through major rivers and mingling with upstream-migrating fluvial stocks, to reach spawning gravels in upper tributary creeks. Age-4 cutthroats, ranging from 14 to 16 inches in length and spawning for the first time, make up about 75 percent of the runs moving up out of Flathead Lake. First-time spawners may be a bit older in some Lake Coeur d'Alene populations, where age-5 or 6 at first maturity appears more common, but elsewhere fish may mature as young as age 3. In the Flathead Lake runs, the percentage of repeat spawners has varied from 19 to 24 percent of the population, with the remainder being sexually immature fish.

11. A meristic, morphometric, and allozyme electrophoresis analysis of cutthroat trout from the Picklejar Lake complex was reported by Carl and Stelfox (1989).

It has been noted that females generally outnumber males in these spawning runs, sometimes by as much as 3 or 4 to 1. One hypothesis advanced to explain this is that males are more vulnerable to angling and thus there are fewer of them remaining at spawning time. Males apparently suffer greater post-spawning mortality than females as well, so there may be fewer of them in the repeat-spawner component of a run.

Each of the female cutthroats carries 1,000 to 1,500 eggs to the spawning gravels. Actual spawning occurs anywhere from March to July and is cued by the temperature of the water, commencing when water temperatures reach 43 to 48 degrees Fahrenheit. Alevins are usually evident after about sixty days, and swimup begins at about day 75.

The young cutthroats will remain in the streams of their birth anywhere from one to four years, but most of them, say 60 percent or so, migrate downstream to the lakes after two years. Most juveniles move downstream in late spring or early to mid summer, akin to the smolt migrations of sea-run cutthroat trout. Indeed, some observers have even referred to these lake-bound juveniles as smolts, even though this is an incorrect application of the term. Lake-bound juveniles are about 7 to 9 inches in fork length at this point, and they grow to the 14- to 16-inch size range over the next couple of years of feeding in the lakes. Clouding

the distinction between lacustrine and fluvial stocks, some juveniles drop down out of natal tributaries but remain in the river, even overwintering there in company with fluvial fish, before completing the migration to their lake the following spring.

Early studies of westslope cutthroats living in the large lakes indicated that their principal feed was zooplankton, supplemented by terrestrial and aquatic insects. In lakes where kokanee have been introduced, such as Flathead and Priest lakes, the planktivorous kokanee virtually monopolize the zooplankton supply, leaving the cutthroats to fend primarily on terrestrial insects. Westslope cutthroat trout show little inclination to prey upon fish, even when abundant forage fishes are present. This sets them apart from other cutthroat subspecies. Dr. Behnke and other scientists attribute this to their having co-evolved with two other predatory fishes, the bull trout and the northern pikeminnow. Over thousands of years of time, a harmonious coexistence evolved in which the westslope cutthroat avoided direct competition and assured its survival in their presence by concentrating on zooplankton and other invertebrates for its food, while the bull trout and northern pikeminnow utilized the forage fish resource.

Lake-dwelling westslope cutthroats seldom live beyond age 7 or 8. At that age, they might run 18 to 20 inches in length or perhaps a bit larger and weigh 2 to 4 pounds. This is about as large as they ever get, a fact that Dr. Behnke and others also attribute to their co-evolved feeding preferences. Old-time angler reports seem to bear out the fact that the lakes within the westslope cutthroat range seldom if ever produced the lunkers that, say, a Pyramid Lake (Chapter 5) or a Bear Lake (Chapter 11), or even a Crescent Lake or a Lake Washington (Chapter 3), would turn out. Yet early on, fishermen the world over were attracted to westslope cutthroat country by the wild, picturesque scenery and the abundance of easy-to-catch, colorful, two-pound trout.[12]

Occasionally, some lake within the westslope cutthroat range will go against what I have just said and will produce a real behemoth of a fish, but invariably these turn out to be hybrids or some other cutthroat subspecies. In 1955, Red Eagle Lake in eastern Glacier National Park produced a 16-pound trout that still stands as the Montana State hook-and-line record cutthroat. But Red Eagle Lake was barren of fish originally, and had been stocked with Yellowstone cutthroats to provide a recreational fishery. So the Montana record cutthroat is probably a Yellowstone cutthroat, not a native westslope. Montana does recognize the rainbow-cutthroat hybrid as a separate category for record purposes. In 1982, a new record was set in this category by a fish taken out of Ashley Lake, near Kalispell, that weighed in at 30 pounds, 4 ounces. That record also still stands. Not only was this fish a rainbow × cutthroat hybrid, but even the cutthroat stock from which it was derived was a cross between westslope and Yellowstone cutthroat bloodlines.

The Fluvial Life-History Strategy

Fluvial westslope cutthroat populations migrate to natal streams to spawn during the period of peak river discharges from March through July, just as lacustrine populations do. Like many lake dwelling stocks, some fluvial stocks also migrate very long distances to reach spawning tributaries. Long-distance spawning migrations are quite common among Salmon River, Idaho stocks, where this behavior may be an adaptive trait to partition spawning niches with sympatric steelhead. Spawning migrations of fluvial stocks may not always be in an upstream direction. By tracking fish with surgically implanted radio transmitters, David Schmetterling of Montana Fish, Wildlife and Parks demonstrated that some stocks in the Blackfoot River (remember the book and movie *A River Runs Through It?*) migrate downstream to reach spawning tributaries. Richard Brown and William Mackay of the University of Alberta earlier showed that some Ram River stocks migrate downstream to spawning tributaries as well.

Flows in most tributaries are high and turbid dur-

12. Wakeman Holberton, an early writer, described the colorful two-pound cutthroats his party found in Lake Pend Oreille: "dark olive green backs shading into deep gold and vermillion on the belly" (Holberton 1890 [1980]). More recent writers have also penned glowing accounts of the bountiful early-day cutthroat fishing in the big lakes, and of the subsequent demise of these fisheries. Claude and Catherine Simpson's book, *North of the Narrows: Men and Women of the Upper Priest Lake Country, Idaho* (Simpson and Simpson 1981), tells of Priest Lake, Idaho—first as a cutthroat fisherman's paradise up through the 1920s, then as this fishery declined in the face of massive egg-takes from its spawning cutthroats for introductions elsewhere coupled with the release of lake trout, which preyed on the cutthroats, then kokanee, which usurped the zooplankton portion of the cutthroats' food base.

ing the spawning migration period, but observers have noted that actual spawning may not commence until flows begin to subside. David Schmetterling, mentioned above, has also studied the spawning behavior of fluvial Blackfoot River stocks. He reported that the fish construct their redds predominately in the tailouts of glides or pools, in gravels up to about 10 percent of their body length in diameter. [13]

Most of the adult trout return to the main river soon after spawning is completed. Those that do not may remain in the spawning tributaries all summer and not drop back until fall. All members of fluvial populations overwinter in the main river channels.

The incubation period for the eggs and alevins of fluvial westslope cutthroat trout is the same as for lacustrine populations, approximately 60 days (about 310 accumulated thermal units) to hatching and 75 to 77 days to swimup, and the young cutthroats rear in their nursery tributaries for the same length of time, two to three years, before migrating downstream to take up their places in the main river.

Life for a trout in the stream reach it uses in summer is based on a linear dominance hierarchy. Here's how it works. In order for a trout to survive, natural selection dictates that it maintain a positive balance in the economics of feeding. In other words, it must minimize the energy cost of capturing food and maximize the energy gain from the food available. The stream reach provides only a finite number of locations, or *territories*, where a trout can maintain that positive balance, and the competition for these territories is intense. Each trout controls the most energy-efficient territory it is capable of defending. The most dominant trout in the reach (i.e., the most aggressive or largest) holds the most profitable position, the next most dominant trout holds the next most profitable position, and so on down the line. Territory defense is usually more ritual and bluff than actual combat because fighting is wasteful of energy, but the trout will do what it takes. If the dominant trout is removed from the reach, everyone moves up a notch in the pecking order. There are usually always a few subordinate trout in the population that cannot defend positions, and one of these so-called "floaters" typically moves into the last position vacated.

Dominance hierarchies do relax from time to time, for example, at the onset of a large hatch of aquatic insects. All sizes and dominance ranks of trout in the stream come near the surface at these times to take advantage of the largess. But when there is no major hatch, the foraging positions of the trout reflect the dominance hierarchy.

Within its territory, an individual trout spends most of its time in a relatively fixed position called the

13. When the adult female cutthroat chooses a spawning site, she is also in effect selecting the incubation environment for her eggs and alevins. During redd construction and spawning, the adult female displaces streambed gravel particles, the spawning pair then deposits eggs and sperm in one or more pockets, then the female covers the pocket with other particles of gravel that she displaces. During this process, fine sediments and organic materials within the substrate are washed away downstream, leaving the redd environment as favorable for incubation as it will ever be. From this point on, water must be able to circulate through the redd as deep as the egg pocket to supply oxygen and carry away waste products. Over the life of a redd, its permeability is inevitably reduced by subsequent deposition of fine sediment carried to it by the stream. But if permeability is reduced too much, or if the redd is dewatered, survival of eggs and alevins diminishes. Thus, the redd must remain permeable enough to support complete development of the eggs and alevins, and it must also be stable enough to resist scour during any high-flow events that might occur during this period. Dissolved oxygen and water temperature are also important.

focal point. Energy-efficient focal points are pockets of quiet water where the velocity is less than 0.5 feet per second, located adjacent to swifter currents that carry food items (referred to as *drift*) past the trout's position. The swifter food-bearing currents may themselves be no more than 1 or 2 feet per second, but salmonids will feed into currents as swift as about 6 feet per second. Rainbow trout are inclined to feed in faster currents than cutthroat trout or bull trout. The trout hold at their focal points, move into the adjacent swifter current to intercept food items in the drift, then return to their focal points. Trout appear to be capable of detecting very small changes in water velocity at their focal points and in amounts of food drifting adjacent to them, and can choose positions that yield the most net energy.

Cover is also important in territory selection, but not so much in the sense of a place where the fish can scurry to escape or hide, although that too is indeed a consideration, but rather to convey a sense of isolation from other fish. Trout seem to feel most secure when they can neither see nor be seen by other fish. An object near to its focal point (a boulder, large cobbles, a chunk of woody debris), an uneven topography of the streambed, or even a blanket of turbulent water or bubbles can provide this security—anything that can screen trout from one another's view. The type of cover sought, and the way it is used, varies among salmonid species and also with the age of the fish. Young trout occupy relatively shallow, slow-moving water in areas closer to overhead cover than do older trout. As they do grow older, individuals of some species, such as bull trout, seek out calmer, deeper water where they orient themselves very closely with the bottom substrate and undercut banks. Cutthroat trout show a preference for pools when they are the sole salmonid in the reach. Rainbow trout seek out swifter areas in which to establish their territories. Rainbow and cutthroat trout, while still orienting with their objects, will assume foraging sites well above the substrate and thus expend more energy for their food-capturing efforts than, say, bull trout. In one comparative study of westslope cutthroat trout and bull trout in a Montana stream, nearly 70 percent of the bull trout did not adopt focal points at all, but moved constantly and captured prey primarily from the streambed. On the other hand, all of the cutthroat trout adopted focal points and fed into the faster currents, capturing their prey from the drift. [14]

Speaking from an angler's perspective, I found that the summer feeding patterns of fluvial westslope cutthroats and the types of water they select for holding stations are not greatly different from the coastal cutthroats I grew up with back home. While working on the first edition of this book, I fished the catch-

Eggs and alevins may survive when dissolved oxygen levels drop below saturation, but their development is retarded. Water temperature acts as a two-edged sword. It affects the rate of egg and alevin development in its own right, and it also affects the solubility of oxygen in water (the higher the temperature, the lower the dissolved oxygen concentration). For more information on the spawning and incubation requirements of trout and salmon, see Bjornn and Reiser (1991). More recently, Kondolf and Wolman (1993) studied the sizes of spawning gravels that trout and salmon are able to utilize, and found that they can build redds in gravels with diameters up to about 10 percent of their body length. For the report on the spawning behavior of fluvial cutthroat trout in the Blackfoot River, see Schmetterling (2000). Brown and Mackay (1995a) published on the spawning ecology of westslope cutthroat trout in the Ram River, Alberta. The book, *Trout: The Wildlife Series* (Stolz and Schnell 1991), is full of excellent color photographs of trout of many species in the act of redd construction and spawning.

14. In footnote 16 of Chapter 1, which appended the discussion of agonistic behavior of trout, papers by Kalleberg (1958) and Jenkins

and-release section of Idaho's Lochsa River with Bill Geoffroy, veteran angler and fishing writer from Troy, Idaho. The Lochsa River is "pull off" fishing. By that I mean that U.S. Highway 12 to Lolo Pass follows the river, and we "pulled off" the highway to wet our lines at all the likely places. What we looked for were the slower, deeper glides or mid-channel pools—water with a depth of four to six feet, smooth, fairly even flow, a liberal studding of boulders, a tongue of turbulent current entering at the head, and broadening into a shallow riffle or rapid at the tailout. Rainbow trout and whitefish inhabit the swifter water, so a cast into the turbulence at the head of the run often produced something other than a cutthroat. But just a few steps downstream, where the water deepened and slowed, and our flies were working through cutthroat water. "There's a distinct difference between cutthroat water and rainbow water," Bill explained. "Fly-fishermen accustomed to rainbows often spend their time working over water that is shunned by the larger cutthroats." It was a lesson I already knew by heart.

Come autumn, a second kind of movement of the fluvial cutthroats commences. This one is confined within the main river channel but can extend over considerable distances, usually (but not always) in a downstream direction. This is an overwintering migration, in which the fish seek better winter refuge cover than their summer foraging territories can provide. The hierarchical pecking order that prevailed in the summer stream breaks down at this time. A trout's metabolism slows down as water temperatures decline. Appetites diminish and behavior has less to do with obtaining food and defending foraging sites than with securing refuge. Now the trout tolerate one another's proximity as they move into deeper waters and into habitats characterized by low water velocities— deeper pools or cutbanks than they used in summer, with boulders or large rubble in the substrate offering deep interstices in which they can take refuge. They also associate with places where springs or upwelling groundwater discharge into the stream. The water here may be several degrees warmer than the temperature of other parts of the stream during the deep of winter in westslope cutthroat country.

Fall and winter movements of trout—and indeed their very survival—can be influenced by ice formation in streams. [15] Three types of ice can form in the winter trout stream. Surface or sheet ice is most common and is the type we're most accustomed to seeing, forming first along the stream edges and growing toward the center. Once a complete layer has formed, surface ice thickens by crystallization on the underside. Frazil ice is small, needle-shaped ice crystals that form within the water column when turbulent water becomes

(1969) were recommended as classic studies of the behavior of trout in streams that should be hunted out and read by anybody interested in learning more about this subject. Chapter 7 of Bill Willers' book, *Trout Biology: a Natural History of Trout and Salmon* (Willers 1991), provides an overview. Two books by Tom Rosenbauer cover the same material from an angler's point of view in an easy-to-read, well-illustrated style; see Rosenbauer (1988, 1993). Kurt D. Fausch of Colorado State University has published on profitable stream positions for trout (see Fausch 1984, 1991), and he also collaborated with Shigeru Nakano of Hokkaido University and others in the study of resource utilization by westslope cutthroat trout and bull trout in Montana (see Nakano et al. 1992).

15. Although it's more than 50 years old now, a good reference on this subject is a paper titled "Ecological Effects of Winter Conditions on Trout and Trout Foods in Convict Creek, California," by Maciolek and Needham (1952). Other more recent papers, specific to fall and winter movements of westslope cutthroat trout and how these trout are affected by stream ice, were published by Brown (1999) and Jakober et al. (1998, 2000).

supercooled. Frazil ice crystals can plug the mouths and gills of trout. These crystals can also be quite adhesive, sticking to each other, to the substrate, and to underwater objects. Frazil ice that has coated onto the substrate or an underwater object becomes the locus for thickening into anchor ice. Anchor ice can build in thickness to the point that it alters water depths and dams streams, leading to stream dewatering at times in reaches below ice dams.

The rising streamflows and warming water temperatures of spring cue an exodus back to summer foraging reaches. Trout that have attained sexual maturity for the first time, which in fluvial westslope cutthroat populations generally occurs in the 3 to 5 age range, will continue the migration up (or in some cases down) to spawning tributaries along with any repeat spawners in the population. Like sea-run cutthroat, even a few fish that have not achieved sexual maturity may join this migration.

Although the age range for first-time spawning in fluvial populations is about the same as in lacustrine populations, the size range runs a bit smaller. Most first-time-spawning fluvial fish will be in the 10- to 14-inch fork length range, where their lacustrine counterparts may average 2 inches or so larger. Repeat spawners may comprise anywhere from 19 to 24 percent of the spawning run, but sometimes fluvial fish may spawn only

every other year. The maximum age and size attained by fluvial fish is about the same as in lacustrine populations: maximum age about 7 or 8 years (maximum age 11 years in upper South Fork Flathead River populations), maximum fork lengths 18 to 20 inches or just a bit larger, and weights of 2 or 3 pounds, the latter two attributes again likely due to their co-evolved feeding preferences to avoid head-on competition with bull trout and northern pikeminnows.

The Stream-Resident Life History Strategy

The life history and habits of westslope cutthroats that live out their lives in small streams are similar in most every respect to the analogous form of coastal cutthroat. It is easiest to understand these behaviors if, again, we accept the concept of two distinct kinds of adfluvial populations. As discussed in Chapter 3, unrestricted resident populations are those that reside in tributaries having no barriers and therefore no physical restrictions on seasonal movement for feeding, growth, and overwintering. Reproductively isolated populations are those in which all life history phases are restricted to reaches above natural barriers and fish movements occur only within these restricted reaches.

Trout in unrestricted resident populations can be quite mobile, with seasonal movements, especially overwintering movements, extending over considerable

Lochsa River, Idaho. Two anglers fish goo water for westslope cutthroats in the specie regulations section of the river. PHOTO BY AUTHO

distances, usually in a downstream direction. Some individuals in unrestricted resident populations may even drop all the way down to the mainstem rivers to find overwintering shelter, but return in the spring to claim summer rearing territories in the small tributaries. On the other hand, the trophic and refuge movements of fish in reproductively isolated populations are necessarily restricted to whatever reach sizes exist above the barriers that isolate them. Even in unrestricted streams, the closer the fish live to the upper extent of fish distribution, the more restricted their year-round range seems to be. There is always some movement from the patches of spawning gravel to the pools and pockets where the feeding stations and refuge habitats are found, but the home territories of these reproductively isolated fishes are generally quite small. [16]

Like lacustrine and fluvial populations, spawning in adfluvial westslope cutthroat populations is cued by water temperature, and commences when the water warms to the 43- to 48-degree Fahrenheit range. In nineteen isolated headwater populations of westslope cutthroat trout in Montana, investigators found that some male fish spawned for the first time at age 2 (average fork length 4.3 inches), but all males sampled were sexually mature by age 4 (average fork length 6.4 inches). Among females, first-time spawning did not occur until age 3, and then only in fish 6 inches in

fork length or larger. Most age-5 females were sexually mature, and all females older than age 5 and greater than 8 inches in fork length were sexually mature. In the same nineteen populations, fecundity of females was size-related and ranged from an average of 227 eggs per female for trout in the 6- to 7-inch fork length range, to 459 eggs per female for trout greater than 8 inches in fork length. The average sex ratio among spawners was 1.3 males per female across all nineteen populations. [17]

Egg incubation and swimup occurs on the same general schedule as for lacustrine and fluvial populations. And, like all trouts, the early weeks of their existence as post-emergent fry are spent in quiet channel-margin habitats. As the season progresses and they grow larger, they move out of the channel margins to more productive summer-rearing habitats in pools and pockets, where they too must find their places in the summer-stream hierarchy. Refuge movements into overwintering habitats commence with the onset of autumn.

As was the case with fluvial westslope cutthroats, the overwinter period can be quite difficult for adfluvial trout. The youngest and therefore smallest trout in the aggregate are at greatest risk during the overwinter period. It is said that this is because smaller trout have a lower lipid content per unit weight (i.e., a lower amount of energy stored as fat) but higher metabolism than larger trout. So, as they burn energy during the

16. At this point, it would be useful to review the evidence for and against the *restricted movement paradigm* discussed in Chapter 3. Richard B. Miller of the University of Alberta published one of the early papers supporting the restricted size of home territory of stream-resident trout based on his experiments with westslope cutthroat trout in Gorge Creek, Alberta (see Miller 1957). His conclusion was that each trout in the creek spent its entire life cycle in a home territory not over 20 yards long. After reviewing his paper and the other references cited in Chapter 3, would you agree with Miller's conclusion?

17. The results reported here are from the Master's thesis studies of Christopher C. Downs at Montana State University (see Downs 1995). They were later published in a joint paper with R.G. White and B.B. Shepard in the journal *North American Journal of Fisheries Management* (see Downs et al. 1997).

transition to winter, their lipid reserves are depleted and overwinter mortality is higher. Because of this so-called *metabolic deficit*, only about half of the trout entering their first winter survive until spring.

Reports of the maximum life span of adfluvial westslope cutthroat trout have ranged from as low as 3 or 4 years to a high of 13 years. Maximum age among the nineteen isolated headwater populations cited above was 8 years. If a life span of 8 years is taken as the norm for adfluvial westslopes, that sets them apart from adfluvial coastal cutthroats, which seldom live more than 4 or 5 years.

STATUS AND FUTURE PROSPECTS

G.O. Shields was another of those late 19th century outdoor writers who chronicled their adventures hunting and fishing in the western mountains. In this passage, his hunting party had been distracted by the fishing available in Montana's Bitterroot River in the vicinity of Missoula:

"The supply of trout in the Bitter Root seems to be almost unlimited, for it has been fished extensively for ten years past and yet a man may catch twenty-five to fifty pounds a day any time during the season, and is almost sure to do so if he is at all skillful or "lucky." I know a native Bitter Rooter who, during the summer and fall of '84 [that's 1884], fished for the market, and averaged thirty pounds a day all through the season, which he sold in Missoula at twenty-five cents a pound. Of course, the majority of the ranchmen along the stream do little or no fishing, but the officers and men at Fort Missoula do an immense amount of it, as do the residents of the town of Missoula; and visiting sportsmen from the East take out hundreds of pounds every season. But the stream is so large and long, and its network of tributaries so vast and furnish such fine spawning and breeding grounds, that it is safe to say there will be trout here a century hence."[18]

Mr. Shields was not much of a prophet. From 1966 to 1973, the U.S. Department of the Interior issued an annual redbook of endangered species. By 1966, two decades shy of Shields' "century hence," the westslope cutthroat had vanished from so much of its historical range that the "Montana westslope cutthroat trout, *Salmo* spp." was included as an endangered species in the first few editions of this redbook. That listing was changed to "status undetermined" in later editions, owing to the confusion in nomenclature and distribution between it and the Yellowstone cutthroat. It was not included at all when the redbooks were replaced by the Endangered Species Act of 1973. Jordan and Evermann had classified the westslope and Yellowstone cutthroats as one and the same subspecies. Since the Yellowstone cutthroat was thought to be faring reasonably well at the time, the decline of the westslope cutthroat as a separate subspecies went unacknowledged and management plans to preserve or restore it were slow to materialize—with one major exception that occurred in the State of Idaho. I will talk about that in a few more paragraphs.

18. Shields, G.O. *Cruising in the Cascades: A Narrative of Travel, Exploration, Amateur Photography, Hunting, and Fishing* (Chicago and New York: Rand McNally & Company, 1889). The passage quoted is in Chapter 26, "Trouting in the Rocky Mountains."

It wasn't until 1984 that management agencies began to quantify how much of the westslope cutthroat subspecies remained and where strongholds of genetically pure populations were located. In that year, George Liknes of the Montana Cooperative Fishery Research Unit, Bozeman, reported on the population status of westslope cutthroat trout in Montana. He stated that genetically pure populations occurred in only 2.5 percent of the total stream length they had occupied in Montana historically. In Montana lakes, 265 lakes contained trout populations that outwardly appeared to be westslopes in 1984, but only 22 of those populations, or 8.3 percent of the occupied lakes, were genetically pure. Nineteen of the 22 genetically pure lake populations occurred in Glacier National Park. The State of Idaho completed an assessment of its westslope cutthroat populations in 1989, which concluded that only 4 percent of the historical range in Idaho still supported strong populations that were not threatened with hybridization. The U.S. Forest Service checked in with status reviews in 1995 and 1996 that essentially reiterated these numbers.

The most recent compilation of westslope cutthroat status information was published in 2003 by a Multi-State Assessment Team composed of state fishery biologists from Montana, Idaho, Washington, and Oregon, and three U.S. Forest Service biologists. That report was at least twice as generous in its assessment as the earlier reports, concluding that trout that would be visually recognized as westslope cutthroats still can be found in 59 percent of the historically occupied habitat, and populations that are genetically pure occur in at least 8 percent and possibly as much as 20 percent of the historically occupied habitat. [19]

The two Forest Service reports and the report by the Multi-State Assessment Team did make the additional important point that remaining pure populations of westslope cutthroat trout occur predominately on public lands in isolated roadless areas, wilderness areas, and in Glacier National Park. However, that alone doesn't guarantee the long-term survival of these populations. In 1997, Bradley Shepard of the Montana Department of Fish, Wildlife and Parks collaborated with Forest Service scientists to assess the risk of extinction of westslope cutthroat populations on federal lands in the upper Missouri River basin. They concluded that these populations are at serious risk of extinction owing to the small habitat fragments they now occupy and the lack of connectivity among them. [20] That assessment should certainly apply as well to small, isolated populations elsewhere in the westslope cutthroat range.

Four primary reasons have been cited for the decline of the westslope cutthroat trout to such low levels: 1) excessive harvest in recreational and, in earlier times, market fisheries, as attested to by the passage I quoted from Shields' book; 2) widespread habitat loss;

19. See Liknes (1984) and Liknes and Graham (1988) for the population status of westslope cutthroat trout in Montana, and Marnell (1988) for the status of the subspecies in Glacier National Park. The State of Idaho assessment was reported by Rieman and Apperson (1989). McIntyre and Rieman (1995) and Van Eimeren (1996) compiled the U.S. Forest Service status reports. Although it omitted some areas of historical distribution that I chose to include in Figure 4-1, the most recent compilation of information on the status of westslope cutthroat trout across the historical range is the Multi-State Assessment Report authored by Shepard et al. (2003), which I referred to in footnote 2 of this Chapter. The information in this report was republished by the same authors in 2005 in the journal *North American Journal of Fisheries Management* (see Shepard et al. 2005).

20. *Population viability analysis* (PVA) is a methodology that has evolved over the last twenty-five years or so to help agencies and managers prioritize and justify conservation and restoration efforts. Several computer software programs, built around models for evaluating the relative risk of extinction of populations, have been developed to help researchers perform the

3) displacement by introduced trout and charr species; and 4) hybridization by introduced rainbow trout and non-native cutthroat trout. In addition to these anthropogenic factors, collectively responsible for historical declines in westslope cutthroat numbers and distribution, global warming looms as a factor that will influence westslope cutthroat numbers and distribution in the relatively near future. [21]

Excessive Harvest

Westslope cutthroat trout are highly vulnerable to angling, as are all other interior cutthroat subspecies. For westslopes, this was first demonstrated back in 1966 by Craig MacPhee, University of Idaho, who showed that westslope cutthroats were twice as easy to catch as brook trout in a previously unfished tributary of the St. Joe River that contained both species. An angling effort of just 12 hours per acre of water (stream width times stream length expressed in acres) was enough to catch 50 percent of all cutthroats in the stream 6 inches in length and over, whereas the same effort caught only 25 percent of the brook trout 6 inches in length and over. Now, in general, the natural mortality rate of catchable-size trout in a stream (i.e., those 6 inches in length and over) will average about 50 percent in a year (with wide fluctuations from stream to stream) even if no angling occurs. Up to that 50 percent level,

computations. The program RAMAS (Akcakaya 1994) is widely used, but is by no means the only one available. The literature on PVA has become quite extensive. A good place to start if you want to learn more is the recent book *Population Viability Analysis* by S.R. Beisinger and D.R. McCollough (2002). The risk of extinction determinations referred to in the text for the upper Missouri River westslope cutthroat populations were carried out by Shepard et al. (1997).

21. The Multi-State Assessment Team's report (see Shepard et al. 2003) also raised the specter of diseases as another potential threat to westslope cutthroat populations. The authors singled out whirling disease, furunculosis, and infectious pancreatic necrosis (IPN) as diseases of particular concern. Here is a brief rundown on each, prepared from the following references that you should consult for more details: Roberts and Shepherd (1974), Wood (1974), Warren (1981), Moring (1991), Watson (1993), and Bartholomew and Wilson (2002).

- *Whirling disease* is caused by the parasitic protozoan *Myxosoma cerebralis*. This organism is not endemic to North America, but was accidentally

introduced from Europe into a hatchery in Pennsylvania in the 1950s, and has spread widely from there. Whirling disease was always thought to be a hatchery problem that posed little risk in the wild. That view changed abruptly in the 1993–1995 period, when the disease was implicated in the severe decimation of naturally-spawning rainbow trout in the Madison River, Montana, and several streams in Colorado. The organism is now present in most, if not all, trout-supporting areas, but is differentially distributed within those areas. *Myxosoma cerebralis* is a shape-changer. It undergoes two spore stages and requires two different hosts to complete its life cycle. One phase is completed inside an aquatic worm, *Tubifex tubifex*, commonly found in fine sediments but in densest numbers in the more eutrophic (i.e., nutrient-rich) waters and in lowest numbers in cold, low-productivity waters, such as high mountain streams. Spores from this phase leave the worm and float in the water. They are not hardy at this stage, and will die in a couple of days if they do not encounter trout, which serve as the second host. The second

phase of the parasite's life cycle is completed in the trout. It does its damage by feeding on the cartilage of very young host fish, before the cartilage hardens into bone. Infected fish swim in a circular motion at the surface, as if chasing their tails, and the tails usually blacken. In trout, cartilage becomes bone when the fish reach about 60 to 80 mm in length, and severe mortality occurs at this stage in heavy infections of the parasite. Fish do survive this infection, but with severely twisted, distorted spines. At this point also, the parasite reverts to its second spore form and simply falls to the streambed. These second-phase spores are very hardy and can survive 10 years or more while waiting to be picked up by tubifex worms to start the cycle anew. Research indicates that rainbow trout and brook trout are highly susceptible, cutthroat trout are moderately susceptible, and brown trout are least susceptible to whirling disease infections.

- *Furunculosis* is a disease caused by the water-borne bacterium *Aeromonas salmonicida*. This organism is evidently another non-native, having arrived from Europe in shipments of brown

or whatever it actually is for the stream in question, angling mortality simply compensates for those that would die of natural causes anyway. But if angling harvest exceeds that compensatory level by a significant amount, then the population will spiral into decline. No wonder Mr. Shields was such a poor prophet. The immense amounts of trout taken out of the Bitterroot and other Montana streams by local residents and market fishermen, and the hundreds of pounds removed annually from these streams by visiting sportsmen, surely exceeded that level.

To further illustrate the vulnerability to angling harvest of westslope cutthroats relative to other trout species, Dr. Behnke performed a couple of calculations that he published in his 1979 monograph. Using angler effort and catch data reported from Dutch Creek, Alberta, he calculated that roughly 150 angling hours per acre of water per year would result in the harvest of 75 percent or more of the catchable-size westslope cutthroats from that stream. For comparison, in the Cache La Poudre River, Colorado, angling pressure of 769 hours per acre per year could not remove more than 35 percent of the catchable-size wild brown trout nor more than 50 percent of the catchable-size wild rainbow trout from the stream.[22]

It may sound paradoxical, but this very vulnerability to overexploitation makes westslope cutthroat

trout. It spread very rapidly across the continent and now affects all species of trout in fresh water, although it is said to be still most common in brown trout. It enters fish through scratches or through the digestive tract, and produces purplish, reddish, or iridescent bluish lesions or blisters, although at the onset of an infection fish can die before these tell-tale lesions become evident. Furunculosis is typically a warm-water disease, and outbreaks usually do not occur until water temperatures consistently exceed 58 to 59 degrees Fahrenheit (15 degrees Celsius). This generally confines the disease to the broader, lower reaches of rivers and/or to summertime conditions. One reference (Roberts and Shepherd 1974) states that the bacterium can only survive and multiply in fish tissues. Between outbreaks, the organism persists in small numbers in the tissues of a few fish in the hatchery or fish farm (most commonly) or in a few wild fish in the watershed. It is treatable with antibiotics in hatcheries and fish farms.

- *IPN virus* is evidently endemic to many of the trout-producing areas of the eastern U.S., Canada, Europe, and Japan, and may have spread to western North America in shipments of brook trout eggs and fry from the east. Brook, brown, rainbow, and cutthroat trout are all susceptible. The disease is most common in hatcheries, where it is treatable, but it can occur in the wild. The virus damages the pancreas and intestines of affected fish. Diseased fish begin whirling, they darken in color, their eyes protrude and their abdomens swell. Hemorrhaging occurs around the vent and often at the bases of fins. Losses have occurred at all water temperatures tested. Fish can be carriers of the virus without suffering ill effects, and can release the virus into the water in feces or other secretions. The virus can also be carried in eggs and milt. Mergansers and seagulls that eat infected fish can carry the virus to distant water courses.

22. The angling effort and catch rate data for Dutch Creek, Alberta, used by Dr. Behnke in his calculations, came from an agency report written by Radford (1977). The Cache La Poudre data were contained in the Ph.D. dissertation of Marshall (1973) and a Colorado Division of Wildlife report authored by Klein (1974).

trout (and the other interior cutthroat subspecies as well) ideal candidates for creating sustainable fisheries based on restricted bag limits, size limits, and restrictions also on the type of terminal tackle anglers can use (e.g., flies or lures as opposed to baited hooks). [23]

In the late 1960s, the State of Idaho estimated that annual mortality rates of catchable-size trout were at or exceeding 75 percent in many of its native westslope cutthroat streams. Up to that point in time, the state had stocked large numbers of catchable-size trout from hatcheries to meet angling pressure (indeed, the most widely stocked cutthroat trout in the state was not even a westslope strain but, rather, Yellowstone cutthroat trout of Henry's Lake origin). But Idaho anglers overwhelmingly supported protection of the native westslopes. In a preference survey, they also signaled strong support for a strategy of reducing harvest, cessation of catchable trout stocking from hatcheries, and limiting angling methods in native cutthroat streams. That came as quite a surprise to state fisheries managers, who had fully expected anglers to go for large numbers of catchable-size trout from hatcheries. So the Idaho Fish and Game Commission responded with an experiment. The Kelly Creek drainage, tributary to the North Fork Clearwater River, was placed under a catch-and-release regulation. The upper St. Joe River was placed under a "trophy fish" regulation (three fish

23. A host of studies carried out over the years (the earliest that I'm aware of was published in 1932) have shown that 30 to 50 percent of the trout caught and released on baited hooks are likely to die of the experience, but that figure drops to 5 percent or considerably less for catch-and-release on lures or artificial flies. Papers by Taylor and White (1992) and Mouneke and Childress (1994) will give you an overall picture of what has been learned from these studies, and will lead you to other references for details. For many cutthroat trout populations, not only is the catch-and-release mortality rate on lures and flies considerably less than 5 percent, it may be considerably less than 1 percent. It has also been shown that individual cutthroat trout may survive many catch-and-release experiences in a season. In one famous study published by Schill et al. (1986), researchers from Idaho State University and the U.S. Fish and Wildlife Service found that each trout in a popular catch-and-release segment of the Yellowstone River was caught and released between nine and ten times on average each season. In another example, reported by Vore (1993), one tagged westslope cutthroat trout in a segment of Rattlesnake Creek near Missoula, previously unfished but newly opened to catch-and-release angling, was caught and released twelve times over a 12-month period.

Lake Coeur d'Alene, Idaho, one of several large lakes within the core historical range that once had magnificent lacustrine populations of westslope cutthroat trout.

bag limit, 13-inch minimum size limit). The drainage of the North Fork Clearwater River upstream from the confluence of Kelly Creek was kept under the standard fishing regulations that allowed liberal harvest to serve as a control until 1972, when a three fish bag limit with no size restriction went into effect there.

The results of this experiment were astounding! When the streams were checked again in 1975, cutthroat abundance had increased by four-fold in the upper St. Joe River and by 13-fold in Kelly Creek. But the cutthroat population in the control portion of the North Fork Clearwater had not increased at all. Furthermore, on the upper St. Joe, the catch of cutthroats per angler-hour had gone up by six-fold, the number of trout in the catch over 10 inches in length had increased by 10-fold, and the number of trout in the catch over 13 inches in length had increased by 30-fold. In Kelly Creek, the fish now averaged 9.6 inches in length, compared to 8.6 inches before the special regulations went into effect. In another experiment, the State of Idaho closed four of the tributaries of the lower St. Joe River to angling from 1973 to 1977. Cutthroat populations in these tributaries increased, and this in turn led to increases in cutthroat abundance in the main St. Joe River and also in Coeur d'Alene Lake, all of which feed the tributaries with spawning fish.[24]

Encouraged by these results, other waters in the westslope cutthroat range in Idaho were brought under special regulation management. Parts of the Coeur d'Alene and North Fork Coeur d'Alene Rivers were added and portions of the middle fork of the Salmon River as well, and in 1976 portions of the Lochsa and Selway Rivers were included. U.S. Highway 12 parallels the Lochsa River from Lolo Pass to its confluence with the Selway. Everywhere along its length, the Lochsa is accessible. The catch-and-release water starts at milepost 123, at the bridge to the Wilderness Gateway Campground, and extends upstream to a few miles beyond the Powell Ranger Station. Below milepost 123, angling is catch-and-keep. Knowledgeable anglers pretty much concede that the catch-and-release section offers the best opportunity for large cutthroats, and overall catch rates have improved each year. When I fished this water with Bill Geoffroy in 1982, I released westslope cutthroats ranging from 12 to 16 inches in length, which is right up there with the largest size this subspecies attains.

In all waters except the Coeur d'Alene River, the result was the same: dramatic increases in the numbers and average size of the native westslopes occurred. Westslope cutthroat populations in the Coeur d'Alene River did not respond to special regulations, perhaps because degraded habitat and water quality in the

24. Results of this Idaho case history were widely published and are still cited by catch-and-release advocates to illustrate how special-regulation fisheries for native cutthroat trout can be viable (and more economically lucrative) alternatives to catchable trout stocking where native trout populations still exist. Four original reports provide the detailed results of this Idaho experience. These are: Gordon et al. (1970), Bjornn (1975), Bjornn and Johnson (1978), and Bjornn and Thurow (1978).

South Fork and main Coeur d'Alene rivers had sup-pressed these populations to the point where no positive response was possible. [25]

Habitat Degradation, Fragmentation, and Loss

Human-caused degradation, fragmentation, and out-right loss of habitat have contributed to the isolation of local populations and restriction of overall westslope cutthroat occurrence to very small portions of the sub-species' historical distribution. Indeed, some biologists consider habitat degradation and loss to be the number one cause of westslope cutthroat decline. Poor grazing practices, historical logging practices, and the linger-ing effects of forest roads, mining, agriculture, and residential developments—all have been blamed for habitat loss in westslope cutthroat country. While each individual habitat alteration or perturbation might not seem so much, the total of all of these, i.e., their *cumu-lative effects*, have contributed greatly to the shrinkage of westslope cutthroat distribution. [26]

One conclusion reached by the various assessments of population status is that stream-resident or adflu-vial populations have been pushed into fragmented headwater habitats where the small sizes of indi-vidual populations and lack of connectivity with other populations increases their risk of extinction. Another conclusion is that fluvial populations and those lacus-

25. In 1885, the Coeur d'Alene Mining District of Idaho became the site of one of the richest silver strikes in the West. Since that time, the South Fork and main Coeur d'Alene rivers from Wallace, Idaho, to Lake Coeur d'Alene have been heavily impacted by mining wastes. From 1885 to 1928, all liquid and solid wastes from milling and smelting opera-tions were dumped directly into the Coeur d'Alene River or its tributaries, and there was a period when U.S. Bureau of Fisheries surveyors could not find a single live fish from Wal-lace to the lake (see Ellis 1940). Even after settling ponds were installed, concentrations of cadmium, lead, and zinc continued to exceed EPA standards for the protection of aquatic life (Hornig et al. 1988). Min-ing was curtailed in the early 1980s, and the old Bunker Hill milling and smelting operation near Kellogg, Idaho, is now a Superfund Site. But even so, toxic metals are still present in stream and lakebed sediments at levels that exceed the EPA criteria, and contaminated sediments have begun to work their way down-stream into the Spokane River and across the border into Washington.

26. Beth Gardner, Swan Lake Ranger District, Flathead National Forest, summarized these threats in a short essay she wrote on the westslope cutthroat trout, one of Montana's fish species of special concern. This essay is available on the Internet at the Montana Chapter American Fisheries Society website at *www.fisheries.org/ AFSmontana/SSCpages/westslope_ cutthroat_trout.htm.*

trine populations of the region's large lakes that make long river migrations to reach spawning tributaries have suffered the greatest declines. Construction of dams, irrigation diversions, and other barriers such as culverts has isolated or eliminated whole reaches that once were available to migratory populations. Entire river basins have been blocked in some cases, e.g., the Pend Oreille and South Fork Flathead rivers.

As is true with coastal cutthroat trout, forest management practices may lead to habitat degradation and loss in westslope cutthroat country through a variety of mechanisms, including disturbance of streambanks and riparian areas, construction of roads and stream crossings, and removal of upland vegetation. In addition, increased levels of fine sediments flushing into westslope cutthroat streams following logging and forest road building have been a primary concern of biologists, particularly in the Idaho Batholith region of Idaho and western Montana. The predominately granitic bedrock within this region weathers easily to coarse and fine sands, silts, and clays. Fine-sediment runoff rates to streams during storm events are intrinsically high in this type of geology, and even higher rates occur when storm events follow man-caused disturbances. Numerous studies have shown that egg-to-fry survival of trout decreases as the percentage of fine sediment in spawning gravel and redds increases. In

addition, fine sediments can fill in stream-edge habitats and pools, making it difficult if not impossible for trout to utilize these important habitat types. Newly-emerged fry depend upon stream-edge habitats for survival during the first few weeks after emergence and, of course, pools provide important summer-rearing and winter-refuge habitats. [27]

Replacement/Displacement by Introduced Species

Introductions of non-native fish species have also contributed to the decline of westslope cutthroat populations. Sometimes, non-native trout were stocked as replacements for the native westslopes in habitats where the natives were already extirpated. But mostly, the decline of the native westslopes has occurred through competition and displacement, and, in the cases of introduced rainbow and non-native cutthroat trout, hybridization. Here, I deal with competition and displacement.

It's kind of hard to know where to start with this topic. Some blame the fish culturists and their colleagues in the early federal and state fish management agencies. Fish culture got off to an early start in the United States; the American Fish Culturist's Association (now American Fisheries Society) was established in 1870 and the U.S. Fish Commission in 1871. Cutthroat trout were propagated early-on in the west

27. Logging and forest road impacts on stream habitats and cutthroat trout populations were touched on in Chapter 3. Footnotes 28 and 41 of that Chapter list references for additional details. Biologists' concerns about the effects of excessive fine sediments in streams of the Idaho Batholith region are reflected in the many reports and papers published on this subject since the late 1960s. Klamt (1976), Bjornn et al. (1977), and Chapman (1988) will give you a sampling of this literature and will lead you to additional references. U.S. Forest Service scientists have developed models for predicting sediment yields and salmonid response (see Cline et al. 1981 and Stowell et al. 1983) and have continued to refine and update these models as the science has improved (see, for example, Megahan and Ketcheson 1996). Still, the strict application of these models for the protection of westslope cutthroat populations rests on federal and state forest management policies, which are being relaxed more and more these days.

(by 1867 or 1868 in California and by 1872 in Utah). Rainbow trout too were propagated as early as 1870 in California, and rainbow trout eggs were being shipped east from there by 1875 or 1876. Brown trout were successfully propagated in the east beginning in 1883.

Others place the onus on the well-heeled 19th century sportsmen who, like G.O. Shields, discovered the vast cutthroat trout populations of Montana and Idaho, fished them down, and clamored for replacements. Being mostly easterners, these people encouraged the stocking of brook, brown, and rainbow trout. Western states, where the tourist dollars were being spent, wanted to keep these people satisfied and the U.S. Fish Commission was happy to oblige.

By the late 19th century, fish culturists had learned how to obtain and propagate millions of trout eggs, and widely distributed both eggs and hatchery-reared trout. The U.S. Fish Commission and some states even had special railroad cars for transporting and distributing fish. Dr. Behnke has called this the "Johnny Appleseed" era of fisheries management. It lasted for roughly 50 years, from the 1890s to World War II. Even in hatcheries propagating cutthroat trout, mixing of various subspecies and hybridization with rainbow trout was common. It didn't seem to matter much what kind of trout were stocked, or where. All one needed to do, Dr. Behnke wrote, was write one's congressman or the U.S.

Fish Commissioner and free fish would be delivered; just bring a bucket or milk can and meet the train at the station. [28]

Regardless of where the blame lay, in westslope cutthroat country the end-result of these introductions was displacement of the native westslopes across much of the subspecies' historical range. Brown trout and rainbow trout became the dominant species in the mainstem rivers and larger tributaries. Brown trout, introduced into all of the blue-ribbon trout streams of the Missouri River drainage, have largely displaced the native cutthroat in these streams. Brook trout displaced westslope cutthroat populations in all but the most remote of the small headwater streams, and is now the most common fish found in the upper creeks and beaver ponds of westslope cutthroat country. In the region's large lakes, westslope cutthroat populations gave way to introduced lake trout, kokanee, and, in some lakes, lake whitefish and yellow perch.

Researchers commonly attribute this displacement to *interspecific competition*, i.e., competition between the native westslope cutthroat and the introduced species of trout for the same set of stream resources, and, more specifically, to *competitive exclusion*, wherein conditions are such that the non-native species inevitably wins the competition. Interspecific competition does not always lead to the loss of native species. There

28. Behnke (1992) gives a good summary of the history of trout-propagation in the United States. Another author who touched briefly on the subject, particularly the role played by late 19th century gentleman-sportsmen, is Bruce A. Staples, who wrote about it in an essay titled "West Yellowstone, Montana is 'Trout Town, USA'" (see Staples 2003).

are still some waters where westslope cutthroat populations have persisted despite large populations of non-native trout. Indeed, some researchers contend that displacement of native cutthroat trout occurred only after degradation of physical habitat or water quality, or overharvest in recreational fisheries, had already dealt a severe blow to the natives. For example, changes in flow regime and deterioration of water quality have been cited as reasons why westslope cutthroat trout declined and non-native brown and rainbow trout (along with native northern pikeminnow populations) became dominant in lower-elevation waters of the Clark Fork and Spokane river drainages. The underlying theory of interspecific competition accommodates each of these observations, but does not provide insights into specific causal mechanisms. That's left for fisheries scientists to work out. Several published studies point to possible causal mechanisms, but to date, consensus among the scientists remains elusive.[29]

Hybridization

Let's begin the discussion of this topic by defining a few terms:

- *Hybridization* is broadly defined as interbreeding between genetically distinct populations. It can occur in nature when individuals of two different species or subspecies come into contact and mate, producing hybrid offspring; or it can be done deliberately, as in a cross between two fish species made in a fish culture facility or laboratory. Some hybrids are infertile and cannot reproduce. Others, e.g., between rainbow and cutthroat trout and between cutthroat trout subspecies, are fertile and can mate and produce offspring with other hybrids or with members of either parental taxon.

- *Introgression* is the movement of genes from one population into another via hybridization. It occurs in nature when hybrid individuals breed with other individuals from either or both of the parental taxa. The term *introgressive hybridization*, which means the same thing, will also be found in the literature. An introgressed population contains both hybrid individuals and pure forms of one or both parental taxa.

- A *hybrid swarm* is a population of individuals that are all hybrids. Hybrid swarms result from multiple generations of mating among hybrids and parental types, and of hybrids among themselves. Eventually, no pure parental types remain in the population.

- *Hybrid zones* are geographic areas (e.g., a reach of stream) where genetically distinct populations come into contact and can hybridize.

Where two trout species have co-evolved as natives of the same body of water or drainage, one or more behavioral differences or the ability to partition available habitat has usually evolved as well, so that reproductive isolation is maintained and mixing of the two species' gene pools either does not occur or is minimized. Short hybrid zones may exist where individuals of the two species do come into persistent contact and may breed with one another, but widespread hybridization does not occur unless natural conditions are altered in some drastic fashion, or relative population sizes change significantly for some reason.

29. Ecologists define competition as the active demand for, pursuit of, or use of a common vital resource (or set of resources) by two or more organisms. Competition among members of the same species is called intraspecific competition. Competition between members of different species is called interspecific competition. When the resource (or set of resources) being competed for is not sufficient to meet the requirements of the organisms seeking it, it becomes a limiting factor in population growth. There is a pair of mathematical equations (called the Lotka-Volterra equations) that allows scientists to model interspecific competition in terms of population growth rates, carrying capacities, and the competition coefficients of the two competing populations (competition coefficients express how much effect each additional member of one population has on the other population relative to a new member of that other population). Out of work with these models comes the competitive exclusion principle (Hardin 1960) which states that two species cannot coexist unless their niches are sufficiently different that each limits its own population growth more than it limits the growth of the other population. If this condition is not met, one of the species will inevitably be

As stated earlier in this Chapter, westslope cutthroat trout co-evolved with the Columbia River form of interior rainbow trout in the Salmon and Clearwater drainages of Idaho, the John Day drainage of Oregon, and in Columbia River tributary drainages of eastern and northeastern Washington. I have already talked about some of the ways reproductive isolation is maintained between the two species in these drainages. Where westslope cutthroat trout have *not* evolved in the presence of rainbow trout, introduction of rainbow trout has almost invariably led to introgressive hybridization of the cutthroat populations.

George Liknes' review of the population status of westslope cutthroat trout in Montana, referred to above, listed introgressive hybridization, particularly by introduced rainbow trout but also by introduced Yellowstone cutthroat trout, as the number one reason for the decline of pure westslope cutthroat populations. Using figures from the Multi-State Assessment Report, also referred to above, we can quantify the severity of this problem. If, as that report concludes, trout that are good visual representatives of the westslope cutthroat subspecies still can be found in 59 percent of the historically occupied habitat but only 8 percent of the historical habitat is occupied by genetically pure populations, then introgressive hybridization has occurred across 51 percent of the subspecies' historical range. [30]

reduced to zero in the competition. What ecological scientists endeavor to do is elucidate the conditions under which one competing species either excludes the other, or the two achieve an accommodation that allows them both to coexist even in disproportionate numbers. My introduction to this subject matter came via two textbooks, one by Wilson and Bossert (1971) and the other by Pielou (1969). I still recommend these for those who would like to dig deeper. Start with Wilson and Bossert (1971), which is a primer, and advance to Pielou (1969). A newer textbook covering the same material is Bulmer (1994). Review papers on interspecific competition and the effects of introduced species on native salmonids, including cutthroat trout, have been published by Hearn (1987), Griffith (1988), Krueger and May (1991), Baltz and Moyle (1993), Northcote (1995), and Dunham et al. (2002).

30. According to Behnke's history of trout propagation (see Behnke 1992), rainbow trout were first introduced into waters outside their native range in 1874. Today, they are found in 39 states in addition to the ones where they are native. They are present in every Canadian province except one, and on every continent except Antarctica. In the interior basins where only cutthroat trout are native, hybridization readily occurred when rainbow trout were introduced. Hybridization also took place when non-native cutthroat trout were introduced; for example, when Yellowstone cutthroats were stocked in waters already inhabited by the westslope subspecies. Widespread stocking of Yellowstone cutthroat trout probably did not commence until the advent of massive egg-taking operations at Yellowstone Lake at the turn of the 20th century, although the U.S. Fish Commission was also taking Yellowstone cutthroat eggs from Henry's Lake in Idaho as early as 1899, according to Van Kirk and Gamblin (2000). For further insights into hybridization and its evolutionary consequences, see Rhymer and Simberloff (1996), Arnold (1997), and Allendorf et al. (2001). For work specifically on hybridization between westslope cutthroat trout and introduced rainbow trout, see Reinitz (1977), Leary et al. (1984), Gyllensten et al. (1985), Rubidge et al. (2001), Hitt et al. (2003), Weigel et al. (2003), Rubidge and Taylor (2005), and Ostberg and Rodriguez (2006).

The Likely Future Impact of Global Climate Change

For decades, observers have pointed out that concentrations of greenhouse gases have been increasing in Earth's atmosphere ever since the advent of the Industrial Revolution in the late 18th century. As I mentioned in Chapter 2, some have even proposed that Earth has entered a new era of geological time, with names such as the Anthropocene Epoch of the Anthropozoic Era, the Age of Humans, being suggested to denote humankind's dominance of the global environment.

Nobody paid much attention to these harbingers until sometime in the late 1970s or early 1980s, when it became clear that under "business as usual" rates of emissions, the concentration of CO_2 in the atmosphere would double in the next century. Furthermore, global climate modelers predicted that if this were allowed to happen, Earth's average surface temperature could increase by anywhere from 2.2 to 10 degrees Fahrenheit, with much faster than average rates of increase in the high and middle latitudes.

Since those first predictions were made, global climate models and the measurements upon which they are based have become ever more sophisticated. Still, a vocal few continue to question if we are really seeing global warming or simply environmental noise. But the trend in global mean surface temperature is sharply up, as illustrated in Fig. 4-3. On a global basis, eleven of the

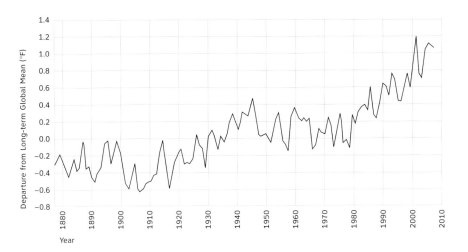

FIGURE 4-3. *Global temperature changes (1880–2004). The zero-line on this chart corresponds to 61.7°F, the global norm for the period of this record.* SOURCE: U.S. NATIONAL CLIMATIC DATA CENTER, NOAA

hottest years on record have occurred in the past twelve years. 2005 finished neck and neck with 1998 as the hottest on record, 2002 and 2003 tied for second-hottest, and 2004 was third hottest since record-keeping began back in 1880. Winter snow cover in the Northern Hemisphere and sea ice cover in the Arctic Ocean have decreased. Sea level has risen 4 to 8 inches over the last century on a global basis, and predictions are that it could rise another 16 inches to 2 feet by the end of the present century. The frequency of extreme weather events has increased throughout much of the U.S., all in keeping with global warming predictions.

But that's not the whole story. The most recent model runs have focused on shorter-term effects at a regional scale. These runs predict prolonged and intensified drought conditions for the American southwest. Winter precipitation is predicted to intensify across the northern latitudes, including the Pacific Northwest, but will come in the form of rain, replacing the winter snow pack. Using thermal tolerance values for cutthroat trout, and excluding the two Canadian provinces, regional projections made in 1996 indicated to fisheries scientists that a doubling of CO_2 concentration in the atmosphere would reduce habitat suitable for cutthroat trout by 1 to 49 percent in the States of Montana, Wyoming, Colorado, and Oregon, and by 50 to 99 percent in Washington, Idaho, Utah, and New Mexico. Model runs made in 2001 that projected changes over shorter time spans were less severe in their projections. These model runs indicate that habitat suitable for all cutthroat subspecies could shrink by 6.8 percent (range of estimates 4-11 percent) by the year 2030; 17.6 percent (range 7–28 percent) by 2060; and 27.4 percent (range 16–40 percent) by 2090.

However, the situation could be exacerbated for westslope cutthroat trout. In a separate analysis, Chris Keleher and Frank Rahel, Department of Zoology and Physiology, University of Wyoming, predicted that the current range of cutthroat trout in Wyoming would decrease by 65 percent with a 5.4-degree Fahrenheit (3-degree Celsius) warming of summer air temperature over that state, and U.S. Forest Service scientists wrote that the same level of warming would result in an equally severe reduction in the distribution of westslope cutthroat trout in Idaho and Montana.

The range of projections published for each run of the global climate models makes for a lot of numbers to absorb, but the bottom line is clear: global warming may further restrict the already reduced distribution of westslope cutthroat populations in the future and severely limit resource managers' options for recovery, because suitably cold aquatic habitat simply won't be available. [31]

Reacting to all this, in June 1997, American Wildlands, an environmental group, along with six other stewardship-minded organizations and individuals, petitioned the U.S. Fish and Wildlife Service to list the westslope cutthroat trout as threatened throughout its range. The Service completed its own status review of the subspecies in 1999, and in 2000 issued a determination that listing was not warranted. Citing deficiencies in the status review, the petitioners promptly challenged this finding in court, where they won, and the judge compelled a second review. That didn't change the outcome in the end, however. The Service corrected the troubling deficiencies, and in 2003 again determined that listing was not warranted owing to the

31. Ever since its inception, the most authoritative assessments of global climate change have been produced by the U.N.-sponsored Intergovernmental Panel on Climate Change (IPCC, website *www.ipcc.ch/*). The United States has also operated its own U.S. Global Change Research Program (USGCRP, website *www. usgcrp.gov*), established by Congress in 1990 to coordinate the resources and research of dozens of federal agencies. The USGCRP provides a major chunk of the research base on which IPCC assessments rely. The U.S. Environmental Protection Agency (EPA) operates a climate change program under this umbrella, and also maintains a website with links to IPCC assessment reports, USGCRP research, and EPA reports on the subject. To reach that site, type in *http://yosemite.epa.gov/ oar/globalwarming.nsf/content/ index.htm*. The most recent IPCC assessments were released in 2007, just as this book was going to press. Copies are available on the IPCC website. IPCC has also published a region-by-region assessment of vulnerability to global warming; see Watson et al. (1998). Other pertinent regional assessments have been published by the EPA (see Eaton and Scheller 1996), by Keleher and Rahel (1996), by Defenders of Wildlife and Natural Resources Defense Council

subspecies' "widespread distribution, available habitat on public lands, and conservation efforts under way by state and federal agencies." [32]

We have already hashed over the first two of these, and you can draw your own conclusions. But what about those conservation efforts under way for this subspecies?

The U.S. Fish and Wildlife Service determination not to list placed quite a bit of emphasis on federal and state administrative programs, strategies, agreements, and plans that, in the course of implementation, should benefit westslope cutthroat trout. At the federal level, the Inland Native Fish Strategy (INFISH), adopted by the U.S. Forest Service and Bureau of Land Management in 1995, is supposed to do for sensitive inland fishes what PACFISH, discussed in Chapter 3, does for sensitive fishes in the Pacific Northwest. INFISH provides riparian management standards, guidelines, and objectives for federal lands in the interior west, just as PACFISH does within the range of the northern spotted owl. Also cited as providing measures of protection were the Clean Water Act, the Federal Land Management Protection Act, the Wilderness Act, the Wild and Scenic Rivers Act, and the National Environmental Policy Act (NEPA), in addition to the various Memos of Understanding and conditions hammered out in formal consultations for the protection of species already listed under the Endangered Species Act whose ranges overlap that of the westslope cutthroat trout. [33] Admin-

(see O'Neal 2002), and by the Pew Center on Global Climate Change (see Poff et al. 2002). The National Geographic Society also carried a big report on the subject in the September 2004 issue of its magazine, *National Geographic* (volume 206, no. 3: pages 2–75), and two books, Tim Flannery's *The Weather Makers: How Man is Changing the Climate and What It Means for Life on Earth* (Atlantic Monthly Press, 2005) and Elizabeth Kolbert's *Field Notes From a Catastrophe: Man, Nature and Climate Change* (Bloomsbury Publishing, 2006), present the world view of global warming. And then there is Al Gore's documentary film and book, *An Inconvenient Truth*, that appeared in 2006. In addition to these authoritative sources, just about everybody who is anybody in the areas of fisheries and watershed management and policy has weighed in on the implications of global warming for fish populations and fisheries. A book edited by Kareiva et al. (1993) included Chapters by experts on aquatic systems and fishes. Cushing (1997) edited another book with Chapters prepared by a panel of experts, and the proceedings of at least two major symposia on the subject have also been published (see Beamish 1995 and McGinn 2002). For our western trout populations, some of the

scariest projections come from even more recent studies of mountain snowpack in western North America, which show that winter/spring snowpack has already declined by an average of 11 percent since 1950 and is likely to decrease by more than 40 percent by the 2050s. The Cascade Mountains of Washington, Oregon, and northern California, where elevations are lower and winter temperatures are milder, may see the greatest decreases if these projections are correct and, indeed, may receive no snow at all in the winter with all precipitation falling as rain by the 2050s. The timing of peak spring snowmelt has also advanced across western North America by anywhere from 10 to 40 days since 1950, and will come even earlier over the next 40 to 50 years. And if all that isn't enough, precipitation patterns across western North America appear to be coupled to Arctic sea ice cover, which is also disappearing rapidly. What this means is that some parts of our cutthroat trout country may receive even less precipitation in the coming years than the already paltry amounts they have received in the most recent drought years. To read some of these projections for yourselves, consult Sewall and Sloan (2004), Stewart et al. (2004), Sewall (2005), and Mote et al. (2005).

32. For this particular petition and associated Fish and Wildlife Service status review, see American Wildlands et al. (1997) and U.S. Fish and Wildlife Service (1999). The reconsidered findings and decision not to list westslope cutthroat trout as threatened are spelled out in a U.S. Federal Register notice published August 7, 2003 (see Williams 2003). While the U.S. chose to take no listing action, the Canadian COSEWIC took the opposite tack for populations of this subspecies in Alberta, placing them on its threatened list in May 2005. However, COSEWIC has indicated it will review the status of these populations once again, along with the status of British Columbia populations, sometime in November 2006. You can check the COSEWIC website at *www.cosewic.gc.ca* for updates.

33. Bull trout (threatened) and the upper Columbia River ESU of steelhead (endangered) are listed species that overlap portions of the westslope cutthroat range. However, like an umbrella caught in a gale, much, if not all, of the protection afforded westslope cutthroat trout by the bull trout listing may have been blown away. In September 2005, the U.S. Fish and Wildlife Service issued a final critical habitat designation for threatened bull

istrative measures cited at the state level included laws and regulations in place in Montana, Idaho, Oregon, and Washington governing forest practices, grazing practices, instream and riparian habitat alterations, and pollutant discharge. Also cited was a formalized four-state coordination process, that also includes the U.S. Forest Service and Bureau of Land Management, that provides for consistency and continuity of goals, objectives, and programs among all these agencies for westslope cutthroat conservation.

If you don't mind my interjecting a note of cynicism here, my problem with administrative measures is that the adequacy of protection they afford depends a lot on who is doing the administrating! For example, at the federal level, both INFISH and PACFISH originated during the Clinton administration. The Bush administration has been working to roll back many of their provisions, as well as those in the Clean Water Act and other federal environmental laws and regulations. The "proof of the pudding," so to speak, is not the administrative measures *per se* but what gets done—or does not get done—via these measures on the ground.

Before going there, however, the Service also cited state actions to stop or curtail stocking of hatchery trout in streams occupied by westslope cutthroat trout. Stocking of rainbow trout has either been stopped entirely or has been redirected to sterile, triploid fish. [34]

trout that reduced critical habitat protection from the 18,450 miles of stream habitat recommended by its own biologists to just 1,748 miles, along with a like reduction in the acreage of lakes and reservoirs, for a total cutback of 82 percent from what scientists believe is needed for the species to recover across its historical range. You can find this plan plus additional information on the Internet at *http://pacific.fws.gov/bulltrout/* or *http://species.fws.gov/bulltrout*, and in a U.S. Federal Resister notice published on September 26, 2005 (see Manson 2005). Meanwhile, the Fish and Wildlife Service also has under way a 5-year review of overall bull trout status to determine if listing is still warranted (this was due out in 2005 as well, but had not appeared as of October 2006). There are also stories circulating that the federal administration may have the endangered listing of upper Columbia River steelhead under review with the intent of dropping that as well. You'll have to stay tuned for the outcomes of these moves.

34. Triploid fish have three sets of chromosomes in their cell nuclei instead of the normal two sets. Triploids are produced artificially by applying a thermal or pressure shock treatment to eggs at a carefully controlled but short time interval after fertilization (e.g., an interval of 10 minutes for rainbow trout). Triploid fish are sterile, which precludes unwanted reproduction. They also have the potential to live longer and grow larger than diploids, an appealing feature to anglers after large trout. However, in keeping with the old axiom in ecology that "you can never do merely one thing," there are a few potential downsides to stocking triploid trout, especially in the presence of or adjacent to native trout populations. For one thing, harking back to our earlier discussion of the competitive exclusion principle, survival and growth

of triploids beyond the normal size and lifespan of diploid natives could tip the advantage to the triploids in competition for stream resources. Or, at a larger size, triploids could also change prey preferences to include otherwise invulnerable size classes; for example, the juveniles of native trout. Also, some male triploids may still be able to produce hormones that trigger participation in courtship and spawning rituals. Even though no progeny would result, too many pairings with wild diploid females could reduce the reproductive capacity of the wild population since the eggs of those females would be wasted. For more on the subject of triploid trout, see Lincoln and Scott (1984), Thorgaard and Allen (1987), Hallerman and Kapuscinski (1993), and Kozfkay et al. (2006).

Montana adopted a policy of no stocking of hatchery fish in streams in 1976. In Idaho, only triploid rainbow trout are now stocked in waters near or connected to reaches occupied by westslope cutthroat. Washington no longer stocks rainbow trout in tributaries with native cutthroat trout, and triploid rainbows are stocked in mainstem rivers, such as the Pend Oreille. Oregon has not stocked streams in the John Day River drainage where that state's westslope cutthroat populations are located since 1997.

The Service's determination not to list also cited more than 700 on-the-ground projects, either completed or ongoing, dealing with physical habitat restoration or westslope cutthroat recovery. Many of these have been collaborative efforts between the U.S. Forest Service or Bureau of Land Management, state fish and game agencies, and sportsmen's groups, such as Trout Unlimited, the Federation of Fly Fishers, or various watershed stewardship organizations. Some of these projects have been funded by matching grants from the Bring Back the Natives program administered by the non-profit National Fish and Wildlife Foundation.[35] From my review of available press releases, progress reports and assessments, on-the-ground projects for westslope cutthroat trout appear to fall into three major categories:

- Projects to restore instream habitat or improve stream corridor structure and function.

- Projects to isolate and insulate remaining genetically pure westslope cutthroat populations from introgression, competition, and disease.

- Projects to remove non-native species either by physical or chemical means, followed by allowing existing genetically pure populations to expand downstream to occupy the vacant habitat, or introduction or reintroduction of genetically pure westslope cutthroats where no existing upstream population is available.

In that third category, even private enterprise is contributing to westslope cutthroat recovery. Cherry Creek, in the Madison River drainage, has approximately 70 miles of trout habitat upstream of a 30-foot waterfall, all located on the Flying D Ranch, a property owned by billionaire Ted Turner's Turner Enterprises, Inc. In 1997, this habitat was occupied by brook, rainbow, and hybridized Yellowstone cutthroat trout, the latter apparently originating from a 1924 fry plant into a headwater lake. Turner Enterprises is working with the Montana Department of Fish, Wildlife and Parks to introduce westslope cutthroat trout into this reach. First the non-native and hybridized trout will be removed using the fish toxicant antimycin, then genetically pure westslope cutthroat trout will be introduced.

Antimycin is an antibiotic produced by the filamentous organism *Streptomyces*. It was first isolated in 1945 by researchers at the University of Wisconsin. In the early 1960s, it was discovered to be toxic to fishes. Its properties and applications in fisheries manage-

35. The National Fish and Wildlife Foundation (website *www.nfwf.org*) was established by Congress in 1984 as a non-profit organization with the mission to partner with federal and state agencies and other non-government organizations to conserve healthy populations of fish, wildlife, and plants. It meets these goals by awarding matching grants to projects benefiting conservation, education, habitat protection and restoration, and natural resource management. Federal partners have included the U.S. Fish and Wildlife Service, National Oceanic and Atmospheric Administration, U.S. Forest Service, Bureau of Land Management, Bureau of Reclamation, and Natural Resources Conservation Service. In 1991, the Foundation, along with the U.S. Forest Service and Bureau of Land Management, launched Bring Back the Natives, a national program to restore the health of riverine systems and their native species. The first grants specifically for this program were awarded in federal Fiscal Year 1992. Trout Unlimited later threw in as a major non-government partner in Bring Back the Natives. Bring Back the Natives awards between 12 and 15 matching grants annually on a competitive basis, average grant size $60,000. Through Fiscal Year 2003 (most recent data available to me), 121 grants had been made to

ment and aquaculture were studied extensively by the U.S. Fish and Wildlife Service. Even though it is expensive, its effectiveness at very low concentrations, its low to zero impact on other aquatic animals, and its rapid breakdown into non-toxic products quickly made it a favorite among biologists who worked in stream systems. It has been used in many projects to remove non-native trout and charr and restore native trout in small headwater streams, including the recovery program for the threatened greenback cutthroat trout (Chapter 13). In practice, after choosing the reach from which non-natives are to be removed, the first step is to locate or create a barrier to prevent re-invasion by non-native fish. A 5- to 10-foot waterfall is usually effective, but where no such feature exists, a barrier can be built. The second step is drip-application of antimycin. This is typically done in late summer, two years in a row, to ensure complete removal of the non-natives. Potassium permanganate, an oxidizing agent that destroys antimycin, can be dripped in at the barrier to limit the treatment to just the designated reach, although, as I said, antimycin does degrade rapidly on its own. Native trout are then introduced into the treated reach the spring following the second application. [36] The plan for the Cherry Creek project is to work downstream in four sections, treating each section twice with antimycin (the first year of this treatment phase got under way

in the summer of 2003). For the introduction phase, fertilized eggs will be placed in streamside incubators so as to incubate and habituate in Cherry Creek water. It's expected that the project will be completed in 2008 or 2009. [37]

The Cherry Creek proposal was first announced in 1997. Also that year, Montana Fish, Wildlife and Parks announced a 10-year plan to restore pure westslope cutthroat trout in 30 additional stream segments in the upper Madison River drainage. In 2001, the department proposed a similar plan for westslope cutthroat trout in the South Fork Flathead watershed. Fish for each of these projects will come from genetically pure broodstocks maintained in state hatcheries. For many years, Montana Fish, Wildlife and Parks operated a large-scale propagation program at its Jocko River Hatchery for stocking into Hungry Horse Reservoir and other waters in the Flathead River drainage. That program was discontinued in 1983, owing to severe inbreeding problems with the broodstock. More recently, trout from South Fork Flathead and Clark Fork tributaries were used to establish a broodstock at the Washoe Park Hatchery near Anaconda. These trout are stocked for recreational fisheries in the northwestern part of the state. Meanwhile, the state has a site at Sekokoni Springs, east of Kalispell, where "nearest neighbor" stocks of westslope cutthroat trout are being

projects benefiting ten of the native cutthroat subspecies recognized in this book. Of these, 35 grants, or 29 percent, went to on-the-ground projects for westslope cutthroat trout.

36. For examples of other projects in which antimycin has been used to remove non-native trout and charr for the purpose of restoring native trout in small headwater streams, see Rinne et al. (1981), Rinne and Turner (1991), Gresswell (1991), and Stefferud et al. (1992). The use of antimycin in the recovery program for greenback cutthroat trout is reviewed in Stuber et al. (1988) and will be touched on again in Chapter 13. Antimycin treatments are not always sure-fire. Examples abound of projects that failed at total removal of non-native fishes. Removal of brook trout may be complicated by their affinity for areas of groundwater upwelling where they may find refuge from the toxicant. For more information on the use of antimycin and rotenone, the only other fish toxicant certified for use in U.S. waters, see Chapter 8 of Sigler and Sigler (1990) and Chapter 10 in Murphy and Willis (1996).

37. Court challenges to the use of antimycin at Cherry Creek (and presumably for other native trout restoration projects as well) were set

developed for recovery efforts in the upper Flathead drainage. The state is also working with another private land owner, the Sun Ranch south of Ennis, to develop a "Madison Valley" westslope cutthroat brood-stock in a small hatchery and rearing pond the ranch owner built on that property.

The State of Idaho has also developed genetically pure westslope cutthroat broodstocks which are stocked to support recreational fisheries. From 2000 to 2005, 449 bodies of water, mostly high lakes but also includ-ing Hayden and Pend Oreille lakes and some streams, have received these trout in the Panhandle, Clearwater, Salmon, and southwest regions of the state. The State of Washington continues to stock westslope cutthroats from its Twin Lakes and Kings Lake broodstocks.

Yellowstone National Park will take a different tack to restore westslope cutthroat trout to the northwest corner of the park where they occurred historically, according to a May 2006 news release quoting Todd Koel, the Park's lead fish biologist. Park biologists, too, will use antimycin to remove non-native and hybrid-ized trout, in this case from East Fork Specimen Creek and from High Lake, where Specimen Creek origi-nates. But to restock these waters, they will not use hatchery trout. Rather, trout will be transferred from a native, genetically pure source population that was recently discovered in an unnamed tributary of nearby Grayling Creek. Barriers will be constructed to prevent non-natives and hybrids from reinvading the restored waters from below.

As always, it seems, in these restoration and recov-ery programs, there are cautionary notes. In Chapter 1, I led you through definitions of gene loci, alleles, and allelic frequency. I explained there that while a diploid individual can have no more than two alleles at a given locus, a population of diploid individuals can have many alleles at that locus. Furthermore, when comparing different populations of individuals, alleles may occur ubiquitously among the populations but at different fre-quencies, or populations may each have its own array of alleles that are common only in that population. These kinds and amounts of genetic difference, toted up across all alleles at all polymorphic loci, are what scientists call *genetic variation* (also sometimes referred to as *genetic divergence* or *genetic differentiation*). To conservation biologists, the primary goal of conservation programs is to ensure that existing genetic variation in a species or a subspecies is maintained. It is this genetic variation, the product of millennia of evolution, that represents the evolutionary legacy of the species or subspecies and is the fuel for future evolution.

The total amount of genetic variation in a species or subspecies is usually structured in a hierarchical geographic order. Thus, a certain portion of the total

aside back in September of 2005 when the Ninth Circuit Court of Appeals ruled that antimycin, when applied to a river or stream for the purpose of native fish restoration, is not a pollutant and does not violate the Clean Water Act. I am grateful to Carter Kruse of Turner Enterprises, Inc. and Pat Clancy of Montana Fish, Wildlife and Parks for details about the Cherry Creek project. Further information about this project can be obtained from the proceedings of a 2001 symposium, "Practi-cal Approaches for Conserving Native Inland Fishes of the West," sponsored by the Montana Chapter, American Fisheries Society (see Shepard 2001). Abstracts of the papers presented at this symposium are also available at the Montana Chapter website at *www.fisheries. org/AFSmontana/Misc/Sympo-sium%20Abstracts/Abstracts.pdf.*

genetic variation may be attributed to genetic differences between population groups at some broad regional level, e.g., between major river basins. The several distinct DPSs and DUs identified for coastal cutthroat trout in Chapter 3, some of which span several river drainages, is another such example. Another portion of the total genetic variation may be due to genetic differences between local populations within regions. And yet another (and typically the largest) portion of the total genetic variation may be due to genetic differences between individuals within local populations. This hierarchical distribution of total genetic variation is commonly referred to as the *population genetic structure* or *genetic population structure* of the species or subspecies. The lowest level of the hierarchy, variation among individuals within a population, is the raw material for breeding programs and aquaculture. It is on the two higher levels where the interests of conservation programs focus.

Fred Allendorf, Robb Leary, and their colleagues at the University of Montana, Missoula, have made extensive studies of the population genetic structure of westslope cutthroat trout using allozyme electrophoresis to detect and obtain estimates of genetic variation. What they found is that the highest level of the hierarchy, i.e., differences among populations from the upper Columbia and upper Missouri river basins, accounts for very little (only 1.29 percent) of the total genetic variation of the subspecies. Based on this result, the U.S. Fish and Wildlife Service decided that the entire westslope subspecies could be encompassed in a single DPS. However, at the second level of the hierarchy, large genetic differences between populations within each drainage were found that account for a very appreciable 33.76 percent of the total genetic variation of the subspecies. Furthermore, these large genetic differences are due primarily to alleles that are narrowly distributed but occur at high frequencies in the local populations where they are found. Only a few variant alleles are shared ubiquitously among westslope cutthroat populations, and even these occur at widely divergent frequencies in different populations. [38]

What all this means is that even over short geographic distances, westslope cutthroat populations can be very different from one another genetically. This suggests that historically there has been very little gene flow among populations, and that extant native populations have probably established local adaptations. Here's the cautionary note: if native fish interbreed with westslope cutthroat trout introduced from broodstocks developed from other localities, even close ones, these local adaptations may be broken down, which could compromise population viability. The potential

38. Full details of the University of Montana group's studies of westslope cutthroat genetic population structure can be found in reports by Phelps and Allendorf (1982), Leary et al. (1985, 1988), Allendorf and Leary (1988), and Leary et al. (1998). For completeness, the remaining 64.95 percent of the total genetic variation not mentioned in the text but found by the University of Montana group occurs at the lowest level of the hierarchy and is due to differences between individuals within local populations. Recent work using microsatellite DNA markers on westslope cutthroat populations from the Pend Oreille River basin of northeastern Washington by researchers from the Washington Department of Fish and Wildlife appears to corroborate the University of Montana findings. Like the Montana group, the Pend Oreille study also found that a high proportion of the total genetic variation was attributable to differences between populations within the drainage. This work is reported in Young et al. (2004).

for this sort of adverse impact needs to be carefully weighed when stocking is proposed.

With genetic variation in westslope cutthroat trout being composed largely of alleles with a narrow geographic distribution but occurring at high frequencies in the local populations where they are found, achieving the goal of maintaining this genetic variation will require the continued preservation of all local populations throughout the subspecies range.

Lahontan Cutthroat Trout, *Oncorhynchus clarkii henshawi*, lake form

CHAPTER 5

Lahontan Cutthroat Trout

Oncorhynchus clarkii henshawi: Chromosomes, $2N = 64$. Scales in lateral series 150–180. Scales above lateral line 33–40. Vertebrae 60–63. Gill rakers 21 to 28, mean value 24, the most gill rakers of any trout of the genus *Oncorhynchus*. Pyloric caeca 40–75, mean value 50 or more, generally higher than in any other subspecies (within the subfamily Salmoninae, high numbers of pyloric caeca tend to occur in the more piscivorous species and subspecies). Spots are medium to large, rounded, and more or less evenly distributed over the sides of the body, the top of the head, and often on the abdomen. This spotting pattern, along with the high numbers of gill rakers and pyloric caeca, serve to set *O. c. henshawi* apart from all other cutthroat subspecies. Colors are generally dull, but the sides and the cheeks often take on pink or red colors.

Lahontan Cutthroat Trout,
Oncorhynchus clarkii henshawi, stream-resident form

THE DESERT LAKE

The People call themselves "Coo-yu-ee Dokado," which means, literally, "cui-ui eaters." The cui-ui (pronounced "kwee-we") is a sucker-like fish peculiar to the large desert lake where The People live. The name they give the lake itself is "Coo-yu-ee Pah." We know it today as Pyramid Lake, and The People are the Pyramid Lake band of Northern Paiute Indians.

These people have their own story of the origin of Pyramid Lake. It is embodied in the story of "Tupep-eaha," the Stone Mother, a large outcropping of tufa that sits on the eastern shoreline, facing south, very near the pyramid that gives the lake its modern name.

The Stone Mother's great face bears an uncanny resemblance to that of a Paiute woman sitting by her basket, weeping. And therein lies the tale.

Long ago in the Great Basin there lived a family known as "Neh-muh" (which means The People). The family had always lived by a credo of kindness to all, but one of the children grew quarrelsome. He persuaded his brothers and sisters to his ill-mannered ways, and soon there was bickering and fighting.

The parents tried to sway their children back to their gentle ways, but to no avail. Finally, with much sadness and reluctance, they decided that the children had to be separated. Thus, the father sent the quarrelsome brother and his family west to the land near the high mountains. These people became the Pit River Indians. Another brother was sent north, where he and his family became the Bannocks. A third brother was sent south to the Owens Valley. His people became the Southern Paiutes. Other brothers and sisters were dispersed elsewhere and formed other bands in their new homes.

The parents now lived in peace and quiet, but they were very sad. Their sorrow mounted when the quarrelsome brother and his Pit River Indians commenced raiding and fighting with the other groups.

Then the father died. There was no consoling the mother. She wailed and mourned. Even when gathering seeds in her basket she wept. Day after day, year after year, her tears flowed, and the water gathered to form a lake. The mother and her burden basket were turned to stone, at a spot where she could look out forevermore over the lake formed by her tears.

The story of the Stone Mother is a beautiful one that explains in a mythical sense the origin of the various Indian tribes, and of Pyramid Lake. The area around the Stone Mother is sacred to the Northern Paiutes. But modern scientists tell us that the lake predates the Paiute people. The Northern Paiute culture goes back only about 500 years, they say, but archaeological evidence indicates that people may have lived in caves along the lakeshore as long as 11,000 years ago, and the lake itself is even older than that.

In fact, the latest work (reviewed in Chapter 2) indicates that pluvial lakes may have existed in the western Great Basin as far back as 650,000 years ago. The ancient lake waxed and waned in accordance with vagaries in climatic conditions as great Ice Age glaciers advanced, retreated, then advanced again, but at its most recent peak, about 25,000 years ago, it was 875 feet deep and covered something like 8,500 square miles. It rivaled present-day Lake Erie in size and was exceeded only by Lake Bonneville, the vast inland sea that lay off to the east in what is now Utah.

Modern scientists call the ancient water body Lake Lahontan, after a French nobleman who wrote fanciful tales about adventures in the American West in the late 17th century, when the French held sway along the Mississippi and sent exploring parties westward.[1] Lake Lahontan drained an area of more than 45,000 square miles. Its longest and most important tributary was the Humboldt River which flowed in from the east. The Truckee, Carson, Walker, and Susan rivers brought in water off the high Sierras to the west. Quinn River and other smaller tributaries drained the lesser ranges to the north.

The first Stone Age people to come upon Lake Lahontan found a sea of blue water that sent long prob-

1. Louis Armand de Lom d'Arce, Baron de Lahontan (or de la Hontan, as some scholars write it), a Frenchman and a pretty imaginative one at that, set out from the French outpost at Michilimackinac in 1688 to explore the Mississippi River and beyond. There, according to his narrative, he discovered a stream, the Riviere Longue (Long River), that flowed in from the west. He ascended the Riviere Longue to its headwaters, so he claimed, where he learned from local Indians that another stream rose on the opposite side of the dividing ridge and flowed westward, eventually emptying into a vast salt lake that in turn drained into the ocean through a large mouth two leagues in breadth. Although historians have widely discredited Lahontan's narrative, I have a somewhat different take. Suppose that what he wrote down in his fanciful account was an impression of western geography passed on in the oral tradition of the American Indians, blurred by the passage of time and the generations of telling and retelling but based on truth nevertheless. It is tempting to conjecture that what the Indians might have conveyed was a vaguely remembered description of an ancient Great Basin lake and its connection to the sea. In any event, Lahontan's name (his title actually)

ing fingers back among forested hills. But about 8,000 years ago, the climate became warmer and drier, and the ancient lake receded. The magnitude of the change is pretty staggering when you stop to think about it. "Where one now sees a world of gray-white alkali dust and sand," as one writer put it, "there [once] had been deep blues, forest greens, and the rich reds, yellows, and violets of wild flowers." [2] It is hard to believe. The only remnants of this once-great body of water are Pyramid and Walker Lakes, the perennial desert lakes of Nevada, the often-dry Honey Lake across the line in California, and the magnificent playas of the Smoke Creek and Black Rock deserts in Nevada.

But trout were already present in Lake Lahontan when the first Stone Age culture settled there. Time has erased many clues as to how the earliest inhabitants lived, but excavations of lakeshore rock shelters have yielded fish remains, fishing lines, and fish-net fragments with dates as old as $9,660 \pm 170$ B.P., leaving little doubt that fishing was practiced in the area almost from the time humans discovered it. [3] People of the Lovelock culture, which settled around Pyramid Lake about 4,000 years ago, lived in community groups of three or four families mostly, and worked together to provide a stable food supply. Remnants of stone piers are thought to have been ancient fish traps. Rock breakwaters may have sheltered rafts used to set nets or to raid the bird rookeries on Anaho Island. Skilled artisans furnished finely chipped spear and arrow points for hunting, large nets for fishing, and baskets for gathering herbs and seeds. Caches of stored food excavated from caves testify to the culture's ability to provide for its members; in fact, they were quite progressive. They traded with other bands—spear points and fish for ocean mollusk-shell jewelry, for example—and exhibited many other elementary characteristics of a modern society.

When the Northern Paiutes settled at Pyramid Lake, they continued to rely on the lake as a source of food. Not only did they utilize the cui-ui, they also netted the huge trout that ascended the Truckee River to spawn. When John C. Fremont came upon the lake in 1844, he found the band in exclusive possession of the fishery. Other tribes had to barter with them for fish. On January 15, 1844, Fremont wrote in his journal:

"An Indian brought in a large fish to trade, which we had the inexpressible satisfaction to find was a salmon-trout; we gathered around him eagerly. The Indians were amused with our delight, and immediately brought in numbers, so that the camp was soon stocked. Their flavor was excellent—superior, in fact, to that of any fish I have ever known. They were of extraordinary size—about as large as Columbia River salmon—generally from two to four feet in length. From the information of Mr. Walker, who passed among some lakes lying more to the eastward, this fish is common to the streams of the inland lakes. He subsequently informed me that he had obtained them weighing six pounds when cleaned and the heads taken off, which corresponds very well with the size of those obtained at this place." [4]

is perpetuated in the Great Basin. To read his narrative in full, search out Volume I of *A General Collection of Voyages and Travels*, John Pinkerton, editor (1808), and turn to "Memoirs of North America." If you can read French, you can also find the complete narrative in Roy (1974).

2. Jackson, D.D. *Sagebrush Country* (New York: Time-Life Books, 1975).

3. Donald R. Tuohy of the Nevada State Museum, Carson City, has summarized these findings; see Tuohy (1990). Primary papers cited by Dr. Tuohy that you may also want to read for yourselves include Heizer (1951), Orr (1956), Follett (1982), Dansie (1987), Rusco and Davis (1987), and Tuohy (1988).

4. See Fremont (1845 [2002]). The Mr. Walker mentioned in this passage was Joseph Reddeford Walker, one of the famous mountain men of the fur-trapping era. He had explored across the Great Basin and on to California with his own party about ten years earlier, and had come upon Walker Lake at that time. Walker was not part of the Fremont expedition that discovered Pyramid Lake, but fell in with the party later, on its return home. Walker's information was added when Fremont wrote up the expedition's official report. Walker did serve as

When you come over the brow of a hill and glimpse one of the big desert lakes or reservoirs, it takes you by surprise. It looks strangely out of place, as though all that water shimmering among the parched brown hills doesn't really belong there. Even in winter, when those hills are cloaked in white, you feel compelled to stop at the first pull-out and just stand there awhile, gazing out over the water.

That's how Fremont must have felt on January 10, 1844, the day of discovery. Leaving the Dalles on the Columbia River, his party of 24 men had followed the eastern base of the Cascade Mountains to Klamath Lake. Their route from there was generally southeast, more or less along a route that immigrant wagons would follow in the reverse direction two or three years hence on their way to the Oregon country. The party found its way through High Rock Canyon, then angled south across a leg of the Black Rock Desert. On January 10, Fremont and Kit Carson, the expedition's guide, were out ahead reconnoitering the route. They came to a good camping spot at the base of a pass, which they marked for the main group, then continued on up to see what was on the other side. Here is what they beheld:

"Beyond, a defile between the mountains descended rapidly about two thousand feet; and, filling up all the lower space, was a sheet of green water, some twenty miles broad. It broke upon our eyes like the ocean. The neighboring peaks rose high above us, and we ascended one of them to obtain a better view. The waves were curling in the breeze, and their dark green colour showed it to be a body of deep water. For a long time we sat enjoying the view, for we had become fatigued with mountains, and the free expanse of moving waves was very grateful."

Fremont stated that Pyramid Lake is "set like a gem in the mountains." A National Park Service committee, taken by its beauty, wrote: "It is no exaggeration to say that Pyramid Lake is the most beautiful desert lake any member of the committee has yet seen—perhaps the most beautiful of its kind in America." [5]

It appeals to the senses in many ways. First there are the sunsets. Miss Julia Hyde was certainly impressed on the evening of her first visit one late fall day in 1888:

"I think I have never in my life seen a more glorious sunset than I witnessed here on the evening of my arrival. The western sky was a mass of purple, vermillion, and gold, resembling painted and gilded towers. From the deep blue of the lake rose the sun-tipped pyramids of the lower world, as though towering to meet those of the more ethereal realms above, while between all the air seemed filled with a glimmering golden haze. All in this wild and treeless region seemed a dazzling picture done in vermillion, blue and gold." [6]

Then there are the nights. Desert nights fall quickly, and they can be incredibly clear and sharp. The stars glitter with a brightness and intensity against the jet-black sky that is almost overpowering.

And, of course, there are the cutthroat trout. Pyramid Lake has produced some monster fish, among the largest in the trout world. The world record for hook-and-line, a 41-pounder, was taken there in 1925 by John

guide for a later Fremont expedition which also crossed the Great Basin and entered California. That one embroiled Fremont in the conspiracy to wrest California away from Mexico and make it part of the United States, which of course is what eventually happened.

5. Cited in Wheeler (1974).

6. Miss Julia Hyde, letter to the editor, *Eureka* (Nevada) *Weekly Sentinel*, November 19, 1888. Cited at page 12 in *Nevada Outdoors and Wildlife Review* 10, no.2 (Summer 1976 issue).

Skimmerhorn, a Paiute Indian. And old-timers swore that even larger fish, trout that sometimes exceeded 60 pounds, were caught on a regular basis in the Indian fish traps and commercial nets that once operated here.

Like the desert itself, Pyramid Lake's cutthroats are somber-hued fish whose colors have been described as earthen, ore-rich glazes, their backs and sides flecked with large, irregular spots. These trout have attracted thousands of anglers over the years. European royalty has fished here; so have the most celebrated of Americans, including Clark Gable and Herbert Hoover. Even now, Pyramid Lake is the focus for individual fly-fishermen and fly-fishing clubs from all over the west who come here year after year in search of trophy trout. As I'll explain later, the original strain of leviathan Pyramid Lake cutthroat has been extinct in the lake since the late 1930s, and the hatchery-reared trout presently swimming there attain only about half or maybe a bit more of the maximum weight of the original strain. But even so, a trout of 20+ pounds is a respectable trophy.

In the summertime, the trout head for the depths. In winter and spring, they roam the shallows. But even then they are hard to come by. You can fish the lake from a boat and with a little care you can wade out to casting positions in the shallow bays and along gently sloping shorelines. Or, if you're particularly venturesome and well insulated against the cold, you can work along parallel to the shoreline in a float tube. The standard fare for fly-fishermen is to cast a big black Woolly Worm or Woolly Bugger (white and fluorescent orange are also good colors) or a black or white marabou streamer, or even a big Zonker pattern, to water that is eight to fifteen feet deep using a sinking line. Work the fly right along the bottom with a variety of retrieves. Spin fishermen sometimes cast a fly-and-bubble rig, but mostly they use large flatfish, wobbling spoons, or deep-running minnow imitations.

Do expect to put in long hours of casting between strikes however. A normal day in the life of a Pyramid Lake angler begins before daylight. If that person's a fly-fisher, dawn will find him or her waist deep in the water, casting away. There might be a stop for breakfast at midmorning and again for a bite of lunch and a short nap in mid afternoon, but otherwise it's cast, cast, and cast again until dark. Many serious anglers fish hard like this for days without a strike. But the chance is always there of hanging onto the trout of a lifetime. The only way to miss it for sure is to not have your fly in the water!

Pyramid Lake, Nevada. John C. Fremont thought this pyramid was "a pretty exact outline of the great pyramid of Cheops," th giving the lake its modern name. PHOTO BY AUT

HISTORICAL RANGE

Figure 5-1 outlines the full extent of the 45,000-square mile Lahontan Basin. This extends down the eastern slopes of the northern Sierra Nevada Mountains of California then eastward across the northern Basin and

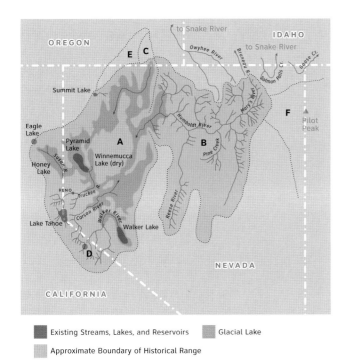

Existing Streams, Lakes, and Reservoirs   **Glacial Lake**

Approximate Boundary of Historical Range

FIGURE 5-1. *Ancient Lake Lahontan and approximate extent of native ranges of* (A) *Lahontan Cutthroat,* (B) *Humboldt Cutthroat,* (C) *Willow/Whitehorse Cutthroat,* (D) *Paiute Cutthroat, and* (E) *Alvord Cutthroat. Also shown: proximity to Snake River tributaries and Bonneville Basin* (F).

Range ecoregion of Nevada. I have designated roughly the western half of this area, including the area inundated by pluvial Lake Lahontan itself, with the letter "A." For the purpose of this book, this area will be considered the historical range of the native cutthroat

trout meeting the description in the chapter header, i.e., Lahontan cutthroat trout *Oncorhynchus clarkii henshawi*.

The cutthroat trout occupying the upper reaches of the Quinn River, which heads up just south of the Coyote Basin marked "C" in Fig. 5-1, do not match this description. Their characters are quite similar to the cutthroat trout of the Humboldt River drainage. I will have more to say about this in Chapter 6.

Strictly speaking, the Lahontan Basin also includes the Humboldt River drainage, designated "B" in Fig. 5-1. The Humboldt River was the major tributary of ancient Lake Lahontan, draining almost the entire eastern half of the basin. This drainage comprises the historical range of a cutthroat trout Dr. Behnke recognizes as distinct at the subspecies level from the cutthroat trout of the western half of the basin.[7] Since I am following Dr. Behnke's classification system in this book, I have set aside the Humboldt River drainage and its native trout for separate discussion in Chapter 6.

I bring all this up here because federal and state management agencies now lump all the cutthroat trouts of the Lahontan Basin together, Humboldt drainage and Quinn River included, as the single subspecies *O. c. henshawi*. The U.S. Fish and Wildlife Service and Nevada Division of Wildlife now designate the trout of the Humboldt River drainage as one of three DPSs of Lahontan cutthroat trout, the others being the Western

7. Dr. Behnke isn't the only scientist who has recognized the distinctiveness of the Humboldt drainage cutthroat trout. Dr. Behnke based his assertion that the Humboldt cutthroat is a distinct subspecies on clear-cut differences in meristic characters that distinguish cutthroat trout from the Humboldt drainage from cutthroat trout from the western portion of the Lahontan Basin 100 percent of the time. Although genetic studies using allozyme electrophoresis have generally failed to show this distinctiveness (Loudenslager and Gall 1980a, Bartley et al. 1987, and Bartley and Gall 1993), studies of variation in mtDNA haplotypes (Williams et al. 1992, 1998) and variation at microsatellite DNA markers (Nielsen and Sage 2002) led investigators employing those techniques to the same conclusion as Behnke's.

Lahontan Basin DPS and the Northwestern Lahontan Basin DPS, which includes the Quinn River drainage. The U.S. Fish and Wildlife Service also includes the cutthroat trout of the adjacent Coyote Basin of Oregon (designated "C" in Fig. 5-1), into which Willow and Whitehorse creeks drain, as a component of the Northwestern Lahontan Basin DPS. Dr. Behnke has regarded the Willow/Whitehorse cutthroat trout as a separate but unnamed subspecies as well, and I will reserve those for discussion in Chapter 7.

The western portion of the Lahontan Basin presents a study in ecological contrasts. [8] The eastern slopes of the Sierra Nevada Mountains of California comprise the western rim of the basin. This is steep, often precipitous mountain terrain creased by many steep-gradient valleys. Mixed conifer forests consisting of white and red firs, Jeffrey pine, Ponderosa pine, and Douglas-fir occur on the lower to mid-elevation slopes, giving way to subalpine species, such as mountain hemlock, red fir, lodgepole pine, western white pine, and whitebark pine, as the altitude increases up to timberline. Alpine glaciation occurred in this ecoregion during the Pleistocene. The Washoe Indians foraged here before the coming of the whites, and logging, mining, and livestock grazing have been predominant land uses since the time of white settlement.

Waters of the Walker, Carson, Truckee, and Susan rivers feed into the Lahontan Basin from off these slopes, beginning as cold mountain streams that drop quickly, but slow occasionally to meander through mountain meadows. Tributaries of the Walker and Carson rivers head up under peaks ranging in elevation from 10,000 to 12,000 feet. The Truckee River originates from deep, oligotrophic Lake Tahoe, elevation 6,226 feet, which in turn is fed by Cascade and Fallen Leaf lakes, two smaller but also deep and oligotrophic bodies of water situated a few hundred feet higher in elevation. Feeder tributaries for the entire Tahoe/Truckee system head up at higher elevations still. The waters of Donner and Independence lakes, two additional mountain lakes in the drainage, join the flow of the Truckee on its run out of the mountains. The Susan River also heads up near the eastern Sierra crest, but closer to its northern terminus where the range is lower in elevation.

These high mountain slopes drop abruptly to the basin floor where the elevation ranges from about 3,600 to maybe 5,400 feet. Ecological conditions shift just as quickly, from alpine/subalpine to high desert. The Sierras form an effective rain shadow, with annual precipitation decreasing from 50 inches at the crest to as little as 5 inches at some places on the basin floor, such as the town of Hawthorne, Nevada, south of Walker Lake. Although this country is called the Great Basin, Basin and Range is a more apt description. The ecoregion is dominated by dozens of north-south

8. More complete descriptions of the topography and natural history of the western portion of the Lahontan Basin can be found in Downs (1966), Bowen (1972), Jackson (1975), Houghton (1976), Wheeler (1974, 1979), Omernik (1987), Trimble (1989), Schoenherr (1992), Grayson (1993), Bailey (1995), and U.S. Environmental Protection Agency (1998).

oriented mountain ranges that rise abruptly, without foothills, to elevations of 7,000 to 10,000 feet. These are separated by broad desert valleys, some of them 20 miles or more across, that form the floor of the basin proper. Saltbrush, rabbitbrush, and sagebrush are the dominant perennial shrubs here, interspersed with bunchgrass in some localities and little more than sand in others; for example, the infamous Forty-Mile Desert that immigrant trains bound for California were obliged to cross. Stands of juniper and pinion pine take over with mountain mahogany on the drier slopes and ridges as one climbs up into the ranges, and these give way to limber pine as one continues even higher. Riparian vegetation consisting of aspen groves, willow thickets, meadows, and willow/meadow complexes are strung out along streams that head up in these ranges.

The four major rivers that drain the Sierra slopes flow out across the basin floor and either empty into expansive, alkaline desert lakes that are remnants of ancient Lake Lahontan, or disappear in marshy sinks. Walker River (or rather, what's left of it after water withdrawals for irrigation and consumptive uses) empties into Walker Lake, whose size and surface elevation have been reduced considerably since the mid-1800s. Were it not for these water diversions, Walker Lake would have maintained a mean surface elevation of about 4,110 feet and spread over an area of more than 69,000 surface acres. Today it stands about 140 feet

lower and its surface area has been reduced by about half. [9] The Carson River flows into a broad, flat, sink area east of Fallon, Nevada. This area, too, has been transformed by the hand of man. In the early 1900s, the Newlands Project, the very first to be completed by the U.S. Bureau of Reclamation, diverted water from the Truckee River into a new storage reservoir in the lower Carson basin, this water to be used to "reclaim" the desert area around Fallon for high-yield crop agriculture. That plan achieved only partial success, and even that success came at a high price to the lower Truckee River/Pyramid Lake system. [10] The Truckee River flows out of the mountains through Truckee Meadows, where the cities of Reno and Sparks now stand, then east and finally north to discharge into Pyramid Lake, present surface elevation around 3,786 feet. Pyramid Lake once had a companion, Lake Winnemucca, which was a shallow body of water that stood at surface elevation 3,862 feet at the level of the slough that joined the two lakes. But the Newlands Project so dewatered the lower Truckee system that the level of Pyramid Lake dropped well below the level of the slough and Lake Winnemucca dried up completely. The Susan River, also much affected by water withdrawals for irrigation, flows into ephemeral Honey Lake southeast of the town of Susanville, California.

In the northeastern corner of the western Lahontan Basin (don't be confused here, this is the area the U.S.

9. In 1987, using records available back to 1870, W. Milne, a graduate student at the Colorado School of Mines, Golden, Colorado, reconstructed pristine water levels for Walker Lake by adjusting annual Walker River discharge records for the amounts of water withdrawn from the river each year for irrigation and consumptive uses. Annualized surface areas of the lake were computed from these adjusted discharge values, and these were corrected for evaporation by applying the mean annual evaporation rate. Data from the U.S. Geological Survey were then used to calculate lake levels from the recomputed annual surface areas. See Milne (1987), Benson and Mifflin (1986), and Benson et al. (1991) for complete details. A study of the limnology of Walker Lake was published by Cooper and Koch in 1984.

10. In my opinion, the best short description of the Newlands Project—its objectives, its politics, the motivations of the various stakeholders, and its consequences—is found in the book *As Long as the River Shall Run* by M.C. Knack and O.C. Stewart (1984). This book presents a frank discussion of the project's glaring downsides as well as its great expectations, a total package you won't find in most other histories of the area. Briefly,

Fish and Wildlife Service considers to be the range of the North*western* Lahontan Basin DPS), the basin floor is dominated by the extensive playa of the Black Rock Desert. Geothermal activity occurs in this area. Numerous hot springs in the Black Rock Desert and its surroundings attracted explorers and immigrant parties, and continue to attract modern recreationists. The Quinn River and its tributaries head up in the ranges that form the northern rim of the basin, and the mainstem Quinn then makes its way in a generally southwesterly course to its sink in the Black Rock Desert. Summit Lake, a shallow 600-acre lake sitting on a plateau at about the 6,000-foot level, lies just inside the divide that separates the western Lahontan Basin from the adjacent Alvord Basin (labeled "E" in Fig. 5-1). Summit Lake is a closed basin of its own at present, but its entire drainage was once tributary to the Lahontan Basin via a stream that flowed southward through Soldier Meadow. In fact, Summit Lake is the remnant of pluvial Lake Parman, a lake that formed after a landslide blocked that Lahontan connection sometime between 9,100 and 19,000 years ago. The drainage has also had hydraulic connections with the Alvord basin in mid- to late-Pleistocene time.[11] Mahogany Creek is Summit Lake's only perennial inlet tributary. Most of its watershed is now set aside in the Lahontan Cutthroat Natural Area administered by the Bureau of

in 1902 Congress passed a bill by Nevada Representative Francis Newlands authorizing the massive federal funding of irrigation projects and establishing the Bureau of Reclamation in the Department of the Interior to administer the funds. Diversion of Truckee River water was the first project to be authorized, and was named for Newlands. Derby Dam, 40 miles upstream from Pyramid Lake, was completed in 1905 to divert 250,000 acre-feet of Truckee River water per year (roughly half of its total annual flow) into the lower Carson drainage. The original scheme was to irrigate 350,000 acres of land in the Fallon and Carson Sink areas for high-yield, high-value crop agriculture. However, once the diversion water was flowing, several unanticipated and unpleasant problems arose. For openers, even though this land looked like desert on the surface, being a river sink meant that the water table was high. Irrigation water met with highly alkaline

ground water, bringing toxic concentrations of salts up to the root zone, killing most of the crops. Even an extensive system of deep drains installed by the government in the years from 1916 to 1920 cured only part of this problem. Then too, much of the topsoil was sandy, infertile, and unsuitable for high-yield crops even when irrigated. Finally, there was huge wastage of water due to an ill-conceived delivery system and poor irrigation practices. This wastage did have one unforeseen benefit: it created even more extensive wetlands than had existed around the original Carson Sink, and these became the Stillwater Wildlife Refuge. But the promise of a booming agricultural economy was largely illusory. Even though the area did become noted for its "heart of gold" cantaloupes, most farmers raised pasturage, hay, and other low-yield crops. No more than 60,000 acres—less than 20 percent of the land originally envisioned— was ever farmed in the project area.

11. For further insights into the ancient drainage patterns of the Summit Lake basin, see Layton (1979), Miflin and Wheat (1979), and Curry and Melhorn (1990).

Land Management to protect the spawning habitat of Summit Lake's trout population. Snow Creek is also an inlet tributary of Summit Lake and historically its drainage was probably used for spawning as well. But Snow Creek was diverted for irrigation by the Summit Lake Paiute Indian Band and its lower stream channel is now desiccated.

In the 1980s, Eric Gerstung, then with the California Department of Fish and Game, delved deep into the historical record to ascertain how and where Lahontan cutthroat trout were distributed when European settlers first arrived on the scene. [12] He found evidence that the trout existed in waters throughout the basin, but mostly, as you would expect, in the cooler perennial streams downstream from impassable falls. The majority of this type of stream habitat occurred in the Walker, Carson, Truckee, and Susan river drainages, and in the drainages encompassing the Smoke Creek and Black Rock deserts. Gerstung estimated that about 1,620 miles of stream habitat held Lahontan cutthroat trout in these five drainage systems, including the Quinn River drainage. In addition, Lahontan cutthroat trout occurred in Tahoe, Cascade, Fallen Leaf, Independence, Donner, Pyramid, Winnemucca, Walker, Upper and Lower Twin, and Summit lakes, which then had a combined surface area of almost 334,000 acres (522 square miles). [13]

Historically, Lahontan cutthroat trout occurred throughout the Walker River drainage from the headwaters in the mountains downstream to Walker Lake. In addition to Walker Lake, Lahontan cutthroats also occurred in Upper and Lower Twin lakes at the head of Robinson Creek, an East Fork Walker River tributary. Great schools of trout migrated upstream from Walker Lake during annual spring spawning runs and were utilized by the Walker Lake Paiutes as a principal source of food. Indeed, the trout were so important to these Native American people that they referred to themselves as "Agai Dokado," meaning "trout eaters," and to Walker Lake as "Agai Pah," the "trout lake." Fremont's explorers observed many Paiute fishing weirs along the lower Walker River, where the Indians speared or dipped fish from the water. As he had done at Pyramid Lake and along the Truckee River, Fremont recorded the fish as "salmon trout." After European settlement of the area had commenced, Walker River and Lake became popular with white sportsmen and market fishermen. The Indians themselves developed a market fishery in the 1880s and continued it until the early 1930s, when it became obvious at last that the last notable runs up the river had occurred. Owing to water withdrawals for irrigation, flows had become too low and too much good spawning habitat had been blocked for the fish to spawn with much success in the river. The trout of the lake

12. Eric Gerstung's review of the Lahontan cutthroat's historical distribution was first used in an abbreviated form in an article he wrote for *Outdoor California*, the news magazine of the California Department of Fish and Game (see Gerstung 1982). It was updated and published in greater detail six years later in a major symposium series on the status and management of interior stocks of cutthroat trout (see Gerstung 1988). Gerstung included the Humboldt River drainage in his original compilation. I have backed that information out of the present discussion for later use in Chapter 6. Other historical information included in this section came from Hutchings (1857 [1962]), Butler (1902), Evermann and Bryant (1919), Shebley (1929), Welch (1929), Sumner (1939), Stollery (1969), Rankel (1977), Tuohy (1978, 1990), Curran (1982), Ono et al. (1983), Coleman and Johnson (1988), and Alexander and Scarpella (1999).

13. Another major lake, Eagle Lake, sits on the western rim of the Lahontan Basin in what is now a disjunct basin north of Susanville, California. Like Summit Lake in Nevada, Eagle Lake and its one inlet stream (Pine Creek) were once tributary to the Lahontan Basin. The native fish fauna of Eagle Lake are Lahontan Basin species—

held on until 1949, when the last few spawners were trapped near the mouth of the river and their eggs used to begin a hatchery broodstock. The present population in Walker Lake is maintained by hatchery propagation.

The extent of historical distribution of Lahontan cutthroat trout in the Carson River drainage is known with more certainty in the upstream direction than in the downstream direction. Trout were present historically in waters upstream to Carson Falls on the East Fork and Faith Valley on the West Fork, but were absent from waters above those points. Downstream, since the Carson River flowed out across the basin floor to disappear in a sink, distribution would have depended on where the water became too warm and habitat conditions too deteriorated to be tolerated by trout. Fairly recent studies indicate that Lahontan cutthroat trout become stressed (decreased appetite and growth and increased mortality) when the maximum water temperature exceeds 72° Fahrenheit (22° Celsius).[14] However, no records or journal entries are available that I am aware of to suggest where along the Carson River this might have occurred. It is known that J. Otterbein Snyder, the ichthyologist, collected Lahontan cutthroat specimens from the Carson River near Genoa in the period just prior to World War I.

No endemic lacustrine populations are known to have existed historically in the Carson River drainage, but trout from the upper West Fork Carson were stocked in the nearby Blue Lakes in 1864 and adapted to a lacustrine life style.[15] Silver King Creek, a tributary of the East Fork Carson, has about 21 stream miles of water, including tributaries, in its upper reaches (designated "D" in Fig. 5-1) that comprise the limited home range of the Paiute cutthroat trout (Chapter 8).

The Truckee River system including Lake Tahoe and its tributary lakes and streams, Donner and Independence lakes and their tributaries, and the mainstem Truckee and its tributaries downstream to and including Pyramid and Winnemucca lakes probably had the greatest abundance of native cutthroat trout, as well as the greatest variety of races, of all the western Lahontan Basin drainages. Cascade and Fallen Leaf lakes, whose outlets drain into Lake Tahoe, had their own populations that spawned in inlet tributaries. From Lake Tahoe itself each spring, thousands of cutthroat trout moved up accessible tributaries to spawn. The more productive tributaries, such as Taylor Creek, the outlet stream from Fallen Leaf Lake, may have supported spawning runs of up to 7,000 fish annually. Other principal spawning tributaries were Meeks, Phipps, McKinney, Blackwood, and Ward creeks in California, and Incline Creek in Nevada. Independence Lake produced spawning runs estimated at 2,000 to 3,000 fish annually.

all, that is, except its native trout, which belong to the rainbow-redband evolutionary lineage. How could that be? The most logical explanation, courtesy of Dr. Behnke (see Behnke 1992), is that the original trout was indeed Lahontan cutthroat, but these were eliminated during a warm, dry period, occurring about 8,000 to 4,000 years ago, that dried up all spawning habitat. Later, about 4,000 to 2,000 years ago when the climate cycle became wetter again, a headwater transfer from the Pit River drainage (since broken) enabled trout of the rainbow-redband lineage to reach Eagle Lake. The trout presently found in Eagle Lake is commonly classified as *Oncorhynchus mykiss aquilarum*.

14. See Vigg and Koch (1980), Dickerson and Vinyard (1999), Meeuwig (2000), and Dunham et al. (2003).

15. Blue Lake (more accurately Blue Lakes, since there are really two of them) is actually located just west of the Sierra crest in the headwaters of the Mokelumne River. In the early 1930s, the California Department of Fish and Game transferred fish from Upper Blue Lake to establish another broodstock in Heenan Lake, which is in the East Fork Carson River drainage. Calhoun (1942, 1944a, 1944b) studied growth rates,

It was from Lake Tahoe that specimens of Lahontan cutthroat trout were first collected and described for science. The name *Salmo henshawi*, in honor of the naturalist H.W. Henshaw, who provided the specimens, was one of two names given these fish by Theodore Gill and David Starr Jordan in 1878 in the 2nd edition of the textbook *Manual of the Vertebrates*.[16] It was generally believed then, and for some years afterward, that two species of trout were present in Lake Tahoe:

- Tahoe trout (initially named *Salmo tsuppitch* by Gill and Jordan)—dark in color, described as a "brown" or "black" form but coppery during the breeding season; boldly and regularly spotted; relatively small in size (2 to 6 lbs mostly in sport and market catches, seldom exceeding 6 lbs; spawners in Taylor Creek averaged 2 lbs); caught chiefly along the western side of the lake and at the south end, from the mid-depths to the surface. A large area extending south a few miles from Observation Point was described as a good feeding ground for Tahoe trout. Ascended the streams to spawn in mid-April; spawning completed by mid-June.

- Silver trout (initially named *Salmo henshawi* by Gill and Jordan)—decidedly silver in color, spots smaller and more elongated than on Tahoe trout; larger, heavier, and deeper-bodied than Tahoe trout (specimens frequently exceeded 6 lbs, the two largest mentioned in the historical record being a fish of 29 lbs and the other the 31 lb 8 oz fish that holds the California hook-and-line record). Caught chiefly in the deeper waters of the lake, especially in the northeastern part, also in Crystal Bay and Sand Harbor. California hatchery workers noted that spawners in the Upper Truckee River and Blackwood Creek were significantly larger (average 10.5 lbs) than spawners in other Tahoe tributaries (average only about 2 lbs), and they believed these to be the large "silver trout."

age at maturity, food habits, and other life history characteristics of these fish in both lakes about a decade after the stock transfer had taken place, and concluded that the fish were then doing somewhat better, condition-wise, in the more food-rich Heenan Lake than they were in Blue Lake, which is a higher elevation, oligotrophic body of water. Unfortunately, prior to the transfer to Heenan Lake, California had also stocked rainbow trout in the Blue Lakes, and it was always suspected that introgression of the cutthroat stock had occurred. Although California did away with the Blue Lakes broodstock at its Heenan Lake operation in 1980 and replaced it with a stock from Independence Lake, the State of Nevada may still maintain a population of the Blue Lakes-derived-Heenan stock at its Marlette Lake facility near Lake Tahoe (King 1982). For a time, the State of Nevada also maintained a stock of these trout at Catnip Reservoir on the Charles M. Sheldon National Antelope Refuge.

16. Henshaw, I believe, was associated with the U.S. Army's Geographical Surveys West of the One-Hundredth Meridian, led by Lt. (later Capt.) George M. Wheeler. In 1876, a contingent of this survey operated in the Lake Tahoe vicinity. I could not locate a copy of the 1878 edition of "Manual of the Vertebrates," but the annual report of the Wheeler survey for the year 1878 contains an appended paper by Jordan and Henshaw describing fishes collected by the survey in 1876 and 1877; see Jordan and Henshaw (1878).

Upper Truckee River a few miles upstream from the town of Truckee, California. Formerly Lahontan cutthroat habitat, brown and rainbow trout live here now.

PHOTO BY AUTHOR

But even the renowned Dr. Jordan vacillated in his opinion about whether there were two species or just one. At first he used the name *Salmo henshawi* specifically to mean the "silver" trout of Lake Tahoe. The Tahoe trout was named as a separate species, *Salmo tsuppitch*. Later, in 1886, he and B.W. Evermann designated the Tahoe trout as *Salmo mykiss henshawi* and withdrew recognition from the silver trout. But two years later, they reinstated the silver trout with the name *Salmo clarkii tahoensis*. By 1914, fishery scientists were coming to the view that only a single subspecies occurred in the lake, and when Dr. Behnke reexamined old museum specimens in 1960, that pretty much cinched it. Behnke studied type specimens of each of these fishes, as well as specimens in other early collections from Tahoe, Pyramid, and the Truckee River. He also looked at specimens from the Carson and Walker drainages. He found no differences in meristic characters between *tahoensis* and *henshawi*, or any of the other pure cutthroat specimens. He concluded, therefore, that only a single subspecies, *O. c. henshawi*, is represented in that part of the Lahontan Basin. [17]

Only a single subspecies may be represented, but to me, the original descriptions of Tahoe trout and silver trout make a compelling case for the existence historically of two distinct morphs, or ecotypes, of *henshawi* in Lake Tahoe. Occurrences of distinct morphs of a species

17. The descriptions summarized here for "Tahoe" and "silver" trout were compiled from reports by Juday (1907a), Snyder (1914, 1917), Evermann and Bryant (1919), and angler accounts published in some of the early books about Lake Tahoe, such as James (1921 [1956]), Scott (1957), and Stollery (1969). Notes on the spawning tributaries used by the two forms are from a recently completed but not yet published manuscript titled "The Trout of Lake Tahoe: History, Biology, and Status" by Almo Cordone, formerly with the California Department of Fish and Game. The back-and-forth opinions about one or two forms and what they should be named are in Jordan and Evermann (1896, 1898), Snyder (1914, 1917), and Behnke (1960, 1972). While we're on the subject of names, there are two other scientific names in the old literature for trouts of the Tahoe-Truckee-Pyramid system that we can dispense with based on Dr. Behnke's examination of museum specimens. *Salmo regalis*, the "royal silver" trout of Lake Tahoe, was described as a new species by J.O. Snyder in 1914, and another new species, *Salmo smaragdus*, the "emerald" trout of Pyramid Lake, was described by Snyder in 1917 (see Snyder 1914, 1917). By these dates, both rainbow trout and hatchery-produced Lahontan cutthroat-rainbow hybrids had been stocked into both lakes and the Truckee River. In his book, *Fishes and Fisheries of Nevada*, LaRivers (1962) wrote that the Nevada State Fish Commission had been quite enthusiastic about these hybrids and stocked them from 1905 through 1912. It is Dr. Behnke's opinion that the specimens of "royal silver" trout were actually these hatchery-produced cutthroat-rainbow hybrids, and the "emerald" trout specimens were introduced rainbow trout.

or subspecies within the same body of water are not that uncommon. There are many examples, invariably occurring in large, deep, oligotrophic lakes. Perhaps the best-studied are the four distinct arctic charr morphs found in Lake Thingvallavatn, Iceland, and the three distinct brown trout morphs that occur in Lough Melvin, Ireland (locals call the Lough Melvin morphs "sonaghen," "ferox," and my favorite name, the "gillaroo"). In each case, the morphs are distinct from one another in phenotype, ecology, life history, and reproductive biology. They differ in color, spotting, size, and body form; they occur in different parts of the lake; they feed on different sorts of prey; and they spawn in different areas and/or at slightly different times. [18] Sounds like "Tahoe" trout and "silver" trout to me!

Meanwhile, downstream at Pyramid and Winnemucca lakes, two distinct forms of trout were suspected as well. This was based on two distinct spawning runs of markedly different-looking trout into the Truckee River each year. The first of these was a run of very large, red-hued trout that the Indians called "tomoo agai" or "winter trout" and the white market and sport fishers referred to as "redfish." This run commenced as early as October in some years but mostly in December, and continued through the winter months. Some of these fish were enormous. The world hook-and-line record cutthroat, that 41-pounder taken

from Pyramid Lake by John Skimmerhorn in 1925, was probably such a fish. And even larger specimens (a 62-pounder is mentioned in early testimony) may have been taken by the market fishermen. [19] In April and early May, after the "redfish" run had drawn to a close, a run of much smaller trout, darker and more heavily spotted, moved up the river from Pyramid and Winnemucca lakes. The Indians called these fish "tama agai" or "spring trout" and the whites referred to them as "tommies." In this case, rather than distinct morphs, the differences in appearance were probably due to intrapopulation differences in size and age between first-time spawners (the "tommies") and older repeat spawners (the "redfish") and the different colors developed by the fish as they aged.

In addition to the abundant winter and spring spawning runs of lake-dwelling trout that utilized the river seasonally, the historical record indicates that the Truckee River was also inhabited by fluvial fish. Indeed, some travelers referred to the stream as Salmon Trout River, as Fremont had done back in 1844. This observation, recorded in the fall of 1849 by a man bound for the California gold fields, serves as an example of what the immigrants saw:

> "Salmon Trout is a most beautiful stream, rushing and roaring over the rocks and its stony bed like a New England mountain river and the waters as clear as crystal.... Saw in our course along the banks of the stream, numbers of those beautiful fish from which the river

18. Descriptions of the arctic charr morphs of Lake Thingvallavatn are in Sandlund et al. (1987), Jonsson et al. (1988), Snorrason et al. (1989), Signurjónsdóttir and Gunnarsson (1989), Skúlason et al. (1989), and Malmquist (1992). For the brown trout morphs of Lough Melvin, see Ferguson and Mason (1981), Cawdrey and Ferguson (1988), Ferguson and Taggart (1991), and McVeigh et al. (1995).

19. In addition to the world hook-and-line record, a Lahontan cutthroat trout also accounts for the State of California hook-and-line record. That fish, weighing 31 lbs 8 oz, was taken off Tallac House, an early-day resort on Lake Tahoe, in 1911 by angler William Pomin. Actually, the large size of the Lahontan cutthroat had attracted the attention of California fish culturists as early as 1870, a year that saw trout from Independence Lake and the Truckee River being propagated in private hatcheries. As I've already explained, it was generally believed at the time that there were two species of trout present in Lake Tahoe. Dr. Behnke related an amusing story about the newly formed California Fish Commission's first attempt to have these two forms classified. In 1870, the commission sent specimens of "brown" trout and "silver" trout from

takes its name; but had no time to spend in catching them.... Some of these fish were two feet long, beautifully spotted like the New England trout."[20]

The Susan River and its tributaries upstream from Honey Lake were also populated with trout. Although the Paiute band that frequented the Honey Lake valley called themselves "Wada Dokado," meaning "wada eaters" (wada are the small black seeds of the seepweed *Suaeda depressa*, a salt-tolerant plant that grows abundantly on the east side of Honey Lake), ethnographers recorded that these Indians also caught suckers and trout from the river. So too did the Maidu, an Indian tribe whose main territory was the higher country west and north of the Honey Lake valley. Early settlers reported that trout were abundant in the Susan River in 1853, long before any stocking, but rainbow trout were introduced there and had become common by 1902. The ichthyologist J. Otterbein Snyder did collect Lahontan cutthroat specimens from Long Valley Creek and from the Susan River near Susanville in the years just prior to World War I, but none are present in the drainage today.

Except for Summit Lake, which has a longer record, the historical record for the Smoke Creek and Black Rock Desert portion of the western Lahontan Basin goes back only to 1935. Between that year and 1960, the Nevada Division of Wildlife performed stream surveys that revealed that 20 streams, most of them in the Quinn River watershed, were occupied by cutthroat trout. These were all small headwater tributaries with stream resident populations, and totaled to perhaps 398 stream miles of habitat at most. The downside to this number as a record of historical occupancy is that even though this is only a sparsely settled region, many of the streams and water sources had already been subjected to withdrawals for irrigation and to heavy riparian grazing by the time the first surveys were made. Fluvial populations may have occupied at least the cooler upper reaches of the Quinn River mainstem and also the upper mainstem of Kings River, a principal tributary, but I have found no historical record or testimony to substantiate this.

The history of the Summit Lake population goes much farther back. According to ethnographers, the Paiute people that inhabited the shores of Summit Lake, like those at Walker Lake, referred to themselves as "Agai ipana Dokado," "trout lake eaters," indicative of a long history of trout in the lake. When spawning migrations to Mahogany Creek, the sole remaining spawning tributary, were first enumerated between 1956 and 1967, annual runs numbered between 785 to 1,200 fish. More recently, they have fluctuated widely, ranging from a high of 5,000 fish in 1974 and 1975, when the run was heavily augmented with hatchery

Lake Tahoe to the noted eastern fish culturist, Seth Green. Mr. Green looked at them and pronounced neither of them to be trout at all, but rather forms of the Sebago or northeastern landlocked salmon!

20. Perkins 1849 [1967]. Also cited in Curran (1982).

fry stocked from Summit Lake eggs, to a low of 472 in 1992. The most recent information I have dates from 1999, when the Summit Lake Paiute Tribe reported a spawning run of 2,400 fish. Presently, the Summit Lake population is the largest self-sustaining lake population of pure Lahontan cutthroat trout remaining in existence.

LIFE HISTORY AND ECOLOGY

Lahontan cutthroat populations exhibit the same three basic life history strategies as westslope cutthroat trout (Chapter 4): [21]

- A lacustrine, or lake-associated, life history, wherein the fish feed and grow in lakes, then migrate to tributary streams to spawn.
- A fluvial, or riverine (potamodromous), life history, wherein the fish feed and grow in main river reaches, then migrate to small tributaries of these rivers where spawning and early rearing occur.
- A stream-resident life history, wherein the trout complete all phases of their life cycle within the streams of their birth, often in a short stream reach and often in the headwaters of the drainage.

The Lacustrine Life History Strategy

My guess is that when anglers think of Lahontan cutthroat trout, their thoughts most often turn to the lunker trout of Pyramid Lake. There is good reason for this: from the late 1970s through the 80s and 90s and even up to the present time, angling magazines—especially the fly-fishing media—have touted the

lake's rejuvenated cutthroat fishery with titles such as "Pyramid Roulette" (*Flyfishing the West*, November-December '78), "Payoff at Pyramid" (*Outdoor Life*, December '82), "Cutthroat Battle at Pyramid Lake" (*Western Outdoors*, October '90), "Pyramid Lake's Giant Lahontan Cutthroats" (*Flyfishing*, March–April '95), and "Pyramid Lake Lahontans" (*Fly Fisherman*, February 2002). Even though the great sizes attained by the original Pyramid Lake "redsides" may be a thing of the past and the present fishery is almost entirely dependent upon hatchery releases, the lake can still kick out some respectable fish.

Then too, some fishery scientists express the view that the large size, longevity, and piscivory (which I shall get to in a few more paragraphs) of Lahontan cutthroat trout in Pyramid Lake are specialized lacustrine life history attributes that the native strain evolved over thousands of years in a continuous lake environment.

Unlike most other trouts, Lahontan cutthroat trout can tolerate unusually high levels of alkalinity and dissolved solids, the latter a surrogate measure of salinity in freshwater systems. Thus they can thrive in environments where alkalinity levels reach 3,000 milligrams per liter, pH of the water reaches 9.4, and dissolved solids exceed 10,650 milligrams per liter. [22] But these trout did (and still do) occur naturally in oligotrophic lakes as well. As a matter of fact, much of what we know about

21. For overall summaries of Lahontan cutthroat life history and ecology, see Behnke's two monographs and his recent book (Behnke 1979 [1981], 1992, 2002); also LaRivers (1962 [1994]), Behnke and Zarn (1976), Sigler and Sigler (1987), and Stolz and Schnell (1991). For details of the lacustrine life history strategy, go to Calhoun (1942, 1944a, 1944b), Lea (1968), Rankel (1977), Vigg and Koch (1980), Johnson et al. (1981), Sigler et al. (1983), and Vinyard and Winzeler (2000). There is nothing published specifically on the fluvial life history strategy of cutthroat trout in the western portion of the Lahontan Basin, but these dissertations and reports give details of the adfluvial or headwater stream-dwelling life history strategy: Dunham (1996), Schroeter (1998), and Dunham et al. (1999, 2003).

22. Studies of the limnology (i.e., the chemical, physical, and biological properties) of Pyramid and Walker lakes have been published by Galat et al. (1981) and Cooper and Koch (1984). Other data on the effect of pH on Lahontan cutthroat trout have been published in Galat et al. (1985) and Wilkie et al. (1993).

the lacustrine life history strategy is based upon studies of these trout in oligotrophic lake environments. We can also guess at what the life-history characteristics of the original populations of the large desert lakes must have been from the historical record and from contemporary studies of the introduced populations now swimming in Pyramid and Walker lakes.

All lake-dwelling populations of Lahontan cutthroat trout appear to be the *lacustrine-adfluvial* type, i.e., they all spawn in inlet streams. Historically, trout from Pyramid and Winnemucca lakes spawned over the entire 120-mile length of the Truckee River, utilizing all suitable spawning areas in the river itself and its tributary streams. Some Pyramid/Winnemucca fish may have passed on through Lake Tahoe and spawned in its tributaries. Francis H. Sumner, then a U.S. Bureau of Fisheries assistant stationed at Stanford University, believed that they did. In a 1939 paper on the decline of the Pyramid Lake fishery, he wrote:

> "Before the coming of the white man, the cutthroat trout was one of the principal staples of the Paiute Indians resident at Pyramid Lake. During the spring spawning migrations up the Truckee River, *as far even as the tributaries of Lake Tahoe*, the thickly-crowded trout were easily captured" [emphasis mine].[23]

If Sumner's observation was true, then these fish would have joined the substantial spawning migration of Lake Tahoe fish, as well as the fish resident in Fallen Leaf

and Cascade lakes. Lacustrine trout from Walker Lake made equally long spawning migrations that extended as far as 125 miles up the Walker River system. Two principal spawning reaches in the Walker River system were in the Bridgeport and Antelope valleys.

Studies indicate that Lahontan cutthroat spawning migrations typically commence when water temperatures rise to 41 degrees Fahrenheit (5 degrees Celsius). In much of the range this occurs in the spring, which is when most historical peak migrations took place. The exception was the late fall and winter migration of the large, older-age Pyramid/Winnemucca "redfish," but this was followed by the run of the younger, smaller "tommies" that occurred in April and early May, consistent with the timing of other lacustrine spawning runs in the basin. Contemporary populations of Pyramid, Walker, Independence, and Summit lakes appear to have the spring spawning migration timing. [24] Egg deposition itself may occur anywhere from April through July, depending on local conditions of stream flow and water temperatures.

In nature, lacustrine Lahontan cutthroats may spawn for the first time at age 3, 4, or 5, but the majority spawn for the first time at age 4. First-time spawning can be pushed to age 2 in hatcheries. Most fish at first spawning are 11 to 14 inches in fork length. These fish can survive to spawn multiple times, but

23. See Sumner (1939). William F. Sigler and his associates, who reported on the life history of Lahontan cutthroat trout in Pyramid Lake in 1983, also stated that historical runs from Pyramid and Winnemucca lakes extended upstream into Lake Tahoe tributaries (see Sigler et al. 1983).

24. In January 1981 and again in January 1982, workers at the Pyramid Lake Paiute tribal hatchery at Sutcliffe found ripe Lahontan cutthroats from the lake that had entered a small artificial stream that returned water to the lake from the hatchery rearing ponds. Eggs were taken from these fish and reared separately, in the hope of eventually re-establishing the fall and winter timing of the original "redfish" spawning run. I received word from Pyramid Lake Fisheries that this program had lapsed back in the 1980s, but might be restarted in 2004. Unfortunately, I have heard nothing since.

repeat spawning in the wild does not always take place in successive years. Rather, it appears common for mature fish to skip a year between spawning episodes. The number of eggs per female carried to the spawning areas typically increases with the size and age of the fish. In a composite of studies at Blue, Independence, Summit, and Pyramid lakes, the number of eggs per female ranged from about 740 in fish of 11 inches in fork length to 7,960 in fish of 26 inches in fork length weighing about 10.5 pounds. Larger fish may carry upwards of 10,000 eggs (what a tremendous amount of reproductive potential those original Pyramid Lake "redfish" must have carried!). Ripe females checked at Summit Lake carried between 2,600 and 2,800 eggs, indicative of fish averaging about 19 inches in fork length and 2.8 pounds in weight. The ratio of male to female fish in those Summit Lake spawning runs was about one to one and a half.

Eggs deposited in the spawning gravels hatch in 4 to 6 weeks, depending on the water temperature. Unlike the lacustrine young-of-the-year of the coastal and westslope cutthroat subspecies, most of the lacustrine Lahontan young-of-the-year trout migrate back to their lakes in their very first summer. Migration takes place throughout the summer and fall, but the highest rate of movement is in July, predominately in the nighttime hours. Some lacustrine juveniles do spend one or more years in natal tributaries however. At Summit Lake, 20 percent of the juveniles did not move down to the lake until the following spring. About 2 percent of the cohort resided in the stream until the second summer, and a handful of fish apparently residualized in the stream.

While in the natal streams, the juvenile trout feed opportunistically, but almost entirely on aquatic insects. Back in the lakes, they continue as opportunists, eating zooplankton, benthic invertebrates, and terrestrial insects. If the lake lacks forage fish, opportunistic feeding on whatever is available continues for the lifespan of the fish. But if indigenous forage fishes are present, the trout profoundly change their feeding behavior when they reach a length of about 12 inches. At that point, they switch to a diet consisting predominately of fish.

Perhaps the most striking characteristic of lacustrine Lahontan cutthroat trout is the propensity to attain great size in certain environments. This came about because *O. c. henshawi* evolved into a very efficient predator on the native forage fishes present in ancient Lake Lahontan. The most abundant of these forage fishes was probably the tui chub, which remains the most abundant fish in Pyramid Lake. [25] The tui chub can itself grow to 15 or 16 inches in length in Pyramid Lake. So, to make the most effective

25. In 1981, Steven Vigg of the Desert Research Institute, University of Nevada Reno, published a census of the species composition and relative abundance of fishes in Pyramid Lake (see Vigg 1981). Five species were found to be present, in this order of numerical abundance: tui chub, Tahoe sucker, Lahontan cutthroat trout, cui-ui, and Sacramento perch. Of these, the tui chub accounted for over 70 percent of the fish in the lake.

use of this prey, the Lahontan cutthroat acquired the genetic basis to attain great size. This carried over to its fullest extent in the original Pyramid Lake strain of cutthroat, where the unique prey-predator relationship with the tui chub continued to exist after the desiccation of Lake Lahontan.[26] In Walker Lake as well, the tui chub comprises the principal prey species of cutthroat trout over 12 inches in length. The same was true in Lake Tahoe when cutthroat trout swam its waters. In Independence Lake, the redside shiner and Paiute sculpin were the principal fish eaten historically, but kokanee were added to the prey selection following their introduction and establishment in the lake. Summit Lake had no native forage fishes, and its trout seldom exceed 20 inches or so in length. It's true that age-6 and 7 fish in Summit Lake can range between 22 and 26 inches, but these are exceedingly rare.[27]

Growth rates of Lahontan cutthroat trout vary. Faster growth occurs in the warmer, more fertile water bodies, especially those where forage fishes are eaten. In Pyramid and Walker lakes, for example, the present populations may reach 16 to 20 inches in fork length by age 4, and 25 or 26 inches in fork length by age 7. In colder, oligotrophic waters, age-4 trout are more typically 11 to 14 inches in fork length, and few live much longer than age 4 or 5.

26. This is an example of evolutionary genetics at work. Evolutionary genetics is the study of the evolution of the diversity of life via natural selection. For more on the general theory and applications of evolutionary genetics, see Ayala (1982), Endler (1986), and Bulmer (1994). Dr. Behnke has elaborated on the example provided by the original Pyramid Lake population of Lahontan cutthroat trout in one of his "About Trout" articles for *Trout Unlimited*; see Behnke (1993). Pyramid Lake is the deepest part of ancient Lake Lahontan and is the only body of water where the native cutthroat trout continued to coexist with the full array of native prey fishes. When Lake Lahontan desiccated, all of the other Lahontan cutthroat populations were isolated in their own portions of the basin and natural selection set them off on different evolutionary paths. Walker Lake, for example, may have desiccated completely in fairly recent geological time, once about 4,500–5,500 years ago and again about 2,000–3,000 years ago (Grayson 1987, Benson et al. 1991). The lacustrine trout of Walker Lake would have been extirpated on these two occasions, and the lake would have had to be recolonized by trout that persisted in the Walker River. Only the Pyramid Lake population remained where its evolutionary genetics could continue to be programmed and fine-tuned for its role as the longest-lived, largest-bodied keystone predator. Because they depend on genes and alleles that are neutral to selection, population genetics studies do not detect important distinctions that arise from natural selection, nor are these distinctions often grasped by fisheries managers and fish culturists who rely solely on the results of population genetics studies for guidance.

27. The largest fish ever reported from Summit Lake was 28 inches in length and weighed 7.5 pounds, but 20 to 21 inches and 3 to 4 pounds is the norm. Summit Lake contained no forage fishes until the early 1970s, when somebody, probably an illegal fisherman, introduced redside shiners. Even so, Summit Lake cutthroats do not utilize these shiners for food to any appreciable extent. Instead, they feed almost exclusively on the dense concentrations of *Gammarus* scuds in the lake, and on midge larvae. On the other hand, Summit Lake cutthroats do develop piscivorous habits when stocked into Pyramid Lake.

Longevity of the trout appears to track these age-length data to some extent, i.e., trout in populations that intrinsically grow largest also intrinsically live longest. Trout in the original populations of Pyramid and Winnemucca lakes may have lived as long as 10 or 11 years. The life span of the trout now present in Pyramid and Walker lakes seldom exceeds 7 years. The maximum age of trout in Summit Lake is 5 years, although the occasional age-6 or 7 fish has been recorded.

The Fluvial Life History Strategy

I know of no specific studies of the life history and ecology of fluvial populations of the western Lahontan Basin. Scientists such as Dr. Behnke believe that an evolutionary programming to specialize as a large lake predator is imprinted in the genetic makeup of the cutthroat trout of the western Lahontan Basin, which would make the fluvial life history a less favored strategy. Even so, history records their presence in the Walker, Carson, Truckee, and Susan rivers, but not for long after white settlement of the region began. Trout do occur in these drainages today, but native cutthroat trout have long since been replaced by introduced rainbow and brown trout.

In all probability, there were no distinctions of life history and ecology between fluvial populations of the western Lahontan Basin and those of other subspe-cies of cutthroat trout. Likely, they responded to the same environmental cues to initiate their spawning migrations, moving to natal tributaries in the spring on rising water temperatures and increasing stream flows. Females in these populations probably spawned for the first time at age 3 or 4 at fork lengths of 10 to 12 inches or so, and these females, along with larger, older, repeat-spawners, would have carried from 600 to 1,800 eggs per female to the spawning tributaries. Incubation would have spanned the same 4- to 6-week period as reported for the lacustrine life history. Also extrapolating from the lacustrine life history, a percentage of the offspring may have moved down to mainstem reaches over the course of their first summer and fall, but the majority probably stayed in the tributaries until the following spring or in a few cases even until their second spring. In the mainstem reaches, they would have been opportunistic feeders and would have made trophic migrations for feeding and shelter. Maximum size and age was probably 16 to 18 inches and 5 or 6 years, although historical accounts already cited indicate that Truckee River fish could grow to at least 24 inches in length.

The Stream-Resident Life History Strategy

According to Eric Gerstung's historical reconstruction, cutthroat populations that I would judge to be adfluvial were widely distributed in the Sierra Nevada streams

Independence Lake in the upper Truckee watershed, California, still supports a small native population of Lahontan cutthroat trout.
PHOTO COURTESY OF ERIC GERSTUNG, FORMERLY WITH
THE CALIFORNIA DEPARTMENT OF FISH AND GAME

along the western rim of the Lahontan Basin, and were quite abundant in some. The original stocks are gone from most of these streams, and the majority of populations now present in the Walker and Carson River drainages are the result of reestablishment efforts, which I will elaborate on in the next section.

The fish assemblages in these eastern Sierra Nevada streams are quite simple. Headwaters usually contain only trout. Historically, these would have been the native cutthroat trout, and this is still the case in the few streams where the natives have persisted or where cutthroat trout have been reintroduced. But nowadays, one finds brook trout most commonly, which give way to rainbow and brown trout at lower elevations. Usually, the first additional native species to appear in the mix in a downstream direction is the Paiute sculpin. Tahoe suckers and speckled dace join in as gradients decrease and pools and glides become more common, and redside shiners appear where deeper pools can form. Further down, where the streams are larger (and the trout are more likely to be fluvial in life-style), the assemblage is joined by mountain suckers, mountain whitefish, and tui chub. [28]

Some of these Sierra Nevada cutthroat streams are quite small indeed, and can be stepped across with ease. Even so, they can produce some surprisingly large trout. Eric Gerstung collected fish from six Sierra Nevada

streams that averaged 3.5, 4.5, 8, and 10.5 inches in fork length at ages 1, 2, 3, and 4, respectively. Fish older than age 4 are uncommon in these adfluvial populations.

These trout spawn for the first time at age 2 for many of the males and age 3 for the females and remainder of the males. Females carry from 100 to 300 eggs. Spring spawning is a relative term in the Sierra Nevada high country and high desert ranges. Actual spawning can take place anywhere from April through July, depending on local conditions of water temperature and stream flow.

The eggs incubate from 4 to 6 weeks, again depending on local water temperatures, and once out of the gravel, the fish adopt the same feeding and trophic movement patterns as adfluvial populations of other cutthroat subspecies. Like all such trouts, the early weeks of their existence as post-emergent fry are spent in quiet channel-margin habitats. As the season progresses and they grow larger, they move out of the channel margins to more productive summer-rearing habitats where they find their places in the summer-stream hierarchy. They are opportunistic feeders on drift organisms, typically insects, during the summer months when most of their growth takes place, with growth rates as indicated by Eric Gerstung's data above. Meadow reaches, where the streams slow and meander, are generally the most productive and

28. This information is derived from Dr. Peter Moyle's revised and expanded *Inland Fishes of California* (Moyle 2002). For more on fish assemblages, distribution patterns, and ecology of Sierra Nevada and Lahontan Basin waters, or for lakes and streams elsewhere in California, this is the place to go.

produce the largest trout. Refuge movements into overwintering habitats commence with the onset of autumn.

STATUS AND FUTURE PROSPECTS

From the seemingly endless abundance of native trout in western Lahontan Basin lakes and streams when European explorers and settlers first arrived, pathetically few endemic populations remained 100 years later. And those that did remain were small in number. When thorough population inventories were finally compiled in the late 1950s, 1960s, and early 1970s, pure, self-sustaining populations were found to occupy less than 2 percent of originally occupied stream miles, and less than 0.5 percent of originally occupied lake acreage.[29]

In the Walker River drainage, nonnative trout first introduced in the late 1800s gradually displaced all fluvial and adfluvial populations of Lahontan cutthroat trout. Only a single endemic stock, By-Day Creek, existed in 1988. The fishery at Walker Lake had slipped into decline by 1911, although the last of the market fishermen didn't give up until 1930. In 1949, the State of Nevada captured the last 39 spawners that attempted to enter the river and took them into a hatchery. Since then, the lake fishery has been entirely dependent on hatchery stocking. All endemic populations had vanished from the Carson River drainage by 1966, although

a couple of small populations, apparently established by "coffee can" transplants, were later discovered in upper tributaries previously thought to be fishless.

At Lake Tahoe, the cutthroat fishery was in decline by the 1880s. Commercial fishing was banned in 1917, but the fishery continued to decline. The last tributary spawning run was observed in 1938, the same year the final spawning run was observed at Pyramid Lake. At Fallen Leaf Lake, the cutthroat trout had disappeared even earlier. The last spawning run was observed at Fallen Leaf in 1920. Elsewhere in the Truckee River, introduced rainbow and brown trout had spread throughout the basin by 1915. By 1960, Pole Creek, an upper tributary, and Independence Lake harbored the only populations of native Lahontan cutthroat trout remaining in the basin. Unfortunately, the Pole Creek population was lost to displacement by introduced brook trout by 1970. And by that same year at Independence Lake, spawning runs of Lahontan cutthroat trout numbered well under 100 fish annually.

Nobody recorded the disappearance of Lahontan cutthroat trout from the Susan River drainage, but one can guess that it probably tracked the dates above. By 1902, only rainbow trout were collected from the Susan River. Only in the Black Rock Desert/Quinn River drainage have cutthroat populations persisted. As I stated earlier, the Summit Lake population is now the

29. Accounts detailing the demise of the native cutthroat trout of the western Lahontan Basin and the reasons for it have been published by Shebley (1929), Welch (1929), Sumner (1939), Scott (1957), McAfee (1966a), Trelease (1969a), Townley (1980), Knack and Stewart (1984), and Gerstung (1988). The first two, by Shebley (1929) and Welch (1929), are first-person accounts by people who were there to witness much of the decline and describe what they saw.

largest remaining self-sustaining lacustrine population in the western Lahontan Basin. In surveys conducted between 1935 and 1960, twenty small, scattered stream-dwelling populations of Lahontan cutthroat trout were discovered in the Quinn River drainage, but this number had dwindled to 11 populations by 1987.

Acting on this information, in 1970 the U.S. Department of the Interior declared the Lahontan cutthroat endangered and added it to its annual redbook of endangered species. That listing carried over when the U.S. Endangered Species Act became law in 1973, but in 1975, the listing was changed to threatened to facilitate recovery and management efforts, and to avoid the problem of enforcing the "no take" provision of the Act at Pyramid and Walker lakes, where by this time stocking of hatchery-reared cutthroats had revived popular sport fisheries.

A litany of reasons has been recited for the Lahontan cutthroat's steep decline:

- Excessive and virtually uncontrolled exploitation by market and sport fishermen in the late 1800s and early 1900s—including massive egg-taking operations by state and federal fisheries agencies.
- Logging, lumbering, and cordwood-cutting operations.
- Mining impacts.
- Water diversions, in particular the Newlands Project's Derby Dam.
- Introduction of nonnative trout and charr species.
- Sheep and cattle grazing.
- Water quality deterioration.

Overexploitation by Market and Sport Fishermen, and by Egg Taking Operations

The Sierra drainages feeding into the western Lahontan Basin lie astraddle the immigrant roads to California. Traffic was heavy, especially after gold was discovered in California in 1849. Soon, the eastern Sierra region experienced a heady boom of its own, fueled by gold and silver strikes at Gold Hill, Virginia City, Aurora, Bodie, Monitor, and Silver Mountain. White settlers and miners provided a ready market for the cutthroat trout that could be caught so easily and in such great numbers from Lake Tahoe, Pyramid and Winnemucca lakes, the Truckee River, and Walker Lake. Huge and almost totally unregulated market fisheries developed at each of these places. As shippers learned how to ice down and preserve their cargoes to reach more distant points, hotels as far away as Chicago featured trout from these sources on their dining room menus. Early fish commission reports noted that rampant poaching and uncontrolled exploitation were adversely affecting the fishery, but little or nothing was done to ramp it down. Where migrating trout massed at the base of water-diversion structures or splash dams, people waded in with pitchforks and hauled fish away by the wagon load.

Even though the historical record indicates that market fishing was already in decline at Pyramid Lake and the Truckee River by the time the following report was written, Nevada Fish commissioner George Mills penned a glowing account in his biannual report for 1889–90:

> "That the magnitude of our fish industry might be generally understood, I have endeavored to collect such data as would furnish reliable information. From October 21, 1888 to April 20, 1889 nearly 100 tons of...trout were shipped through Wells Cargo and Company and by railroad freight, selling in markets of destination at twenty cents and upward per pound and from October 21, 1899 to November 27, 1890 nearly the same amount. This does not include...trout sold in our home markets."[30]

One hundred tons of trout: if the average fish weighed 10 pounds (a not unlikely figure at the time), that's 20,000 trout per season, even in a declining fishery. If visiting anglers couldn't catch 100 pounds of trout in a couple of hours, they were rated unlucky, unskilled, or both.

From Lake Tahoe as well, shipments of trout averaged upwards of 70,000 pounds a season through the 1870s. Permanent traps established by market fishermen in the more productive tributaries often captured entire annual spawning runs. Trout were also taken in traps, in gill nets, and by beach seining near the mouths of spawning tributaries. Beach seines up to one-half mile in length were reportedly used in these operations. Even though the Lake Tahoe fishery had declined by the turn of the century, market shipments of 158,667; 11,981; 7,982; 13,977; and 22,730 pounds

were reported in the years 1900, 1901, 1902, 1903, and 1904, respectively.

In addition to market demand for the fish themselves, there also developed a lucrative market among commercial and sport fishermen for trout roe to be used for bait to catch more trout. Tons of roe were also frozen and shipped to other markets. The demand was so great that countless spawners were killed for the roe alone and the carcasses discarded.

From 1882 to 1938, when the spawning runs finally failed at Lake Tahoe, fish culturists from the State of California also took cutthroat eggs from adults trapped in tributary streams. Sometimes the workers found themselves competing with market fishermen for the trout entering spawning streams. Even so, as many as two to four million eggs per season were taken in good years. Although some of the progeny were returned to the lake or its tributaries and are credited with maintaining the runs for a short while, the majority of the progeny were stocked into other waters of the state. This, in effect, mined Lake Tahoe of its native trout during a period when its population was already in decline.

In Nevada, fish culturists from the State of Nevada and the U.S. Bureau of Fisheries were taking eggs from trout on their spawning run up the Truckee River at least as early as 1889. As in California, the net result was fish-mining. For the six years between 1919 and

30. This quote, copied faithfully from the actual report, is from Nevada State Fish Commissioner George T. Mills' Biennial report of the Fish Commissioner, 1889–1890. I emphasize the faithfulness of the copy because everybody else who has used this passage has substituted Wells Fargo for the "Wells Cargo and Company" mentioned in the original document. I'm not sure which is correct. There is a town of Wells, Nevada that dates back to that period, and I found websites on the Internet for two contemporary Wells Cargo Companies, one of which is located in Nevada. I sent inquiries to both contemporary companies to see if there was a link, but they must have thought it was some kind of joke because neither replied.

1925 (the only period for which records are available), 6.4 million eggs were taken but only 1.7 million progeny, less than 30 percent of the egg take, were released back into the stream. The remainder were either used for stocking other waters or were shipped to other states, often in trade for exotic species that were in turn released into the Truckee or Pyramid Lake or other waters of the state. [31]

Logging, Lumbering, and Cordwood-Cutting Operations

Logging, lumbering, and woodcutting were other activities fueled by the region's mining boom. Lumber and shingles were needed for building and construction. Timbers were needed to shore up and support mine shafts. Ties were needed for rail lines. Cordwood was needed to make the charcoal used in the smelters. Cordwood was also needed to produce the steam that powered the machinery, not to mention for heating the homes, boarding houses, hotels, and other businesses in winter. From 1853 to 1914, no less than 64 sawmills are known to have operated in the Walker, Carson, and Truckee river watersheds, and there were probably others that the record doesn't mention. The towns and mining operations of the Comstock Lode alone (i.e., the area around Virginia City and Gold Hill, Nevada) consumed 73 million board-feet of timber annually,

plus another 250,000 cords of cordwood. Other major mining centers sprang up around Aurora, Bodie, Monitor, and Silver Mountain with demands for wood of their own. [32]

All the timber that could be profitably reached was cut. Logging crews took all the larger trees, primarily Jeffrey and Ponderosa pine, red fir, and hemlock, and the woodcutters took all the smaller stuff that could be cut into cordwood. As they progressed, the country was all but denuded. What visitors and even most locals don't appreciate these days is the extent of the deforestation that took place in the Walker, Carson, and Truckee river watersheds during that period. The trees one sees in these watersheds today are almost all second-growth—or even third-rotation in areas where modern-day logging has or is occurring.

Of course, no riparian buffers were left in those days, so an inevitable result of stripping the valleys and hill slopes bare was tremendous deluges of sediment and debris-laden runoff water sluicing down the streams following each good rainstorm. Stream habitat in the tributaries must have suffered heavily, as did the trout trying to eke out a living in the impacted streams.

Some logs were moved to the banks of larger streams by dragging or by horse- or oxen-drawn wagons, but the more efficient way was to build flumes and let water carry them down. Some flumes extended

31. The Sacramento perch came to Pyramid Lake in this way in 1889 or 1890. Rainbow trout, brown trout, brook trout, lake trout, chinook and coho salmon, largemouth bass, brown bullhead, and common carp were also acquired via these egg trades and introduced into Nevada waters, including the Truckee River and Pyramid Lake.

32. In 1941, the editors of *The Timber-man*, a trade journal for the logging and lumbering industry, published a history of lumbering in western Nevada which also took in the east slopes of the Sierra Nevada in California, including the Walker, Carson, and Tahoe/Truckee drainages (see Editors 1941). In addition, Eric Gerstung, formerly with the California Department of Fish and Game, provided me with a set of notes he compiled on the history of logging, lumbering, and woodcutting principally in the Carson River drainage. Among Eric Gerstung's primary sources were books by Bruns (undated), Maule (1938), Howatt (1968), and Dangberg (1975). Another book on the history of logging and lumbering in the Truckee Basin is Dick Wilson's *Sawdust Trails in the Truckee Basin* (Wilson 1992). W. Storer Lee's book, *The Sierra* (Lee 1962) has a particularly colorful chapter on this era, including

up into the mountains for miles. Where streams were large enough to build up good heads of water, splash dams were constructed. Logs were sent downriver to the sawmills in huge drives in the spring of the year, usually just after the first rush of snowmelt water occurred. Smaller bolts and cordwood, as well as lumber, ties, and mine timbers that were cut in the woods, were often splashed down even the smaller tributaries.

The timing of these drives coincided with the spawning of cutthroat trout. Spawning migrations were disrupted when the fish encountered splash dams or streams chock-full of logs. River beds and spawning riffles were gouged and scoured. Redds already constructed were destroyed. Many natural habitat features that provided cover for fish were deliberately removed, channels were straightened, and side-channels were blocked, all to make the rivers snag-free for floating logs. Log drives could impact miles-long reaches of a river. One of the main retaining ponds on the Carson River, for example, was located at the head of Carson Valley on the East Fork. When this dam was opened and the logs released, wood was floated all the way to Empire, Nevada, east of Carson City. The final river drive on the Carson River took place in 1896.

Many of the sawmills were small and relatively portable, and followed the logging camps up into the mountains. The larger and more permanent mills were built at centers along the main rivers. Antelope City on the West Fork Walker River; Bridgeport on the East Fork Walker; Centerville, Markleeville, Woodford, Genoa, and Empire on the Carson; and Lake Tahoe (several mills), Truckee, Verdi, and Truckee Meadow on the Truckee River were among those highlighted in the historical record as centers for sawmilling. The earlier mills were mostly water powered, but these gave way to steam driven operations as the logging and lumbering era extended. Either way, they required water to operate, and this was acquired by building diversion structures in the streams and taking what they needed. The sawdust generated by these mills was most often disposed of by dumping it into the streams. An eye-witness "...saw the Truckee River in the summer months, practically colored white by the immense amount of sawdust coming down from the large mills operating on the river." [33] Sometimes the sawdust was thick as mush in the Truckee, coating spawning beds, smothering eggs, fry, and even larger fish, and sometimes extending all the way downstream to Pyramid Lake.

Many of the loggers, woodcutters, and workers who manned the river drives of the logging and lumbering era came from French Canada. An interesting side-note to the history of this period is that these men may be responsible for "milk can transfers" of trout that helped preserve indigenous populations. This

the account of a ride down one of the log flumes by a couple of well-heeled gents who should have known better; see Lee's Chapter 9, "They Thought the Trees Would Last Forever." For those of you who might not be familiar with the terms, the board foot is a measure of wood volume used in the logging and lumber industries and is equivalent to a board measuring 12 inches by 12 inches by one inch thick. Twelve board-feet equal 1 cubic foot. The cord, another measure of wood volume, is a stack measuring 4 ft. by 4 ft. by 8 ft.

33. Reported by Shebley (1929).

may have occurred in the Silver King Creek drainage (Chapter 8). Also, some historians speculate that names for features in the East Fork Carson drainage, such as Poison Creek, Poison Lake, and Poison Flat, may be anglicized versions of the French "poisson," meaning fish. If so, this would be additional testimony to the historical presence of fish in that part of the watershed.

Mining Impacts

Curiously, even though mining and mineral extraction was a booming part of the history and economic development of the western Lahontan Basin, there is hardly a mention of any more direct impacts that this industry might have had on the native cutthroat trout than those discussed above. Indeed, of all the authorities who have written about the decline of western Lahontan Basin cutthroat trout, only a single observer made reference to "…depleted waters of the foothill streams that had been destroyed by mining debris." [34] And that's all that particular writer had to say on the subject. In general, however, the following impacts on rivers, streams, and trout are attributed to mining activities.

Placer mining, where the miners work directly in the stream itself or in alluvial deposits of the stream, does not appear much in the historical record for eastern Sierra streams, but it could have been practiced in operations too small to have been mentioned. Placer operations not only gouge out and destroy the stream-bed in reaches where they actually take place, they also impact reaches further downstream by generating large amounts of sediment. For underground or hard rock mining, which was widely practiced in the region, the commonly cited impacts are: 1) acid drainage from tailing piles and old mine shafts; 2) toxic metal inputs into streams; and 3) sedimentation either from tailing pile runoff or from direct dumping of tailings into the stream.

Acid drainage can occur anywhere there is iron pyrite ("fool's gold," a sulfide of iron) or sulfides of other heavy metals, such as copper, zinc, lead, or cadmium. These oxidize to produce acids that lower the pH of the water and can be directly toxic to most forms of aquatic life. The heavy metals themselves, when solubilized in any form and leached into the stream, can also be toxic to fish. Fine sediment carried in from tailing pile runoff can fill pools, depriving fish of rearing habitat and cover, and can embed spawning gravels. In addition, liquid and solid wastes from milling and smelting operations were frequently dumped directly into the streams. But even when they weren't, nothing much was done to keep leachates from reaching the streams. These materials too are laced with toxic heavy metals such as lead, cadmium, and zinc. The effects of runoff from these old mines and milling operations can persist for decades, even up to the present time.

34. Reported by Shebley (1929). For a more thorough review of the impacts of mining and mineral extraction on trout and their habitats, see Nelson et al. (1991).

Water Diversions

I have already mentioned the water diversion structures and splash dams built by logging and lumbering crews and how they could interfere with Lahontan cutthroat spawning activities. Those of the region's settlers who devoted themselves to farming and ranching rather than gold-seeking also erected diversion dams to obtain irrigation water. Many of the earliest ones were low structures built of rubble. On high flows, when water typically overtopped these structures, the trout could pass them easily. But when the rivers ran low, water percolated through the structures and the fish were blocked, stalling spawning migrations and making easy pickings for the market hunters.

Another problem with these water diversions was that few, if any, were screened. Countless numbers of young trout, and adults as well, perished after being shunted into farmers' fields via unscreened diversion ditches and pipes.

As problematic as these early diversion structures were for the trout, it was the construction of higher, permanent dams made of concrete that ultimately spelled extirpation for the Truckee River spawning runs and Pyramid Lake's cutthroat trout, and near-extirpation for the cutthroat trout of Walker Lake as well. Many of these concrete structures were built without fish ladders. Even when ladders were provided,

many failed to enable passage and apparently little, if any, effort was ever made to fix these problems.

The first of these concrete monsters was the U.S. Bureau of Reclamation's Derby Dam on the Truckee River, the capstone of the Newlands Project. Derby Dam was completed in 1905 approximately 40 miles upstream from Pyramid Lake, with the objective of diverting 250,000 acre-feet per year of Truckee River water into the lower Carson basin. As the water began to flow through the diversion canal, the level of Pyramid Lake began to drop. Winnemucca Lake, Pyramid's shallow companion that had shared the natural flow of the Truckee, dried up completely in 1938. As the level of Pyramid Lake receded, the river began to downcut through ancient lake deposits, transporting them down to the new river mouth and depositing them as a wide, shallow sand bar or delta across the river mouth. The river channel braided and anastomosed as it worked its way across this delta in rivulets that in many years were too shallow for spawning trout to ascend.

Even for the cutthroat trout that could ascend into the Truckee, the impact of Derby Dam was noticed almost right away. The dam had been built with a fishway and the trout did try to use it, but it was never adequate. In 1912, Nevada Fish commissioners Mills, Yerington, and Clarke wrote in frustration:

> "The erection of Derby Dam on the Truckee River by the government has occasioned the Nevada Fish Commission more

Mahogany Creek, inlet tributary of Summit Lake, Nevada, is the sole spawning tributary for the largest remaining endemic population of lacustrine Lahontan cutthroat trout.
PHOTO COURTESY OF U.S. FISH AND WILDLIFE SERVICE FISHERY ASSISTANCE OFFICE, RENO, NEVADA

correspondence and absorbed more time and explanation than any one subject or object in its history. The dam is a government construction with a view to supplying the Truckee-Carson Reclamation Project with water. The Nevada Fish Commission has no fight along this score, but it has, since the first, demanded of the government that it build proper fish ladders that the spawning trout may come up the river during the season, not alone for the engagement of taking eggs, both by the government and the state, but that the trout might distribute along the best of the interstate mountain streams in the west. It is an admitted fact by the California Commissioners, the government experts, and the home people that this poorly provided fish ladder or runway, installed by the government, is, in many ways, inadequate. Tons of fish have died at Derby Dam. Millions of eggs have been lost to the Nevada hatchery, the government, and to the natural spawn beds of the river, all because of the inadequate provisions made by the government." [35]

Finally, that "poorly provided" fish ladder washed out completely and was never replaced. Surprisingly enough, the trout held on for several more years by utilizing river reaches below the dam for spawning when they could ascend past the delta, but that wasn't enough. Fishing in Pyramid Lake tailed off. By 1934, it was considered poor. After about 1928, fewer and fewer trout ascended into the river and none came at all after 1938. The last cutthroat was seen in the lake in 1940. The cutthroat population of Pyramid Lake had finally been snuffed out.

Walker Lake also contained large Lahontan cutthroats that supported a commercial fishery until about 1930. Irrigation withdrawals for agricultural development had been taking their toll on the fishery even before the turn of the 20th century, but it wasn't until the 1920s that the last good spawning runs occurred, pretty much coincident with a series of concrete dams built along the Walker River system in the 1920s and 1930s. Bridgeport Dam, completed in 1924, blocked spawning migrations up the East Walker River. Bridgeport Dam had a fish ladder, but this was another of those cases where it never worked and was never improved. Topaz Diversion was another project built in this period. Weber Dam, built in 1935 without a fish ladder, was sited below the Walker River forks and ended all cutthroat migrations beyond that point. Again, trout persisted by utilizing the less than adequate spawning habitat in the river below Weber Dam, but could not maintain the population. In 1949, perhaps mindful of the extinction episode at Pyramid Lake a decade earlier, the State of Nevada captured the last 39 trout attempting to ascend the river and took them into a hatchery. The present Walker Lake sport fishery is maintained by stocking hatchery-reared progeny from this broodstock.

Introduction of Non-Native Trout and Charr Species

The cutthroat trout of the western Lahontan Basin evolved in the absence of other trout or charr species

35. Cited at pages 9–10 in *Nevada Outdoors and Wildlife Review* 10, no.2 (Summer 1976 issue).

and do not compete well with them when they come into contact. Having examined the historical record, Eric Gerstung pointed out that in stream environments of the western Lahontan Basin, native cutthroat trout were seldom able to co-exist with non-native trout for more than a decade. A case in point is Pole Creek, an upper Truckee River tributary, which had native cutthroat trout in 1960 but only brook trout ten years later. Lahontan cutthroat trout also hybridize with rainbow trout. Although a few introgressed populations of Lahontan cutthroat trout have maintained themselves—for example, in upper tributaries of the Quinn River drainage—by far the more usual outcome in this region has been displacement of the native cutthroats when rainbow trout have been introduced. [36]

Rainbow trout, brown trout, and brook trout were first introduced into the western Lahontan Basin prior to the turn of the 20th century. Rainbow and brown trout displaced Lahontan cutthroats in all mainstem waters and larger tributaries of the Walker, Carson, and Truckee river drainages, and brook trout occupied most of the upper reaches and smaller headwater tributaries once occupied by native cutthroat trout. In the Susan River, only rainbow trout were captured in 1902. Small adfluvial populations of Lahontan cutthroat trout have persisted in upper tributaries of the Quinn River system in the face of non-native trout stocking, but hybridization and displacement have occurred widely in that drainage as well.

Rainbow and brown trout were also introduced into the large oligotrophic lakes of the western Lahontan Basin prior to the turn of the 20th century. Also introduced early-on were lake trout, often referred to as "mackinaws" by local anglers. These large, primarily benthic predators quickly became established in Tahoe, Fallen Leaf, Donner, and Independence lakes (but not in Cascade Lake, where only rainbow and brown trout were introduced), and soon became popular with local and visiting sport anglers who quickly forgot about the original native "Tahoe trout" and "silver trout." Only in Independence Lake did native cutthroats manage to persist in the presence of lake trout.

Between 1956 and 1962, the California Department of Fish and Game planted nearly 1 million fingerling and yearling Lahontan cutthroats in Lake Tahoe in an effort to reestablish a cutthroat fishery. This, and a later effort involving the release of Yellowstone cutthroat trout, failed utterly. Nearly all of the stocked cutthroats became fodder for the lake trout. Recent studies by a research group from the University of California, Davis, showed that the food web of Lake Tahoe has been altered dramatically since the establishment of lake trout, as well as the introduction of the freshwater shrimp *Mysis relicta* into the lake. Where the histori-

36. I introduced you to some of the underlying theory of displacement of native species by non-natives in Chapter 4. The broader field of inquiry of which this science is a part is known as invasion biology, which examines how species become invasive and the resulting effects of invasion. Introduced species that became invasive are one of the primary causes of endangerment and extinction of just about all of our native cutthroat subspecies. The following references will help familiarize you with this field: Lodge (1993), Tilman and Kareiva (1997), Kolar and Lodge (2001), Dunham et al. (2002), and Peterson et al. (2004).

cal food web favored the production of pelagic fishes, such as the native cutthroat trout, the present food web favors the production of benthic fishes, such as the lake trout that has now assumed the top-predator niche. The present food web structure of Lake Tahoe actually functions as a barrier to the restoration of the native fish community. [37]

Sheep and Cattle Grazing

As the logging and woodcutting era drew to a close in the eastern Sierra watersheds, it was replaced by an era of livestock grazing in the revegetating mountain meadows. Basque herders with flocks of sheep and stockmen with cattle replaced the French-Canadian loggers and woodcutters. Livestock grazing continues to this day as a principal land use in the eastern Sierra back country. In the Soldier Meadow area and Mahogany Creek drainage at Summit Lake, and in the Quinn River watershed including Kings River, a major tributary of the Quinn, livestock grazing became the principal land use soon after the start of white incursion into the area. [38]

Although grazing cattle clearly have become the icon of the open western range, stockmen have always run both sheep and cattle on western Lahontan Basin grazing lands. According to range scientists, sheep and cattle have different impacts in a watershed. Sheep

37. For the University of California, Davis, report showing how the food web structure of Lake Tahoe has changed over time, and a discussion of its implications for the restoration of Lahontan cutthroat trout to the basin, see Vander Zanden et al. (2003). Cordone and Frantz (1968) have a brief account of the failed early efforts to reestablish a cutthroat trout fishery in Lake Tahoe.

38. Sessions S. Wheeler's book, *The Black Rock Desert* (Wheeler 1979), contains an interesting account of the early days of livestock grazing in the Soldier Meadow, Summit Lake, and Quinn River country. Dahlem (1979) and Myers and Swanson (1996a, 1996b) describe the more recent history of the Mahogany Creek watershed as a case study in aquatic habitat improvement through grazing management. Behnke (1979 [1981]) has an excellent and succinct discussion of livestock-fisheries interactions (at pages 24–26 in my copy of the 1979 edition), and for broader discussions see Cheney et al. (1991) and Platts (1991). Livestock grazing on public lands has become one of those polarizing, lightning-rod issues with much written and much posted on the Internet by those both for and against. This is another topic I will revisit several times in this book. For additional reviews of the science behind this issue, I recommend Platts (1981a, 1981b), Kauffman et al. (1984), Belsky et al. (1999), Jones (2000), and Van Vuren (2001).

are typically grazed in flocks and are actively herded, meaning they are moved along as forage is grazed. In this regard, their use of the range is like the American bison that once grazed the great North American grasslands in such large numbers. Bison, too, moved steadily along as forage was grazed, spending the majority of their time in the uplands and moving to a water source only for short intervals, perhaps only once a day. Grazers of this type convert forage to meat with relatively little direct impact on streamside habitat.

Cattle, on the other hand, are allowed to range freely, but they prefer to graze streamside environments. This is due to the evolutionary biology of our domestic breeds of cattle. Cattle evolved in relatively moist Eurasian climes, and require the equivalent of 40-plus inches of rain a year. Consequently, and especially in the semi-arid and arid regions that define most of the interior west, where 10 to 20 inches of rain a year may be a lot, riparian areas (where the most succulent plants grow, and of course where the water is located) receive disproportionately heavy use when cattle are on an allotment. Cattle will bunch up along a stream and remain, consuming all the vegetation, trampling the banks, and even fouling the stream with feces, until something—or someone—comes and drives them along. However, it has been found that yearling cattle tend to spend less time on stream bottoms than cows and calves.

The damage from cattle is most acute in semi-arid and arid regions where climate conditions may indeed produce forage plants, but not in sufficient quantity or in vigorous enough growth to support high grazing levels. It is very easy to overgraze these areas—and it's easy to recognize the symptoms of overgrazing. Riparian areas are reduced to bare, compacted soil. As soil is compacted and favorable ground cover is reduced, infiltration of rainwater lessens and surface runoff increases, accelerating erosion, and increasing peak flows and the flashiness of streams receiving this runoff. Streambanks are destabilized. Fine sediment smothers spawning and rearing habitats of trout and other fishes in the stream. Topsoil is lost and the water table lowers. Streams respond to these altered hydraulic conditions in either of two ways. The stream may widen out to produce a shallower, higher-velocity channel, or it may trench downward, deepening the channel as the stream seeks to accommodate to the lowered water table. Either way, the resulting channel conditions are of only poor value as trout habitat.

Water Quality Deterioration

I have already mentioned the discharges of sawdust from sawmills, the likely inputs of acid drainage and toxic metals from mine tailings and ore-processing, and the fouling of streams by cattle that overstay their time along riparian reaches as factors that would have impaired water quality of western Lahontan Basin streams. The record also lists another type of industrial discharge, namely effluent from a paper mill that operated for two decades until about 1930 at Floriston, California. Paper mills have always been large users of water, which would have meant another major water withdrawal from the Truckee River in the early 20th century. But aside from that, the effluent from this mill was discharged directly into the river. Although the nature and composition of this particular effluent is not recorded (other paper mills of the period used a variety of chemicals and their effluents also contained suspended materials, such as cellulose fibers and clays), what is known is that a sample of the Floriston effluent taken in 1913 killed trout in an experimental hatchery. [39]

Also, the various towns and mining camps ran their sewage directly into the rivers. This included the cities of Reno and Sparks until about 1930, when those towns constructed sewage disposal plants. Taken together with all the other factors, pollution of the rivers and streams was another condition that worked against the survival of the native cutthroat trout and led to their demise in the western Lahontan Basin.

Ironically, even as man was working to bring about the demise of all these remarkable fisheries, the large size of the Lahontan cutthroat had attracted interest in propagating it for stocking in other waters. I have already mentioned how trout from Independence Lake were being propagated as early as 1870 in private hatcheries in California. Egg-taking from Lake Tahoe fish commenced in 1882, and from Pyramid Lake fish by at least 1889. Record-keeping to identify where hatchery-reared progeny were stocked was often pretty slipshod in those days. Sometimes records were forgotten altogether. Later, when odd forms of trout were discovered in unlikely places, they could become real mystery fish.

Such was the case in 1907, when Joseph Grinnell, then a Stanford University student (he later became Director of the Museum of Vertebrate Zoology, University of California), discovered trout in the upper reaches of the South Fork Santa Ana River, a stream that flows out of southern California's San Gorgonio Mountains. Grinnell's professor, David Starr Jordan, described these specimens as a new species, *Salmo evermanni*, the San Gorgonio trout. For 40-odd years, the San Gorgonio trout was considered a valid species, but no scientists and few anglers ever visited its waters to take new samples. It wasn't until 1950 that Robert Rush Miller, another scientist examining old museum specimens, called the identity of *Salmo evermanni* into question. Dr. Miller discovered the specimens were not a distinct species at all, but rather, were cutthroat trout!

39. Reported in Sumner (1939).

So how did cutthroat trout get into the mountains of southern California? It took another ten years for the story to unfold. The answer had been there all along, in the Fourteenth Biennial Report of the Fish Commissioner of the State of California for the Years 1895–96. That report revealed that cutthroat trout, reared at the Sisson Hatchery from eggs taken in tributaries of Lake Tahoe, had been stocked in the Santa Ana River and nearby streams back in 1895 and 1896. However, by 1960 when this report came to light, so many rainbow trout had been stocked in the Santa Ana drainage that the original population sampled by Grinnell was extinct. R.R. Miller had not classified the San Gorgonio cutthroat specimens as to subspecies, so out came those old museum specimens once again, this time to be compared with *O. c. henshawi* specimens. And sure enough, they were identical. [40]

Macklin Creek, a small tributary of the Middle Fork Yuba River west of the Sierra crest, harbors another of those Lahontan cutthroat populations originating from hatchery operations at Lake Tahoe. These trout have remained free of introgression since being stocked in the 1920s, and have been used by the State of California in recovery efforts in the upper Truckee drainage. Pole Creek, mentioned earlier, where the original Lahontan cutthroat population had been replaced by brook trout, was chemically treated in 1975, 1976, and 1977, then stocked with Macklin Creek trout. By 1982, Lahontan cutthroats were once again abundant in Pole Creek and have remained so.

Another successful restoration using Macklin Creek trout was made in the upper Truckee upstream from Lake Tahoe in the late 1980s. Four miles of mainstem Upper Truckee, along with three tributaries upstream from the Four Lakes outlet stream, 10-acre Meiss Lake, and its outlet stream, were all chemically treated in 1988, 1989, and 1990. A barrier was also built to prevent upstream recolonization by brook trout. Streams were stocked with Macklin Creek trout, and Meiss Lake was planted with trout originating from Independence Lake.

The original Pyramid Lake strain, once thought to be extinct, also may have been preserved by one of those old transplants. In April of 1977, Kent Sumners, Utah Division of Wildlife Resources, discovered a small population of cutthroat trout in a tiny creek on BLM land on the slopes of Pilot Peak, which sits on the Utah-Nevada border north of Wendover (Fig. 5-1). Pilot Peak had been a landmark for the old immigrant trains and it is clearly visible from Interstate 80, but it lies in the Bonneville Basin. Terry Hickman, then a graduate student at Colorado State University, had been studying Bonneville Basin cutthroat populations. Hickman collected specimens from the Pilot Peak population in June of 1977 and determined that they were

40. For more of the story of the San Gorgonio trout, see Jordan and Grinnell (1908), Miller (1950), and Benson and Behnke (1960).

Lahontan cutthroat trout, not the expected Bonneville Basin subspecies. Dr. Behnke, Hickman's thesis advisor, determined that the creek (named Donner Creek by Hickman) had not been stocked since 1930, and before that, all Lahontan cutthroats stocked in either Nevada or Utah had come from eggs taken from Pyramid Lake trout. The Donner Creek cutthroats thus appear to be the original Pyramid Lake genotype—or as much of it as remains after many generations of reselection in a harsh, small-stream environment.

You would have thought that the discovery of pure-strain descendents of original Pyramid Lake cutthroats would generate at least a little bit of excitement, and initially it did. Hickman and Behnke published a paper on the discovery in a fisheries journal,[41] news items and at least two full articles appeared in the angling media, and even *Sports Illustrated* carried a piece titled "The Fish that Wouldn't Die" in its March 17, 1980 issue. U.S. Fish and Wildlife Service biologists did take eggs from Donner Creek cutthroats in 1980 to develop a broodstock at the National Fish Hatchery in Hotchkiss, Colorado. However, this attempt failed and federal interest in working with this stock apparently waned. The State of Utah assumed the role of managing and monitoring this population through the 1980s and 1990s.[42] In the process, the name Donner Creek morphed into Morrison Creek. Trout from Donner/Morrison Creek were even-

tually transplanted into two additional creeks to increase the number of stream-dwelling populations and provide a hedge against extirpation. Trout were also stocked into a couple of small reservoirs on cooperating ranches in the area in order to observe the behavior and growth of this strain in a lacustrine environment. Individual trout were said to have grown to 5.5 pounds in weight after five years in one of these ponds, which was taken as a good sign that lacustrine adaptation had not been selected out of this population.

In 1996, the U.S. Fish and Wildlife Service renewed work to develop a broodstock from the Pilot Peak stock, this time working at the Lahontan National Fish Hatchery near Gardnerville, Nevada. According to Fish and Wildlife Service fish culturist Jay Bigelow, enough success has now been achieved, despite setbacks, that progeny from this stock have been released into Walker Lake and into Fallen Leaf and Cascade lakes, the latter introductions part of a new recovery plan for the Tahoe Basin, which I will talk about again a bit later.

Between 1938 and the early 1950s, little was done to preserve population of Lahontan cutthroat trout within the historical range. The State of Nevada did succeed in maintaining a cutthroat fishery at Walker Lake using hatchery-reared fish from a broodstock captured at the lake in 1949. Based on meristic characters, Dr. Behnke had called the genetic purity of this stock into

41. Hickman, T.J. and R.J. Behnke. "Probable Discovery of the Original Pyramid Lake Cutthroat Trout," *Progressive Fish-Culturist* 41, no. 3 (1979): 135–137. See also: Hickman (1978a) and Hickman and Duff (1978).

42. Details of the State of Utah's work with the Pilot Peak Lahontan cutthroat population can be found in the report, "Native Cutthroat Trout Management Plan" (Utah Division of Wildlife Resources, 1993).

question at one point. The opportunity for spawning cutthroats from Walker Lake to hybridize with rainbow trout planted in the Walker River had existed for more than half a century prior to the time the broodstock was established. However, the fish have always been good representatives of the Lahontan cutthroat trout appearance-wise, and subsequent genetic testing has not detected evidence of introgression.

In the early 1950s, State of Nevada biologists conducted experiments that indicated trout could still survive in the increasingly more alkaline and saline waters of Pyramid Lake. The state, in cooperation with the Pyramid Lake Indian Tribe, attempted to resurrect a sport fishery there, first by introducing rainbow trout, then cutthroats derived from the Walker Lake, Heenan Lake, and Summit Lake broodstocks. [43] The broodstock then present in Heenan Lake, although another good Lahontan cutthroat appearance-wise, had originated from a transplant from California's Blue Lakes, a stock suspected of being introgressed by rainbow trout. But the Summit Lake broodstock, as indicated above, is a pure-strain stock.

Another development that eventually aided in the restoration of the Pyramid Lake fishery occurred in 1956, when Congress passed the Washoe Act. This Act provided for increased water flows to Pyramid Lake and for facilities to help restore the fishery. However, it wasn't until eighteen years later, in 1974, that money to implement this Act finally came through. Even so, it was better late than never, and by then it had become apparent that the cutthroat fishing could once again be phenomenal. The median weight of trout caught by anglers throughout the 1960s was 6 pounds, with fish of 10 pounds or more becoming an ever more common occurrence.

Up until the mid-1970s, the State of Nevada had been the main contributor to the restoration effort at Pyramid Lake. With Washoe Act funds, the Pyramid Lake Paiute Tribe staked out a larger role. Assisted initially by the U.S. Fish and Wildlife Service, the Tribe established its own cutthroat hatchery at Sutcliffe (a second tribal hatchery on the Truckee River near Wadsworth went into operation in 1981), initially using a broodstock obtained from Summit Lake. As their own hatcheries came on line, the Tribe assumed more and more of the responsibility for stocking the lake and now handles it all. The present operation relies only on spawners that have reared and matured in the lake.

Meanwhile, in 1976, also with funds from the Washoe Act, the Fish and Wildlife Service built a dam at Marble Bluff, near the former mouth of the Truckee River. This also included a fishway that extends 3.5 miles to the lake to provide passage across the delta to the dam. A second fishway was built at Numana Dam about 13 miles up the river. Fish ascending to the Mar-

43. A short account of the early efforts to revitalize sport fishing at Pyramid Lake (and the surprising event that led to these efforts) was written by Thomas J. Trelease, then-Chief of Fisheries for the Nevada Division of Wildlife. See Trelease (1969b).

ble Bluff Dam are trapped and can either be passed into the river above to spawn if and where they can, or be taken into the hatcheries. Natural reproduction in the lower river had been assumed negligible, owing to low flows, low dissolved oxygen levels, and high spring and summer water temperatures that frequently exceed 68 degrees Fahrenheit.

Also in 1976, water in Stampede Reservoir, a water storage reservoir on the Little Truckee River in California, was dedicated to cui-ui and Lahontan cutthroat trout. It took a court fight, because western water law generally favors the "first in time" and does not recognize the needs of fish as a beneficial use. [44] But in the end, the courts affirmed a strategy to prioritize this water for the benefit of the Pyramid Lake fishery until the cui-ui and Lahontan cutthroat trout are no longer classified as endangered or threatened, or until sufficient water becomes available from other sources to conserve these fishes. [45]

All of these activities now meld with a series of recovery and management plans developed by the federal, tribal, and state fishery management agencies following the listing of Lahontan cutthroat trout under the Endangered Species Act (endangered in 1973, down-listed to threatened in 1975). [46] The most recent overall recovery plan was issued by the U.S. Fish and Wildlife Service in 1995. This Plan includes both the

44. Western water law is based on the prior appropriation doctrine. This allows water to be withdrawn from streams and lakes for beneficial purposes based on date of application for withdrawal. It is a "first in time, first in line" doctrine. In other words, if your water-right was filed first, before somebody else's, you get your full quota of water before they get any of theirs, and so on down the line. The prior appropriation doctrine thus favors long-time water users. It does not allocate water based on social or economic value of uses or products. Nor does it consider the value of leaving water in the streams. The prior appropriation doctrine has been affirmed by the courts many times. Water users throughout the American West have learned to live with this system—but not without conflict. There is, perhaps, no single natural resource issue that has been so divisive. Friends, even family members, have become bitter enemies. And yes, there have even been gunfights. Some western states have modified their water laws to maintain water levels in natural lakes and streams to support public uses (for the good of fish, for pollution abatement, and other like uses) while still preserving the prior appropriation doctrine. Other states have rejected such attempts. But even where instream

flows are mandated for these public uses, conflict resolution depends upon creative dialog and accommodation among stakeholders. Since water rights underlie so many trout conservation and recovery issues, it behooves everyone who is interested in native trout, from nature lovers to conservationists and anglers to professional managers and scientists, to become at least acquainted, if not fully knowledgeable, about western water law and how it works. This becomes doubly important as global warming portends even further reductions in fresh water supplies. The National Geographic Society devoted a special issue of its magazine to the broader issues of fresh water supply and demand across North America (see National Geographic Society 1993). Two good books that bring the issues closer to home are Marc Reisner's *Cadillac Desert: The American West and its Disappearing Waters* (Reisner 1986) and Robert G. Dunbar's *Forging New Rights in Western Waters* (Dunbar 1983). You might also want to check out David Boling's *How to Save a River: A Handbook for Citizen Action* (Boling 1994).

45. It was mainly concern for the cui-ui (federally listed as endangered), not cutthroat trout, that prompted the change in Truckee River water

allocation. Even so, native trout conservationists owe a debt to the cui-ui. For those not acquainted with this fish, here is a species profile compiled from information in Koch (1973), Sigler et al. (1985), Scoppettone et al. (1986), Strekal et al. (1992), and Emlin et al. (1993). This large, omnivorous sucker (family Catostomidae, genus and species name *Chasmistes cujus*) is found only in Pyramid Lake. It also inhabited Lake Winnemucca when that lake existed. It is a long-lived fish (life span 41 years) and matures at 8–12 years of age. Females can carry from 40,000 to 180,000 eggs, depending upon the size and age of the fish. Cui-ui are obligate stream spawners, and do so in the lower 12 miles or so of the Truckee River. Each spring, adults gather in a pre-spawning aggregate at the mouth of the river. If stream flows and water temperatures are suitable and access is possible across the delta (for some reason, cui-ui shun the fishway and use what's left of the river channel), the fish ascend to the base of Marble Bluff Dam, where they are lifted into the river. A portion may also be taken here to supply eggs for a Pyramid Lake Paiute tribal hatchery dedicated to their propagation. When spawning is completed, the fish drop back down to the lake. Historical spawning runs

Humboldt River DPS and the Northwestern Lahontan Basin DPS in addition to the Western Lahontan Basin DPS. In the Western Lahontan Basin DPS, the Recovery Plan identified 18 self-sustaining stream-dwelling populations, including seven populations in the Truckee basin, six in the Carson basin, and five in the Walker River basin, with lacustrine populations present in Pyramid and Walker lakes (both populations maintained by hatchery stocking) and Independence Lake (a small, naturally spawning population). In the Northwestern Lahontan Basin DPS, excluding the Coyote Basin, 15 stream-dwelling populations were identified in 1995, including 11 populations in the Quinn River drainage and four populations in other small Black Rock Desert drainages. Also included here was the Summit Lake population, the largest remaining self-reproducing lacustrine stock within the historical range. These populations represent the starting point for future recovery of the Lahontan cutthroat subspecies.

The 1995, U.S. Fish and Wildlife Service Recovery Plan also acknowledged the existence of nine stream-dwelling populations of Lahontan cutthroat trout in California, two in Nevada, and three in Utah, plus one pond-dwelling population in Utah, that are all outside the historical range but originated from stocking with western Lahontan Basin fish. Much of the stocking that led to these populations occurred during the 1893

may have easily numbered in the hundreds of thousands, prompting Snyder (1917) to write:

> "...at times cui-ui appeared in such large and densely packed schools that considerable numbers were crowded out of the water in shallow places...."

Although adults aggregate at the river mouth every spring, and despite the federal allocation of water for their benefit, under present conditions they cannot always ascend to be lifted into the river. Successful spawning may now occur only one to three times a decade, and even then, numbers of adults are an order of magnitude lower than in Snyder's time. In wet years (for example, 1980 to 1986), an average of 65,000 adults a year spawned in the lower Truckee. But in dry years (e.g., 1988 through 1992), no spawning occurred. Three successive drought years also occurred in 2000, 2001, and 2002. No cui-ui spawned in 2001, but a release of water from storage, allocated for this purpose, allowed 40,000 cui-ui to spawn in the river in 2002. I have no more recent data.

46. Management and recovery plans for Lahontan cutthroat trout, as I have defined the subspecies in this book, include:

1) Fisheries Management Plan, Summit Lake Indian Reservation (Rankel 1977).

2) Fishery Management Plan for Lahontan Cutthroat Trout (*Salmo clarkii henshawi*) in California and Western Nevada Waters (Gerstung 1986).

3) Walker Lake Fisheries Management Plan (Sevon 1988).

4) Pyramid Lake Fishery Conservation Plan (Pyramid Lake Fisheries 1992).

5) Recovery Plan for the Lahontan Cutthroat Trout (Coffin and Cowan 1995).

6) Lahontan Cutthroat Species Management Plan for the Quinn River/Black Rock Basin and North Fork Little Humboldt River Sub-Basin (Sevon et al. 1999).

7) Short-Term Action Plan for Lahontan Cutthroat Trout (*Oncorhynchus clarkii henshawi*) in the Truckee River Basin (Truckee River Basin Recovery Implementation Team, 2003).

8) Short-Term Action Plan for Lahontan Cutthroat Trout (*Oncorhynchus clarkii henshawi*) in the Walker River Basin (Walker River Basin Recovery Implementation Team, 2003).

to 1938 period, when eggs were obtained from Lake Tahoe's spawning tributaries and from the Pyramid Lake spawning runs, and hatchery-reared progeny were stocked widely in California, Nevada, and Utah. Most of these populations were displaced by subsequent introductions of other trout, but those noted in the 1995 Recovery Plan are genetically pure and self-sustaining.

Not included in the 1995 Recovery Plan's list of out-of-basin stocks are several artificially maintained lacustrine populations outside the historical range. The Lahontan cutthroat's great tolerance to high alkalinity did not escape the attention of fishery managers in California and other states who have waters of this type that will not support other trout. In California, Indian Tom Lake in Siskiyou County is so alkaline that repeated plants of rainbow trout and warmwater species failed to establish a fishery. But Lahontan cutthroat fingerlings survived and grew well. The lake now treats anglers to 16- to 20-inch trout.

In Oregon, Lahontan cutthroat trout based on the old and possibly slightly introgressed Heenan Lake broodstock were planted in Mann Lake, a shallow desert sump of 275 surface acres near the upper end of the Alvord Basin. Mann Lake has an inlet tributary, Trail Creek, which mature trout attempt to enter during the spring spawning period. However, Trail Creek is a seasonal stream that dries up after snow runoff is complete, and even when the stream is flowing, its

water is diverted for irrigation purposes. It is my understanding that mature trout are captured and spawned at the lake each spring, and the eggs are flown to the State of Oregon Klamath Falls Hatchery for rearing. The progeny are then released back into Mann Lake as fingerlings the following spring.

The State of Washington also has many saline and highly alkaline lakes in the dry, barren Columbia Basin country in the eastern part of the state. The largest of these, Omak Lake on the Colville Indian Reservation in north-central Washington, covers 3,244 surface acres and has mean and maximum depths of 98 and 319 feet respectively. It sits in a closed basin. All of its inlet streams are small and not suitable for spawning, although there is a plan afoot to build a spawning channel on one of these tributaries to provide natural recruitment and perhaps make the lake a self-sustaining fishery.

Omak Lake was stocked with Lahontan cutthroat trout from the Summit Lake strain in 1968, following failures of stocked rainbow and brook trout. In 1969 and 1970, trout from the old, possibly introgressed Heenan Lake strain were added. Subsequently, the broodstock has consisted of mature fish taken from the lake itself. Progeny for restocking in the lake were first reared at the Winthrop National Fish Hatchery, but now this is done at the tribal hatchery at Bridgeport, Washington. The lake provides an excellent sport fishery for trout averaging 4 to 5 pounds, but occasionally a

Derby Dam on the Truckee River 40 miles upstream from Pyramid Lake diverts Truckee water into the lower Carson basin via the canal in the lower left foreground. This project doomed the fabled giants of Pyramid Lake. PHOTO COURTESY OF ALAN RUGER, FORMERLY WITH PYRAMID LAKE INDIAN TRIBAL ENTERPRISES

much larger one is taken. The current state hook-and-line angling record for Lahontan cutthroat trout is a fish from Omak Lake that weighed in at 18.04 pounds.

Lake Lenore, situated at the extreme lower end of Grand Coulee, the ancient channel of the Columbia River in Washington, is so highly alkaline that it was once thought to be good only for mineral baths. The first impressions you get of this long, narrow 1,300-acre lake are of barren gray bluffs dropping into chalky-green water; of rock strata in the exposed cliff faces frozen in grotesque shapes, some resembling ancient sea-serpents; of basalt columns stacked one atop the other; of stark-white alkali deposits lining a shore bereft of vegetation; of the hundreds of seagulls riding choppy waters awaiting a meal of *Gammarus* scuds. But then you see there's a boat with a party of fishermen out there. And up at the north end you spot a couple of anglers wading the edges casting their flies to big Lahontan cutthroats that ghost across the shallows. These anglers are looking for trout originally introduced by state biologists in 1979, after live-box experiments showed that they would survive even after rainbow and brown trout, chinook salmon, and kokanee had failed. Washington's initial Lahontan cutthroat eggs came from Summit Lake stock. In 1981, age-2 cutthroats were observed trying to enter a tributary creek presumably to spawn, so in 1982 the state began trapping fish to develop its own egg source with a view to expanding the Lahontan cutthroat program to other eastern Washington lakes that had not supported fisheries in the past. This program now encompasses a half-dozen or so lakes that are stocked on a routine basis.

Lake Lenore proved to be bountiful habitat for the introduced cutthroats with fish stocked as three-inch fingerlings in the fall growing to 18 to 25 inches two summers hence. After the state began recognizing Lahontan cutthroat trout in its hook-and-line record program, Lake Lenore fish regularly challenged for that distinction. But a glitch occurred in the summer of 1998 when long-lasting, record heat in eastern Washington produced surface water temperatures at Lenore in excess of 80 degrees Fahrenheit. A fish kill estimated at 75 percent of the population occurred. Lenore was restocked with trout from Omak Lake, but it was a couple of years before the lake again produced consistently. [47]

But getting back to the 1995 Recovery Plan, that plan listed four major strategies for recovery: 1) population management, 2) habitat management, 3) research, and 4) updating and revising the recovery plan. I'll hit the highlights of the first two of these here.

Population Management

Here the Recovery Plan emphasized monitoring and maintaining the 33 viable stream-dwelling populations and four lacustrine populations that existed in the western Lahontan Basin in 1995, and reintroducing

47. The experience with Lahontan cutthroat trout at Omak Lake, Washington has been written up by Kucera et al. (1985). The information about Mann Lake, Oregon and Lake Lenore, Washington came from various news releases and personal communications with biologists from the Oregon and Washington fish and wildlife departments. I have also visited the Oregon and Washington lakes myself many times to fish for their cutthroat trout. Exotic goldfish have recently turned up in Mann Lake, according to a report attributed to the Oregon Department of Fish and Wildlife. It's uncertain at this point just how they will affect the Lahontan cutthroat fishery, but in other locales, goldfish have reproduced so prolifically that they compete for food with game-fish and can crowd them out. They are omnivorous feeders, and can eliminate aquatic plants and greatly increase turbidity. Although not usually thought of as a coldwater fish, they can survive and even thrive in cold waters if the littoral zone is large enough and warm enough for breeding. In captivity, they can live up to 25 years.

additional populations using endemic stocks where they are available or "genetically matched" stocks where they are not. [48] The plan endorsed efforts to remove non-native trout when necessary to make way for these reintroductions. The desire here is to reconnect reaches of stream and tributary habitat into networks populated throughout with Lahontan cutthroats, i.e., to reestablish *metapopulation* structure wherever possible within the major drainages in order to increase the odds of long-term survival of these populations. [49]

However, the plan essentially concedes that existing habitat conditions, land ownership patterns, competing uses for water, and the like, along with the current status of existing Lahontan cutthroat populations, limits opportunities to reestablish and maintain meta-populations within western Lahontan Basin drainages. Therefore, where metapopulation structure cannot be developed, the plan encourages reintroduction and maintenance of new isolated populations, both to hedge against extinction from catastrophic events and to serve as genetic repositories for local stocks.

Habitat Management

Much of what the 1995 Recovery Plan calls for in the way of population management is predicated on having healthy (and connected) stream habitat in which to work. On federal lands, habitat management strategies are supposed to follow the provisions of four federal

48. To this end, a number of population genetics studies have been carried out on western Lahontan Basin cut-throat stocks. For details and results, see Loudenslager and Gall (1980a), Gall and Loudenslager (1981), Bartley et al. (1987), Cowan (1988), Williams and Shiozawa (1989), Bartley and Gall (1989, 1993), Williams et al. (1992, 1998), Nielsen and Sage (2002), and Peacock and Kirchoff (2004). See also Appendix H of the Short-Term Action Plan for Lahontan Cutthroat Trout (*Oncorhynchus clarkii henshawi*) in the Truckee River Basin (Truckee River Basin Recovery Implementation Team, 2003).

49. In 1970, Richard Levins coined the term *metapopulation* to refer to a collection of local populations that are connected to one another by avenues of dispersal. Metapopulation structure has a couple of advantages. First, it has been shown that migration of as few as one breeding individual per population per generation among populations in the network maintains genetic variation and prevents local decreases in fitness due to inbreeding depression. Second, and most importantly for this discussion, the risk of extinction of a connected array of populations is lower than for that same array when they exist as disconnected local popula-

tions. If a local population exists in isolation, as when its habitat has become fragmented, it is at great risk of extinction simply from chance decreases in its spawning success or chance increases in its mortality rate or random variation in sex ratio; and from chance changes in weather, food supply, predation, or diseases (scientists refer to chance changes such as these as *demographic* and *environmental stochasticity*; see Lande 1993). Furthermore, the smaller the local population, the greater its risk. Such a population is also at great risk of extinction from any catastrophic event that might befall it—a flash flood, a debris torrent, or wildfire, for example. Among interconnected populations, local extinctions may still occur, but when they do, vacant habitat may be reoccupied by dispersal from one or more of the other local populations in the network. This notion that regional populations of a species, subspecies, or DPS may persist long-term if the habitats of local populations can be connected up is especially appealing to biologists and managers charged with recovery and management of rare, threatened and endangered salmonids. Gilpin and Soulé (1986), Gilpin (1987), Harrison (1991), Hanski and Gilpin (1996), Hanski (1998), and McCullough (1996) are good places to learn more

about the theory of metapopulation dynamics; and Franklin (1980), Varvio et al. (1986), and Mills and Allendorf (1996) provide insights into metapopulation genetics. Hilderbrand and Kirschner (2000a) and Rieman and Dunham (2000) discuss the application of metapopulation theory to salmon and trout, and Dunham et al. (1997) address fragmented habitat and the extinction risk of Lahontan cutthroat populations.

laws: the National Environmental Policy Act (NEPA), adopted in 1969; the Endangered Species Act of 1973; the National Forest Management Act of 1976; and the Federal Land Policy and Management Act, also adopted in 1976. In addition to these, the Inland Native Fish Strategy (INFISH), which I discussed in Chapter 4, was adopted by the U.S. Forest Service and Bureau of Land Management in 1995. INFISH provides riparian management standards, guidelines, and objectives for sensitive inland fishes.

One habitat suitability index model already exists for cutthroat trout,[50] but it was designed to assess minimum instream flow requirements, not local habitat conditions as defined by the interactions of climate, land type and condition, and hydrologic processes on a stream channel and the way it functions. The federal guidelines laid out in the legislation and policies listed above focus on managing degraded watersheds to achieve a desired future condition that is healthy and properly functioning, and to preclude further degradation. The measure of success here is the extent to which all these standards and guidelines can be met.

As a step toward properly functioning condition, the 1995 Recovery Plan calls for designating streamside management zones that include the stream, riparian and streambank vegetation, and adjacent areas that would be managed to improve water quality, streambank stability, instream habitat condition, fish, and other aquatic resources. In this regard, recent land acquisitions by non-profits, such as The Nature Conservancy and the Trust for Public Land in the Truckee, Carson, and Walker river drainages, have made many formerly privately-held stream miles in these watersheds available for streamside management zones. For example, on the Truckee River downstream from the Reno/Sparks area, The Nature Conservancy and BLM have acquired approximately nine stream miles (the McCarran Ranch and Mustang Ranch parcels) through which the Army Corps of Engineers had channelized the river and disconnected it from its floodplain. The plan here is to reintroduce natural meanders and reconnect the river with its natural, wider floodplain, and to replant the area with cottonwoods and willows to recreate the original floodplain vegetation cover. One ironic twist to this venture, according to news reports, is that the Army Corps of Engineers will contribute $5 million of the $7 million project cost.

Timetable for Recovery

The objective of the 1995 Recovery Plan is to delist Lahontan cutthroat trout in all three of the DPSs defined by the U.S. Fish and Wildlife Service. The plan described a timetable of actions to be carried out through 2018. At the end of this period actions should have been implemented that would:

1. enhance and protect the habitat necessary to sustain as viable the 18 stream and 3 lacustrine populations that were present in

50. Hickman, T. and R.F. Raleigh. *Habitat suitability index models: cutthroat trout* (Washington, D.C.: U.S. Fish and Wildlife Service, Western Energy and Land Use Team, Office of Biological Services, Report FWS/OBS-82/10.5, 1982).

the Truckee, Carson, and Walker river basins in 1995, and the 15 stream and 1 lacustrine population that were present in the Black Rock Desert/Quinn River basin in 1995;

2. enhance and protect the habitat necessary to sustain as viable the 13 populations recognized by the plan outside the historical range (these would be considered refugial sources of Lahontan cutthroat trout); and

3. enhance and protect the habitat necessary to sustain as viable 6 reintroduced stream populations in each of the Truckee, Carson, and Walker river basins, and 12 reintroduced stream populations in the Black Rock Desert/Quinn River basin.

A viable population, per this plan, is one that has been established for five or more years and has three or more age classes of self-sustaining trout.

These goals, if actually achieved, would approximately double the number of viable Lahontan cutthroat populations within the historical range by 2018. Still, this seems a modest accomplishment when compared with the historical abundance and wide distribution of Lahontan cutthroat trout in western and northwestern Lahontan Basin drainages.

The prospects for even greater strides in recovery got a big boost following a 100-year flood event that occurred in the region in 1997. This galvanized local and regional stakeholders to collaborate to modify and improve watershed management in the Truckee system. At the federal level, the many water rights conflicts between California and Nevada, between rural and urban users, and between the Pyramid Lake Paiute

Tribe and its many adversaries were addressed by Congress in the Truckee-Carson-Pyramid Lake Water Rights Settlement Act, signed into law in 2000. A new Truckee River Operating Agreement was proposed to cover storage and allocation of water from the four federal reservoirs in the upper Truckee system, including spring and summer flows for the spawning and rearing needs of cui-ui and cutthroat trout. At the regional and local levels, the Truckee River Habitat Restoration Group was founded to promote watershed restoration in the upper Truckee between Lake Tahoe and the Nevada line. In the lower river, the Pyramid Lake Paiute Tribe, The Nature Conservancy, and scientists from state and federal agencies formed the Lower Truckee River Restoration Steering Committee for the same purpose. Water quality issues in the lower Truckee River are also being addressed. Perhaps the single most polluted tributary in the entire Truckee watershed is Steamboat Creek, which drains approximately 200 square miles of urban developments, small ranches, and farms immediately south of Reno and Sparks. Although there is still plenty of heartburn as contentious stakeholders work through the details of these collaborations, any and all successes that are achieved should help Lahontan cutthroat trout attempting to spawn and rear in the river. [51]

In that regard, in 1998, with much fanfare capped by a visit by then-Secretary of the Interior Bruce Bab-

51. John Cobourn of the University of Nevada Cooperative Extension wrote up a good summary of recent restoration and watershed management collaborations in the Truckee River basin; see Cobourn (1999).

bitt, Trout Unlimited and the Pyramid Lake Paiute Tribe began an egg-incubator program on tribal land along the lower river beginning with two incubators, capacity 90,000 eggs each, with the eggs supplied by one of the tribal hatcheries. In 2002, the State of Nevada joined with the Tribe to continue stocking the river. Some, if not all, of the fish being released by the state are 24 inches in length and should be capable of spawning in the lower Truckee if indeed spawning there is yet possible. The last information I had on this effort was in May, 2003, so stay tuned.

Meanwhile, perhaps the biggest breakthrough of all is occurring at Derby Dam. There, in 2002, a 930-foot bypass channel was constructed that should allow fish to pass to and from the river upstream of the dam for the first time in almost a century. The diversion canal that carries Truckee River water to the lower Carson basin is also being screened to ensure that Truckee trout stay in the Truckee.

Restoring connectivity of the Truckee River will be a key achievement of the Short-Term Action Plan written for that watershed in 2003, and it would appear from the discussion above that the elements needed to achieve that are falling into place. In contrast, a Short-Term Action Plan for recovery of Lahontan cutthroat trout in the Walker River Basin, also written in 2003, has started off much more tentatively. That's probably because a series of public meetings held in

Walker Basin communities prior to writing the plan revealed many apprehensions and outright opposition to Lahontan cutthroat restoration. Concerns about potentially negative effects on individual livelihoods and community economies ran high. So, the Walker River plan proposes to essentially ease its way into the recovery process. While acknowledging that the presence of fish passage barriers is a significant recovery issue for the basin and promising to address these over time, the plan also states that the principal short-term activities on the ground will continue to be the stocking of Lahontan cutthroats into selected headwater reaches and managing some tributaries strictly for Lahontan cutthroat trout. Meanwhile, for the first five-year period of the plan, which may be close to expiration by the time you read this, the following dozen tasks received highest-priority for accomplishment:

1. Draw together all existing information on fish management, water management, and habitat conditions in the basin.

2. Develop an education and outreach program.

3. Identify native and non-native salmonid populations in the basin that are maintained by natural production.

4. Identify the role to be played by hatcheries in the recovery program for Walker Basin Lahontan cutthroat trout.

5. Complete the genetic work on Lahontan cutthroat stocks in the Walker Basin and decide which should be used in recovery efforts.

6. Conduct a watershed analysis of the physical components of the Walker River Basin.

Pilot Peak on the Utah-Nevada border, once a landmark for California-bound immigrants. A transplant of trout into a small stream on its northeastern slope may have preserved the original Pyramid Lake strain of Lahontan cutthroat trout. PHOTO BY AUTHOR

7. Develop and carry out hydrological studies of the basin.

8. Identify where Lahontan cutthroat trout existed in the watershed in the past and what species assemblages exist there now.

9. Develop, implement, and monitor a management plan for wild Lahontan cutthroat trout that will not impact donor or newly established populations.

10. Develop native and non-native fish-distribution overlays for the basin's geographic information system (GIS)-based data system.

11. Complete a management plan for the U.S. Forest Service's Rosa-schi Ranch segment of the East Fork Walker River.

12. Evaluate the potential of Lahontan cutthroat recovery in the basin as a recreational fishing opportunity.

Mighty powerful steps, those.

CHAPTER 6

Humboldt Cutthroat Trout

Unnamed. Chromosomes, $2N=64$. Scales in the lateral series 125–150 (fewer than in *O. c. henshawi*, as described in Chapter 5); scales above the lateral line 26–40 (fewer than in *henshawi*); and most significantly, fewer gill rakers (19–23, mean of 21 in the Humboldt cutthroat vs. 21–28, mean of 24 for *henshawi*). Pyloric caeca 50–60 (same as *henshawi*, except that trout native to Hanks Creek, a Mary's River tributary, average 66 pyloric caeca, the most of any form of cutthroat or rainbow trout). Pyloric caeca 50–60 (same as *henshawi*, except that trout native to Hanks Creek, a Mary's River tributary, average 66 pyloric caeca, the most of any form of cutthroat or rainbow trout). With respect to lateral series scale counts, specimens from Ruby Mountain streams consistently have higher counts than specimens from other parts of the Humboldt drainage, and the trout of Gance Creek, a tributary of the North Fork Humboldt, consistently have the lowest counts. Like *henshawi*, the colors are generally dull, typically brassy, coppery, or burnished silver with some tendency toward yellow. Sometimes pink tints appear on the sides. The ventral region is white to gray and the lower fins are brownish with sometimes pinkish tints. Spotting varies, but the spots on the Humboldt cutthroat are typically fewer than on *henshawi*, and tend to be concentrated more on the posterior part of the body. Only rarely are spots found on the abdomen of the Humboldt cutthroat, whereas they are often found on the abdomen of *henshawi*.

Humboldt Cutthroat Trout, *Oncorhynchus clarkii* subspecies

BASIN AND RANGE COUNTRY— THE HUMBOLDT RIVER DRAINAGE

The first white men to see the Humboldt River country were the Hudson's Bay Company fur trappers, led by Peter Skene Ogden, who came this way during the period between 1825 and 1829 (John Work led a trapping brigade through the area again in 1831). Ogden first named the river the Unknown River because he did not know where it started or where it ended. Later he called it Paul's River, then Swampy River, and finally Mary's River. The Bonneville expedition of 1833–34 called it the Barren River for the lack of trees along its banks. It was John C. Fremont who gave it the name that finally stuck: the Humboldt River, after the Prussian explorer, Baron Alexander von Humboldt. Although Humboldt never laid eyes on his namesake river, he secured a place in Fremont's esteem by preparing one of the earliest maps of the Rocky Mountain region from data gathered by the Escalante expedition of 1776.

Fremont painted a rather glowing picture of the Humboldt River country. He described the river as:

> "...arising in two streams in the mountains west of the Great Salt Lake, which unite after some fifty miles and bears westwardly along the northern side of the basin. The mountains in which it rises are rounded and handsome in their outline, capped with snow the greater part of the year, well clothed in grass and wood, and abundance of water. The stream is a narrow line, without affluents, losing by absorption and evaporation as it goes, and terminating in a marshy lake, with low shores, fringed with bulrushes, and whitened with saline incrustations. It has a moderate current, is from two to six feet in depth in the dry season, and probably not fordable anywhere below the junction of the forks during the time of snow melt, when both lake and river are considerably enlarged."

He referred to the Humboldt valley itself as "a rich alluvian beautifully covered with blue grass, herd grass, clover, and other nutritious grasses." [1]

Immigrants who later followed the Humboldt on their way to California saw it differently. Being pretty gritty and travel-worn already by the time they entered the Humboldt basin, they found the country to be dry, dusty, and inhospitable during dry years and something of a quagmire and equally inhospitable during wet ones. They labeled the river "the Humbug" and derided Fremont and the others who had come before as "scribbling asses describing nutritious grasses." [2]

"Sickly rivulets," Mark Twain disdainfully called the rivers of the Great Basin in *Roughing It*, an account of his journey to Nevada in 1861. As for the Humboldt, he wrote:

> "One of the pleasantest and most invigorating exercises one can contrive of is to run and jump across the Humboldt River until he is overheated and then drink it dry."

The truth about the Humboldt River country lay somewhere in between the glowing accounts of Fremont and Mark Twain's derisive report. What the

1. Fremont's exploration down the Humboldt occurred during his third expedition of 1845–46, chronicled in Spence and Jackson (1973), which took to the field the year following his coming upon Pyramid Lake, its Paiute people, and its large "salmon trout." His description and remarks about the Humboldt are also quoted in LaRivers (1962).

2. As an illustration of the immigrants' sometimes grim humor regarding what they encountered along the Humboldt, one of them, a John Grantham, wrote this poem (cited in Jackson 1980; the original is in the Palmer C. Tiffany manuscript at Beinecke Library, Yale University):

> "From all the books that we have read
> And all that travelers had said
> We most implicitly believed
> Nor dreamed that we should be deceived,
> That when the mountains we should pass
> We'd find on Humboldt fine blue grass;
> Nay, that's not all, we learned moreover
> That we'd get in the midst of clover.
> Nay, more yet, these scribbling asses
> Told of "other nutritious grasses;"
> But great indeed was our surprise
> To find it all a pack of lies...."

immigrants actually found was a relatively easy arterial across the Great Basin—easy, that is, once they learned how to avoid the Salt Lake desert. The Salt Lake desert guards the eastern approach to the Humboldt basin. The present Interstate freeway bores straight across it, and this is the route some of the earliest immigrant parties tried to follow. The modern traveler has a far easier time!

The Bartleson-Bidwell party was the first to attempt the crossing in the year 1841. They succeeded in reaching California, but not without paying a terrific price. Dry as the climate is in the Great Basin, the Salt Lake desert can become a horrible quagmire when it does rain. The party lost time and wagons. It wasn't until September that they emerged from the desert, and then they had to swing far south to avoid the Ruby Mountains. They crossed the Rubies at Harrison Pass, then followed the South Fork Humboldt back to the main river. From there the route was broad and fairly flat for a time, but then, beyond the Humboldt Sink, they came to the Forty-Mile Desert. The Forty-Mile Desert could be hell on travelers too, and still there remained the mountain passes of the Sierra Nevada to be crossed. It was dangerously late in the year when the bone-weary party finally made it through.

Lansford Hastings tried the route again in 1846. He, too, made it to California, but suffered the same delays and hardships as his predecessors. And when the Donner-Reed party followed a bit later that same year, they were trapped by heavy snows in the Sierras with tragic results.

After that, the trail was rerouted. Subsequent wagon trains went north from Fort Bridger, following the Oregon Trail to Fort Hall on the Snake River. From there they angled southwesterly along Goose Creek, entering Nevada in the northeast corner, then followed Tabor Creek down to the Humboldt. A favored camping spot was at Humboldt Wells, where the town of Wells, Nevada, is located today.

The upper Humboldt drainage is typical high desert, basin and range country. Sagebrush dominates the valley floors, which lie at an elevation of 5,000 to 6,000 feet. Juniper and pinion pine dot the approaches to the ranges and mountain mahogany grows on the ridges. Sixty-five percent of the basin gets less than ten inches of moisture per year, and most of that falls as snow in the winter and early spring. The rest comes in the form of violent thunderstorms that often vent their fury in severe flash floods. But Fremont was not entirely wrong in his description of lush grasses. The valleys through which the Humboldt and its tributaries flow did support palatable grasses amongst the shrubs in Fremont's day, and wild rye grew in the alkali bottoms in almost pure stands. A traveler entering the basin in the 1840s

Other historical references to the Humboldt River and to the immigrants' attitudes toward it are from Morgan (1943), Stewart (1962), Cline (1963), Jackson (1975, 1980), and Curran (1982).

would have found much more grass and much less sagebrush than one encounters today. But what water there was to support the wild grasses has long since been diverted into cultivated fields, and the sagebrush has filled in the gap.

Across the sage-studded distance lie the mountains, north to south oriented ranges with names such as East Humboldt, Ruby, Snake, Jarbidge, Independence, Toiyabe, and Tuscarora. They appear to be low, rounded, and hazy-blue off there in the distance. But up close, they rise abruptly from alluvial slopes that grade gently out onto the basin floor. Some of the peaks reach heights of 11,000 to 12,000 feet. Their flanks are timbered, and some wear mantles of snow for most if not all of the year. Up in these mountains you find life zones more in keeping with the Rockies or Sierra Nevada than with the basin floor.

The ranges receive more precipitation than the valleys, and this water feeds into small headwater streams via groundwater seeps and springs. Many of these upper stream reaches flow perennially and cool, even in the hottest part of the summer, as they tumble down out of the ranges in small, swift, high-gradient channels bedded with gravel, cobble, and boulders. Yet when they exit their canyons and flow out onto the valley floor, they often become intermittent or peter

out altogether, simply disappearing in the alluvial gravel. There is water enough, excepting possibly in extreme drought years, to fill the lower channels and form continuous networks during the spring runoff, which usually commences in March, but for seven months of the year, from August until the following March, there is often no visible flow at all between isolated pools.

Meager as these waters are now, the streams of the upper Humboldt drainage were clear and trout-filled when the trappers, immigrants, and settlers first saw them. Peter Skene Ogden's men were the first to sample the trout; they called them "salmon fish." Diaries kept by members of the Bartleson-Bidwell party spoke about eating trout they had caught as they moved along the Humboldt. Early newspaper accounts testified to the excellent fishing and the quality of the habitat, prompting some to call the Humboldt drainage "a delightful region, represented as the paradise of Nevada" (shades of John C. Fremont!). [3]

HISTORICAL RANGE

I outlined the full extent of the 45,000-square mile Lahontan Basin in Fig. 5-1. The Humboldt River drainage, comprising almost the entire eastern half of the basin, is designated "B" in that figure, and is remapped

3. The quote used here was cited by Patrick Coffin in an early report he prepared on the distribution and life history of the Humboldt cutthroat trout (see Coffin 1981).

Humboldt River between Elko and Carlin, Nevada. PHOTO BY AUTHOR

in greater detail in Fig. 6-1. The U.S. Fish and Wildlife Service, Nevada Division of Wildlife, and other participants in the recovery program for threatened Lahontan cutthroat trout now designate the native trout of this drainage as the Humboldt River Basin DPS of *Oncorhynchus clarkii henshawi.* Dr. Behnke, on the other hand, has always regarded it as distinct at the subspecies level from the cutthroat trout of the western half of the Lahontan Basin (excepting those of the Quinn River drainage, which are similar in characters to the Humboldt form), although he has never offered a subspecies name to go along with his formal description.

Dr. Behnke has written that he first became aware of the distinction between Humboldt cutthroats and other Lahontan basin populations when he examined the museum specimens collected by J. Otterbein Snyder back in 1915. Snyder himself had not examined any of the trout he collected from the Humboldt drainage; thus it was left to Dr. Behnke to point out that there were consistently lower gill raker counts in the Humboldt drainage trout, as well as lower scale counts. Additional specimens collected from small streams in the Humboldt drainage, first in the early 1960s then again in 1972, also showed this distinction. To add more weight to the argument, Terry Hickman, one of Dr. Behnke's graduate students, set up a discriminate function computer analysis of sixteen meristic characters

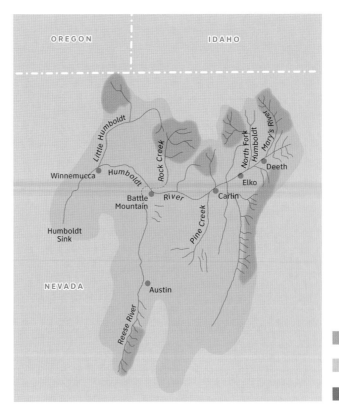

FIGURE 6-1. *Humboldt River drainage*

Basin and Range country. Humboldt River drainage, Nevada. PHOTO BY AUTHOR

▇ Areas Where Humboldt Cutthroat Populations Persist

▇ Approximate Boundary of Historical Range

▇ Existing Streams, Lakes, and Reservoirs

that separated Humboldt drainage trout from those of the Walker, Carson, Truckee/Pyramid, and Summit Lake drainages with 100 percent accuracy. Allozyme electrophoresis studies of Nevada cutthroat trout populations do not show this sharp distinction; so, based on those studies, no subspecies delineation would be warranted. However, more recent genetic studies using mtDNA and microsatellite methods appear to support Dr. Behnke's view. [4]

Even though they differ in characters, the evolution of the Humboldt drainage cutthroat is linked to the history of ancient Lake Lahontan and its fishery, a subject I dealt with in Chapters 2 and 5. The Lahontan basin may have been invaded by a $2N=64$ ancestral cutthroat very early on in the lake's history. According to Dr. Marith Reheis, U.S. Geological Survey, and her colleagues, the highest-ever stand of an ancient Lake Lahontan occurred about 650,000–700,000 years ago, possibly with overflow into the Alvord Basin and adjacent Coyote Basin, and from there perhaps discharging into the Snake River. [5] Noting that Humboldt drainage cutthroats consistently have 21 gill rakers while all populations on the west side of the Lahontan Basin have 24 (except those in the Quinn River drainage, which also have 21), Dr. Behnke suggested that sometime after ancestral trout got into the basin, a lowering of the ancient lake isolated the cutthroats that had spread into the ancient Humboldt drainage, kicking off the process of genetic differentiation at a stage where the trout had an average of 21 gill rakers. Contact between the isolated populations would have undoubtedly

been reestablished during later high stands of Lake Lahontan, but by then, selective pressures to maintain a river-specialized life history form with 21 gill rakers in the Humboldt drainage would have been too great to overcome. Meanwhile, cutthroats remaining in the Lake Lahontan system evolved into the 24-gill raker form, a lake adaptation that remains a characteristic of the native trout of the west side of the basin.

As for the Quinn River populations, I mentioned in Chapter 5 that these populations differ significantly from Lahontan cutthroat trout of the Walker, Carson, Truckee/Pyramid, and Summit Lake drainages but are essentially similar to the cutthroat trout of the Humboldt drainage in meristic character counts. Evidence now available indicates that from mid-Pleistocene time, 600,000 to 700,000 years ago, to the end of the Pleistocene Epoch, about 10,000 years ago, the Humboldt River flowed northward from a point between present-day Winnemucca and Rye Patch Reservoir rather than follow its present course southwestward to the Humboldt Sink. At those times, the Quinn River was a tributary of this northern meander of the ancient Humboldt, which then emptied into the Black Rock Desert basin or the Black Rock arm of Lake Lahontan if the ancient lake was at a high enough stand to fill that arm. Dr. Behnke has expressed the belief that Quinn River and Humboldt

4. To review the literature on this matter, see Behnke (1979 [1981], 1992), Loudenslager and Gall (1980), Bartley et al. (1987), Bartley and Gall (1993), Williams et al. (1992, 1998), Nielsen and Sage (2002), Peacock et al. (2003), and Peacock and Kirchoff (2004).

5. See Reheis and Morrison (1997), Reheis (1999), and Reheis et al. (2002).

cutthroat trout share a common origin, and this connection of ancient drainages may explain how that came about. [6]

Eric Gerstung's historical reconstruction, published in 1988, indicated that cutthroat trout occupied approximately 2,175 miles of stream habitat within the Humboldt River drainage when the trappers, explorers, and immigrants first arrived on the scene. Trout occupied all sections of the Humboldt River itself, from approximately Battle Mountain upstream, and all of the Reese River, the Humboldt's longest tributary, upstream from Reese River Canyon. It is probable that the only waters that were *not* cutthroat habitat were the lower Humboldt below Battle Mountain; the lower mainstem of Rock Creek, a Humboldt tributary that enters from the north near Battle Mountain; and the lower reaches of the Reese and Little Humboldt rivers. No lake populations of cutthroat trout occurred historically in the Humboldt drainage. Even though pluvial lakes existed there in Pleistocene time, they had long since vanished by historical time, and the small lakes that are present today in the high ranges were all originally fishless, probably owing to alpine glaciation during the late-Pleistocene. [7]

The main Humboldt River in the Elko area became widely known for its trout fishing, as did many of its tributaries such as Bishop Creek, Mary's River, the North Fork Humboldt, and Pine Creek. The July 4th, 1869 issue of the *Elko Independent* carried an item noting that:

> "At Carlin,...boys fish in Maggie, Susie, and Mary's River...and come in with long strings of fine delicious trout, which are sold so cheaply in the streets that hotel keepers can't compete. Trout, fresh from the limpid waters of the Humboldt, weighing 1–3 pounds are plentiful in the market at fifty cents each. Not unusual for 5–6 pound trout to be caught; one recent one weighed eight pounds; these command higher prices. The fish are caught with hook and line, using grasshoppers for bait."

Early accounts of Pine Valley and Pine Creek describe a long, grassy valley with a clear, silvery stream of water that was known for its splendid trout fishing. Likewise, the North Fork Humboldt and its tributaries were described as clear, trout-filled streams surrounded by range lands clothed with luxuriant grasses. [8]

In the Reese River near present-day Austin, Nevada, trout of 2 and 3 pounds were common. The Indians' name for this stream was "Pang-que-o-whop-pe," meaning "Fish Creek" in the Shoshonian language, according to Capt. James Simpson of the U.S. Army Corps of Topographical Engineers, who renamed the stream for his guide, a Mr. Reese, on May 28, 1859. Capt. Simpson wrote:

> "Reese River is 10 feet wide, 1½ deep; current moderate; water good, though slightly milky color from sediment, runs northwardly, and is the largest stream we have seen this side of the Jordan [in Utah]. Trout weighing 2½ pounds are found in it."

6. Evidence for the northward loop of the Humboldt River and its connection with the Quinn River during Pleistocene time is presented in Davis (1982, 1990), Benson and Peterman (1995), and Adams et al. (1999), and is also discussed in Reheis and Morrison (1997). Dr. Behnke's views on the relatedness of Humboldt and Quinn river cutthroat trout are expressed in Behnke (1992) and Behnke (2002).

7. The papers by Reheis and colleagues cited in footnote 5 also list the pluvial lakes that existed in the Humboldt drainage during Pleistocene time. See also Morrison (1965) and Hubbs et al. (1974). Records of late-Pleistocene alpine glaciers that existed in the high ranges are found in Blackwelder (1934), Sharp (1938), Morrison (1965), Wayne (1984), Thompson (1992), and Osborn and Bevis (2001)..

8. These historical accounts are cited in Coffin (1981).

In 1899, an Austin boy caught a 12-pound trout from the Reese River that was over 30 inches long. [9]

The main Humboldt was still cutthroat habitat at least as far downstream as the town of Palisade (now a ghost town), about 10 miles downstream from Carlin, as recently as 1915 when ichthyologist J. Otterbein Snyder completed his four-year collection of native fishes of the Lahontan Basin. Snyder collected native trout from Starr Creek and Mary's River near Deeth, from Pine Creek near Palisade, from the Humboldt River near Palisade, and from the Humboldt near Carlin. Today, these sites are barren of trout and the once-abundant habitat is highly degraded. The lush wild range grasses have been supplanted by hay fields irrigated by the waters that once fed the tributaries and the main river. The native cutthroats of the Humboldt drainage have largely retreated to the remote, often intermittent headwaters. By 1995, it was estimated that only about 14 percent of the original historic range remained occupied by native Humboldt cutthroat trout.

LIFE HISTORY AND ECOLOGY

Unlike the cutthroat trout of the western portion of the Lahontan Basin that evolved a lake-adapted life history, the Humboldt cutthroat became a river-specialized or small stream-specialized form, and these are the two life histories that have predominated in the drainage. [10]

No lake populations occurred in the drainage historically; however, the Humboldt cutthroat has shown in recent years that it does possess an ability to adapt to harsh lacustrine environments. In the early 1970s, Humboldt cutthroats were found in a warm, turbid, eutrophic irrigation reservoir near Jiggs, south of Elko, which they evidently entered in high runoff years. And even earlier, Humboldt cutthroats took up residence in Willow Creek Reservoir between Tuscarora and Midas, northwest of Elko, and made annual spawning runs into upper Willow Creek. Thus, individuals and populations of Humboldt cutthroat trout may display one of three basic life history strategies, although the first two are the predominate ones and the latter one must be considered provisional:

- The fluvial, or riverine (potamodromous), life history, wherein the fish feed and grow in main river reaches, then migrate to small tributaries of these rivers where spawning and early rearing occur.
- A stream-resident life history, wherein the trout complete all phases of their life cycle within the streams of their birth, often in a short stream reach and often in the headwaters of the drainage.
- A possible reservoir-adapted (lacustrine) life history.

The Fluvial, or Riverine, Life History

Truly riverine populations, those that feed and grow in the mainstems of the larger streams, then migrate to small tributaries to spawn, may be few and far between in the Humboldt drainage today, even though these

9. Capt. James Simpson's U.S. Army Corps of Topographical Engineers survey for a wagon road from Camp Floyd, Utah to Genoa, California made special note of the trout of the Reese River, and contains the cited information about the local Indians' knowledge of this resource; see Simpson (1876 [1983]). The guide for whom Simpson named the river, identified only as "Mr. Reese" in the report, was most probably John Reese of Salt Lake City. Reese owned and operated the trading post at Mormon Station, later called Genoa, from 1850 to 1856. The report of the 12-pound trout from Reese River near Austin is cited in Coffin (1981).

10. Most of the life history information specific to fluvial and adfluvial Humboldt drainage cutthroat trout was compiled and reported by Coffin (1981, 1983). Gerstung (1988) also has a summary, but most of that came from Coffin's reports. Other published studies of Lahontan Basin cutthroat trout that contain life history information derived from Humboldt drainage populations include Platts and Nelson (1983), Nelson et al. (1987, 1992), Dunham (1996), and Dunham et al. (1999, 2000, 2003). Two consultant's reports, prepared in 1997 and 1998 by AATA International of Fort Collins,

were the trout that generated all the early-day public-ity. However, a few riverine fish can still be found in the Mary's River, a principal Humboldt tributary that drains the Jarbidge Range north of Deeth and Wells.

On the bright, shirtsleeve-weather day in late August 1982 that Patrick Coffin took me up to the Jar-bidge Wilderness country to show me Mary's River, the water was extremely low. We had stopped at one small tributary where only a trickle of water flowed between tiny pools—no, puddles would be a more apt descrip-tion. Yet everywhere there was cover, the least little bit of cut-bank or the smallest tuft of grass, we observed scores of young-of-the-year trout. Adult cutthroats had come up this little tributary to spawn back in May or June, during the spring runoff when the stream was carrying maybe 100 times more water than it was on this day. The spawners dropped back down before the runoff receded. The little trout we were seeing would survive here and overwinter, then migrate downstream themselves on the next spring runoff.

The high desert country appeared to be barren at first, but there was, and is, plenty of life here. Elko County is a high-use area during hunting season, accounting for 50 percent of the state's deer kill and 30 percent of the sage grouse, not to mention ante-lope. At one point on our tour, we came around a bend to confront a band of eight antelope, including one

good buck. As one, they wheeled away, loping grace-fully down a draw, and disappeared from view. Further along, we spotted a mule deer standing on a side hill. It looked to me to be about twice the size of one of our Pacific coastal blacktails, but Patrick dismissed it with a shrug. "A baby," was his only comment.

At Mary's River, we found the cutthroats lying in the shallow, shaded pools. Patrick had told me that the Mary's River produces some big cutthroats, and there were three or four good-size trout lying together in the darkest part of one pool that was deep in the shade of an aspen grove. We had startled them when we approached. We watched their wakes as they scur-ried up and down a few times, but they never did leave the shaded area, even though they could have found shelter further away had they only been willing to pass through the open, sunlit water.

Nor did our disturbance keep them from rising to my fly. I figured that once we had spooked them, there would be no chance. The cast I made into the shaded run was strictly to work out the kinks after a long, hot, bone-rattling ride across the desert. The serious fish-ing would begin at a spot further along, preceded by a stealthier approach.

Thus it was that the strike of a trout at my size 16 Adams caught me completely off guard. Needless to say, I missed it. Quickly as I could, I got the line straight-

Colorado, and provided to me by Louis A. Schack, Newmont Mining Corporation, Carlin, Nevada, also contained good information on the summer and winter ecology of adfluvial populations in tributaries of Maggie Creek. However, because of their proprietary nature and limited availability, I have not included them in the bibliography. Their titles are "Lahontan Cutthroat Trout Baseline Summary Report" (1997) and "Lahontan Cutthroat Trout Winter Habitat Survey" (1998).

ened out and the fly back out on the water. It bounced along into the deep shade—and this time the strike was true! The water erupted in a shower of spray as the surprisingly powerful trout surged back and forth. This time, thoroughly alarmed, the other trout abandoned the pool, flashing and thrashing through the skimpy, sunlit water to other hiding places further upstream.

My first Humboldt cutthroat, and what a battle it put up against my wispy fly-weight rod. But finally the fish wore down and I tailed it onto the gravel at the stream edge to be measured and photographed before release—a fat male, 15 inches in fork length, the colors of its speckled sides a deep, dull reddish bronze.

The similarities in life history and ecology between fluvial populations of the Humboldt River drainage and those of other subspecies of cutthroat trout are undoubtedly many. They respond to the same environmental cues to initiate their spawning migrations, moving to natal tributaries in the spring, usually May or June according to Patrick Coffin, on rising water temperatures and increasing stream flows. Based on the historical record, trout could grow to 2 to 3 pounds or more in weight in Humboldt drainage mainstem reaches. Such fish in good condition would have been 17 to 19 inches or more in fork length, and their life span would probably have been somewhere in the order of 6 to 8 years. My guess is that females in these populations probably spawned for the first time at age 3 or 4 at fork lengths of 10 to 12 inches or so, and those females, along with larger, older, repeat-spawners, probably carried from 600 to upward of 2,500 eggs per female to the spawning tributaries.

However, recent information indicates that the size, and in all probability fecundity and maximum age as well, has decreased considerably in the fluvial populations that remain. As populations have retreated to remote, possibly less productive parts of their range, size has decreased to the point where very few trout now attain or exceed 12 inches in fork length. Indeed, up to 90 percent of the fish in some recent surveys have been 6 inches in fork length or less. As size has declined, so too, evidently, has age at first spawning, which now appears to be age 2 among most males and age 3 for females. These first-time spawners may be only 5 to 6 or 7 inches in fork length and may produce only 100 to 300 eggs per female.

One distinctive feature of fluvial Humboldt cutthroat trout is their ability to utilize the smallest, most marginal of tributaries as spawning and rearing habitat—mere trickles of stream where often in summer there is no surface flow at all between tiny, isolated pools. Patrick Coffin calls these waters "streamlets." Yet they appear to be the important spawning and early-rearing habitats for fluvial populations. They

This view looks out over the Mary's River valley from the slopes of the Jarbidge Range. Cool, perennial, high-gradient spring- and seep-fed tributaries head up here and flow down steep canyons to the basin floor. PHOTO BY AUTHOR

have adequate water for spawning and incubation during the spring runoff, and the mature cutthroats have learned to time their spawning runs to coincide with this period. When spawning is complete, the adult fish drop back downstream to mainstem reaches where there is suitable year-round habitat. If they didn't, they would be stranded.

In July, stream flows begin to decrease rapidly. By August and early September, pools in the spawning streamlets would not float even the smallest adult cutthroat. But the young-of-the-year trout remain in these streamlets, hovering in small schools in pools that frequently have no other cover than the small rubble on the bottom, into which they scurry when danger threatens. Sub-surface flow must occur through the gravel, otherwise the pools would stagnate. But with no cover, water temperatures sometimes reach or exceed 80 degrees Fahrenheit (26.7 degrees Celsius). [11]

Come the following spring, when the runoff begins again and the adult cutthroats are ascending to spawn once more, the now-yearling survivors from the previous year's spawning move downstream to areas where they will feed, grow, and mature. Patrick Coffin described the typical Humboldt-drainage mainstem habitat where this feeding and growth to maturity now occurs as ranging from 3 to 23 feet in width but only 1 ½ inches to 8 inches in average depth. He cited a typical pool-to-riffle ratio of 58 percent, which is a very decent number in any trout stream, but the pools are small, shallow, and poorly shaded. The trout occupying these mainstem reaches become opportunistic feeders and, like other fluvial populations, probably make trophic migrations for feeding and shelter. The streams of the Humboldt drainage support an array of aquatic insects, including caddisflies and mayflies (but few, if any, stoneflies, according to some recent studies), and dipteran insects, such as chironomids and blackflies, and the cutthroats feed on the nymphs and adults of these insects as they become available. To a lesser extent, such terrestrial insects as ants, beetles, and grasshoppers are also utilized.

In the western part of the Lahontan Basin, when cutthroats reach a length of 12 inches or thereabout, they convert to a diet of fish if forage fishes are available in the food chain. In the Humboldt drainage, speckled dace are sometimes found in perennial upper tributaries, and suckers, dace, and redside shiners enter the mix as one moves further downstream. Sculpins are also sometimes found in the middle reaches of Humboldt-drainage streams. In historical times, when the range of fluvial Humboldt cutthroat populations extended much further downstream, forage fishes may indeed have been part of the diet and could have accounted for the remarkable sizes

11. Water that flows underground beneath the stream bed and then re-emerges in a small pool can provide an effective cool-water refuge for trout. Bilby (1984) found that water flowing beneath the bed emerged 7 to 8 degrees Fahrenheit cooler than water flowing down the channel above-ground in a small, warm stream in the Pacific Northwest.

of trout reported in the historical record. But recent studies have found few forage fishes in the stomachs of Humboldt cutthroat trout, so it isn't surprising that present-day maximum sizes rarely match those reported in the early days.

The Adfluvial, or Stream-Resident, Life History

In terms of numbers of extant populations, the principal remaining life history form of Humboldt cutthroat trout may be that which inhabits the small, unstable, flood-and-drought cycle streams that head up in the high ranges. Genotypes have evolved that can persist in the harsh, changeable, small-stream environments that characterize these headwaters. [12]

It is well that this has occurred, because the native cutthroat has long since been displaced from all of the better-quality, more stable trout streams by brook trout, rainbows, and brown trout. For example, the streams along the west sides of the East Humboldt Range and the breathtakingly beautiful Ruby Mountains to the south are some of the highest-quality trout waters in all of the Humboldt drainage. Non-native species took hold very rapidly once they were planted, and displaced the native cutthroat from more than 95 percent of these streams. Only in the most marginal streams do the native cutthroats still persist in the East Humboldt and Ruby ranges.

On the other hand, in the headwater streams of the Rock Creek, Maggie Creek, Mary's River, and North Fork Humboldt sub-basins, where stream flows are often smaller and less stable and streambeds may be more exposed to the sunlight, here it is the native cutthroat trout that holds sway. Dr. Behnke has reported taking Humboldt cutthroat specimens in late summer from small streams with no flowing water, only intermittent pools, but with flood debris from the spring runoff littering the banks some five to six feet above the streambed, indicative of the massive volumes of water flowing down these small channels at that time of year. He has also reported Humboldt cutthroats thriving in small pockets where the water temperature was 78 degrees Fahrenheit (25.6 degrees Celsius). In laboratory studies of the thermal tolerance of Lahontan cutthroat trout, the fish show signs of stress (decreased appetite and growth and increased mortality) when water temperature exceeds 71.6 degrees Fahrenheit (22 degrees Celsius) for even short periods. [13] Rainbow trout were stocked repeatedly in some of these streams in years past, but they always disappeared quickly between stocking intervals, leaving only the native cutthroats as pure populations showing little or no evidence of introgression.

The environment in which these populations exist is highly changeable. Years of extended drought will

12. The tolerance of populations to the patterns of disturbance their habitats experience has become a major field of study in ecological research. Perhaps the defining paper on this subject as it applies to stream systems is N. LeRoy Poff and J.V. Ward's "Physical Habitat Template of Lotic Systems: Recovery in the Context of Historical Pattern of Spatiotemporal Heterogeneity," published in the journal *Environmental Management* in 1990 (see Poff and Ward 1990). Don't be put off by the title. What it says, basically, is that the pattern of natural disturbance experienced by the stream system (drought cycles, flood frequency, recurrence intervals for natural wildfires, and the like) will have selected for a corresponding set of traits in the native population (in our case, Humboldt cutthroat trout) that enables it to endure under that particular pattern. Since I live and work in the forested region of the Pacific Northwest, the research I am most familiar with focuses on wildfire and debris torrents. On one of my trips to Nevada not long ago, I arrived just after a large range fire had burned in the Mahogany Creek watershed. I asked Roger Bryan, a supervising biologist at the BLM Winnemucca field office, about the recurrence interval for wildfire and flash floods in Lahontan Basin drainages. He

be broken by good water years and even flooding, as happened in 1997 when many Lahontan Basin stream systems experienced 100-year flood events, then more years of drought may set in (drought conditions have prevailed throughout much of the west since 1999, and as I write this, 2004 is shaping up to be yet another dry year). In addition, when high water years do occur, the spring runoff can sometimes come in a torrent, often with more than one peak. These capricious shifts, not only in overall climate but also in runoff volume in wet years, are accompanied by equally dramatic fluctuations in weights, lengths, condition, and numbers in the stream populations of native Humboldt cutthroat trout.

But here may be one of the unique genotypic adaptations exhibited by these trout. Although their numbers and physical condition decline sharply when adverse conditions prevail, as in periods of drought, they have demonstrated an ability to rebound quickly in good water years even after flood flows. William Platts and Roger Nelson of the U.S. Forest Service observed this in the population of Gance Creek, a small North Fork Humboldt tributary, over a string of three good water years. In their test section of Gance Creek, the population increased by three-fold in the first good water year, then increased by three-fold again in the second year before declining (but

only slightly) in the third year. In Chimney Creek, a Mary's River tributary, Nelson and Platts along with Osborne Casey, a colleague with the Bureau of Land Management, observed a similarly large and rapid population rebound over four successive years in which spring runoff was abnormally high. Here again, population numbers increased dramatically in each of the first two years then tailed off slightly in the last two years of the cycle. [14]

Nelson, Platts, and Casey explained their observations as resulting from plasticity of reproductive behavior. These observations also suggest that adfluvial Humboldt cutthroat populations (and fluvial populations as well, since Chimney Creek is also a spawning and nursery stream for the Mary's River fluvial population) can express very high levels of *intrinsic productivity*. Intrinsic productivity is a measure of the number of surviving offspring per female a parent stock is capable of producing when parent stock numbers are very low. [15] The higher this value, the quicker the population can rebound. Intrinsic productivity is a product of the natural fecundity of the females and the ability of the progeny to survive to reproducing age. Patrick Coffin has reported that adfluvial Humboldt cutthroat spawners carry fewer than 100 to as many as 300 eggs per female, and that males mature at age 2 and females at age 3 in these populations. Thus, for

told me that in untrammeled native vegetation, a major natural-caused wildfire would happen on average about once every 75 years, but on disturbed land, where cheat grass and native but normally suppressed tumbleweed has taken over (both are more fire-prone than the usually dominant native vegetation types), the wildfire interval reduces to about once in 10 years. Flash flooding, he said, has a 15 to 20-year recurrence interval, but streams also experience heavy debris-laden runoff events following the first good rainstorms after wildfires.

13. Lab results on thermal tolerance limits for Lahontan cutthroat trout are reported in Vigg and Koch (1980), Dickerson and Vinyard (1999), Meeuwig (2000), and Dunham et al. (2003).

14. See Platts and Nelson (1983) and Nelson et al. (1987).

15. The concept of intrinsic productivity of a fish stock derives from *stock-recruitment* theory, which predicts that under certain conditions, fish stocks will produce a surplus of recruits over and above the number needed to replace the parent stock. This has become the prevailing paradigm for managing harvest fisheries worldwide. The literature on this subject would fill volumes,

high intrinsic productivity to manifest itself in these populations, a very high percentage of the eggs must survive, and so must the juvenile trout for the two to three years it takes them to reach reproductive age.

Rapid embryo development may be another genotypic adaptation displayed by these trout. In snow-melt-fed streams that go intermittent by mid-summer, rapid embryo development should be an advantage for survival. The only information available on embryo development in Humboldt cutthroat trout comes from a migratory population that makes spawning runs from a reservoir (next section) into snowmelt-fed tributaries of upper Willow Creek. Fertilized eggs from these fish taken to a hatchery hatched out in 23 days at 11 degrees Celsius (254 thermal units), which is an unusually short time for cutthroat trout.

The ability of the juveniles to survive the harsh and variable conditions of the natal streams for the two years (if male) to three years (if female) it takes to reach reproducing age may be yet another genotypic trait of these populations. Growth of Humboldt cutthroats apparently occurs in early-season spurts. That is, in the spring and early summer months, when aquatic and terrestrial insects are available, the trout can feed adequately and growth occurs. But from mid-summer on, when the waters warm and flows drop to base levels and in some reaches become intermittent,

and again in the winter months, when there is snow cover and even the perennial reaches may be iced over, there is little feed in the streams. The trout can survive for long periods without food in cold water, but they do not grow at these times.

A Possible Reservoir-Adapted (Lacustrine) Life History

Willow Creek Reservoir is a 640-acre irrigation impoundment that sits alongside the long gravel road between Tuscarora and Midas (lots of mines around those two places). The reservoir lies unprotected, no shade, in a region of low, rolling, clayey hills and its water temperature can ratchet up 30 degrees or more between sunup and sundown. Its waters are almost always cloudy, owing to the ultra-fine suspension of clay particles carried in from the surrounding hills that never quite settle out.

It's one of those bodies of water that is seemingly totally unsuited for trout. Indeed, brook and rainbow trout stocked there years ago exhibited very poor growth and survival, and recent management emphasis has shifted to warmwater species: white crappie, largemouth bass, channel catfish and white catfish. But at some point in the past, cutthroat trout from the upper Willow Creek drainage also took up residence in the impoundment, and some of these fish lived to attain weights of 5 to 7 pounds.

but the two classic treatises are William E. Ricker's "Stock and Recruitment" (Ricker 1954) and "On the Dynamics of Exploited Fish Populations" by R.J.H. Beverton and S.J. Holt (see Beverton and Holt 1957 [1993]). I have also obtained useful insights from a paper on the subject by Solomon (1985), and from the book *Quantitative Fisheries Stock Assessment: Choice, Dynamics and Uncertainty* by Hilborn and Walters (1992). Stock-recruitment theory also predicts the tail-off in population numbers, as observed by Platts, Nelson, and Casey in Gance and Chimney creeks, which comes about following population buildup when density-dependent factors curtail the number of surviving offspring produced per parent fish, thus guiding population numbers toward a level that matches the carrying capacity of the stream.

Not that much is known about the specifics of this population's life history. Patrick Coffin did report on its spawning runs, noting that trout ascended into upper Willow Creek and its tributaries from about the first of April through the end of May, on increasing stream flows and rising water temperatures. He also noted that Willow Creek water temperature at a Nevada Division of Wildlife trap a mile above the reservoir was 54 degrees Fahrenheit (12.2 degrees Celsius) at the beginning of the spawning run, 60 degrees Fahrenheit (15.5 degrees Celsius) when the run peaked around the 7th of May, and near 80 degrees Fahrenheit (26.6 degrees Celsius) when the last of the fish passed at the end of May. This is a much warmer temperature regime than reported for any other spring-spawning salmonid. Although, as I noted, a few fish attained larger size, most of the mature trout ranged from 13 to 17 inches in fork length, and were quite fecund. Three females in this size range, trapped for an experiment in hatchery incubation and rearing, yielded 7,055 eggs, or approximately 2,350 eggs per female.

Willow Creek Reservoir, along with a couple of the upstream tributaries of Willow Creek, continue to be listed in guidebooks as likely destinations for anglers seeking Humboldt cutthroat trout, although one well-respected Nevada fishing writer has stated that anglers encounter cutthroats only occasionally nowadays in the impoundment itself. However, for as long as this population does persist, it represents what may be the first and only lacustrine population to occur in the Humboldt River drainage since the pluvial lakes of the Pleistocene Epoch. [16]

STATUS AND FUTURE PROSPECTS

When the wagon trains first appeared in the Humboldt River country, the native Humboldt cutthroat was the only trout in the drainage. In those days, it may have occupied as much as 90 percent of the stream miles in the basin. Today, one would be hard-pressed to find any stream that has not been altered directly or indirectly by human activities, and today the Humboldt cutthroat can be found in only a few small upper reaches and headwater tributaries. As recently as 1915, the ichthyologist J. Otterbein Snyder collected native Humboldt cutthroat trout from Starr Creek and Mary's River near Deeth, from Pine Creek near Palisade, from the Humboldt River near Palisade, and from the Humboldt near Carlin. None of Snyder's collection sites are cutthroat habitat today.

Three principal reasons are cited for the reduction in range and population numbers of Humboldt cutthroat trout: 1) intensive livestock grazing; 2) water diversions and withdrawals for irrigation; and 3) introductions of non-native trout and char. A fourth

16. The life history information presented here is from Behnke (1979 [1981]) and Coffin (1981, 1983). Richard Dickerson's *Nevada Angler's Guide: Fish Tails in the Sagebrush* (Dickerson 1997), although several years old now, provides reasonably contemporary information about fishing prospects at Willow Creek Reservoir.

Mary's River near the Jarbidge Wilderness boundary. Mary's River is the only Humboldt Basin stream with any remaining semblance of metapopulation structure. The person standing is biologist Patrick Coffin, formerly with Nevada Division of Wildlife. PHOTO BY AUTHOR

factor, mining impacts, is listed as well, but more in the context of an area of present and future concern than a cause of decline.

Intensive Livestock Grazing

Livestock grazing has been practiced for 125 to 130 years in the Humboldt River basin and continues to be the number one land use, and most of the tributary watersheds are severely overgrazed. A 1980 survey of 23 tributary watersheds on BLM-administered lands concluded that "priority habitat factors are limiting for most of the streams"—a bureaucratic way of saying that range and riparian habitat conditions were poor in most watersheds and only fair in the others surveyed. [17] Even though a few riparian livestock exclosures were established (indeed, Nevada's first-ever livestock exclosure was built by the BLM in 1968 on a 40-acre site along Tabor Creek) and habitat restoration projects were undertaken, a resurvey of BLM and U.S. Forest Service lands taken in 1987 concluded that overall conditions had not improved measurably. Streambank trampling, the loss of riparian cover, and associated erosion have been major causes of habitat deterioration. [18] These factors, along with the seasonal extremes of water flow and the occasional flash flood, have combined to make life for most Humboldt cutthroat populations precarious at best.

17. Two Nevada Department of Wildlife reports prepared by Patrick Coffin contain detailed tables of habitat conditions in Humboldt drainage tributary watersheds as they occurred in 1980; see Coffin (1981, 1983). I lifted the quote I used to summarize these conditions from Coffin's 1981 report. Gerstung (1988) cited overall results of the 1987 surveys, although he presented no tables of data. The U.S. General Accounting Office issued two reports in 1988 and two more in 1991 that alerted federal and state governments and the U.S. public to the poor condition of western public rangeland and riparian areas and highlighted needed improvements in management (I've included citations for these reports in the bibliography). New federal rules for grazing on public lands were issued in 1995 to address these problems, but late in 2003, under a different federal administration, a still newer set of rules was proposed, which critics from several environmental organizations say will end public lands protection, including the requirement to take prompt action to address harmful grazing practices. According to the Associated Press, the Bureau of Land Management acknowledges that the new rules could have "some short-term adverse effects," meaning that some rangeland, already in poor health, might suffer even more initially, but would "result in long-term positive effects on rangeland." How far off in the future those promised "long-term positive effects" might be was left unsaid. You can access the new rules and read them for yourselves in the U.S. Federal Register for Monday, December 8, 2003 (see Griles 2003). I also found them, along with news releases and a draft environmental impact statement, on the BLM website at *www.blm.gov/grazing/*.

18. Footnote 38 of Chapter 5 directs you to several authoritative references on the effects of livestock grazing on riparian and stream habitat, which would be worth your time to review again here. Of related interest, many fisheries biologists implicate beaver activity as exacerbating the problem of livestock abuse of riparian areas. I heard this often during my field excursions, especially in Great Basin locales. Although these same biologists will concede that active beaver ponds may provide productive feeding and refuge areas in streams that might otherwise offer little to the trout in summer low-flow periods or in winter, they also point out that the beavers themselves eliminate riparian cover, exposing the expanded water surface area of the pond to sunlight and warming already temperature-limited waters even more. Furthermore, they contend, when the beavers abandon an area as they typically do after just a few seasons of use, their dams deteriorate and wash away during high flow events, distributing silt from the ponds downstream and generally leaving the stream in *less* favorable condition for trout than before the beavers came. Other biologists in other areas of North America do not share these concerns about beavers. In 1983 the Wisconsin Department of Natural Resources compiled a bibliography of more than 445 references on beaver, trout, wildlife, and forest relationships with special emphasis on beavers and trout. To locate this, see Avery (1983). Even more information is available in two really excellent reviews on the general ecology of beavers and their influence on stream systems, riparian areas, and trout. These were published by Collen and Gibson (2001) and Pollock et al. (2003).

Water Diversions and Withdrawals

Although livestock grazing is the number one land use, many managers believe that the main causes of decline of the Humboldt cutthroat trout have been water diversions to irrigate hay meadows and pasture, withdrawals for mining and municipal use, and associated alterations of the stream courses. Many small streams were completely dewatered for these purposes, fragmenting any metapopulation structure that might have existed among the native trout stocks of the affected drainages. Flow depletion made other streams warm and silty. Channelization and alteration of water-flow patterns and the associated soil erosion that typically occurs as channels try to adjust to these changes, along with the return of poor-quality irrigation waters, all contributed to the deterioration of Humboldt basin stream reaches as native cutthroat habitat.

Displacement by Non-Native Species

In the 1860s and 1870s, it was not uncommon for ranchers or settlers to transplant trout from one stream where they were native to another where they were not, in order to have fishing closer to home. In this way, for example, streams on the east side of the Toiyabe Range, formerly devoid of trout, were stocked with Humboldt cutthroats from the Reese River system located on the west side of the Toiyabes. The distribu-

tion of the Humboldt cutthroat was thus expanded somewhat initially, but then followed the inexorable process of wiping them out.

In 1877, the Office of Nevada State Fish Commissioner was created, and stocking of non-native species into Nevada waters began in earnest. Rainbow trout and eastern brook trout were introduced into waters of the Humboldt drainage early on, and brown trout and Yellowstone cutthroats were planted heavily during the period between 1929 and 1941. Hybridization with rainbow trout and Yellowstone cutthroats appears to have been only a minor occurrence, [19] but extensive displacement of the native cutthroat populations did take place. The native cutthroats persisted only in the upper reaches of the more marginal trout waters, which were beyond the abilities of the non-native species to survive.

Mining Activities

As noted above, federal and state fisheries managers mention mining activities more in the context of an area of present and future concern than a cause of decline of Humboldt Basin native cutthroats. Still, Nevada is a heavily mineralized state, and prospectors and miners combed its ranges in days of yore. Old ghost towns, most of them mining centers in their heydays, dot the landscape of the Humboldt drainage. Most of the early-day mines were underground, hard rock

19. Regarding hybridization and introgression, the 1995 U.S. Fish and Wildlife Service Recovery Plan (Coffin and Cowan 1995) noted that 10 of 23 Humboldt drainage streams surveyed for a late-1970s study of population genetics were occupied by mixed populations of Humboldt cutthroat and rainbow trout. Of these, only three streams showed any evidence of introgression of the cutthroat populations. In the other seven streams, the natives and non-natives co-occurred without interbreeding. See Loudenslager and Gall (1980a) for the complete results of this study.

View of Chimney Creek, a Mary's River tributary, looking upstream in late August. Trout from the river move into small streamlets like this when the water is high in the spring. They spawn, then retreat to the mainstem to live out the summer and winter.
PHOTO BY AUTHOR

operations that were simply abandoned when the ore veins played out.

In Chapter 5, I wrote about the adverse impacts that sediment and runoff from these old-time mining and milling operations could have on nearby streams and their trout populations. The mines themselves and associated milling operations drew water from the streams, as did the old towns, and probably affected nearby streams with their sediments, tailing pile runoff, and sewage discharge. I also pointed out how sediment and runoff effects from old mines can persist for decades, even up to the present time.

What I didn't say in Chapter 5 is that in 1996 the U.S. Geological Survey mapped abandoned mine sites as well as tributaries and streams tainted with runoff from hard rock mining. The Truckee, Carson, and Walker rivers in the western Lahontan Basin were among the streams mapped as tainted. A more thorough survey was undertaken in the years between 1997 and 2001 as part of the U.S. Geological Survey's Abandoned Mine Lands Initiative, but, fearing damage to development and tourism in their states, state officials and members of Congress have teamed to thus far block release of any new maps. The State of Nevada lists 3,077 old mine sites located throughout the counties of Elko, Eureka, Humboldt, Lander, and Nye, which together encompass the historical range of the Humboldt cutthroat trout.[20] Not all of these old sites are actually in the Humboldt watershed, of course, but you can see the magnitude of the potential problem. But none of this is mentioned by fisheries managers as a cause for decline of the Humboldt cutthroat, nor, evidently, are leachates and sediments from abandoned mine sites considered threats to the persistence or recovery of the subspecies.

Mining is still going strong in the Humboldt River Basin, especially in the middle part of the basin around Carlin, where major gold deposits occur. But the scale of operation has changed dramatically from the early days. Large-scale open pit mining, a relatively inexpensive way to extract large ore bodies containing gold particles so tiny the early-day miners would never have seen them, is now the norm in Nevada and is a land use that is raising increasing levels of concern.

There are two, make that *three* major areas of concern. First of all, the mining industry has a terrible history of polluting while mines are producing and then callously walking away, leaving local citizens and the taxpayers to clean up the mess. The companies operating in today's industry bend over backwards to overcome this image and convince us that they are better corporate citizens, but winning trust even with the promise of jobs, prosperity, and appeals to national security is an uphill battle, and the watchdogs are exceedingly skeptical.

20. Back in 2001, one of my home town newspapers, *The Seattle Post-Intelligencer*, published a four-part series, "The Mining of the West: Profit and Pollution on Public Lands," by Robert McClure and Andrew Schneider (with Lise Olsen for Part 4). The 1996 U.S. Geological Survey maps were reprinted in that series. You can access the series (at least it was still possible to do so as of October 2006) on the Internet at *http://seattlepi.nwsource.com/specials/mining/*. I also found a link to it on the Trout Unlimited website, *www.tu.org*. The U.S. Geological Survey's Abandoned Mine Lands Initiative also has a website, *http://amli.usgs.gov*. When I checked that site, also in October 2006, I found several progress reports, but no maps. Closer to the subject matter of this Chapter, the State of Nevada keeps track of its own abandoned mines through its Abandoned Mine Lands Program, administered by the Division of Minerals in Carson City. Its website at *http://minerals.state.nv.us/programs/aml.htm* had the numbers I cited in the text, and these were current as of April 7, 2006.

The second concern is related to the first, in that it becomes a problem when the mines stop producing and are closed. Open pit mining gouges out immense chunks of real estate. The pits are both staggeringly large and deep, often sinking well below the water table. Even though reclamation and restoration technology has made great strides,[21] water will inevitably flow into those holes, creating pit lakes that often become highly acidic. Contaminants in some pit lakes threaten to pollute surrounding ground water, and in some cases defy efforts at neutralization. According to a 1997 statement by Dr. Gary Vinyard, University of Nevada Reno, the water in the Sleeper Mine pit lake (located, I believe, in the Quinn River drainage) remained at pH 3 even after treatment with 100,000 pounds of lime.

The third, and in my view most vital, concern is the fate of the water table. Groundwater is the source of the springs and seeps that feed those upper stream reaches where the native cutthroats live. Lose those springs and seeps and you lose those upper reaches. Lose those upper reaches and you lose their native trout populations. Of particular concern are the known populations in tributaries of Maggie Creek. These tributaries drain from the area where the major gold deposits occur, and therefore stand to be within the radius of water table reductions due to mining.

Here's the problem: As the levels of operating pit mines descend below the water table, mine operators must pump water out of the pits. Most of this water is discharged onto the ground somewhere outside the pits, and so recycles into local aquifers. But in a couple of cases at least, mines are permitted to discharge the water into the Humboldt or its tributaries. This has increased the flow in the Humboldt downstream from the Carlin area, much to the benefit of downstream irrigators. But it has also lowered the water table in places in the mid-Humboldt basin, especially in the vicinity of the mines. It is also expected that in 2008 and 2011, when the mines now discharging pit water into the Humboldt and its tributaries close and cease pumping, the flow in the Humboldt will decline. Concern has also been expressed about the impacts of pulling down the water table on stream flows when drought conditions return, as they have across much of the interior west in the years since 1999.

In 1995, to get a better handle on the cumulative effects of groundwater withdrawals for mining use, dewatering, and other water uses on the groundwater and surface water resources of the basin, the U.S. Geological Survey launched a multi-year Humboldt River Basin Assessment, focusing initially on the middle basin from Carlin to a point about 5 miles upstream from the town of Golconda. The first report, issued

21. One overview of ecology and contemporary methods for mineral extraction that you might find useful for context is the book *Environmental Effects of Mining* by E.A. Ripley, R.C. Redmann, and A.A. Crowder (Ripley et al. 1995). Although the data and case studies used are mostly from Canada, this 11-chapter book describes mineral extraction methods, ecosystem effects, environmental impacts, and reclamation and rehabilitation methodologies. Two additional books on the subject are Duane Smith's *Mining America: the Industry and the Environment* (Smith 1993) and Jerrold Marcus, editor, *Mining Environment Handbook: Effects of Mining on the Environment and American Environmental Control on Mining* (Marcus 1997).

in 1999, determined that agricultural withdrawals had reduced groundwater levels as much as 70 feet in irrigated areas of the middle basin over the previous 30 to 40 years of record. But much more severe declines in the water table, from 200 to 1,000 feet, had occurred in areas around open pit mines, and these reductions had occurred within just a 10-year period from 1988 to 1998. As of 1999, these water level declines had not interfered with other water uses in the middle Humboldt basin. But most of that 10-year span was a wet period. How this increased demand for and use of groundwater resources will play out in the future, especially now that a drought cycle has returned to the interior west, remains a question for serious study. [22]

Figure 6–1, redrawn from a map originally produced by Jason B. Dunham of the U.S. Forest Service, shows the Humboldt River drainage and the areas of the basin where Humboldt cutthroat populations persist. Within these areas, currently occupied habitat comprises about 318 stream miles in total, or 14 percent of the originally occupied habitat. [23] The U.S. Fish and Wildlife Service Recovery Plan for the Lahontan cutthroat trout, written in 1995, tallied 93 populations in the Humboldt River Basin DPS distributed among seven sub-basins and two localized areas as follows (number of populations in parentheses): Mary's River sub-basin (17); North Fork Humboldt sub-basin (12);

South Fork Humboldt sub-basin (20); Maggie Creek sub-basin (7); Rock Creek sub-basin (6); Reese River sub-basin (9); Little Humboldt River sub-basin (15); East Humboldt River area (6); and Lower Humboldt River area (1). Of these, only the populations of the Mary's River sub-basin appear connected to any extent as a metapopulation.

The 1995 Recovery Plan emphasized population management and habitat management as major strategies for recovery of the Humboldt River cutthroat DPS (research, and updating and revision of the recovery plan were also listed as major strategies but will not be discussed here).

Population Management

Here the federal Recovery Plan emphasized monitoring and maintaining the 93 populations that existed in the western Lahontan Basin in 1995. No objectives for reintroducing additional populations were specified, although the plan did endorse efforts to remove non-native trout when necessary to make way for reintroductions. A Species Management Plan for the North Fork Little Humboldt sub-basin written by Nevada Division of Wildlife in 1999 went a step further, however. The state plan identified four tributaries with genetically pure populations of native cutthroat trout that were deemed below habitat carrying capacity,

22. The Humboldt River Basin Assessment project (website http://nevada.usgs.gov/humb/) was still ongoing as of October 2006. Its first report, issued in 1999 (see Plume and Ponce 1999), drew considerable attention from the news media, but I haven't seen any media coverage since. Four additional reports, all in the U.S. Geological Survey's Water-Resources Investigations series, have been issued through 2002, and two others were promised for 2004, but had not yet been posted when I checked the website in October 2006. However, the site does offer a good overview of the project, and also has instructions for obtaining your own copies of the reports that have been issued.

23. These figures for presently occupied stream miles come from the most recent U.S. Fish and Wildlife Service Recovery Plan for Lahontan cutthroat trout; see Coffin and Cowan (1995). Other status, management, and recovery plans that include Humboldt cutthroat trout populations were written by Coffin (1983) for the subspecies as a whole, and by Sevon et al. (1999) for North Fork Little Humboldt populations. At the time I was working on this Chapter (in May 2004), the Nevada Division of Wildlife had just completed a brand-new species management plan for the upper Humboldt Basin, but had not yet

and four additional tributaries with good habitat and populations at carrying capacity, but those populations were introgressed. Under this plan, the pure populations were to be augmented with trout from other nearby streams to bring them up to carrying capacity, and the introgressed populations were to be eradicated and then restocked with genetically pure trout. I presume these projects were accomplished. Nevada Division of Wildlife proposed to complete a number of similar supplementation and eradication/reintroduction projects in other Humboldt cutthroat sub-basins in a Fishery Management Plan written in 1983, but the only such project that I know was completed for sure was in Illinois Creek, an upper tributary of the Reese River, in 1987.[24]

The 1995 federal Recovery Plan essentially concedes that existing habitat conditions, land ownership patterns, competing uses for water, and the like, along with the current status of existing cutthroat populations, limits opportunities to reestablish and maintain metapopulations in any but the Mary's River sub-basin, where some semblance of metapopulation structure already exists. But, having said that, the Plan also expressed the desire to reconnect reaches of stream and tributary habitat into metapopulation networks in the Maggie Creek, Rock Creek, North Fork Humboldt, and Little Humboldt sub-basins as an objective for sometime off in the future.

Habitat Management

Much of what the 1995 Recovery Plan calls for in the way of population management depends on having healthy stream habitat in which to work. On federal lands, habitat management strategies are supposed to follow the provisions of four federal laws: the National Environmental Policy Act (NEPA), adopted in 1969; the Endangered Species Act of 1973; the National Forest Management Act of 1976; and the Federal Land Policy and Management Act, also adopted in 1976. In addition to these, the Inland Native Fish Strategy (INFISH), which I discussed in Chapter 4, was adopted by the U.S. Forest Service and Bureau of Land Management in 1995. INFISH provides riparian management standards, guidelines, and objectives for sensitive inland fishes. If these provisions are followed, degraded watersheds would be managed to achieve a desired future condition that is healthy and properly functioning, and further degradation would be precluded. As a step toward achieving the properly functioning condition part of this mandate, the 1995 Recovery Plan calls for designating streamside management zones that include the stream, riparian and streambank vegetation, and adjacent areas that would be managed to improve water quality, streambank stability, instream habitat condition, fish, and other aquatic resources.

In this regard, action commenced in the Mary's

released it to the public. That plan was finally made available in March 2005, on the Division's website at *www.ndow.org/wild/conservation/ fish/lct_smp.pdf.* See Elliott and Layton (2004) for the citation.

24. The two State of Nevada plans referred to in this paragraph are "Lahontan Cutthroat Trout Species Management Plan for the Quinn River/Black Rock Basins and North Fork Little Humboldt River Sub-Basin" (see Sevon et al. 1999) and "Lahontan Cutthroat Trout Fishery Management Plan for the Humboldt River Drainage Basin" (see Coffin 1983). *Augmentation* is defined as the act of supplementing an existing wild population where it is determined that that population is below habitat carrying capacity. In this case, supplementation is done during years of average or above average production in streams that are at carrying capacity, by collecting trout from these donor populations and releasing them into the below-capacity populations. Trout from these donor populations are also the source fish for introductions into streams where introgressed or non-native fish have been eradicated. The Illinois Creek project was announced in a Nevada Division of Wildlife news release in the winter of 1986.

River sub-basin long before the Recovery Plan was ever written. In 1988, a collaboration among federal and state land management and fisheries agencies, local ranchers, and sportsmen kicked off to restore trout habitat along the Mary's River. A big donation from Barrick Goldstrike Mines, Inc. was matched by federal Sport Fish Restoration dollars. Land swaps and conservation easements secured most of Mary's River's 128 stream miles. Then, in 1992, through the Bring Back the Natives program that I introduced you to in Chapter 4, the National Fish and Wildlife Foundation provided six grants for additional fencing and habitat improvement projects, as well as monitoring and research studies. Much progress was made in the Mary's River sub-basin as a result of this work, but much remains to be done. A trip through the sub-basin still reveals grazing damage that needs mitigating, and blocking culverts and water diversions that need attention if improved metapopulation connectivity is to be realized. [25]

Elsewhere in the Humboldt River Basin, I've already mentioned that the first livestock grazing exclosure in Nevada was built by the Bureau of Land Management on Tabor Creek, in the Humboldt River headwaters, in 1968. Between then and 1982, the BLM built livestock grazing exclosures on 580 acres along five native cutthroat streams in Elko County. Instream habitat improvement projects were also initiated as early as 1969 on U.S. Forest Service land in the North Fork Humboldt sub-basin, and on BLM land along Sherman Creek, a Humboldt River tributary near Elko. Since the 1995 Recovery Plan was written, consideration of the habitat needs of native cutthroat trout has increased in BLM and U.S. Forest Service land-use planning and management. Grazing allotment management plans have or are being updated to improve stream, riparian, and watershed conditions. Management strategies to effect these improvements include more exclosure fencing, changes in season of livestock use, rest-rotation grazing cycles, and herding as opposed to free-range grazing.

Timetable for Recovery

The objective of the 1995 federal Recovery Plan is to delist Lahontan cutthroat trout in all three of the DPSs defined by the U.S. Fish and Wildlife Service. No target date for delisting is given; the plan describes merely a timetable of recovery actions to be carried out through 2018. Whether or not even this timetable will be met is uncertain. The path to recovery has not been a smooth one in Nevada.

Some outside observers looking in on the situation have attributed much, if not all, of what has happened to the frustration people who have long used public

25. The Mary's River restoration project has received several glowing write-ups in the conservation columns of national fly fishing magazines, and in the journals of such organizations as Trout Unlimited, the Federation of Fly Fishers, and the Isaac Walton League. It is also described from the U.S. Fish and Wildlife Service and Bureau of Land Management perspectives in the book *Watershed Restoration: Principles and Practices* (American Fisheries Society, 1997); see Gutzwiller et al. (1997).

lands for their own benefit must feel in having to deal with a federal government that owns 87 percent of the land in Nevada. The State itself has found this to be an annoyance. In 1979, the Nevada legislature passed a law claiming title to all federal lands in the state. Later, when the "county supremacy" movement emerged, it took hold with greatest vigor in Nevada. All but one of Nevada's counties passed resolutions to assume control of federal lands. Some launched grand jury investigations of federal land management officials, and some took stronger law enforcement actions against BLM and Forest Service employees. In 1993, a group calling itself the Elko County Federal Land Use Planning Commission petitioned the U.S. Fish and Wildlife Service to delist the cutthroat trout of the Humboldt River drainage basin. That petition was rejected on the ground that it did not present any new evidence that the DPS was recovered. In 1994, a Nye County Commissioner bulldozed a closed road in the Toiyabe National Forest. There were also confrontations in Elko County. Three bombings were directed at federal land management offices in Reno.

That sort of extremism diminished following the Oklahoma City bombing that so shocked the nation on April 19, 1995. And later in 1995 and continuing through 1996, the county supremacists lost a string of key court cases that took the oomph out of the move-ment. Even that 17-year-old Nevada law claiming title to federal lands was struck down as unconstitutional. Some semblance of civility returned to the scene—at least for a while.

In 1998, the Nevada Wildlife Commission changed its fishing regulations for the Mary's River from an allowable harvest of 10 trout per day to catch-and-release only. This was hailed as a big step forward in the recovery process for the Mary's River cutthroat population. Sadly however, in 2001, in what appeared to be another clash over local vs. federal control, Nevada wildlife officials withdrew from the U.S. Fish and Wildlife Service recovery effort for Lahontan cutthroat trout, and since then have gone their own way, although they still cooperate with the other entities participating in the recovery plan. One rather sad action they took, in my view, as part of this maneuvering was to reopen the Mary's River to harvest angling, although they did set the allowable harvest at 5 trout per day, half the original level. Nevada Division of Wildlife deputy administrator Gene Weller defended this action. "If we felt the population would be impacted by the regulation there is no way we would have changed it," he asserted. Noting that his department had recorded only 50 to 60 angling days on the Mary's River annually, he stated, "We felt it [catch-and-release] was unnecessarily restrictive." [26]

26. Deputy administrator Gene Weller's quotes are from an Associated Press newspaper article titled "NEVADA FOCUS: Cutthroat Politics Fuel State-Federal Feud over Nevada Trout" in the November 9, 2001 issue of *The Las Vegas SUN.*

Patrick Coffin, formerly Nevada Division of Wildlife, on the bank of a perennial stream high in the North Fork Humboldt sub-basin. Stream-resident Humboldt cutthroat trout are present in this stream. PHOTO BY AUTHOR

CHAPTER 7

Willow/Whitehorse Cutthroat Trout

Unnamed. Chromosomes, 2n=64. Scales in the lateral series 131 to 164 (means of 150 in Willow Creek and 147 in Whitehorse Creek specimens). Scales above lateral line 34 to 45 (means of 40 in Willow Creek and 38 in Whitehorse Creek specimens). Gill rakers 18 to 24 (means of 21 in Willow Creek and 22 in Whitehorse Creek specimens). Pyloric caeca 35 to 55 (means of 44 and 43 in Willow Creek and Whitehorse Creek specimens, respectively). Vertebrae 59 to 64. These meristic character values are only slightly differentiated from the Quinn River and Humboldt forms (Chapter 6) of cutthroat trout (except for pyloric caeca counts which are much lower in Willow/Whitehorse cutthroats) and may represent a transitional gradient of characters. Fairly large spots sparsely distributed but with a tendency to concentrate on the posterior and above the lateral line on the anterior. Coloration, as in the Lahontan cutthroat, runs to a dull brassy shade with often a rose-colored tint on the sides.

Whitehorse Basin Cutthroat Trout, *Oncorhynchus clarkii* subspecies, Willow Creek form

BASIN AND RANGE COUNTRY—COYOTE BASIN AND TROUT CREEK MOUNTAINS

The record breaking heat wave was in its fifth day with temperatures in the high 90s. The breeze flowing through the opened vents and windows of my old station wagon provided no relief as I drove along U.S. 20 toward the town of Burns, Oregon. The high desert country was dry as tinder and, frankly, I was worried.

The voice crackling through the static on the car radio had told about the 2,000-acre range fire burning out of control on BLM land on the west slope of Steens Mountain. And now, as I looked off to the south, I could see the dirty white smoke blurring an otherwise clear blue sky.

What worried me was the field trip I was supposed to take the next day with BLM fishery biologist Mike Crouse and Steve Pribyl, his counterpart with the Oregon Department of Fish and Wildlife. I had driven over 500 miles to visit Willow and Whitehorse creeks, home of a rare subspecies of cutthroat trout in the Whitehorse Basin (aka Coyote Basin) of southeastern Oregon.[1] Although the fire was 60 to 70 miles off to the west and no direct threat whatsoever, combating the blaze was a BLM responsibility. I was concerned that somehow that priority would preempt my plans.

But I needn't have fretted. Next morning, right on schedule, a four-seater Cessna touched down on the BLM landing strip at Burns Junction, discharging Mike and Steve, and before long we were rolling out across the desert in a 4WD truck.

We were headed for the Trout Creek Mountains. The Trout Creeks are just one of many fault-block mountain ranges that rise up above the basin floor every 30 miles or so throughout the Great Basin. The Trout Creeks rise gradually on their western flanks and fall away more steeply on their eastern slopes, but nowhere near as steeply as the sheared-off east face of Steens Mountain, visible across the basin to the west.

Three separate desert basins receive the meager waters that flow down out of the Trout Creek Mountains. Trout Creek and Cottonwood Creek rise on the western slopes and flow into the Alvord Basin. A second group of streams that includes McDermitt Creek, Oregon Canyon Creek, and Kings River arises on the eastern and southern slopes and flows toward the Quinn River, part of the Lahontan Basin. Coming off the northern slopes and flowing into the totally enclosed Coyote Basin are the streams I had come to see: Willow Creek and Whitehorse Creek, along with major Whitehorse tributaries, such as Little Whitehorse, Fifteenmile, Cottonwood, and Doolittle creeks. These creeks begin at an altitude of nearly 8,000 feet, then flow down through steep, narrow canyons. As they near the valley floor, elevation about 4,400 feet, the

1. Whitehorse Basin is the name given to the basin by geologist I.C. Russell during his Great Basin surveys back at the turn of the 20th century (see Russell 1903a, 1903b). The basin went by that name as the more or less "official" designation until fairly recently, when Coyote Basin, after the ancient lake (now a playa) that existed there, came into common use.

slopes flatten out, the canyons open up, and finally they blend into the floor of the basin. The streams meander out into shallow marshy washes, then disappear altogether. Further out on the floor of the basin lies Coyote Lake, a large, flat, shallow playa.

The Coyote Basin as it exists today is an internally draining basin lying immediately north of the Lahontan Basin. It is contiguous along a relatively low rim with the Alvord Basin to the west, and with the Snake/Columbia Basin to the north and east. Like the Alvord Basin to the west, where a Pleistocene lake occupied a trough some 75 miles long and 5 to 10 miles wide, recent geological evidence indicates that the Coyote Basin was occupied by a shallow Pleistocene lake as well, at times. Overflow, passing from Lake Alvord through Coyote Lake and into Crooked Creek, a tributary of the Owyhee River of the Snake River drainage, may have occurred at least twice in the basin's prehistory, once in mid-Pleistocene time and again in the late-Pleistocene. [2] We pulled off at a vantage point where I could look out over the dry, glistening-white bed of Coyote Lake, and Mike pointed out the very low, almost imperceptible divide that separates the basin from the Crooked Creek/Owyhee drainage. It was certainly easy for someone new to the country to miss. A water level of 60 to 90 feet in ancient Coyote Lake would have been more than enough to spill it into Crooked Creek. Mike told me in addition that when one flies over this country, one can see faint channels where underground seeps still follow the ancient waterway.

The first leg of our trip on this day had taken us south on a paved road, then we angled off across the desert on a graded dirt road, a thick cloud of dust billowing out behind us. Perhaps it was getting off the pavement that did it, I'm not really sure. But suddenly I was struck by the sheer expanse of this dry, dun-colored land. E.R. Jackman and R.A. Long tried to describe it in their book, *Oregon Desert*: "The bigness is almost like a noise." [3]

Even though this is an arid land, it is rich in bird and animal life. A burrowing owl stood atop its mound and watched us pass. Ravens, doves, shrikes, and red-tailed hawks flew overhead. In a marshy place where one of the creeks bottomed out, we saw a mule deer doe with twin fawns. In the same spot, a marsh hawk winged its way past, intent on its hunt. Twice we saw families of chukars scurrying up the canyons. Jackrabbits were literally everywhere. Antelope roam the countryside and it is also one of the region's major wild horse ranges, but we saw neither on this trip.

We crossed the low divide between Crooked Creek and the Coyote Basin and skirted the basin edge. Presently, we came to the irrigated fields and painted buildings of the Whitehorse Ranch, a cattle operation

2. According to Dr. Marith Reheis, U.S. Geological Survey, and her colleagues (see Reheis 1999, Reheis and Morrison 1997, and Reheis et al. 2002), the highest-ever stand of an ancient Lake Lahontan occurred about 650,000 to 700,000 years ago, possibly with overflow into the Alvord Basin and adjacent Coyote Basin, and from there perhaps discharging into the Owyhee River of the Snake River drainage. Recent work by Deron Carter, Central Washington University, and his colleagues provides evidence of a late-Pleistocene catastrophic flood from Lake Alvord and Coyote Lake into the Owyhee River (see Carter et al. 2003, 2004). The timing of the late-Pleistocene event corresponds roughly with the overflow of pluvial Lake Parman (the precursor of Summit Lake) into the Alvord Basin that I discussed in Chapter 5 (see Layton 1979).

3. Jackman, E.R. and R.A. Long. *Oregon Desert* (Caldwell, Idaho: The Caxton Printers, Ltd., 1964).

situated where Willow and Whitehorse Creeks flow down out of the hills.

A few cattle had been trailed through this country since the 1840s, when immigrants passed this way. But 1869 marked the beginning of a real range cattle industry, with big ranches that eventually controlled thousands of acres of land. It began in the late summer of that year, when thirty year old John Devine, a large, imperious man who liked to dress the part of a Spanish don, trailed a herd of 2,500 head into the basin from California. He also brought a cavalcade of horses and a full compliment of Mexican vaqueros to ride them. He acquired an abandoned military post, Camp C.F. Smith, built three years before and occupied briefly during a period of Indian unrest, and on this site he established the Whitehorse Ranch, thereby becoming the region's first permanent white settler. From here he extended his range, preempting all of the country east of Steen's Mountain and becoming undisputed king of the Coyote and Alvord basins.

For twenty years the "reign" lasted. It was ended when a hard winter followed by two dry summers proved too much for Devine's finances, already strained to the limit by the demands of his aristocratic lifestyle. The Whitehorse Ranch and its once vast range fell into other hands. Devine retained control of only the Alvord Ranch, which he had established during

the good times at the foot of Steens Mountain, and there he lived out the rest of his days. The Whitehorse Ranch has endured, and now, as if time has come full-circle, it is owned by another Californian, businessman Ted Naftzger.

John Devine may have been the first permanent white settler, but the first white men to actually glimpse southeast Oregon were undoubtedly the Hudson's Bay Company fur trappers of Peter Skene Ogden and John Work. Ogden's men passed through in the winter of 1828–29. Work's party made the transect two winters later.

Gradually then had come the immigrants. Gold was discovered over in Idaho. Wagon roads and supply routes were pushed through the basin. More cattle and horses arrived. Sheep were brought in. As the Indians felt the pressure, there were the inevitable raids and uprisings, which in turn brought in the military. Outposts such as Camp C.F. Smith were established. Here, as elsewhere in the American West, disputes arose over land and water rights, disputes that were, as often as not, settled by force. "There's a lot of history in this region," Mike commented as we turned up Willow Creek and followed it back into the hills. [4]

Old-time naturalists, following C. Hart Merriam, called the life zone characterized by the valleys and lower slopes of the high desert country the "Upper

4. You can read about this history and more in French (1964), Jackman and Long (1964), Jackman and Scharff (1967), and Boreson et al. (1979).

Looking down from the rimrock into a "beaver park" on Little Whitehorse Creek, 1981. You can't see them in the photo, but cattle were grazing in this "park."
PHOTO BY AUTHOR

Sonoran" zone. Here the gray-green sagebrush, rabbitbrush, and native bunchgrasses hold sway. Climb higher and you enter what they called the "Arid Transition" zone, where a few scattered junipers and mountain mahoganies begin to appear. The few perennial streams flow in narrow canyons, protected there by willows, wild rose thickets and sage, and occasional stands of aspen.

But perennial streams are few, and even those are quite small. A typical one might be a couple of feet wide and only six inches to a foot deep in its deepest pools. In their lower-gradient, lower-elevation reaches, these streams may reach or exceed 80 degrees Fahrenheit during daytime in midsummer, but cool to 55 to 60 degrees Fahrenheit at night. Yet here, in this harsh, hardscrabble environment where conditions are seemingly stacked against them, the Willow/Whitehorse cutthroats have managed to survive, and at times even thrive.

HISTORICAL RANGE

The field trip I described above took place in August 1981 as part of a broad sweep across the western U.S. that I made that year—the years 1979, 1980, 1982, and 1983 were likewise spent in the field—to observe native cutthroat populations in their native habitats. In Fig. 5-1, the Coyote Basin is the area designated "C." The expanded view of the basin in Fig. 7-1 shows the lay of the land in a bit more detail. Of particular importance is the relationship of the source areas of the various creeks to one another as they flow off the Trout Creek Mountains into their respective drainage basins.

Although the ranchers and other inhabitants of the Coyote Basin and its surrounding area surely knew about and undoubtedly fished for the trout that occupied the Willow and Whitehorse creek drainages, the first specimens to be described for science were collected by Carl Hubbs, University of Michigan, and his family in the summer of 1934. At that time, cutthroat trout occurred only in those two creek drainages. In 1971, trout were transferred to Antelope Creek, another Coyote Basin stream, which had been without fish prior to that introduction. Shortly after my field trip in 1981, I was told that Willow/Whitehorse cutthroats had also been transplanted into fishless streams draining into the Alvord Basin off the east side of Steens Mountain and the Pueblo Mountains, and that populations had become established in Van Horn and Mosquito creeks. But whether or not they still exist, I have not been able to confirm. I also learned recently from Dr. Behnke that, in 1957, trout from the Coyote Basin were stocked in Guano Creek, a stream that flows into the Catlow Basin of southeastern Oregon, but later stockings of

The combined effects of beavers and cattle Little Whitehorse Creek, 1981. PHOTO BY AUTHO

Heenan Lake-strain Lahontan cutthroats and rainbow trout make it unlikely they still exist in pure form.

Cutthroat trout were, and still are, the only fish species found in the Coyote Basin. Dr. Behnke has written that this argues for a long period of isolation of the Coyote Basin from all surrounding basins. It also argues for original occupancy by trout having occurred via headwater stream transfer from another basin, from an uppermost reach inhabited only by cutthroat trout. Most probably, this ancestral trout was transferred from a headwater tributary of the Quinn River drainage of the Lahontan Basin.

What is the evidence for this? Had the trout been introduced from Lake Alvord by the overflow events described in the previous section, or had they come via upstream migration from the Snake River drainage while those floodways were still open, then other fish fauna should have been introduced along with the trout, and none were. For example, the Alvord Basin has an indigenous species of chub in addition to cutthroat trout. You'd think that a transfer of cutthroat trout from Lake Alvord would have transferred the chub as well, but chubs are not found in Coyote Basin waters.

Then there is the evidence of the trout themselves. Meristic characters of Willow/Whitehorse cutthroat trout are only slightly differentiated from the Quinn River and Humboldt forms, and what slight differ-

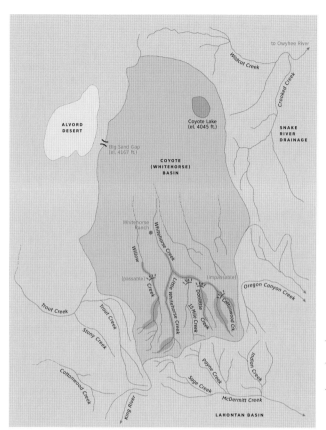

Exposed pool on Little Whitehorse Creek, 1981. The water temperature exceeded 82 degrees, yet we spooked a cutthroat trout from this pool. PHOTO BY AUTHOR

Core Population Reaches

Occupied Habitat
(Based on 1994 survey)

Barriers or Impediments
to Instream Migration

Approximate Boundary
of Historical Range

FIGURE 7-1. *Drainages in the Coyote (Whitehorse) Basin and proximity of Alvord, Lahontan, and Snake River Basins*

ences there are may be explained by the divergence of characters that one would expect after long isolation. Of particular significance here are the gill raker counts of these forms (21 and 22 gill rakers respectively for Willow Creek and Whitehorse Creek specimens; 21 gill rakers for Quinn River and Humboldt drainage specimens). Alvord Basin cutthroat trout, which I will discuss in more detail in Chapter 15, had closer affinities to the Lahontan cutthroat trout of the western Lahontan drainages and Summit Lake, including a higher number of gill rakers (24 gill rakers in all of these forms). This again rules against a transfer from the Alvord Basin via overflow from Lake Alvord, which would have introduced a 24-gillraker form. It also rules out an origin via headwater stream transfer from the Trout Creek drainage of the Alvord Basin, even though the headwaters of Trout Creek intertwine with those of Willow Creek.

Thus, Dr. Behnke reasons, the cutthroat trout native to the Coyote Basin must have been derived from the Quinn River drainage of the Lahontan Basin. I recall that in 1981, as a prelude to my field trip, Neil Armantrout of the Bureau of Land Management had made this very same suggestion. His candidate for the actual point of transfer was a series of washes near the head of Oregon Canyon Creek, a Quinn River tributary, that are, as Mike Crouse pointed out while we were out in the field that year, quite close to the headwaters of the Whitehorse Creek drainage. In fact, from its source, Oregon Canyon Creek takes a short northwesterly course directly toward the Whitehorse Creek drainage, but then loops east, then south, away from the Coyote Basin as it flows out of the mountains and off to the Quinn River. It is not hard to envision a direct connection in the headwaters in cooler, wetter climes. We didn't examine any of these closely, but several McDermitt Creek headwater tributaries also originate in proximity to headwaters of both Willow and Whitehorse creeks. Willow Creek and the Whitehorse Creek drainage are completely isolated from one another at present, but again, it isn't hard to envision connecting flowages during pluvial climate periods.

In his book, *Trout and Salmon of North America* (Behnke 2002), Dr. Behnke took his line of reasoning a step further. "The cutthroat trout native to the Whitehorse basin was derived from the Quinn River drainage of the Lahontan Basin," he wrote. The minor differentiation of characters between Coyote Basin, Quinn River, and Humboldt drainage forms may represent a transitional gradient of characters. Thus, another option, aside from recognizing the Willow/Whitehorse cutthroat trout as a separate subspecies, "would be to combine the Willow/Whitehorse and Quinn River drainage cutthroats with the Humboldt cutthroat as a single subspecies that is, as yet, unnamed." [5]

5. In late 1991, following a taxonomic review that included genetic evidence mainly from two studies by Williams and Shiozawa (1989) and Williams (1991), the U.S. Fish and Wildlife Service concluded that Willow/Whitehorse cutthroat trout were indistinguishable from Lahontan cutthroat trout and, therefore, would be designated as part of the Northwestern Lahontan Basin DPS of *O. c. henshawi*. In these two studies, nine specimens of Willow Creek cutthroat and four specimens of Whitehorse drainage cutthroat had mtDNA patterns typical of Lahontan cutthroat trout, but five Whitehorse drainage specimens had mtDNA patterns that were unique and not found in any other cutthroat trout collection. More recent genetic work by Williams et al. (1998) indicates that Willow/Whitehorse, Quinn River, Humboldt, lower Lahontan Basin (i.e., Walker, Carson, and Tahoe/Truckee drainages), and Summit Lake forms all exhibit genetic differentiation from one another, but not enough, evidently, for the U.S. Fish and Wildlife Service to change its designation. Dr. Behnke's summation of the evidence for the origin of Willow/Whitehorse cutthroat trout and how they should be classified can be read in his 1992 monograph, "Native Trout of Western North America" (American Fisheries

The U.S. Fish and Wildlife Service sees this classification question differently. In 1991, the Service opted to classify the Willow/Whitehorse cutthroat trout as part of the Northwestern Lahontan Basin DPS of *O. c. henshawi*, the Lahontan cutthroat trout.

LIFE HISTORY AND ECOLOGY

There has never been a formal study made of the life history of the Willow/Whitehorse Creek cutthroat, although graduate student Andrew Talabere, Oregon State University, did study the influence of water temperature and beaver ponds on these trout back in 2002.[6] Everything else we know is based on observations passed on by the Bureau of Land Management and Oregon Department of Fish and Wildlife biologists who oversee them. These trout exhibit a stream-resident life history and, like the Humboldt (and Quinn River) forms, are remarkable for their ability to live under extreme conditions in a fluctuating, arid-land environment. They likely reach sexual maturity at age 2, with the females carrying from fewer than 100 to 300 or more eggs depending on their size and age, and embryo development probably occurs rapidly. And, like the adfluvial populations of the upper tributaries of the Humboldt and Quinn River drainages, they have undoubtedly evolved other coping mechanisms and adaptations that have enabled them to persist. For example, Andrew Talabere observed that large numbers of age-2 and -3 trout congregated in beaver ponds in summer, even when water temperatures exceeded 75 degrees Fahrenheit (24 degrees Celsius), a temperature considered lethal for cutthroat trout.

During my 1981 field trip with Mike Crouse and Steve Pribyl, we stopped the truck at one exposed pool on Little Whitehorse Creek where the mid-afternoon water temperature topped 82 degrees Fahrenheit (27.8 degrees Celsius). Very unlikely trout habitat, yet all three of us saw the cutthroat, spooked by our approach, flash out of the pool where it had been holding and off to shelter downstream. Later, we electrofished a section of Willow Creek. I was appalled at the volumes of silt that billowed out of the gravel with every step we took. Just about everything the books say could be wrong with this water as trout habitat *was* wrong, yet the trout not only succeeded in spawning here, they reared in this segment to a fairly respectable size as well. Six to 7 inches in fork length is about tops, I'd guess, for fish in their third or fourth season of life, although one fish we electroshocked did reach the 10-inch mark on the measuring board. Later, I heard of other credible observers who reported fish up to 14 inches in fork length from beaver ponds in the system. We found our fish holding under the undercut banks, in the deeper parts of the sheltered pools, and especially under the

Society Monograph 6, 1992) at pages 130–131, and in his recent book, *Trout and Salmon of North America* (Behnke 2002) at pages 226–227.

6. Talabere, A.G. 2002. Influence of Water Temperature and Beaver Ponds on Lahontan Cutthroat Trout in a High-Desert Stream, Southeastern Oregon. Master's Thesis, Oregon State University, Corvallis.

Little Whitehorse creek, 1981, showing down-cutting and erosion of the stream bank and lack of cover.
PHOTO BY AUTHOR

outfalls of beaver dams, where we presumed they could find the coolest, best-oxygenated water in these warm, summer-lit streams. Feeding occurs opportunistically on whatever aquatic or terrestrial insect life the streams brought along.

In 1985, 1989, 1994, 1999, and 2005, the Oregon Department of Fish and Wildlife conducted population surveys of Willow/Whitehorse cutthroat trout.[7] The 1994 survey, conducted in the month of October, not only estimated population size but also determined the age class structure and something about the dynamics of these populations. In 1994, an estimated 39,500 cutthroat trout occupied Coyote Basin streams including Willow Creek, Little Whitehorse Creek, and Whitehorse Creek itself plus its tributaries, Fifteen-mile, Doolittle, and Cottonwood creeks. Of these, an estimated 44.3 percent were age 0; 40.7 percent were age 1; 11.1 percent were age 2; and 3.9 percent were age 3 or older. Densities of trout, both age-0 and older-age classes, were highest in the headwater reaches of all streams (i.e., reaches higher than 5,900 feet in elevation), where water flows were coolest and most consistent during the summer. Trout densities were lowest in the low-elevation reaches (those below 4,920 feet). Trout densities ranged from 0.21 fish per sq. yd (0.25 fish per sq. m) or less in the low-elevation reaches, to 0.45–0.91 fish per sq. yd (0.54–1.09 fish per sq. m) in pools in the high-elevation reaches. These

high-elevation reach densities are well within the range typically observed in other eastern Oregon and Nevada trout streams.[8]

Interestingly though, young-of-the-year trout rearing in the warmer, lower-elevation reaches grew better than those in the higher-elevation reaches. Median lengths of age-0 trout in October were 2.83 inches (72 mm) in the reaches below 4,920 feet; 2.52 inches (64 mm) in the 4,920–5,900 feet sites; and 2.01 inches (51 mm) at elevations above 5,900 feet. This could be a result of the warmer water spurring faster growth in the low-elevation reaches, or the higher trout densities suppressing growth in the higher-elevation sites, or perhaps both.

When the 1994 survey team compared their results with the surveys completed in 1985 and 1989, they noted that trout densities increased downstream in mid- and lower-elevation sites during wet years when there was more water in the stream channels longer into the summer, but retreated to a few headwater reaches with cool, dependable summer flows during dry periods. The team identified six high-elevation stream reaches (mapped in Fig. 7-1) where these populations persist year-round, during both favorable conditions (wet years) and unfavorable conditions (dry years). These *core populations* expand downstream during favorable conditions (as in 1985) and contract to the core reaches during unfavorable conditions (as in 1989 and 1994).[9]

7. See Buckman (1989), Perkins et al. (1991), Jones et al. (1998), and Gunckel and Jacobs (2006) for reports covering the 1985, 1989, 1994, and 2005 surveys. I have never been able to find a report for the 1999 survey.

8. A good reference to have handy for making comparisons like this is the report "Density and Biomass of Trout and Char in Western Streams," USDA Forest Service General Technical Report INT-241 (see Platts and McHenry 1988). As I noted in footnote 7 above, I have never been able to find a report for the 1999 survey, but a reference to it in Gunckel and Jacobs (2006) indicates that even more trout were present in 1999, perhaps as many as 57,400 if the young-of-the-year were included, as they were in the 1994 count. That would make 1999 the highest abundance recorded in all of the five surveys to date.

9. Metapopulation connectivity can be described by at least five different models of ways individuals can move among the aggregate of populations. In one of these, called the *mainland/island* or *source/ sink* or *core-and-satellite* model, dispersal from a large "core" (or "mainland" or "source") population results in "satellite" (or "island" or

However, the team did note that the core populations in Cottonwood and Doolittle creeks have waterfalls near their stream mouths. Downstream migration from these populations is possible, especially in high stream flows, but upstream migration is blocked by these falls. Therefore, metapopulation structure in the Whitehorse Creek system is limited. Should either or both of these core populations wink out, their core habitats cannot be recolonized by immigration from other populations.

STATUS AND FUTURE PROSPECTS

In 1994, an estimated 39,500 cutthroat trout occupied Coyote Basin streams including Willow Creek (13 miles of occupied stream habitat), Little Whitehorse Creek (8.7 miles of occupied habitat), and Whitehorse Creek plus its tributaries, Fifteenmile, Doolittle, and Cottonwood creeks (13 miles total of occupied stream habitat). Six core populations were identified in high-elevation reaches of these streams with a limited amount of metapopulation connectivity. A large total number of trout, fish densities (in the core population areas) well within a healthy range for arid-western trout streams, an equally healthy-looking distribution of age classes—all pretty respectable indicators for a component of a DPS that the U.S. Fish and Wildlife Service lists as threatened under the Endangered Species Act. But it hadn't always been that way.

Old creel records and angler checks do testify to an abundance of trout in the Willow and Whitehorse creek drainages prior to 1960. Fishing was good enough to draw anglers despite the area's rugged remoteness, especially in upper Willow Creek where biologists have subsequently learned the core populations of that drainage are located. To illustrate how numerous the trout were in upper Willow Creek at the time, here is an excerpt from an Oregon Department of Fish and Wildlife report dated 1955:

> "On April 26, 1955 a trip was made into upper Willow Creek for the purpose of obtaining a number of native cutthroat trout.... The trout were found to be quite plentiful and little difficulty was experienced in obtaining 650 trout.
>
> Several methods were employed to secure the trout; however, the most effective means was the hook and line with a fly and piece of worm attached. Several beaver ponds were drained down to a point where a seine could be used and thus a large number of trout were secured by this method."[10]

Fifteen years later, however, that easy angling was only a memory, as the following quote from an agency report dated 1970 attests:

> "In past years, Willow Creek was considered to have been the more popular angling stream in the Whitehorse/Willow Creek drainage. Today, however, angler use on Willow Creek is considered to be much lower than it was prior to 1960. Angler use since that date has been very light. In past years, Willow Creek was considered to be one of the better small early summer fishing streams in the region."

"sink") populations at or beyond the periphery of the core. These can retract to the core during adverse times. This model seems to best describe what Oregon Department of Fish and Wildlife biologists have observed in the Coyote Basin cutthroat populations, where trout distribute downstream from core areas during wet years and retreat, if they can, during periods of drought. To learn more about metapopulation models, see Harrison (1991). Schlosser and Angermeier (1995) discuss applications to stream fishes.

10. The passages quoted here, as well as the numbers of trout cited, came from an assortment of Oregon Department of Fish and Wildlife and Bureau of Land Management monthly reports, memos, and creel census notes compiled and provided to me by the BLM biologist Mike Crouse, as "homework" for my 1981 field trip.

These are stream systems that had been exposed to livestock grazing for almost 90 years prior to 1955, yet they continued to provide good fishing. But after 1960 they did not. What changed so abruptly in these watersheds in that five-year period that had such an adverse effect?

I found the answer in an Oregon Department of Fish and Wildlife stream survey report, also dated 1970, which said:

> "In 1954, much of the cover along the stream and canyon bottom [in both the Willow Creek and Whitehorse Creek drainages] was destroyed by fires set intentionally to clear out the brush in order to provide more grass for cattle. This resulted in heavy erosion and siltation during the following years. Many of the active beaver dams found on Willow Creek...were silted in and washed out causing frequent changes in the stream course. With the elimination of the beaver ponds, a majority of the best trout habitat was lost. Today, pools or good holding water is scarce on Willow Creek."

In other words, in the years following the intentionally set fires, the stream ecosystems simply unraveled and continued grazing never gave them a chance to recover. In 1980, only 170 trout were collected by electrofishing that same core-population segment that had yielded the 650 trout in 1955. This is the same segment where Mike Crouse, Steve Pribyl, and I stopped to do our electrofishing in 1981. While we did find trout, we didn't do a population estimate, so I can't say how we stacked up against the 1980 survey. But I can say that

we did find plenty of fine sediment in the substrate, severely trampled stream banks, and ample evidence of instream and riparian habitat damage, all symptoms of over use by grazing cattle.

Both the agencies and the Whitehorse Ranch, which is the principal grazing lessee and whose owner is a fisherman who was concerned about the trout, had taken steps early on that they hoped would improve habitat conditions for the trout and reverse population declines. In 1962, the Bureau of Land Management's $14 million Vale Project brought almost 2,000 miles of fences to divide the public range into pastures, 574 small reservoirs, and over 200,000 acres of seeding using the Eurasian native, crested wheatgrass—not all of this in Coyote Basin drainages of course, but all intended to improve the range. The cows were happy and so were the ranchers, but the streams continued to unravel and the water table continued to decline. So, in 1971, a habitat management plan was initiated for Willow and Whitehorse creeks in which rim-and-gap cattle exclosures were constructed,[11] 49 trash-catcher dams were built in the streams to create pool habitat, and about 40,000 willow cuttings were planted along 7 miles of stream bank. But aside from exclusion of cattle from riparian areas within the exclosures, nothing much was done to alter grazing intensity elsewhere along the streams, or in the uplands.

11. Rim-and-gap exclosures work as follows. Steep canyon walls generally prevent access to the stream except at major side drainages. Some of these major draws are fenced off at the top to prevent access, but others are left open so that livestock can come down to water. But the stream is fenced above and below these water gaps to prevent cattle from moving up or downstream. This is done by extending the fences upslope to tie into the existing rimrock. Gates in the fences allow cattle to be herded through the exclosure to and from allotted grazing areas.

Subsequent inspection revealed that the willow plantings either had not taken hold or had themselves been removed by grazing. Only about 20 of the trash-catcher dams still remained when we toured the area in 1981, and less than half of those had created the hoped-for pool habitat. The rest had silted in or else the water had channeled around them.

But the cattle exclosures were remarkable successes. Regrowth of streamside vegetation was excellent within them. The dense new growth stabilized banks and provided shade and cover, which in turn cooled the water. Shaded pools within exclosures were as much as 9 degrees Fahrenheit cooler than exposed pools just outside the exclosures in the hottest part of the day in late summer. Within the exclosures, where logs had fallen or boulders had tumbled into the streams, good natural instream pools, pockets, and cover had been created.

The habitat management plan was updated in 1980 to provide for additional exclosures and more instream habitat structures (log structures and boulder placements to emulate what nature was doing, rather than the ineffective trash-catcher dams) to control erosion and create pools. A land exchange was also executed with the Whitehorse Ranch that added four additional stream miles of stream habitat to Bureau of Land Management ownership, bringing the total to 36 of the 40 stream miles known to be occupied by trout in 1980.

One key objective of the 1980 habitat management plan was to improve riparian vegetation and cover along the perennial portions of the drainages, even in reaches outside of exclosures. This was to be done by managing grazing so that riparian vegetation could establish itself and be maintained along these reaches, and so that adjoining slopes would also establish improved surface cover. Not only would riparian and instream habitat improve, re-established vegetation would also help control erosion and increase the water table. The new proposals for grazing management were:

- A shift in the rest-rotation system, from the standard 2-year rotation to one that would allow at least 3 years of rest from grazing for important riparian areas not within exclosures.

- A change in season of use for grazing in important riparian areas, from spring-through-fall to winter-spring only, to permit vegetation regrowth; or, grazing of these areas by yearling cattle only, because yearlings spend less time on stream bottoms than cows and calves.

- A shift of grazing pressure away from riparian areas by developing additional seedings and water sources at other locations.

Looking back, this is about when the dissention started. Ranchers, faced with grazing fewer cows or otherwise grazing less intensely on their allotments, became more recalcitrant and testy. Implementation of proposed changes was delayed. Environmental organizations and conservation groups, becoming more involved in the management of federal lands, threat-

Willow Creek has cut down to bedrock at this location. The resulting high-gradient cascade is not considered a barrier to fish passage. PHOTO BY AUTHOR

ened lawsuits. Meanwhile, trout habitat remained in a seriously degraded condition and trout populations remained depressed.

By the summer of 1988, the situation had deteriorated to the point that ranchers were on the verge of having grazing permits lifted. That's when the equivalent of a miracle occurred. That summer, the fractious parties came together and commenced to work their way toward long-term solutions to the problems that could be applied on the ground and would be satisfactory to all. It wasn't easy, but the Trout Creek Mountain Working Group persevered and succeeded. [12] In 1989, the Whitehorse Ranch agreed to a 3-year rest of high elevation pastures in its Whitehorse Butte Allotment. The new Whitehorse Butte Allotment Management Plan, in place since 1992, permits grazing for two months, down from four months in the mountain pastures, plus a reduction in the number of cows in these pastures and in lower elevation pastures. A number of other range improvements were constructed as well, to provide forage and water for the cattle.

Grazing continues in the Trout Creek Mountains, but at a lighter, ecologically sound level. Trout habitat has shown a remarkable rebound, as have the trout populations of the Willow and Whitehorse creek drainages. In 1991, following the designation of the Willow/ Whitehorse cutthroat as part of the Northwestern

12. The Trout Creek Mountain Working Group consisted of representatives from the Oregon Cattleman's Association, the five ranches that operate in the Trout Creeks, the Oregon Environmental Council, Oregon Trout, the Isaac Walton League, the Oregon Department of Fish and Wildlife, Oregon State University, the U.S. Fish and Wildlife Service, and the Bureau of Land Management. Much of the credit for getting these people together and keeping them working together goes to Connie and Doc Hatfield of Brothers, Oregon, a ranching couple who had the vision, along with prior experience getting fractious parties working together when they'd just as soon be punching one another out. I learned more about the Trout Creek Mountain Working Group at the website *www.mtnvisions.com/ Aurora/tcmtinfo.html*. The experience of this group is also documented as a case study by the organization Sustainable Oregon, *www.sustain-ableoregon.net/casestudies/more. cfm?caseID=37*, and in an article by Pulitzer Prize-winner Tom Knudson in *High Country News* back in 1999. I found this at *www.hcn.org/servlets/ hcn.PrintableArticle?article_id=4810*. All of these websites were still up when I last checked in January 2007, but nothing regarding more recent activities of the Working Group was posted.

A trash-catcher structure constructed by t BLM in 1979 to catch debris and create poo habitat behind the resulting dam. These u mostly ineffective because, as in this one, the stream tended to channel around then But exclosures, glimpsed in the backgroun were quite successful, allowing healing of riparian zone that in turn improved the stream habitat. PHOTO BY AUTHOR

Lahontan Basin DPS of the threatened Lahontan cutthroat trout, the U.S. Fish and Wildlife Service examined the new Allotment Management Plan as required by law and issued a "No Jeopardy" opinion. Riparian and instream habitat recovered nicely, and the increased numbers of trout in the streams in 1994 and 1999 attested to the success of the Trout Creek Mountain Working Group.

But alas, the trout have not been able to sustain those levels of gain. The 2005 survey found a population estimated at only about 14,000 age-1 and older trout (young-of-the-year were not counted in the 2005 survey), which is the second lowest level since the 1989 survey that estimated that only about 9,000 age-1 and older trout were present in basin streams. The steep decline in abundance seen in 2005 was attributed to the effects of prolonged drought. Five years in a row of lower-than-normal streamflows had impacted the Coyote Basin. The trout had retreated to their core areas, and even in those reaches, productivity and recruitment was low. Clearly, trout abundance in Coyote Basin streams is highest during higher flow regimes, which may bode ill for these populations if, as expected, global warming and climate change brings intensified drought conditions to this region.

Paiute Cutthroat Trout, *Oncorhynchus clarkii seleniris*

CHAPTER 8

Paiute Cutthroat Trout

Oncorhynchus clarkii seleniris: Chromosomes, 2N = 64. All meristic characters are similar to Lahontan cutthroat (Chapter 5). Scales in lateral series 150–180; scales above lateral line 33–40; vertebrae 60–63; pyloric caeca 50–70; gill rakers 21–27. The only trait distinguishing *seleniris* from *henshawi* is the consistent absence of spots on the body of *seleniris*. But even this is not a sure-fire distinction. Dr. Behnke wrote of

reexamining the original specimens used to name *seleniris* and finding as many as nine body spots. He has also written about observing specimens of a remnant population of *henshawi* isolated in the very headwaters of the East Fork Carson River which had almost no body spots and which he said he would have classified as *seleniris* had they been found where Paiute cutthroat trout occur. The California Department of Fish and Game now uses a five-spot criterion to separate unhybridized from hybridized trout. *Seleniris* retains its parr marks to maturity. The coloration most often seen on specimens today is olive-bronze on the back shading to light lemon-yellow on the sides, with white undersides and often a faint rose band along the lateral line. Ichthyologist J.O. Snyder's original description of coloration read as follows: "The color is pale, the whole body much suffused with yellow. The upper surface is pale yellowish olive or greenish olive in some lights; lateral stripe light coral red; region below rich ivory; ventral surface clear white; head light brownish above; the cheeks red like the lateral stripe; dorsal fins and caudal suffused with yellow and pink; lower fins pink. The entire body exhibits evanescent opaline reflections, and the skin is translucent, so much so that the dorsal cranial bones are partly outlined through the overlying tissue." [1]

1. J.O. Snyder's original description of the Paiute cutthroat trout was published in two papers (see Snyder 1933 and 1934) and a summary description was published later in *The Trouts of California* (see Snyder 1940). The 1934 and 1940 publications include a color plate painted by artist Chloe Lesley Starks from living specimens that were displayed at the Steinhart Aquarium in San Francisco. Snyder stated that this painting well portrayed the characteristic colors of this trout, whose evanescent tints bore a fanciful resemblance to the rainbow-like halo that sometimes forms around the moon; hence the name *seleniris*.

Silver King Creek heads up at an elevation of 9,600 feet in a saddle high in the Sierra Nevada Mountains of Alpine County, California, in the southernmost portion of the East Fork Carson River drainage. Off to the west lies the Sierra crest. Over to the east a few miles, the slanting border of California butts up against the State of Nevada. Jedediah Smith, the mountain man, crossed the Sierras just north of here in 1827, and Joseph Walker passed through south of the saddle as he worked his way up the West Walker River and into California via Sonora Pass. Immigrant parties bound for California later used Walker's route as one of their main arterials. John C. Fremont led his exploring expedition over the mountains a little to the east and south of the saddle in the winter snows of 1844. While struggling over one particularly difficult peak, Fremont abandoned a brass cannon. Folks have been hunting that cannon on the slopes of "Lost Cannon Peak" ever since.

The Creek drops quickly out of the saddle, then slows its descent as it flows through timber and on past sagebrush covered slopes. Glacier-strewn boulders litter the landscape. Here and there, water trickles in from nearby soda springs. Presently, the creek enters an area where dried-up channels and the skeletons of trees speak of washed-out beaver ponds.[2] This is the confluence of Fly Valley Creek. A little further down, Four

Mile Canyon Creek comes in off the slopes to the east. Now the land opens out into a mile-long valley, the first of three glacier-sculpted valleys that Silver King Creek will encounter on its fourteen-mile journey to the East Fork Carson River. This one is Upper Fish Valley. The creek becomes a meandering meadow brook here, digging deep undercuts into the grassy banks.

A low, sparsely timbered moraine guards the lower end of Upper Fish Valley. Over the years, the creek has eroded through this moraine and cut its way down to bedrock. Here, it cascades down to the lip of Llewellyn Falls, where it drops into Lower Fish Valley. The elevation change here is only about twenty feet, and Llewellyn Falls itself can't be more than a six or seven foot drop; nevertheless, it provides an effective barrier to the upstream movement of fish.

In Lower Fish Valley, Silver King Creek meanders again. A couple more tributaries join the mainstream here, then the creek descends into Long Valley. Behind the low ridge that forms Long Valley's west flank lies forested, seldom-visited Tamarack Creek. This creek joins the mainstream about a half-mile below Long Valley. Over behind the east ridge lies Coyote Valley Creek. This creek loses its identity where it merges with Corral Valley Creek, and the latter flows into Silver King about a half-mile below Tamarack's junction. A hiker following Silver King Creek would have

2. The native distribution of beavers in California includes the river drainages of the Great Central Valley and the Modoc Plateau, but not the eastern Sierra Nevada. Where they have been introduced into the Sierra Nevada, they fall aspen and willows, often removing aspen faster than it can regenerate. The dams they build can also flood meadows and cause sedimentation that covers gravel used by spawning trout. This has been the case in the upper Silver King drainage, particularly along Fly Valley Creek. See Ingles (1965), James and Peeters (1988), and Schoenherr (1992) for more information.

easy, pleasant going for about 2 miles below the mouth of Tamarack Creek. But then he or she would come to a steep canyon where the creek descends rapidly. This is Silver King Canyon. There are waterfalls located within this canyon that, like Llewellyn Falls, are barriers to upstream migration of fish. Silver King emerges from its canyon in the vicinity of Snodgrass Creek. From there it flows through Silver King Valley for a shade over 3 miles to its confluence with the East Fork Carson River.

Aside from the grassy valley floors, this is a zone of Jeffrey and lodgepole pine, red fir, and dense stands of aspen. Junipers are also found in the drier places. Near the bottom of one steep, eroded gully on the trail into Upper Fish Valley stands a particularly massive juniper. The trail guides say it is one of the largest and perhaps oldest in the Sierras. But there are a few places where it is too dry even for juniper, and these spots are covered with pinion pine and mountain mahogany. Sagebrush is common along the outer periphery of the undisturbed mainstem meadows. Willows and sedges dominate in the riparian zones.

But this country is high and fairly dry, and the sagebrush can creep out onto the valley floor when the meadows are too heavily grazed, as they have been in the past. Early one morning, out there among that sagebrush, I witnessed one of the most remarkable displays of animal teamwork I have ever seen. As I crawled out of my tent, I spotted a pair of coyotes working up the valley hunting ground squirrels for breakfast. One coyote trailed the other by about thirty feet. The lead coyote would spot a ground squirrel and chase it into its burrow. This coyote would dig excitedly at the opening, then sit back and watch anxiously for a moment. Then more digging and another spell of watching, each time less intently than the time before, until finally it would lope off up the valley as if discouraged. Meanwhile, coyote number two would creep stealthily up to the opening. The ground squirrel, unaware of the second coyote and thinking the coast was clear, would pop out of his hole, and POUNCE— breakfast! The coyotes would then repeat the whole process as they worked on up the valley. [3]

Before the mining boom that kicked off in the region following the discovery of the Comstock Lode in 1858, the only visitors to the valleys and canyons of Silver King Creek were Washoe and Paiute Indians, who would move in when the seasons permitted to gather food. Pine nuts were a staple. But when the white man came, the area became sort of a "working wilderness"—although "wilderness" might not be quite the right word. A bustling silver mining center developed at Kongsberg, later renamed Silver Mountain City, a few miles northwest of Silver King Creek,

3. I have visited the meadows of upper Silver King Creek three times, first in 1980, then in 1984, and again in 1997. In 1980 and 1984, cattle were grazing the meadows. In 1997, cattle had been off the allotment for two years. The difference was astounding. The ecology of these meadows must change profoundly when they are grazed. I recall that sagebrush, normally present only along the periphery of the meadows at the edges of the slopes, was distributed well out across the valley floor in 1980 and 1984. What grass there was grew in sparse, close-cropped clumps and patches, and ground squirrel burrows were everywhere. In contrast, in 1997 after just two seasons of rest, the grass was thick and lush, up to our knees in some places, the sagebrush had retreated to the base of the slopes, and there wasn't a ground squirrel burrow to be seen. Another thing that stands out in my mind was the proliferation of tiny green Pacific treefrogs in those lush, grassy meadows in 1997. They were literally everywhere. "Millions and billions of them," my wife exclaimed. I don't recall seeing any during my visits in 1980 and 1984. I observed the ground squirrel-hunting coyotes in 1980. I thought what I saw was unique until my wife made me a present of the book *The Voice of the Coyote* by J. Frank

and a smaller, shorter-lived mining center with a cluster of buildings, a sawmill, and 15 mining claims developed at Silver King City in the lower part of the Silver King Creek drainage. [4]

The miners themselves did not work the upper part of the Silver King drainage, but the loggers and woodcutters did, in order to supply timbers for the mines, lumber for the flumes, trestles, and buildings, and cordwood for the fuel that the mining camps required. And when the loggers and woodcutters pulled out, livestock grazing took over. Members of the ethnic group known as the Basques were among the earliest to graze sheep in the high meadows. To pass the time, lonely Basque herders built rock cairns, called "arri mutillaks" or "stone boys," and carved their names on aspen trees for other passersby to read. [5]

For years, through the summer of 1994, cattle grazed the valleys of the Silver King drainage. When I first visited the drainage, the authorities didn't feel that grazing posed much of a problem, although the meadows were covered with more sage and sand than with grass, and stream bank erosion was glaringly evident. But later, there was a change of heart. In the summer of 1986, volunteers from Trout Unlimited joined U.S. Forest Service personnel to build several fenced cattle exclosures, and in subsequent years, other work was done to repair and stabilize damaged stream banks.

In 1995, the Silver King grazing allotment was placed under administrative rest. It is now vacant and the plan is for it to remain so.

HISTORICAL RANGE

Silver King Creek and its tributaries, identified with the letter "D" in Fig. 5-1, comprise the southernmost drainage of the East Fork Carson River, on the very fringe of the natural range of the Lahontan cutthroat trout. Figure 8–1 shows an enlarged view of the Silver King drainage including all the landmark features I refer to in this Chapter.

Sources differ as to the timing, but anywhere from 60,000 to 12,000 years ago, late-Pleistocene alpine glaciers occupied the drainages of the East Fork Carson River headwaters, including the drainage of Silver King Creek. The Silver King glacier was a relatively short one, extending down-drainage only about 8 miles and leaving its terminal deposits not far from the mouth of Corral Valley Creek. Other recessional moraines were deposited further up the drainage as the glacier retreated, including the one at the lower end of Upper Fish Valley where the creek cut down to form Llewellyn Falls.

When these glaciers retreated for the last time, Lahontan cutthroat trout moved up into the newly accessible waters. It may have happened quickly, or it

Dobie (2001), where I found similar cooperative behaviors described. No wonder Native Americans featured Coyote as the trickster in so much of their lore.

4. I have in my possession a copy of an 1864 "Map of the Silver Mountain Mining District including the Territory of the Proposed new County of Alpine," compiled by Theron Reed and published by H.H. Bancroft & Company, San Francisco. The original is in the Bancroft Library, University of California, Berkeley. This map shows Silver King City as a cluster of buildings and 15 associated mining claims spreading up the slopes on both sides of Silver King Creek, which goes by the name "East Fork of East Carson River" on this map. The Silver King Mill and Lumber Company sawmill, a water-powered operation, started up in 1865 about a mile and a half downstream from the town (see Editors 1941).

5. The old aspens are decaying and falling now, and it won't be long until this record of the Basques' presence will be gone. Scholars studying Basque culture have visited the Silver King drainage from time to time to transcribe the old names. You can read one treatise on these Basque tree carvers in Mallea-Olaetxe (2000).

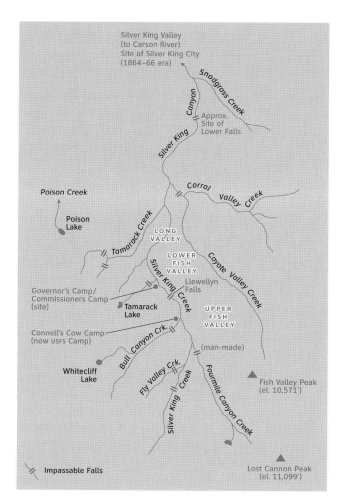

FIGURE 8-1. *Silver King Creek and its tributaries, Alpine County, California. Historic and present range of* Oncorhynchus clarkii seleneris, *the Paiute Cutthroat.*

may be that many generations of Lahontan cutthroats lived out their lives in these newly populated waters. But at some point, as the creek cut down to bedrock and formed barrier falls, populations became isolated from fresh infusions of Lahontan stock from below. On Silver King Creek, such a barrier falls developed in the Silver King Canyon reach upstream of Snodgrass Creek. And, of course, another barrier falls developed further upstream at Llewellyn Falls. As the millennia passed, populations isolated above or between these barriers in Silver King Creek adapted and evolved into the distinctive form that we recognize today as the Paiute cutthroat trout.

The present distribution of Paiute cutthroat trout in the Silver King drainage is in Silver King Creek itself upstream from Llewellyn Falls, in Four Mile Canyon and Fly Valley creeks, both tributaries of Silver King upstream from Llewellyn Falls, and in Corral Valley and Coyote Valley creeks upstream of a barrier in lower Corral Valley Creek. With the exception of Fly Valley Creek, which had its own barrier falls and was fishless at the time, this was the distribution occupied by the Paiute cutthroat when J.O. Snyder was informed of its presence and first described the trout for science in 1933. Silver King Creek upstream from Llewellyn Falls became the type locality for the newly described trout, which Snyder initially named as a full species, *Salmo seleneris.* [6]

6. See Snyder (1933, 1934).

Silver King Creek, California, view upstream toward headwaters. PHOTO BY AUTHOR

Snyder had learned of this unique trout from Thomas Hanna, a mine owner, who was also the son-in-law of the famed naturalist and conservationist John Muir. Hanna was an old hand in the eastern Sierra Nevada, and had spent a period early in the 20th century working as a surveyor in the Silver King drainage. He had returned many times to fish for these trout in the waters above Llewellyn Falls. Mrs. Lynn Llewellyn, owner of the pack station at Leavitt Meadows on the West Walker River, was another who had known about these trout for many years. Mrs. Llewellyn outfitted the men who collected Snyder's specimens. She called these fish "Paiute trout," because "these trout and the Paiute Indians are the old settlers of these mountains," and that was the common name Snyder chose to adopt. [7]

However, it is possible that *none* of the stream reaches and tributaries where Paiute cutthroat trout are now found, including the type locality reach of Silver King Creek above Llewellyn Falls, may actually be the native range of this subspecies. A fortuitous series of transplants may have established populations in new sections of the Silver King drainage that had formerly been barren of fish, just in time to keep the subspecies from going extinct and preserving it for Snyder to collect and describe for science in 1933. It is possible that the true native distribution may have been that part of the Silver King drainage located between Llewellyn

Falls and the downstream barrier falls in Silver King Canyon. There is still some lingering uncertainty about all this, which I'll touch on later, but the accepted version of what transpired follows the recollections of Virgil Connell, an early stockman. Here is that story. [8]

The first white men to encounter the Paiute cutthroat trout were probably the miners or woodcutters who worked in the drainage in the 1860s. Virgil Connell believed it was these men who transplanted fish into previously barren Corral Valley and Coyote Valley Creeks. Connell himself first visited Lower and Upper Fish Valleys with Harry Powell, a boyhood chum, in 1890. They found Paiute trout in abundance, and returned home after a three-day trip with 1,500 trout. All of the trout caught on that trip and several others over the next few years were caught *below* Llewellyn Falls. Not a fish did they find in the waters above Llewellyn Falls.

Connell ran sheep in the upper valleys of the Silver King drainage from 1893 until at least 1944. In 1912, one of his herders, a young Basque named Joe Jaunsaras, put some Paiute cutthroats in a can and carried them above Llewellyn Falls. By 1924, according to Connell, the stream above the falls was so well populated that fishing in Upper Fish Valley was actually better than in Lower Fish Valley. It's a good thing, too, because, beginning about 1925, rainbow

7. I am grateful to Eric Gerstung, California Department of Fish and Game, for providing the biographical sketch of Thomas Hanna, and also the information about Mrs. Lynn Llewellyn and the naming of the Paiute trout.

8. Virgil Connell's recollections were contained in a letter he wrote to Brian Curtis, California Department of Fish and Game, dated August 6, 1944. This letter is included as Appendix B in the report, "Status of the Paiute Cutthroat Trout, *Salmo clarkii seleniris* Snyder, in California" (California Department of Fish and Game, 1976); see Ryan and Nicola (1976).

trout were stocked repeatedly, and probably Lahontan cutthroats as well, into the stream below Llewellyn Falls and the fish in Lower Fish Valley were becoming increasingly introgressed.

When J.O. Snyder published the first scientific description of the Paiute cutthroat trout in 1933, his specimens were collected from above Llewellyn Falls. The fish below Llewellyn Falls, in the Paiute cutthroat's presumed original territory, were by then thoroughly introgressed. These Snyder described as being "well covered with large and very conspicuous roundish black spots," which he compared to Lahontan cutthroats from Lake Tahoe, Truckee River, and Pyramid Lake. He further described the trout from below Llewellyn Falls in 1933 as being considerably deeper in the body and more robust in appearance, and not nearly so bright in coloration as the slender-bodied, evanescently hued Paiute trout from above Llewellyn Falls. [9]

That fortuitous stocking of Paiute cutthroats in Upper Fish Valley above Llewellyn Falls by the young Basque sheepherder may have saved the subspecies from extinction. But had Upper Fish Valley always been barren of fish? In 1991, Eric Gerstung of the California Department of Fish and Game compiled information on native trout distribution in the Silver King drainage based on letters, interviews with long-time residents, published articles, and notes in California Department of Fish and Game files. Although Upper Fish Valley may have had no trout in 1912, Thomas Hanna reported catching trout there in 1900. Also, Gerstung found an old hand-drawn map of the area prepared by the California Geological Survey in 1866, which suggests fish may have been present. On this map, Gerstung noted (compare landmarks with Fig. 8-1), "Corral Valley Creek is labeled as 'Poisoned Carson Creek' and Silver King Creek is called 'Silver King Fork.' The outlet of Tamarack Lake is called 'Poisoned Valley Creek.' Upper Fish Valley is called 'Fish Lake Valley.' The map shows two cabins in Lower Fish Valley and a trail paralleling Silver King Creek." The "Fish Lake Valley" name used by the old mapmaker implies the presence of fish in 1866.

Gerstung also found evidence that a splash dam existed at the lower end of Upper Fish Valley during the logging and woodcutting days. "A breached dam about 15 feet high exists near the valley outlet," he wrote. "This splash dam was periodically breached during the spring to flush cordwood downstream.... By 1866 the lake may have been allowed to remain full thus becoming good fish habitat, therefore the name." French-Canadian loggers and woodcutters could have provided the word "poisson" for fish, later anglicized to "poisoned" on the 1866 map features. [10]

If one assumes the truth of Hanna's testimony and

9. See Snyder (1934).

10. Eric Gerstung compiled the information cited here in a 1991 "memo to file" titled "Some Notes on the State of Confusion Regarding the Historic Distribution of Paiute Cutthroat Trout in the Silver King Drainage," which he updated in 1997. Regarding the 1866 hand-drawn map prepared by the California Geological Survey, Eric noted that the original from which his notes were taken is in the Bancroft Library, University of California, Berkeley. I was not able to locate this particular map, even with the aid of the Bancroft Library staff. We did find three other maps of Alpine County dating from the 1864–66 period, including the Theron Reed map I cited in footnote 4 above, but none of these contained the details Eric described in his memo.

the implications of the 1866 map, then the historical range of the Paiute cutthroat trout would have included the Upper Fish Valley reach of Silver King Creek downstream to the barrier falls in Silver King Canyon, and all accessible tributaries within that span. Four Mile Canyon Creek was probably fish-bearing water as well in this scenario, because the mapped barrier near its lower end was originally a beaver dam that washed out occasionally, allowing fish passage until a permanent barrier was erected by Cal Fish and Game. The accessible range within this span, both above and below Llewellyn Falls, totals to about 14.6 stream miles, the lowest amount of historical range of any of the cutthroat trout subspecies.

Through the supposed early introductions by loggers and woodcutters, populations of Paiute cutthroat trout exist in Corral Valley and Coyote Creeks, and through the efforts of the California Department of Fish and Game, populations are present in Silver King Creek upstream from Llewellyn Falls, in Four Mile Canyon Creek, and in Fly Valley Creek. Beginning about 1937, Paiute cutthroats were introduced into waters in other drainages in order to establish refugial populations. Of these, only North Fork Cottonwood Creek and Cabin Creek in the Inyo National Forest, Mono County, and Stairway and Sharktooth creeks in the Sierra National Forest, Madera and Fresno counties, have self-reproducing populations at this time. A 1937 transplant into upper and lower Leland Lakes, El Dorado County, California, failed. A 1957 transplant into Birchim Lake, Inyo County, succeeded and the population became self-sustaining, but was lost to introgression by rainbow trout that had been stocked upstream of the lake and later escaped and interbred with the Paiute cutthroats.

LIFE HISTORY AND ECOLOGY

Paiute cutthroat trout populations exhibit a stream-resident life history, with trophic movements within the stream for spawning and summer rearing, and refuge movements in winter to find shelter from stream ice. [11] In these regards, their life history and habitat requirements are little different from those reported for other western stream-dwelling salmonids. Paiute cutthroats can survive and grow well in lakes and ponds (the introduced population in Birchim Lake produced fish up to 18 inches in length [12]), but the only hint of the existence of a naturally-occurring lake population within the historical range is the cryptic evidence of that 1866 map of "Fish Lake Valley."

Paiute cutthroats are spring spawners; that means late June or early July at the altitude of the Silver King drainage. Sexual maturity is reached at age 2, and the fecundity of females is consistent with what has been

11. Life history and ecology summaries of Paiute cutthroat trout can be found in McAfee (1966b), Behnke and Zarn (1976), Moyle (2002), and U.S. Fish and Wildlife Service (2004). Detailed studies of the introduced population in North Fork Cottonwood Creek are reported in Diana (1975), Wong (1975), and Diana and Lane (1978).

12. See McAfee (1966b).

reported for other stream-resident trout, i.e., from 50 to 300 eggs per female, depending upon size and age. Like other stream trout, Paiutes require clean gravel and cold, flowing, unpolluted water for spawning.

The eggs hatch in 6 to 8 weeks, and it may be another 2 to 3 weeks before fry emerge from the gravel. Early on, young-of-the-year trout rear in channel margins or they may even enter intermittent tributaries. But when they reach a size of about 50 mm (2 inches), they seek out runs or riffles where they can find more productive feeding habitats. As they grow, they will move into pools and slow, meandering meadow sections with undercut banks and healthy riparian vegetation. They set up dominance hierarchies in summer-rearing stream habitats, and will defend their territories.

In the environment of Silver King Creek and its tributaries, Paiute cutthroat trout are feeders of opportunity, the fare being aquatic and terrestrial insects. The fish attain lengths of 6 to 8 inches after three years of growth. We did electroshock one fish 11.5 inches in fork length out of Silver King Creek during one of my week-long stays with the California Department of Fish and Game, and tiny Corral Valley Creek yielded a trout of 10.5 inches in fork length. Bill Somers, the Cal Fish and Game crew leader on more recent surveys, lists 13.5 inches in fork length as the maximum size he has observed in Silver King Creek. But these sizes are

the exceptions. Only the occasional fish ever exceeds 8 inches in fork length in the stream environment.

Perhaps owing to their Lahontan cutthroat origin, Paiute cutthroat trout can survive in lakes and can attain their greatest lengths in such environments. Although the fish also do well growth-wise in beaver ponds, the overall population benefits are only temporary, according to Cal Fish and Game biologists. When the beavers exhaust available food supplies, they abandon their ponds and move on, leaving washed out dams, mud flats, and eroding channels. Thus, in the Silver King drainage, where beavers are not native, they have brought about a net *loss* of suitable habitat. Graphic evidence of this could be seen on Fly Valley Creek just above its confluence with Silver King Creek. Downed trees and material from washed out beaver dams had clogged the little stream and diverted it from its main channel. Beaver dams also blocked spawning movements, inundated or silted up spawning gravels, and warmed up the water to the detriment of the trout. Jim Ryan, one of the California Department of Fish and Game biologists who guided me on my first visit to the Silver King drainage, told me that beaver-caused habitat effects led to a 10-fold reduction of the populations in the Four Mile Canyon Creek and Fly Valley Creek populations. Although it's hard to imagine, the authorities once considered habitat damage from bea-

Silver King Creek, Upper Fish Valley, as the author found it in 1997. PHOTO BY AUTHOR

ver colonies in the Silver King drainage to be a much more serious threat to the existence of Paiute cutthroat populations than the streambank erosion and riparian damage caused by grazing cattle.

Paiute cutthroat trout do not coexist with other trouts. Rainbow and Lahontan cutthroat trout hybridize readily with them, and non-native species, such as brook trout and brown trout, are capable of outcompeting them, as they seem to do with all other subspecies of cutthroat trout, for all but the most marginal of habitats.

STATUS AND FUTURE PROSPECTS

In terms of historical range, the Paiute cutthroat trout is the rarest of all the existing cutthroat subspecies. If its native range was indeed that section of Silver King Creek and its tributaries between Llewellyn Falls and that second impassable falls downstream in Silver King Canyon, as Virgil Connell's testimony suggests, then never in its entire existence up to Connell's time did the Paiute cutthroat subspecies ever occupy more than about 10 stream miles of habitat. And it's thankfully that man intervened, however inadvertently, or even that small population would have been lost.

Although man may indeed have helped preserve the Paiute cutthroat trout in pure form, after 1933 he seemingly tried his best to blot it out again. Poaching was heavy, even though Silver King Creek was closed to angling above Llewellyn Falls from 1934 to 1952 and again from 1965 to the present. Add to this a couple of unintentional stockings, one of rainbow trout in 1949 and the other of Lahontan cutthroats in 1955, and the stage was set for introgression once again. Sure enough, hybrids began showing up in 1957 and by 1963, pure-strain Paiutes no longer existed in Silver King Creek above Llewellyn Falls. And somewhere along the line, somebody had also put rainbow trout into Corral Valley and Coyote Valley creeks. By 1963, these populations too were thoroughly introgressed.

Fortunately, pure Paiutes had been stocked in Fly Valley and Four Mile Canyon creeks, tributaries of Silver King Creek above Llewellyn Falls, and the Fly Valley Creek population remained untainted. Paiute cutthroat populations had also been introduced into other California waters outside the Silver King drainage. So a few pure stocks did remain from which to try to recoup the setbacks suffered in the Silver King drainage.

The first attempt at this was undertaken in 1964 by treating Silver King Creek above Llewellyn Falls and the Corral Valley/Coyote Creek system with the fish toxicant rotenone, followed by restocking with pure Paiute cutthroats from Fly Valley Creek. Alas, that attempt failed. Spotted trout had reappeared in the treated reaches by 1968. For several years after that, electroshocking was used to remove any spotted trout encountered. Corral Valley and Coyote creeks were rotenone-treated again in 1977, again unsuccessfully, and yet again in 1987 and 1988 in back-to-back applications of rotenone prior to restocking. That seemed to do the trick. Pure-strain trout stocked following that treatment have remained pure, and these trout were used to restock Silver King above Llewellyn Falls and Four Mile Canyon Creek following rotenone treatments of those waters in 1991, 1992, and 1993. These treatments, too, apparently succeeded in removing all introgressed fish, as none have appeared in the treated reaches since.[13]

Meanwhile, in 1970, the Paiute cutthroat trout was added to the list of endangered species kept by the U.S. Department of the Interior. When the U.S. Endangered Species Act became law in 1973, the Paiute cutthroat was one of the three cutthroat subspecies that carried over with a listing of endangered. In 1975, its status was changed to threatened to enable the California Department of Fish and Game to carry out its monitoring and recovery efforts. An additional measure of protection was afforded in 1971, when the U.S. Forest Service completed a land exchange with the Sierra Pacific Power Company in which the Forest Service acquired 640 acres of Power Company ownership in Upper Fish Valley. Then, in 1984, the California Wilderness Act placed the entire Silver King drainage, along with the rest of the East Fork Carson River drainage upstream of Silver King Valley, in the Carson-Iceberg Wilderness.

In 1985, the U.S. Fish and Wildlife Service wrote up its first formal recovery plan for the Paiute cutthroat subspecies.[14] Since then, as noted, Cal Fish and Game has reintroduced and maintained pure stocks of Paiute cutthroat trout in Corral Valley and Coyote Valley creeks, in Four Mile Canyon and Fly Valley creeks, and in Silver King Creek above Llewellyn Falls. The out-of-basin populations in North Fork Cottonwood, Cabin, Stairway, and Sharktooth creeks also have been maintained. These accomplishments fulfilled the objectives of that first formal recovery plan. In 2004, a revised recovery plan was issued that acknowledged these achievements and set forth one additional major objective: to re-establish pure Paiute cutthroats in that original segment of historical habitat downstream from Llewellyn Falls to the impassable falls in Silver King Canyon. When that is done, and long-term conservation plans and conservation agreements are in place to maintain the viability of all Paiute cutthroat populations, then the subspecies will have met the criteria for delisting.[15]

As outlined in the 2004 recovery plan, the program

13. Full details of the history of these recovery efforts can be found in Vestal (1947), Ryan and Nicola (1976), and U.S. Fish and Wildlife Service (1985, 2004). Results of studies to test genetic purity of stocks are reported by Busack and Gall (1981) and Israel et al. (2002).

14. See U.S. Fish and Wildlife Service (1985).

15. See U.S. Fish and Wildlife Service (2004).

to reintroduce Paiute cutthroat trout into the historical habitat segment will include three successive years of rotenone treatment to remove all non-native fish from Silver King Creek between Llewellyn Falls and the lower impassable falls in Silver King Canyon. Tamarack Lake and its outlet stream will also be treated to remove any trout remaining from a release that was made there in 1991. Within one year of the final treatment, pure Paiute cutthroats will be released into the treated reach of Silver King Creek. Restocking may need to continue for several years to enhance recolonization and build population numbers to a level that will maintain adequate long-term genetic variation and buffer against catastrophic events.

There are two amphibian species, the mountain yellow-legged frog and Yosemite toad, whose native ranges also encompass the Silver King drainage. Both are candidates for listing under the U.S. Endangered Species Act. Adults and tadpoles of the yellow-legged frog have been found in the past in Upper Fish Valley, the lower portion of Fly Valley Creek, and at Whitecliff Lake, but evidently not the Yosemite toad. Neither species has been observed in extensive recent surveys of the stream segments to be treated with rotenone, but prior to treatment a resurvey will be made to capture and relocate any amphibians found. Whitecliff Lake, Tamarack Lake, and their outlet

streams will be maintained as fishless waters for the benefit of amphibians. Removing the non-native fishes from the treated segment of Silver King Creek and replacing them with a subspecies of trout that was historically present should also benefit yellow-legged frogs and Yosemite toads. Since these amphibians co-evolved with Paiute cutthroat trout in the Silver King drainage, reintroduction of Paiute cutthroats to their historical habitat downstream from Llewellyn Falls should have no long-term adverse effect on either amphibian species.

The plan to expand the distribution of Paiute cutthroat trout downstream into its historical habitat below Llewellyn Falls has been a dream of Cal Fish and Game biologists for many years. I first heard Eric Gerstung talk about it as a distant possibility way back in 1980. But discussions for actually carrying out this plan didn't begin among the responsible agencies until 1999. Treatment was originally scheduled to commence in the summer of 2002, but there was a glitch. This arose from a complaint filed in U.S. District Court by the Center for Biological Diversity, an environmental advocacy group, against the use of rotenone for the project.

Sadly, this complaint contained a number of inaccurate descriptions of the impacts of rotenone on aquatic animals, assertions that have been tested many times

Beaver-damaged habitat, Fly Valley Creek. Beavers are not native to the eastern Sierra Nevada, and fish-habitat biologists in Paiute cutthroat territory consider them a detriment to recovery. PHOTO BY AUTHOR

over and found to be wanting. [16] It also alleged a link between the use of rotenone and Parkinson's disease in humans, based on a study published in 2000 by a research team from Emory University. [17] The Emory University team worked with lab rats by injecting high concentrations of rotenone (about 20 times higher than are used in fisheries applications) directly and continuously into their jugular veins for 1 to 5 weeks straight. Their objective was to produce a set of molecular, biochemical, and neurological effects that model the way Parkinson's disease is thought to develop. Twelve of the 25 lab animals treated in this way developed the desired symptoms; the others did not. And the researchers themselves stated that lab animals fed the same amounts of rotenone by mouth produced none of the effects they were looking for.

So the alleged link with Parkinson's disease appears to be a spurious claim as well. But the complaint did have the effect of postponing the proposed reintroduction of Paiute cutthroat trout into the subspecies' historical range. 2002 and 2003 passed without any action, as did 2004, when the Lahontan Region Water Quality Control Board declined to vote on a permit for the project even though the State Water Resources Control Board, which oversees the regional boards, had assured Cal Fish and Game that the permit would be issued. The required permit finally came through in

16. Back in Chapter 4, I discussed in some detail how antimycin, a fish toxicant, is used to remove non-native fishes. Rotenone, sometimes called "derris" or "cube root," is another naturally-occurring fish toxicant, this one found in the roots of several Malaysian and South American plants in the bean family. Indigenous peoples of South America have used rotenone for centuries to capture fish for food. Rotenone has been used for fish management applications in the United States since 1934, and is also used as an organic garden insecticide against chewing insects, and in pet and livestock dips to control external parasites. Rotenone has been scrutinized extensively by public health agencies and environmental regulators at both the federal and state levels. Rotenone is insoluble in water, so is usually formulated for ease of dispersal. Commercial rotenone products come in 2.5 to 5 percent concentrations and are typically applied at 0.5 to 1.0 ppm (parts per million) of formulated product, which translates to 0.025 to 0.100 ppm of actual rotenone in the water. Unlike antimycin, which breaks down rapidly, rotenone may persist in water for several weeks. Therefore, when applied in streams,

it is neutralized at the downstream end of the treated reach with potassium permanganate. The National Academy of Sciences has suggested a safe level of rotenone in drinking water of 0.014 ppm; the State of California has suggested 0.004 ppm. "Safe" here is relative to a lifetime of exposure to these levels. In a study done for the U.S. Fish and Wildlife Service, Marking (1988) found that laboratory rats fed doses of 75 ppm of rotenone per day for 24 months suffered no physical or neurological maladies whatsoever. For the facts on rotenone's effects on other aquatic species, see Hamilton (1941), Binns (1967), Cook and Moore (1969), Meadows (1973), Haley (1978), Chandler and Marking (1982), Gilderhus (1982), Fontenot et al. (1994), and Mangum and Madrigal (1999). For more information on the practical applications of rotenone in fisheries management, see Chapter 8 of Sigler and Sigler (1990) and Chapter 10 of Murphy and Willis (1996). Another good source of information is the manual *Rotenone Use in Fisheries Management: Administrative and Technical Guidelines Manual*, published by the American Fisheries Society. You can find this on the Internet at *www.fisheries.org/rotenone*.

17. To read and review the Emory University study for yourselves, see Betarbet et al. (2000).

the summer of 2005, and work was set to begin. But at just the eleventh hour, the federal court stepped in once again, stopping the work with an injunction. Following that, Cal Fish and Game announced that it was giving up on the project.

As I write this, early in 2007, the Paiute cutthroat recovery program remains in limbo. In stopping the restoration work in 2005, the federal court had also ordered that a full Environmental Impact Statement (EIS) be prepared for the project under the provisions of the National Environmental Policy Act (NEPA). That process was started by the U.S. Fish and Wildlife Service in June of 2006. I'm told that the draft EIS should be completed and ready for public comment in the summer of 2007, with the final document to be issued in the fall or early winter. That would clear the road to restarting the restoration work during the field season of 2008. But frankly, having worked with the NEPA process many times myself, I think that's way too optimistic a projection. Given the intransigence of the opponents, there are bound to be additional challenges and delays. Meanwhile, the trout hold out as best they can in their limited sanctuary stream reaches. [18]

18. The Reno, Nevada, office of the U.S. Fish and Wildlife Service has the lead in preparing the Environmental Impact Statement. A call to that office in February 2007 also elicited the information that the Service has a five-year status review of the Paiute cutthroat trout under way that may be completed by summer 2007 as well. That review should provide the latest information on the strength and well-being of existing populations, and whether or not the present listing as threatened under the Endangered Species Act is still appropriate for protecting these populations. Stay tuned.

Llewellyn Falls on Silver King Creek. The water above now contains pure Paiute cutthroat trout. The water below, formerly the historical range of the Paiute cutthroat, now contains rainbow trout and hybrids. PHOTO COURTESY OF RENA LANGILLE, SEATTLE, WASHINGTON

CHAPTER 9

Yellowstone Cutthroat Trout

Oncorhynchus clarkii bouvieri: Chromosomes, 2N=64. Scales in lateral series 150–200, typically 165-180. Pyloric caeca 25–50, typically 35–42, except for a stream-resident population in Sedge Creek, a tributary of Yellowstone Lake but cut off from the lake by geothermal features, which has 52–63 (mean 58) pyloric caeca (Yellowstone Lake and Henry's Lake specimens typically have 40–42 pyloric caeca). Gill rakers 17–23, typically 19–20. Trout from Yellowstone Lake have gill raker counts that can be used as diagnostic characters for those populations, there being typically 20–22 gill rakers on the first gill arch (vs. 17–20 for other Yellowstone cutthroat populations and other cutthroat subspecies) and 5–15 gill rakers on the posterior side of the first gill arch (vs. 0–3 in other populations and subspecies). Trout from Yellowstone Lake typically have more basibranchial teeth than Yellowstone cutthroats from other sources or other subspecies, averaging 20 or more in a sample vs. 5–10 in other cutthroats. Vertebrae 60–63, typically 61 or 62. Spots are typically medium-large, rounded, and concentrated on the caudal peduncle except in Yellowstone Lake specimens, which have spots more evenly distributed on the body. Typical coloration is dull, running to yellowish brown, brassy, or golden-olive tones. Sometimes rosy tints appear on mature fish especially on the cheeks and along the lateral line, but bright golden-yellow, orange, or red colors noted on some of the other subspecies are typically absent from *O. c. bouvieri*.

Yellowstone Cutthroat Trout, *Oncorhynchus clarkii bouvieri*

Yellowstone Cutthroat Trout,
Oncorhynchus clarkii bouvieri, stream-resident form

A TRIP TO YELLOWSTONE MEADOW

The Yellowstone was roily from snow runoff. You could barely see the tops of boulders only eight inches beneath the surface, yet trout were rising here and there to a hatch of tan-colored mayflies. One fish was working steadily not a foot out from an undercut bank just below where I was standing. I floated my fly down to it, and it took—fast! It zipped out into the main current, gave a couple of quick, powerful tugs, and broke me off! The whole thing had taken no more than a few seconds. It was the only fish I was to hook that evening.

It was the 8th of July, 1980. [1] We had set up our base camp in a grove of lodgepole pines at the foot of an imposing dome called Hawk's Rest, after a 17-mile ride across Two Ocean Pass to Yellowstone Meadow. There were nine of us in the party: Fred and Barbara Washburn and their three children from Illinois, ten-year-old P.J. Korn, son of Box K Ranch owner Walter Korn, who had outfitted the expedition, Sue Roberts, the party's cook, Roger Zander, the friendly but quiet wrangler, and myself. The day had been hot, with thunderheads building most of the afternoon. And now, as I reeled in my line and broken leader, lightning flashed at the end of the meadow and heavy drops of rain splattered my face and clothing. That would be all for the fishing this evening.

1. Another, more recent account of a pack trip across Two Ocean Pass to fish for trout in the upper Yellowstone country can be found in Dave Hughes' book, *The Yellowstone River and its Angling* (Hughes 1992). Hughes and other angling writers have also contributed pieces to outdoor magazines about their experiences in this area. One of the first I saw, after my own adventure, was Gerald Almy's "Horsepacking for Teton Trout" in the February 1985 issue of *Sports Afield* magazine. Hughes' piece, "Upper Yellowstone, Horse Packing into the Headwaters of the West's Most Famous Trout Stream," appeared in the November–December 1992 issue of *Flyfishing.* An article by Rich Osthoff titled "A Chance Encounter" appeared in the September/October 1995 issue of *American Angler,* and another, "Upper Yellowstone River Cutthroat Migration," was published in the summer 2001 issue of *Flyfishing and Tying Journal.* For a contemporary description of the Hawk's Rest country, pick up a copy of *Hawk's Rest: A Season in the Remote Heart of Yellowstone* by Gary Ferguson (2003). Ferguson spent the summer of 2002 as a seasonal ranger stationed at the Hawk's Rest Guard Station.

Two days before, after a hearty ranch breakfast, we had watched as Roger and Walter Korn diamond-hitched our gear onto the pack horses. Then we were on the trail, climbing steadily through stands of quaking aspen, pine, and Douglas-fir. At one point, I reined in my horse to look back. Spread out below was the whole of the Buffalo Fork Valley with a view all the way down to the Snake. The Tetons stood like sentinels on the far horizon.

We dropped over into Lava Creek, where mule deer watched us from low willow bushes, and followed up that drainage to our first night's camp at the edge of a meadow 12 miles out from the ranch. After dinner that evening, I strolled out onto the meadow to a point where I could look off to the east toward Two Ocean Pass. Enos Lake was just visible in the distance. On the way back, as I came around the base of a low side hill, I confronted a feeding moose not thirty yards away. More curious than startled, he eyed me, then loped off into the timber.

July 7th. This was to have been a traveling day, but Barb Washburn was under the weather, so we decided to spare her another day on a horse. Roger, Fred Washburn, the youngsters, and I took a day trip to Enos Lake instead.

Enos Lake is populated with Yellowstone cutthroat trout. Fish were working all along a narrow shelf that looked as if it extended all the way around the lake. We could see the fish clearly in the shallow water. They were taking something right in the surface film but we couldn't make out just what.

The trout weren't big, but they were bright and feisty, and since they responded well to a Royal Coachman bucktail cast out and worked back over the edge of the drop-off, I didn't bother to ascertain what they were eating. A chance to do that came later. The kids had kept a few of their trout for dinner that evening, so to get a look at the stomach contents I volunteered to demonstrate my fish-cleaning prowess. Mostly, the trout were filled with olive-colored chironomid pupae. I might have guessed at something like that from the way they were working the surface film. But one trout had also taken some kind of small fish and another had swallowed an iridescent green beetle, so they hadn't been completely selective.

Clouds had moved in and the sky had become overcast as we rode into the lake that morning, and that had undoubtedly helped the fishing. But now, as we headed back to camp, a few drops of rain began to fall. I went to bed early and drifted off to sleep to the sound of a breeze blowing gently through the evergreens and the light spackle of raindrops on the tent.

July 8th. Barb Washburn was much improved this morning, so we broke camp and rode on. We still had 17 miles to go before reaching our base camp in Yellowstone Meadow. Our route took us past Enos Lake once again, then over a low divide to Pacific Creek. We would follow Pacific Creek up to Two Ocean Pass, stopping for lunch where the creek flows out of the meadow, then continue on through the pass, following Atlantic Creek down to its confluence with the Yellowstone River. We would then turn north, following the Yellowstone to our campsite at the foot of Hawk's Rest.

But Enos Lake had one more treat for us as we passed: a glimpse of the rare trumpeter swans that nest there. They had kept well away from us when we were fishing yesterday, and now we could see why: they had cygnets! I counted three young swans swimming in train with the two adult birds.

Ominous banks of clouds were building as we rode up Pacific Creek. Roger and Sue, experienced in mountain travel, kept glancing up at the sky and I found myself wondering if we would make the full 17 miles before the first thunderstorm caught us. But nothing could dampen my interest in the surroundings as we rode into Two Ocean Pass. This was one of several routes used by the old mountain men to get in and out of Jackson Hole. In the fur trappers' day, they followed Pacific Creek all the way from its confluence with the Snake, whereas we had "ridden over the mountain," so to speak, to hit Pacific Creek further up. But the pass itself still looks pretty much

the same as it did when Osborn Russell first described it back in 1836. [2]

Two Ocean Pass is a flat, narrow meadow some 2 miles long and maybe a quarter-mile or so wide. Bluffs rear up to flank it on both the north and south sides, and a little less than halfway through, the line of bluffs on the south side is broken by a snowy peak that juts into the air. The trail had been following the south side of the meadow, but when we had gone about a third of the way through, we swung across to the north side. Here a fork in the trail leads back into the timber and up a gentle slope to the base of the bluff. North Two Ocean Creek comes down off the plateau at this point and divides into two almost perfectly equal branches. One branch becomes Pacific Creek, along which we had just been riding. The other branch becomes Atlantic Creek, which we shall follow the rest of the way across the pass. We each stood beneath the wooden sign proclaiming "Parting of the Waters" and had our pictures taken, one arm pointing to the Atlantic, the other pointing to the Pacific. As in the trappers' day, Two Ocean Pass proved a popular thoroughfare. Three other horse parties were camped at the edge of the meadow and other groups passed us as we paused at Parting of the Waters.

Atlantic Creek took on the character of a rich meadow stream as we rode on, with deep pools, riffles,

2. Osborne Russell's description of Two Ocean Pass (as well as his other adventures in the Jackson Hole, Yellowstone, and upper Snake River regions) can be read in his book, *Journal of a Trapper, or, Nine Years in the Rocky Mountains, 1834–43* (see Russell 1986).

and undercut banks. It took Polaroid sunglasses for best viewing, but everywhere the trail came close to the bank, we could see trout holding in the current. More fishermen were working the lower end of Atlantic Creek where it flows across the broad expanse of Yellowstone Meadow. This is the time of year when the cutthroats move up out of Yellowstone Lake to spawn. We could see dozens of trout on every riffle, trailed by schools of what looked like suckers. The fishermen we talked to, though, said the trout were lethargic and not hitting well, and that fishing was much better at Bridger Lake. We would be camped less than a mile from there.

We heard the first roll of thunder as we descended toward Yellowstone Meadow. Fortunately, it was a long way off, and even though the storm clouds continued to threaten, we rode along in bright sunlight. The ride along the Yellowstone was uneventful, and we soon arrived at our campsite just a short walk from the Hawk's Rest Patrol Cabin. Gordon Reese, a school teacher from Salt Lake City, his wife Judy, and their two daughters had occupied the rustic cabin at the base of the towering cave-studded bluff for the past three summers. Their front window looks out across the meadow to the river and beyond. Nature had certainly provided them with one beautiful front yard. And I learned that real happiness, when you're hot, tired, and

grimy from riding, is a long drink of water from their incredibly cold, bracing spring.

All of us adults were saddle-weary, but not so the youngsters. They grabbed their fishing rods and headed for the river. When they promptly landed three hefty cutthroats, I decided I'd better rig up my fly rod and join them. That's what led to my unceremonious encounter with the trout, and that's when the threatening thunderstorms finally overtook us. There would be no fishing the evening rise tonight. Dinner was accompanied by the play of lightning in the gathering clouds and the sound of thunder rolling and reverberating down the valley.

July 9th. We awoke to a heavy mist over Yellowstone Meadow and the sight of two mule deer grazing with the horses. The sun burned through about 10 AM and the rest of the day was absolutely glorious.

Today we would fish Bridger Lake, which is surrounded by timber except at the west end, where it outlets through a low, marshy opening to the Yellowstone River. You can wade the shallows, which are mostly rock and rubble, and as I did so I was amazed! Big cutthroats cruised virtually at my feet, unmindful of my presence as long as I was reasonably careful. Unlike the trout in Enos Lake, these fish were not rising, but instead were working leisurely right along the bottom, sometimes in water only inches deep. I fished a

A horse party follows Atlantic Creek across Two Ocean Pass to the upper Yellowstone River. PHOTO BY AUTHOR

bright bucktail wet fly, a fly whose combination of red tail, deep-orange body, badger hackle, and white wing is a favorite of mine for sea-run cutthroats back home. The Bridger Lake trout responded with enthusiasm. In two hours' fishing, I hooked and landed eight trout ranging in size from 13 to 17 inches, and lost many others. They were all plump, robust, and full of fight. One fish that I measured scaled 16 inches in length and 8 inches in girth—a solid pound and a half by the length-girth formula.

Again I got to examine the stomach contents of the fish we kept for supper. They were filled with masses of scuds, all about a size 12 and all colored in a subtle orange-tan-olive hue. I had seen some large creamy-brown sedges on the water as well, but the scuds explained why there was no sign of surface feeding: just too much rich food in the shallows. Why then did they respond so well to my bright orange and white bucktail? Who knows, that's how trout fishing goes sometimes.

July 10th. Our last full day in "permanent" camp. Even in July, the high country mornings can be crisp, and so it was today. It warmed up once the sun topped the ridges, but with the sun came the wind. A stiff breeze blew all day long, making it tough to fly fish, but at least it kept the mosquitoes at bay.

This morning, while Roger took the kids across Thorofare Creek into Yellowstone Park, Fred and Barbara Washburn and I went back up the meadow to fish Atlantic Creek. Although the wind was bothersome, I chose to work with dry flies. It was, to borrow a corny phrase, like shooting fish in a barrel. Cutthroats in the 15- to 17-inch class were on every riffle and at the head of every pool. I hooked and released fish after fish, maybe ten or so in all, before I got disgusted with myself. They were all males, dark, sluggish, and close to spawning, and I'm sure they struck at the flies more in defense of their spawning territories than from any other instinct.

It's curious to me that the upper Yellowstone River and its tributaries within Yellowstone National Park are closed to fishing at this time of year to protect the cutthroats on their spawning runs. But much of the actual spawning takes place in waters outside of the park in Wyoming's jurisdiction, where fishing remains open. The State of Wyoming has always treated its "sports" generously, and has evidently never seen fit to close these upper waters. In 1986, the state did reduce the daily bag limit from twelve fish, which was the statewide allowance in 1980, when I made the trip just described, to two fish. That keeps the harvest under control, but the spawning trout must still put up with the harassment.

HISTORICAL RANGE

Figure 9–1 shows the historical distribution of the Yellowstone cutthroat subspecies. Every time I look at this map, I think of one of those old gerrymandered political districts, boundaries manipulated to consolidate or isolate voting blocs. It includes the Snake River drainage from Shoshone Falls upstream to its headwaters at the Continental Divide—except for a segment of the South Fork Snake mainstem extending from Palisades Reservoir on the Wyoming-Idaho border upstream through Jackson Hole to Jackson Lake. That segment, including all but the upper tributaries of the Gros Ventre (pronounced "Gro Vont") River, is the range of the finespotted Snake River cutthroat trout (Chapter 10).

Nobody knows how far down the main Snake River the finespotted subspecies extended before Palisades Reservoir was constructed (it was authorized prior to World War II but not completed until 1957), but we do know that all of the Snake River tributaries between Palisades Reservoir and Shoshone Falls—this includes Henry's Fork and all of its tributaries, Willow Creek, the Blackfoot, Portneuf, and Raft rivers, Goose Creek, and even Palisades Creek, a tributary of the South Fork Snake that enters just a bit over 3 miles downstream from Palisades Reservoir—have the Yellowstone cutthroat as their native trout. So, too, do the uppermost tributaries of the Gros Ventre River, even though the lower reaches of this stream have the finespotted trout. Yellowstone cutthroats also occupy all of the Buffalo River, Pacific Creek, and Spread Creek, each of which joins the Snake River below its discharge from Jackson Lake but upstream from the mouth of the Gros Ventre.

Prior to 1911, when a dam was erected at the discharge of Jackson Lake, there was nothing to bar free movement of trout up and down the Snake River into the reaches above Jackson Lake. Upstream from Jackson Lake, David Starr Jordan reconnoitered the streams and lakes of Yellowstone National Park on behalf of the U.S. Fish Commission, and reported his findings in 1891. Jordan noted a difference in spot size and pattern between trout in Heart Lake in the headwaters of the Snake River and the trout of Yellowstone Lake, but he did not identify the Heart Lake trout as a fine-spotted form. Neither did Barton W. Evermann of the U.S. Bureau of Fisheries, who caught trout from the Snake River 12 miles upstream from Jackson Lake a year or two after Jordan's reconnaissance. [3] The cutthroat trout one finds in Heart Lake today are clearly the Yellowstone subspecies. So I, like most others, show the distribution of the finespotted Snake River cutthroat trout extending only up to Jackson Lake.

The Goose Creek drainage, which heads up in Idaho but loops south into northern Nevada then back again into Idaho as it flows toward its junction with the Snake

3. Jordan's record of the difference in spotting between Heart Lake and Yellowstone Lake trout is reported in "A Reconnaissance of the Streams and Lakes of the Yellowstone National Park, Wyoming, in the Interest of the U.S. Fish Commission" (see Jordan 1891a). Evermann's report is in "A Reconnaissance of the Streams and Lakes of western Montana and northwestern Wyoming" (see Evermann 1893).

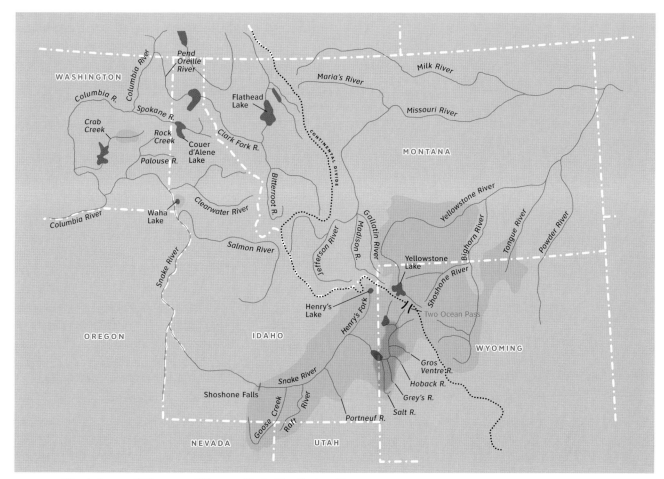

FIGURE 9-1. *Historical range of Yellowstone and Finespotted Snake River Cutthroat Trouts*

Yellowstone Cutthroat Historical Range; Also Waha Lake and Crab Creek

Finespotted Snake River Cutthroat Range

River, was Yellowstone cutthroat habitat in its entirety and is Nevada's only chunk of the subspecies' historical range. Likewise, upper tributaries of Raft River originating just across the border in Utah are that state's only portion of the native distribution.

The distribution of the Yellowstone cutthroat trout also extends across the Continental Divide into the Yellowstone River drainage. The Shoshone Indians have a story that explains this. According to their version, an old woman carrying fish in a basket accidentally spilled the contents into the waters on the east side of the Great Divide. Our version is less poetic but just as intriguing, in my view, and very likely involved a crossover at Two Ocean Pass, as I discussed in Chapter 2. This could not have happened until the Yellowstone Plateau country was free of ice at the end of the Pleistocene. Even though the great continental glaciers themselves never reached the Yellowstone country, the Yellowstone Plateau was under the influence of alpine glaciers during several of the major ice advances of the Pleistocene.[4] It was probably not until this alpine glacial ice receded at the end of the last major advance (called the Pinedale glaciation in this region) some 10,000 to 12,000 years ago that the present distribution of cutthroat trout was established in the upper Yellowstone River and Yellowstone Lake. Once that happened, cutthroat trout spread downstream in the Yellowstone

River, colonizing eastward as far as the Tongue River drainage. The Powder River drainage, the next Yellowstone tributary system downstream from the Tongue River, never had native trout.

Perhaps the best-known and most abundant populations of Yellowstone cutthroat trout are those occurring in Yellowstone Lake and the Yellowstone River above the falls in Yellowstone National Park. Here the trout are so abundant and pure that they have been called museum specimens in a museum setting. Since 1872, when the Park was established, these trout have been widely proclaimed in both popular angling literature and scientific accounts. Even titled European sportsmen were attracted to the region early-on. In more recent times, before restrictive regulations were imposed, there were some years when anglers took more than 300,000 of these trout. Given that abundance, what many anglers find hard to realize when it's first brought to their attention is that fully 40 percent of Yellowstone National Park was originally barren of trout. The barren portion included almost all of the Firehole River system now famous for its trout fishing, much of the Gibbon River system, both Lewis and Shoshone Lakes, and many other lakes and streams in the western part of the Park. Only a bit of the northwest corner, where westslope cutthroat trout (Chapter 3) were native in the Gallatin and Madison rivers, and

4. In addition to alpine glaciation, the Yellowstone region has suffered episodes of cataclysmic volcanism. One violent explosion occurred about 2 million years ago, and another about 600,000 years ago. That 600,000-year event occurred around the edges of a 1,000 square mile ellipse that takes in the present-day Yellowstone Lake and Hayden Valley areas. That entire area collapsed into the empty chamber, forming a caldera that in time held an immense lake. The present Yellowstone Lake is a relict of that 600,000-year volcanic event. You can read more about the geology of the Yellowstone region in Good and Pierce (1993) and Smith and Siegel (2000). References to alpine glaciation of the region during the late-Pleistocene include Keefer (1972), Porter et al. (1983), and Barnosky et al. (1987).

of course the Yellowstone River/Yellowstone Lake corridor and associated tributaries where Yellowstone cutthroat trout abounded, originally had native trout.[5]

In the eastern portion of the range, it has long been assumed that Yellowstone cutthroat trout extended down the Yellowstone River as far as the Tongue River, colonizing all tributary systems they came to including the Tongue. Distribution maps in Dr. Behnke's writings, for example, show this as the eastern extent of historical distribution. However, the historical record itself indicates a greatly truncated distribution in this area at the time of first white contact. William Clark of the Lewis and Clark expedition explored down the Yellowstone River in 1806. His description of both streams at the confluence of the Yellowstone and Tongue rivers suggests habitat more suited to warm water fishes, an observation that was confirmed when the party caught only catfish from these waters.

Bruce E. May, a forest fisheries biologist for the U.S. Forest Service, examined the historical record in detail. His sources included accounts of early settlers and road builders, records of General Crook's military expedition against the Sioux and Cheyenne Indians in 1876, and U.S. Bureau of Fisheries surveys of the region in the early 1890s. Writing in the U.S. Forest Service "Conservation Assessment for Inland Cutthroat Trout" in 1996, May pegged the historical

5. John Varley, formerly chief of research for Yellowstone National Park, and Paul Schullery, formerly Park historian, co-authored two excellent books on the distribution and biology of the fishes and the history of fish management and fisheries in the Park: *Freshwater Wilderness: Yellowstone Fishes and Their World* (Varley 1983), and *Yellowstone Fishes: Ecology, History, and Angling in the Park* (Varley 1998). Although the later book is an updated second edition of the first, I own copies of both and value each of them. Early expeditions sent to check on the truth of the "fantastic" sights reported from the Yellowstone region were the Washburn-Doane exploration of 1870 (see Langford 1905, Bonney and Bonney 1970, and Schullery 2006) and the Hayden survey of 1871 (see Hayden 1872). These parties also reported on the trout fishing they encountered; especially Doane, whose daily journal was published in "Senate letter from the Secretary of War communicating the Yellowstone Expedition of 1870," 41st Congress, 3rd session, executive document 51 (1871), but a more accessible version was reprinted in Bonney and Bonney (1970). Exploits of the titled English sportsmen who entered the Yellowstone region are recounted in Merritt (1985) and the memoirs of the Fourth Earl of Dunraven; see Dunraven (1876 [1917]).

Bridger Lake looking west toward its outlet to the Yellowstone River. PHOTO BY AUTHOR

downstream distribution of trout in the Yellowstone River as a point just upstream from the mouth of the Bighorn River. In the Bighorn River system, cutthroat trout extended downstream only to a point between the towns of Worland and Thermopolis, Wyoming. Trout were present in abundance in upper tributaries and reaches of the Tongue River, but not in downstream reaches. I chose to go with May's distribution limits in Fig. 9-1. Dr. Behnke's distribution may reflect more the prehistoric extent of downstream colonization in these river systems. [6]

Yellowstone cutthroat trout in upper tributaries of the Tongue River may have helped shape events of the great military conflict of 1876 that occurred between U.S. Army columns under generals Terry, Custer, and Crook and hostile Sioux and Cheyenne Indians led by Sitting Bull, Crazy Horse, and Gall. At least, historian John H. Monnett suggested they might have.

According to the Army strategy, Terry and Custer's troops would move south from the Yellowstone and Crook's troops would move north and west from the Powder River, catching the hostiles between them. General George Crook, an avid sportsman who always had his hunting and fishing gear with him, even on campaign, had already found good trout fishing in the Bighorn Mountains where his troops had crossed over from the Powder River drainage to a camp on Goose

Creek, an upper tributary of the Tongue River. But as they advanced north and westward out of the mountains to Rosebud Creek, they encountered the hostiles and a pitched battle ensued. Even though Crook's forces held the field when the fighting was done, rather than press his advance, pushing the hostiles toward Terry's column as originally ordered, he withdrew to the upper Tongue and sent for more supplies and reinforcements.

For two weeks Crook tarried, hunting and fishing all the while along various upper tributaries of the Tongue River. Capt. John G. Bourke, a staff officer who chronicled Crook's campaigns, had this to say about that interlude:

> "The credulity of the reader will be taxed to the utmost limit if he follow my record of the catches of trout made in all these streams...but the hundreds and thousands of fine fish taken from that set of creeks by officers and soldiers, who had nothing but the rudest of appliances, speaks to the wonderful resources of the country...at that time."

Crook was still encamped on the upper Tongue, and still hunting and fishing, when dispatches arrived telling of the disaster that had befallen the impetuous Custer and his troops on the Little Bighorn. Historian Monnett, writing about this campaign in a 1993 essay that appeared in *The American Fly Fisher* (the journal of the American Museum of Fly Fishing, Manchester, Vermont), wondered if Crook could have altered the

6. May's interpretation of the historical record for the extent of downstream distribution of Yellowstone cutthroat trout in the Yellowstone, Bighorn, and Tongue rivers is in the U.S. Forest Service report, "Conservation Assessment for Inland Cutthroat Trout" (see May 1996). May's sources included accounts of early settlers of the Bighorn basin on file at his U.S. Forest Service office. May also cited Major John Owen's 1864 journal of a trip over Jim Bridger's road from the Oregon Trail to Montana (see Dunbar 1927). Owen's party traveled down the Bighorn River, then over to the Greybull and Shoshone rivers, but didn't encounter trout until they reached those latter two streams. William Clark's description of the Yellowstone and Tongue rivers at the latter's confluence with the Yellowstone, along with his note on the type of fish the party caught there, is at page 248, Volume 8 of the Moulton edition of the Lewis and Clark journals (see Moulton 1986–97). The U.S. Bureau of Fisheries survey cited by May is in Evermann and Cox (1896).

events at Little Bighorn if he had continued to push north as originally ordered—or did the lure of trout fishing help seal Custer's fate? [7]

In addition to the core distribution in the upper Snake and upper Yellowstone river drainages, Fig. 9-1 also shows the historical presence of two outlier populations that have been presumed to be Yellowstone cutthroat trout. Some elements of these populations' descriptions support this subspecies assignment (others do not), but mostly, I think, it's because one population occurred in the Snake River drainage, albeit the lower part of the drainage well downstream from the present limit of Yellowstone cutthroat distribution at Shoshone Falls, and the other occurred in a creek whose headwaters lie in close association with the headwaters of another lower Snake River tributary (as I pointed out in Chapter 2, it has long been the belief that the 64-chromosome ancestor of the Yellowstone cutthroat trout originally colonized the entire Snake River drainage). These outlier populations, both now extinct, were present in Waha Lake, Idaho, and upper Crab Creek, Washington. Indeed, Waha Lake is the type locality for *bouvieri*, the formal subspecies name given the Yellowstone cutthroat trout.

Waha Lake is situated on a shoulder of Craig Mountain approximately 20 miles southeast of Lewiston, Idaho. Lapwai, the Nez Perce tribal headquarters and site of an old military post, is 14 miles to the east. The lake has no surface outlet and appears to be spring fed, but it is said that at some time back in prehistory, a landslide blocked a canyon in the Waha Creek (now called Sweetwater Creek) drainage, forming the basin where the lake lies. This drainage is tributary to the Clearwater River, which is probably how trout populated the lake originally. Descriptions of Waha Lake differ. Army Capt. Charles Bendire, who first visited in 1869, described a lake "3 miles long and a mile and a half wide [approximately 2,880 surface acres], very deep, with very cold water." A description from the late 1930s is of a lake of 180 surface acres and a depth of 60 feet. However, by then, Waha Lake had been tapped for irrigation water for Lewiston Orchards, a Bureau of Reclamation project for agricultural development in the valley below. In Bendire's time, there were two other small but fishless lakes in a chain with Waha but at lower elevations. Both have long since passed out of existence.

The record of trout in Waha Lake goes well back in time. H.J. Spinden, an anthropologist writing about the Nez Perce Indians in 1908, said families from that tribe had visited the lake on a regular basis for generations to capture "Waha Lake trout" for part of their annual food supply. Later, in the 1860s and 1870s, soldiers from Fort Lapwai fished the lake for trout. In 1882, a hotel called Lake House was established there and became

7. Capt. Bourke's chronicle of these events can be found in his book, *On the Border with Crook* (Bourke 1891 [reprinted by University of Nebraska Press, 1971]). Historian Monnett's essay, "Mystery of the Bighorns: Did a Fishing Trip Seal Custer's Fate?," appeared in the Fall, 1993 issue of *The American Fly Fisher*, the journal of the American Museum of Fly Fishing, Manchester, Vermont (see Monnett 1993).

famous in the region for its venison, bear, chicken, and, of course, trout dinners. Lake House became a destination resort for citizens of the Lewiston area who would travel by horse and buggy or stagecoach. Many built summer homes on the lake. A post office opened in 1890 and served the community until 1965.[8]

Capt. Charles Bendire, a U.S. Army officer with a keen interest in natural history, fished Waha Lake many times, beginning in 1869 when he served at old Fort Lapwai. In 1880, he collected specimens that were sent to the U.S. National Museum. Here is his description of these trout as published in 1882 [italics highlighting a couple of key points about belly color and spots are mine]:

> "All the fish caught there are about the same size, from 6 to 10 inches long. I believe it is a new species or variety.... Back bluish green, olive color, sides silvery; *in some instances the whole belly is red, in others the sides only*; a few show a yellowish tinge; no red on the sides. *There are round black spots near the tail*, and fainter ones on the flanks. A few have an occasional spot on the head; two vermillion-colored stripes on each side of the under jaw; fins edged with brick red, ranging from this color to an orange."

David Starr Jordan provided a summary description of Capt. Bendire's preserved type specimen, which he wrote was: "Similar to typical *Salmo clarkii*, but with dark spots only on the dorsal, caudal and adipose fins, and on the caudal peduncle behind front of anal, where the spots are very profuse, *smaller than the pupil*"

[again, the italics highlighting the point about spot size are mine].[9] In Fig. 9-2a, I have reproduced an illustration of the Waha Lake trout that accompanied this description in one of Jordan's publications. Capt. Bendire may have used the name *Salmo bouvieri* for these trout in honor of Capt. Bouvier, presumably an army colleague. Jordan and his associates used this name as well, but also used *Salmo purpuratus* variety Bouvieri and *Salmo mykiss bouvieri* in their various publications.

As I tried to explain earlier, because of the spotting but primarily because of the location of Waha Lake in the presumed path of colonization by the 64-chromosome large-spotted ancestor of the Yellowstone cutthroat trout, it has long been thought that Waha Lake trout were Yellowstone cutthroats. Dr. Behnke held this view, and selected *bouvieri* as the earliest-used scientific name for any large-spotted cutthroat trout in either the Snake or Yellowstone river basins. But the Clearwater River drainage has westslope cutthroat trout as its historically indigenous cutthroat, and referring

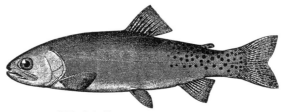

FIGURE 9-2A. *Waha Lake Trout*

8. The background information regarding Waha Lake came from the following sources: Spinden (1908), Bailey (1935 [1947]), Federal Writers' Project (1938), and Boone (1988). There is also a bit of contemporary information on the Bureau of Reclamation's recreation website at *www.recreation.gov*. Fishing at Waha Lake today is for introduced smallmouth bass and stocked rainbow trout.

9. Capt. Bendire's original description of the Waha Lake trout was published in the Proceedings of the United States National Museum in 1882; see Bendire (1882). A summary description and illustration of Bendire's type specimen subsequently appeared in several publications by David Starr Jordan and his colleagues; see Jordan and Gilbert (1883) and Jordan and Evermann (1896, 1902). Jordan, by the way, was able to move the collection of Waha Lake trout with him when he went to Indiana and later to Stanford University. The specimens are now part of the California Academy of Sciences' ichthyology collection.

back to Chapter 2 again, there is an alternative explanation for inland colonization by ancestral cutthroat trout that accounts for this directly. In some of his most recent writings, Dr. Behnke has taken note of the red colors on the bellies and sides described by Capt. Bendire, which are not typical of Yellowstone cutthroat trout but do appear on westslope cutthroat trout. Dr. Behnke appears to accept the possibility that rather than being an outlier population of Yellowstone cutthroats, the Waha Lake trout may have been westslope cutthroats with unusually large spots, as are found in some populations of westslope cutthroat in eastern Washington.[10]

The second outlier population, originally named *Salmo eremogenes*, was reported from upper Crab Creek, eastern Washington, in 1909 by U.S. Bureau of Fisheries workers who visited there the year before. Crab Creek drains a substantial portion of the Columbia Plateau region, a land that was once cloaked in bunch grass and sage, but now supports wheat fields as far as the eye can see. The creek heads up just south of the town of Reardan, in proximity to headwater tributaries of the Palouse River, a tributary of the lower Snake River. Crab Creek follows a westerly course to within a few miles of Soap Lake, then turns south to flow into Moses Lake. Below Moses Lake, it continues south for a few miles before turning west again to join the Columbia River downstream from the town of Vantage. Observers have called it a "lost creek" because it sinks entirely into the porous basalt underlayment in places, emerging each time several miles downstream with an often greater flow, having received the cold waters of many underground water courses and springs. Even more curious, Crab Creek flows through the channeled scablands scoured by the flood waters from Glacial Lake Missoula.

As was the case for Waha Lake, the record of trout in Crab Creek goes far back in time. The earliest written record in my own file dates back to 1870, but if you believe the Native Americans who lived in the area, the trout were there long before that. Thomas Nelson Strong, a local historian, knew the Crab Creek of 1870 well. He wrote:

> "Crab Creek, on the great plains of the Columbia, in Eastern Washington, is one of the most remarkable streams in the Northwest. At its source near Medical Lake it is a mere brook, and here, in 1870, there were trout, little fingerlings, by the hundreds. A few miles to the westward the stream disappeared in sand and basaltic rock. Again a few miles below it came to the surface a larger stream than at first, and with larger trout. For 100 miles went this peculiar stream in this way, now sinking and now rising, every reach of open water stocked with trout of appropriate size, until at a point a little below Moses Lake, south of the Grand Coulee and 20 or 30 miles from the Columbia River, it finally disappeared in a waste of sand and rock."[11]

Strong asked old Chief Moses, head of a local Indian band, how the trout got into Crab Creek. Here was Moses' reply:

10. See Behnke (1992) at pages 91–93 and Behnke (2002) at page 171 for his discussions of this possibility.

11. These quotes came from Thomas Nelson Strong's book, *Cathlamet on the Columbia*, originally published in 1906 as a history of the lower Columbia River region, but with an added informative Chapter on Crab Creek; see Strong (1906 [1930]).

"You want to know how the little salmon got into the little creek? No, no, they didn't get in. My people know, and I know, that they have always been there."

Crab Creek was not the only stream in the region known to be inhabited by trout. So was nearby upper Cow Creek, a tributary of the Palouse River. These trout were highlighted in some of the literature used by the railroads and land boosters to promote the area to settlers throughout the 1880s and early 1890s. Here is one typical item: [12]

> "Crab Creek is about 15 miles north of Ritzville and Cow Creek ten miles east—both flowing a large body of water the year around, and both stocked with trout."

I should point out here that when these early writers used words such as "stocked with trout," they were not referring to hatchery stocking but to naturally reproducing populations. Hatchery stocking of trout didn't get started in these parts until 1903 and 1904, when two releases of eastern brook trout were made into headwater reaches of Crab Creek. Another release of hatchery trout was made in 1907, but more about that one in a minute.

Because of the railroads' land promotions, these and other reports of trout in Crab Creek had been reaching the U.S. Bureau of Fisheries for years, but it wasn't until July 29, 1908 that the Bureau finally followed up. On that date two Bureau of Fisheries workers arrived at Crab Creek at a road crossing called Rocky Ford, [13] 12 miles north of the town of Ritzville, to make a study of local conditions and collect specimens of the trout and other fishes inhabiting the stream. They spent three days there. Their report was published the following year, 1909, in the *Proceedings of the Biological Society of Washington*, along with photos of the site and a drawing of their type specimen, a trout that was 10 inches long (I have reproduced that drawing in Fig. 9-2b). They described this trout as a new species, *Salmo eremogenes*. Following is an excerpt from their report: [14]

> "Trout were found in considerable abundance but, probably owing to an abundant food supply, they did not take the fly or baited hook with any avidity. Young trout two to three inches long were abundant in the creek and some were found in an irrigation ditch which received its waters from the creek."

They were able to capture only four specimens including the 10-inch type specimen, which they described as follows [italics highlighting flank color and spot size are, again, mine]:

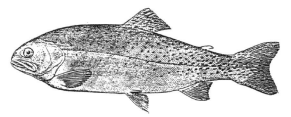

FIGURE 9-2B. Salmo eremogenes *from Crab Creek, Washington*

12. In addition to Thomas Nelson Strong's book, other historical notes on Crab Creek and its trout came from *An Illustrated History of the Big Bend Country embracing Lincoln, Douglas, Adams, and Franklin Counties, State of Washington* (Anonymous 1904) and *As I Remember* (Phillippay 1970), a history of the Ritzville area written by Minnie Phillippay.

13. Anglers familiar with fly-fishing in Washington State should not confuse the Crab Creek Rocky Ford with the blue-ribbon trout fishery on Rocky Ford Creek, a tributary of Moses Lake near the town of Ephrata. Rocky Ford Creek is miles to the west of the Crab Creek Rocky Ford, which is located about 12 miles due north of the town of Ritzville. The Rocky Ford on Crab Creek is an easy place to find and it looks pretty much the same as it did back in 1908, except for the bridge that now takes you across the creek instead of a wet crossing. There is also a rattlesnake or two in the vicinity, so if you go there, watch where you step.

14. Evermann, B.W. and J.T. Nichols. 1909. "Notes on the Fishes of Crab Creek, Washington, with Description of a New Species of Trout." *Proceedings of the Biological Society of Washington* 22: 91–94, with accompanying Plate II. The Biological Society of

"Color (in life)—Above dark olive; caudal peduncle with numerous close-set *roundish spots of moderate size*, these spots becoming less numerous anteriorly, there being only a few in front of dorsal and none on head; dorsal and caudal fins with black spots, other fins immaculate, the spots on the anterior part of body more nearly round than those on caudal peduncle; cheeks and opercle olive yellowish, tinged with pink; *lower part of side from base of pectoral to anal more or less pink*, the color showing a little on pectoral and more on ventrals; ventral surface whitish; pectoral green; anal olive with more or less reddish tint; ventrals more or less olive, tinged to a considerable extent with the pink of the flanks; red on lower jaw quite distinct."

Continuing from their report:

"The Crab Creek trout is evidently a species of the cut-throat series. Red marks on throat are very distinct, and the scales are small, there being 165 to 175 in a longitudinal series.... As regards coloration, *the Crab Creek fish have the spots, which are large*, and vary in abundance, much the more abundant caudally. The spots are mostly on the caudal peduncle, the back as far forward as the dorsal fin, and the dorsal and caudal fins. In each specimen, however, a few spots occur in the front part of the body.... In the number, size and arrangement of the spots, this species most resembles *Salmo stomias*, the trout of headwaters of the Platte and Arkansas. It differs from that species, however, in the shorter snout, larger eye and the somewhat larger scales."

The same ambiguity clouds the identification of these trout as Yellowstone cutthroats as obscures the true identity of the Waha Lake trout, with two added complications. There is no doubt, based on the historical record, that trout were present in upper Crab Creek long before the U.S. Bureau of Fisheries workers arrived on the scene. But, being in the path of the Lake Missoula floods, one would think that had Crab Creek been colonized by trout flushed down by those waters, they would be westslope cutthroats, not Yellowstones. That's one complication. Then there is that stocking record from 1907, the year *before* the U.S. Bureau of Fisheries visit. Those trout, 28,000 in number, originated from the State of Washington egg-taking and hatchery operation at Lake Chelan, where we learned in Chapter 4 that the westslope cutthroat trout is native. [15]

So, which form of cutthroat trout did the Bureau of Fisheries workers actually collect that summer of 1908 at Rocky Ford crossing—the native or the hatchery fish? And if the natives, were they really Yellowstone cutthroats or westslopes with an unusual spotting pattern?

As early as 1992, Dr. Behnke proposed that these questions might be answered for both Waha Lake and Crab Creek by analyzing DNA extracted from preserved museum specimens. DNA from preserved and ancient specimens was a hot topic about 10 to 15 years ago. Researchers competed to publish in high-profile journals, each claiming to have recovered DNA from ever more ancient specimens and giving rise to such pop-culture favorites as *Jurassic Park* before the hype subsided. Although there was some good, careful work done on ancient DNA during that period, some of the most spectacular claims could not be substanti-

Washington was a Washington, D.C. institution, not one associated with the State of Washington.

15. The record of stocking of Lake Chelan-origin trout in Crab Creek is documented in the 18th and 19th Annual Reports of the Washington State Fish Commissioner for the Years 1907 and 1908; see Riseland (1909). The 14th and 15th Annual Reports for 1903 and 1904 have the records of the two earlier stockings of brook trout. There is nothing in Evermann and Nichols' paper to indicate that the U.S. Bureau of Fisheries was ever made aware of this stocking activity before their visit in 1908.

ated. DNA degrades over time and the problems with contamination are enormous, so it's very easy to be led astray. Even so, with care and exacting technique, it is sometimes possible to extract usable DNA from preserved museum specimens including fish. [16] Preserved specimens of the original Waha Lake trout are housed in the California Academy of Sciences ichthyology collection, and a while back, Dr. Behnke told me that the Field Museum in Chicago may have the specimens from Crab Creek.

Although the U.S. Bureau of Fisheries workers wrote that trout were present "in considerable abundance" in Crab Creek in 1908, they were able to collect only four specimens by hook-and-line angling, "probably owing to an abundant food supply" in the creek. To my knowledge, no cutthroat trout have been collected there since, either from Crab Creek or any other nearby stream, and it is assumed that the original populations are extinct.

LIFE HISTORY AND ECOLOGY

Populations of Yellowstone cutthroat trout exhibit three life history strategies:

- A lacustrine, or lake-associated, life history, wherein the fish feed and grow in lakes, then migrate to tributary streams to spawn.
- A fluvial, or riverine (potamodromous), life history, wherein the fish feed and grow in main river reaches, then migrate to spawning reaches that may be elsewhere in the mainstem or in small tributaries of these rivers where spawning and early rearing occur.

- A stream-resident life, history wherein the trout complete all phases of their life cycle within a small stream, often in a short stream reach and often in the headwaters of the drainage.

The Lacustrine Life History Strategy

Lanky mountain man Jim Bridger earned his reputation as a teller of tall tales with his yarns about the wonders of Yellowstone country. He kept listeners entertained with stories about petrified trees with petrified birds singing petrified songs; of a creek so full of alum it shrank everything including time and distance; of a canyon so broad and deep he could yell "Wake up, Jim" before bedding down at night and be awakened next morning by his echo; of catching trout from Yellowstone Lake, then immediately cooking them in a handy hot spring without even taking them off the hook.

But wait. That last story may not have been so far-fetched after all. In August 1923, *National Geographic* ran an article titled "Our Heritage of the Fresh Waters," in which a photograph, taken in 1919 and supplied to National Geographic by the National Park Service, shows two anglers cooking a trout in Fishing Cone, a hot spring located on the shore of Yellowstone Lake. [17] By 1919, when that photo was taken, so many visitors to the Park had tried to duplicate the feat that finally the American Humane Society objected and an edict was issued forbid-

16. For those interested in learning more about extracting DNA from preserved museum specimens, here are some places to start: Pääbo (1989), Pääbo et al. (1989), Shiozawa et al. (1992), Lindahl (1993a, 1993b), Poinar et al. (1996), Shedlock et al. (1997), Wirgin et al. (1997), Butler and Bowers (1998), Rivers and Ardren (1998), and Cooper and Poinar (2000).

17. This same photo also appeared at page 101 in the book *Freshwater Wilderness: Yellowstone Fishes and Their World* by John Varley and Paul Schullery (Yellowstone Library and Museum Association, 1983). I tried to obtain a copy to run in the first edition of this book, but was told it could not be located. But it wasn't just at Fishing Cone where fresh-caught trout could be cooked in a geyser. Henry J. Winser wrote several early travelers' guidebooks for the Northern Pacific Railroad. In 1883, he described how he had accomplished the same feat on the banks of the Gardiner River near the Park headquarters at Mammoth Hot Springs. Here's Winser's story as reprinted from an essay in the Spring, 1980 issue of *The American Fly Fisher*, the journal of the American Museum of Fly Fishing, Manchester, Vermont (see Schullery 1980):

"...this extraordinary feat may certainly be accomplished, not only at the Yellowstone Lake, but also

ding the practice. Fishing Cone and other geothermal features along the lakeshore are now off-limits to visitors.

But tall tales have never been needed to describe the outstanding trout fishing to be had at Yellowstone Lake. Various late-20th century estimates of the standing population of cutthroat trout in the lake have ranged as high as 7.5 to 10 million, and there may have been even more back in the 1890s. The potential of this resource was not lost on the fish culturists of that era either. Between 1899 and 1957, when operations ceased, more than 818 million eggs were taken from cutthroat trout captured at Yellowstone Lake. Shipments were distributed to 22 states, most of the Canadian Provinces, and to several countries around the world. [18]

For these reasons, perhaps most of what we know about the life history and ecology of the subspecies as a whole is based on the trout of Yellowstone Lake. [19] The lake itself sits at an elevation of 7,733 feet in country where winters can be severe but summers, while sometimes variable, are usually mild and pleasant. Wintertime snows often exceed six feet on the level and the lake usually freezes over by late December, not thawing again until May or early June. Therefore, its growing season is short. The lake is large (surface area close to 139 square miles with a shoreline that extends nearly 149 miles around), deep (mean and maximum depths estimated at 159 feet and 351 feet respectively),

on the Gardiner River, below the Mammoth Hot Springs. The writer performed it at the latter place, in the presence of nine witnesses.... Selecting a likely pool of the ice-cold stream, with a boiling spring fifteen feet distant from the bank, he stood upon a projecting rock and made a cast. His flies soon tempted a trout to his doom. The fish was small enough to be lifted out of the water without the aid of a landing net, and it was quite easy to drop him into the bubbling hot spring behind. His life must have been extinguished instantly. This procedure was repeated several times, and each of the spectators who had purposely assembled to test the truth of the strange assertation, partook of the fish thus caught and boiled. It required from three to five minutes to thoroughly cook the victims of the experiment, and it was the general verdict that they only needed a little salt to make them quite palatable."

18. A record of shipments from the Yellowstone Lake egg-taking and hatchery operations can be found in National Park Service Information Paper 34 (see Varley 1979). For additional information about the history of fisheries management in the Park, see Gresswell (1980), Gresswell and Varley (1988), and Varley and Schullery (1983, 1996, 1998).

19. The reference string on life history and ecology of the cutthroat trout of Yellowstone Lake is a long one, and includes: Muttkowski (1925), Welsh (1952), Ball (1955), Cope (1956, 1957a, 1957b, 1957c), Benson (1960, 1961), Ball and Cope (1961), Beisinger (1961), Bulkley (1961, 1963), Bulkley and Benson (1962), Benson and Bulkley (1963), McCleave (1967), Jahn (1969), LaBar (1971), Raleigh (1971), Raleigh and Chapman (1971), Bowler (1975), Kelly (1993), Gresswell et al. (1994, 1997), and Kaeding and Boltz (2001). There are summaries in Varley and Schullery (1983, 1998), Gresswell and Varley (1988), and Behnke (1992, 2002).

cold (summer surface water temperatures rarely exceed 64.4 degrees Fahrenheit [18 degrees Celsius]) and oligotrophic, which means, simply, that it is not a rich producer of food organisms. Amphipods (fly-fishers call these scuds), chironomids, and various kinds of plankton comprise the principal invertebrate fauna along with some *Callibaetis* mayflies and caddisflies. An algae bloom occurs each August, after the lake stratifies. [20]

Besides the cutthroat, Yellowstone Lake has only one other native fish, the longnose dace, whose numbers are, and apparently always were, low. The longnose sucker was introduced in 1923 and spread throughout the lake. The redside shiner appeared in 1957 and is now abundant as well, but is restricted to littoral areas. Lake chub appeared around the same time as the redside shiner, but never became abundant. Those that are found come mostly from lagoons in West Thumb and the northern part of the lake. Lake trout were discovered in 1994, but based on the age structure of specimens captured since, they have probably been present in the lake since the mid 1970s. Lake trout have become well established and pose a serious threat to the lake's cutthroat trout, a subject I'll return to later.

Yellowstone Lake has 124 identified tributary streams. The largest, of course, is the Yellowstone River. Most of the tributaries are clear, cold alpine streams, but a few receive the drainage from geothermal features. Lake-dwelling Yellowstone cutthroat trout have been observed spawning in 68 of these streams, but the principal spawning tributaries are Pelican, Clear, and Cub creeks in the northern section of the lake; Columbine and Beaverdam creeks and the Yellowstone River in the Southeast Arm; Chipmunk and Grouse creeks in the South Arm; and Arnica and Solution creeks in West Thumb. These are all inlet tributaries that host *lacustrine-adfluvial* spawning populations. One *allacustrine* population of the northern section of the lake spawns in the Yellowstone River outlet stream. For many years, it was thought that a fluvial population also inhabited the Yellowstone River upstream of the Upper Falls and used the same spawning areas as the allacustrine trout, with the two populations maintaining reproductive isolation by occupying the spawning areas at slightly different times. But work conducted in the 1993–95 spawning seasons indicates this is not so: that the two populations are not reproductively isolated, and, furthermore, the fluvial component is quite small if it exists at all, comprising no more than 10 percent of all trout spawning in the Yellowstone River between the lake and the Upper Falls. The large number of trout that makes this segment of the river so popular with anglers during the summer fishing season are allacustrine fish from Yellowstone Lake that hold in the river

20. Additional information about the environmental features of Yellowstone Lake can be found in Benson (1961), Kaplinski (1991), Gresswell et al. (1997), and Krajick (2001).

throughout the summer months, then return to the lake in the fall. [21]

Spawning runs into Yellowstone Lake tributaries take place in the spring and summer months, between April and August, with run timing and duration in individual tributaries being a function of the size of the tributary basin, its elevation, and its aspect. In Clear Creek, where the spawning run has been particularly well studied, 92 percent of the cutthroat spawners moved upstream after the date of the peak discharge from snow runoff as flows receded and water temperature increased, a pattern that is consistent for Yellowstone cutthroat trout throughout their range.

Homing of first-time spawners to their natal tributaries is high, as is fidelity of repeat spawners to these same streams. The rate of straying is 3 percent or less. Evidence from tagging studies also suggests that mature cutthroats, once they have completed spawning, exhibit fidelity to those areas of the lake where their spawning tributaries enter. For example, about 75 percent of the trout tagged in spawning tributaries of West Thumb and the South Arm were recaptured later in the summer in those areas of the lake. Post-spawning movements of the Clear Creek population are a bit more complex; about half of the tagged trout from Clear Creek were later recaptured in the South Arm, although they, like the other tagged groups, appar-

ently stayed together all summer. Thus, several distinct subpopulations, or races, of Yellowstone cutthroat trout may share Yellowstone Lake. [22]

Spawning cutthroats in Yellowstone Lake range from 3 to 9 years of age, but ages 4, 5, and 6 predominate. Studies carried out in the early 1950s found that almost all of the fish were first-time spawners, and mortality in the spawning streams was very high (up to 48 percent), so that only about 2 percent of the run were repeat spawners. It was also reported that most repeat spawners skipped a year and came back in alternate years. However, those results appear to have been artifacts of the fisheries management policies in place at the time, which allowed high levels of angler harvest and compensated with hatchery supplementation. Hatchery stocking stopped in the mid-1950s and harvest has been restricted since the early 1970s. The trout population now shows a greater proportion of older, larger fish, with mature fish typically repeat-spawning on an annual basis rather than every other year as reported earlier. Also, spawning mortality levels are now down in the 7 to 19 percent range.

As a general rule, adult fish do not remain long in the spawning tributaries, but drop back into the lake after just two to three weeks. I recall Gordon Reese, the seasonal ranger who occupied the Hawk's Rest Guard Station on the upper Yellowstone, telling me that if he

21. The early work that led to the conclusion that reproductively isolated fluvial and allacustrine populations spawned in the Yellowstone River outlet stream was that of Ball and Cope (1961). The 1993–95 studies that concluded that these populations are not reproductively isolated, and, furthermore, that the fluvial component is very small if it exists at all, were reported by Kaeding and Boltz (2001).

22. For a deeper look at the evidence for subpopulation structure in the cutthroat trout of Yellowstone Lake, see Cope (1957), LaBar (1971), Raleigh and Chapman (1971), Bowler (1975), and Gresswell et al. (1994, 1997).

wanted to get any trout for his family to eat, he'd have to get them before the last week of July, because after that the river in front of his cabin would be empty of cutthroats. But this pattern doesn't hold in all tributaries. Marking studies of Pelican Creek spawners indicated that a few of these trout not only remain in the stream after spawning, but overwinter there as well. And lake-dwelling trout that spawn in the Yellowstone River outlet stream between the lake and the Upper Falls remain in the river all summer, not returning to the lake until fall.

Nor is it common for the newly emerged fry to spend much time in their natal streams. Eggs hatch in about 30 days (requiring 278–365 thermal units) and the fry emerge from the gravel about two weeks later. About two weeks after that, the majority of these fry head down to the lake. Those that do not migrate to the lake at this time may remain and overwinter, then drop down to the lake as one-year-olds. Some may even stay for another year and migrate to the lake as two-year-olds, but this is rare. [23]

Once in the lake, the young fish tend to segregate from the adults. The nursery area of Yellowstone Lake is its vast limnetic zone, i.e., the deeper, open-water area where zooplankton are most abundant. This is where young-of-the-year, yearling, and age-2 trout feed and grow. The older, larger trout are found in inshore waters.

Zooplankton sustain the juvenile trout until they reach a length of about 9 inches. At about that size, they move into inshore waters and begin to feed more heavily on the organisms associated with the littoral zone. Freshwater scuds of the *Gammarus* and *Hyallela* genera (fly-fishers take note) are important food items, as are chironomid pupae and adults and the nymphs and adults of *Callibaetis* mayflies. The scuds et al. decline in importance in August, when *Daphnia* begin to swarm. *Daphnia* swarms continue through late summer and fall and can drive fly-fishermen to distraction as the trout cruise the surface, inhaling the pinhead-size organisms while totally ignoring the smallest of fur-and-feather offerings.

The fly rodder has a few more things to imitate in the West Thumb area of Yellowstone Lake. Here, the feeding habits of the trout are influenced by the prevailing south and southwesterly winds that blow off the nearby lodgepole pine forests. In nearshore waters of this area of the lake in particular, wind-blown wasps, bark beetles, and other terrestrial insects augment the food supply of the trout.

Possibly because their evolutionary programming occurred in the absence of abundant forage fishes (only longnose dace are native but were never known to be numerous), the cutthroat trout of Yellowstone Lake did not evolve into piscivores, as did the Lahon-

23. Pelican Creek, where we have already seen that some mature trout hold over in the stream after spawning, may also be an exception as regards juvenile rearing. In the study cited (see Gresswell et al. 1994 for details), all trout migrating upstream during the 1980 spawning run were marked with a fin-clip. Subsequently, all downstream-migrating fish were checked for the mark. About half of the fish moving down to the lake did not have the mark, which suggests that these trout had reared in the stream for at least a year before migrating (no young-of-the-year trout were counted in this study). From among the unmarked fish, a group of trout ranging in size from 4 to 19 inches was checked for previous spawning. Half of those had never spawned before. To attain these sizes, at least some of the trout that had never spawned must have reared in Pelican Creek for multiple years before migrating to Yellowstone Lake.

tan cutthroat trout of Pyramid Lake. Although they are not completely puritan in this regard (indeed, they will take a fish or two on occasion), they do not feed to any appreciable extent on their own young, or on the other introduced species, such as redside shiner, lake chub, or longnose sucker. Subsisting as they do primarily on scuds, aquatic or terrestrial insects, and *Daphnia*, the cutthroat trout of Yellowstone Lake seldom exceed 2 to 3 pounds in weight, although occasionally a 4-pounder will be reported. Life span in Yellowstone Lake is 7 to 9 years. [24]

No treatise on the ecology of the cutthroat trout of Yellowstone Lake would be complete without mentioning the role they play—possibly a *keystone* role—in the functioning of their ecosystem. [25] At least 15 species of fish-eating birds, including white pelicans, ospreys, bald eagles, eared and western grebes, California gulls, Caspian terns, Barrow's goldeneye, loons, mergansers, cormorants, kingfishers, and the like live at or near Yellowstone Lake and its tributaries and depend on the trout for most, if not all, of their food for about 135 days each year. Of these, the white pelican is the most important avian predator on adult cutthroats in the lake. The osprey diet consists of 93 percent cutthroat trout, mostly fish smaller than 13 inches. The lake's eagles eat many fish too, but they prey more on ducks than they do on fish. Even so, there's still a significant

24. The life span of these trout depends much upon the environment in which they live. In South Gap Lake, Wyoming, Dr. Behnke examined a population of cutthroat trout that had been introduced from Yellowstone Lake, but here the trout lived to 10–11 years of age. South Gap Lake sits at an elevation of 11,155 feet, where it probably experiences no more than 90 ice-free days per year. Metabolism and growth processes are slowed in such an environment, extending the life span of the trout. Although the trout in South Gap Lake lived longer, they grew only slowly, attaining maximum lengths of only 11–13 inches. This report is in Behnke (1992) at page 95.

25. A *keystone species* is one whose direct and/or indirect effects on other species or on ecosystem function is disproportionally large, and without which other species in its community would disappear or the ecosystem would unravel. The term itself was first used by ecological scientist Robert T. Paine in 1969 to describe the role of a predator species in an intertidal community. Paine's keystone predator, a *Piaster* sea star, kept the density of a mussel population well enough in check for an array of other intertidal species to persist in the community, where they would have otherwise been pushed out by the mussels. Since Paine's time, other scientists have expanded the concept to include other keystone roles. In addition to the keystone predator role, a keystone species may be a mutualist, such as a pollinator; a habitat-modifier, such as the beaver; or it may be a keystone food resource. That is the role of the cutthroat trout of Yellowstone Lake in supporting the piscivorous bird community, and in providing protein and energy for a component of the Park's grizzly bear population at a crucial time of year. To read more about keystone species, see Paine (1969), Soulé (1986), Mills et al. (1993), and Meffe and Carroll (1994).

connection because the ducks themselves are primarily fish-eating varieties. Overall, fish-eating birds are believed to consume about 200,000 lbs of Yellowstone Lake cutthroat trout annually. Across the spectrum of trout sizes consumed by the birds, the average trout would probably weigh 0.3 lb, which calculates out to 666,667 trout per year.

In addition to the birds, it's estimated that a minimum of 84 members of the Park's grizzly bear population, including females with cubs, concentrate their activities around Yellowstone Lake's spawning tributaries for about 47 days each year during the spawning period of the trout. Trout appear to be an important source of energy and protein for these bears in the spring and early summer, especially for the reproductive female bears. I extrapolated from earlier data (when it was thought that there were fewer bears working the streams) to estimate that this grizzly bear contingent consumes 39,920 cutthroat spawners annually. Black bears, rivers otters, mink, and other animals also utilize the spawning trout, but no information is available about their levels of consumption.

To put these numbers in perspective, let's compare them with estimates of the standing population size of cutthroat trout in Yellowstone Lake, along with estimates of the total annual mortality in this population. Various estimates of the standing population size have yielded values ranging from 7.5 to 10 million trout, with about 70 percent of these (5.25 to 7 million trout) being age-3 and younger juveniles and 30 percent (2.25 to 3 million trout) being age-4 and older adults. Annual mortality rates have been estimated at 63 percent for the juvenile trout (3.3 to 4.4 million fish per year) and 52 percent for the adults (1.2 to 1.6 million fish per year).[26] If these estimates are anywhere close to being right, then total annual mortality is somewhere between 4.5 to 6 million trout per year. Fish-eating birds account for only 11–15 percent of this amount, and the grizzly bears less than 1 percent.

At this point it is also instructive to see how much human angling contributed to total annual mortality during the period when these population estimates applied. From the early 1970s until 2001, when a catch-and-release regulation was adopted, anglers were allowed to harvest two trout per day under 13 inches in total length, but only during certain periods and in certain waters. According to a 1996 National Park Service creel survey regarded as typical for that period, anglers captured 218,915 trout that year from Yellowstone Lake, but kept only 21,359. The rest, including almost all of the fish under 10 inches, were released (in effect, most anglers were practicing a *de facto* slot limit, keeping trout only between 10 and 13 inches in length). If the released fish suffered a 10 percent hooking

26. These numbers were gleaned from various National Park Service and U.S. Fish and Wildlife Service sources, and from publications of university scientists commissioned by these agencies. Pertinent references include Kaeding et al. (1994), Varley and Schullery (1995, 1996, 1998), and Ruzycki et al. (2003). These sources also include references to the primary studies of bird and grizzly bear predation on Yellowstone Lake cutthroat trout, which I urge you to seek out and read as well. Rather than total annual mortality, some of these reports talk in terms of annual production of the cutthroat population. These numbers should be equivalent. In other words, if total annual mortality is 4.5 to 6 million trout per year, then the population must produce 4.5 to 6 million new trout per year to maintain the total standing population size.

mortality rate, an additional 21,892 trout would have succumbed for a total angling mortality of 43,251 trout. So, like the grizzly bears, the angling contribution to total annual mortality was less than 1 percent. [27]

As "ballpark" as these comparisons might be, it should be evident that the abundance of cutthroat trout in Yellowstone Lake has not been limited by indigenous natural predators, such as fish-eating birds and bears, or, under recent fishing regulations, by human anglers. Nor has the number of cutthroat trout limited the number of indigenous predators the ecosystem could support. But, should the cutthroat trout decline significantly in abundance and biomass, and should their spawning runs falter as a result, the grizzly bears and bird predators would likely decline as well, or maybe even disappear altogether.

The establishment of a new predator, the lake trout, not discovered in Yellowstone Lake until 1994 but introduced illegally perhaps as early as the mid-1970s (and recent testing suggests that illegal releases of Lake Trout from nearby Lewis Lake may also have been made in 1989 and 1996), has the potential to bring about just such a collapse. [28] Yellowstone Lake's cutthroat trout evolved in the absence of any large predatory fish, so there is no natural selection for coexistence with such a predator. By the time Park biologists and managers became aware of the problem, the

27. Prior to 1970, when restrictions were placed on when and where one could fish and the harvest limit of two-fish per day under 13 inches was imposed on angling in Yellowstone Lake and its tributaries, human anglers accounted for a very high percentage of the total annual mortality of the trout—high enough to virtually eliminate the older age classes from the population, and to send the population as a whole spiraling into decline. This is chronicled in the references already cited on the history of fisheries management and angling in the Park; see Gresswell (1980), Gresswell and Varley (1988), and Varley and Schullery (1983, 1996, 1998). I also refer you back to Chapter 4, where I explained how angling mortality, if high enough, becomes additive rather than compensatory with other forms of mortality, causing trout populations to decline.

28. The lake trout *Salvelinus namaycush* is a large-bodied, long-lived, piscivorous cousin of the brook trout. Its native range extends from the northern areas of New York and New England northward and westward across boreal and subarctic Canada to Alaska. Within its native range, it lives in coldwater lakes and ponds, and is well adapted to deep, cold, oligotrophic lakes. Lake trout have been widely introduced into such waters outside the species' native range, including many of the large lakes within the native distribution of the cutthroat trout, where in most cases they either brought about themselves or contributed to the demise of the native cutthroats. Pend Oreille, Coeur d'Alene, and Flathead lakes in westslope cutthroat country and Lake Tahoe, where the Lahontan cutthroat was native, are examples of water bodies where this happened. Lake trout were introduced into previously fishless Lewis and Shoshone lakes in the upper Snake River drainage of Yellowstone National Park in 1890, from which they quickly spread downstream over waterfalls that are impassable to upstream migration and had kept these two lakes from being colonized by cutthroat trout at the end of the Pleistocene. Once lake trout had passed over these waterfalls, they spread downstream as far as Jackson Lake, where a population became established, and upstream into Heart Lake in Yellowstone National Park, where they also became established. And now, they are established in Yellowstone Lake as well. The history of their discovery in Yellowstone Lake, and the alarm this set off among fisheries managers and biologists, can be found in publications by Kaeding et al. (1994), Varley and Schullery (1995, 1998), and Koel et al. (2005). Lake trout have been captured throughout Yellowstone Lake, but adults are most abundant in West Thumb, where four major spawning areas have been located near Carrington Island, Geyser Basin and Solution Creek, and in Breeze Channel. They spawn in gravel and cobble shoal areas, and spend their entire life cycle within the lake itself. Lake trout smaller than about 12 inches feed primarily on invertebrate organisms, but as they grow, their diet shifts to greater proportions of fish. The diet of age-9 and older lake trout in Yellowstone Lake (these are fish 24 inches in total length and larger) is greater than 95 percent fish. It had been hoped that the abundant longnose sucker would act as a buffer to diminish lake trout predation on Yellowstone Lake's cutthroat trout, but that has not happened. Studies to date indicate that 99 percent of the fish consumed by lake trout in Yellowstone Lake are cutthroats. Lake trout can live for 20+ years (the oldest caught from Yellowstone Lake was 23 years of age) and attain lengths of 34 inches or more, and are capable of consuming fish more than half their body length in size. In Yellowstone

population of lake trout old enough to be preying on cutthroat trout had already grown to more than 11,000 fish. In 1996, it was estimated that lake trout consumed 522,000 cutthroats. Modeling studies commissioned that year projected that by 1999, if the lake trout population were allowed to grow unchecked, cutthroat consumption would have increased to 684,000, about the same proportion of the total annual mortality as attributed to the indigenous fish-eating birds and grizzly bears. But here's where things got scary. Projecting farther into the future, the modelers concluded that by 2025, without control, the lake trout population would grow large enough to alone consume the entire total annual production of the cutthroat population, and by 2033, lake trout predation would exceed the cutthroats' standing population size. In other words, the cutthroat population of Yellowstone Lake would be extirpated. [29]

Fortunately, Park biologists and managers didn't wait for the results of these studies to be in before taking action. In 1996, they commenced an intensive gillnetting effort that became even more focused once lake trout spawning areas were pinpointed. In addition, anglers on Yellowstone Lake were required to kill all lake trout brought to hand, regardless of size. This effort, though expensive (it costs around $300,000 per year), has held the lake trout population in check and actually reduced the number of lake trout old enough

to be eating cutthroats. That portion of the lake trout population in Yellowstone Lake now stands at about 3,500 fish.

I will revisit the lake trout situation one more time in the last section of this Chapter. But now it's time to move on.

Dr. Behnke and others have expressed concern that all this focus on the life history and ecology of cutthroat trout of Yellowstone Lake has produced misconceptions about the ecological variability of the subspecies as a whole. Evolution in the relatively stable, oligotrophic environment of Yellowstone Lake, free from competition from other fish species, may have resulted in an innate lack of ability of these trout to adapt to other environments. Of the hundreds, perhaps thousands of transplants made from the extensive Yellowstone Lake egg-taking operations of the 1899–1957 period, most failed to establish any new populations, let alone thriving ones. [30] Lack of adaptability in the parent Yellowstone Lake population may be one reason why.

But other lake-dwelling populations exhibit greater ecological variability and adaptability. In the northeast corner of Yellowstone Park, in the Slough Creek drainage, lies 23-acre McBride Lake, elevation 6,558 feet. McBride Lake supports an indigenous, self-reproducing population of Yellowstone cutthroat trout that is genetically pure despite a 1936 plant of rainbow trout.

Lake, the lake trout seem to focus on cutthroats between 3 and 14.5 to 15 inches in length. For more information, see Ruzycki et al. (2003).

29. See Ruzycki et al. (2003) for the details of these modeling studies.

30. A few of those transplants did indeed succeed, establishing populations of pure Yellowstone cutthroat trout outside their historical range in some cases, and in other cases resulting in introgression of native cutthroat subspecies already present in the stocked waters. Both outcomes occurred from stocking in Glacier National Park, within the range of the westslope cutthroat trout, where pure populations of Yellowstone cutthroat trout now occur in some Glacier Park lakes and hybrid populations are documented in others. Reports by Marnell (1980) and Marnell et al. (1987) give details.

These trout, too, evolved in a simple ecosystem with no competition from other native fishes. Yet, in contrast to Yellowstone Lake fish that performed poorly for State of Montana biologists when stocked in other waters, trout of the McBride Lake strain adapted quickly to other lacustrine ecosystems, even lakes as high as 10,869-foot Marker Lake, the highest-elevation fishery in Montana. In most alpine lakes, growth of McBride Lake trout to 10 to 12 inches is achieved in 3 to 4 years, but in productive waters, lengths up to 24 inches and weights of 6.6 lbs have been documented. Large *Daphnia* species and chironomid pupae and adults comprise the major portion of the diet for trout in most mountain lakes, augmented seasonally by terrestrial insects, but in one lake inhabited by mottled sculpins, trout of the McBride Lake strain preyed extensively on those fishes. Versatility in their spawning behavior is also a characteristic; trout of the McBride Lake strain reproduce equally well in inlet or outlet streams, whichever is accessible. [31]

The Yellowstone cutthroats of 2,150-acre Heart Lake, only about 8–10 miles south of West Thumb as the crow flies but across the Continental Divide in the upper Snake River drainage, provide another example of the ecological diversity of this subspecies. Heart Lake, elevation 7,469 feet, sits in a depression at the eastern foot of Mt. Sheridan. At the upper end of the lake, and in the lake itself, are the numerous geysers and hot springs of the Heart Lake geyser basin. Witch Creek, the principal inlet tributary, also heads up amid geysers and hot springs that make its upper waters scalding hot, but the water temperature moderates over its 2–3 mile run to the lake so that the temperature at the inlet averages about 75 degrees Fahrenheit (24 degrees Celsius). Probably because of these high inlet temperatures, Heart Lake cutthroats evolved to be outlet spawners. They move downstream into the accommodating waters of Heart River, the lake's outlet stream, that flow into the upper Snake River after a run of 4 miles.

Heart Lake is shallow and wadable around the edges, but drops off quickly into deeper water. Its water is slightly saline, like very weak sea water. It is not a particularly cold lake, and there is ample food to support an array of native fishes that includes, of course, the cutthroat trout and also mountain whitefish, mountain sucker, Utah sucker, Utah chub, redside shiner, speckled dace, and mottled sculpin. In this environment, the Yellowstone cutthroat trout did evolve to be piscivores, with the Utah chub being a preferred prey. The switchover from invertebrates to fish occurs when the trout reach 9 to 12 inches in length, and after that they can grow to impressive sizes. It is well known among the region's anglers that even though it's a long hike in, Heart Lake

31. More on the State of Montana's experience with the McBride Lake strain of Yellowstone cutthroat trout can be read in "Use of McBride Lake Strain Yellowstone Cutthroat Trout for Lake and Reservoir Management in Montana" (McMullin and Dotson 1988).

is the place to go for lunker Yellowstone cutthroats. Trout weighing up to 8 pounds have been reported.

Lake trout invaded Heart Lake in the early 1890s, following the stocking of Lewis and Shoshone lakes. No longer at the top of the aquatic food chain after lake trout became established, the Heart Lake cutthroat population took a big hit. But cutthroat trout have persisted in Heart Lake because here, in contrast to the situation in Yellowstone Lake, the indigenous forage fishes do provide a buffer against lake trout predation. Lake trout consume large numbers of suckers and Utah chub in Heart Lake in addition to their take of cutthroat trout. [32]

The final example I'll mention of ecological diversity among lake-dwelling Yellowstone cutthroat trout is the population of 6,500-acre Henry's Lake. Henry's Lake is located in the northernmost corner of the Henry's Fork watershed in Idaho, in an area that anglers have called the "golden triangle" of western trout fishing. It's only about an 18-mile drive across the Continental Divide at Targhee Pass on U.S. Highway 20 from the town of West Yellowstone, and a run of about the same distance or a bit more north across Raynolds Pass on State Route 87 to the famous Madison River fishery below Hebgen Lake.

Henry's Lake has long been noted for its robust trout. [33] Originally, these were Yellowstone cutthroat trout exclusively, but brook trout appeared early on

32. Jordan (1891), Smith and Kendall (1921), and Varley and Schullery (1983) provide useful descriptions of Heart Lake, its fishes, and its fishing. Fly-fishers with Internet access may also want to check out *www.yellowstonenationalpark.com/ffsoutheast.htm*. In addition to the lunker cutthroat trout, a lake trout of 42 pounds, the largest ever caught in Yellowstone National Park, was taken by an angler at Heart Lake in 1931.

33. E.N. "Uncle Nick" Wilson, for whom the town of Wilson, Wyoming, is named, ran away from home at age 12 and spent two years with Chief Washakie's band of Shoshone Indians. For generations, the Shoshone had made seasonal rounds that included subsistence fishing. Henry's Lake was one such stop for the band. Wilson described the lake as "...fairly alive with fish. Oh, how I did catch them" (see Wilson 1919 [1991]). The fur trappers had also passed this way from time to time beginning with Andrew Henry, for whom the Henry's Fork and Henry's Lake are named, in 1811. Trapper Osborne Russell described a rather expansive body of water "about 30 miles in circumference" when he camped at Henry's Lake on June 6th, 1838 (see

Russell 1986). He also described the surrounding area as being covered in "pine woods" except for "a small prairie about a mile wide" on the southeast side of the lake. However, by wealthy New York sportsman George W. Wingate's time in 1885 (see Wingate 1886 [1999]), those "pine woods" had vanished and the entire valley eastward from Henry's Lake was covered with a thick growth of bunch grass. The first permanent settler at Henry's Lake was Gilman Sawtell, who arrived in 1868. Sawtell ran cattle, and he also fished the lake commercially, selling the trout he caught to the "sporting houses" in the gold mining camps at Virginia City, Montana. Sawtell is supposed to have told Ferdinand Hayden, the explorer and surveyor whose government party came this way in 1872, that "The trout cost the customers almost as much as the girls." Later settlers also exploited Henry's Lake trout for commercial purposes, selling to markets as far away as Butte and Salt Lake City. By 1900, there were 37 commercial fish operations that harvested each winter between 50,000 and 100,000 pounds of trout. The U.S. Commission of Fish and Fisheries also

used the lake as a source of trout eggs for distribution to other parts of the country, and may have been responsible for at least some of the non-native trout that now prevail in the Henry's Fork watershed. By 1912, though, Henry's Lake itself was approaching extinction. Although still populated with cutthroat trout, a U.S. Bureau of Fisheries survey conducted that year found a lake with a maximum depth of only 6.5 feet and an average depth of only 4.8 feet, its waters in imminent danger of being completely choked with aquatic vegetation (see Kemmerer et al. 1924). Then, in 1922 or 1923, a group of farmers organized as the North Fork Reservoir Company built a dam across the outlet stream that raised the level of the lake and increased its surface area and volume to more or less their present levels. For more on the history (and prehistory) of Henry's Lake and its fishery, see Christiansen (1982), Brooks (1986), Green (1990), Van Kirk and Benjamin (2000), and Van Kirk and Gamblin (2000). The Sawtell quote relayed by Fredinand Hayden is in Sprague (1964).

and are still present in the lake. Rainbow trout were also stocked, first in 1890 and then sporadically and in small numbers over about a 50-year span, from the late 1920s or early 1930s to 1982. Although rainbow trout never took hold in the lake, some spawning and limited interbreeding with the native cutthroats apparently did occur, as a few rainbow × cutthroat hybrids were reported by R.B. Irving of the Idaho Department of Fish and Game in 1954. In 1960, the State of Idaho began deliberately producing and stocking first-generation rainbow × cutthroat hybrids in Henry's Lake, a move that proved very popular with anglers because these hybrids grow larger and more vigorous than either parent. The agency now stocks about 200,000 of these hybrids annually. Since 1998, to prevent introgression of the remaining pure Yellowstone cutthroat population, these have been heat-shock treated to produce sterile triploids. [34]

Since the early 1920s, in addition to its trout fishing, Henry's Lake has also served as a storage reservoir for irrigators. It is shallow (average depth 12 feet, maximum depth 21 feet) and nutrient-rich, with a luxuriant growth of aquatic vegetation. Although it freezes over in winter, it warms quickly in spring and summer and its growing season is long. In this environment, the trout grow quickly, feeding on a rich mix of invertebrates consisting of *Gammarus* scuds, damselflies, and chironomids, augmented by hatches of large *Siphlonu-*

34. This brief review of Henry's Lake fisheries management using rainbow × cutthroat hybrids was taken from papers by Rohrer (1983), Rohrer and Thorgaard (1986), and Campbell et al. (2002). Geneticists and breeders use the term *heterosis* or *hybrid vigor* to describe the tendency of crossbred individuals to exhibit qualities superior to those of their parents. This phenomenon has been exploited by plant breeders and agriculturists for years to produce important and productive crop varieties; hybrid corn, for example. The "cuttbows" released into Henry's Lake express hybrid vigor in growth rate and size. Cuttbows creeled by anglers in the 1982 fishing season averaged 17.8 inches in total length vs. 16.4 inches for the average pure cutthroat; and cuttbows returning to the hatchery averaged even larger, at 23.2 inches for the cuttbows vs. 18.5 inches for the pure Yellowstones (Rohrer 1983). A few years later, in a test of sterile triploid hybrids, the cuttbows grew larger still compared to the pure cutthroats. At age 4, the sterile triploid cuttbows had attained total lengths up to 29.5 inches and weights up to 13.2 lbs, while pure age-4 Yellowstone cutthroats had attained only 20.7 inches and 4.4 lbs (Rohrer and Thorgaard 1986). Needless to say, Henry's Lake anglers are delighted with these cuttbows. However, hybrid vigor is a one-generation-only phenomenon. Diploid cuttbows are fertile and capable of reproduction, but if mating does occur, either with other hybrids or with parental stocks, the superior traits diminish in subsequent generations. Thus, just as corn farmers have to go back to the seed companies each year for new hybrid seed in order to maintain their high crop yields, so fishery managers must go back to the fish-culture facility each year for a new batch of first-generation cuttbows in order to continue reaping the benefits of hybrid vigor.

Waha Lake, Idaho, type locality for the subspecies name bouvieri. PHOTO BY AUTHOR

rus mayflies and plenty of fat leeches. With regard to the native Yellowstone cutthroats, these trout reach first sexual maturity at age 2 or 3 for the males and age 3 or 4 for the females, and nowadays ascend the inlet tributaries for spawning from mid-April through mid-June (those ascending Hatchery Creek run up earlier, from the first of March through April, but this is a result of hatchery selection for early run timing). As in Yellowstone Lake, fidelity to their natal tributaries is strong among the naturally-spawned cutthroats and the rate of straying is low. Evidently, in the years before the dam, a component of ripe cutthroat trout also swam down the Henry's Lake Outlet stream and spawned in that stream or its tributaries. Cutthroat trout still spawn in these tributaries, but these must be fluvial fish since there is no passage at the Henry's Lake dam.

Owing to the rich food supply and long growing season in Henry's Lake, its Yellowstone cutthroat trout grow more rapidly than the trout of Yellowstone Lake. They attain a larger maximum size, reaching 24 inches and 6 lbs as a general maximum length and weight (vs. 21 inches and 4 lbs for the cutthroat trout of Yellowstone Lake), but their life span is shorter. Where the cutthroat trout of Yellowstone Lake can reach 7 to 9 years of age, those in Henry's Lake do not live beyond 6 or 7 years. [35]

The Fluvial, or Riverine, Life History Strategy

The riverine life history strategy was once widespread in rivers and streams throughout the historical range of the Yellowstone cutthroat subspecies. This life history pattern prevailed in the Yellowstone River downstream from the Lower Falls to the lowest extent of historical occupancy by trout near the mouth of the Bighorn River; in the Lamar River in Yellowstone National Park; in historically occupied reaches of the mainstem Tongue, Bighorn, Clark Fork, Greybull, and Shoshone rivers; in Henry's Fork of the Snake River and all of its main tributaries; in the South Fork Snake (except for the reach between Jackson Lake and Palisades Reservoir); and in the main Snake and its major tributaries downstream from the Henry's Fork confluence to Shoshone Falls. However, little was published concerning the life history and ecology of riverine populations prior to the mid-1980s. Since then, a great deal of information has been acquired. [36]

Trout exhibiting this life history strategy typically make spawning runs to natal tributaries when runoff flows begin to subside from peak levels and water temperatures warm up to around 41 degrees Fahrenheit (5 degrees Celsius). This could occur anytime between March and August in different parts of the range, but in both the Yellowstone River drainage downstream from Yellowstone National Park and the South Fork

35. Information on the life history and ecology of pure Yellowstone cutthroat trout in Henry's Lake was compiled from Irving (1954), Adriano (1956), Thurow et al. (1988), and Campbell et al. (2002).

36. Details or summaries of the life history and ecology of riverine populations of Yellowstone cutthroat trout are given in Behnke (1979, 1992, 2002), Thurow (1982), Moore and Schill (1984), Clancy (1988), Corsi (1988), Thurow et al. (1988), Varley and Gresswell (1988), Byorth (1990), Griffith and Smith (1993), Thurow and King (1994), Baxter and Stone (1995), Gresswell (1995), Henderson et al. (2000), and Meyer et al. (2003b).

Snake River drainage downstream from Palisades Reservoir, actual spawning of fluvial trout takes place from May through early July. In the South Fork Snake River downstream from Palisades Reservoir, spawning of fluvial Yellowstone cutthroat trout is not restricted to tributary creeks. In one segment of the river where the floodplain is broad and a network of side channels has formed, trout move into and spawn in these side channels as well.

Although spawning tributaries and side channels typically flow perennially, use of intermittent or seasonal streams for spawning is also known. Some of these go dry in July or early August. Trout that utilize these seasonal streams may enter and spawn earlier, during the height of spring runoff rather than after the peak. This gives the embryos a bit more time to incubate, hatch, and gain a modicum of growth before exiting to the main river when the natal tributary goes dry.

Spawning runs of fluvial trout consist mostly of first-time spawners. In the upper Snake River system of Idaho—this includes Henry's Fork and its major tributaries, the South Fork Snake and its major tributaries, and the main Snake and its major tributaries downstream from the forks—the trout spawn for the first time at age 4 or 5. These trout would be about 12 inches in length at age 4. Fluvial trout in the Yellowstone River system downstream from Yellowstone National Park are about the same size and age at first spawning. Repeat spawning is common in fluvial Yellowstone cutthroat trout, but not always in consecutive years. This reduces the percentage of repeat spawners in a given year's spawning run to numbers as low as 15 percent. What's more, some spawning runs can have really skewed sex ratios among the repeat spawners. For example, some years ago the Idaho Department of Fish and Game reported that 93 percent of the repeat spawners in the Blackfoot River fluvial spawning migration were females. [37]

Females tend to outnumber males in fluvial spawning runs sampled in Idaho, sometimes by big ratios. Female-to-male ratios ranging from 2.7 to 1 to 5.9 to 1 have been reported for Idaho's Blackfoot River, but more balanced ratios nearer to 1.2 to 1 may be the norm in the South Fork Snake system. The sexes are also more nearly equal in the Yellowstone River system in Montana, where, if anything, males may slightly outnumber females. Fecundity of the females in these spawning runs varies with fish size. The average-size female spawner in the South Fork Snake River is just under 15 inches in length. Such a trout would carry about 1,400 eggs.

As in lake-dwelling populations, homing of fluvial spawners to their natal tributaries is high and straying rates are low. The distance traveled to reach spawn-

37. This report, authored by Russell F. Thurow, was issued by the Idaho Department of Fish and Game in 1982 under the title, "Blackfoot River Fishery Investigations" (Boise, Idaho Department of Fish and Game, Job Completion Report, Project F-73-R-3).

ing sites varies by population. Four of the six major spawning tributaries in an 83-mile segment of the Yellowstone River from Yellowstone National Park to Springdale, Montana, are located near the upper end of the segment, but account for most of the recruitment to the middle portion of the segment where no spawning tributaries exist. Fluvial trout inhabiting that portion of the river must make upstream spawning runs of half that distance or more to account for this recruitment. In Pine Creek, a tributary of the South Fork Snake River, fluvial spawners enter the creek and migrate 10 to 15 miles upstream to reach spawning sites, but in Palisades Creek, another South Fork Snake tributary, all fluvial trout spawn in the lower mile and a quarter of the stream. [38]

Water depth, velocity, reach gradient, and substrate size attributes of spawning sites used by fluvial Yellowstone cutthroat trout are within the range of values published for other cutthroat subspecies, i.e., water depths ranging from 4 to 12 inches, velocities in the 0.8 to 2 feet per second range, reach gradients not exceeding 3 percent, and gravel substrate ranging from less than a quarter-inch to 4 inches in diameter but most of the particles between 0.6 and 2.5 inches in diameter for trout ranging from 12 to 20 inches in total length. Having deposited and fertilized their eggs, the adult trout linger in the tributaries from two to four weeks as they recover their condition before dropping back down to the main rivers.

Eggs deposited in the spawning gravels hatch out when they have accumulated about 310 thermal units, which typically takes about 25 to 30 days, and swimup fry appear in the natal streams about two weeks later. Fry emergence typically commences about mid-July and may continue into the fall months depending on location, climate conditions, and just when spawning actually took place.

In this life history strategy, juveniles may rear in natal tributaries for 1 to 3 years, or they may return to the mainstem rivers as age-0 fry at lengths of no more than an inch and a half to two and three-quarters inches. In intermittent spawning tributaries, this is, of course, the norm for juvenile emigration. However, at this small a size, the fry are not capable of swimming in the main river currents. They congregate instead in shallow water along the river shorelines, and there they remain until they have attained enough size and swimming stamina to join the summer rearing hierarchy. In winter, when water temperatures drop below about 45 degrees Fahrenheit (7 degrees Celsius), these age-0 juveniles will again be found in the shallows, in water less than 18 inches deep within 3 feet or so of the stream bank, now concealing themselves in interstices within the substrate during daylight hours and not emerging from cover until night.

38. An irrigation diversion is in place in Palisades Creek 0.8 miles up from its mouth. Fluvial trout can evidently pass this diversion but, according to the telemetry data of Henderson et al. (2000), do not migrate beyond stream mile 1.24 for spawning. High-quality spawning habitat is abundant in upstream reaches of Palisades Creek, but is evidently left to the use of the creek's population of stream-resident cutthroat trout, and to the lacustrine cutthroats that feed and grow in two small lakes in the upper part of the drainage.

Turning now to feeding behavior, the larger sub-adult and adult cutthroats are feeders of opportunity during the summer feeding and growth period, consuming whatever items in the food chain that become available in their particular environments. Fluvial Yellowstone cutthroat trout are not noted as fish eaters, but even so, they may be more innately piscivorous than, say, westslope cutthroat trout (Chapter 4). In 1992, Dr. Behnke wrote about examining Yellowstone cutthroat trout in Goose Creek, Nevada, in which fish remains made up almost 100 percent of the stomach contents of all cutthroats over 12 inches in length. [39]

Waters of the upper Snake River drainage tend to be alkaline and are generally quite productive of aquatic food organisms. So, too, are the waters of the Yellowstone River and certain of its spring-fed tributaries in the reach between Yellowstone National Park and the town of Livingston in Paradise Valley, especially near Livingston where several limestone spring creeks provide world-class trout fishing. [40] In these productive waters, fluvial Yellowstone cutthroats can grow to large sizes. Trout in the Blackfoot River and Willow Creek, Idaho, have attained total lengths of 20 to 24 inches and weights of 4.5 to 9 pounds. They can also enjoy fairly long life spans in these waters, with some trout living to 8 or 9 years of age.

39. This observation is in Behnke (1992) at page 94.

40. What makes some streams more productive of fish food and fish biomass than others? Several factors come into play: the water temperature regime, the amount of sunlight reaching the stream through the canopy, and the concentrations of inorganic nutrients in the water are examples. Productive sites have moderate water temperatures that seldom, if ever, fluctuate to extreme highs or lows (this is especially true of spring-fed systems where temperatures are buffered by groundwater inputs the year round). Their canopies allow ample sunlight to penetrate into the water to generate autotrophic production, and their waters have relatively high concentrations of inorganic nutrients. But above all, it may be the geology. The most productive sites have hard waters, i.e., alkaline pHs resulting from the solubilization of calcium compounds from sedimentary rocks, such as limestone and dolomite, or from metamorphic rock types, such as gypsum or marble. The bicarbonate generated when these calcium compounds go into solution raises the pH of the water. The dissolved calcium encourages the growth of the algae, diatoms, insects, crustaceans, and rooted aquatic plants that constitute the food chain supporting the trout. Rock types composed mostly of insoluble silica, such as gneiss, quartzite, granite, or schist, contribute none of these productivity-enhancing components. Streams flowing through geological settings with silica rock types have soft waters that are generally low in productivity. Anglers (and some stream ecologists) have taken to calling these "freestone" streams as opposed to the more productive "limestone" streams. To learn more about rich and poor trout streams, the place I'd start is Tom Rosenbauer's *Prospecting for Trout: Fly Fishing Secrets from a Streamside Observer* (1993). His Chapter 2 is the most readable presentation of this subject that I have ever come across. For the more technical details, consult any good book on stream ecology. Here are four that I can recommend: *The Ecology of Running Waters* by H.B.N. Hynes (1970); *Stream Ecology: Structure and Function of Running Waters* by J.D. Allan (1995); *River Ecology and Management: Lessons from the Pacific Coastal Ecoregion*, R.J. Naiman and R.E. Bilby, editors (1998); and *Wildstream: A Natural History of the Free-Flowing River* by T.F. Waters (2000). Another useful volume is *Methods in Stream Ecology*, edited by F.R. Hauer and G.A. Lamberti (1996 [2006]), which is, as the title suggests, a methods manual for students and practitioners of the science of stream ecology. Finally, you might also want to look up U.S. Forest Service General Technical Report INT-241 by W.S. Platts and M.L. McHenry (Ogden, Utah: USDA Forest Service Intermountain Region, 1988), which lists the measured productivities of various western streams in terms of the densities and biomass of trout they produce.

The Stream-Resident Life History Strategy

Nowadays, most genetically pure populations of Yellowstone cutthroat trout outside of the Yellowstone Lake drainage occur as stream-resident populations in remote, high-elevation headwater areas. Here they persist, many doing so even in the face of encroachment by introduced species, such as brook, brown, and rainbow trout, which have displaced native cutthroats in downstream reaches. Some scientists suggest that stream-resident cutthroats enjoy a home-field advantage in high-elevation headwater reaches by being able to function better in cold environments than non-native species. [41]

To further characterize the physical attributes of high-elevation sites occupied by Yellowstone cutthroat trout, scientists from the Wyoming Cooperative Fish and Wildlife Research Unit and University of Wyoming examined 151 sites in 56 perennial streams ranging in elevation from 7,546 feet to 10,663 feet in the drainage of the Greybull River and its principal tributary, the Wood River in Wyoming. They found that trout were absent from all sites above 10,440 feet, and from all but one site with a channel slope greater than 10 percent (the one exception that held trout had a channel slope of 17 percent). Also, no Yellowstone cutthroats occurred naturally in sites upstream from waterfalls or other impassable geological barriers,

regardless of channel slope or site elevation. However, at four sites above such barriers, Yellowstone cutthroat trout had been stocked from hatcheries and these populations were doing well. [42]

Stream resident Yellowstone cutthroat populations have been found in streams as small as 3 feet in wetted width with flows as low as 2 cfs. Viable and even robust populations occur in streams with summer maximum temperatures of only 41 to 46 degrees Fahrenheit [5 to 8 degrees Celsius], as exemplified by Gregg Fork of Bechler River, Sedge Creek, and Bear Creek, all in Yellowstone National Park. At the other extreme, there are reports that stream resident cutthroats can survive in Yellowstone Park streams, such as Alum and Witch creeks, that get as hot as 81 degrees Fahrenheit [27 degrees Celsius] in summer owing to geothermal inputs, but these trout do so by seeking out and holding in cooler refugia.

Adult trout belonging to stream resident populations disperse locally for spawning in their home streams. Depending on altitude, water temperature, and runoff conditions, spawning could occur anywhere from late April or early May into August. In the cold, high-altitude, headwater settings that characterize the territories of most stream resident populations, growing seasons are short and growth is slow. Fish may become sexually mature at only 4 to 5 inches in

41. I touched on this in Chapter 4 with regard to the persistence of westslope cutthroat populations in the cold, upper reaches of streams where only rainbow trout now occur in downstream reaches. In footnote 5 of that Chapter, I also referred to the work of Vincent and Miller (1965), Fausch (1989), De Staso and Rahel (1994), and Taniguchi et al. (1998) that collectively describes the ability of other cutthroat subspecies to outperform brook trout and other non-native salmonids at the cold water temperatures one would typically find in these upper stream reaches during key life-stages of the fish.

42. Complete details of the research summarized here can be found in Kruse et al. (1997a). Sources for the remainder of this section on life history and ecology of stream-resident Yellowstone cutthroats include Jones et al. (1979), Yekel (1980), Remmick (1981), Kent (1984), Gulley and Hubert (1985), Skinner (1985), Varley and Gresswell (1988), Gresswell (1995), Kruse (1995), and Kruse et al. (1997b).

total length and the number of eggs per female is correspondingly low. But life spans may be extended in these conditions, affording the trout the opportunity to spawn multiple times.

Fry may move upstream or down following emergence, or remain in the vicinity of the redds as they seek to avail themselves of what's left of the summer growing season. This is a critical period, especially for age-0 trout in alpine and higher subalpine settings, because overwinter conditions can be especially harsh in these streams, with low temperatures and extreme ice conditions that last up to 8 or 9 months. Those that survive into the second and subsequent growing seasons feed on an array of items: the nymphs and adults of aquatic insects, such as caddisflies, stoneflies, mayflies, and two-winged insects, such as chironomids and blackflies, and lots of terrestrial insects, such as ants and beetles. In a study of the food and feeding of resident trout in Palisades Creek, Idaho, the fish were even found to take large quantities of the small whitish berries of red-osier dogwood, a riparian shrub common in places along the creek. [43]

STATUS AND FUTURE PROSPECTS

The Yellowstone cutthroat trout, like other interior cutthroat subspecies, has experienced major reductions in overall distribution and abundance during the last

43. William D. Skinner, Department of Biology, Idaho State University, Pocatello, collected the data for his feeding study of Palisades Creek trout in August 1983, and published his results in 1985; see Skinner (1985). Palisades Creek is a good-sized tributary of the South Fork Snake River with stream widths in late summer of 40 to 45 feet, and a reputation for turning out the occasional trout of 16 to 18 inches (although those that yours truly encountered were all between 7 and 12 inches). The best fishing requires a bit of a hike. As I recall, the first 4 miles or so are a steady but moderate climb, and then comes a short but steep ascent to Lower Palisades Lake. The creek pours out of this lake in a falling cascade of water. But above this, between Lower and Upper Palisades lakes, is flat terrain bordered by hill slopes, where the creek meanders among willows. This is your destination. This is mountainous country, and it's not uncommon to see mountain goats up on the slopes on the way in. Moose may also be seen, especially in the willow flats between the two lakes. Be careful here; momma moose are skittish and unpredictable about their calves. Deer and elk are also present and so are bears, although I never saw any.

Don't be surprised if, in addition to the cutthroats, you also hook a rainbow trout or a cuttbow up here. The rainbow trout have increased to the point of being a threat to the native cutthroat population. The State of Idaho encourages anglers to keep the rainbow trout they catch everywhere in the South Fork Snake River drainage, which is now being managed strictly for the native Yellowstone cutthroat. One of Skinner's findings should be of particular interest if you're a fly-fisher. He noted that the larger the trout, the more bottom-oriented its feeding. Skinner found that one of the principal prey items of these larger trout was the nymph of the *Ephemerella doddsi* mayfly, one of the two mayflies responsible for hatches of what fly-fishers know as the Western Green Drake (the other and better-known Western Green Drake is *Ephemerella grandis*). *E. doddsi* nymphs are bottom dwellers that seldom appear in the drift, so trout have to root around and scarf them off the rocks (some angling writers describe this as "tailing"). The larger the trout, the more of these nymphs they consumed in Palisades Creek. Charles E. Brooks' Ida May Nymph in size 8 or 10 is a good imitation. Fish it deep, right along the bottom.

The popular Buffalo Ford reach of the Yellowstone River in Yellowstone National Park. PHOTO BY AUTHOR

century—reductions in the order of 70 to 90 percent, depending on who's compiling the estimates. These reductions have been masked, until fairly recently, by the large numbers of Yellowstone cutthroat trout that exist in the Yellowstone Lake/Yellowstone River drainage upstream from the falls in Yellowstone National Park. As many as 7.5–10 million trout (juveniles and adults combined; predominately the lacustrine stocks of Yellowstone Lake itself) may be present in this drainage. This area represents the greatest concentration of genetically pure interior cutthroat trout of any subspecies still existing in a natural environment. But the picture for the Yellowstone cutthroat subspecies as a whole is not that bright. Some say that more than 90 percent of the present distribution of genetically pure Yellowstone cutthroat trout now lies inside the boundaries of Yellowstone National Park. [44]

Boosted by the numbers of trout in Yellowstone Lake, lacustrine Yellowstone cutthroat trout could be said to be doing rather well. In the other large lakes within the historical range, an abundant population remains in Henry's Lake, although that lake is now being managed for rainbow × cutthroat hybrids along with its pure cutthroat population. Yellowstone cutthroat trout also persist in Heart Lake and in Jackson Lake, but those populations declined significantly after lake trout became established. A population that apparently has adapted successfully to a lacustrine life

history is present in Buffalo Bill Reservoir on the North Fork Shoshone River, Wyoming, but shares the reservoir with other gamefish species. These trout migrate for spawning to several tributaries of the North Fork Shoshone River. In Idaho, Lower and Upper Palisades Lakes in the Palisades Creek drainage contain Yellowstone cutthroat trout, but these populations are under threat of introgression and displacement by rainbow trout. In Montana, 143 mountain lakes within the historical range support populations of Yellowstone cutthroat trout, many (or perhaps most) of these established by stocking with hatchery-reared trout of the McBride Lake strain. Overall, it is estimated that genetically pure populations of Yellowstone cutthroat trout still occupy about 85 percent of their historical lacustrine habitat.

But this picture is offset by losses of fluvial populations from the large low-elevation rivers, and by losses of stream-resident populations from all but small, isolated headwater tributaries where they are able to withstand encroachment by non-native species. Large metapopulations of Yellowstone cutthroat trout that once encompassed several connected watersheds and included both fluvial and stream-resident populations have been largely reduced to the few stream-resident populations that can persist in spatially disjunct headwater streams.

About 1986, the State of Montana began to accurately define where genetically pure Yellowstone

44. A slew of status reports have been written for Yellowstone cutthroat trout dating all the way back to 1959. These include: Varley and Gresswell (1988), Gresswell (1995), May (1996), and May et al. (2003) regarding status across the entire historical range; Thurow et al. (1988), Van Kirk et al. (1997), Jaeger et al. (2000), and Meyer et al. (2003a, 2006) for status in Idaho's portion of the range; Dufek et al. (1999) and Kruse et al. (2000) for status in Wyoming; and Hanzel (1959), Hadley (1984), and Montana Department of Fish, Wildlife and Parks (2000) for status in Montana. With regard to the Yellowstone cutthroat trout in Montana, Michael K. Young of the U.S. Forest Service compiled an overview document that discusses current status, historical and current distribution, ecology, threats, and management opportunities. This document, "Montana's Fish Species of Special Concern: Yellowstone Cutthroat Trout," is on the Montana Chapter American Fisheries Society website at *www.fisheries.org/ AFSmontana/SSCpages/yellowstone_cutthroat_trout.htm*.

cutthroat trout still occur within the historical range in that state, which includes the Yellowstone River and its tributary sub-basins downstream from Yellowstone National Park. In the most recent compilation, submitted to the U.S. Fish and Wildlife Service in 2000, Montana biologists identified just 40 genetically pure Yellowstone cutthroat populations still present in 12 of the 13 historically occupied sub-basins in that state, altogether occupying less than 10 percent of the subspecies' historical stream miles. Furthermore, brook, brown, and/or rainbow trout occur in all of the sub-basins surveyed. These non-native species pose a high risk of displacing or introgressing the remaining Yellowstone cutthroat populations in all but one of the sub-basins surveyed.

In Montana, fluvial cutthroat populations no longer exist in the larger Yellowstone tributaries, having been replaced by rainbow trout, rainbow × cutthroat hybrid populations, and brown trout. In the mainstem Yellowstone itself, riverine cutthroats shared the river with rainbow and brown trout in a segment extending about 125 miles downstream from Yellowstone National Park, but were still relatively common in this segment up until the early to mid-1990s. The cutthroats maintained at least some semblance of reproductive isolation from rainbow trout by spawning in tributaries at slightly different times.[45] But recent reports indicate that the rainbows are increasingly displacing the cut-

45. Over the years, introduced rainbow trout have become well established in many of the river systems once occupied solely by fluvial Yellowstone cutthroat trout. Where native cutthroat trout have persisted, they have maintained at least some degree of reproductive isolation by spawning at slightly different times. For example, in lower Palisades Creek and the various side channels of the South Fork Snake River, the same spawning areas are used by all forms of fluvial trout, but the median spawning time of the cutthroat trout is about three weeks later than that of the rainbow trout and hybrids. This also appears to be true in Yellowstone River tributaries downstream from Yellowstone National Park. However, reproductive isolation is not complete in these locations because the durations of the spawning periods overlap, allowing some interbreeding to take place. In other locations, as in Idaho's Blackfoot River and Pine Creek, Yellowstone cutthroats spawn much higher up in the system than rainbow trout, and reproductive isolation is maintained. Thurow (1982), Clancy (1988), and Henderson (2000) have more detailed information.

throats from this segment of the river, as rainbow and brown trout have already done further downstream. Angler-author Dave Hughes wrote of catching about an equal mix of cutthroat, rainbow, and brown trout in the Paradise Valley reach in 1991. Dr. Behnke wrote that when he visited the river in 2000, his catches in the Livingston area (about 50 miles downstream from the park at the downstream end of Paradise Valley) were all either typical rainbow trout or hybrids that expressed a strong rainbow-like phenotype. He had to fish up near Gardiner, near the park boundary, to catch a typical cutthroat. And he caught hybrids even there, albeit ones that were more cutthroat-like in appearance.[46]

The picture is similar in Wyoming, where it's estimated that Yellowstone cutthroat trout have been extirpated from all but about 30 percent of their historical stream habitat. Fluvial Yellowstone cut-throats are still relatively common in the upper Snake River drainage above Jackson Lake, but in Wyoming's Yellowstone River tributaries, fluvial populations have been replaced by non-native species. Genetically pure stream-resident populations persist in a few upper tributaries of the Greybull, Wood, and South Fork Shoshone River watersheds in the Absaroka Mountains along the eastern boundary of Yellowstone National Park, in high elevation, in isolated streams in the Teton Range along the west boundary of Grand Teton

National Park, and also in a few upper tributaries in the Bighorn Mountains. But hatchery cutthroats have been stocked in the Bighorn Mountain drainages, which may account for the present populations. Also, hatchery-reared finespotted Snake River cutthroat trout were extensively stocked in the Greybull River watershed from 1972 to 1975, and may have interbred with some of the Yellowstone cutthroat stocks that occurred there. Only a single genetically pure Yellowstone cutthroat population is present in the Clark's Fork watershed, and that one, too, originated from stocked trout. Other stream-resident populations in Wyoming occur in upper tributaries of the Gros Ventre River, where large-spotted Yellowstone cutthroat trout provide an upper bound to the distribution of the fine-spotted Snake River cutthroat (Chapter 10).

The State of Idaho gives itself more credit, saying that genetically pure Yellowstone cutthroat trout may still be found in about 43 percent of their historical stream habitat in Idaho. Between Palisades Reservoir and Shoshone Falls in the Snake River drainage, genetically pure Yellowstone cutthroats still are present in the South Fork Snake and its tributaries, which may be the subspecies' strongest remaining enclave in Idaho, but these waters have become increasingly populated with rainbow trout and rainbow × cutthroat hybrids. Brown trout are also present. In the Henry's

46. These reports are in Hughes (1992) at pages 67–77 and in Behnke (2002) at page 167.

Fork, rainbow trout have long since replaced the native cutthroat trout as the dominant species everywhere but in the Teton River system, but there, too, rainbow trout and rainbow × cutthroat hybrids appear to be quickly gaining the upper hand. Late in 2004, results of a State of Idaho population survey of two segments of the upper Teton River came to hand. During this survey, actually conducted in 2003, state biologists counted only 50 Yellowstone cutthroat trout, a decrease in cutthroat numbers of nearly 95 percent from a previous state survey of the same segments in 1999. In contrast, the numbers of rainbow trout and brook trout in these segments had almost tripled. The state estimated that today, Yellowstone cutthroat trout comprise about 5 percent of the upper Teton trout fishery, down from 50 percent in 1999.

Downstream, from the forks of the Snake to Shoshone Falls, fluvial Yellowstone cutthroat trout remain in the upper Blackfoot River and Willow Creek, but again, rainbow trout and rainbow × cutthroat hybrids are reported to be encroaching in increasing numbers in the upper Blackfoot. Yellowstone cutthroats have given way in the mainstem Snake and most all the other tributaries to rainbow trout, rainbow × cutthroat hybrids, and brown trout. Extant stream-resident populations in Idaho include Eight Mile Creek, a tributary of the Portneuf River, and Palisades, Rainey,

Pine, Prichard, Falls, and Burns creeks, all tributaries of the South Fork Snake River. The State of Idaho also credits populations in McCoy, Big Elk, and Bear creeks, which flow into Palisades Reservoir, but these may be finespotted Snake River cutthroats which I deal with in Chapter 10. Yellowstone cutthroat trout still persist in about 40 stream miles in the upper Goose Creek drainage in Nevada, although these fish are under threat of displacement by brook trout, and recent surveys have also found Yellowstone cutthroat trout in upper Raft River tributaries in Utah occupying about 42 miles of stream.

Stream-resident populations of Yellowstone cutthroat trout that remain within the historical range of the subspecies are those that persist in small, isolated headwater drainages where conditions are harsh enough to inhibit encroachment by nonnative brook, brown, and rainbow trout that have pretty much completely displaced the native cutthroats from the more habitable trout waters downstream. Examples of historical drainages where Yellowstone cutthroat trout have disappeared altogether, having long since been replaced by brook, brown, and rainbow trout, include the Tongue River drainage, where the trout fishing so occupied General Crook and his troops in 1876; the Little Bighorn drainage, where Crook's advance, had he pressed it, might have averted Custer's debacle; and the Rosebud Creek drainage, where, as disclosed in Chapter 1, Charles Hallock caught the trout he named "cut-throat" back in 1884.

As I've recited this litany, I've made it sound as if hybridization with or displacement by non-native trout has been the number-one cause of decline of the Yellowstone cutthroat trout. Most of the status reviewers agree that this is indeed the major contributing factor. Furthermore, they also agree that few of the remaining pure Yellowstone cutthroat populations are secure from this threat. In 2002 and 2003, volunteers working on summer surveys of Yellowstone National Park waters even captured rainbow trout in the First Meadow section of Slough Creek, long a bastion of Yellowstone cutthroat trout within the Park, much to the consternation of Park fisheries managers. Reviewers of Wyoming's Yellowstone cutthroat population status have gone so far as to conclude that without intense management to control exotic salmonids and to reestablish large, genetically pure, allopatric populations, the few remaining genetically pure enclaves found in Wyoming will be eliminated within decades. [47]

But displacement and introgression by non-native species is not alone on the list of threats. Past overharvest in sport fisheries is also listed as a factor in the decline of the Yellowstone cutthroat, as are habitat alterations and losses attributable to water storage projects, diversions for irrigation, grazing, mineral extraction, and timber harvest. I have written about most of these in earlier Chapters. Within the historical range of the Yellowstone cutthroat trout, water diversions and withdrawals have been significant in the decline of the subspecies in Montana, especially as they impact spawning tributaries in that segment of the Yellowstone River downstream from the Park. The same is true in Idaho where populations of the Henry's Fork, Teton River, the main Snake River downstream from the forks, Willow Creek, the Blackfoot River, the Portneuf River, and Raft River are all seriously affected by water withdrawals. Intensive grazing has caused degradation of riparian areas and subsequent sloughing of stream banks, channel instability, erosion, and siltation throughout the upper Snake River basin in Idaho and Wyoming. In Montana, livestock grazing is considered a lesser threat than water withdrawals. Examples of the effects of mineral extraction include old placer mining operations in the headwaters of McCoy Creek, around the old mining camp of Caribou City, Idaho, that destroyed much instream habitat along with adjacent riparian and upslope areas. Contemporary visitors find these areas still scarred and in the process of recovery even after more than 100 years since the operations ceased. An abandoned gold mine in the headwaters of Soda Butte Creek near Cooke

47. This conclusion is stated in Kruse et al. (2000).

City, Montana, caused extensive pollution in the 1960s, resulting in low fish numbers downstream, even within Yellowstone National Park. The tailing pile was finally stabilized and reclaimed and the trout population rebuilt itself, but it has taken three decades. The memory of all this led to recent concerns and downright opposition to the planned expansion of gold mining in the same region. The most recent indictments of mineral extraction have centered on the Blackfoot River drainage in Idaho, where open-pit phosphate mines have been linked to increased sediment levels in streams and also to sharply elevated levels of selenium, an element that is toxic in high levels to some fishes and may lead to recruitment failure and extirpation of trout populations. [48]

To the above list, which, except for the specific examples, are threats common to most interior cutthroat subspecies, I would add the specter of global warming, which, as I mentioned in Chapter 4, may hit Idaho, Montana, and Wyoming trout habitat particularly hard. Lake trout predation on the cutthroat trout of Yellowstone Lake, while in check for the present, has the potential to extirpate that entire population should the expensive gillnetting program presently used for lake trout control be curtailed or stopped, for lack of funding, say. This is not just idle supposition, either. Supporters of the program have had to go to the

48. Selenium is a naturally occurring element that is both an essential dietary trace element and a toxic substance, with a very narrow range between levels that are required in the diet and toxic threshold concentrations. High selenium levels have killed fishes in standing, warmwater environments. In one well-studied case, Belews Lake, South Carolina, selenium-contaminated waste water discharge from a coal-fired power plant resulted in extirpation of 19 of the 20 fish species present. While no outright fish kills have been reported in streams supporting any of the salmonid species studied (these include chinook salmon, rainbow trout, cutthroat trout, and brook trout), selenium does bioaccumulate up the food chain into the bodies of adult salmonids in high levels, and is passed on to their eggs when they spawn. As the embryos develop into fry, skeletal deformities, edema, and deformities to the head and face appear that can impair further development and increase their mortality rate to the point that recruitment into the next generation of breeders either does not happen at all or there are too few new breeders to sustain the population. Recruitment failure has serious consequences for a trout population. Over the course of one to a few generations, the population may simply disappear. A recent modeling study of the population-level effects of chronic selenium exposure on Yellowstone cutthroat trout, performed by Dr. Rob Van Kirk of Idaho State University, has shown that population extinction can happen very quickly. Kendell Creek in the Blackfoot River drainage, in the phosphate mining region of southeastern Idaho, may provide a real-life case in point. Specimens of Yellowstone cutthroat trout were captured there in the summer of 2001 for an Idaho Department of Environmental Quality assessment of high selenium levels on human and ecological health. But when Forest Service biologists returned to the creek just one year later, in the summer of 2002, they found no fish at all. You can learn more about the effects of selenium in aquatic systems, and also about the studies and examples I've cited here, by consulting the following references: Hodson et al. (1980), Hamilton et al. (1990), Lemly (1997, 2002), Kennedy et al. (2000), Hamilton and Palace (2001), Christensen (2002), Holm et al. (2003), Hamilton (2004), and Van Kirk (2006).

mat already, in 2002, to get funding for the program restored after it had been cut by the administration from the federal budget. They might not be so successful in future budget battles.

Additional threats have emerged in the form of whirling disease (see footnote 21, Chapter 4 for a rundown on whirling disease) and a new exotic invader, the tiny New Zealand mud snail. Whirling disease was discovered in cutthroat trout taken from Yellowstone Lake near the mouth of Clear Creek in 1998, and subsequently around Fishing Bridge and in Pelican Creek, the lake's second-largest spawning tributary. In Pelican Creek, the numbers of surviving juveniles and, consequently, the numbers of spawners returning to the creek have fallen so dramatically that Park officials closed the creek indefinitely beginning with the 2004 angling season to protect the remaining trout and slow the spread of the organism.

As for the New Zealand mud snail, this insidious little creature was first detected in the middle Snake River, Idaho, in 1987, and is rapidly expanding its range in the western U.S. [49] Within the Yellowstone cutthroat range, mud snails are now present in the upper Snake River in the Flagg Ranch area and at Polecat Creek, both sites upstream from Jackson Lake. They are also present in the lower Gardner River in Yellowstone National Park; in the Yellowstone River

49. The New Zealand mud snail *Potamopyrgus antipodarum* is a tiny, prolific snail that lives in a variety of habitats in its native New Zealand, including estuaries, lakes, large rivers, and small streams. It can colonize all substrate types, including silt, sand, gravel, cobble, and vegetation, where it occupies a scraper/grazer feeding niche consuming diatoms, plant and animal detritus, and periphyton. These snails can reproduce both sexually and clonally (the strain or strains that have invaded the western U.S. appear to be clonal), and can multiply quickly and in great numbers. In New Zealand, young are born every three months, but in the western U.S., new broods appear most often in summer and autumn. Baby snails are about the size of the period at the end of this sentence. They grow to about a quarter-inch in length over their life span of just a bit more than a year, but within that life span they can produce very high-density colonies and can expand their range quite rapidly. Densities are usually highest in systems with high primary productivity, constant temperatures, and constant flow—limestone spring creeks, for example. In New Zealand, several species of trematodes keep mud snail populations in check, but there are no known natural predators or parasites in the western U.S. Mud snails have been found in the stomachs of mountain whitefish, but they evidently provide little nutrition and may pass though unscathed. New Zealand mud snails can completely carpet a stream bed (densities of 46,000–47,000 snails per square foot have been found in some reaches of the Madison River) and, since they eat the same foods, they can eliminate the native aquatic invertebrates in a colonized reach. They have been implicated in recent reductions and loss of mayfly hatches in the Madison and Firehole rivers and DePuy's Spring Creek. Stonefly and caddisfly populations suffer the same fate, except that rushing mountain streams where these insects predominate do not seem to be suited for the strain of mud snail that has invaded the western U.S. Another bright spot is that some invertebrates, such as scuds and some species of chironomids, can survive in colonized reaches by feeding on the mud snails' nitrogen-rich feces. For more information about New Zealand mud snails, including distribution maps and pinpointed collection sites, check out these two websites maintained by researchers at the University of Montana. One is at *www.esg.montana.edu/aim/mollusca/nzms/*. The other is at www2.montana.edu/nzms/. Both were active when I last checked on September 13, 2004. Ralph Cutter, a noted fly-fishing writer and operator of the California School of Fly Fishing, *www.flyline.com/environmental/nzms*, also has information on mud snails as well as the latest method for disinfecting your wading gear using a 5-minute treatment with a 50 percent solution of the household disinfectant Formula 409® to prevent transporting the critters. This procedure is based on a recent study made by the California Department of Fish and Game; see Hosea and Finlayson (2005). There's a link to that report on the Cutter website.

from the Park downstream to the vicinity of Springdale, including the world-famous Nelson's Spring and DePuy's Spring creeks; in the Henry's Fork near Last Chance and Box Canyon Camp; and in the Blackfoot River near the head of Blackfoot Reservoir. So small that a dozen can fit on a dime and a singleton can fit comfortably on the head of a kitchen match, New Zealand mud snails can multiply prolifically enough to blanket a stream bed, eliminating the native invertebrates by preempting their food and living space and thereby eliminating the aquatic food supply of the trout. They evidently supply little in the way of nutrition themselves if eaten by fish. Although they have been found in mountain whitefish stomachs, trout apparently shun them.

Much of the foregoing information about range shrinkage, reasons for decline, and ongoing threats was available to the public by 1998. That year, a coalition of environmental organizations and a concerned citizen petitioned to list the Yellowstone cutthroat trout as threatened under the U.S. Endangered Species Act. Having been relegated to a second tier of priority activities, the petition languished in federal offices for three years. Finally, in 2001, after a threat of court action, the U.S. Fish and Wildlife Service issued a 90-Day Finding rejecting the petition and declining to list the subspecies. [50]

As a first order of business in arriving at its 90-Day Finding, the Service considered the genetic population structure of Yellowstone cutthroat trout, and decided that the entire subspecies comprises but a single DPS (distinct population segment) as it defines that term. Even though, as we have seen, Yellowstone cutthroat trout display as much diversity among populations as any other subspecies in terms of adaptation to different environments and biotic communities, none of this variation is detectable using available methods of allozyme electrophoresis, mitochondrial DNA analysis, or nuclear DNA analysis. By these methods, each population of Yellowstone cutthroat trout appears pretty much the same as every other population. Each population has the same array of detectable alleles at the same frequencies as all the other populations, and no population displays any rare or unique alleles that set it apart from other populations. [51] In other words, using available allozyme and DNA methods, there is no detectable population subdivision within the Yellowstone cutthroat subspecies. That doesn't mean it does not exist, just that available allozyme and DNA methods don't detect it. But the Fish and Wildlife Service relied only on those methods in arriving at its DPS determination.

The 90-Day Finding went on to say that even though the Yellowstone cutthroat trout is much

50. For the petition to list the Yellowstone cutthroat trout as threatened, and for the 90-Day Finding not to list, see Biodiversity Legal Foundation et al. (1998) and Kaeding (2001), respectively. You can download your own copies of the petition, the 90-Day Finding, a list of questions and answers about the Finding, and other information regarding Yellowstone cutthroat trout from a U.S. Fish and Wildlife Service website at *http://mountain-prairie.fws.gov/ species/fish/yct/*. When you get there, click on Yellowstone Cutthroat Trout Archives.

51. Contrast this with the westslope cutthroat trout (Chapter 4), where each population carries a different set of the subspecies' total detectable alleles, and, in cases where populations do share one or more alleles, the frequencies of these alleles are significantly different in each population. For detailed studies of the population genetic structure of Yellowstone cutthroat trout, see Loudenslager and Kitchen (1979), Loudenslager and Gall (1980a), Wishard et al. (1980), Allendorf and Leary (1988), Toline et al. (1999), Novak et al. (2005), and Cegelski et al. (2006).

reduced from its historical distribution, the petition-ers provided no evidence that the subspecies as a whole is declining toward extinction in the foreseeable future or that the probability of extinction is high. The 90-Day Finding essentially dismissed concerns about the important threats being faced by the subspecies, and instead made the argument that most of the strongholds of extant Yellowstone cutthroat stocks occur within isolated roadless or wilderness areas of the National Forest system and in Yellowstone National Park, where considerable protection is afforded to Yellowstone cutthroat trout. In addition, it stated:

1) Numerous federal and state regulatory mechanisms are in place that, *if properly administered and implemented* [italics mine], will protect Yellowstone cutthroat trout and their habitats;

2) each of the principal state and federal agencies responsible for Yellowstone cutthroat management has a long history of working to conserve the subspecies; and

3) federal and state agencies have numerous ongoing projects directed toward protection and restoration of Yellowstone cutthroat trout and their habitats.

With regard to the first point, it's true that each of the seven National Forests that contain historical Yellowstone cutthroat habitat does treat it as a Sensitive Species. This means that biological evaluations of all land management activities that may affect the subspecies' habitat are supposed to be prepared to ensure that these activities do not result in loss of species viability or increase the likelihood of federal listing under the Endangered Species Act. Compliance with the National Environmental Policy Act (NEPA), the Clean Water Act, the Federal Land Management Protection Act, the Wilderness Act, and the Wild and Scenic Rivers Act, where any of these apply, is also required. The Sensitive Species designation also directs the U.S. Forest Service to cooperate with and assist the states in achieving their conservation goals for Yellowstone cutthroat trout.

Yellowstone National Park, of course, sits right in the middle of the Yellowstone cutthroat range and this is fortunate because its waters largely escaped the exploitation for water, minerals, energy, timber, and grazing that prevailed in other areas of the west. True, some unfortunate fish introductions were made early-on that resulted in exotic species becoming established in many park waters. But in 1936, a formal stocking policy was established that discouraged further stocking of non-native fish in waters containing native trout, and encouraged the propagation and stocking of native species instead. This policy has been followed ever since. In addition, in 1949, the National Park Service began to worry about the impact of the massive Yellowstone Lake egg-taking operation on the ecological balance of the lake. Studies showed that even though some trout reared from these eggs were returned to the lake each year, the removal of so many

Fly-fishermen working Slough Creek in First Meadow, Yellowstone National Park

eggs for shipment elsewhere was detrimental to natural reproduction. So, in the mid-1950s, the egg-taking and hatchery operations were shut down.

That helped, but it wasn't the whole answer. Anglers were also accounting for a very high percentage of the total annual mortality of the trout—high enough to virtually eliminate the older age classes and send the population as a whole spiraling into decline. The trout were being pounded at such popular destinations as Fishing Bridge at the outlet of Yellowstone Lake. I first visited Yellowstone Park in the summer of 1957, on my way east for graduate school. I can still remember the crowds of anglers at Fishing Bridge, lining the rails and the edges of the roadway, dangling their lines off both sides of the bridge and crossing haphazardly from one side to the other, seemingly oblivious to the motorists trying to inch their way across without hitting anybody.

During that period, fishery management at Yellowstone Lake followed the *maximum sustained yield* (MSY) paradigm [52] in which escapement to the spawning tributaries was managed to maximize a so-called harvestable surplus of trout. But that "harvestable surplus" never materialized; instead, the population continued to decline. It wasn't until the MSY paradigm was abandoned, beginning in 1969, that the Yellowstone cutthroat population began to recover.

52. Fishery managers have long observed that in fish stocks there appears to be a relationship between the number of spawners in an adult generation and the number of surviving recruits these spawners will produce for the next reproductive cycle. The form of this relationship is such that everywhere below a certain number of spawners, more recruits are produced, on average, than are needed to replace the parents. Furthermore, for each stock, the relationship predicts that there should be an optimum number of spawners that maximizes the production of surplus recruits. Therefore, fishery managers reasoned, if a stock could be managed so that just that number of spawners was allowed to escape to the spawning grounds each season, then all the surplus recruits could be harvested and the cycle should sustain itself indefinitely. They referred to this optimum number of spawners as the *escapement goal*, and called the resulting management paradigm *maximum sustained yield* (MSY). In practice, fishery managers arrived at their MSY escapement goals by applying a model of the spawner-recruit relationship for their particular stock. Not all spawner-recruit relationships have the same shape. Two models were developed that fit the most common shapes, the Ricker model (see Ricker 1954) and the Beverton-Holt model (see Beverton and Holt 1957). Both models assume that the form of the relationship is determined by density-dependent interactions that occur in the stream, but they differ in how they assume these interactions come into play. You can read more about these models and the implications of their different assumptions in the original publications or in Hilborn and Walters (1992). The Ricker model generates a dome-shaped curve in which the number of recruits increases to a maximum and then declines as the number of spawners continues to increase. The Beverton-Holt model generates a curve in which the number of recruits also increases, but then levels off as it approaches some limiting number of recruits. In both models, the number of recruits produced per spawner rises to a maximum that may be many times greater than 1 (that being the number of recruits per spawner needed to just replace the parent stock), but then drops off to less than 1 as the number of spawners increases. The MSY level is the point where the number of recruits per spawner is maximum. Despite its promise of sustainability, the history of fishery management under the MSY paradigm has been a sorry one. Intrinsic problems with this approach, largely unrecognized by those who developed the concept as a management tool, have contributed to declines rather than sustainability in the majority of fish stocks being managed worldwide. As early as 1977, scientists who had initially championed MSY urged that it be abandoned as a management tool (see Larkin 1977). Shortcomings of the MSY paradigm were recognized at Yellowstone Lake as well, and, as noted, management of the lake's fishery quickly changed. You can read more about this in Robert Gresswell's well-articulated case study, "Yellowstone Lake—A Lesson in Fishery Management," presented at the Wild Trout II Symposium in 1979 (proceedings published 1980; see Gresswell 1980). For more information about the history of fisheries management in Yellowstone National Park, see Gresswell and Varley (1988) and Varley and Schullery (1983, 1996, 1998).

Fishing with bait was banned in 1969. In 1970, a 14 inch minimum size limit was imposed. In 1973, a 2-fish creel limit was imposed, and Fishing Bridge and the Yellowstone Lake outlet area was closed to angling. Those steps actually increased angler catch rates, and, since more of these trout were being released than ever before, reduced the annual angler harvest. The trout population started building in numbers, but the older age classes were still absent from the age structure. That was rectified in 1975, when the size limit for the lake and its tributaries was changed to a 13-inch *maximum*. The proportion of older and larger trout quickly increased in spawning streams and in the angler catch—indeed, a common angler lament was that it was getting too hard to catch a fish *small* enough to keep—and the trout population rebounded.

Other steps taken subsequently were to place Yellowstone Lake and other Park waters under catch-and-release regulations for Yellowstone cutthroat trout and to close many of them to angling until mid-July to protect spawners. Park fishery managers were also proactive in dealing with the threat of lake trout predation in Yellowstone Lake, instituting the intensive gillnetting program (although that program may yet be vulnerable to administration budget writers) and requiring anglers to keep every lake trout they catch. [53] They have also reacted to the whirling disease threat by closing Pelican Creek to protect the remaining fish in that population and prevent further spread of the organism.

The State of Montana is another jurisdiction that took forward-looking steps (for the time) that benefited its Yellowstone cutthroat populations, even if that wasn't necessarily the purpose of the actions. As early as 1967, the state began studying the effects of stocking catchable-size rainbow trout on wild populations of rainbow and brown trout. Based on results showing that such stocking actually depressed the abundance and biomass of the wild populations, the state curtailed stocking of hatchery trout in its streams. In 1987, the state also took another unprecedented step, this time one that was aimed directly at protecting a Yellowstone cutthroat population, by closing the world-famous Nelson's Spring Creek to wading anglers. Anglers could fish from the banks (with permission from property owners, of course), but they couldn't wade in the creek from June 15th to September 15th. Nelson's Spring Creek is the principal spawning tributary for Yellowstone cutthroat trout in the adjacent segment of the Yellowstone River, and that's the period the spawning population's eggs and alevins are in the gravel. This closure was taken because a first-of-its-kind study by Montana State University graduate student Bruce C. Roberts had shown

53. As the 2006 fishing season rolled around, a number of articles appeared in angling magazines exhorting anglers to catch and remove as many lake trout as possible from Yellowstone Lake. One in particular caught my eye: a piece titled "Catch and Fillet: Eat a Lake Trout and Save a Cutthroat," that appeared in the July/October, 2006 issue of *Fly Rod & Reel* magazine. Author Wayne Phillips not only outlined where, when, and how to fish the lake for lake trout, he also offered tempting recipes for preparing your catch, among them baked lake trout with roasted red pepper sauce, baked lake trout with mango salsa, pan-fried lake trout with pesto over pasta, and a hearty lake trout chowder. Bon Appétit!

that daily wading could kill over 80 percent of the eggs and alevins in the spawning areas. [54]

Montana took other steps to curtail angler harvest of Yellowstone cutthroat trout. Between 1973 and 1983, Montana's fishing regulations allowed a harvest of five trout per day (only one of which could be over 18 inches in total length, however). During that period, reflecting what had been observed at Yellowstone Lake, the mean length of spawning cutthroats in the 50 miles of the Yellowstone River downstream from Yellowstone National Park to the town of Livingston had decreased and the proportion of spawners greater than 15 inches total length had also declined—both observations indicating that angling mortality had become high enough to be additive with other forms of mortality, sending the population spiraling into decline. Montana responded by going to a catch-and-release regulation for Yellowstone cutthroat trout in that segment of the river in 1984, which helped to improve the proportion of larger-size spawners in the population. The catch-and-release regulation was extended to the 75 miles of river downstream from Livingston to Springdale in 1994 to encompass the entire remaining portion of the Yellowstone cutthroat's distribution in the Yellowstone River mainstem (rainbow and brown trout may be kept, but Yellowstone cutthroat trout must be released). These moves seem to have arrested the angling

mortality impacts on the riverine cutthroats of the Yellowstone River, but they haven't affected encroachment and displacement of the cutthroat population by rainbow trout. As I pointed out earlier, rainbow trout have gradually become the predominant species even in this, the last-bastion segment of fluvial cutthroat range in the Yellowstone River downstream from Yellowstone National Park.

In Idaho, various combinations of special regulations have been tried in different waters in an effort to conserve Yellowstone cutthroat trout and still provide for angling recreation. On the South Fork Snake River, a two-fish harvest limit plus a slot limit on Yellowstone cutthroat trout (all cutthroats between 10 and 16 inches had to be released) was effective in increasing the proportion of older and larger trout in that population. But again, that didn't arrest the encroachment of rainbow trout in the South Fork Snake. To combat that, Idaho has now gone to a catch-and-release regulation for cutthroat trout and has removed its limits on the harvest of rainbow trout. The state encourages anglers to keep the rainbow trout they catch from its Yellowstone cutthroat waters. In addition, the state has operated traps in many spawning tributaries of the South Fork Snake to remove rainbow trout before they can spawn. Idaho's goal is to reduce the proportion of rainbow trout in the South Fork Snake system to about 10 to 20 percent

54. The studies referred to in this paragraph on the effects of catchable-trout stocking on wild trout populations were conducted on a segment of the Madison River and one of its tributaries over the period 1967–76. They were reported in a series of agency reports in 1969 and 1970, and then in a landmark paper that appeared in the *North American Journal of Fisheries Management* in 1987 (see Vincent 1987). The work on the effects of angler wading on the mortality of eggs and fry first appeared in Bruce C. Roberts' Master's thesis (see Roberts 1988), and later was published in *North American Journal of Fisheries Management* (see Roberts and White 1992).

of the overall trout population. Meanwhile, despite the encroachment of rainbow trout in the South Fork Snake and other streams, a recent survey published by Idaho Department of Fish and Game biologists concluded that the abundance and size structure of Yellowstone cutthroat trout has remained relatively stable over the last 10 to 20 years in 77 stream sites across southeastern Idaho. [55]

Wyoming, as I've said before, has always been generous to its sports. In 1980, when I made my first trip across Two Ocean Pass, the general statewide bag limit was twelve fish per day with only one larger than 20 inches. The state still tries to be as generous as it can, but it too has made changes to conserve its remaining Yellowstone cutthroat populations. Wyoming operates under a plethora of angling regulations, with the specifics depending on which of the state's several management areas you find yourself in. For example, the general statewide daily bag limit now in place is six fish with only one over 20 inches. In the upper Shoshone River drainage, that bag limit has been halved again to three fish per day with only one over 20 inches. In the Clark Fork drainage, you may harvest six trout on flies or artificial lures only, but you must release all trout over 8 inches—unless you are fishing in what they call the Canyon of the Clark Fork, where the daily bag limit is three fish but only one can exceed 12

inches. Wyoming also now manages the upper Greybull River drainage for native Yellowstone cutthroat trout, meaning no stocking with exotics, but no special regulations govern angling in that drainage.

The upper Yellowstone drainage in Wyoming, including Thorofare and Atlantic creeks, has been managed for wild Yellowstone cutthroat trout since 1986 with a two-trout daily bag limit. Still, the bulk of the angling that occurs in this drainage is for trout on their spawning runs. In the Jackson management area, in waters within designated Wilderness areas, that two-trout bag limit has been in place since the early 1970s. Plus, anglers are given a bonus allowance of brook trout as a way to hopefully curb the encroachment of that exotic species. But outside of designated Wilderness areas, the general statewide bag limit applies. [56]

Several of these state and federal jurisdictions have joined in agreements to coordinate efforts aimed at conserving Yellowstone cutthroat populations. In 1994, for example, concern for the fluvial and stream-resident populations of the Yellowstone River basin, which the two states share downstream from Yellowstone National Park, prompted the Montana Department of Fish, Wildlife and Parks and Wyoming Game and Fish Department to form a working group that drafted a joint management guide for future protection and conservation of Yellowstone cutthroat trout in that

55. This survey was published in 2003 in the *North American Journal of Fisheries Management* (see Meyer et al. 2003a), and in 2006 as a status review of Yellowstone cutthroat trout in Idaho in the journal *Transactions of the American Fisheries Society* (see Meyer et al. 2006).

56. The notes on bag limits and angling regulations presented here came from the report, "Status and Management of Yellowstone Cutthroat Trout (*Oncorhynchus clarkii bouvieri*)" by D. Dufek and 6 co-authors (1999), compiled by the Wyoming Game and Fish Department as input to the U.S. Fish and Wildlife Service 90-Day Finding on the petition to list the subspecies under the Endangered Species Act. In a couple of places, this report promised a review of management options and regulations to be completed in time for the 2002–03 seasons, but I have received no word that this review was completed or that any changes from the 1999 report were made.

part of the subspecies range. Wyoming has also stocked several headwater streams in its portion of this basin with genetically pure Yellowstone cutthroat trout of the Paintrock Creek strain (origin an isolated tributary of the Bighorn River drainage) in the hope of reestablishing self-sustaining populations in streams formerly occupied by this subspecies.

Another working agreement, this one springing from the U.S. Forest Service Sensitive Species mandate to assist states in achieving their conservation goals for the subspecies, was signed by the Forest Service and Montana Department of Fish, Wildlife and Parks for managing Yellowstone cutthroat trout and their habitats on Forest Service lands in Montana. I haven't heard of similar agreements in Idaho or Wyoming, but I'm sure they must exist.

The broadest collaborative agreement of all for the Yellowstone cutthroat trout was signed in 2000 between the states of Montana, Idaho, Wyoming, Nevada, and Utah, the U.S. Forest Service, and Yellowstone and Grand Teton National Parks. That collaboration produced much of the information on present distribution, genetic purity, and current status across the subspecies' historical range that I drew upon for this Chapter. [57]

Although I've devoted many words in this Chapter to the Yellowstone River drainage upstream from Yellowstone Lake, and I've even visited that area a time or two myself in the past, I was surprised to learn from Dr. Todd Koel, Supervisory Fish Biologist at Yellowstone National Park, that there had never been a complete fisheries survey of this segment of the river or its vast array of tributaries. Completing such a survey was the objective of the most recent joint agreement to come to my attention, one signed in 2003 between Yellowstone National Park and Wyoming Game and Fish Department. In this program, the movements of radio-tagged adult trout were to be followed during their spawning migrations, and other work was to be done to determine if and where stream-resident populations are present in the upper Yellowstone drainage. This project was scheduled to run for two years, so results should be available by the time you read this.

Since many of these joint agreements involve land-management agencies, the way is open for collaboration on habitat restoration and protection projects. Indeed, many such projects are under way or have already been completed. Some of these have been funded in part by the Bring Back the Natives Program of the National Fish and Wildlife Foundation, a program I told you about in Chapter 4. Between 1991 and 2003 (last information available to me), 12 percent of the Bring Back the Natives funding went to on-the-ground projects benefiting Yellowstone cutthroat trout.

57. See May et al. (2003).

It's a good thing, too, that these collaborations are in place and apparently functioning, because all the U.S. Fish and Wildlife Service appears willing to offer is moral support. In a Questions and Answers release posted to explain its 90-Day Finding not to list the subspecies as threatened back in 2001, the Service offered only this regarding the future management of Yellowstone cutthroat trout:

> "The Service strongly recommends that state fish and game departments, federal land-management agencies, tribal governments, private groups, and other concerned entities continue to work individually and cooperatively to develop and implement programs to protect and restore stocks of Yellowstone cutthroat trout throughout the historical range of the subspecies."

This wasn't quite the end of the story, however. Early in 2004, environmental groups filed suit in federal court alleging that the Fish and Wildlife Service illegally denied listing Yellowstone cutthroat trout as threatened. I'm unclear why it took so long after the 90-Day Finding for this action to be taken, but nevertheless it was effective. Just before Christmas in 2004, a federal judge in Denver upheld the challenge and ordered the Fish and Wildlife Service to undertake a full 12-month status review and reconsider its listing decision. The Service did that, but didn't change its mind. So for now, the Yellowstone cutthroat trout remains off the list. [58]

58. For the full details of the 12-Month Finding not to list, see Hall (2006). You can download your own copies of the 12-Month Finding, a list of questions and answers about the Finding, and other information regarding Yellowstone cutthroat trout from the U.S. Fish and Wildlife Service website at *http://mountain-prairie.fws.gov/species/fish/yct/*.

CHAPTER 10

Finespotted Snake River Cutthroat Trout

Oncorhynchus clarkii behnkei: Chromosomes, 2N=64. Scales in lateral series 136 to 188 with mean values of 153 to 176 in populations from different spawning tributaries. Pyloric caeca 32 to 51 with mean values of 39 to 46 in different populations. Vertebrae range from 60 to 65. Meristic counts differ among populations of this subspecies from different spawning tributaries, indicating that individual spawning populations maintain a high level of reproductive isolation. But the range of counts for this subspecies overlaps that of the Yellowstone cutthroat subspecies (Chapter 9), so meristic characters cannot distinguish one from the other. Nor can the two subspecies be distinguished by allozyme electrophoresis or, so far, by methods employing DNA analysis. What clearly sets this subspecies apart is its spotting. The spots themselves are very small, the smallest of any native western trout, and are profusely distributed over the body but with a heavier concentration on the caudal peduncle area and above the lateral line anterior to the dorsal fin. This spotting pattern has been likened to a heavy sprinkling of ground pepper. Coloration is predominately yellowish brown, similar to the Yellowstone cutthroat, or sometimes silvery or even faintly purplish. The lower fins are often orange or red.

Snake River Finespotted Cutthroat Trout, *Oncorhynchus clarkii benkei*

BOB CARMICHAEL AND THE FINESPOTTED CUTTHROAT OF JACKSON HOLE

Bob Carmichael came to Jackson Hole a sick man. Born Robert Whittier Carmichael on March 19, 1899 in Wellsburg, West Virginia, he had served in the U.S. Marine Corps and had been wounded in World War I. He worked for the Senate Press in Washington, D.C. for a number of years, but a steadily worsening case of asthma forced him to give it up and head west. That was in 1930, when he was 31 years old.

The clear, robust mountain air proved just the ticket. Carmichael's health improved steadily as he lived the life of a fishing guide at Jenny Lake. He also used his time to learn the other productive waters of the Jackson Hole and Yellowstone country, and there were plenty of waters to learn. One of Carmichael's contemporaries once wrote, "One thoroughly familiar with the country could fish with flies every single day of the open season, get good to excellent fishing every day, and *never fish the same water twice*."

One of Carmichael's favorite rivers was the Gros Ventre, a tributary of the Snake, which flows westerly into Jackson Hole from the Gros Ventre Mountains. It is easy to see why one would like this river. Its valley is one of those intriguingly beautiful, comparatively remote places that is well worth the drive over the Forest Service road just for the scenery.

But to fly fish in a setting like this in Carmichael's day was an experience in itself.

And Bob Carmichael *was* a fly-fisherman, and a dry fly man first and foremost. He held strong opinions about this; when his wife insisted that they add a line of lures to the tackle shop they operated in Moose, Wyoming, he drew a chalk line down the center of the floor and growled to customers, "The hardware department is hers!"

Sometime around 1935, Carmichael began experimenting with the Adams dry-fly pattern, finding it to be a real taker in the Jackson Hole area. His experiments soon led to other good patterns based on the mixed Adams hackle and hackle point wings. The Carmichael's Indispensable, developed for him by fly tyer Don Martinez, became a favorite. Later, Carmichael formed a business association with Roy Donnelly, another noted fly tyer, who originated two additional Carmichael favorites, the Donnelly Variant Light and Donnelly Variant Dark.

Over the years, he guided many famous people, among them former President Herbert Hoover and his son, Allen, famous *Life* magazine photographer George Silk, U.S. Air Force General Curtis LeMay, and many members of the Rockefeller family. Being a former journalist himself, he wrote his own copy for the Carmichael Tackle Shop ads published weekly in *The*

Jackson Hole News. The following is an example of his pithy commentary:

"FISHING TIP OF THE WEEK. Anglers have enjoyed excellent fishing this week to compensate for the poor period caused by the PM storms and accompanying north winds. Barometric pressures have maintained a higher level than average, but somehow anglers have been plagued with poor fishing as soon as one of these storms hits. However, dropping the crying towel temporarily, AM fishing has been best 8:45 to 1:30. On days when no disturbance was registered, many fly fishermen scored heavily, some claiming the best fishing of their lives. Our top anglers claim fish can be taken almost everywhere, but cut banks seem to produce the heavier fish. Follow this tip to the letter: fish these cut banks carefully from water's edge rather than from the top, in order not to be seen by the feeding trout. Often these fish will take a fly when presented only two feet or less from the shore. These trout, obviously, are taking large dries (Donnelly Dark, Irresistible, or Humpies) for grasshoppers. Hoppers often misjudge their landings or are blown off course by the wind. Fishing conditions should remain as good for the coming week as it was this past, and even improve if the PM thunderstorms slack off. Best dry, Irresistible; best wet, badger Woolly Worm."

About the time World War II broke out, Carmichael got a fine location for his tackle sales and guide service at Moose, Wyoming. His business grew and prospered but his health did not. The asthma came back and grew steadily worse, and this, accompanied by problems with his heart, finally forced him off the rivers. In the final years of his life, he held forth from his tackle shop, presiding over the frequent gatherings of fly-fishermen from a seat near his horseshoe-shaped mahogany fly cabinet, still drawing detailed and unerringly accurate maps of favorite pools and runs, and still leading the salty, good-natured, give-and-take discourse, answering frequent questions with a peppery "Bet your sweet ass!"

Bob Carmichael was a unique individual with strong opinions about the trout and the fishing in the Jackson Hole country, and he could be cantankerous and temperamental about it. But he loved the Jackson Hole cutthroat trout. There's a passage in one of Ernest Schwiebert's books telling how Carmichael set the record straight when Schwiebert, then a young man new to Jackson Hole, voiced a derogatory opinion about the fighting qualities of cutthroat trout. "Young man," Carmichael growled in annoyance, "when you know enough about this part of the country to have an opinion about the fishing, you'll know there's cutthroats and there's cutthroats!" When Schwiebert asked if the local fish were somehow different from other cutthroat trout, Carmichael shot back, "Bet your sweet ass! These fish ain't no pantywaists. They're Jackson Hole cutthroats!"

Carmichael recognized that the native Jackson Hole cutthroat was a fish of the main rivers and heavy currents of such streams as the Gros Ventre and the Snake. He took a dim view of the introduction of exotic species into the Snake. In a letter to J. Edson Leonard dated April 1949 he wrote, "...we are alarmed for fear the rainbow will take over Snake River. 1948 catches on the Snake during September and October ran in excess of 20 percent rainbow." Later in the same letter, he wrote, "Fortunately the Loch Levens [brown trout were planted in Lewis Lake near the headwaters of the Snake River in 1889] seldom if ever come down the Snake below Jackson Lake dam so there is no apparent danger of these Lochs taking over our Snake."

Bob Carmichael passed away in 1959. In his two-volume book *Trout*, published nineteen years later, Ernest Schwiebert included a painting of the fine-spotted, current-loving Jackson Hole cutthroat and labeled it *Salmo carmichaeli*—a fitting tribute to Bob Carmichael perhaps, but not an acceptable procedure under the rules of zoological nomenclature. Getting the naming procedure down correctly—well, almost correctly—fell to another, more recent author, M.R. Montgomery, who wrote a more proper description in his book, *Many Rivers to Cross: of Good Running Water, Native Trout, and the Remains of Wilderness*. The name Montgomery chose to bestow is *Oncorhynchus clarkii behnkei*, in honor of Dr. Behnke, not only to recognize his expertise "on salmonids in general and trout in particular," but also because "he is a great crusader for habitat protection and the preservation of subtly different genetic stocks of native American trout and salmon." [1]

1. The story of Bob Carmichael of Jackson Hole was pieced together from a correspondence with his widow, Mrs. Julius Mosley of Moose, Wyoming, that took place in 1982 and 1983. Other material came from his obituary, as published in *The Jackson Hole News* of April 30, 1959, and from the writings of others who knew him. For example, the young Ernest Schwiebert's indoctrination into the fighting qualities of Jackson Hole cutthroat trout can be found in richer detail at pages 287–290 in Volume 1 of Schwiebert's two-volume opus titled *Trout* (1978). Schwiebert's painting of the Jackson Hole cutthroat is in a color place between pages 174 and 175 of the same volume. There is no mistaking the fine-spotted pattern in this depiction of the Jackson Hole cutthroat. Carmichael's letter to J. Edson Leonard was published in Leonard's book, *Flies* (1950). As for the formal naming of the subspecies, it's not the usual thing for this to happen in a piece of popular writing, because most popular writers are not aware of the standards for formal naming in the International Code of Zoological Nomenclature. Schwiebert's *Salmo carmichaeli*, associated only with the painting of the trout, didn't make the grade. But when author M.R. Montgomery bestowed the name *Oncorhynchus clarkii behnkei* on the finespotted Snake River cutthroat in his book *Many Rivers to Cross: of Good Running Water, Native Trout, and the Remains of Wilderness* (Montgomery 1995) at page 127, he got it close to right, formal description of specimens and all. However, some still do not accept *behnkei* as a proper scientific name because Montgomery did not designate any one particular specimen as the type specimen for the subspecies. Indeed, agencies have been reluctant to recognize the finespotted Snake River cutthroat as a subspecies at all, preferring instead to lump it in with the Yellowstone cutthroat for status reviews and management purposes. I'll say more about this later in the Chapter.

Snake River from the Snake River Overlook on U.S. Highway 26/89/191, showing layers of glacial outwash and Pleistocene lake sediments. PHOTO BY AUTHOR

HISTORICAL RANGE

The known historical range of the finespotted Snake
River cutthroat is the South Fork Snake River from
Palisades Reservoir on the Wyoming-Idaho border
upstream through Jackson Hole to Jackson Lake, and
all of the major tributaries of this segment from the
Salt River upstream to the Gros Ventre. This distribution was mapped in Fig. 9-1, and I have enlarged it for
greater detail in Fig. 10-1.

Nobody knows for sure how far down the South
Fork Snake River the finespotted subspecies actually
extended before Palisades Reservoir was constructed (it
was authorized prior to World War II but not completed
until 1957) because nobody wrote any descriptions of
the trout they caught from this reach prior to that time.
But we do know, from Chapter 9, that all of the Snake
River tributaries between Palisades Reservoir and
Shoshone Falls have the Yellowstone cutthroat as their
native trout. In Palisades Reservoir itself, the finespotted
cutthroat once comprised about 97 to 99 percent of the
trout caught by anglers. But by then, large numbers of
finespotted trout were being stocked from the Jackson
National Fish Hatchery, built to mitigate for impacts
on fisheries caused by construction of the dam, and
from State hatcheries. Finespotted trout still swim in
the reservoir, and apparently remain the predominant
gamefish, but brown trout are also present and now comprise a significant proportion of the catch.

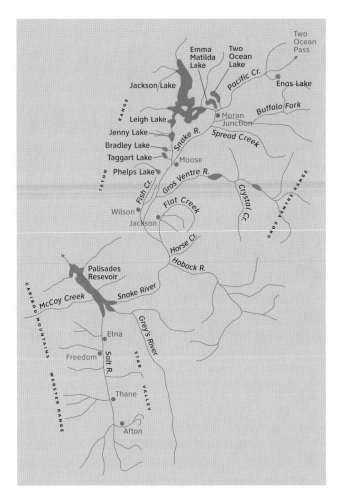

FIGURE 10-1. *Snake River drainages associated with the historical
distribution of the Finespotted Snake River Cutthroat Trout*

At the upper end of the distribution, at Jackson Lake, it was also possible for fish to move freely up and down the river and in and out of the lake prior to construction of the dam at the Jackson Lake outlet. One of the earliest historical records of trout dates back to trapper Osborne Russell's journal for the years 1836–39, where at least twice he mentioned the abundant trout in Jackson Lake and the nearby Snake River. Russell and his companions even enjoyed a sumptuous dinner there on July 4, 1839, feasting on the twenty fine trout he had caught at the Jackson Lake outlet, but he identified the fish only as salmon trout. English sportsman William Baillie-Grohman also wrote about fine dining on trout that he caught at the Jenny Lake outlet stream in the early 1880s, but he too described them only as salmon trout. Old photos and occasional anecdotes passed on by anglers do suggest that the finespotted subspecies may have inhabited Jackson Lake in the years prior to 1911, when the dam was built, and perhaps for a time thereafter. [2] But the invasion of Jackson Lake by lake trout following their introduction into Lewis and Shoshone lakes in the 1890s reduced the cutthroat population, and, again, finespotted cutthroat trout from hatcheries were among the trout frequently stocked in Jackson Lake to maintain the fishery.

If finespotted cutthroat trout did occur historically in Jackson Lake, I think it likely that they migrated to tributaries downstream from the lake or utilized small tributaries of the lake itself for spawning, and did not migrate to upstream reaches. That's because, in the Snake River above Jackson Lake, the historical record is a bit more convincing that the Yellowstone cutthroat, not the finespotted form, was the native trout. David Starr Jordan examined the streams and lakes of the upper Snake River drainage in Yellowstone National Park for the U.S. Fish Commission, reporting his findings in 1891. Barton W. Evermann of the U.S. Bureau of Fisheries also caught trout from the Snake River 12 miles upstream from Jackson Lake a year or two after Jordan's reconnaissance. Both of these men were quick to note new fishes when they found them, but neither man marked the existence of a finespotted cutthroat trout upstream from the lake. [3]

Downstream from Jackson Lake, Yellowstone cutthroats occupy all of the Pacific Creek, Buffalo Fork, and Spread Creek drainages, the first three tributaries of the Snake River below the Jackson Lake outlet before you come to the Gros Ventre River. Headwater tributaries of the Gros Ventre are also inhabited by Yellowstone cutthroat trout even though the finespotted trout occupies the remainder of the Gros Ventre system. Finespotted trout also occupy all of the Hoback, Greys, and Salt river drainages, although Yellowstone cutthroat trout do occur in the headwater tributaries of

2. Osborne Russell was far from the first to fish for subsistence at Jackson Lake. Paleo-Indians left clues in the form of stone weights for fishing nets at ancient campsites near the upper end of Jackson Lake (Crockett 1999). But Russell may have been the first to leave a written record of his experiences. His accounts can be found in *Journal of a Trapper, or Nine Years Residence Among the Rocky Mountains Between the Years of 1834 and 1843* (Russell 1986). William Baillie-Grohman wrote up his account in *Camps in the Rockies* (Baillie-Grohman 1884). Photos of finespotted Snake River cutthroat trout from the Jackson Lake area, believed taken around 1900, were published by Paul Schullery in a piece titled "Snake River Cutthroats" that appeared in *The American Fly Fisher*, the journal of the American Museum of Fly Fishing, Manchester, Vermont (see Schullery 1982). Simon (1946), in his *Wyoming Fishes*, reported a 14-pound, 10-ounce finespotted cutthroat trout taken from Jackson Lake in 1932.

3. See Jordan (1891) and Evermann (1893) for these accounts.

these drainages. The present picture of finespotted cut-throat distribution in the Salt River drainage is clouded because Yellowstone cutthroat trout from state hatcheries in Idaho have been stocked in Idaho tributaries of this system for many years. On the west side of Jackson Hole, tributaries flowing off the Teton Range into Leigh Lake, Jenny Lake, and Phelps Lake have Yellowstone cutthroat trout, but early stocking, presumably of Yellowstone cutthroat trout, also occurred in some of these streams as well.

The known historical distribution of the finespotted Snake River cutthroat is remarkable in two respects: 1) it lies totally enclosed within the territory of another subspecies, the Yellowstone cutthroat trout; and 2) the finespotted trout historically resisted introgression and displacement by the Yellowstone cutthroat and rainbow trout that were stocked frequently within this enclave. The finespotted trout also resisted hybridization with Yellowstone cutthroat trout at the boundaries of the distribution where hybrid zones typically occur, although occasionally "intergrades" were reported; for example, from the upper Gros Ventre River. [4]

Another curious aspect of this distribution is the fine-pepper spotting pattern itself and how this might have originated, given that finespotted Snake River and Yellowstone cutthroat trout are all but indistinguishable using meristic counts, allozyme electrophoresis,

and DNA methodologies. A clue to this came from, out of all places, Norway, in 1987 and 1988. In 1987, researchers from the University of Bergen and Norwegian Directorate of Fisheries described a finespotted population of brown trout that coexists with normally spotted brown trout in a lake in Norway's Hardangervidda National Park. From photos and the published description, the spot size and pattern of these finespotted Norwegian brown trout appear identical to the spotting on finespotted Snake River cutthroat trout. In 1988, the Norwegian researchers published results of breeding experiments that showed that inheritance of this spotting pattern in the Hardangervidda brown trout is controlled by a single gene locus with two codominant alleles. [5]

Um, OK, but what exactly does that mean? First, let's review. As I outlined back in Chapter 1, each diploid individual has two copies of each of its nuclear genes, one copy inherited from its mother and one from its father. These copies, called alleles, may be exactly the same, in which case the individual is *homozygous* at that gene locus, or they might differ slightly, in which case the individual is *heterozygous* at that gene locus. It is these allele pairs that determine hereditary traits in individuals. Many inherited traits are determined by only a single allele pair, i.e., the pair of alleles at a single gene locus. Such traits are called *qualitative* traits or

4. T.C. Murphy, a graduate student from Colorado State University who collected Snake River cutthroat trout for taxonomic study back in the early 1970s, reported "intergrades" from the Gros Ventre River; see Murphy (1974). These were trout with spots of intermediate size and somewhat more profuse distribution than on normal Yellowstone cut-throats, but distinctly less profuse than on true finespotted trout.

5. See Skaala and Jørstad (1987, 1988).

Mendelian traits, because all of the traits that Augustinian friar Gregor Mendel studied when he postulated the existence of the factors that would later be called genes, as well as his principles of inheritance, were traits of this sort—pea color either green or yellow, peas with either wrinkled or smooth skins, pods either pinched or swollen, and so on. These traits typically manifest themselves in an either-or fashion, producing clearly one phenotype or another, although sometimes a clearly discernable intermediate phenotype may also be expressed. [6]

It should go without saying that individuals homozygous for a Mendelian trait will exhibit only one phenotype because the two alleles governing that trait are identical. But in heterozygous individuals, the two alleles can interact in different ways. One allele may be *dominant* and the other allele *recessive*, in which case only the dominant allele's phenotype is expressed. In other cases, neither allele can dominate the other and both contribute equally to the observed phenotype. Such alleles are called *codominant* and their expression is often referred to as *additive* gene action in textbooks. There are also cases of *incomplete dominance*, in which an intermediate phenotype is expressed, but it resembles one parent more than the other.

An often-cited textbook example of how dominant and recessive alleles interact is the albino trait in

6. Gregor Mendel published his original paper in 1865, but his work created no stir in the scientific community of his time and was forgotten for nearly 40 years. For you history-of-science buffs, I've cited the original publication and an English translation that was published after the work was "rediscovered;" see Mendel (1865 [1901]). To learn more about Mendelian genetics from modern texts, see Ayala (1982), Tave (1986), or Allendorf and Ferguson (1990). Bill Willers' book, *Trout Biology: A Natural History of Salmon and Trout* (1991), also has a Chapter on inheritance in trout that deals mostly with qualitative traits. But qualitative traits are only part of the story. Most inherited traits are determined by allele pairs at many gene loci, not just one, which leads to a distribution of "values" of the trait, some low and some high, but most clustering around an average value for the group. Traits controlled by many genes are called *quantitative* (or *polygenic*) traits. Body length at a given age in a population of trout may be an example. Individuals from the high end of the distribution can be chosen to mate for the next generation, and the next, etc., in hopes of increasing the average value of the trait in succeeding groups. Principles and methods for working with quantitative traits are covered in an excellent textbook by Falconer (1989). Ayala (1982), Tave (1986), and Allendorf and Ferguson (1990) also have sections on quantitative genetics.

Snake River below Wilson, Wyoming.

rainbow trout. In rainbow trout, an allele designated *A* (the dominant allele) is associated with normal skin color whereas the *a*-allele (the recessive) produces no color (note that dominant alleles are designated with italicized capital letters and recessives with the same lowercase letter, also italicized). [7] Individuals that inherit the *AA* allele pair have normal skin color, but so do those that inherit the *Aa* allele pair, because the expression of the dominant allele prevails. Only individuals that inherit the recessive *aa* allele pair express the albino phenotype.

The inheritance of the finespotted pattern in Norwegian brown trout is an example of codominant allele action. Here I've labeled the alleles involved as *S* and *S'*. The homozygous *SS* phenotype is the normal large-spotted form. The homozygous *S'S'* phenotype is the finespotted form. But in this case, the heterozygous *SS'* individuals express an intermediate spotting pattern that is visually about halfway between what one sees as the large-spotted (i.e., "normal" brown trout) and finespotted forms.

If this same codominant allele mechanism also governs inheritance of the fine-pepper spotting phenotype in our finespotted Snake River cutthroat trout (and I hasten to add this has *not* been established), [8] it means that everywhere within its enclave of distribution the trout are homozygous (i.e., *S'S'*) at that gene

7. A standard nomenclature for protein-coding loci in fish was published by Shaklee et al. (1990), and has been adopted by most scientists in the field and their professional societies. It names gene loci for the proteins they encode, if known, or designates them with letters as I have done here when the proteins are not known. Enzyme names used in this nomenclature are those recognized by the International Union of Biochemistry (see IUBNC 1984).

8. A letter from Kerry Grande, manager of the U.S. Fish and Wildlife Service's Jackson National Fish Hatchery, received October 18, 2004, states: "We have not performed any mating experiments to determine the mode of inheritance" in finespotted Snake River cutthroat trout. As far as I have been able to determine, neither have fish culturists employed by the State of Wyoming or any other state that has imported finespotted Snake River cutthroats for stocking purposes. To verify this mode of inheritance, controlled mating experiments would have to be performed involving the following crosses, with the expected results as indicated. If the expected results are not obtained, then some other mode of inheritance governs the trait in this subspecies.

PARENTS	PROGENY
finespotted × Yellowstone	all intermediate
intermediate (from 1st cross) × finespotted	half finespotted, half intermediate
intermediate (from 1st cross) × Yellowstone	half Yellowstone, half intermediate
intermediate × intermediate (both from 1st cross)	finespotted, intermediate, and Yellowstone in proportions of 1:2:1

locus. In other words, they are fixed for the allele pair that produces the fine-spotted pattern and only that pattern. This, in turn, implies that the Yellowstone cutthroat subspecies does not have the *S'* allele. Yellowstone cutthroat trout would either be fixed for the *SS* allele pair that expresses the large-spotted pattern, or have the *S* allele paired with other alleles at this gene locus that interact to produce other large-spotted pattern variations. [9]

Assuming for the moment that the single locus, codominant allele mechanism does govern inheritance of the finespotting pattern in the Snake River enclave, when and how could it have originated and been confined to just this area of distribution? We have better information about the "when" part of this question than we do about the "hows." In 1979, Eric Loudenslager and Robert Kitchin, two scientists then at the University of Wyoming Department of Zoology and Physiology, published the first genetic comparison of finespotted Snake River cutthroat with populations of Yellowstone cutthroat from both the Snake River and Yellowstone Lake systems. Using allozyme electrophoresis, they found very little genetic variability among any of the populations, but what little they did find led them to a molecular clock estimate of 20,000 years for the time since divergence of the fine-spotted and large-spotted forms. [10]

9. A brand-new wrinkle to this story of spot size and pattern in cutthroat trout associated with Jackson Hole drainages appeared in 2005, when Mark Novak and Jeffrey Kershner of the U.S. Forest Service and Karen Mock of Utah State University reported the discovery of Yellowstone cutthroat trout with ultra-large, disperse spots on their bodies in isolated, above-barrier, canyon reaches of four streams, all flowing out of the Teton Mountains on the west side of Jackson Hole, namely Cascade Creek, Death Canyon, Leigh Canyon, and Paintbrush Canyon. To my knowledge, trout with spots this large (almost twice the diameter of "normal" Yellowstone cutthroat spots and roughly five times the diameter of spots on finespotted trout) have never been observed anywhere else in the Yellowstone or finespotted cutthroat range. These ultra-large spotted trout all carried a single mtDNA haplotype, one of four haplotypes found to be predominate in the upper Snake River headwaters area of Wyoming, but that same haplotype was also found in normal-spotted Yellowstone cutthroats, finespotted Snake River cutthroats, and intermediates between the two, especially in the Jackson Hole portion of these Snake River headwaters. For more details, see Novak et al. (2005).

10. See Loudenslager and Kitchin (1979).

Twenty thousand years. In the grand sense of geological time, that's a very recent date. Even so, it takes us back to late-Pleistocene time, to a period when at least a part of Jackson Hole was covered with glacial ice.

The Jackson Hole area has been shaped by repeated episodes of faulting, warping, uplifting and downdropping, bouts of volcanism, inundation by lakes, and sculpting by glacial ice. [11] The Teton Range that forms the west side of the valley began pushing up about 9 million years ago, in Miocene time. The Gros Ventre Mountains on the east side are much, much older. In Pliocene time, Lake Teewinot, a large, shallow body of water, formed by uplifting of the valley floor at about the southern boundary of present-day Grand Teton National Park, filled a part of Jackson Hole. The Pleistocene Epoch opened with downdropping between faults that extended along the east and west sides of the valley. This was followed by a nearly 500,000-year period of volcanism and extrusions of lava, with eruptive centers just north and west of the town of Jackson. About 1.3 million years ago, downdropping of southern Jackson Hole blocked the Snake River drainage for a time, impounding another large lake that extended upvalley and overlapped at least part of the site of Pliocene Lake Teewinot. This was a deep lake, and it lasted long enough to deposit about 200 feet of lakebed sediment before subsequent faulting and warping allowed it to drain away. Around 700,000 years ago, another downdrop of the valley in the same general area impounded a second Quaternary lake that also endured for a time before it, too, was destroyed by the same faulting and warping mechanism that drained its predecessor.

Then came the ice. The most recent interpretations of the evidence indicate that two major advances of glacial ice occurred in the Jackson Hole area, although the younger of the two may have itself been a three-phase episode. Each of these advances occurred in the late-Pleistocene. The first and by far the most extensive, now referred to as the Munger glaciation (although timing-wise, this event is associated with the ice advance called the Bull Lake glaciation in other parts of the Rocky Mountain west), commenced about 170,000 years ago. [12] Ice from the Beartooth and Absaroka ice centers converged on Jackson Hole from the north and northeast with lobes also flowing down the Pacific Creek and Buffalo Fork drainages, and advanced south along the Teton Range. This was met by ice flowing down the Gros Ventre valley from the Wind River and Gros Ventre ranges. Glaciers also flowed down from the canyons of the Teton Range. Jackson Hole was buried under about 2,000 feet of ice, and some believe that the glacier extended all the way down through the Snake River canyon past the mouth of the Salt River into eastern Idaho. If trout were pres-

11. The information presented here was distilled from publications dealing with the geology and pre- and post-glacial vegetation and climate conditions of Jackson Hole and Grand Teton National Park by Love and Reed (1968 [1971, 1984]), Pierce and Good (1992), Whitlock (1993), Love (1994), Elias (1996), and Good and Pierce (1996). Another summary, updated in 2003, can be found on the website for Grand Teton National Park at *www.nps.gov/grte/ nat/glac.htm.*

12. Some of the authorities cited in footnote 11 have assigned different dates and used different names for the glacial advances discussed here, especially in the earlier references, but the dates and names I have used are the ones that are recognized currently. If earlier ice advances occurred in this area, evidence of their occurrence was wiped out by subsequent events.

ent in this segment of the Snake River drainage prior to the Munger glaciation, they would certainly have been expelled by the ice, unless there were refugia we don't know about.

Munger ice receded about 130,000 years ago, and the area apparently remained ice-free until the advent of the Pinedale glaciation, about 70,000 years ago. This was a three-phase event in the Jackson Hole area. The main center of ice for the Pinedale advance was the Absaroka Range. In its initial advance, called the Burned Ridge phase in this area, lobes flowed down the Pacific Creek and Buffalo Fork drainages, with the Buffalo Fork lobe being the major player, to join ice flowing down canyons from the Teton Range. Ice associated with the Burned Ridge phase advanced about halfway down Jackson Hole, leaving the large moraine known today as Burned Ridge. This ice also excavated the basins that hold Emma Matilda and Two Ocean lakes.

The phase-two advance of the Pinedale glaciation, called the Hedrick Pond phase, may have begun around 30,000 years ago. This time, ice advanced down the Snake River drainage, extending south of the Jackson Lake site to join with a major lobe advancing down Pacific Creek. Ice also pushed down the Buffalo Fork drainage, but not as extensively as it had during phase one. There is evidence that a lake formed in a depres-

sion left by the Hedrick Pond ice south of the Jackson Lake site, with an outlet near the present Snake River Overlook on U.S. Highway 26/89/191. Some authorities call this the Triangle-X Lake for the Triangle-X Ranch that now occupies its former basin. There is also evidence for glacial lakes formed by ice dams that, like other glacial lakes, failed repeatedly when the waters rose to 90 percent of the height of the dams. One such lake occupied the Buffalo Fork valley during this and the final phase of Pinedale glaciation. Another glacial lake occupied a part of the Gros Ventre valley. Flood debris from these breaches covers much of the area downstream from the Snake River Overlook.

The final advance of Pinedale ice, called the Jackson Lake phase, came hard on the heels of the second phase, commencing perhaps as much as 25,000 years ago. This is the event that left behind the lakes we presently see in and around Jackson Hole: Jackson Lake, Jenny and Leigh lakes, and Bradley, Taggart, and Phelps lakes. Glaciers flowing down upper canyons of the Teton Range combined with the once-again advancing Yellowstone ice mass to cover the present site of Jackson Lake. Other glaciers flowing down Teton canyons south of Jackson Lake left the moraines that encircle Jenny Lake and the others. Ice from this advance began to recede perhaps 15,000 to 17,000 years ago, but remnants of glacial ice may have still been

A section of the lower Gros Ventre River under a threatening sky. The Gros Ventre was one of Bob Carmichael's favorite streams.
PHOTO BY AUTHOR

present on the floor of Jackson Hole when the first humans began using the area.

The maximum advance of the Pinedale ice pretty much delineates the upstream boundary of the fine-spotted Snake River cutthroat distribution, and the molecular clock evidence, if it can be relied on, dates its divergence from the large-spotted Yellowstone cutthroat form to a period when this ice mass was still in place. But how did the divergence actually come about? Could it have been a spontaneous mutation, perhaps, at the gene controlling spotting pattern? DNA sequences are normally copied exactly during the process of gene replication, but on very rare occasions an error will occur that gives rise to a change in the sequence. That's all a mutation is. If it happens to occur in a cell that divides to become a sperm or an egg cell, it will be inherited by the next generation. Such mutations are the source of the genetic variation that underlies molecular clock calculations. Could such a mutation have produced an allele that then, owing to its codominant nature, spread through the cutthroat population occupying the Jackson Hole enclave between the Pinedale ice and the Palisades—a population that was probably small and bottlenecked anyway, owing to the marginal conditions of existence so near to the ice? Perhaps, but it's all speculation at this point.

How the two distinct forms were able to maintain themselves in this region without apparent hybridiza-tion for so long after the glacial ice receded also remains a mystery. As long as we're speculating, perhaps the fine-spotted pattern is discernable by the trout and forms a basis for mate selection when spawning. That would forestall hybridization. Whatever the mechanism, somehow the two forms were able to partition this portion of the Snake River drainage and maintain this partitioning, even though no physical barriers prevented them from intermingling and interbreeding.

Alas, however, this historical resistance to introgression may now be eroding. A recent letter from Kerry Grande, manager of the Jackson National Fish Hatchery, states that the spotting pattern they now see in the hatchery trout is highly variable: "We have fish with very few spots; some that look like Yellowstone cutts (large spots typically posterior to the dorsal fin); some that have very large spots all over; and some with the fine pepper spotting that you would expect to see in the Snake River fish. Our fish are infused with wild fish every three years. Even the fish taken from the river are variable in the spotting pattern."[13]

LIFE HISTORY AND ECOLOGY

Like many of the other interior cutthroat subspecies discussed already, the finespotted Snake River cutthroat can exhibit three basic life history strategies:[14]

- The fluvial, or riverine (potamodromous), life history, wherein the fish feed and grow in main river reaches, then migrate to tributar-

13. Kerry Grande, Manager, Jackson National Fish Hatchery, U.S. Fish and Wildlife Service, Jackson Wyoming. Personal communication received October 18, 2004.

14. In 1964, in view of then-perceived threats from increasing fishing pressure, demand for irrigation water, proposed dams, and continued loss of habitat due to siltation of spawning sites and construction of flood control structures, the Wyoming Game and Fish Department kicked off a comprehensive investigation into the biology and ecology of the finespotted Snake River cutthroat trout. The results of this investigation were written up by John W. Kiefling in 1978 in what may remain the most thorough compilation of biological information for this subspecies; see Kiefling (1978). Other sources used here were Hayden (1968), Novak (1989), Stonecypher et al. (1994), Joyce (2001), and Joyce and Hubert (2004) on spawning ecology; Hazzard and Madsen (1933), Hagenbuck (1970), Foster (1978), and Isaak and Hubert (2004) on feeding, age and growth, and summer rearing ecology; Harper and Farag (2004) on winter rearing; and Wiley (1969), Trojnar and Behnke (1974), Sekulich (1974), Mullan (1975), and Hazzard and McDonald (1981) regarding adaptability to other environments and response of populations to angler harvest.

ies (primarily spring creeks, in this case) where spawning and early rearing take place.

- A stream-resident life history, wherein the trout complete all phases of their life cycle within the streams of their birth, often in a short stream reach and often in the headwaters of the drainage.
- A lacustrine, or lake-associated, life history, wherein the fish feed and grow in lakes, ponds, or reservoirs, then migrate to tributary streams to spawn.

The Fluvial, or Riverine, Life History Strategy

The principal and most prevalent life history form of the finespotted Snake River cutthroat is the fluvial, or riverine, trout of the mainstem Snake, Gros Ventre, Hoback, Greys, and Salt rivers. These were the trout, husky inhabitants of the robust currents, that provided the tremendous sport fishery in Bob Carmichael's day. Some mighty prodigious trout were reported back then. A 1946 publication of the Wyoming Game and Fish Department mentioned trout weighing over 10 pounds taken from the Snake River and its tributaries in the early 1940s. M.D. Rollefson, also with the Wyoming Game and Fish Department, told of a 12-pounder taken from the Snake in 1967, and said that 6- to 8-pound trout were taken annually.[15] Certainly in Bob Carmichael's time and right up to the present, anglers have regarded these trout as the brawniest, most acrobatic fighters of all the cutthroats, although today's anglers may have to probe waters further downstream—say, the canyon reach below Astoria Hot Springs—to consistently find the larger trout.

The finespotted Snake River cutthroat does not follow the custom of other cutthroat subspecies in seeking out the slower, quieter places in a river. Instead, it is a current lover, a seeker of the swift-flowing runs. These cutthroats hold in water where you'd normally expect to find rainbow trout. If you happen to live in the Pacific Northwest, as I do, or make regular visits to this region to fish in summer-run steelhead streams, you'll recognize the kinds of places to cast your flies or lures; they're the same lies that would hold a summer steelhead in a Pacific Northwest stream. Bob Carmichael would urge his angler clients to fish the long, bouncy runs with water maybe waist-deep or a bit more, with a heavier rapid or chute emptying into the head end. The cutthroats will sometimes hold pretty high up in such water, he would advise, close in along the edges of the chutes or up in the throats of the pools. Carmichael himself, while he was able, would fish large, buoyant dry flies in such waters: a Humpy, an Irresistible, a heavily dressed Wulff pattern perhaps, or, more likely, a Donnelly Light or Dark Variant, which were his favorite patterns for the Gros Ventre River. Today's anglers might use the same patterns or opt for a lime-green Trude, a Jay-Dave's Hopper, a Turck's Tarantula, or a foam-bodied Chernobyl Ant or Orange Crush.

15. M.D. Rollefson's reports of the 12-pound finespotted cutthroat trout and the annual capture of trout weighing 6–8 pounds were cited by Mullan (1975) in a presentation to the 1975 annual meeting of the Western Association of State Fish and Game Commissioners. The mention of the 10-pound trout of the early 1940s is in Simon (1946).

When spawning time draws near, these trout migrate into the area's spring creeks to mate and deposit their eggs. There is a distinction among spawning tributaries based on the timing of the spawning period and the reaches of the river from which the spawners are drawn and to which their progeny recruit, with the line of demarcation being the Wilson highway bridge. In general, spawning occurs in tributaries north of this bridge in May and June, and the progeny recruit to the fishery in reaches north of the bridge. In tributaries south of the bridge, spawning is about a month earlier, in April and May, and the progeny recruit to reaches south of the bridge. Fish Creek, which joins the Snake River south of the Wilson highway bridge, is an exception. Snake River fish entering Fish Creek to spawn do so in May and June, but their progeny recruit to the mainstem population south of the bridge. Homing of trout to their natal spring creek tributaries is precise.

Spawning runs are generally composed of about equal numbers of males and females, with the majority of fish being first-time spawners in age classes 3 (mostly males) and 4 (mostly females). About 5 percent of a typical run are first-time spawners of age 2, usually males. Repeat spawners, mostly ages 5 and 6, make up only about 10 percent of the run. The oldest spawner ever checked by a biologist was an 8-year-old

male, but the normal life span of these trout is 6 to 7 years.

Sexually mature finespotted females range from about 15.5 inches up to 20–22 inches in length, and average about 2,300 eggs per female. They deposit their eggs in redds dug in clean gravel in water that may average less than 12 inches deep. Usually they choose quieter areas near the creek bank, or midstream below a bend. Redds may harbor two or three egg pockets each, and the fish may deposit eggs in more than one redd.

Post-spawning mortality is high in finespotted Snake River cutthroat trout. Field observations indicate that many of the spawners are in poor condition following spawning and are quite susceptible to debilitating fungal attack. They are also particularly vulnerable to such predators as great blue heron and ospreys at this time. All in all, in a given year, about half of the run may succumb to the rigors of spawning. This, added to the natural mortality and whatever man-caused mortality may occur in the mainstem rivers between spawning runs, accounts for the low percentage of repeat spawners in subsequent runs.

High annual adult mortality is offset by high survival of eggs to the swim-up stage for this subspecies. If the female trout can find clean, round, silt-free gravel in which to prepare their egg pockets in these spring

A panorama of the upper Gros Ventre River looking upstream (east) from a point near the Goosewing Ranger Station. PHOTO BY AUTHOR

creek tributaries, egg mortality is typically about 11 to 12 percent, a value that is considered quite low. The apparently high survival of juveniles over the time they spend in these natal tributaries may also offset high annual adult mortality. [16]

After swim-up, the young-of-the-year-trout remain in the natal spring creeks for the remainder of the summer, fall, and early winter. Come January and February, just short of their first birthdays, about half of them move down to the main rivers. Among this group, for reasons unknown, the greatest daily downstream movement comes in periods of extreme cold. Of the remainder, some will rear in the natal tributary until spring or summer of their first year, migrating to their river as age-1 juveniles. Others will rear for another year and recruit to the river as age-2 juveniles.

The rate of growth for the finespotted subspecies is quite good, considering that it lives in a northern temperate riverine environment. A typical finespotted Snake River trout is about 4 inches long at age 1. By age 2, it will have grown to 8–10.5 inches in length; to 11–14 inches at age 3; to 14–15.5 inches at age 4; and to 17–18.5 inches at age 5. This is about 20 percent faster than the growth of a typical cutthroat trout in Yellowstone Lake. Trout up to 22 inches in length have been observed by biologists in the Salt River, and even larger ones, fish 27 to 28 inches or more weighing 10 pounds

or better, were reported in the 1940s, 50s, and 60s from the Snake River, but these fish were not aged. [17]

And what do they use to fuel this growth? For the youngest age classes, the dominant food items during the spring and early summer rearing seasons are the nymphs, larvae, and pupal forms of aquatic insects along with the first of the adult hatches, whereas late summer and fall brings a shift to the emerger and adult forms. Terrestrial insects, carpenter ants early and grasshoppers later on, also become important food items during summer and fall. The older age classes utilize these food items as well, but trout greater than 11 inches in length feed heavily on fish throughout the spring and summer rearing period. Two species of dace, two of chub, one shiner, three species of sucker, and the mountain whitefish inhabit this portion of the Snake River and its tributaries. Both the Paiute and mottled sculpins inhabit these streams as well, and are important food items of the larger trout, as are gastropods, worms and leeches, nymphs of the large stonefly species, and adult mayflies. Even such items as mice and voles are often found in trout of 15 inches and larger. [18]

As is the case with other salmonids, the onset of winter brings on changes in behavior and habitat use by finespotted Snake River cutthroat trout. Radio telemetry was used to study winter habitat use by these trout in the Snake River between Moose, Wyoming,

16. This may also account for a common lament, heard especially among visiting anglers, about the large number of "snits," i.e., undersize trout, encountered on fishing excursions along the Snake. The guides, fly shop owners, and local anglers don't have much patience with these complaints, because the larger trout, the ones that will put a real bend in your sturdiest rod, are indeed out there. But it's true, according to the Wyoming Game and Fish Department, that somewhere between 70 and 86 percent of the trout now caught by anglers on the Snake are 2- and 3-year-old fish that are 11 inches and smaller.

17. Simon (1946) and Mullan (1975) refer to the 10+ pound trout of the 1940s through the 1960s. Sizes of trout observed in the Salt River fishery are reported in Gelwicks et al. (2002).

18. For you fly-fishers in the audience, here is some information that may be of interest, even if the work is nearly 40 years old now. In 1968, the Wyoming Game and Fish Department initiated a study of the food habits of cutthroat trout between Jackson Lake and Palisades Reservoir. University of Wyoming graduate student L.E. Foster examined the collected stomach samples and reported his findings in 1978 (see

and the South Park Bridge on Highway 26/89/191, a distance of about 30 stream miles, during the winters of 1998–99, 1999–2000, and 2000–01 . River habitat in this reach is dominated by riffles and deep runs, with only infrequent pools and backwater habitat. The channel is moderately to extensively braided in places, but rip-rap flood control levees have reduced the number and complexity of side channels, large woody debris, and off-channel pools along this reach. [19]

The telemetry study found that in winter, the trout move into the deep runs, all but abandoning the riffles and few mainstem pools that exist in the reach. The trout also utilize off-channel pools with groundwater infiltration, especially when water temperatures drop to 32 degrees Fahrenheit (zero degrees Celsius) or below in the mainstem, which occurs frequently in Jackson Hole. Large numbers of trout, and mountain whitefish too, congregate in these off-channel pools at these times. Here they find water temperatures maintaining a comfortable 39 to 46 degrees Fahrenheit (4 to 8 degrees Celsius), owing to the modulating effect of the infiltrating groundwater, even as ice is forming in the mainstem runs.

The Stream-Resident Life History Strategy

Resident, non-migratory stocks are found in most, if not all, of the spring creek tributaries of the Snake and

Foster 1978; also summarized in Kiefling 1978). Foster discovered that there are some differences in feeding habits in different sections of the river. Just below the Jackson Lake dam, for example, the trout feed less on aquatic nymphs and larvae and more on emergers and adult insects. Fishes are also a less common item in the diet of trout in this section. The *Hydropsyche* caddisfly (common name spotted sedge), whose larvae are net-spinners, is the most commonly eaten caddisfly in most of the study area, but in the section from the mouth of the Buffalo Fork downstream to Wilson, the *Brachycentrus* caddis, commonly known as the Grannom, becomes nearly as important. Grannom larvae are case-builders, and the trout will eat them case and all. Stoneflies supplant the caddisflies as the most important food item in the section from Wilson to Astoria Hot Springs. The mottled sculpin is taken by trout the entire length of the Snake, but the Paiute sculpin appears in trout stomachs only in the section from Astoria Hot Springs to Palisades Reservoir. In that lower section of the Snake, from Astoria Hot Springs to Palisades, stomach contents also reveal some differences between morning and evening feeding. In the morning hours, the trout feed along the bottom, taking mostly stonefly and mayfly nymphs and caddis larvae,

while in the evenings they feed at or near the surface on drifting stoneflies, emerging mayflies, and chironomid pupae. Forage fish are eaten during all feeding hours.

19. Studies by the U.S. Army Corps of Engineers have confirmed these losses of floodplain habitat complexity. Furthermore, modeling done by the Corps indicates that channelization may result in continuing loss of habitat complexity within this reach in the future. This is important because habitat complexity has been linked to population abundance in other salmonid species; see Swales et al. (1985), Cunjak and Power (1986, 1987), and Quinn and Peterson (1996) for more details. The Corps of Engineers' report on habitat conditions on the Snake River in Jackson Hole and the feasibility of restoration is cited in the bibliography under U.S. Army Corps of Engineers (2000), and is available on the Internet at *www.nww.usacoe. army.mil/reports/jackson/report.htm.* The radio telemetry study of winter habitat use by finespotted Snake River cutthroat trout is in Harper and Farag (2004). Also, I refer you back to my discussion in Chapter 4 on winter stream conditions and habitat use by trout, and to the references cited in footnote 15 of that Chapter. See also Cunjak (1996).

This is Crystal Creek, a tributary of the Gro Ventre River. PHOTO BY AUTHOR

Gros Ventre rivers in Jackson Hole, and resident populations have been identified in spring creek tributaries of the Salt River as well. To my knowledge, no formal study has been made of any of these populations, but we do have a few tidbits of information about them from biologists working on other things, and from anglers who have written down what they observed. [20]

Adult trout belonging to stream resident populations disperse locally for spawning in their home streams. In Salt River spring creeks, the spawning period extends from early April to about the second week of June, with the peak in late May. This timing coincides with the spawning migration of the riverine trout into these same tributaries. The size of spawning adults is also about the same for both the resident and riverine populations—which raises the question: with this apparent overlap in size of adults and spawning time, how do the trout maintain their separate identities as discrete stream-resident and riverine-migratory stocks? Don't know, maybe they don't. Maybe the population structure and interactions between the trout in these streams are more complex than we comprehend. All we do know is that more adult cutthroats (sometimes many more) spawn in these tributaries than are accounted for by movements past weirs designed to trap, count, and tag the riverine migrants. The extra spawners must have reared within the tributaries. Redd counts have been recorded for these tributaries over many years, but no attempt has been made to determine if the two forms use different spawning areas.

Conditions for growth and development of the resident trout are evidently quite good in these spring creek tributaries, every bit as good, perhaps, as in the mainstem rivers. Author M.R. Montgomery described finespotted cutthroat trout observed in September in Fish Creek, a spring creek tributary of the Snake River that flows through the town of Wilson, Wyoming, as long as his arm and so well-fed that they were shaped more like footballs than fish. And noted angler and writer Ernest Schwiebert once said that for many years his best cutthroat was an 8-pound finespotted trout taken from Blacktail Spring Creek near Moose, Wyoming, on a carpenter ant imitation fished during an early mating flight of those insects.

Although they may thrive in the spring creeks, this particular subspecies may not be as well adapted as other cutthroat subspecies to life in small, swift-flowing, freestone mountain streams. Those who studied the Salt River fishery noted that in mountain tributaries of the upper Salt River, brook trout introduced many years ago have become the dominant species and have displaced the native cutthroats completely from some tributaries. Nor have finespotted Snake River cutthroats done well when stocked into small, swift, high-gradient streams

20. These include observations made by State of Wyoming biologists engaged in a comprehensive study of the Salt River trout fishery from 1995 through 1999 (see Gelwicks et al. 2002), and by researchers from the Wyoming Cooperative Fish and Wildlife Research Unit working on the spawning ecology of trout in Salt River spring creek tributaries (see Joyce and Hubert 2004). Angling writers who have reported their observations of finespotted Snake River cutthroat trout living in spring creek tributaries include M.R. Montgomery (see Montgomery 1995, pages 171–181) and Ernest Schwiebert (see Schwiebert 1978 at page 305).

outside their native range. Following up on angler complaints about the skinny, "snakey-looking" Snake River cutthroats in three such streams where they had been stocked in Utah's Uinta Mountains, James Mullan of the U.S. Fish and Wildlife Service found that the trout were indeed in poor condition when he checked them in the fall. He expressed shock over the "skin and bones," emaciated appearance of the trout from these streams, and the year following his investigation, he recommended to the Western Association of State Fish and Game Commissioners that finespotted Snake River cutthroat trout not be used to stock such streams. [21]

The Lacustrine Life History Strategy

Many lakes and ponds are found within the distributional enclave of the finespotted Snake River cutthroat trout, including several good-sized ones. There is Jackson Lake itself, and Jenny and Leigh lakes, and Bradley, Taggart, and Phelps lakes among others. All of these would certainly have hosted populations of trout adapted to a lake-dwelling life history, but again, no formal study has been made of any of these populations.

The similarities in life history and ecology between lacustrine populations of the finespotted Snake River cutthroat and those of other cutthroat subspecies are undoubtedly many. They likely respond to the same environmental cues to initiate their spawning migra-tions, moving to natal tributaries in the spring, from early April to about the second week of June in the Jackson Hole country. Females probably spawn for the first time at age 3 or 4 in these populations, the same as the fish from the mainstem rivers, and these females, along with larger, older, repeat-spawners, probably carry from 600 to upward of 2,500 eggs per female to the spawning tributaries. Their life span is probably in the order of 6 to 8 years, again similar to the riverine trout, and in this span of time in some of these lakes, the historical record indicates trout could grow to 14 pounds or more. These lakes still have cutthroat trout, but now they are augmented by hatchery stocking and are also populated with brown trout and lake trout.

Even riverine trout of this subspecies adapt well to lacustrine environments. Most, if not all, of the hatchery stocks of finespotted cutthroat trout originated from Snake River fish. They have been widely stocked in lakes and reservoirs, including the large Jackson Hole lakes and Palisades Reservoir. It is said that before Palisades Reservoir was completed, McCoy, Big Elk, and Bear creeks, now tributaries of today's reservoir, hosted significant runs of spawning cutthroats from the Snake River. Reservoir cutthroats still run up these tributaries to spawn, McCoy Creek in particular. McCoy Creek may be the major spawning tributary of the reservoir's hatchery-augmented finespotted cutthroat population. [22]

21. Mullan's report, "Condition (K) as Indicative of Non-Suitability of Snake River Cutthroat Trout in the Management of High Gradient, Low Diversity Streams," was presented to the Western Association of State Fish and Game Commissioners in 1975; see Mullan (1975).

22. My source for the little note on pre-reservoir spawning history of the streams now tributary to Palisades Reservoir is Bruce Staples' book, *Snake River Country: Flies and Waters* (1991). Gelwicks et al. (2002) also highlighted McCoy Creek as a principal spawning tributary for the cutthroats populating the reservoir today. However, brown trout numbers have grown in Palisades Reservoir, and lake trout are also present. Brown trout and lake trout are also present, along with hatchery-augmented populations of finespotted cutthroat trout, in Jackson, Jenny, Leigh, and the other large lakes in Jackson Hole.

Finespotted Snake River cutthroat trout of hatchery origin have been widely stocked in waters outside the historical range of the subspecies, including lakes and reservoirs of all different types, elevations, and trophic states, even warm, shallow, eutrophic desert impoundments seemingly more suitable for warmwater species than trout. They have adapted well to all of these environments, learning quickly to exploit whatever major food source is available, and shifting readily from bottom-oriented to open-water to surface-feeding as conditions warrant. At 12 to 14 inches in length, they focus on large organisms, such as fish and crayfish, if available, and growth rates increase rapidly thereafter on such fare. Dr. Behnke reported having raised finespotted cutthroat trout in a small pond on his property near Fort Collins, Colorado, where, on a diet containing abundant crayfish, they reached almost 22 inches in length and 4 pounds in weight by age 4.[23]

Another interesting aspect of the ecological adaptability of finespotted Snake River cutthroat trout stocked into lakes and reservoirs is their apparent ability to adjust their feeding niche to avoid direct competition with other populations. John R. Trojnar, one of Dr. Behnke's graduate students, investigated an example of this in a reservoir on a North Platte River tributary, Colorado, where finespotted cutthroat, greenback cutthroat, and brook trout had all been introduced (the North Platte drainage never had native trout). Trojnar found that the three populations partitioned the available food resources, with the brook trout utilizing bottom organisms, the greenback cutthroats feeding on benthic organisms, primarily *Daphnia* (exclusively *Daphnia* in August and September), and the finespotted Snake River cutthroats on surface prey items, mainly terrestrial insects. Dr. Behnke himself identified another example of this in Clinton Reservoir, near the Continental Divide separating the Colorado River drainage from the headwaters of the Arkansas River. In Clinton Reservoir, finespotted Snake River cutthroat trout had been stocked in a body of water already occupied by a native population of Colorado River cutthroat trout (Chapter 12). After only a year or two of sympatric existence, the two subspecies had partitioned the feed so that, again, the finespotted Snake River cutthroats were utilizing predominately surface items and the native Colorado River cutthroats were obtaining most of their food from the bottom.[24]

STATUS AND FUTURE PROSPECTS

Because of its ready adaptation to new environments and its ability to produce lunker-size trout that anglers flock to catch, the finespotted Snake River cutthroat became something of a favorite among recreational fishery managers in some states. It is now the most

23. One of the major studies of the feeding and growth of finespotted Snake River cutthroats in reservoir environments was carried out by P.T. Sekulich, one of Dr. Behnke's graduate students, and written up as a Master's thesis in 1974 (see Sekulich 1974). Dr. Behnke's own experience raising finespotted cutthroat trout in his home pond is reported in Behnke (1992) at page 102.

24. Trojnar's work was written up in a joint paper with Dr. Behnke in the journal *Transactions of the American Fisheries Society* (see Trojnar and Behnke 1974). Dr. Behnke's observations at Clinton Reservoir were included in a consulting report he prepared for the Climax Molybdenum Company (see Behnke 1979).

widely propagated of all the cutthroat subspecies, and the most widely stocked in waters outside its historical range. I have not been able to get a handle on just how widely, but it is present in the fabled fly-fishing waters of the Green River below Flaming Gorge Dam in Utah, and in many of the other lakes and impoundments in the states of Wyoming, Utah, and Colorado. Even the State of Maryland has finespotted Snake River cutthroats, and lists a hook-and-line angling record for cutthroat trout. The current Maryland record holder is a 6-pound, 9-ounce finespotted cutthroat taken from the North Branch of the Potomac River in 2000.

The finespotted Snake River cutthroat trout is also the only cutthroat subspecies that still dominates within its historical range—although all is not completely well in that world either, as I shall explain. It has not been replaced by non-native species, even though brown trout have become well established in some waters within the distributional enclave. The Hoback, Greys, and Salt rivers have established brown trout populations, as have Jackson, Jenny, Leigh, and the other large lakes of Jackson Hole. Rainbow trout, too, are present in some of these waters, including Bob Carmichael's favorite, the Gros Ventre. Also, I pointed out earlier that brook trout, stocked years ago, now are either dominant or have replaced the native cutthroats in mountain tributaries in the Salt River headwaters.

There are signs as well that this subspecies' historical resistance to introgression may be eroding. Increasing variability in spotting pattern, with the occurrence of more and more large-spotted individuals, is now being seen in the hatchery stock at the Jackson National Fish Hatchery, and also in trout taken into the hatchery from the Snake River to periodically infuse the hatchery stock with wild genes. Recent collections from the upper Snake River headwaters in Wyoming for population genetics studies have also turned up many large-spotted and intermediate-spotted individuals in waters that once were solidly within the finespotted enclave. [25]

The specter of introgression is troubling enough, but in addition, recent agency status reviews have lumped the finespotted Snake River cutthroat together with Yellowstone cutthroat as merely a finespotted form of the Yellowstone subspecies. [26] Since the U.S. Fish and Wildlife Service has already decreed that the entire Yellowstone subspecies comprises but a single DPS, effective actions to preserve the finespotted form may be slow to be taken, or not taken at all. Under these policies, the finespotted Snake River cutthroat could eventually be homogenized out of existence. Cutthroat trout would remain in the Snake River and the other

25. See Novak et al. (2005).

26. See May (1996), Dufek et al. (1999), May et al. (2003), and Hall (2006) for recent agency status reviews. Kaeding (2001, 2006) summarized the current U.S. Fish and Wildlife Service position that Yellowstone cutthroat trout comprises but a single DPS everywhere across the range of the subspecies, including the finespotted Snake River enclave. Kaeding (2006) also emphasized the position that taxonomic validation of the finespotted Snake River cutthroat as a separate subspecies remains the responsibility of taxonomists, geneticists, and other qualified scientists, not the U.S. Fish and Wildlife Service. To grapple with that taxonomic validity question, The Idaho Chapter American Fisheries Society sponsored a symposium early in 2006. Although no consensus was reached on whether the finespotted Snake River cutthroat trout deserves recognition as a distinct subspecies, everyone agreed that it "represents an evolutionarily and ecologically important group of individuals....and therefore is deserving of management and protection aimed at preserving its distinct morphological and ecological features." For the proceedings of this symposium (actually, only the extended abstracts of the talks presented, including Kaeding's) see Van Kirk et al. (2006).

waters of the distributional enclave, but they would no longer be the unique finespotted form.

In the previous section, I highlighted the several years of investigations sponsored by the Wyoming Game and Fish Department into the life history and ecology of finespotted Snake River cutthroat trout. These investigations revealed two significant habitat features that may be limiting factors for overall survival and abundance: 1) spawning habitat in spring creek tributaries, and 2) river habitat complexity, especially an abundance of off-channel pools with groundwater infiltration for overwintering during periods of extreme cold and ice formation.

The vital importance of the spring creek spawning tributaries was brought to light in the late 1960s, a period when spawner densities dropped to low levels. After increased stocking of tributaries with hatchery trout failed to improve these spawning runs, the Department concluded that the underlying problem was the loss of spawning habitat due to heavy siltation. Most of this was blamed on erosion caused by wildlife and livestock use of riparian areas, and the return of silted runoff water from irrigated fields. A program was initiated to rehabilitate the spring creeks by excavating resting pools, cleaning old gravel riffles, constructing new ones, trucking in clean gravels when necessary, and constructing catch basins for silt. Eggs were also placed in hatching baskets to augment natural spawning. This program succeeded. Spawning runs increased by two to almost five and one-half times, and in two cases, spawning runs were initiated in spring creeks that had never had them before. [27]

But sustaining these gains requires periodic maintenance, which is expensive. Most of the spring creeks remain vulnerable, and need additional rehabilitation from time to time to maintain their productivity. The Teton Conservation District and non-government organizations, such as Trout Unlimited, Federation of Fly Fishers, and the One Fly Foundation [28], have stepped up to help. However, the increasing popularity of the Jackson Hole area, with attendant increases in economic and commercial development, adds to the burden and to the need for continued attention to the spring creek spawning tributaries.

The second limiting factor for overall survival and abundance, particularly of those trout populations inhabiting the Snake River itself, is associated with these populations' need for an abundance of complex features, such as side channels, large woody debris, and, especially, off-channel pools with groundwater infiltration where the trout can congregate during critically cold, ice-forming conditions. What has happened along a lengthy reach of the Snake River extending from Moose, Wyoming, downstream to

27. John W. Kiefling of the Wyoming Game and Fish Department prepared a comprehensive report on the rehabilitation program for Snake River spring creek spawning tributaries (see Kiefling 1997).

28. A word or two about the One Fly Foundation and the very popular Jackson Hole One Fly competition: I've never been a fan of competition fly-fishing, preferring for myself the more contemplative aspects of the sport. But I must say that proceeds from the Jackson Hole One Fly, distributed by the One Fly Foundation, do go to help fund restoration and management programs that benefit coldwater fisheries, primarily in the domains of the finespotted Snake River and Yellowstone cutthroat subspecies. Here's how it works: The Jackson Hole One Fly is a two-day competitive event held the second week of September each year, in which contestants may use just one fly each day. If that fly is lost, the contestant is through scoring for the day. Forty teams of four persons each pay a $4,000 entry fee per team. Since its inception back in 1986, the Jackson Hole One Fly has been a major source of grant money for projects. In 2003, the One Fly Foundation funded more than $270,000 worth of projects. I don't have the figures for the

South Park, is that rip-rap levees built for flood control have tended to channelize the river, reducing the amount and complexity of these vital overwintering habitats. Furthermore, modeling studies conducted by the U.S. Army Corps of Engineers indicate that if left alone, the levees will continue to nudge the channel into a less complex condition, reducing overwintering habitat even more.

In 1990, Congress authorized the Corps of Engineers to study the feasibility of restoring habitat complexity and actually increasing overwintering habitat along this reach of the Snake River while still maintaining flood protection. Preliminary findings available by 1998 indicated that restoration was indeed feasible, so in 1998 and 1999, the Teton Conservation District (with the National Fish and Wildlife Foundation and Wyoming Game and Fish Department as partners) completed a demonstration project of restoration methods in a short reach of the Snake River just below the Wilson Bridge. With those results in hand, the Corps finalized its feasibility study in 2000, and recommended a comprehensive $52.3 million program (in fiscal year 1999 dollars) to complete the overall restoration. This program, to span 14 years of work to completion and be funded jointly by the feds (65 percent of the cost) and Teton County (35 percent of the cost), was slated to kick off in 2002. But that

years since then, but I'm sure they were equally substantial. In 2003, the One Fly Foundation threw in with the National Fish and Wildlife Foundation (I introduced you to that organization back in Chapter 4) to further leverage their dollars for stream improvement and fish conservation projects.

A view of the Teton Range and Jenny Lake, Grand Teton National Park. PHOTO BY AUTHOR

didn't happen. The Administration has not yet seen fit to include this program in the federal budget, and the small amount of funding that the Corps has been able to assemble has been only enough to make additional improvements in the demonstration project reach.

The Snake River within the distributional enclave of the finespotted Snake River cutthroat trout is rated as a Class 1 or blue-ribbon trout stream by the State of Wyoming, which designates the river as a stream of national importance deserving of the highest priority for protection. [29] Of its finespotted cutthroat trout, Fred Beal, former head of the Wyoming Game and Fish Division, once wrote "…we feel that this fish is part of Wyoming and should be held in as true a strain as is possible for coming generations." [30]

Palisades Reservoir on the Wyoming-Idaho border is the downstream distribution limit of finespotted Snake River cutthroat trout.
PHOTO BY AUTHOR

29. Concerned citizens from the angling, boating, and conservation communities are marshalling forces to help attain this protection. Early in 2006, I heard about the Campaign for the Snake Headwaters, whose goal is to persuade Congress to pass Wild and Scenic Rivers Act legislation for 38 segments of the upper Snake, Lewis, Buffalo Fork, Gros Ventre, Hoback, Greys, and Salt rivers, as well as several tributaries including Pacific, Crystal, Granite, and Cliff creeks. Although this effort is not just about the finespotted Snake River cutthroat, it would certainly benefit this subspecies and its habitat. To learn more about this campaign, or to throw in with them, go to the organization's website at *www.snakeheadwaters.org.*

30. Beal, F.W. 1959. "For the Cutthroat, a Brighter Outlook." *Wyoming Wildlife* 23, no. 8: 7–9. Before leaving this Chapter, I want to mention two additional programs that were started to encourage anglers to learn more and gain a better appreciation for native cutthroat trout. One is the State of Wyoming's Cutt-Slam program. What you do is catch each of Wyoming's four extant cutthroat subspecies in its native range in Wyoming (those would be the finespotted Snake River cutthroat, the Yellowstone cutthroat,

the Bonneville cutthroat, and the Colorado River cutthroat); submit a clear photo of each to the Wyoming Game and Fish Department for verification; and provide the Department with information on the date and location of each catch. When the "slam" is completed, the State issues you a certificate. For more details and an application, consult the Department's website at *http://gf.state.wy.us.* Navigate to Fish & Fishing, then to Cutt-Slam Program. The Federation of Fly Fishers started a similar program called Project Cuttcatch to raise awareness and appreciation of *all* of our native cutthroat subspecies. To claim this award, catch any four subspecies on artificial flies within the trouts' native range, provide a photo of the trout raised just far enough out of the water for it to be identified, then release the trout unharmed. Rules for Project Cuttcatch and other necessary paperwork can be found on the Federation's website at *www.fedflyfishers.org/Conserve/projects/Cuttcatch/Apply.htm.*

Bonneville Cutthroat Trout, *Oncorhynchus clarkii utah*, stream-resident form

CHAPTER 11

Bonneville Cutthroat Trout

Oncorhynchus clarkii utah: Chromosomes, 2N=64. There appear to be three slightly differentiated groups of *O. c. utah*: one associated with the Bonneville basin itself (Jordan, Provo, Weber, Ogden, Beaver, and Sevier river drainages); one associated with the Bear River drainage which includes the Logan River and Bear Lake; and one associated with the Snake Valley region on the extreme west side of the basin. Overall,

meristic character counts are similar to the Yellowstone cutthroat (Chapter 9). Scales in lateral series 135–190 with a mean value around 160 (lowest counts in Snake Valley trout and highest counts in specimens of the Bear River drainage). Scales above lateral line 33–46, mean value 38. Pyloric caeca 25 to 60, mean value 35 except for Bear River fish that average more than 40. Gill rakers 16–24, mean values 18 or 19 (Snake Valley fish have 18–24 gill rakers with means of 20–22). Vertebrae counts typically run 62–63 which is slightly higher than other cutthroat subspecies at 61–62. Snake Valley trout have many basibranchial teeth, averaging 20–28 but with some specimens having as many as 90, whereas other Bonneville cutthroats average 5–10 basibranchial teeth. Body

colors are generally dull, a trait shared with all other Great Basin cutthroat subspecies as well as with Yellowstone cutthroat trout, except that trout in streams of the Deep Creek range often develop reddish hues on the body. The major difference from the Yellowstone subspecies is in the tendency to develop larger spots that are more evenly distributed over the body in Bonneville cutthroats, whereas the spots tend to concentrate on the posterior in most Yellowstone cutthroat population. Snake Valley cutthroats have more spots and smaller ones than other Bonneville cutthroats, and present something of a stocky appearance with large but blunt-looking heads, thick bodies, shortened caudal peduncles, and dorsal and anal fins that appear oversized for the body.

Bear Lake Cutthroat Trout, *Oncorhynchus clarkii utah*

THE BONNEVILLE BASIN

Driving west along Interstate 80 across southwestern Wyoming, you are tracing a route that was a major arterial long before recorded history. Ancient peoples, and later the American Indians, used it in going from the high plains of Wyoming to the Great Basin. The Indians told the fur trappers, and they, too, came this way. So did the first wagon trains to Oregon and California. It was used by Brigham Young's Mormons in 1847, and later by the Pony Express. The overland stage followed it and so, too, did the transcontinental railroad. Today's travelers zip along on a modern highway.

From Fort Bridger you cross the Bear River divide near present-day Evanston, Wyoming. You can turn

north here, generally following the Bear River along the original route of the old Oregon Trail. The evidence of energy exploration and production is clearly visible in the dry hills of the Bear River divide as you leave Evanston, but presently you are driving through beautifully-kept hay fields and meadows full of grazing sheep and cattle. You climb a bit, and then, as you round a bend, comes the first startling glimpse of Bear Lake. It spreads out before you, its waters a distinctive robin's egg blue.

At Garden City, about halfway up the west side of Bear Lake, you intersect U.S. Highway 89. This leads you west again, over a shoulder of the Wasatch Range, to the headwaters of the Logan River. "Cache Valley is one of the most extensive and beautiful vales of the Rocky Mountain Range," wrote Warren Angus Ferris in July 1830 when he emerged from the mouth of Logan Canyon.[1] The mountain men first wintered here in 1824–25, and they held their yearly rendezvous in Cache Valley in 1826. They were all here, the whole colorful band: Jim Bridger, only 21 that summer, Tom Fitzpatrick, Bill Sublette, Jim Beckwourth, William Ashley, General David Jackson, Jedediah Smith. They loved this valley. It made a man seem welcome, they said.

If you elected to stay on the Interstate instead of turning off back at Evanston, you would continue west through Echo Canyon into Utah, then come to a point where the trail branches again. Bearing north here sends you through the Wasatch Range along the Weber River, exiting at the city of Ogden. Ogden was named after Peter Skene Ogden, whose Hudson's Bay Company brigade, along with the other fur trappers and mountain men, were the first whites to explore this area.

The south branch of the trail begins at the sleepy little town of Henefer, Utah. It takes you up East Canyon, then up and over the oak-canopied crest of the Wasatch Range. This is the way the Mormons came. If you know where to look and you are a careful observer, you can still see in places the ruts of the wagon wheels and hand carts. An advance party of the Brethren

emerged from Emigration Canyon to a breathtaking view of the Great Salt Lake and the spreading Salt Lake valley on July 19, 1847. Brigham Young, traveling with the main party, did not arrive until July 24th, by which time a few members of the scout group were enjoying a well-earned dip in the buoyant waters of the Great Salt Lake.

Emerging from the canyons along the west face of the Wasatch Range is a lot like stepping off the edge of the world. You are, literally, on the very brink of the Great Basin. There, far below you, is the basin floor with the waters of the Great Salt Lake stretching off to the horizon. When Jim Bridger saw this, and especially when he tasted the water, he thought he had discovered the Pacific Ocean.

What he had really found was but a dwindling remnant of an even more awesome phenomenon: a vast inland sea that had existed here many thousands of years before. It had lapped high up against the Wasatch Range and covered much of western Utah. You can still see the terraces that mark the ancient shorelines etched into the sides of the mountains.

Ancient Lake Bonneville was by far the largest of the Ice Age lakes of western North America. At its highest level, it had a maximum depth of about 1,000 feet and covered close to 20,000 square miles, making it comparable in size to present-day Lake Michigan.

1. Warren Angus Ferris was a trapper with the American Fur Company during the years when the fur trade was at its peak. He left an account of his adventures, called *Life in the Rocky Mountains*, which contains this quote (see Ferris 1940 [1983]). I also found the quote inscribed on a historical marker overlooking the Cache Valley, beside the highway leading to Logan, Utah.

The main body of the lake encompassed the Great Salt Lake, of course, and all of the Great Salt Lake Desert. It extended south to cover the Sevier Lake area, and a fingerlike bay extended southwesterly, filling what is now the Escalante Desert. Other major bays existed in the Snake Valley on the west, the Utah Lake area where the city of Provo now stands, and the Cache Valley, so beloved by the mountain men. Other Pleistocene lakes may have occupied this basin prior to the filling of Lake Bonneville, and it is said that Lake Bonneville itself went through at least four periods of high-low fluctuation during its existence. Final desiccation occurred about 8,000 years ago. [2]

Originally, the major tributaries of ancient Lake Bonneville were the Logan, Ogden, Weber, Provo, and Sevier Rivers. The Bear River, a tributary of the Snake River throughout much of the Pleistocene, was occasionally diverted into the Bonneville Basin as well. Then, a little over 34,000 years ago, a lava intrusion in a canyon of the Bear River diverted that stream southward into Lake Bonneville for a final time. The Bear River supplied as much water, perhaps even a bit more, than all the other tributaries put together. The ancient lake rose steadily until its surface stood at 5,100 feet above sea level. At this point, it found a weak spot in its northern rim at a place we now call Red Rock Pass. It breached the rim catastrophically, overflowing into the Snake River via the Portneuf channel. That ancient flood has been studied quite extensively by geologists. It had an impact on the topography far downstream, and the date of that event provides an important clue to the timing of the inland migration of the cutthroat trout.

Today, much of the floor of ancient Lake Bonneville is blistering desert, characterized by such harsh terrain as the Great Salt Lake Desert and the Bonneville salt flats. This country didn't exactly stop the first wagon trains bound for California, but it did hamper them so severely that after 1846, most detoured around it. Instead of setting out straight across the desert from Salt Lake City, the trail looped north via the Oregon Trail to Fort Hall on the Snake River, then south again to hit the Humboldt River well to the west of the bad country. The Pony Express, and later the overland stage route, skirted the worst of the desert by cutting south.

But the Great Basin is really basin-and-range country. The basin floor may be dominated by glistening, gray-white playas or dotted with sagebrush and juniper, but the ranges are something else again. Rising more or less gradually from the west, the highest peaks in these ranges often tower to 11,000 or 12,000 feet or more before falling away abruptly on their eastern faces. The geologists call these tilted fault block ranges. They rear up out of the valley, their ridge lines running generally north to south, separated from one another

2. For an excellent summary of the chronology of Pleistocene lakes of the Bonneville Basin, see Oviatt and McCoy (1992). Oviatt et al. (1987) and Scott et al. (1982) give additional details.

by maybe 30, 40, 50, or more miles of dry basin floor. Up in these ranges are worlds of oak brush, choke-cherry, mountain mahogany, aspen, and pine. Up here, too, are the habitats of blue grouse, bobcat, cougar, and mule deer, and in some of the ranges, bighorn sheep.

Stone Age people were able to eke out an existence along the shores of Lake Bonneville and in its basin after the lake dwindled away. These ancient peoples were displaced by other nomadic cultures. When the Spanish explorers and, a bit later, the fur trappers arrived, the dominant tribes of the Great Basin were the Utes in the Rocky Mountain areas along the eastern rim, the Shoshone, who generally inhabited the northern rim, the Goshutes (who call themselves "Kusiutta," a term some translate as "dry earth people"), who lived in the region southwest of the Great Salt Lake, and the Paiutes, who lived in loose, scattered bands throughout the basin. Today, more than 77 percent of the people are concentrated in a corridor along the western base of the Wasatch Range within 50 miles of downtown Salt Lake City. [3]

When those ancient Stone Age people occupied the territory, the ancestral large-spotted cutthroat trout swam the waters of the Lake Bonneville system. When the white explorers, trappers and Mormon settlers arrived, the Bonneville cutthroat was still plentiful in places such as Bear Lake, Utah Lake, and Panguitch

3. Beck and Jones (2000) summarize the evidence for early human occupation and subsistence in the Great Basin. Two good overviews of the Ute and Paiute Indians of the region are books by Holt (1992) and Simmons (2000). Information about the modern valley of the Great Salt Lake was drawn in part from an article titled "Salt Lake Valley's Leap of Faith" by L.M. La Roe, a senior editor for *National Geographic* magazine, that appeared in 2002. See La Roe (2002) for the complete citation.

Echo Canyon, an ancient Indian trail from the high plains of Wyoming to the Great Basin. Used by the ill-fated Donner-Reed party, later by the Mormons, and later still by the transcontinental railroad, it is now the route of Interstate 80. PHOTO BY AUTHOR

Lake; the Bear, Logan, Weber, Provo, Jordan, Beaver, and Sevier rivers; and in streams that fed off of the ranges around the perimeter of the Bonneville basin. But up until recently, you would have had to probe far back into the headwaters to find even traces of the remnant populations.

HISTORICAL RANGE

Figure 11-1 shows the Bonneville Basin and its major tributary river systems. I have also dotted in the shoreline of ancient Lake Bonneville. Also shown is what is currently believed to be the distribution of Bonneville cutthroat trout at the time of first contact with explorers, trappers, emigrants, and settlers. [4]

In Dr. Behnke's view, movement of ancestral $2N=64$ cutthroat trout into the Bonneville Basin is intimately linked with the pattern of drainage of the Bear River during Pleistocene time. [5] During much of the Pleistocene, the Bear River followed the Portneuf channel and was a Snake River tributary, so in due course the ancestral large-spotted cutthroats would have found their way into the Bear River system. And indeed, back in Chapter 2, I included the Bear River drainage as part of my upper Yellowstone distributional enclave for ancestral $2N=64$ cutthroat trout. However, the evidence indicates that the Bear River was diverted into the Bonneville Basin on more than one occasion, each

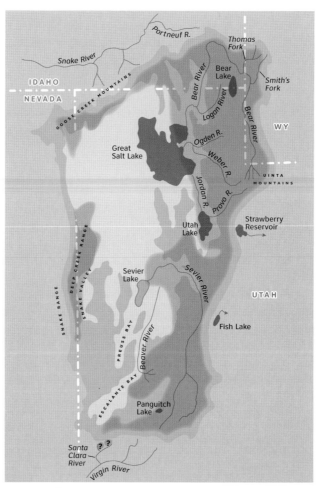

■	Existing Streams, Lakes, and Reservoirs
■	Ephemeral Lakes
■	Historical Distribution of Cutthroat
■	Bonneville Basin Boundary
■	Shoreline of Pleistocene Lake Bonneville at Its Maximum Size

FIGURE 11-1. *Bonneville drainage basin and probable historical distribution of Bonneville Cutthroat Trout*

4. The historical distribution shown in this map was taken from two recent status reviews of the Bonneville cutthroat trout, one compiled in 1996 by the U.S. Forest Service (see Duff 1996) and the other by the U.S. Fish and Wildlife Service in 2001 (see U.S. Fish and Wildlife Service (2001). Historical accounts of trout in this region include Townsend (1839 [1905, 1978]), Fremont (1845 [2002]), Stansbury (1852 [1988]), Burton (1862), Suckley (1874), Yarrow (1874), Cope and Yarrow (1875), Robinson (1950), Pratt (1970), Hickman (1978b), Cleland and Brooks (1983), Rawley (1985), and Chavez and Warner (1995). But this interpretation of the historical range of the subspecies also includes waters that were deemed suitable habitat for trout historically, even if not mentioned in early accounts of fish. Many were dewatered for irrigation by early settlers. The Mormons, who settled most all of this territory, kept excellent records. These were tapped for compilations of early irrigation diversions and water withdrawals by Kendrick (1984) and Utah Division of Wildlife Resources (1991).

5. Back in Chapter 2, I pointed out that an alternative view of ancestral cutthroat dispersal and subspeciation has been

diversion probably enabling a fresh invasion of ancestral cutthroat trout. The final diversion of the Bear River, setting up the present drainage pattern, occurred at little more than 34,000 years ago.

The trout found the vast and abundant waters of Lake Bonneville much to their liking. They spread, occupying all adjoining watersheds on every side of the basin. The Sevier River at the south end of the lake, the streams coming off the Snake Range and Deep Creek Mountains to the west, the many streams flowing off the Wasatch Range on the east—all of these waters teemed with cutthroats.

In the Bear River system, the fish occurred in every tributary stream from the Salt River range to the Uintas. Bathtub-shaped Bear Lake, which sits astraddle the Utah-Idaho border, was formed about the time that Lake Bonneville was at its height (not by intrusion of water from Lake Bonneville, but by separate geological events), and trout from the Bear River became established there too. [6]

As previously mentioned, scientists believe Lake Bonneville underwent at least four periods of high-low fluctuation during its existence, with final desiccation occurring about 8,000 years ago. During these fluctuations and later, as desiccation progressed, the various drainages in the basin were separated from one another, isolating their cutthroat populations.

These isolated populations dwindled in numbers and retreated to the higher mountain headwaters as a natural consequence of the climate becoming drier. But even so, when the white man came to the basin, the fish were still widely distributed, and in good numbers. They still inhabited the Logan, Weber, Jordan, Provo, and Sevier River watersheds. They were still present in Bear Lake and throughout the Bear River drainage. Populations of trout thrived in Utah and Panguich Lakes. Trout were also present in the Beaver River, which flows out into the Escalante Desert, and in the streams of the Snake Valley drainage.

The Indians had relied heavily on the native trout of the basin for food. South and east of here, both the Navajo and Apache peoples had taboos against eating fish. But the Paiutes and the Utes, now, they had no such taboos. Take Panguitch Lake, for example, near the southern end of the historical distribution. The name Panguitch for the lake, its creek, the range in which it lies, and the Mormon settlement 19 miles east of the lake is said to come from a Paiute word meaning "big fish." Trout averaging 3 pounds in weight but with larger ones reaching several times that were so plentiful that the Indians could obtain an ample supply by simply walking the shore and spearing them. [7]

When the settlers arrived, they continued to seine the trout out of the lakes and streams, for their own use

expressed by the University of Michigan's Gerald. R. Smith and his colleagues. This group has proposed that cutthroat trout may have originated in the interior and dispersed from the Bonneville Basin into all corners of the species' historical distribution. See footnote 25 of Chapter 2 for the key reference.

6. The earliest geologist to examine this region was A.C. Peale, who served with the Hayden survey in 1877. In his report (see Peale 1879), he concluded that the Bear Lake valley had been occupied by three earlier lakes, one in Pliocene time, one in the early Pleistocene, and one in the late-Pleistocene. Mansfield (1927) and Richardson (1941) describe just a single Pleistocene-age lake in the Bear Lake valley.

7. Two good overviews of the Indians of this region and how they subsisted can be found in Virginia McConnnel Simmons' book on the Ute Indians of Utah, Colorado, and New Mexico (see Simmons 2000) and R.L. Holt's book on the ethnohistory of Utah's Paiutes (see Holt 1992). Mentions of Apache and Navajo taboos against eating fish are in Miller (1961), Reichard (1974 [1983]), and Matthews (1994). But it wasn't just the Paiute and Ute Indians who caught large numbers of

at first, but later to supply mining and railroad camps. They also established extensive irrigation systems that diverted the water and disconnected and altered the habitat. The result, as might be expected, was a severe decline in the native trout populations to the point where they continued to exist only in the most remote mountain headwaters. As the native trout populations declined, hatchery operations commenced, often aided by the U.S. Commission of Fish and Fisheries. Exotic species, including rainbow, brook, and brown trout as well as other subspecies of cutthroat trout, were introduced, which hastened the replacement and introgression of pure populations of native trout. When the Mormons arrived in the Bonneville basin in 1847, native trout were bountiful. Just a little over one hundred years later, they were believed to be extinct in pure form. [8]

Fortunately, that was not the case. Pure populations of Bonneville cutthroat trout were indeed few and far between, but when the meristic characters of populations collected in the 1950s and 1960s were compared with authentic Bonneville Basin specimens that had been preserved in museums, a handful of apparently pure populations were discovered. Diligent searching of the tiny, remote streams around the periphery of the historical distribution has been ongoing, with the result that more and more pure populations have been found—not great numbers of fish, mind you, but

trout from Panguitch Lake. Mormon pioneer John D. Lee recorded in his diary that he "caught some 300 trought" [Lee's spelling, not mine] at the lake during a one-day trip in September, 1872 (see Cleland and Brooks 1983). And H.C. Yarrow, a surgeon-naturalist serving with the Wheeler surveys west of the 100th meridian, stated that he himself had no trouble taking "from thirty to forty pounds weight in a single hour's fishing" (see Yarrow 1874). Cutthroat trout are still listed as being present in Panguitch Lake, but nowadays anglers are attracted to the lake for its rainbow, brook, and brown trout.

8. The decline in native trout populations was well documented, as were the advent of hatchery operations and the introduction of non-native species. You can trace this history for yourselves in Yarrow (1874), Stone (1874), Siler (1884), Woodruff (1892), Popov and Low (1950), and Sigler and Sigler (1986). A note of possible historical interest is that the Bonneville cutthroat may well have been the first of the species to be reared in a public hatchery. In 1872, Dr. Livingston Stone of the U.S. Commission of Fish and Fisheries visited the Salt Lake City municipal trout hatchery where he discussed trout propagation with the superintendent, Mr. A.P. Rockford. The trout being propagated came from Bear Lake, the Bear River, Utah Lake, and various streams in the mountains near Salt Lake City. The view that the Bonneville Basin native cutthroat had gone extinct was first expressed in a paper by Cope (1955) and then in publications by Platts (1957) and Sigler and Miller (1963).

A roily Bear River flows into Cache Valley, a place much loved by the mountain men. The old Oregon Trail followed Bear River north to Soda Springs, Idaho, where the river loops south for its run into Cache Valley and on to the Great Salt Lake. PHOTO BY AUTHOR

enough here and there to nurture the hope of rebuilding and maintaining the subspecies.

Based on meristic counts, allozyme electrophoresis, and mitochondrial DNA analysis of populations discovered within the historical range,[9] Dr. Behnke reckons that the native cutthroat trout of the Bonneville Basin comprises three slightly differentiated groups associated with:

1) The Bear River drainage, which now includes the Logan River, and Bear Lake.

2) The eastern Bonneville Basin, including the Ogden, Weber, Provo, Jordan, Beaver, and Sevier river drainages.

3) The western basin, which includes the Snake Range and Deep Creek Mountain populations.

The group associated with the Bear River drainage and Bear Lake is very little different from Yellowstone cutthroat trout. This may reflect the most recent (approximately 34,000 years) diversion of the Bear River from the Snake River drainage into the Bonneville Basin. The other groups may reflect more ancient invasions. The eastern Bonneville Basin group shares a protein allele (but no mitochondrial DNA features) with the Colorado River cutthroat trout (Chapter 12), with which it also shares a distributional boundary along the crest of the Wasatch Range.[10]

Two headwater populations of Bonneville cutthroat trout occur outside the Bonneville Basin in tributaries

9. Dr. Behnke made the initial comparisons between authentic museum specimens and cutthroat populations collected in 1959 using meristic character analysis. Subsequent work with additional specimens was done by his students. The progression of this work can be tracked in Behnke (1970), Murphy (1974), and Hickman (1978b). More recently, population genetics studies using allozyme electrophoresis and mitochondrial DNA analysis have been carried out by Klar and Stalnaker (1979), Loudenslager and Gall (1980b), Martin et al. (1985), and a group led by Dennis Shiozawa at Brigham Young University; see Shiozawa et al. (1993) and Shiozawa and Evans (1994a, 1994b, 1995).

10. The U.S. Forest Service and U.S. Fish and Wildlife Service have opted to delineate the Bonneville Basin cutthroat subspecies into four groups based on what they call *hydrogeographic areas* or *geographic management units* (GMUs). These seem to be nothing more than convenient groupings of contiguous watersheds. Their Bear River GMU encompasses all waters within the Bear River drainage in Utah, Idaho, and Wyoming from the Bear River headwaters to its confluence with the Great Salt Lake. Their Northern Bonneville GMU includes the waters of the Ogden, Weber, Jordan, and Spanish Fork rivers and the Utah Lake sub-basins. Their Southern Bonneville GMU includes the Sevier River, Sevier Lake, and Escalante Desert sub-basins as well as the upper Virgin River sub-basin. Their Western Bonneville GMU includes the western drainages along the Utah-Nevada border from the Raft River Range in the north to the Snake Valley area in the south, including several westward-draining stream systems in Nevada where Bonneville cutthroat have been transplanted. For more details on these delineations, see Duff (1996) and U.S. Fish and Wildlife Service (2001).

Bear Lake on the Utah-Idaho border.
PHOTO BY AUTHOR

of the Santa Clara River of the Virgin River drainage. It's not known for sure if these populations are native or introduced, so I have designated them with question marks in Fig. 11-1. However, early accounts have been cited to the effect that cutthroat trout were present in the Santa Clara River in Pine Valley and Grass Valley in 1863. Only a low and gentle divide separates Pine Valley from the Bonneville Basin, so headwater stream transfer during a long-ago wetter clime could account for these out-of-basin populations. Another early account suggests that trout may have inhabited upper tributaries of the North Fork Virgin River as well, upstream from Zion National Park. [11]

Another handful of Bonneville cutthroat populations occur just outside the Bonneville Basin in streams flowing into Nevada off the western slopes of the Snake Range. I didn't show these in Fig. 11-1 because these populations, unlike the populations in the Santa Clara drainage, are known with certainty to have been introduced. At least one of these introductions was accidental. Pine Creek, on the west side of the Snake Range, in Nevada, was most likely invaded by trout after completion of the Osceola Ditch, a waterway built by miners to draw water from Lehman Creek, also a Snake Range stream but flowing east off the slopes of Mt. Wheeler into the Snake Valley. In other instances, ranchers, desirous of having trout to be caught a little closer to home, may have moved fish deliberately.

At any rate, the Pine Creek population persisted, despite the fact that the creek itself is small and offers very limited trout habitat, not to mention being vulnerable to flash flooding. In 1953 and again in 1960, the Nevada Fish and Game Department transplanted trout from Pine Creek into other small streams, namely Hampton Creek, which is a Snake Valley tributary, and Goshute Creek, which flows off of the Cherry Creek Mountains into a separate basin north of Ely. But Goshute and Hampton Creeks are themselves vulnerable to flash flooding, so in 1977, the Nevada authorities stocked cutthroats from Goshute Creek into several other small streams in the same area for extra insurance.

LIFE HISTORY AND ECOLOGY

Based on the old reports and historical accounts, the native cutthroat of the Bonneville basin must have existed in all three of the interior life history forms:

- The lacustrine, or lake-associated, life history, wherein the fish feed and grow in lakes, then migrate to tributary streams to spawn.

- The fluvial, or riverine (potamodromous), life history, wherein the fish feed and grow in main river reaches, then migrate to small tributaries of these rivers where spawning and early rearing occur.

- A stream-resident life history, wherein the fish complete all phases of their life cycle within the streams of their birth, often in a short stream reach and often in the headwaters of the drainage.

11. These early-day accounts were cited by Robert Rush Miller in a paper he wrote about man's alterations of the native fish fauna of the American southwest; see Miller (1961).

The Lacustrine Life History Strategy

There is little question that this was a predominant life history form of Bonneville cutthroat, even in historical times. Utah Lake, which remained in the Bonneville basin after the desiccation of ancient Lake Bonneville and Bear Lake, which is not a remnant of Lake Bonneville but was formed separately during the Pleistocene, continued to support abundant cutthroat populations when the first settlers arrived. Panguitch Lake, so often mentioned in the historical record, is a mountain lake whose drainage is tributary to the Sevier River. It, too, had an abundant cutthroat population. Trout from these lakes were large and so plentiful that they seemed inexhaustible. The Utah Lake population was such an important food supply for the settlers that one early legislator proposed that it be revered along with the seagull in the historical record of Utah. [12]

Unfortunately, the supply of these lake dwelling cutthroats was not inexhaustible. As early as the 1870s, concerned writers were reporting on the alarming decline of the Utah Lake population. Thus it is that the only information available on the life history and ecology of this important trout population is what can be inferred from the early descriptions of scientific workers, such as George Suckley, who first used the name *Salmo utah* for the trout of Utah Lake in 1859, and H.C. Yarrow who reported in 1874. Suckley, a fly-fisherman as well as a naturalist, reported that the trout ran up the Timpanogus River (now the Provo) to spawn. "They are said to grow occasionally to 30 inches in length and are an active, fine fish," he added, "affording much sport to the fly-fisher, and a delicacy to the epicure." [13]

Suckley, in naming these trout *Salmo utah*, was merely trying to differentiate the trout of Utah Lake from cutthroat trout he had observed elsewhere in the Bonneville basin. The trout of Utah Lake were silvery in color, possibly the result of their living in an alkaline lake environment, and the spotting pattern was affected as well. The lake specimens evidently had only a few small speckle-like spots rather than the large, rather pronounced spots of trout found in the mountain streams. These factors, along with the great size of the trout from Utah Lake, made the latter seem quite distinct from other trout populations in the basin. All the other cutthroats were called, collectively, *Salmo virginalis* at the time. [14]

In his report to the U.S. Fish Commissioner in 1874, H.C. Yarrow described the Utah Lake cutthroat as voracious and non-selective in its feeding habits. Trout stomachs contained an eclectic array of items including invertebrates, snakes, frogs, and small fishes. Elaborating on the fishes eaten, Yarrow wrote:

> "The trout is very voracious devouring other fish smaller than itself, particularly a species locally known as silver-sides, of from two to six inches in length; on dissection, I have found the stomach of the trout crammed with these little fish."

12. This was cited in Hickman (1978b) from an essay titled "The Grasshopper Famine, the Mullet and the Trout" by P. Madsen, that appeared in a 1910 volume of *Improvement Era*, a publication of the Young Men's Improvement Association, Salt Lake City. The Bonneville cutthroat trout eventually did receive its recognition, in 1997, when it was designated the Utah State Fish to replace the rainbow trout, a non-native species. Utah Lake has an additional distinction. After the desiccation of Lake Bonneville, it became the sole remaining refuge of a Miocene-age species of sucker, *Chasmistes liorus*, commonly known as the June sucker, a sister species to the cui-ui of Pyramid Lake (Chapter 5). Another sister species, the shortnose sucker *C. brevirostris*, holds out in Upper Klamath Lake, Oregon. All three are listed as endangered under the U.S. Endangered Species Act. But the June sucker of Utah Lake may no longer exist in pure form, having hybridized with the more common Utah sucker *Catostomus ardens*.

13. In Suckley (1874).

14. Girard (1856) assigned the name *Salar virginalis* to trout specimens collected by a railroad survey party from Ute Creek near Fort

Although Yarrow's report didn't identify the "silver-sides," it was probably the Utah chub.

As an historical note, it is recorded that in 1864, one haul of a commercial fishing net in Utah Lake came up with somewhere between 3,500 and 3,700 pounds of trout. In 1930, only one solitary cutthroat was caught in the lake during the entire fishing season, and none were seen after that. Today, Utah Lake is not even trout habitat. Walleyes, largemouth bass, white bass, channel catfish, and black bullheads now swim its depths, and when you see the lake, it's hard to believe that those murky, chalk-green waters could ever have been clear and blue.

To learn about the lacustrine life history of Bonneville cutthroat today, one has to look to Bear Lake on the Utah-Idaho border; to an out-of-basin introduction of Bear Lake cutthroats into Strawberry Reservoir on the Colorado River side of the Wasatch crest; and to Lake Alice, a mountain lake in the Smith's Fork drainage of Wyoming where a population of Bonneville cutthroat trout isolated by a landslide has adapted to a lacustrine lifestyle.

As indicated a couple of times already, Bear Lake was not part of ancient Lake Bonneville, but was formed by separate geologic events during the Pleistocene. It, too, has had its ups and downs in water level, with direct connection with the Bear River during some periods of its existence but disconnected during other periods, and evidence of the different ancient shorelines still present around the borders of the lake.

The history of the Bear Lake fishery from the time of the first records until, say, the World War I era, parallels that of Utah Lake. "Bear Lake is alive with natives—redthroats," proclaimed an article about angling in the State of Utah that appeared in 1906. These trout were also netted for the commercial market in quantities of 500 to 2,000 pounds of fish per day during the season, which coincided with the annual spawning run. Utah legislated against commercial netting in 1897, but Idaho did not. Utah Fish Commissioner Sharp, writing in 1906 about the continuing netting on the Idaho side of the border during the spawning run, minced no words about his feelings:

"Though these trout live in the lake, they seek the streams for spawning. They spawn in June, seeking the Lake shores and following them till they are caught in the nets or escape them and find their spawning beds up the river. Of course these vandals with their nets take good care that few, if any, shall escape. For a mile on both sides of the lake, leading to the river, nets are set, and then, to make assurance doubly sure, someone has staked a battery of nets about 200 yds above the mouth of the river. None but the ignorant or criminal would do this—it is the same kind of a fellow who would kill a setting hen and sell her, or slaughter a cow with a calf." [15]

The spawning streams used by these Bear Lake cutthroats are not really rivers, as Commissioner Sharp's

Massachusetts in Colorado's San Luis Valley. This is in the Rio Grande drainage (Chapter 14). For years it was mistakenly assumed that Ute Creek was in the Bonneville basin, so the name *virginalis* was used for the Bonneville cutthroat trout as well. This usage continued in the literature until the error in the location of Ute Creek was pointed out by J.O. Snyder in 1919, in a footnote to a paper on another fishery topic. But Suckley, writing in the 1859–61 period (although his monograph was not published until 1874) and not knowing of this error, continued to think of all the trouts of the Bonneville basin as *virginalis* except for the trout of Utah Lake, which he named separately as *Salmo utah*.

15. Commissioner Sharp's acerbic comments were quoted by a correspondent who signed himself only as "Gage" in an article titled "Utah's Trout Waters" in the July 14, 1906 issue *of Forest and Stream* magazine. "Gage" is also responsible for the earlier quote about the native "red-throats" in Bear Lake. See "Gage" (1906) for the full article.

diatribe indicated, but creeks. St. Charles Creek in Idaho and Swan Creek and Big Spring Creek in Utah are the principal ones. Commercial netting and diversion of these tributaries for irrigation by the early Mormon settlers dealt severe blows to the cutthroat population of Bear Lake. Egg-taking operations for stocking into other Utah and Idaho waters were also conducted at Bear Lake for many years, further mining the native trout population. But despite all this, angling for cutthroat trout held up fairly well in Bear Lake until after World War I, sustained mainly, it would appear, by the escapement into Swan and Big Spring creeks where netting was prohibited. The average weight of the trout taken from the lake was 4 pounds in those years, but 10- to 19-pounders were not uncommon in commercial nets and trout of 22 and 23-¾ pounds were reported in 1913.

But after that, things went downhill in a hurry. The native cutthroat population fell away to almost nothing, and despite the stocking of a wide array of other salmonids, including four species of Pacific salmon, landlocked Atlantic salmon, lake trout, brook trout, brown trout, rainbow trout, and Yellowstone cutthroats, by the 1940s the lake had a poor reputation among anglers as a fish producer. Rainbow trout did reasonably well in the lake for a time and did contribute to the sport fishery, as did the lake trout, but none of the other

stocked species contributed much. And even the lake trout, which spawns on shoals in the lake, cannot sustain a population through natural spawning; augmentation by stocking from hatcheries is required. The stocking of rainbow trout and non-native cutthroats also raised the specter of introgression of any of the native Bear Lake cutthroats that might still have survived.

By all rights, the native Bear Lake cutthroat should have gone the way of the Utah Lake cutthroat. And so indeed did it seem. In 1939, 1940, and 1941, aquatic biologists set nets in Bear Lake in all areas and at all depths and never captured a single native cutthroat. But occasionally a fisherman would take one, and finally the biologists' nets began to pick up a few. The natives were easy enough to recognize. They were silvery in overall color, except that the backs and especially the tops of the heads often took on a distinctive robin's egg blue color that earned them the nickname "bluenoses." Spotting was typically sparse, with the few spots they did have being round in shape and concentrated above the lateral line and behind the dorsal fin. Males developed a rosy-orange hue at spawning time.

It wasn't until the early 1970s that the State of Utah (this time with Idaho's cooperation) decided to act to repopulate Bear Lake with the native "bluenoses." State biologists set up a program to take spawn from the best phenotypical representatives of the native cut-

Bear Lake cutthroat trout spawn in St. Charles Creek, Idaho, shown here, and in Swan Creek over the line in Utah. A stream-resident population of Bonneville cutthroat trout occurs in St. Charles Creek in its upper reaches.
PHOTO BY AUTHOR

throat they could find in the spawning streams. They did allow some spawners to escape to take advantage of whatever natural production the spawning streams could provide, and they also undertook habitat restoration projects to improve the tributaries for spawning. But the main emphasis was placed on hatchery production. The little fish would be hatched and reared in a hatchery, then restocked into the lake.

Utah's Bear Lake Cutthroat Trout Enhancement Program has been in place since 1973, and fishing in the lake made a spectacular comeback (although it has suffered again in recent years, owing to the severe drought conditions that have plagued the region). The lake hook-and-line record, a cutthroat trout of 18 pounds, 15 ounces, was actually taken in 1970, prior to the inception of the program, by an Idaho angler. But since the program started, the percentage of fish that would rival that record had increased steadily until the recent drought, not to mention the increased numbers of smaller cutthroats, fish in the 2- to 5-pound range, available to anglers. [16]

Bear Lake is a sparkling, azure-blue jewel set in a field of sage-gray. The lake itself is unique in several respects. It is an oligotrophic body of water about 20 miles long and 8 wide, sitting at an elevation of 5,922 feet. Average depth is 92 feet and maximum depth is 208 feet. It is ice-covered about four winters out of five.

It became detached from the Bear River at some point back in the late-Pleistocene, perhaps as long as 28,000 years ago, and has since depended on precipitation and inflow from several small tributaries and groundwater sources to maintain its level. [17]

Its water chemistry is unique as well. It is high in dissolved solids (456 ppm on the average) and very alkaline (pH 8.4 to 8.6). But unlike most alkaline lakes, which derive their alkalinity from high levels of calcium, Bear Lake's is the result of high levels of magnesium.

Shortly after the turn of the century, the Telluride Power Company (now PacifiCorp) reconnected the Bear River with Bear Lake by diverting the river into Mud Lake, aka Dingle Marsh, a wetland area at the north end of Bear Lake, and from there into Bear Lake itself. A pumping station was constructed and an outlet canal was dug to carry discharge water from Bear Lake back to the Bear River. This diversion has had some impacts, mainly on water quality. In 1993, one of the old dams collapsed and was rebuilt in such a way that silt and sediment from the Bear River could flow directly into Bear Lake, rather than filtering through the Mud Lake wetland as the diversion had been set up previously. Also, fishes not originally found in Bear Lake have invaded from the river, such as the Utah chub, common carp, yellow perch, green sunfish, speckled dace, and redside shiner. Historically, only six

16. I have on my desk a stack of Utah Division of Wildlife Resources reports on the Bear Lake Cutthroat Trout Enhancement Program covering every year from 1975, the program's third year, to 1996. These were written by Bryce R. Nielson, the project leader, either as sole author or jointly with one or more of the project biologists. These make for interesting but detailed and protracted reading, so rather than citing them all, I'll direct you to an excellent summary of the program prepared by Nielson and Leo Lentsch for the American Fisheries Society Symposium, "Status and Management of Interior Stocks of Cutthroat Trout," proceedings published in 1988; see Nielson and Lentsch (1988) for the citation. This paper also has good information about the life history of the Bear Lake cutthroat population. Kershner (1995) has a discussion of life history characteristics that includes information about spawning in St. Charles Creek, Idaho. Other information about the life history and ecology of Bear Lake cutthroat trout can be found in papers by Ruzycki et al. (2001) and Wagner et al. (2001).

17. More on the geological history of Bear Lake and on the hydrology and present trophic state of the lake can be found in Robertson (1978) and Lamarra et al. (1986).

species inhabited the lake: Bonneville cutthroat trout, Utah sucker, Bonneville cisco, Bonneville whitefish, Bear Lake whitefish, and Bear Lake sculpin. The last four are found nowhere else. The Bonneville cutthroat trout evolved to be the lake's top predator. [18]

Spawning runs from Bear Lake begin around the middle of April, shortly after ice-out. The run peaks in May, then continues on a diminishing basis through the month of June. In St. Charles Creek, the lacustrine trout spawn in the lower one-third of the stream where the channel is characterized by high sinuosity and gradients of less than 1 percent, typically choosing pool tailouts to dig their redds. It has also been discovered that St. Charles Creek has a population of stream-resident Bonneville cutthroats that spawn in upstream reaches where gradients run between 1.5 and 3.5 percent and the channel is less sinuous. The stream-resident trout also spawn somewhat earlier, in April and May.

Bear Lake cutthroats evidently mature late. Some reach sexual maturity at age 4 or 5, but most do not spawn for the first time until they are 6 years of age or older. The age range of first-time spawners can extend to age 11 (work carried out in the 1990s by scientists from Utah State University pegged the maximum age for these trout at 14 years). Spawning runs are dominated by first-time spawners 21 to 24 inches in total length and averaging 4.5 pounds in weight, with each female carrying about 2,600 eggs. Repeat spawners typically account for less than 4 percent of the run.

Although there is a limited amount of natural production from the spawning tributaries (perhaps as much as 10 percent of the angler catch in some years, according to the Utah Division of Wildlife Resources), no research has been done on the early life history of naturally spawned trout at Bear Lake. Because the lake is oligotrophic and the young trout would have to compete with all the other fishes for the limited available invertebrate food if they recruited to the lake as young-of-the-year, it is assumed that they rear for 1 or 2 years in the natal streams before migrating to the lake. For the hatchery-produced trout, as I understand the current process, the hatcheries push to raise their trout to a total length of 175 mm (just under 7 inches) by the time they are yearlings, for release into the lake in the spring. But sometimes, if hatchery space and costs permit, trout are held over until fall and released as advanced yearlings at an average size of 200 mm (just under 8 inches). The trade-off here is between survival of the juveniles in the lake and the cost of rearing in the hatcheries; the larger juveniles survive better.

Once in the lake, the little Bear Lake cutthroats start off mainly (but not exclusively) on a diet of insects. Terrestrial insects are available to trout feeding on

18. In company with many of the world's large, deep, oligotrophic lakes, Bear Lake has one other curious distinction and that is the story of a serpent-like creature that lurks in its depths but occasionally allows itself to be seen on the surface. The stories of the Bear Lake monster go back to 1868 and a series of articles written by Joseph C. Rich, a Mormon pioneer of the Bear Lake settlement period. These articles, appearing in Salt Lake City's *Deseret Evening News*, claimed several upstanding citizens (but not Rich himself) had seen the creature. Sightings have been reported off and on ever since. It has now become something of a tourist attraction. In the summer of 2004, one enterprising concessionaire even operated a serpent-shaped pontoon boat for tours of Bear Lake.

or near the surface during the warm months, but the principal aquatic insects present in the lake are bottom-dwelling larvae. The little trout utilize both, but show their flexibility by also preying on small specimens of Bonneville cisco and Bear Lake sculpin.

The trout grow reasonably well on this fare, typically reaching 11 to 12 inches in total length at age 2. At about this size, a major shift in their diet occurs. Now they become primarily fish eaters. They will take sculpins in large numbers during the spring when the latter are spawning, and they will also prey on newly stocked cutthroat fingerlings, lake trout juveniles, and any of the other fish species present in the lake. But their major prey item from this time forward is the Bonneville cisco.

The Bonneville cisco is a schooling fish. The cisco schools move freely about the open waters of the lake for most of the year, and the Bear Lake cutthroats follow these schools, feeding and putting on weight even in the winter months. Knowledgeable Bear Lake anglers have learned this, and find the cutthroats by tracking the cisco schools with depth finders and fish locators. Most boat anglers fish the lake by trolling Countdown Rapalas and the like. Fly-fishers don't fish the lake in any great numbers, although those that do have reported good success at times by fishing big streamers from float tubes or kick-boats. Of course,

anybody who is going to venture out on Bear Lake is well advised to use caution. The lake can get pretty rough on occasion, and there are few natural bays, points, or other structures to offer shelter.

Strawberry Reservoir sits at an elevation of 7,602 feet in the headwaters of the Strawberry River east of the Wasatch crest, just outside the Bonneville Basin in the Colorado River drainage. Although there may have been a dam and reservoir here as early as 1904, the official version is that the existing reservoir was built in 1922 to store and deliver water to the Bonneville Basin for irrigation and for industrial and domestic uses. The reservoir was 8,400 acres in size initially, but was enlarged to 17,164 acres in 1985. In the 1930s, the reservoir was stocked with Yellowstone cutthroat trout. Rainbow trout were also introduced. These trout spawned in the tributaries and interbred with native Colorado River cutthroat trout to produce a hybrid strain, the "Strawberry cutthroat," that prospered in the relatively shallow, fertile waters of the original reservoir. Indeed, the Utah state hook-and-line record cutthroat, a trout of 26 pounds, was captured there, and for many years this strain was the primary source of cutthroat eggs for Utah's statewide fishery management programs. But Utah chubs, Utah suckers, carp, yellow perch, and redside shiners also got into the reservoir, and by the 1950s were so numerous that neither the hybrid cutthroats nor the rainbow trout could compete. A rotenone treatment followed by restocking with Strawberry cutthroats and rainbow trout improved the situation for awhile, but the chubs, suckers, and shiners were back by 1978 and expanded rapidly through the 1980s. By 1990, they comprised more than 98 percent of the fish biomass in the now expanded reservoir.

This set the stage, in 1990, for the largest rotenone treatment of a water body ever attempted. After that was completed, the reservoir was again stocked with game fish, but this time with Bear Lake cutthroat trout (a trout that authorities hoped would grow to trophy size and compete with, and eat, the chubs, which they fully expected to return) and kokanee (another popular game fish expected to hold the chubs in check by competing with them for the rich zooplankton supply in the reservoir). Rainbow trout, popular with the reservoir's anglers but sterilized so as not to interbreed with the cutthroats, were stocked as well, but this was stopped after the 1995 release when the federal Food and Drug Administration restricted the use of the chemical agent used to sterilize the fish.

Because Bear Lake cutthroat trout coexist with Utah chub in Bear Lake, it was expected that these trout could compete successfully with the prolific chub in Strawberry Reservoir, whereas the old Strawberry cutthroat strain could not. Also, because of its reputa-tion as a fish-eater, it was expected that the Bear Lake cutthroat would keep the chub population in balance by utilizing them for food. Initially, it didn't look like that was going to happen. The chubs had reappeared in Strawberry Reservoir by 1993, but studies carried out in the 1996 and 1997 seasons by researchers from the Utah Cooperative Fish and Wildlife Research Unit, Utah State University, indicated that even the larger cutthroats were ignoring them. In fact, the only time these trout exhibited piscivorous behavior at all was when the annual release of hatchery-reared cutthroat fingerlings took place. Otherwise, even the larger trout rearing in the reservoir were doing quite well, thank you, on the rich supply of aquatic invertebrates.

But biologists working on the reservoir have continued to monitor chub numbers as well as the feeding behavior of the Bear Lake cutthroats, and it now appears that the cutthroats are behaving as was originally expected. The most recent results, obtained from the Utah Division of Wildlife Resources for the 2003 season, indicate that the predatory behavior of the cutthroats has increased. Forty-seven percent of the cutthroats over 18 inches in total length were found to be feeding primarily on chubs in 2003, as were 64 percent of the cutthroats over 20 inches, whereas few, if any, trout of any size had been doing so in 1996 and 1997. Along with this change in feeding behavior of

Utah Lake, shown here with Mt. Timpanogus in the background. Trout from this lake once so sustained the settlers that one early legislator proposed it for the historical record along with the seagull. The lake is not even trout habitat today. PHOTO BY AUTHOR

the trout, the numbers of age-1 and age-2 chubs were down significantly from earlier years, and this is the preferred prey size range for the cutthroats. In the 1996 and 1997 studies, Bear Lake cutthroats in Strawberry Reservoir consumed juvenile trout up to 39 percent of their body length in size. That's about an 8-inch trout for a 20-inch Bear Lake cutthroat. But chubs appear to outgrow susceptibility to predation by cutthroats when they reach about 6 inches in length.

Although hatchery-reared Bear Lake cutthroats continue to be stocked, either as 2- to 4-inch 1-year old fingerlings released in the spring, or as advanced fingerlings 4 to 6 inches in length released later in the season, adult trout also spawn in Strawberry Reservoir inlet tributaries, especially in Indian and Trout creeks. The early life history of emergent juveniles was studied in these creeks by Curtis Knight, as part of his graduate studies at Utah State University. He found that emergence of fry from the gravels commenced about the second week in August and extended for a period of five weeks, with most of the emergence occurring at night. He also observed that a fraction of these emergent fry left the tributaries immediately, apparently the very night they emerged, migrating to the reservoir at a length of only 26 mm (about an inch). The majority of the fry remained in the streams, however, where they lingered in quiet channel-margin habitats until, by late September, they had approximately doubled in size. At that point, they took up residence out in the main channels. These fry then remained in the tributaries for 1 to 2 years before migrating to the reservoir.

The very early and immediate exit from the natal streams of a portion of each year's naturally-spawned fry production prompted Knight to perform some additional experiments. Early fry migration is usually the result of displacement as the young-of-the-year trout compete for suitable rearing territories. But movement by displacement is usually more protracted. Inferior competitors are forced downstream, but usually only as far as is necessary for them to attain suitable territories. Some may be continually displaced until they are either stressed to death, succumb to predators, or finally leave the streams altogether. But still, competition for suitable rearing space is the governing mechanism. To test if this was the case in the Strawberry tributaries, Knight caught groups of the tiny immediate migrants in traps and placed them in enclosures, together with fry of equal size that had chosen the channel margins. He found that both groups competed equally well and both groups grew at the same rate, regardless of how densely they were packed together. Since no competitive differences were evident among the migrant and non-migrant groups, Knight concluded that the immediate migration behavior of a fraction of the Bear Lake cutthroat fry is an innate response of the population and part of its evolutionary programming. [19] The nature of this programming and its evolutionary significance remain to be ascertained.

Strawberry Reservoir has been, and remains, one of Utah's most heavily fished waters, receiving as many as 1.5 million angler-hours of use each season. Fly-fishers visit it frequently, using float tubes and kick-boats to work the shallow areas and downwind bays. A favored method is to probe the submerged vegetation with a size 6 Woolly Bugger or leech pattern fished on a sinking line. Damsel nymph imitations are also effective, especially in July when these nymphs are swimming to the areas where they will hatch. A couple of "Strawberry Special" streamer patterns are local favorites, and flies tied to imitate small chubs or cutthroat fingerlings ought to work too.

Lake Alice is a 230-acre mountain lake, surface elevation 7,745 feet, located on a tributary of Hobble Creek in the Smith's Fork drainage of Wyoming. It is the largest natural lake in the southern portion of Bridger-Teton National Forest, and was created in fairly recent Holocene time when a massive landslide filled this tributary watercourse with a plug of rock and debris more than a mile in length. The lake is 3 miles long and averages 46 feet in depth with a maximum

19. The spawning behavior and early life history of Bear Lake cutthroat trout in Strawberry Reservoir tributaries was studied by Knight (1997) and reported in his Master's thesis and in a paper by Knight et al. (1999). Papers by Beauchamp et al. (1999) and Baldwin et al. (2000, 2002), as well as Baldwin's Master's thesis from Utah State University (see Baldwin 1998), have the details of the Utah Cooperative Fish and Wildlife Research Unit's investigations of the movements and feeding behavior of Bear Lake cutthroats in the reservoir itself. My source for other information about Strawberry Reservoir was the Utah Division of Wildlife Resources. This agency has an excellent webpage devoted to the reservoir, its history, and the current status of its fishery at *www.wildlife.utah.gov/strawberry/*. For those of you who may be interested in the old "Strawberry cutthroat" hybrid strain, W.S. Platts published a short summary of age and growth data for these trout in Strawberry Reservoir in the Proceedings of the Utah Academy of Sciences, Arts, and Letters back in 1958 (see Platts 1958). Hepworth et al. (1999) give a more recent evaluation of the performance of this strain in put-grow-and-take fisheries elsewhere in Utah.

Logan River, Utah, provides the type of rearing habitat used by fluvial populations of Bonneville cutthroat trout. PHOTO BY AUTHOR

depth of 174 feet. Its outlet is subsurface, through the dammed-up debris, and this acts as a total barrier to fish movement to or from the stream below. Therefore, trout that were isolated upstream from the debris dam have been left to adapt to this mountain lake environment for at least a few thousand years. The only known glitches in this otherwise uninterrupted history are records of deliberate stocking, the first a single release of fingerling Yellowstone cutthroat trout that occurred in the early 1930s, and releases of cutthroat trout of unknown or unreported origin in 1939, 1940, and 1941. However, Wyoming biologists believe that none of these fingerlings survived to reach adulthood, let alone reproduce.

The population of Bonneville cutthroat trout in Lake Alice is large and apparently well established, but in this particular environment their growth, reproduction, and life span are more like a population of stream-resident trout than the large, long-lived denizens of Bear Lake.[20] Cutthroat trout and sculpins are evidently the only fishes present in Lake Alice, and the cutthroats subsist primarily on a diet of aquatic and terrestrial invertebrates.[21] Growth is rapid for the younger trout on this fare, but the rate of growth tails off for the older age classes. Based on ages obtained by counting annuli on scales, the trout attain a total length of about 3 inches at age 1, 7 inches at age 2, 9 inches at age 3, and 10.5 inches at age 4. The largest trout

captured in gillnet sets at Lake Alice was 13.5 inches in total length, but no trout older than age 4 have been captured. Adults ready for spawning enter the inlet tributaries from about the third week of May, when water temperatures are estimated to reach about 45 degrees Fahrenheit, to the middle of June. Spawning activity peaks around the 1st or 2nd of June. Based on fecundity measurements made back in the 1950s, adult females carry an average of 474 eggs each.

Despite the fact that the trout are small and Lake Alice is a hike-in fishery over rocky terrain, the walk is short (about a mile and a half from the Hobble Creek Campground) and the lake is visited often by anglers. No motorized vehicles are allowed on any of the trails in the area, and no motorized floating devices are allowed on the lake.

The Fluvial Life History Strategy

There is plenty of evidence, in the form of references in early settlers' journals and biological reports, to indicate that the Bonneville cutthroat also existed in a form adapted to life in the main rivers of the Bonneville Basin. These include the Bear River and its main tributaries, the Ogden and Weber rivers, the Jordan River, the Provo River, and perhaps even the Sevier and Beaver rivers. George Suckley, writing in his journal in 1859 (not published until 1874), mentioned trout taken

20. The Wyoming Game and Fish Department collected information about age, growth, and reproduction of Lake Alice trout during spawning operations conducted in 1951 and from gillnet and creel surveys in 1960, 1973, 1977, and 1979. This was all included in a Fisheries Technical Bulletin, "Bonneville Cutthroat Trout *Salmo clarkii utah* in Wyoming," authored by Dr. Niles Allen Binns and issued by the Department in 1981. See Binns (1981) for the complete citation.

21. These trout will feed on small fishes if given the opportunity. The Binns report, cited in footnote 20 above, contains an observation by Wyoming biologists Orv Landen and Earle Wilde that when Yellowstone cutthroat fingerlings were released in the 1930s, Lake Alice trout appeared quickly at the release site and were seen feeding on the fingerlings. According to a Department report by G.O. Hagen (1951) that is also cited in the Binns report, biologists conducting spawning operations at the lake in 1951 also noted that a trout or two had eaten sculpins, and a few had also consumed loose trout eggs.

from the Weber and Provo rivers. It is known, of course, that lake-dwelling trout migrated up the Provo River from Utah Lake to spawn. But even in the late summer and fall, after all of the lake-dwelling spawners should have returned to the lake, feisty cutthroats could be found in the pools of the Provo. Regarding his experience fishing the Provo River, Suckley wrote:

> "About the 1st of September last, we caught three trout from the same stream.... They rose freely to large, dark hackles, but refused gaudy or light-colored flies. Owing to poor flies, which had been in our possession for several years, the whipping of the hooks having shrunk so that they were easily pulled off, we caught but three of the many fish that jumped at them."

Even earlier, during an 1848 survey of the Salt Lake Valley, Capt. Howard Stansbury of the U.S. Army Corps of Topographical Engineers remarked on the abundance and size of the "speckled trout" in the Bear River and the streams of Cache Valley. And the anonymous correspondent, "Gage," whom we met in our discussion of Bear Lake, highlighted the native red-throats in the Provo, the Bear River, the upper portals of the Ogden, and the Weber River in his 1906 piece on Utah's trout waters. [22]

By the early 1950s, however, it was believed that these trout had long since gone the way of the Utah Lake and Panguitch populations. Cutthroat trout still could be found in some rivers, for example, the upper reaches of the Logan River where George G. Fleener,

Utah State Agricultural College (now Utah State University) studied their life history characteristics in 1948 and 1949. [23] But even Fleener believed the original Bonneville cutthroat trout was extinct in this drainage owing to hybridization with Yellowstone cutthroat and rainbow trout, both of which had been planted heavily for years prior to his study.

Despite the dubious pedigree of the present Logan River cutthroats, their life history characteristics probably reflect those of the pure-strain, original inhabitants. Subsequent work in the Logan River drainage has shown that its cutthroat trout populations can exhibit both fluvial and stream-resident life history strategies, and, like such trout elsewhere, stream-resident individuals can exhibit either mobile or relatively sedentary life history patterns. [24]

In the years 1973, 1974, and 1975, David Bernard and Eugene Israelsen, both from Utah State University, focused on a fluvial population that spends its summer-rearing and overwintering periods in a segment of the Logan River near the confluence of the Temple Fork, upstream from the city of Logan. Migration distances for these trout are relatively short. For spawning, they pass upstream through about a mile-long segment of Temple Fork and into Spawn Creek, a spring-fed Temple Fork tributary that is a little under 3 miles long. Brook trout populate the upper mile or two of

22. The passages cited here are from Suckley (1874), Stansbury (1852 [1988]), and "Gage" (1906).

23. See Fleener (1951).

24. This work is detailed in Bernard (1976), Bernard and Israelsen (1982), and Hilderbrand and Keirshner (2000b, 2004a, 2004b). Two even more recent papers have examined how cutthroat trout and brown trout distribute themselves along an altitudinal gradient in the Logan River; see de la Hoz Franco and Budy (2005) and McHugh and Budy (2005). McHugh and Budy (2006) have also studied the effects of brown trout on the individual and population-level performance of Bonneville cutthroat trout in the Logan River.

Spawn Creek, and brown trout from the Logan River also spawn there in the fall. Rainbow trout had been stocked in the upper reaches of Temple Fork prior to their work, but Bernard and Israelsen indicated that no cutthroat trout reside the year round in either Spawn Creek or Temple Fork.

In Fleener's study of Logan River cutthroat trout, he had found that all fish age 2 or older were sexually mature. Bernard and Israelsen reported that only a portion of the trout in their study population matured at age 2, and all of these trout were males. Females, as well as the rest of the males in the population, did not mature until age 3. At any rate, the first of the mature cutthroats entered Spawn Creek in early April. The peak of the migration varied in different years from the first week to the third week in May, but it always occurred just after the peak in snowmelt runoff had passed. Male and female spawning migrations were about a week out of phase at the Spawn Creek migrant trap, with the peak in migration of male trout occurring first. The same was true of the return migration of adults to the Logan River. The peak in downstream movement of spawned-out males occurred about a week ahead of the peak downstream movement of females. Little spawning activity took place after mid-June, and by the end of July all adult cutthroat trout had dropped out of Spawn Creek and back to the Logan River.

Spawning migrations of sexually mature cutthroat trout often attract a few younger sub-adult trout that join in the migration and linger in the spawning tributaries throughout the spawning period. This is true at Spawn Creek as well, with the sub-adults in this case being age-1 and age-2 trout. There is also a small migration of sub-adult cutthroats into Spawn Creek in the fall, which Bernard thought might be due to the attraction of the brown trout spawning run.

Following emergence, young-of-the-year cutthroat trout produced in Spawn Creek move into quiet, channel-margin areas associated with streamside vegetation. About the last week in September, almost all of these young-of-the-year trout leave Spawn Creek and move down to the Logan River. Young cutthroat trout from Spawn Creek overwinter in the Logan River, as do the sub-adult and adult trout.

During the summer rearing period, the trout feed primarily on terrestrial and aquatic insects. Ants, beetles, grasshoppers, wasps, and, yes, yellowjackets comprise the terrestrial fare while mayflies and caddisflies make up the bulk of the aquatic insects eaten. [25] Occasionally one of the larger cutthroats will capture a juvenile salmonid, but piscivory is evidently rare.

Like the lake-adapted trout of Lake Alice, the Logan River population is short lived, and the growth rate tails off in older trout. Fleener's data are the only

25. The Logan River produces mayflies and caddisflies, which its trout eat, but you'll notice that stoneflies are not listed. Up through the 1950s, the large *Pteronarcys californica* and *Pteronarcella badia* stoneflies, known collectively to Utah anglers as salmonflies, were among the most abundant aquatic insects in the river. But they disappeared completely sometime in the early to mid-1960s. The last one known to have been collected was taken in 1966. The reason for their demise in the Logan River is a mystery. They are abundant in the nearby Blacksmith Fork River, which is a tributary of the Logan, but they have never recolonized the Logan on their own. In an attempt to kick-start a recolonization, scientists from Utah State University joined with volunteers from the Cache Valley Anglers Chapter of Trout Unlimited and elsewhere to collect salmonfly nymphs and adults from the Blacksmith Fork and release them into the Logan River. This project has been active since 2004, and has yielded at least some evidence of success: in mid-June 2005, split-open nymphal cases were found on streamside bushes in a reach upstream from Birch Glen, indicating that at least a few adults had completed emergence. For updates on this project, go to *www.usu.edu/buglab* and click on Projects and Research.

published size-at-age figures, and these extend only to age 3. He found that age-1 trout attained an average total length of 3.9 inches, age-2 trout 6.8 inches, and age-3 trout 9.2 inches. Maximum age in the Logan River population appears to be age 4.

More recently, researchers from both Utah State University and the University of Wyoming have been studying a fluvial population of Bonneville cutthroat trout that inhabits the Thomas Fork of Bear River in Wyoming. In contrast to the smallish trout of the Logan River, spawning adults in the Thomas Fork population attain sizes ranging from 9.3 to 19.8 inches in total length. That's more in keeping with historical reports, e.g., John Kirk Townsend, who wrote of the abundance of trout in the 15- to 16-inch size range captured by members of his party from a branch of the Bear River on July 4th, 1834. [26] Life-span data have not yet been reported for these Thomas Fork trout, but for trout of this size, I imagine it extends to at least age 5 or 6.

The trout of the Thomas Fork ascend into four upper tributaries, Water Canyon, Huff, Little Muddy, and Coal creeks, where they spawn from about the middle of May to the middle of June. They do not linger long after spawning; within 30 days, most will have departed from the tributaries for their return to mainstem reaches of the Thomas Fork where they take up habitats for summer feeding and growth.

University of Wyoming researchers studying the post-spawning movements of these fluvial trout found that the distance of downstream travel was correlated with fish size, with the largest trout migrating farthest. The longest travel distance recorded, by the two largest trout tagged for this particular study, was 51 miles. These two trout passed all the way down the Thomas Fork and into the mainstem Bear River before they took up summer residence. Once they had found suitable summer habitats, the trout moved very little.

Following emergence, young-of-the-year trout remain in the natal tributaries, at least through the summer months. No young-of-the-year trout were ever observed by the University of Wyoming researchers in the mainstem Thomas Fork in the summer months, but many were observed in all four of the natal tributaries. It is possible, even probable, that the young-of-the-year trout may exit the tributaries later in the fall, as they do at Spawn Creek in the Logan River drainage, but this remains to be confirmed.

The Stream-Resident Life History Strategy

The present bastions of pure-strain stream-resident Bonneville cutthroat trout are small, sometimes intermittent headwater streams located in the high ranges around the perimeter of the Bonneville Basin.

26. Townsend's observation on the size of the trout in the Bear River drainage in 1839 is in Townsend (1839 [1905, 1978]). The most recent studies of the Thomas Fork population have now begun to appear in thesis form and in journal publications. For what's available to date, see Colyer (2002), White (2003), and Schrank and Rahel (2004, 2006).

Fig. 11-1 shows the principal watersheds where these populations have been discovered or have been reintroduced. Reading clockwise from the top, these are the Bear River drainage; the Jordan River drainage; the Sevier River drainage; the Beaver River drainage; and the streams draining into Snake Valley. Those along the east side of the basin, including the Bear, Jordan, Beaver, and Sevier river drainages, originate in the habitat province ecologists call the Wasatch and Uinta Mountains ecoregion. The streams draining into Snake Valley originate in mountain ranges of the Northern Basin and Range ecoregion. [27]

Up in the high country of the Wasatch and Uinta Mountains ecoregion, conifer forests alternate with aspen groves and open, grassy meadows. Sagebrush occurs in the high meadows, but, where untrammeled, there is more grass and wet meadow than sage. There is also more rainfall up here, along the western edge of the Rocky Mountains, and stream flows are more dependable. Beavers work the willow-bordered streams, and up in the headwaters of Wyoming tributaries of the Bear River drainage, moose wade the muddy fringes to feed on tender shoots. Brook trout are present in many of the headwater stream reaches where cutthroats might otherwise be found, and rainbow and brown trout are also present, but usually in lower stream reaches. Finespotted Snake River cutthroat trout (Chapter 10) were also introduced some years ago in Wyoming's Smith's Fork, drainage.

In Utah's portion of the Northern Basin and Range ecoregion, much of the basin floor is dry desert country—so dry that according to one old cowboy tale, "the boys would ride two days out of their way just to see a mirage." The annual rainfall is only about 8.5 inches, and most of that falls as snow in the winter. The rest comes in swift, violent summer thunderstorms that sweep over the country, often sending walls of water surging down the normally parched washes in devastating flash floods. In the summertime, the floor of the basin can be bake-oven hot. Only in the evenings, when the air flows gently down off the ranges, bringing with it a cool scent of juniper and pine, do you sense that there's something more to be found up there among the peaks.

And it truly is a different world up there. The peaks, some of them soaring to more than 12,000 feet, receive a bit more rain than the valley floor, and the vegetation reflects it. There is chokecherry, mountain mahogany, aspen, spruce, Douglas-fir, and pine. Some areas support stands of limber pine and the venerable bristlecone pine. The timber is more scattered and sparse than it is in the Wasatch and Uinta Mountains ecoregion, but it is there.

And so, too, are the tiny streams where the cutthroats are found. Subject to high flows during spring

27. U.S. Environmental Protection Agency (1998).

runoff and extremely low flows in late summer and fall, with pools no more than a foot or so deep, prone to flash flooding from any sudden cloudburst, and vulnerable to habitat degradation from grazing, mining, and irrigation diversion—all this notwithstanding, the headwater reaches of a handful of these little creeks harbor relict populations of genetically pure Bonneville cutthroats that have adapted very well to their marginal environments. Rainbow and brook trout are also found in many of the small streams of the Snake Valley drainage, but these species never seem to grow quite as large nor fare quite as well as the native cutthroats in adjacent streams. Possibly it is the poor showing of the non-native trouts in these waters that has enabled the Bonneville cutthroat to resist replacement and maintain a high level of genetic purity.

The Bonneville cutthroats in these small, isolated streams spawn on the declining limb of the hydrograph, with spawning activity beginning just after the peak of the spring snow runoff. In a normal year, spawning activity would peak in May but extend through about the middle of June. Water temperatures during the peak spawning period range from 44 degrees to about 48 degrees Fahrenheit. Incubation extends to the end of July, which is about when the first of the emergent young-of-the-year fry are observed in channel margin habitats. As they gain size and swim-

ming ability, the young-of-the-year trout move out of the channel margins to join the mainstream hierarchy, assuming the most profitable positions in pools and pockets that they can defend.

Aquatic and terrestrial insects provide the major food items for these trout during the summer rearing period. As you might expect, the greatest variety comes in the early summer when the "cycle of the season" aquatic insect hatches are at their peak. Terrestrial insects, primarily ants in headwater reaches of Snake Valley streams, dominate the food chain later in the summer and fall. A late-summer study of the feeding habits of stream-resident cutthroat trout in upper Beaver Creek, a mountain tributary of the Logan River, found that in this stream, diet varies in different habitat types. In a forested, high gradient segment with step pools and channel slopes of 3 to 5 percent, terrestrial invertebrates dominated the diet. In a low gradient, meandering, meadow segment with channel slope less than 1.5 percent, aquatic Diptera (chironomids and the like) and terrestrials shared top billing as the dominant food items. Trout habituating in beaver ponds fed predominately on Diptera. Although these were the dominant food items, the trout did exhibit selectivity in all segments by consistently taking terrestrials and caddisflies whenever they were present.

Although they do appear to fare better than

Provo River, Utah. Bonneville cutthroat trout once ascended this river from Utah Lake to spawn. Now noted for its brown trout fishery.
PHOTO BY AUTHOR

introduced rainbow and brook trout in streams of
the Northern Basin and Range ecoregion, the native
stream-dwelling cutthroat trout do not always attain
large sizes on average. As an example, trout from
Birch Creek, a tributary of the Beaver River, average
3.3 inches in total length at age 1, 4.7 inches at age
2, 6.2 inches at age 3, and 7.8 inches at age 4. On the
other hand, in Coantag and Giraffe creeks, mountain
tributaries of Wyoming's Smith's Fork and Thomas
Fork respectively, trout in the older age classes may
grow more rapidly and attain larger sizes. In these
tributaries, the native cutthroats average 3 to 3.5
inches in total length at age 1, 6.2 to 6.9 inches at age
2, 8 to 10.7 inches at age 3, and 11.1 to 12.6 inches at age
4. No trout older than age 4 have been recorded in any
of these studies.

Based on observations of specimens from Birch
Creek, stream-resident Bonneville cutthroats attain
sexual maturity and spawn for the first time at age 2
for males and age 3 for females. Birch Creek females
ranging from 5.8 to 6.4 inches in total length produced
an average of 112 eggs, but a group of larger females
(6.7 to 7.6 inches in total length) from Trout Creek, a
Snake Valley stream, carried an average of 180 eggs.
Females ranging from 4.9 to 9.8 inches in total length
from Raymond Creek, Wyoming, a tributary of Smith's
Fork, carried an average of 165 eggs. [28]

STATUS AND FUTURE PROSPECTS

I mentioned earlier that by the mid-1950s many
authorities believed the Bonneville cutthroat trout
to be extinct as a subspecies in genetically pure form.
They blamed the usual suspects: overharvest, alteration
and destruction of habitat by man and livestock, and
widespread introductions of non-native fishes. But it
was also in 1950 that a relict population was discovered
in Pine Creek in the Snake Range, and transplants
from that population in 1953 and 1960 boosted the
number of known populations to three. Still later, the
populations in Reservoir Canyon and Water Canyon
creeks in the headwaters of the Santa Clara River
drainage were recognized. A couple of additional relict
populations were turned up in the Deep Creek Range
in the early 1970s by biologists associated with the
Bureau of Land Management. But despite the fact that
the total number of known populations was small and
likely restricted to less than 5 stream miles of habitat in all, and despite the fact that they all existed in
precarious circumstances with leading experts urging
all possible protection, the Bonneville cutthroat was
not included when the U.S. Endangered Species Act
became law in 1973. [29]

By 1978, the number of known stream populations that had either been discovered or established by
transplants had been boosted to 11. By that date too, the

28. The discovery of more and more
populations of stream-resident
Bonneville cutthroat trout has been
accompanied by an increasing
number of investigations into their
life history and ecology. You can
find details of these investigations
in Hickman (1977, 1978b), May et
al. (1978), Binns (1981), Buys et al.
(1995), Hilderbrand and Kershner
(2000b, 2004a, 2004b), Buys (2002),
Johnstone and Rahel (2003), Schrank
et al. (2003), de la Hoz Franco and
Budy (2005), and McHugh and Budy
(2005). Good overall summaries are
provided in Kershner (1995) and
Benhke (1992, 2002).

29. The view that the Bonneville Basin
native cutthroat had gone extinct
as a pure subspecies was expressed
in papers by Cope (1955), Platts
(1957), and Sigler and Miller (1963).
Although the newly discovered
relict populations in Snake Valley
headwater streams were known
to these authors, many thought
they were distinct enough to be
classified as a separate subspecies.
But those views gradually changed,
and now they are all recognized
as *bona fide* Bonneville cutthroat
trout. Discoveries of these and
other relict populations, and early
efforts to establish new populations
based on transplants from these,
are chronicled in Hickman and

populations of Bear Lake and Lake Alice had also been recognized as likely pure Bonneville cutthroat stocks. But none of the stream populations held abundant numbers of trout, and none led an existence any less precarious than it had been in 1973. The American Fisheries Society considered the subspecies endangered, and in 1979, the Society's Bonneville Chapter joined with the Desert Fishes Council in a petition to list it under the Endangered Species Act.

Prompted by this petition, the U.S. Fish and Wildlife Service conducted its own status review, eventually concluding that listing was indeed warranted but was precluded by other, higher priority endangered species activities. In 1982, the Service placed the subspecies on its Category 2 priority list, a place-holder for species and subspecies deemed warranted for listing but needed more information to support an actual listing proposal. In 1985, that priority level was raised to Category 1, the highest level for candidates for listing. [30]

By 1988, the continuing program of searching remote headwaters, transplanting from known populations, and, increasingly, genetic testing of populations heretofore thought to be introgressed, had increased the number of known populations of genetically pure Bonneville cutthroat trout to 41, including the Bear Lake and Lake Alice populations. But still, the 39 known stream populations occupied only 43 to 44 miles

Duff (1978), Behnke (1979 [1981]), and Hepworth et al. (1997, 2002). Dr. Behnke, writing in a Rare and Endangered Species Report to the U.S. Fish and Wildlife Service (see Behnke 1970), pointed out that pure, native populations of Bonneville cutthroat trout "are extremely rare and should receive all possible protection." Two prestigious groups of scientists, the American Fisheries Society's Committee on Endangered Species and the Conservation Committee of the American Society of Ichthyologists and Herpetologists, included the Snake Valley form in a list of threatened fishes of the United States (see Miller 1972). Holden et al. (1974) considered the entire subspecies endangered in the State of Utah.

30. The Service's conclusion that the subspecies was warranted but precluded was announced in a Federal Register notice issued in January 1984 (see Chambers 1984). For the actual status review document, see Hickman (1984). The earlier placement of the subspecies on the Category-2 candidate list was in a Federal Register notice issued in December 1982 (see Potter 1982), and the later elevation to Category 1 was announced in the Federal Register in September 1985 (see Dodd 1985).

Smith's Fork, a Bear River tributary. Typical habitat of Bonneville cutthroat trout in Wyoming. PHOTO BY AUTHOR

of total stream habitat and still, the Fish and Wildlife Service had not moved toward listing the subspecies. In 1992, impatient with the lack of action, the Desert Fishes Council petitioned again to list the subspecies as threatened, this time joined by the Utah Wilderness Alliance. This second petition presented no new information, stated the Service, so again no action was taken. The Bonneville cutthroat remained on the warranted but precluded list where it had languished for the last ten years, only now, in a little-noticed administrative move announced in 1991, it was back at the Category 2 priority level.

What followed next was not directed specifically at Bonneville cutthroat trout, but it does serve to illustrate how a subspecies can be pushed off the table by administrative maneuvers alone. In 1996, the U.S. Fish and Wildlife Service announced a major change in policy regarding listing priorities and categories in which all Category 2 species were summarily removed from the candidate list. Henceforth, only Category 1 species would be considered candidates for listing. Having been bumped back to Category 2 status in 1991, the Bonneville cutthroat was now dropped altogether from the listing queue. [31]

That triggered yet another petition, filed in 1998 by the Biodiversity Legal Foundation, which the U.S. Fish and Wildlife Service followed with its second formal status review of the subspecies. But this time, the Service concluded that listing was no longer warranted. By 2001, the Service asserted, 291 Bonneville cutthroat populations were known to occur in an estimated 852 miles of stream habitat and more than 70,000 acres of lake habitat, which amounts to between 5 and 17 percent of the subspecies' historical range. [32]

While that doesn't sound like much on a percentage basis, and these findings *are* being challenged on the grounds that most of the existing populations are very small and remain below the population size or occupied stream length required for long-term survival, [33] the Service contends that additional populations continue to be identified as waters previously thought to be occupied by introgressed populations are found to be pure. Furthermore, conservation and restoration actions for Bonneville cutthroat trout have become cornerstones of state and tribal fishery management plans and in the land management plans of federal agencies, such as the U.S. Forest Service, Bureau of Land Management, National Park Service, and Bureau of Reclamation.

OK, so what has been happening in those venues? And just what are those conservation and restoration actions that have become cornerstones of agency plans?

I've mentioned a time or two already that three principal activities have accounted for the progressively increasing number of pure Bonneville cutthroat

31. The history of these moves as they affected the Bonneville cutthroat trout was summarized by two U.S. Fish and Wildlife Service biologists, Yvette Converse and Janet Mizzi, in a piece that appeared in the journal *Endangered Species Bulletin* (see Converse and Mizzi 1999). This article was also posted on the Internet at *www.nativefish.org/Articles/UBC_Trout.htm* (accessed 25 November, 2004). The pertinent Federal Register notices are: Drewery (1991), which moved the subspecies back to Category 2 on the candidate list, and Drewery and Sayers (1996), which dropped all but the Category 1 species.

32. For the latest petition to list the Bonneville cutthroat trout as threatened, see Biodiversity Legal Foundation (1998). For the U.S. Fish and Wildlife Service's decision not to list, and its latest status review on which that decision is based, see Converse (2001) and U.S. Fish and Wildlife Service (2001), respectively. You can download your own copies of these documents, a list of questions and answers about the decision, and other information regarding Bonneville cutthroat trout from a U.S. Fish and Wildlife Service website at *http://mountain-prairie.fws.gov/species/fish/bct/*.

33. The challengers are evidently basing their case on studies by Hilderbrand and Kershner (2000a), who examined the prognosis for long-term persistence

populations known to exist in the historical range: 1) diligent searching of remote headwater streams to ferret out relict populations; 2) translocations of trout from discovered populations to establish additional populations in other historically occupied waters, often after removing non-native trout; and 3) genetic testing of populations that were already known but were thought to be introgressed. [34]

I believe the Bureau of Land Management deserves credit for initiating the search for relict populations of Bonneville cutthroat trout in remote headwater streams. In 1973, this agency began surveys in the Deep Creek Range that led to the discovery of a pure population in the extreme headwaters of Trout Creek in 1974 and other relict populations in subsequent years. Also in 1974, a relict population was discovered in Salt Lake County, Utah. Dr. Behnke called attention to the pure populations in the Santa Clara River drainage and in Birch Creek of the Beaver River drainage, southern Utah, in 1976, and other relict populations were verified later in this same general area by the Utah Division of Wildlife Resources. The Wyoming Game and Fish Department kicked off surveys of its waters in the Bear River drainage in 1977.

Also in 1977, the Utah Division of Wildlife Resources began efforts to increase the occupied range of Bonneville cutthroat trout in southern Utah by

of isolated cutthroat trout populations. To support a population of 2,500 trout 3 inches in length or larger at an abundance level of 480 fish per mile, a stream length of 5.2 miles of suitable habitat is needed. To maintain that same 2,500 trout at an abundance level of 160 trout per mile would require 15.5 miles of suitable stream length. Hilderbrand and Kershner chose the population level of 2,500 trout to correspond to an *effective population size* of 500 individuals. Effective population size refers to the number of breeding individuals that actually contribute their genes to the next generation. Effective population size is typically only a small fraction of the actual population size. Too low an effective population size leads to a loss of genetic variation in a population, which may in turn lead to lower overall fitness and a decrease in reproductive performance, all of which bode ill for long-term viability. An effective population size of 500 is generally accepted as the minimum level needed to offset these effects. The challengers contend that few of the remnant populations of Bonneville cutthroat trout, or even populations established by transplanting from these remnant populations, are large enough to meet the minimum effective population size of 500, nor do the majority

of them occupy the prerequisite number of stream miles. However, Hilderbrand and Kershner were quick to point out that insufficient stream length to maintain the 2,500 individuals needed for an effective population size of 500 does not necessarily mean that a population *will* go extinct. Some small, isolated populations have persisted for centuries and may have adapted to restricted space. But their *risk* of extinction is greater nevertheless, because their limited space makes them more vulnerable to localized disturbances than populations with a wider distribution. For example, they remain vulnerable to drought, or, when the heavy rains do finally come, to flash floods that could sweep away whole populations at once. Metapopulation structure, or the lack of it, comes into play here. To refresh yourselves on that subject, I refer you back to the discussion and references in footnote 49 of Chapter 5. To read more about effective population size and its implications for genetic structure and long-term survival of small populations, see Franklin (1980), Gilpin and Soulé (1986), Frankham (1995), and Nunney (1995).

34. You can track the progress of these programs for yourselves by reading the array of status reports that have been written for Bonneville cutthroat trout by the various management agencies since the late 1970s. In addition to the two status reviews published by the U.S. Fish and Wildlife Service in 1984 and 2001 (cited earlier), the list includes:

1) Current Status of Cutthroat Trout Subspecies in the Western Bonneville Basin (Hickman and Duff 1978).

2) Distribution, Systematics and Biology of the Bonneville Cutthroat Trout, *Salmo clarkii utah* (May et al. 1978).

3) Bonneville Cutthroat Trout *Salmo clarkii utah* in Wyoming (Binns 1981).

4) Bonneville Cutthroat Trout: Current Status and Management (Duff 1988).

5) Management Plan and Status Report: Native Cutthroat Trout of the Dixie National Forest (Duffield 1990).

6) Current Status of Bonneville Cutthroat Trout in Nevada (Haskins 1993).

7) Status of the Bonneville Cutthroat Trout in the Bridger-Teton National Forest, Wyoming (Nelson 1993).

translocating trout collected from the Birch Creek population into other waters. Over time, additional relict populations were discovered and translocations were made from these as well, when suitable reaches of water could be found and non-native trout removed using rotenone. The agency adopted a "nearest neighbor" policy in which the closest available source of native trout within the same drainage is used for translocations. Since so many of Utah's wild populations are introgressed, the agency also adopted some management designations based on the level of introgression. Only the purest populations, with less than a 1 percent level of introgression, are used for translocations. These so-called "core" populations are included in a broader category of "conservation" populations, which are those with less than a 10 percent level of introgression. Conservation populations receive no stocking and maintain themselves strictly by spawning in the wild. Populations that are more than 10 percent introgressed, as well as populations established and maintained by stocking from state hatcheries, are designated "sport fish" populations. Sport angling is generally legal for all population categories including conservation populations, but may be more restrictive for the latter. [35]

Fish-stocking policies were changed as well—not just in southern Utah, but state-wide. Stocking of catchable-size rainbow trout, prior to this a staple of Utah sport fish management, was discontinued in most rivers and streams. Non-native subspecies of cutthroat trout were discontinued altogether. Where it is still deemed desirable to stock non-natives, the state now experiments with sterile hybrids or triploids.

I don't know why this still surprises me, I've heard it so often before, but Utah fisheries managers say that one of the biggest roadblocks they face in reintroducing native cutthroats is resistance from sport anglers fearing changes to popular fisheries, even when those fisheries are based on non-native, stocked trout. To defuse some of this resistance in southern Utah, state biologists established a wild broodstock of Bonneville cutthroat trout at Manning Meadow Reservoir using a blend of pure-strain trout from several southern Utah streams. More and more in southern Utah, it is hatchery-reared trout from the Manning Meadow broodstock that are utilized when it's necessary to establish and maintain sport fish programs by stocking. Management plans and interagency conservation agreements call for additional broodstocks to be developed in the other areas around the Bonneville Basin where the native cutthroat was historically present. [36]

Among the parties that have signed onto these interagency conservation agreements is the Confederated Tribes of the Goshute Reservation, which takes us out of southern Utah and back up the west side of

8) Bonneville Cutthroat Trout in Idaho: 1993 Status (Scully 1993).

9) Bonneville Cutthroat Trout (Kershner 1995).

10) Bonneville Cutthroat Trout *Oncorhynchus clarkii utah* (Duff 1996).

11) Distribution and Abundance of Native Bonneville Cutthroat Trout (*Oncorhynchus clarkii utah*) in Southwestern Utah (Hepworth et al. 1997).

12) Abundance of Bonneville Cutthroat Trout in Southern Utah, 2001–02 , Compared to Previous Surveys (Hepworth et al. 2003).

35. Dale Hepworth and two colleagues, Utah Division of Wildlife Resources, wrote up an excellent review of the progress and problems of the native trout conservation program in southern Utah, which they published in the *Intermountain Journal of Science* in 2002. Consult the bibliography entry under Hepworth et al. (2002) for the complete citation.

36. Interagency conservation agreements have helped greatly to alleviate concerns among stake holders over Endangered Species Act and state and federal sensitive species listings. One such agreement, for the Bonneville cutthroat

the Bonneville Basin to the Deep Creek Range. You'll recall that here is where the first organized searches by biologists associated with the Bureau of Land Management found relict populations of Bonneville cutthroat trout in the uppermost headwaters of Trout and Birch creeks, two streams that flow off the east slopes of the range. The Goshute Reservation extends across the Utah-Nevada border and encompasses streams flowing off the west slopes of the range. Genetically pure populations have since been found in some of those west-slope streams as well.

For several years now, the Goshute Indians have worked with Trout Unlimited and the U.S. Fish and Wildlife Service to protect the relict populations that occupy their streams, and to restore the native trout to other formerly occupied Reservation waters. Beginning with four genetically pure populations occupying 22.2 stream miles of habitat, the Goshutes and their collaborators added two brood ponds and two spawning channels to produce trout both for fishing and for additional introductions. They also use streamside incubators to increase fry production and help their brood program along.

Something unique and significant to the tribal culture has happened already as a result of this program. Fish Creek (known as "Painkwi Okwai" to the Goshutes), which joins with Fifteenmile Creek to form

trout in Utah, was finalized in 1997 between the Utah Division of Wildlife Resources, Utah Department of Natural Resources, Utah Reclamation Mitigation and Conservation Commission, U.S. Fish and Wildlife Service, U.S. Forest Service, Bureau of Land Management, Bureau of Reclamation, and the Confederated Tribes of the Goshute Reservation. See Lentsch et al. (1997) for the complete citation. A broader-reaching range-wide conservation agreement was signed in 2000 by each of the above agencies plus the National Park Service and the fish and wildlife agencies of Wyoming, Idaho, and Nevada. For that one, see

Lentsch et al. (2000). These agreements are typically made for five years, with provisions for extending if need be, and for writing new ones. Some of these agreements have now reached the end of their initial terms, and the first completion reports have begun to emerge. For example, the Utah Bonneville Cutthroat Trout Conservation Team (2004) has written a completion report for the 1997 agreement covering Bonneville cutthroat trout in Utah. With these reports in hand, one can match progress and accomplishments against the original conservation goals.

Deep Creek, the namesake of the range, had held trout historically. The Goshute elders referred to these trout as "ainka painkwi," which translates to "red fish." Initially, this struck the volunteers working with the tribe as odd, because Bonneville cutthroat trout are generally dull in coloration. But, according to Tim Zink, writing in the May 2005 issue of *Trout*, the coldwater fish conservation journal of Trout Unlimited, within the first year of being back in Fish Creek, the trout were sporting a reddish coloration, much to the joy of tribal elders who still recalled the trout of their youth. [37]

Meanwhile, over on the east side of the Deep Creek, Trout Unlimited, Utah Division of Wildlife Resources, the Bureau of Land Management, and rancher Buck Douglass of the Deep Creek Mountain Ranch have similarly collaborated to establish and maintain six genetically pure populations of Bonneville cutthroat trout in 31.7 miles of stream habitat. Four brood ponds and two spawning channels have also been established, and, again, the brood program is helped along by the use of streamside incubators to boost fry production.

Logging, grazing, and mining have all taken place in the Deep Creek, and most of the perennial streams are diverted for irrigation. Although these activities, along with the stocking of non-native gamefish species, no doubt did contribute to the demise of the native Bonneville cutthroat trout in all but a handful of the uppermost, remote headwaters, they are con-

sidered only minimal threats to existing populations and conservation programs. Even so, clouds of uncertainty still float across the Deep Creek from time to time. As an example, in the late 1980s or early 1990s, word came that a private developer had been granted a preliminary license by the Federal Energy Regulatory Commission for a small hydropower facility that would have affected both Trout and Birch creeks. This license was challenged by just about everybody involved in Bonneville cutthroat conservation, and the latest word is that the project is off the table. But vigilance against such threats must be constant in order to protect the gains that have been made. [38]

A traverse south and west from the Deep Creek takes us across the state line into Nevada and onto the slopes of the Snake Range. It was up in this range, on the west flank, in 1950, that a Nevada Department of Wildlife biologist made the initial discovery of a relict Bonneville cutthroat population in Pine Creek. As I mentioned earlier, Pine Creek is actually outside the Bonneville Basin (the Snake Range crest marks the Basin boundary). The trout were probably transferred incidentally back in the 1800s, when miners diverted water from Lehman Creek, a Bonneville Basin stream on the east flank of 13,063-foot Wheeler Peak, into Pine Creek via the Osceola Ditch. This was fortunate, because the Lehman Creek population was eventually extirpated. Later, after they

37. A good review of the Goshute tribal program, including the story of the "ainka painkwi," can be found in Tim Zink's article, "Restoration on the Reservation: Protecting Native Fish on Native Lands" in the Spring 2005 issue of *Trout* magazine; see Zink (2005). The figures I've cited here for the number of genetically pure populations and stream miles of habitat occupied in the Deep Creek Range are from tables in the range-wide conservation agreement that was signed in 2000 (see Lentsch et al. 2000 for the complete citation). News releases, mostly from Trout Unlimited, have also highlighted these collaborative programs. A good source for these (as of June 2005) is the Utah Council Trout Unlimited website at *www.tuutah. org/*. You can also find information there on the streamside incubators used in these projects.

38. A large chunk of Bureau of Land Management land in the Deep Creek Range was set aside as a Wilderness Study Area some years ago, and was included in a citizens' proposed Wilderness Bill that was first introduced in Congress back in 1989. That bill, now called the Redrock Wilderness Act, has gotten nowhere, and probably won't for a while, given the anti-Wilderness bias in both Congress and the pres-

had been recognized as a unique subspecies, trout from Pine Creek were stocked into Hampton Creek and other remote streams to increase the number of known populations.

U.S. Highway 50 ("The Loneliest Highway in America"), which crosses at Sacramento Pass, bisects the Snake Range into roughly equal-sized northern and southern segments. Mt. Moriah (elevation 12,050 feet) and Hampton Creek, an east-flowing Bonneville Basin stream, lie in the northern segment. Wheeler Peak, Pine Creek, and Lehman Creek are in the southern segment. When I first visited this region for the first edition of this book, most of the Snake Range was administered by the U.S. Forest Service. What wasn't was under Bureau of Land Management jurisdiction or was held privately. Water diversions, mining, and domestic livestock grazing were common, and these practices had altered the streams by reducing stream flows, increasing sediment, and decreasing streamside cover. Non-native rainbow, brown, and brook trout had been introduced and were still being stocked, and these species provided most of the available angling opportunities. In 1989, land was set aside in the northern segment of the range for the Mt. Moriah Wilderness. Three years earlier, in 1986, a big piece of the federal land in the southern segment, including Wheeler Peak and its environs, became the Great Basin National Park.

The National Park Service is another party to the interagency conservation agreement for Bonneville cutthroat trout. Early in the park's history, the only known pure populations were in Pine, Willard, and Ridge creeks on the west side of the park. None were known to exist in the eastside streams of the Bonneville Basin *per se*, although Mill Creek on the east side had a population that was thought to be introgressed. Streams in the park that were identified as potential reintroduction sites were the South Fork of Big Wash, Strawberry Creek, upper Snake Creek, and South Fork Baker Creek. Upper Lehman Creek, the original source of the Pine Creek population, was also mentioned as a reintroduction candidate in 1989, but does not appear on recent lists.

The park's original objective under the interagency conservation agreement was to reintroduce and maintain Bonneville cutthroat trout into 13 miles of historically occupied stream habitat in the four park watersheds mentioned above. Subsequent genetic tests of the Mill Creek population revealed that it is pure Bonneville cutthroat rather than introgressed, as had been believed, which put the project ahead of the game. Rather than 13 stream miles of occupied habitat, the park and its environs will have 21 miles when the current agreement is fulfilled. As I understand the current status, South Fork Big Wash and Strawberry Creek have received their reintroduced populations and are

ent administration. Meanwhile, in hopes of resolving the continuing battle between proponents and opponents of Wilderness in Utah, the State has embarked on a new initiative based on Wilderness Working Groups in counties across the state. Each group, comprising government representatives, ranchers and farmers, industry officials, and environmentalists, is charged with hashing out land management strategies for each local area. As these plans are agreed on and completed, they will be presented to Congress. Meanwhile, the sponsors of the Redrock Wilderness Act have not given up. While fully intending to participate in the county-by-county process, they say, they also intend to reintroduce their bill into each new session of Congress.

being monitored to ensure their stability. Snake Creek and South Fork Baker Creek will be ready to receive Bonneville cutthroat trout in 2005. When all these populations have stabilized, recreational angling under close regulation will be allowed. [39]

Combined with the progress in the Deep Creek, these reports of progress on Bonneville cutthroat reintroductions into Snake Range streams give plenty of reason for optimism regarding the viability of the native cutthroat trout in historical habitat on the west side of the Bonneville Basin. Ah, but also like the Deep Creek, an occasional cloud can darken the horizon. One such cloud was a mining venture that focused on Hampton Creek in the northern segment of the Snake Range. It turns out that there are garnet deposits in the Hampton Creek canyon, not gemstone quality, but industrial-grade material used in abrasives, filtration applications, and as proppant for oil and gas wells (a proppant is a grain-like or bead-like material that is suspended in hydraulic fluid and injected into oil or gas wells under very high pressure to fracture the formation and hold the cracks and crevices open so the oil or gas will flow more freely). Thirty-five claims were on file for those garnet deposits both inside and outside the Mt. Moriah Wilderness, with Hampton Creek running through the middle. Some mining was done at these claims in the 1960s and again in 1994 on claims on

Bureau of Land Management land downstream from the Bonneville cutthroat habitat. In 1996, the Forest Service received a proposal to dig test trenches as a prelude to mining on claims just outside the Wilderness boundary, along the Hampton Creek reach where the Bonneville cutthroats do reside. An Environmental Assessment was completed the same year, and even though opposition was voiced by Utah and Nevada Trout Unlimited chapters, several Native American tribes, and a group calling itself the Moriah Defense Fund, I believe the test trenches were dug. But in this case, the proponents eventually decided not to mine in the area and the project was dropped. [40]

Ok, let's go back across the Bonneville Basin now, to the streams flowing out of the Wasatch Front in what the interagency conservation agreement calls the North Bonneville Management Unit. This unit encompasses four drainages: the Ogden, Weber, Jordan, and Utah Lake/Provo River drainages. It is at the feet of these drainages where the bulk of Utah's human population is concentrated, and here is where the most extensive water withdrawals and stream alterations have taken place to accommodate these people. The interagency conservation agreement set forth two goals for this unit: 1) to maintain fifteen conservation populations in 119.4 occupied stream miles and 700 surface acres of lake or reservoir waters across all the

39. My sources for the reintroduction program for Bonneville cutthroat trout in Great Basin National Park are personal communications from William S. Brock, Resource Management Specialist in 1989; Albert A. Hendricks, Park Superintendent in 1992 and 1994; and Gretchen Baker, Park Ecologist in 2004. There is also a nice summary of the project and its progress on the park website at *www.nps.gov/grba/bct.htm*.

40. Patricia N. Irwin, Ely District Ranger, Humboldt-Toiyabe National Forest, provided me with the latest information that I have on the status of this mining proposal in a personal communication, February 2, 2005. I also found information about the garnet deposits themselves at a U.S. Geological Survey website at *http://minerals.usgs.gov/minerals/pubs/commodity/gemstones/sp14-95/garnet.html*.

four drainages, and 2) to establish and maintain two sport fish populations in 30.2 occupied stream miles and 350 surface acres of lake or reservoir waters in the Jordan River drainage and two sport fish populations in 33 occupied stream miles in the Utah Lake/Provo River drainage.

Because of the greater numbers of rivers and headwater streams in these drainages that could harbor unrecognized populations of pure Bonneville cutthroat trout, considerable emphasis was placed on searching for such populations and on genetic analysis. The most exciting outcome to date has been the discovery of two functioning metapopulations, both in the Weber River drainage. The most extensive of these is in the Chalk Creek sub-basin where pure-strain Bonneville cutthroat trout were found in 15 interconnected tributaries totaling about 103 stream miles of occupied habitat. The other is in the Hardscrabble Creek sub-basin, where the native trout occupy five interconnected tributaries and approximately 20 stream miles of habitat. Meanwhile, three wild broodstocks were developed at Red Butte Reservoir, Mountain Dell Reservoir, and Little Dell Reservoir, the latter located in the Salt Lake City municipal watershed. Trout from these sources have been used to meet the conservation agreement's objectives for establishing sport fish populations.

Using the same combination of diligent searching and genetic testing that paid off in the North Bonneville Management Unit, the number of known "core" and "conservation" populations in Utah's portion of the Bear River drainage has been increased from 14 populations in 56 miles of occupied stream habitat in 1997 to 23 populations in 166 miles of occupied stream habitat in 2004. In addition, one functioning metapopulation was documented in the upper Logan River watershed in 13 interconnected tributaries totaling 60 occupied stream miles, and another was found in the Woodruff Creek watershed in seven interconnected tributaries totaling 37 occupied stream miles. No sport fish objectives had been set in the conservation agreement for Utah's Bear River waters. Even so, a Bear River broodstock established at the Mantua State Fish Hatchery has produced more than 120,000 cutthroat trout for stocking in these waters.

The Wyoming Game and Fish Department likewise developed a captive Bonneville cutthroat broodstock at its Daniel Hatchery for stocking in waters of Wyoming's portion of the Bear River drainage, but did so many years prior to any of the interagency conservation agreements. The source population for this operation was originally the pure-strain population of Raymond Creek, a Thomas Fork tributary, but the plan was to also incorporate trout from other tributaries of the Thomas Fork and Smith's Fork as well, to broaden

The Osceola Ditch is dry now, but it once carried water from Lehman Creek on the east slope of the Snake Range to the Osceola mining district on the west slope, and incidentally transported Bonneville cutthroat trout into Pine Creek, where they persisted.
PHOTO BY AUTHOR

the genetic base. While in Wyoming, I have fished often for the pure-strain native trout of the Bear River drainage using very careful catch-and-release methods and tiny barbless-hook dry flies. My recollection of the trout I have brought to hand is of beautiful, delicately colored, opalescent little creatures, overall color pale fawn, darker on the back and washed with the faintest shade of pastel pink on the sides, spots relatively few in number but uniformly distributed, standing out against the pale backdrop like intense black dots of India ink.

One cloud hanging over the future of Wyoming's Bear River cutthroats is energy exploration and development. I will get into this subject more fully in the next Chapter, because more of the range of the Colorado River cutthroat subspecies is affected by these activities. Suffice it to say here that the Thomas Fork and Smith's Fork drainages lie along the western edge of what geologists call the *overthrust belt*. There's oil and gas down there. When I visited these drainages in 1982 for the first edition of this book, the "boom, boom, boom" of seismologists' charges reverberated through the forests, and signs proclaiming oil and gas leases appeared seemingly everywhere. The next time I visited, four years later, the country was experiencing an oil glut and the exploration frenzy had dropped to next to nothing. But I imagine that intensity has turned

up again by now, owing to the present administration's pro-extraction energy policy.

Even so, I am optimistic about the outlook for the Bonneville cutthroat trout. Getting back to Utah for a minute, progress on native cutthroat conservation has been accompanied by positive steps by other elements of Utah's populace. First, of course, the legislature named the Bonneville cutthroat trout the official State Fish of Utah in 1997, replacing the rainbow trout, a non-native species. Second, Salt Lake City supported the restoration of Bonneville cutthroat trout within its municipal watershed, including Little Dell Creek and Little Dell Reservoir. Even the organizing committee for the XIX Winter Olympic Games, held in Utah in 2002, stepped up, agreeing to a change of venue for the Cross Country and Biathlon ski events to prevent harm to the newly established trout in the Little Dell watershed. The organizing committee also mandated changing the location of the public access road to the Super G and Giant Slalom site to protect the Bonneville cutthroat population of Wheeler Creek. Also in 2002, Salt Lake City's Hogle Zoo installed a native trout interpretive exhibit and did habitat restoration work on Emigration Creek, which flows through the zoo grounds.

On a larger scale, the Utah Reclamation Mitigation and Conservation Commission, another of the agencies that signed the interagency conservation agreements for Bonneville cutthroat trout, is restoring a 10-mile reach of the middle Provo River, which had been channelized and diked in the late 1950s and early 1960s. While the river's brown trout fishery will benefit most from this for now, it does signal willingness to improve stream habitat in the mainstem rivers that once supported native cutthroats. Along this same line, PacifiCorp announced it will remove the American Fork Dam in 2006. Again, brown and rainbow trout will benefit most initially in this Utah Lake tributary, but the State of Utah does plan to restore Bonneville cutthroat trout to the American Fork.

There is hope in these reports, enough to get some conservationists thinking ahead to a day when Bonneville cutthroat trout might again swim in Utah Lake and the other locations where they once thrived. Noah Greenwald, a conservation biologist with the Center for Biological Diversity, has said, "We would like to bring back the historic populations. That is something we think would benefit the people of Utah. The Bonneville cutthroat trout is the State Fish of Utah and is part of what makes Utah unique." [41]

41. The reports cited in the last three paragraphs came from an assortment of news releases and articles. You can learn more about the Utah Reclamation Mitigation and Conservation Commission's middle Provo River program as well as its other activities at the agency's website, *www.mitigationcommission.gov.* The final quote by Noah Greenwald, Center for Biological Diversity, appeared in an interview in the Provo, Utah *Daily Herald,* September 19, 2004.

Colorado River Cutthroat Trout, *Oncorhynchus clarkii pleuriticus*

CHAPTER 12

Colorado River Cutthroat Trout

Oncorhynchus clarkii pleuriticus: Chromosomes, 2N=64. Scales in lateral series 170–200+. Scales above lateral line 38-48 (this subspecies, along with the greenback cutthroat (Chapter 13), exhibit the highest scale counts of any of the cutthroat subspecies). Vertebrae 60–63. Gill rakers 17–21. Pyloric caeca 25–45. Spotting varies. Populations native to the upper Green River basin north from LaBarge Creek have small to moderate-size spots, typically smaller than the pupil of the eye, concentrated mainly on the caudal peduncle. Spots on those specimens that do appear toward the front of the body are mainly above the lateral line. Specimens from the Little Snake River drainage (tributary of the Yampa River) have much larger spots, more like those on greenback trout, but all degrees of spotting in between these extremes are also found. This subspecies, along with the greenback cutthroat, is the most colorful of all the recognized cutthroat subspecies. Although coloration is highly variable, there may be a diffuse band of pink or red along the lateral line and the sides may be brassy yellow to bronze-gold. The ventral region and belly may become bright crimson, especially on sexually mature males.

TRAPPER'S LAKE AND THE COLORADO HIGH COUNTRY

"Trapper's Lake. Make your first stop Trapper's Lake," urged a correspondent when I was laying plans for my initial trip to Colorado for the first edition of this book. "It's Colorado's Yellowstone Lake."

That was an apt description back then, in the sense that it was a high altitude, self-reproducing cutthroat fishery that the State of Colorado tapped for eggs to produce trout to stock in other waters in the north central region of the state. The Colorado Division of Wildlife informed me that spawn taking operations had begun at Trapper's Lake about 1914. The annual take at the time of my visit was around 350,000 eggs, but had been boosted to 600,000 at times. Eggs were collected in June and July, when the trout ascended the inlet creeks to spawn. They were hatched at the Glenwood Springs Hatchery, and the trout were then stocked into back country waters by truck, plane, or horseback. [1]

The setting of Trapper's Lake nearly defies description. It sits in a large, natural amphitheater at the edge of Colorado's Flattop Wilderness, surrounded on three sides by massive, flat-topped mountains. Dark evergreen forests march up the flanks of these mountains, but soon give way to towering walls of gray rock. The outlet of the lake, on the north side, is the White River, which soon turns west on its journey into Utah and its juncture with the Green River. The Yampa River heads up east of the lake, flowing north initially, then it too swings around to the west to also flow into Utah and join the Green River.

The name, Trapper's Lake, evidently goes back to the time of the Taos trappers, mountain men who ranged northward during the 1820s and 1830s to trap the streams of the Yampa, White, and Green river drainages for beaver. The Utes who inhabited the Yampa and White River country would certainly have known of the lake and utilized its trout for food, just as the Ute and Paiute bands that lived around Fish Lake and Panguitch Lake in Utah used the trout in their home waters. So too, no doubt, did the trappers and the white settlers and sportsmen who followed them. But the first actual mention of trout and angling at Trapper's Lake that I have been able to track down is in the memoirs of Luther S. "Yellowstone" Kelly, a scout for the army, who led a party of officers there in the spring of 1880:

> "On the north fork [of White River] near the head of Bill Williams' fork of Bear River [an early name for the Yampa], I found a beautiful little lake hidden in the forest, called Trapper's Lake, that was swarming with trout. To this spot I led a party of officers from the cantonment [which later became the town of Meeker, Colorado]. There was no trail, so I took a straight course through the woods, crossing hills and ravines. We found a little spot at the head of the lake where we might camp and have scant picking for the animals for at least one day and night. We made a pleasant camp in the open and caught with ease all the trout needful for our use and to take back to friends at the cantonment." [2]

1. Clee Sealing, Colorado Division of Wildlife, personal communication, March 12, 1982. Three agency reports and two papers published in fisheries journals provided me with additional details about Trapper's Lake and its cutthroat trout; see Snyder and Tanner (1960), Drummond and McKinney (1965), Colborn (1966), Drummond (1966), and Babcock (1971). And Bill Haggerty, regional information officer with the Colorado Division of Wildlife in Grand Junction, wrote a piece on the broodstock and stocking operation titled "A Trappers Lake Tradition" for the September/October 1987 issue of the Division of Wildlife's public magazine, *Colorado Outdoors*, which I obtained shortly after the first edition of this book came out.

2. See Kelly (1926). For you history buffs interested in the Taos trappers, Weber (1971) and Utley (1997) have additional details and references.

By five or six years later, Trapper's Lake had been "discovered" by others. The first permanent structures, a cluster of three log cabins intended to serve as a hunting and fishing lodge for the industrialist John Cleveland Osgood, owner of Colorado Fuel and Iron Company, were constructed there in 1886. [3]

David Starr Jordan and his students packed in and collected trout from Trapper's Lake in 1889. It was Jordan who assigned the name *pleuriticus* solely to the cutthroat trout of the Colorado River basin. Before that, the name had been applied at various times to trout from the Yellowstone drainage, the South Platte River, the Rio Grande River, and the Bonneville Basin. [4]

This garbling of nomenclature may have been due to confusion among the eastern ichthyologists who examined the specimens as to where they had actually been collected. Or maybe the combination of preservative and long-distance shipment muted features that would have distinguished them. Otherwise, it's hard to imagine anyone, let alone a scientist of that period, when every little nuance of difference was described as a new species, getting native trout from the Colorado basin mixed up, especially with Yellowstone or Bonneville cutthroats. The change in coloration that occurs between the native cutthroats of these basins is quite striking. Dr. Behnke once wrote, "although some of the headwater tributary streams may be only a few miles apart, it would be possible to know that one had crossed a divide when the somber, dull hues of *S. c. utah* or *S. c. bouvieri* gave way to the brilliant coloration of *S. c. pleuriticus*." [5]

Trapper's Lake has managed to escape man's heaviest hand. I suspect it's the wilderness aura as much as anything that drew the notable sportsmen of the world to visit here. Zane Grey came in on horseback prior to World War I, approaching overland from the east. He wrote a glowing, four-part series of articles on his adventures in Colorado for *Outdoor Life* magazine. His articles on Trapper's Lake appeared in the March and April 1918 issues. Years later, Ray Bergman fished here and wrote of his experiences in his immortal book, *Trout*. Bergman's party came in by car, over a road which by that time had been pushed up the White River, and they stayed at Trapper's Lake Lodge, a rustic camp for anglers and hunters that overlooks the lake and is still in operation today. [6]

I stayed at Trapper's Lake Lodge, too, on that first visit, sleeping in one of the cabins and sharing my meals with then-owners Red and Mary Gulliford and their crew of wranglers. "Contact Red Gulliford at Trapper's Lake Lodge," my correspondent had advised. "He knows all the cutthroats in that whole area by their first names, 'cause he's packed most of them in on his horses."

So that's what I had done, and in response Red wrote, "Bring along plenty of number 10 Cowdungs."

3. The story of the first hunting and fishing lodge at Trapper's Lake was taken from a side bar piece by Dave Weber, Colorado Division of Wildlife, that was included in the *Colorado Outdoors* article cited in footnote 1 above.

4. See Jordan (1891b).

5. See Behnke (1979). Dr. Behnke's comments were written back in the days when all the western trouts were classified in the genus *Salmo*, hence the abbreviation for *Salmo* rather than *Oncorhynchus* in this quote.

6. Zane Grey's Trapper's Lake account also appeared in abridged and edited form in two books under the Chapter title, "Colorado Trails." See Reiger (1972) and Rae (2001) for these references. Ray Bergman's book was published in 1938, with a second edition in 1964. In the 1964 edition, his Trapper's Lake adventure appears at pages 257–258. That Trapper's Lake remained as free of man's imprint as it has is credited to Arthur H. Carhart, a landscape architect hired by the Forest Service in 1919 to lay out a plan for roads and lakeshore home sites. This he did, but recommended that the plan *not* be implemented in order to preserve the natural beauty of the area. Surprisingly, the Forest Service complied and left the area alone. But that didn't stop the hatchery trucks. According to Anita

Red's version of the Cowdung turned out to be quite different from the standard ginger-winged, olive-bodied wet fly I had stocked in my fly box. Instead, he proffered a plain brown hackle pattern with a body of fluorescent green yarn and a brown hackle tail. I had good luck with both versions fishing the shoreline of Trapper's Lake for an hour or so each evening. Other trout flies that work there still are midge patterns in black, pale green, and cream colors, *Gammarus* scud imitations, and something tied on a size 12 or 14 hook to imitate the flying ants that appear in late summer. The ones I saw were a deep brown color. Red also told me there was a nice "Adams hatch" (likely the *Callibaetis* mayfly) and lots of caddisflies earlier in the season.

There are dozens of other lakes (ponds, Red called them back then) in the high country around Trapper's Lake. I fished as many as my limited stay would permit on that first visit. Wherever Red told me that "natives" occurred or had been stocked, I tried to go. "There are natives in some of those ponds as long as your arm," he asserted. I never connected with any of the real lunkers Red told me about, although nearby Coffin Lake did yield one nice fat one that measured 18 inches in total length.

Later, I visited a high alpine basin on Climax Molybdenum Company land just a couple of miles from Fremont Pass, elevation 11,316 feet. When you stand there looking up toward the head of that basin,

Martinez, Colorado Division of Wildlife, who first called attention to the hybridized condition of the native cutthroats of Trapper's Lake (see Martinez 1988), Yellowstone cutthroat trout were released into the lake from 1943 through 1950, and trout labeled "black spotted trout" were released in 1952 and 1965. Little Trapper's Lake, which drains into Trapper's Lake via one of the main spawning tributaries, was also stocked with "black spotted trout" in ten years between 1954 and 1970. Rainbow trout were stocked in Little Trapper's Lake in 1970, and catchable-size rainbows were stocked often in the North Fork White River, the Trapper's Lake outlet stream. Even though a barrier was constructed at the lake outlet in 1960 to keep rainbows from the North Fork White River out of Trapper's Lake, obvious cutthroat-rainbow hybrids, some of considerable size and weight, were taken from nearby

waters. A mounted specimen, a trout that weighed 10 pounds, 4 ounces, hung in Trapper's Lake Lodge at the time of my stay there, and just a week before my visit, another angler fly-fishing a pond out in front of the Lodge caught one that weighed 7 pounds, 12 ounces. Brook trout, also stocked in nearby waters in years past, have also appeared in Trapper's Lake, in particularly large numbers during the early 1990s. Efforts have been made to remove or at least reduce the number of these brookies, which spawn in the lake itself, by netting the spawning adults and by allowing anglers a liberal harvest of brookies when they fish Trapper's Lake. Trapper's Lake Lodge has changed hands a time or two as well, since my visit. During the devastating forest fire season of 2002, the main lodge and several other buildings were torched by the Big Fish Fire that burned more than 17,000 acres. The present

owner rebuilt and reopened in 2004, and I'm told that the affected area is also recovering nicely. Meanwhile, also in 2004, the Colorado Division of Wildlife mounted a project to resurrect the genetic purity of the Trapper's Lake cutthroat population by introducing genetically pure trout from Lake Nanita in Rocky Mountain National Park. Lake Nanita's cutthroat population originated from Trapper's Lake trout that had been released there in 1931, prior to the stocking of non-native subspecies and hybridization at Trapper's Lake. Prior to that 1931 release, Lake Nanita was fishless. Once in Trapper's Lake, the marked Nanita-origin trout will be allowed to pass into the spawning tributaries while the unmarked hybrid trout will be trapped and removed. Anglers may also be allowed to harvest any unmarked hybrid cutthroats they catch. For other details of this project, see Rogers and Wangnild (2004 [2005]).

you are viewing a wild mountain setting. But just behind you is an altogether different scene: a manmade reservoir and cars whizzing by on the highway to Leadville. And just over the hill to the south is the big Climax molybdenum mine, where, at the time of my visit, machinery was chewing away at the ridgeline of the Continental Divide.

What is unique about this basin and the tiny creek meandering through it is that it too contains a population of pure Colorado River cutthroat trout. These are small fish, 6 to 8 inches on the average, but colorful specimens of pure *pleuriticus* nevertheless, maintaining their headwater stream existence even though civilization is only moments away. [7]

HISTORICAL RANGE

At the point in geological time when the $2N=64$ ancestral cutthroat trout found their way across the divides separating the upper Snake/Portneuf/Bear River enclave I outlined in Chapter 2, there were several routes they could have taken to gain access to the Colorado River basin. I mentioned in Chapter 2 that Wallace R. Hansen of the U.S. Geological Survey postulated an ancient diversion of the upper Bear River into Muddy Creek, a tributary of Black's Fork of the Green River. A crossing from the Grays River to LaBarge Creek, a trout-filled tributary of the Green River, is another possibility. The Grays

River divide is up on the eastern flanks of the Salt River Range, but here, in an open grassy meadow separated by only a very gentle rise, the very headwaters of the Grays lie within about a quarter-mile of the head of LaBarge Creek. This is an easy crossing as evidenced by the fact that the Lander cutoff of the Oregon Trail passed through here. [8] The divide between the Hoback River and the Green River is easy as well, save for a short, fairly steep climb on the Hoback side just short of the divide itself, which is called The Rim, elevation 7,921 feet. The Green River side of The Rim is one of those long, high valleys where the descent is so gradual you hardly notice you're losing altitude.

Other possible avenues by which the large spotted cutthroat might have entered the Green/Colorado system were from the Bonneville basin via headwater transfers in the Wasatch Range. Rivers such as the Heber and Provo head up in topography where headwater stream transfers could have easily taken place to headwater streams of the Colorado system. But here, the fish would have had to await the melt-off of ice from the divides, because this region was subjected to alpine glaciation during major glacial maxima.

Thousands of years ago, when this invasion of the Colorado basin occurred, the waters of the basin were cold enough to support trout at least as far south as the beginnings of the Grand Canyon. On the west side of

7. The Climax Molybdenum Company abruptly ceased mining at this site in 1982 and mothballed the entire operation. The water stored in Clinton Reservoir, where I stood to look into the headwater basin of Clinton Creek, now provides an allocation for snowmaking at several of the area's ski resorts. Finespotted Snake River cutthroat trout were stocked in Clinton Reservoir in 1976 and 1977, where they co-existed with the native Colorado River cutthroat trout at least through 1981 or 1982 before they all were caught or died out, but no hybridization was ever detected in the Colorado River cutthroat population. Today, according to Tom Kroening, Colorado Division of Wildlife (personal communication, February 2005), Colorado River cutthroats alone swim the waters of Clinton Reservoir, some reaching sizes of 15 to 17 inches. There's no need to stock the reservoir anymore, Kroening says; the population can fully support itself through natural spawning in the creek upstream.

8. Frederick W. Lander, a government surveyor and engineer, built what we now call the Lander Cutoff, but known officially at the time as the Fort Kearney, South Pass, and Honey Lake Wagon Road, between 1857 and 1860 with a $300,000 appropriation from Congress. He was an

the basin, the trout occupied the Escalante and Dirty Devil rivers of Utah. On the east side, they occupied waters at least as far south as the San Juan River drainage of Arizona, New Mexico, and Colorado.

As the climate warmed, the mainstem waters of the Green and Colorado Rivers and their major tributaries below about 5,000 feet elevation grew too warm to maintain trout populations. Today, the towns of Green River, Wyoming, and Rifle, Colorado, mark the approximate downstream limits even for brown and rainbow trout in these rivers—except, of course, for the spectacular tailwater fisheries that have developed for browns, rainbows, and sometimes introduced finespotted Snake River cutthroats (Chapter 10) downstream of such major dams as Flaming Gorge Dam on the Green River and Glen Canyon Dam on the Colorado. The native Colorado River cutthroats long ago retreated to the colder tributary systems at the higher elevations. The distribution of the native cutthroats thus became a discontinuous one, as I have illustrated in Fig. 12-1. [9]

Even so, when the trappers and explorers came, and later the settlers and miners, the cutthroat was the only trout that occurred anywhere in the upper Colorado River basin. [10] All the great trout waters of the upper basin—the Green, the upper Colorado, the upper Yampa, the Gunnison, the Roaring Fork, the upper San Juan—all were cutthroat waters. Witness this report by Wilford Woodruff, a Mormon who

enthusiastic promoter of his new route, as this excerpt from his report testifies (see Lander 1861):

> "The route is but a few days shorter in travel than the present emigrant roads, but is so abundantly furnished with grass, timber, and pure water, with mountain streams abounding with fish, plains thronged with game, and so avoids the deleterious alkaline deposits of the south that it may be described as furnishing all that has been so long sought for through this section of the country—an excellent and healthy emigrant road, over which individuals of small means may move their families and herds of stock to the Pacific Coast in a single season, without loss."

9. Some distribution maps for this subspecies, notably in Young et al. (1996) and Behnke (2002), show the historical distribution as connected in a strip extending diagonally across the top of the basin, roughly from the Little Snake River to Green River, Wyoming. However, this strip is arid, so was unlikely to have provided trout habitat historically. Indeed, some emigrants described the route across this strip from South Pass to the Green River as mile after mile of alkaline desert. Avoiding this difficult area was one of Frederick W. Lander's objectives (footnote 8 above) in looping his route northwesterly from South Pass to an upstream crossing of Big Sandy Creek, then west to the Green River near the present-day town of Big Piney. Back in 1980, Behnke and Benson (1980) mapped the historical distribution of Colorado River cutthroat trout as discontinuous in this region, and I follow their lead in this book.

10. While it's true that the Colorado River cutthroat trout is the only trout to occur historically in the *upper* Colorado River basin, the Gila and Apache trouts *Oncorhynchus gilae gilae* and *O. gilae apache* (Dr. Behnke has long considered these to be two subspecies of just one species) occur in the upper Salt and Gila river sub-basins in Arizona and New Mexico, and once occurred in the Verde River sub-basin of Arizona, all tributary to the Gila River in the *lower* Colorado River basin.

was with the first group of pioneers led by Brigham Young and who later became President of the Church himself. Early in July 1847, Woodruff's party camped near Jim Bridger's fort, situated in the broad valley of Black's Fork, a tributary of the Green River, and it was there, on July 7th, that Woodruff recorded in his daily journal what may well be the first reference to fly-fishing in the American west.

Woodruff reported that trout inhabited more than a dozen "trout brooks" in the vicinity of the fort in numbers plentiful enough so that a good many of the brethren baited up with fresh meat or grasshoppers to try their luck. Woodruff had evidently learned fly-fishing in England while on a mission for the Church, but he had never tried it in America. He wrote:

"I threw my fly into the water and it being the first time that I ever tried the artificial fly in America or ever saw it tried, I watched it as it floated upon the water with as much interest as Franklin did his kite when he was experimenting in drawing lightning from the sky and as he (Franklin) received great joy when he saw the electricity descend on his kite string, so was I highly gratified when I saw the nimble trout dart at my fly hook, and run away with the line. I soon worried him out and drew him to shore.... I fished two or three hours during the morning and evening and caught twelve in all. One half of them would weigh three-fourths of a pound each...." [11]

These trout were, of course, Colorado River cutthroats. The populations Woodruff encountered were the same ones referred to twelve years later by Dr. George Suckley, a surgeon and naturalist who stopped at Fort

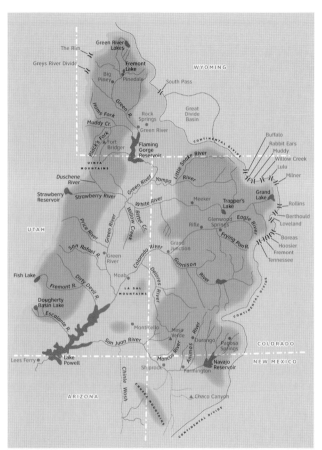

FIGURE 12-1. *Upper Colorado River basin and historical distribution of Colorado River Cutthroat Trout*

Map legend:
- Existing Streams, Lakes, and Reservoirs
- Historical Distribution of Cutthroat
- Upper Colorado Basin Boundary
-)(Passes
- Existing Towns and Cities
- ▲ Historical and Prehistoric Sites
- Continental Divide

11. See Cowley (1916 [1964]); also cited in Moon (1982).

Upper LaBarge Creek, Commissary Ridge in the left background. The LaBarge Colorado River Cutthroat Restoration Project is the largest attempt at a meta-population restoration ever to be undertaken in Wyoming. PHOTO BY AUTHOR

Bridger in August of 1859. Suckley, too, was a fly-fisherman and he, too, had his gear along. Like Woodruff, he, too, found that the trout, though not large, were plentiful and would take a fly readily:

> "It [the trout] is abundant in Black's Fork, from which on the 25th of August we caught half a dozen, and on the following day about forty, with the artificial fly, to which they rose exactly in the manner of their more easterly relatives, and greedily seized, like unsophisticated fish, as they were, scarcely learning caution or timidity until pricked once or twice by the alluring and deceitful bait. Probably but few artificial flies, if any, have ever before been cast on these waters. One specimen, about ten inches in length, caught with a red-hackle, was selected for examination and description…. Upon inquiry at Fort Bridger, we learned that 17 or 18 inches might be considered the maximum size in those waters, and out of forty or fifty fish it is rare to find one over a foot in length." [12]

Not all emigrant parties were as well equipped with fishing tackle as Woodruff and Suckley, but still, they managed to catch their share of the trout. Catherine Sager Pringle came west to Oregon in 1844, and wrote this about how her party went about it:

> "At Fort Bridger the stream was full of fish, and we made nets of wagon sheets to catch them." [13]

Another early angler was Lewis B. France, whose book, *With Rod and Line in Colorado Waters*, published in 1884, may have been the first by a national writer to promote fly-fishing in Colorado. France wrote an earlier article for the August 19, 1882 issue of *The American Angler* about a week on the Grand (the Grand was the early name for the Colorado River upstream from the Green River confluence). Here's his report of the week's catch by himself and one other angler:

> "We have not kept close count, but it is safe to say we have caught one hundred and fifty pounds of mountain trout, no fingerlings, and none wasted…. On Friday we brought in twenty-one trout; nine fell to my score; dressed, they weighed thirteen pounds, that catch we considered a pretty fair average, and we spent about an hour on the stream where we had been fishing all the week." [14]

In the southeastern corner of the range, the upper San Juan River and its tributaries drain out of the San Juan Mountains of Colorado. Most authorities map the historical distribution of Colorado River cutthroat trout to encompass upper San Juan River tributaries in New Mexico and extend down the San Juan River itself, at least as far as the present site of Navajo Dam. New Mexico authorities do not dispute that this area was inhabited by Colorado River cutthroat trout historically, but they do note that no museum records from New Mexico exist for this subspecies, and the trout are no longer present in this portion of the historical range. [15]

The historical record leaves no doubt that trout did populate the tributaries of the upper San Juan on the Colorado side of the border. Benjamin Alfred Wetherill, who, with his brother, discovered the awe-inspiring Anasazi cliff houses of Mesa Verde, lived on a ranch established on his family's homestead on the Mancos River a few miles downstream from the town of Man-

12. See Suckley (1874).

13. See Pringle (1905 [1989, 1993]). Other early journals and records that help to define the historical distribution of Colorado River cutthroat trout include Palmer (1847), Bryant, (1848 [2000]), Burton (1862), Wislizenus (1912), Jackson (1929), Brewerton (1930 [1993]), Arms (1938), Field (1957), Hafen and Hafen (1959), Stewart (1961), Porter and Davenport (1963), The Fitzpatricks (1966), Gowan and Campbell (1975), and Fletcher (1977). Reports of the Hayden surveys of 1873–75 in Colorado also contain references to where the surveyors observed trout or caught them themselves; see Annual Reports (1868–83).

14. See France (1882 [1980]). Lewis B. France came to Colorado Territory in 1861. He was a lawyer by profession and, after statehood had been granted in 1876, was appointed clerk of the Colorado State Supreme Court. He was also an avid fly-fisherman who became noted for his stories about fly-fishing and other outdoor adventures that appeared often in national sporting journals from the 1880s until his death in 1907. During the winter of 1862, he built what may have been the first fly rod ever to be constructed in Colorado—a 10-footer with a butt section of pine wood, the

cos and not far from Mesa Verde itself. Writing of this period (the late 1870s and early 1880s) in his autobiography, Wetherill said:

> "Between stretches of work building fences and digging ditches for irrigation, we ranged around the valleys and over the fields to fish for the beautiful trout."

The Wetherills took a deep interest in archaeology and in the ancient people who had lived in the cliff houses and, they asserted, in other houses built not among the cliffs where they were hidden from direct view but out in the open, along the Mancos and other nearby streams. Speculating about the foods these ancient peoples might have utilized, Wetherill wrote:

> "For meat, they had the usual wild game of all kinds—sage hens, grouse, ducks, prairie dogs, squirrels, beaver, rabbit, deer, and the like, as well as the ferocious animals such as bear, mountain lion, wolves and cats. The mountain streams abound in trout, so that [the trout] could be used for meat, although they could have had the same superstitions regarding the fish tribe that the Navajos have."[16]

I touched briefly on the Navajo and Apache superstition against eating fish in Chapter 11, where I contrasted it with the Utes, who regularly used trout for food wherever they could obtain them. Obviously, the Wetherills knew of this tradition as well.

In addition to the two large areas of historical distribution sweeping down along either side of the upper Colorado River basin, Fig. 12-1 also shows four smaller areas of disjunct historical distribution within the basin.

Reading from north to south, these are: 1) the highlands of the Book Cliffs region encompassing perennial tributaries of the Bitter and Willow creek drainages; 2) the La Sal Mountains east of the town of Moab; 3) the Abajo Mountains west of Monticello, Utah; and 4) the Chuska Mountains that straddle the Arizona-New Mexico border within the Navajo Reservation. Each of these areas has its share of anecdotes passed down from local inhabitants or "old-timers," whether written down or not, that tell of trout in the perennial streams long before hatchery trout were ever stocked in these waters. In addition, and even more indicative of their historical presence, relict specimens have been collected from each area that have proved to be either pure or introgressed Colorado River cutthroat trout.[17]

The Book Cliffs rise up out of the East Tavaputs Plateau that begins near Green River, Utah, and extends eastward into Colorado. From an elevation of about 4,100 feet near Green River, the area rears abruptly upward to nearly 8,500 feet before falling away gradually to the north to a base altitude of about 5,500 feet. Stands of aspen and fir near the top give way to pinion, juniper, sage, and desert shrub as the elevation decreases. Looking northward from up in the heights, broad pinion- and juniper-studded vistas are broken by deep valleys, some containing perennial streams, including Bitter and Willow creeks and their tributaries. This is an area rich in birds and wildlife, including

middle and tip sections of cedar, and a buggy whip handle for the grip. He had a local watch repairman make the ferrules and guides for him for the princely sum of $7 in gold dust. His wife pronounced the finished product "just perfect, the finest rod in Colorado Territory." France told the story of how he built that rod, and its eventual fate, in an article titled "Memories of an Angler" which appeared in *Western World*, a Colorado magazine, in September 1907. You can also read about it in historian John H. Monnett's book, *Cutthroat and Campfire Tales: the Fly-Fishing Heritage of the West* (1988).

15. See Sublette, Hatch, and Sublette (1990).

16. The Wetherill autobiography was written up originally as a report to the U.S. Department of the Interior on the brothers' activities in and around Mesa Verde. It was later published in book form; see Fletcher (1977).

17. News releases from The Nature Conservancy, who acquired lands in the Book Cliffs in the early 1990s, and later descriptions of the area published by the Bureau of Land Management (see, for example, *www.blm.gov/utah/vernal/rec/bcrec.html*) alluded to the historical occurrence of Colorado River cutthroat trout in the area's perennial streams. Surveys of streams in the Book Cliffs, the

waterfowl, shorebirds, blue and sage grouse, golden eagles and other birds of prey, and mule deer, elk, antelope, cougars, bears, and the occasional bighorn sheep. It's also rich in natural gas, oil shale, and tar sands, with many wells already producing and many more oil and gas leases on file. Livestock grazing has been a major land use over the years and there has also been some logging, although the timber resources aren't substantial. Approximately 70 percent of the area is administered by the Bureau of Land Management, 25 percent by the School and Institutional Trust Land Administration (a State of Utah entity), and 5 percent is private. The Hill Creek Extension of the Uintah and Ouray Indian Reservation also extends into the area along the west side.

The La Sal Mountains are what geologists call a *laccolith*, which is an upthrust of igneous rock through a layer of sedimentary deposits. The range towers more than 7,000 feet above the surrounding slickrock canyons and mesas, with Mt. Peale, the highest peak, reaching 12,721 feet in elevation. In this dry region, the range is a weather-interceptor. The slopes of the La Sals receive nearly three times more rainfall than nearby areas, and temperatures are cooler as well. Forests with stands of ponderosa pine, aspen, spruce, and fir border meadows rich with grasses and forbs. And above these, there is alpine tundra, a real rarity

in this region. Humans have been here off and on for 11,000 to 12,000 years. Clovis, Folsom, and Plano points have been found on the benches and mesas. The Anasazi occupied the area for a time, and later, the Utes. Navajos visited as well for food and medicinal plants. The Mormons tried to establish a settlement at the site of Moab in 1855, but were driven out by the Utes. It wasn't until the late 1870s that white settlers finally gained a foothold. Although there were some attempts at mining in the La Sals, livestock grazing and logging have been the mainstay land uses. Early on, these activities went totally unrestrained and the streams, valleys, and hillslopes suffered greatly from erosion and flooding. Today, much of the area is within the bounds of the Manti-La Sal National Forest.

A unit of the Manti-La Sal National Forest also embraces the Abajo Mountains. This range rises to 11,360 feet just west of Monticello, Utah, but drops off farther to the west into redrock plateaus and canyons. The Dark Canyon Wilderness abuts on the west, and Canyonlands National Park lies off to the northwest. The Abajos are weather-interceptors as well, and are forested with aspen and spruce, earning them the local name Blue Mountains, for the color of the forest cover when seen from a distance. This area, too, has a long history of human use. In my one and only excursion into the Abajos some years ago, I remember pass-

La Sal Mountains, and the Abajo Mountains, compiled for a Conservation Agreement signed in 1997 (see Lentsch and Converse 1997), reported Colorado River cutthroat trout introgressed with rainbow trout in several small Willow Creek tributaries and two short segments of Bitter Creek in the Book Cliffs area, and in one small segment of Indian Creek in the Abajo Mountains. One genetically pure population of Colorado River cutthroat trout was found in Beaver Creek in the La Sal Mountains during the same surveys.

ing several mines, numerous stock ponds, and roads leading to private ranch inholdings. I also remember passing a sawmill on the outskirts of Blanding, Utah, just south of the National Forest boundary. So, like the La Sal Mountains, a list that includes livestock grazing, mining, and logging would probably cover the principal land uses.

The Chuska Mountains, in the Navajo Reservation, rise to nearly 10,000 feet in elevation. Their roughly southeast-to-northwest orientation extends them across the line from New Mexico into Arizona, although in this territory it's only a line on a map. The Chuskas are part of an upward warp in Earth's crust that geologists call the Defiance Uplift. The adjacent and more gentle terrain of the Defiance Plateau, separated from the Chuskas by the narrow valley of Black Creek, is also part of this uplift, but rises to only 7,000 to 8,000 feet. The combined uplift area is the wettest portion of the entire Navajo Reservation, generating close to two-thirds of its average annual surface water runoff. Much of this runoff drains west off the Chuskas into Canyon del Muerto and Canyon de Chelly, and thence to the San Juan River via Chinle Wash. The east slopes drain to Peña Blanca Creek and to the Chaco River system that also drains the Chaco Canyon Anasazi site, and thence again to the San Juan River. Forest mantles the Chuskas, with stands of spruce, fir, Douglas-fir, and aspen on the north-facing slopes and ponderosa pine along the crest and the east slopes. Some of those ponderosa stands were just magnificent, with old-growth trees that traditional Navajos revered as "grandfather trees." [18]

But other, more pragmatic tribal interests had eyes on those stands as well. Between 1960 and 1992, Navajo Forest Products Industries, an industrial forestry enterprise created by the tribe and backed by the Bureau of Indian Affairs, cut and processed an average of 40 million board feet of timber a year from Chuska and Defiance Plateau forests, with little, if any, regard for traditional uses or values, and, evidently, an equal disregard for any obligation to reforest the tracts that were cut. In the early 1990s, as the cutters began removing the "grandfather trees," an intense intertribal conflict erupted in which one forest activist died under troubling circumstances. Navajo Forest Products Industries ended up a casualty as well, declaring bankruptcy in 1995 and closing down its sawmill. Things quieted in the forests after that, at least for a time, as the Navajo Nation's Department of Forestry took on more responsibility for managing and preserving their cultural values as well as their natural resource values. [19]

If the historical distribution of Colorado River cutthroat trout ever extended into Arizona, it was certainly in the streams of the Chuska Mountains. In his 1979

18. The Chuska Mountains themselves may have great cultural significance to the Navajo people. One story that I heard, or maybe read somewhere, says that the Fifth World (our present one, according to Navajo tradition) will end when evil causes Father Sun to depart, thereby making the world cold. At the onset of that period, Beautiful Mountain in the Chuskas will open up so that the People may find refuge. Later, when the time is right, they will call back Father Sun, who will return the warmth and the People will emerge into the Sixth World.

monograph, Dr. Behnke told of receiving a single specimen from one of these streams that had been collected by U.S. Fish and Wildlife Service biologists. "Although hybridized with rainbow trout, it still retained some of the strong coloration of *pleuriticus*," he wrote. [20]

Which brings me to another early break with traditional Navajo taboos that occurred in the Chuska Mountains, namely the stocking of trout for fishing and consumption. Small, shallow, sometimes intermittent lakes occur in numerous depressions along the crest of the southern Chuskas where the terrain is relatively flat. At least one of those lakes, Whiskey Lake, was dammed to raise its level—not to store water for irrigation, as one would expect in this country, but for fishing, which was maintained by a stocking program. Trout were stocked in streams as well, including Tsaile Creek, which flows for several miles through Chuska forestland before entering Canyon del Muerto. Upper Tsaile Creek held rainbow trout and brook trout as recently as 1999, according to a guidebook to Arizona trout fishing that was published that year, [21] and I imagine they're still there.

As for the Book Cliffs, La Sal Mountains, and Abajo Mountains, maps included in a State of Utah publication on fish, game, and fur animals introduced into the state prior to 1950 showed both the La Sal and Abajo mountains as areas of rainbow trout and brook trout distribution, but not the Book Cliffs. [22] Introductions of rainbow trout into Book Cliff streams must have occurred later, because surveys of these streams, compiled in 1996 for a Conservation Agreement for the Colorado River cutthroat subspecies, revealed small populations of Colorado River cutthroat trout introgressed with rainbow trout in several small tributaries of Willow Creek and in two short segments of Bitter Creek. Those same surveys revealed an additional introgressed Colorado River cutthroat population in a short segment of Indian Creek in the Abajo Mountains, and one genetically pure population occupying 11 miles of stream habitat in Beaver Creek in the La Sal Mountains. [23]

LIFE HISTORY AND ECOLOGY

In general, populations of this subspecies exhibit the same three life history strategies that are common to the other interior subspecies: [24]

- A lacustrine, or lake-associated, life history, wherein the fish feed and grow in lakes, then migrate to tributary streams to spawn.
- A fluvial, or riverine (potamodromous), life history, wherein the fish feed and grow in main river reaches, then migrate to small tributaries of these rivers where spawning and early rearing occur.
- A stream-resident life history, wherein the trout complete all phases of their life cycle within the streams of their birth, often in a short stream reach and often in the headwaters of the drainage.

However, the Colorado River cutthroat does appear to enjoy a broad-based, rather generalized ecology, and individuals that exhibit one life history strategy

19. Actually, the Navajos were not the first to log in the Chuskas. Archaeological evidence shows that the Anasazis of Chaco Canyon cut and transported logs from the Chuskas, for use as construction beams and probably also for fuel, from about A.D. 1000 to A.D. 1170, which is about when their tenure at Chaco Canyon came to an end. Jared Diamond's recent book, *Collapse: How Societies Choose to Fail or Succeed* (2005), has a discussion of the Chaco Canyon Anasazi society and reasons for its collapse at pages 136–156, including the key references to evidence of their logging in the Chuskas. The story of the puzzling death of Navajo environmental activist Leroy Jackson can be found in Alex Shoumatoff's *Legends of the American Desert* (1997, [1999]). For more on the logging and forestry operations of the Navajo Nation and their cultural and economic consequences, see Pynes (2000, 2002).

20. See Behnke (1979) at page 117.

21. See Johnson (1999). The information about the lakes of the southern Chuskas came from Wright (1964).

22. See Popov and Low (1950).

23. See Lentsch and Converse (1997).

24. Good overviews of Colorado River cutthroat life history and ecology can be found in Dr. Behnke's mono-

in their native habitat can adapt to other habitats and establish populations that exhibit a different life history strategy. This is an attribute fishery managers are exploiting as they work toward increasing the number of populations within the historical range. Most of these projects involve the transfer of trout from stream-resident populations into ponds, small lakes, or reservoirs to develop broodstocks that exhibit a lacustrine life history. Often the progeny are reintroduction into streams, where they revert back to a stream-resident life history.

The Lacustrine Life History Strategy

Along with its use as a broodstock lake, another point that's been made in referring to Trapper's Lake as the Yellowstone Lake of Colorado is that, until recently, just about everything we knew about the lacustrine life history and ecology of Colorado River cutthroat trout was based on observations of the Trapper's Lake population. To be sure, Colorado River cutthroat trout originally existed in all natural lakes and ponds capable of supporting trout that were accessible to them following retreat of the alpine glaciers that covered much of the high country during Pleistocene ice advances. Finis Mitchell, the famous 1930s-era packer and guide of the Wind River Range, noted the absence of trout in many high lakes when he set up his first camp at Mud Lake, where the Big Sandy River exits

the mountains. Only five of the hundreds of lakes in the high country accessible from that camp held native trout. Most all of the others were in glacier-carved cirques with steep waterfalls below that had prevented trout from getting to them. [25]

Among the larger lakes within the historical range that did hold notable and abundant native cutthroat populations, in addition to Trapper's Lake, were Fremont Lake in Wyoming (near Finis Mitchell's territory), Grand Lake in Colorado (Colorado's largest and deepest natural lake), and Fish Lake in Utah.

Fremont Lake, at 10, 11, or 12 miles long and one-half or one mile wide (you'll find that even contemporary descriptions vary), is Wyoming's second largest natural lake and also, at 609 feet, its deepest. It lies in a glacier-gouged trough that was dammed by the glacier's terminal moraine. Actually, its original name was Stewart Lake, after Sir William Drummond Stewart, a Scottish soldier and nobleman, who accompanied supply caravans to the annual rendezvous of the fur trappers each year beginning in 1833. In 1837, with the artist Alfred Jacob Miller, he visited the lake, where Miller "painted scene after scene." Stewart returned in 1843, this time with newspaper correspondent Matthew C. Field, and they fished at the lake for several days. Meanwhile, John C. Fremont had come in 1842 on his first exploring expedition. His main party camped at the lake while Fremont and several others climbed nearby

graphs and book (see Behnke 1979 [1981], 1992, 2002) and in Behnke and Zarn (1976), Binns (1977), Behnke and Benson (1980), Speas et al. (1994), and Young (1995).

25. See Mitchell (1975). Mitchell identified the original five fish-bearing lakes in the area in which he operated as Boulder, Dad's, Donald, Big Sandy, and Black Joe lakes. All were inhabited by native Colorado River cutthroats. In the seven years he operated (from 1930 to 1937), Mitchell used his pack trains to stock 314 of the barren lakes with trout, but he planted whatever species he could obtain from fish-culture operations. These included rainbow, cutthroat (probably Yellowstones), and golden trout, the latter obtained from California; also brook trout and brown trout.

Fremont Peak. Fremont called the lake Mountain Lake, but evidently his own name was inserted in subsequent mapping. As for the native trout, Matthew C. Field's dispatches are the only ones that actually described these fish. Although abundant and easily caught, he wrote, they were uniformly about a foot in length. Fremont Lake still attracts anglers even though the once abundant native cutthroats are gone. Today's attractions are rainbow trout, brown trout, kokanee, and lake trout, some of the latter said to approach 40 pounds. [26]

Grand Lake, near the headwaters of the Colorado River, has been attracting tourists for many thousands of years, it would appear. There is evidence that Clovis hunters passed this way, no doubt in search of the mammoths that once ranged in the area. Later paleo-Indian hunter-gatherers also visited the area seasonally, as did the Arapaho and their allies, the Cheyenne and Sioux, in more recent times. Ute bands lived throughout Middle Park and fought with the Arapaho and their allies, but the Utes had a tradition that Grand Lake itself was evil, so they may have avoided it. [27] John C. Fremont passed through in 1844, but hastened on owing to the belligerence of the Arapahos he met, and apparently missed Grand Lake. Sir St. George Gore, an Irish nobleman notorious for the amount of game and fish he slaughtered on a three-year junket through the western plains and Rocky Mountains, brought a caravan of servants, baggage, and wagons through here in 1855,

26. The town of Pinedale, only 3 miles from Fremont Lake, lends its name as the "type locality" for deposits from the major Pleistocene ice advance that formed the lake (i.e., the Pinedale glaciation), because other nearby deposits from the same glaciation were the first of their type to be examined by earth scientists. For accounts of Sir William Drummond Stewart's adventures in the area and Matthew Field's descriptions of the fishing, see Porter and Davenport (1963) and Field (1957). Fremont's sojourn at the lake and its environs is described in his report of the exploring expedition to the Rocky Mountains in the year 1842 (see Fremont 1845 [2002]).

27. The Ute name for Grand Lake is Spirit Lake, and its story goes like this. One day a band of Ute hunters, accompanied by their wives and children, were camped on the shore of the lake when they were attacked by Arapaho and Cheyenne warriors. The Ute men engaged the attackers while the women and children took to rafts on the lake for safety. As the battle raged, a thunderstorm blew in over the mountains and whipped the waters into treacherous waves. The Ute men eventually beat off the attackers, only to learn that the stormy waters of the lake had torn the rafts apart and swallowed up their families. Next morning, as tendrils of mist rose off the now-calm waters, the spirits of the vanished women and children were seen to rise up with them. The Utes thereafter avoided the lake, because it is haunted by the spirits of their dead. The Arapaho have a couple of Spirit Lake stories in their tradition as well, one of which is similar to the Ute story, so the battle must have taken place at some point in time. But in the Arapaho version it was they who won, killing many of the Ute men and driving the rest up Shadow Mountain. The different versions of this story can be found on the Internet at *www.grandlakechamber.com/activities-glhistory.htm* for the Ute version and *www.colorado.edu/csilw/arapahoproject/* for the Arapaho version (both accessed March 11, 2005).

guided by none other than mountain man Jim Bridger. Sir St. George even had one retainer whose only job was to keep him supplied with fresh-tied fishing flies.

Joseph Westcott, the first white settler at Grand Lake in 1867, subdivided his holdings and sold lots. Some of the state's most prominent families built summer homes, coming in by wagon or stagecoach from Denver and Fort Collins. During the mining boom years, the community became a supply hub for the various towns and mining camps that sprang up first along Clear Creek on the east side of the Continental Divide, then later in areas north and west of Grand Lake. This included market fishing to help feed the hungry miners. It is said that native trout were seined out of Grand Lake by the tubful and the wagon load, with the occasional specimen scaling 20 pounds in weight.

Water development for diversion to the east side of the Continental Divide for power generation and irrigation has also been big around Grand Lake. Shadow Mountain Reservoir just downstream from Grand Lake, and Lake Granby downstream from that, were constructed to store water for transport through the mountains to the east side. These three lakes continue to draw tourists and anglers, but, like Fremont Lake in Wyoming, the native cutthroats are gone. Today a mix of rainbow, brook, brown trout, and kokanee provide the bulk of the catch, with lake trout to about 25 pounds also adding to the sport. [28]

Fish Lake, in Utah, sits at an elevation of 8,800 feet in the headwaters of the Fremont River. It's a little over 5 miles long and about three-quarters of a mile wide. Even though its maximum depth is 117 feet, it has extensive shallow areas and abundant growth of aquatic vegetation, and has always been a rich producer of trout food. Early researchers recorded snails, caddisflies, mayflies, midges, leeches, *Gammarus*, and *Daphnia* among the aquatic trout foods along with ants, wasps, and beetles blown in from the surrounding landscape. The mottled sculpin was indigenous to the lake along with the native Colorado River cutthroat trout. The trout were said to have reached weights of 5 to 6 pounds in this environment, and would appear each year by the thousands in the six spring-fed inlet streams where they spawned. These trout in turn attracted the bands of Paiute and Ute Indians that inhabited the area. The Indians would come every year to take the trout by netting, trapping, spearing, shooting them with arrows, or sometimes by simply clubbing them in the spawning tributaries. Clubbing—that's how Kit Carson's men got them in 1848 when they stopped briefly at what was probably Fish Lake on their way from California to Taos. This fishery was so attractive that in 1889, a group of Mormons organized as the "Fremont Eragation Companys" [their spelling, not mine] took the unprecedented step of actually purchasing fishing rights from one of the Indian bands. The price they paid? Nine

28. This exposition on the pre-history and history of Grand Lake was synthesized from several sources including Fremont (1845 [2002]), Lavender (1968 [1981]), Merritt (1985), Benedict (1992), Utley (1997), and Geary (1999). More information on water diversion from the Grand Lake area can be found on the Northern Colorado Water Conservancy District website at *www.ncwcd.org/projects/features/East_Portal.asp*, which I accessed March 11, 2005.

horses, 500 pounds of flour, one beef steer, and one suit of clothes (the latter no doubt for the chief). Although the new owners called themselves an irrigation company, their real purpose was market fishing to supply Mormon settlements and towns. [29]

As the native cutthroat population dwindled under all this pressure, Utah authorities and the U.S. Bureau of Fisheries began introducing exotics, which of course just hastened the demise of the native cutthroats. Among the species these agencies released in Fish Lake were chum and coho salmon, landlocked (Atlantic) salmon, rainbow, brook, and brown trout, lake trout, and splake (a male brook trout × female lake trout hybrid). There have been some illegal or accidental releases as well, among them Utah chub, Utah sucker, redside shiner, and yellow perch. These days, rainbow trout and splake comprise the bulk of the angler catch of salmonids, with lake trout available for the trophy hunters. Utah chub are also abundant, as are yellow perch, which are attracting a following of anglers. The yellow perch was illegally introduced around 1970 and may now be the most abundant fish in the lake. [30]

The native cutthroat trout did hold out for a while in Fish Lake. A U.S. Bureau of Fisheries paper listed "about twenty" cutthroat spawners in 1933 and anglers would occasionally catch one over the next few years, but then they were gone. [31]

29. Early investigations of trout food in Fish Lake were carried out by the U.S. Bureau of Fisheries (see Hildebrand and Towers 1927 and Hazzard 1935) and Utah State Agricultural College (see Madsen 1942). Two good overviews of the Indians of this region and how they subsisted can be found in Virginia McConnnel Simmons' book on the Ute Indians of Utah, Colorado, and New Mexico (see Simmons 2000) and R.L. Holt's book on the ethnohistory of Utah's Paiutes (see Holt 1992). Holt's book has the story of the Mormon purchase of fishing rights on Fish Lake for the purpose of market fishing. For an account of Kit Carson's 1848 journey from California to Taos, including the stop at what may have been Fish Lake, see Brewerton (1930 [1993]).

30. Hazzard (1935), along with Popov and Low (1950) and Sigler (1953), listed early records of non-native fish introductions at Fish Lake. For more on the yellow perch situation, including tips and encouragement on how to catch them, see the Utah Division of Wildlife Resources website at *www.wildlife.utah.gov/fishing/fish_lake/yellow_perch.html* (accessed March 14, 2005).

31. Hazzard (1935), in reporting the depleted number of native cutthroat spawners in the 1933 spawning run, commented further that "the native cutthroat trout, *Salmo pleuriticus*, has been virtually displaced by plantings of rainbow, mackinaw and eastern brook trout."

So, of necessity, to understand the life history and ecology of the lake-dwelling form, it's back to Trapper's Lake. Until fairly recently, just about everything we knew about the lacustrine life history and ecology of these trout was based on focused studies made of the Trapper's Lake population by Colorado Division of Wildlife biologists in the years between 1958 and 1968. [32]

At Trapper's Lake, the cutthroats spawn in three inlet creeks. Migration first begins in Cabin Creek (which receives about 75 percent of the spawners) soon after the ice recedes from the lake. At the 9,604-foot elevation of Trapper's Lake, this usually happens in early June, when the water temperature in Cabin Creek is about 45 degrees Fahrenheit. First entry into Heberton and Frasier creeks, which are somewhat colder tributaries, usually lags Cabin Creek by one to two weeks. The peak of the Trapper's Lake spawning run occurs about the last week of June, when water temperatures in each of the tributaries have reached about 50 degrees Fahrenheit, and then dwindles to an end about the middle of July. If homing fidelity to natal tributaries is as true in Trapper's Lake as it is in other cutthroat subspecies, then this population may be structured into three subpopulations or races based on their tributary of origin.

The commencement of spawning at Trapper's Lake even as ice may still be receding from the lake is con-sistent with a handful of observations made elsewhere. For example, Lt. George Brewerton, narrator of Kit Carson's journey from Los Angeles to Taos in 1848, including the party's stop at what may have been Fish Lake sometime in the last week of May or first week in June, wrote: "In this same section of country, we encamped one evening upon a beautiful little lake situated in a hollow among the mountains, but at so great an elevation that it was, even in summer, surrounded by snow, and partially covered with ice." Yet its spawning tributaries were "swarming with fish," according to Brewerton. [33] At the Williamson Lakes, California (three lakes ranging in elevation from 11,200 feet to 11,760 feet, well above timberline in the Sierra Nevada Mountains), where Colorado River cutthroats from Trapper's Lake were stocked in 1931, spawning does not commence until midsummer. [34]

Like the trout that Matthew C. Field found so easy to catch out of Fremont Lake in 1843, the Colorado River cutthroat trout of Trapper's Lake run to the small side. Adults in the spawning runs studied by the Colorado Division of Wildlife averaged 10.4 inches in total length at age 2 (comprising about 18 percent of the spawning run), 11.9 inches in total length at age 3, 12.5 inches in total length at age 4 (ages 3 and 4 together comprised about 80 percent of the spawning run), and 14.5 inches in total length at age 5 (comprising

32. Three Colorado Division of Wildlife agency reports and two papers published in fisheries journals give details about Trapper's Lake and the life history of its cutthroat trout; see Snyder and Tanner (1960), Drummond and McKinney (1965), Colborn (1966), Drummond (1966), and Babcock (1971).

33. See Brewerton (1930 [1993]).

34. The Williamson Lakes are a group of five to seven lakes in high, difficult terrain in the Sierra Nevada Mountains of Inyo County, California. These lakes are in a portion of the John Muir Wilderness that is additionally managed as the Big Horn Sheep Zoological Area where public use, though not forbidden, is also not encouraged. The lower three of the Williamson Lakes were stocked with trout from eggs taken at Trapper's Lake in 1931, in exchange for a shipment of golden trout eggs from California. The State of California has managed this population strictly as a gene bank for future restocking of native waters. In 1987, when introgression of the Trapper's Lake population was pretty much acknowledged, approximately 300 of these trout were collected and transported back to Colorado. There they were stocked in Bench Lake in Rocky Mountain National

2 percent of the spawning run). Several much larger trout were captured, including one age-5 individual that measured 20.3 inches in total length, but, owing to their dull coloration, these trout were deemed to be Yellowstone cutthroats, descended no doubt from a group that had been released in the lake some years earlier. The ominous thing about these trout is that they were trapped on spawning runs into Trapper's Lake tributaries together with the Colorado River cutthroat trout, which doesn't speak well for the genetic purity of the latter population.

Another notable aspect of the spawning runs of Colorado River cutthroats at Trapper's Lake was the sex ratio of the trout, which was skewed toward females. The ratios measured over six seasons of study were 1 male to 2.5 females. Fecundity of females at Trapper's Lake ranged from 670 to 850 eggs per female for trout in the 11.9- to 12.5-inch size range. At Dougherty Basin Lake, Utah, where, in the 1990s, a wild broodstock was established by transplants of trout from a stream-dwelling population, females averaging from 11.1 to 12.8 inches in length in succeeding spawning seasons produced 483 to 644 eggs per female. [35]

After completion of the spawning act, most of the adult cutthroats remain in the spawning tributaries for about three weeks, then drop back down to the lake. A small but unreported percentage stays longer, possibly remaining in the streams until after the downstream migration of fry runs its course, before returning to the lake in the fall. Adult cutthroats that survive in the lake over the next winter period will spawn again the following year.

The first appearance of swimup fry in the Trapper's Lake natal tributaries usually occurs about a month and a half after the first spawning activity occurs (in late July if first spawning took place in early June). These fry begin moving downstream to the lake very soon after emergence, at a size of only 24 to 34 mm (from a little less to a little more than an inch as the migration progresses), with downstream movement peaking in mid-August and dwindling to a close in early September. All of this downstream movement of fry was observed to take place in the evening hours, between 6 PM and midnight, with peak downstream movement coming earlier as daylength shortened. A very small percentage of the young-of-the-year fry (less than 1 percent in the studies I'm citing) evidently do remain in the natal tributaries after downstream movement ceases, presumably to overwinter there and migrate to the lake the following spring.

I indicated earlier that Trapper's Lake produces invertebrate aquatic trout foods including midges, mayflies, caddisflies, and *Gammarus*, and the surrounding landscape contributes terrestrial insects

Park, in hopes of developing a new broodstock for additional introductions. For more on the Williamson Lakes population, see Gerstung (1979), Gold et al. (1978), Martinez (1988), and Pister (1990). Another out-of-basin lacustrine population of Colorado River cutthroat trout may still occur in Ypsilon Lake, also in Rocky Mountain National Park but in the South Platte drainage on the east side of the Continental Divide. This population apparently originated from a release of trout that took place in 1915, perhaps from Trapper's Lake or maybe from some closer population on the west side of the Continental Divide. My information on this population goes back to Dr. Behnke, who told me about it in 1979. I have nothing new to add.

35. The fecundity data for the Dougherty Basin Lake broodstock are in Hepworth et al. (2004).

including an abundant "hatch" of flying ants in late summer. First-year growth of the trout is rapid on this fare, with age-1 trout attaining an average total length of 8.9 inches. But growth tapers off in this environment: age-2 trout averaged 10.4 inches in total length, age-3 trout 11.9 inches in total length, age-4 trout 12.5 inches in total length, and the few trout that attained age 5 averaged 14.5 inches in total length.

Evidently, the trout of Trapper's Lake do not become piscivorous as they grow larger (although it was suspected by Colorado Division of Wildlife biologists that those adults that did not return to Trapper's Lake after spawning but over-summered in the natal streams may have preyed on young-of-the-year trout). It is likely, though, that Colorado River cutthroat trout inhabiting other lake environments did become piscivorous and thereby attained much larger sizes. The historical record indicates that the native cutthroat trout of Fish Lake, Utah reached weights of 5 to 6 pounds, and if you believe the anecdotes, native trout in Grand Lake may have reached 20 pounds. Studies of the feeding habits of trout introduced into Fish Lake following the demise of the native cutthroats showed that rainbow trout over 2 pounds in weight fed predominately on small fish.[36] I think it's safe to assume that Colorado River cutthroat trout over 2 pounds in weight did so as well when they occupied the lake.

The Fluvial Life History Strategy

All the major rivers of the upper Colorado River basin, including the mainstem Green, the upper Colorado itself (or Grand River, as it was called in earlier times), the upper Yampa, the Gunnison, the Roaring Fork, and the upper San Juan, were populated with Colorado River cutthroat trout that almost certainly would have exhibited fluvial life history characteristics. But fluvial populations (and most stream-resident populations for that matter) disappeared rapidly after white settlement and exploitation of the region's resources began. Mining, logging, and water diversions took their toll on the habitat of these trout, eliminating populations from some streams and making life difficult at best in others. Market fishing to supply the mining camps and the growing permanent towns and cities also took its toll, as did sport angling, which early on in the west placed great emphasis on numbers and size of trout caught— the larger the number and heavier the creel, the better.

Those fluvial populations that did persist in the face of these pressures were quickly displaced after non-native trouts were introduced. Dr. Behnke, and, later, William Wiltzius of the Colorado Division of Wildlife, used the example of the Gunnison River to illustrate just how rapidly displacement occurred. Rainbow trout were first stocked in the Gunnison River in 1888. By 1897, just nine years later, the Gunnison was famed

36. See Madsen (1942).

A 1981 view of Trapper's Lake looking past the Colorado Division of Wildlife station near Cabin Creek, the principal spawning tributary. PHOTO BY AUTHOR

for its rainbow trout fishery (and heavily promoted by the Denver and Rio Grand Railroad as an angling destination to boot), and the native cutthroat trout had all but disappeared. The upper Yampa River may have escaped stocking for just a little longer, but its native cutthroat population was doomed to displacement as well. There's a story that one angler in the early 1900s even wrote a plea to *Outdoor Life* magazine, which was published in Denver at the time, that authorities not stock rainbow trout in the Yampa, because that was the last place in Colorado he could catch native trout. His plea, of course, was ignored. [37]

Although no fluvial populations of Colorado River cutthroat trout are available to examine first hand, I expect that the similarities in life history and ecology between historical populations and those of extant interior cutthroat subspecies were many. It's likely that they responded to the same environmental cues to initiate their spawning migrations, moving to natal tributaries just after the peak of the snow runoff had passed and water temperatures in the tributaries had started to rise. Based on historical accounts and old photos, trout in some of those historical populations could grow to 5 to 7 pounds or more in weight and attain lengths that truly were as long as your arm. [38] Their life span would probably have been somewhere in the order of 6 to 8 years. My guess is that females in these

37. The Gunnison example was first cited in Behnke (1979) at page 117, and later in Wiltzius (1985) as part of a history of fish propagation and stocking in Colorado. The plea to spare the Yampa River (called Bear River at the time) was supposed to have appeared in *Outdoor Life* in 1905. The library where I found early issues of this magazine had them only on microfiche. I looked closely at every issue from 1900 through 1906, poring over them until my vision swam, but I never found this reference.

38. Writer Charles Hallock published a *Sportsman's Gazetteer and General Guide* in 1877 (although the only copy I was able to locate was published later, in 1883), in which he contrasted the trout in Twin Lakes in the Arkansas River drainage east of the Continental Divide near Leadville with the trout in the Eagle River drainage on the west side of the Continental Divide. He noted that Twin Lakes trout (the greenback subspecies, Chapter 13) were small, but anglers could hike over Tennessee Pass to the Eagle River drainage for larger trout. A year later, an article in *Forest and Stream* magazine [Anonymous, "Trout in the Rocky Mountains." 9, no.25: 268–269 (1878)] made the same assertion, stating firmly that although the greenback trout was abundant on the east slope of the Rockies, anglers would travel to the west slope, the Colorado River side, if they wanted large trout.

Clinton Reservoir, Colorado, view toward head of reservoir. Clinton Creek, entering from the basin above the reservoir, held genetically pure Colorado River cutthroat trout in 1981, and these trout now populate the reservoir as well. The ridgeline in the background is the Continental Divide.
PHOTO BY AUTHOR

populations spawned for the first time at age 3 or 4 at fork lengths of 12 to 14 inches or so, and those females probably carried from 600 to 1,000 eggs each to the spawning tributaries. The larger, older, repeat-spawners would have carried more eggs, perhaps upward of 3,000 eggs per female.

The Stream-Resident Life History Strategy

The rapid depletion and displacement of Colorado River cutthroat trout was not limited to the fluvial populations of the large rivers, but affected the resident populations in the smaller tributary streams as well. By the early 1970s, populations resembling pure Colorado River cutthroat trout existed in only a handful of the most remote, oftentimes smallest headwater streams.[39] Some believed that no populations existed at all that had not been exposed to some level of introgression. However, several populations did remain that were phenotypically good representatives of the subspecies, and as more attention has focused on the subspecies, many additional headwater stream populations have been discovered, a number of them genetically pure.

Serious studies of the life history and ecology of these relict stream-resident populations began in the late 1970s with the State of Wyoming, whose biologists, directed by Dr. Niles Allen Binns, examined life history and stream habitat attributes of populations inhabiting four enclaves: 1) westside tributaries of the upper Green River, 2) upper Black's Fork drainage, 3) Big Sandstone (a Little Snake River tributary), and 4) North Fork Little Snake River. More recent studies have broadened our knowledge base into Colorado and Utah, but still, a majority of even the recent publications have focused on the North Fork Little Snake enclave which may have the largest and least fragmented of existing stream populations.[40]

Stream reaches where these populations are found are mostly upward of 8,000 feet in elevation (Big Gulch Creek, Wyoming, at 6,760 feet, is the lowest I have found among the listings), the waters are cold and well-oxygenated, and gradients are fairly steep, 4 percent or greater in most cases. Despite these high gradients, the better reaches have a good balance of pools to riffle areas and substrates consisting of cobbles, boulders, and gravels. Pools formed by beaver dams, large wood, boulders, scour against or under root wads, or other structural elements are the habitats used most by the trout, especially the larger individuals. The deeper pools assume particular importance to all sizes of trout in late summer and fall, when water levels decline, and also during the overwintering period, when the trout seek out the deepest pools for shelter.

Spawning takes place when peak flows begin to diminish and water temperatures rise to about 45

39. Dr. Robert Rush Miller, writing for the American Fisheries Society's Committee on Endangered Species and the Conservation Committee of the American Society of Ichthyologists and Herpetologists in 1972, included the Colorado River cutthroat trout in his list of threatened freshwater fishes of the United States (see Miller (1972). Behnke and Zarn (1976) stated that only two populations of the many they had examined up to that point appeared to be wholly pure based on meristic characters (those populations were found in the upper Green River drainage of Wyoming, one population in an isolated segment of a LaBarge Creek tributary and the other in an isolated tributary of Piney Creek). The population of Clinton Creek, Colorado, was determined to be genetically pure in 1978. Five additional populations considered typical of pure Colorado River cutthroat trout had also become known by 1978; these included populations in the headwaters of the Colorado River, Rocky Fork, Cunningham Creek, Northwater Creek, and Rock Creek (see Knox et al. 1980).

40. The list of publications covering stream-resident life history and ecology of Colorado River cutthroat trout includes Binns (1977), Quinlan (1980), Jespersen (1981), Scarnec-

degrees Fahrenheit (7 degrees Celsius). This may occur anywhere from late May to mid-July, depending on location and local conditions. The redds are typically small (because the spawning trout are themselves typically small), and are located in gravel patches where the water may range from only 2 to 3 inches to about a foot in depth, and water velocities are 1 to 2 feet per second. Swimup fry appear in the streams in late July and early August in locations where spawning occurs early, but not until late August or early September where conditions make spawning a later event. Spawning has been observed in reaches that are intermittent and therefore not suitable as rearing habitat, in which cases the adults move out soon after spawning, and so do the fry soon after swimup.

The trout assume their positions in dominance hierarchies during what's left of the summer rearing period, and feed opportunistically on the usual array of aquatic invertebrates, such as midges, mayflies, caddisflies, and stoneflies, plus any hapless terrestrial insects that fall into the streams. But growing seasons are short at these high elevations and the streams are not productive as a general rule, so the trout do not typically attain large sizes. At age two, when most reach sexual maturity, they may be no more than 5 inches in total length and not much larger than 8 inches at age 4. Trout in this 5 to 8 inch size range typically carry 100 to 200 eggs per female to the spawning gravels. Larger trout have been reported from some headwater reaches, but a 10- or 12-inch trout from one of these enclaves of stream-resident distribution would be a real trophy.

STATUS AND FUTURE PROSPECTS

The history of the depletion of the Colorado River cutthroat trout and of recent efforts to restore the subspecies and keep it off the Endangered Species list parallels, in many respects, the story of the Bonneville cutthroat trout as told in the last Chapter. The points of similarity are these:

1) From the time of first settlement through the mining camp years, to well into the 20th century, Colorado River cutthroat populations are lost to the destructive effects of mining, logging, water diversions, livestock grazing, heavy harvest by market fishermen and sportsmen, and displacement and introgression owing to the widespread stocking of non-native species, such as rainbow, brown, and brook trout.

2) By the early 1970s, only a small handful of genetically pure populations are known to exist.

3) In 1972, the two major professional societies representing U.S. fishery scientists and managers place the Colorado River cutthroat trout on their joint list of threatened freshwater fishes of the United States—but the subspecies is *not* included as threatened or endangered when the U.S. Endangered Species Act becomes law in 1973. The U.S. Fish and Wildlife Service does eventually add the subspecies to its Category 2 candidate list but takes no further action, and in 1996, a change in administrative rules drops that category altogether. [41]

chia and Bergersen (1986), Bozek and Rahel (1991, 1992), Bozek et al. (1994), Herger et al. (1996), Young (1996, 1998), Kershner et al. (1997), Young et al. (1997, 1998), and Horan et al. (2000).

41. I discussed this administrative rule change in Chapter 11, and listed references to the pertinent Federal Register notices in footnotes 30 and 31 of that Chapter.

4) State fish and wildlife agencies step into the breach, beginning with Wyoming in the late 1970s and then Colorado and Utah. They undertake surveys of their respective waters that reveal the existence of additional pure or virtually pure populations. They also enter into joint agreements for conservation and improvement of habitat with individual national forests and Bureau of Land Management resource areas. These activities continue through the 1980s and into the 1990s. [42]

5) The Inland Native Fish Strategy (acronym INFISH) hits the streets in 1995. Federal land management agencies become more pro-active in Colorado River cutthroat trout conservation, including two major conservation assessments by the U.S. Forest Service, published in 1995 and 1996. [43]

6) In 1999, state and federal agencies sign onto a major new Conservation Agreement and Strategy that sets more comprehensive goals for improving habitat, protection of core populations, and restoring Colorado River cutthroat populations to additional historical waters. [44]

7) Also in 1999, a group of nonprofit conservation organizations, together with a concerned individual, petitions to list the subspecies under the U.S. Endangered Species Act. The U.S. Fish and Wildlife Service rejects this petition in 2004, citing positive results achieved in items 4, 5, and 6, as well as the strong commitment of all agencies to further conservation and restoration actions per the 1999 Conservation Agreement and Strategy. The petitioners sue, and in 2006, a federal judge gives the Fish and Wildlife Service nine months to review again the status of the subspecies and reconsider its decision not to list. [45]

Let's go back and examine the status of the subspecies as it stood in 1999, when that Conservation Agreement and Strategy document was signed. As of July 1, 1998, according to that document, there were 161 so-called

42. A list of early conservation agreements includes, for Colorado, Sealing et al. (1992) and Langlois et al. (1994); for Wyoming, Green River Westside Working Group (1993) and Little Snake River Working Group (1994); and for Utah, Lentsch and Converse (1997). Reports on accomplishments under these agreements have been produced for the Green River westside tributary enclave (see Green River Westside Working Group 1998) and for southern Utah (see Hepworth et al. 2002, 2004).

43. These are in Young (1995) and Young et al. (1996).

44. The multi-agency Conservation Agreement and Strategy, signed in 1999 and updated in 2001, sets goals and objectives for conservation and restoration of Colorado River cutthroat populations across all of the subspecies range in Utah, Colorado, and Wyoming. See CRCT Task Force (1999 [2001]) for the complete citation, and download your own copies from the U.S. Fish and Wildlife Service website at *http://mountain-prairie.fws.gov/species/fish/crct/* for the original 1999 version and *http://wildlife.state.co.us/Research/aquatic/CutthroatTrout/* for the 2001 update. The latter site also has a 5-year progress report that highlights management activities and accomplishments under this agreement that was issued in 2004. In 2006, a new Conservation Strategy for Colorado River Cutthroat Trout in the States of Colorado, Utah, and Wyoming was issued, which I also downloaded from the State of Colorado site along with a new range-wide status assessment that was completed in 2005. Be warned, however, that the status assessment report is a very large file (25.9 MB) and could take awhile to download. See CRCT Coordination Team (2006) and Hirsch et al. (2006) in the bibliography for citations for these two items.

45. See Center for Biological Diversity et al. (1999) for the petition to list the Colorado River cutthroat trout as threatened or endangered, and Gelatt (2004) for the 90-Day Finding that rejected this petition. You can download copies of both, plus a list of questions and answers regarding the Service's decision and other information about Colorado River cutthroat trout, from the mountain-prairie website highlighted in footnote 44 above. The federal court's order throwing out the Fish and Wildlife Service decision and ordering the agency to prepare a new status review was announced in news releases dated September 7, 2006. You can find stories about the court's action at both the Earthjustice and Environmental News Network websites at *www.earthjustice.org* and *www.enn.com*, respectively. The Fish and Wildlife Service commenced its new review in November 2006, so look for it to be completed in the fall of 2007.

conservation populations of Colorado River cutthroat trout, i.e., populations of 90 percent genetic purity or better, known to exist in 524 miles of stream habitat, and 12 such populations in 601 acres of lakes within the historical range in Colorado, Wyoming, and Utah. The agreement divided the historical range into 14 (later consolidated into 8) geographic management units (GMUs), which the language of the document called common-sense groupings of river drainages (the intent was to morph these into distinct population segments, the U.S. Fish and Wildlife Service terminology, when enough was learned about how the subspecies genetic variation is distributed). Going in, the large majority of existing conservation populations were in GMUs in Colorado and Wyoming (for example, 47 individual populations in the originally defined Colorado River GMU and 18 in the originally defined Yampa GMU in Colorado, and 32 individual populations in the originally defined Little Snake GMU in Wyoming). Several GMUs had no populations in lakes, but no GMU had fewer than two known stream populations.

The agencies set themselves the objective of increasing the number of individual conservation populations, and the amount of stream habitat and lake acreage occupied, to 383 populations in 1,754 miles of stream habitat and 18 populations in 652 acres of lakes. Furthermore, they set themselves a long-term goal of establishing two self-sustaining metapopulations, each

consisting of at least five viable connected populations, in each GMU within the historical range, and a short-term goal of establishing one such metapopulation in each GMU. "Long-term" and "short-term" were not specifically defined, but the agencies did state that they believed all of their goals and objectives could be achieved in 10 years (although this initial Agreement only extended for five years). The Agreement also noted that drainages in some GMUs are very simple with few habitable tributaries, and so may never have metapopulation structures more extensive than five individual interconnected populations.

So, how have the agencies done so far? Very well, actually, especially in regard to the restoration of individual conservation populations (those of 90 percent genetic purity or better). As of June 2005 (see footnote 44), which marked the end of five years of work, 326 individual conservation populations of Colorado River cutthroat trout occupied approximately 1,796 miles of stream habitat (285 individual populations) and 1,123 acres of lakes (41 individual populations). That's already 85 percent of the objective originally set for number of restored stream populations and more than 100 percent of the original stream mileage target. For lakes, the objective has already been exceeded by more than twice for number of populations and not quite twice for occupied lake acreage. Another highlight item is that 153 of the individual stream populations enumerated above

*Colorado River near Rifle, Colorado. Belo[w]
here, warm silt-laden waters probably lim[it]
the historical distribution of Colorado Riv[er]
cutthroat trout to colder, clearer tributarie[s].*
PHOTO BY AUTHOR

are genetically pure and have never been influenced by genetic alteration linked to human intervention. These are designated as *core conservation populations.*

On the down side, only about half of the individual core conservation populations occupying streams occur in secure reaches above barriers that prevent intrusion of non-native species capable of displacing or introgressing them. Also, most of the existing populations are very small and remain below the population size or occupied stream length required for long-term survival. [46] Forty percent of the core conservation populations occur in stream segments 2 miles long or less, and, based on abundance data available in 2004, 82 percent had adult populations of less than 1,000 individuals and 64 percent had fewer than 500 individuals, so there is much room for improvement here.

Progress has been mixed as well on the metapopulation front. Based on data reported in 2004 (see footnote 44), six of the eight GMUs had at least one metapopulation that meets the original goal's criterion of 5 viable interconnected populations, which is 75 percent of the original short-term target, and three GMUs had at least two such metapopulations, which is not quite 40 percent of the long-term target for metapopulations. Also on the plus side, the LaBarge CRCT Restoration Project, started in Wyoming's Upper Green GMU in 2000, is on target for completion in 2007, which will accomplish

the largest and most extensive Colorado River cutthroat metapopulation restoration in Wyoming. [47] But two GMUs, namely the Dolores and San Juan GMUs, have no metapopulations meeting the criterion of five interconnected viable populations, and the Dolores GMU has no metapopulation in which even two to four viable populations were connected up.

So the work continues, with much already accomplished but much remaining to be done. In 2006, a new Conservation Strategy for Colorado River Cutthroat Trout in the States of Colorado, Utah, and Wyoming was issued (see footnote 44), which extends out for another five years. This new Strategy sets goals of 36 new conservation populations in 310.5 additional stream miles of habitat and four new conservation populations in 65 additional acres of lakes in the next five years. But for metapopulations, it hedges, mandating only that metapopulations should be created where possible, but setting no numerical goals for any GMU.

The colorful Colorado River cutthroat, like almost all the other interior subspecies, has found it difficult to coexist with man and his ever-increasing activities. Although progress on conservation and restoration is indeed being made, this subspecies still faces an uphill struggle. Trout habitat and populations can be lost to more than just the effects of mining, logging,

46. I reviewed the scientific basis for this claim in footnote 33 of Chapter 11.

47. I have many pleasant memories of the LaBarge Creek drainage. That's where I got my first look at a stream-dwelling specimen of the upper Green River form of Colorado River cutthroat trout. It's also where I had one of those exhilarating encounters that lend spice to the outdoor experience. I had waded a little too deep into the backwater of a beaver pond—just an inch or two of freeboard at the top of my waders and a gooey-soft, uncertain bottom. I worked carefully toward shore and slightly shallower water, but was blocked from the stream bank itself by a wall of willows. I was just edging around one of those willow clumps when suddenly, right before me, loomed this huge dark head. A cow moose—a BIG cow moose—not five feet away! We stared at one another for what seemed the longest time, then she broke the spell, turning quickly away and splashing heavily across the pond to the opposite side. Then I saw the calf she had in tow as it followed her, half swimming and half running, across and out of the stream. There I was, the alien intruder in her natural habitat and maybe a threat to the calf I hadn't seen. Had she reached that conclusion, she could have stomped me

irrigation, and livestock grazing, which are the culprits usually discussed in this context and do indeed get a lot of ink in the new Conservation Strategy. But demand has arisen again for extraction of the region's rich energy resources. What will this hold in store? And what about the massive diversions of water from the river drainages on the west side of the Continental Divide to supply the insatiable water needs of burgeoning east-slope cities? Let's talk energy exploration and development first.

Some of you may be old enough to remember the year 1973, when the organization of the world's major oil producing and exporting countries (acronym OPEC) choked off our foreign oil supply and triggered our first major energy crisis. It certainly brought the American dependence on fossil fuel, and especially our dependence on foreign oil, home to roost.

Now, geologists had known for years that tremendous pools of oil and gas lay deep beneath the Overthrust Belt, a wide geological zone that stretches through the Rocky Mountains from Mexico to Canada that was formed when two plates of Earth's crust slid one over the other. The trouble was, these reserves lay 4 to 5 miles deep, and it wasn't until that 1973 oil boycott that it became economical to explore for and drill down after them. Nowhere was the oil and gas exploration so intense as it was in the upper Green River country of

southwestern Wyoming, northwestern Colorado, and northeastern Utah, the historical range of the Colorado River cutthroat trout. These same sections of country are layered with a limestone rock called marl, aka oil shale, though it is not really shale and there is no actual oil. Marl does contain a fossil material known as kerogen, which is sort of a precursor to oil, and can, with very complex and expensive processing, be converted into a low-grade of oil. Deposits of tar sand, so named because the sand contains 10–15 percent of bitumen, occur in northeastern Utah. When heated to over 500 degrees Celsius, bitumen will also convert into a crude oil. The crisis made it feasible to look closely at these as possibilities for supplying our needs as well. Small, quiet towns such as Pinedale, Big Piney, Green River, and Rock Springs, Wyoming; Vernal, Utah; and Grand Junction, Rangely, and Rifle, Colorado, took on boomtown proportions as they became staging areas for the frenzied effort to find and extract the crude.

I got a chance to observe some of this activity in the early 1980s, when I made my first extended trip through the region for the first edition of this book, and frankly it wasn't pretty. One only had to spend a day or two in one of the towns mentioned above, or, better yet, take a few trips into the surrounding country, to see that the rush had far outpaced all ability to mitigate its effects. Populations of the small towns doubled

into the mud. But enough about that. There have been a number of projects in the LaBarge Creek drainage over the years devoted to habitat improvement and protection of the relict Colorado River cutthroat populations that survived in the upper watershed. Some have been funded by the Bring Back the Natives program I told you about in an earlier Chapter. Some have involved volunteer workers from Trout Unlimited and other organizations. The present and most ambitious project, started in 2000, includes the completion of a migration barrier in lower LaBarge Creek in 2001, construction of temporary barriers on 6 tributaries, and chemical treatment of 13 tributaries totaling 31 stream miles plus 27 miles of mainstem LaBarge to remove all non-native salmonids. When all this work is completed, in 2007, the temporary barriers will be removed to reopen connection between the tributaries and LaBarge Creek itself. In addition, beginning in 2007, the entire project area will be stocked with pure Colorado River cutthroat trout to hasten repopulation of the mainstem and augment the populations already present in the tributaries.

and tripled. Trailer parks sprang up all around. Those who couldn't get in simply pitched tents and squatted wherever they could along nearby streams, some even sleeping in the backs of station wagons. Overcrowding, litter, and pollution became common. Out among the hills, acres of habitat gave way to oil and gas rigs, test wells, new roads and utility rights-of-way, processing plants, and the like. And even more acres where the seismographers had been at work looked like bombed-out targets of the Strategic Air Command. Most of the concern you saw in print in those years was over the effects on big game animals and the loss of big game habitat. But disturbance of the trout populations and habitat in the small headwater streams couldn't help but result as well.

Field biologists I conferred with at the time told me that what they noticed most was greatly increased fishing pressure and poaching, attributable to the huge influx of people. But subsequently, some acknowledged that local impacts on streams and trout populations had occurred from the exploration and production activities too, most associated with sediment problems from road building and maintenance, impedance or sometimes total blockage of instream movement of trout owing to poor culvert installations at road crossings, siting of wells and ancillary facilities too close to streams, and water depletion for well drilling and dust control. [48]

Producing oil and gas wells also bring up huge amounts of briny water (called *produced water* in industry parlance), which is loaded with chlorides, sulfates, and dissolved solids, and also contains small amounts of dissolved and suspended petroleum that remain after the oil and gas is separated. Produced water is typically disposed of by either reinjecting it into the well, atomizing it by blowing it into the air to hasten evaporation, or discharging it into closed containment ponds. But sometimes it is disposed of by discharging directly into surface waters or onto areas where it can get into surface waters. In the early 1980s and even now, the states of Wyoming and Colorado allow a concentration of 10 parts per million of oil in produced water that is discharged into surface waters, which is about the level at which an oil sheen can be discerned. But testing has shown that this is much too high a level for fish health. The flow-through 96-hour LC50 for salmonids exposed to oil, including cutthroat trout exposed to Wyoming crude, is 2.4 parts per million (the 96-hour LC50 value is the concentration that kills half the exposed population in 96 hours, and the term flow-through refers to how the test is run), and sublethal effects such as impaired growth and survival, fin-tissue loss, and development of lesions on gills and eyes occur at much lower concentrations. In 1976, the U.S. Environmental Protection Agency proposed that

48. One retrospective acknowledge-ment of local impacts from energy exploration and development can be found in the Conservation Agree-ment and Strategy for Colorado River Cutthroat Trout in the State of Utah, prepared by Lentsch and Converse (1997). The proceedings of a symposium on present and future prospects for the fishes of the upper Colorado River basin, published back in 1982, has a paper by R.D. Jacobsen titled "New Impacts by Man in the Upper Colorado River Basin" with maps of some of the fossil fuel deposits in the basin; see Jacobsen (1982) for the complete citation. A recent article by John G. Mitchell (with photos by Joel Sartore) in *National Geographic* magazine has a rundown on the natural gas boom that is sweeping the basin, the conflicts it is stirring, and how public lands are being transformed by the accelerated pace of this development; see Mitchell (2005).

for oil, effluents into surface waters should contain no more than one-hundredth of the flow-through 96-hour LC50 value, or no more than 0.024 part per million of oil, to avoid these effects. But neither state has chosen to adopt this more stringent standard for produced water discharges into its trout bearing waters. [49]

Energy developers always insist that it is possible to get the oil and gas out of the ground without destroying habitat. Advances in exploration and production technology have been made so that a facility's footprint on the ground is small. Improvements in the way at least some companies in the oil and natural gas industries approach sensitive areas have been made as well, as they strive to keep intact their "social license to operate," as Jared Diamond put it. [50] I hope so. But having said that, the industry itself often takes positions that make it hard for me to be optimistic.

Here's an example: One of the major new areas recently opened by the BLM for oil and gas drilling is the Roan Plateau in Colorado's Upper Colorado GMU—this despite strong local sentiment *not* to drill at all on the Roan Plateau. There are but five remaining populations of Colorado River cutthroat trout on the Roan Plateau, of which just two are genetically pure core conservation populations. The BLM plan, six years in the making, opens half of the plateau's public lands for drilling, allowing nearly 90 percent of its estimated 4.2 trillion cubic feet of natural gas to be extracted—but under a carefully phased development timeline and other restrictions designed to preserve wildlife and environmental values. But the plan had hardly been presented before energy industry spokespersons and company executives were grousing about how unattractive it would be to do business under these carefully crafted restrictions. [51]

Tapping our country's energy resources is essential to our nation and our economy. It would be nice, though, if this time around the frenzied "gold rush" approach of the last energy exploration boom could be tempered so that something of value remains after the oil and gas are gone, and all those advances in Colorado River cutthroat restoration will not be undone.

The other element of uncertainty I want to touch on here is the effect of the massive water diversions from the Colorado River side of the Continental Divide to quench the thirst of the cities and agricultural enterprises on the east side of the Front Range in Colorado, where most of that state's population originally concentrated and where it continues to expand, and also farther north in Wyoming. The City of Denver built the first transmountain water pipeline shortly after the railroads showed that large-scale tunneling through the Rockies was feasible in 1927. Denver now draws water from multiple transmountain pipelines.

49. Woodward, Mehrle, and Mauck (1981) published the tests of Wyoming crude oil on cutthroat trout, and provided references to other work on the effects of oil on other species of trout and salmon. The U.S. Environmental Protection Agency criterion for oil in water is from its 1976 publication, "Quality Criteria for Water," cited under U.S. Environmental Protection Agency (1976) in the bibliography. Check the EPA website at *www.epa.gov* for anything more recent. Websites for information on water quality criteria in Wyoming and Colorado are *http://deq.state.wy.us/wqd/ WQDrules/index.asp* and *www. cdphe.state.co.us/wq/wghom.asp*, respectively.

50. See Diamond (2005). His Chapter 15, titled "Big Business and the Environment: Different Conditions, Different Outcomes," provides good discussions not only of the oil industry, but the hardrock mining industry, the logging industry, and the fishing industry as well.

51. Details of the BLM plan for tapping the natural gas resources of the Roan Plateau can be obtained from the BLM website at *www.blm.gov/ rmp/co/roanplateau*. See also Connell (2006), which is a Federal Register notice of the availability

The City of Pueblo has been ditching water across the Continental Divide near Tennessee Pass since 1932. Colorado Springs and the City of Aurora have followed suit. Water from the upper Colorado River watershed supplies needs in the Fort Collins area. The Eagle River is tapped, and so is the Roaring Fork drainage, the Frying Pan, and all the other major tributaries, with the demand showing no sign of slackening.

One of the more recent trans-Divide diversions, the City of Cheyenne Water Project in Wyoming, points up the problems these diversions can pose for the Colorado River cutthroat trout and for those who are working to restore it. The Cheyenne Water Project first became operational in 1967 and was expanded in the mid-1980s to draw water from the North Fork Little Snake River drainage, right in the heart of the North Fork Little Snake enclave of Colorado River cutthroat distribution. The purpose of this diversion is at least in part to satisfy a water-right obligation. As I understand it, the City draws the water it actually uses from the North Platte drainage, but its water right is held in the North Fork Little Snake drainage, so it must pay back the North Platte water. [52]

The Cheyenne Water Project diverts water from every tributary containing Colorado River cutthroat trout in the North Fork Little Snake drainage. Diversion structures are constructed on every perennial tributary and on the mainstem North Fork itself at about the 8,700-foot level of elevation. These structures block fish passage in both directions in the streams where they are located. Aside from the drain of water from the streams (which is mitigated for by instream flow agreements) and any direct loss of trout into the diversion pipelines that may occur, the presence of these structures impairs instream trophic movements of the trout. For example, some of the best spawning habitat and highest fry survival rates in the enclave occur in the headwater areas, but as the fry grow to the point where they need habitat of greater depth in order to survive, they move downstream. This sort of instream movement is inhibited by the water diversions. The diversion structures thus may limit or prevent upstream spawning migrations as well as recruitment of trout to downstream populations, which in turn may limit population numbers and viability in some parts of the system. In addition, these structures limit the development of metapopulation structure in the system. The largest present uninterrupted stream segment remaining in the project area is a 17.3-mile segment of North Fork Little Snake mainstem and connected tributaries downstream from the City of Cheyenne mainstem water diversion structure (another constructed barrier in the mainstem at the lower end of this segment prevents the Colorado River cutthroat enclave from being invaded by brown and

of the plan and its final environmental impact statement. For news commentary on the BLM plan, I ran a Google search for Roan Plateau Energy Development that quickly turned up articles published in *The Denver Post*, *The Rocky Mountain News*, and *The Grand Junction Sentinel*, among others. I'll leave it to you to do your own search if you want to read these articles and more.

52. I introduced the subject of western water law in Chapter 5 and footnote 44 of that Chapter.

Green River, Wyoming, upstream from the town of Pinedale. Colorado River cutthroat trout inhabited this reach historically. Wind River Range in the far distance. PHOTO BY AUTHOR

rainbow trout from below). The next-largest is 1.4 stream miles in the North Fork Little Snake and its Rhodine Creek tributary upstream from that main-stem diversion structure. These represent the limits of metapopulation size and structure that can be attained in this drainage. [53]

53. My information about the Cheyenne Water Project came from the City of Cheyenne Board of Public Utilities, website *www.cheyennecity.org/bpu.htm*, and the Water Resources Data System Library at the University of Wyoming, website *http://library.wrds.uwyo.edu/wwdcrept.html*. Instream flow mitigation information is from Wolff et al. (1986). Effects on the Colorado River cutthroat trout and metapopulation connectivity in the North Fork Little Snake River are reported in Speas et al. (1994) and Young (1996).

CHAPTER 13

Greenback Cutthroat Trout

Oncorhynchus clarkii stomias: Chromosomes, 2N=64. Typical characters similar to Colorado River cutthroat trout (Chapter 12) with perhaps a tendency toward more scales and larger spots. Scales in lateral series 175–220+, typically more than 185. Scales above lateral line 42–60, typically more than 45. Vertebrae 59–63. Pyloric caeca 24–46. Spots are typically large, the largest and most pronounced of any cutthroat, and round to oblong in shape, tending to concentrate on the caudal peduncle. Coloration, especially the tendency to develop intense colors, is also similar to *O. c. pleuriticus.* Mature males can develop brilliant red over the lower sides and ventral region.

The common name, greenback, is something of a misnomer; the backs of these trout are no more green than the backs of any of the other cutthroat subspecies.

Greenback Cutthroat Trout, *Oncorhynchus clarkii stomias,* stream-resident form

Greenback Cutthroat Trout, *Oncorhynchus clarkii stomias*, Bear Lake spawning male

THE UPPER ARKANSAS, BAYOU SALADO, AND THE UPPER SOUTH PLATTE

Dr. Behnke wrote, some years ago, that based purely on the characters listed above, the greenback cut-throat and the Colorado River cutthroat might very well be considered one and the same subspecies. That they are not is more because the subspecies names *pleuriticus* and *stomias* have long been rec-ognized to designate the cutthroats native to specific drainage basins (the upper Colorado River basin for *pleuriticus* and the upper Arkansas and South Platte drainages for *stomias*) than for any real distinctions in meristic characters.

In fact, it is not even clear where the type locality for *stomias* might be. *Salmo stomias* was named by E.D. Cope from two specimens that had been sent to the Academy of Natural Sciences, Philadelphia, from Fort Riley, Kansas, back in 1856. Those specimens had been sent by the post surgeon, W.A. Hammond, but they had actually been collected somewhere along the route of an army surveying expedition from Fort Riley to Bridger's Pass, Wyoming, and back in the summer of that same year. The expedi-tion passed through parts of the Kansas, South Platte, North Platte, and Green River drainages, but cut-throats could have been collected from only two, the

South Platte or Green River systems, since neither the Kansas nor the North Platte ever had native trout. The Green, of course, is *pleuriticus* territory. Unfortunately, when the specimens were shipped to Philadelphia, where Cope later inspected them, they were labeled only with the point of shipment, Fort Riley, Kansas, not the water from which they had come. It was David Starr Jordan who redefined *stomias* and limited the name to the cutthroat trout native to the South Platte and Arkansas river drainages, which lie east of the Continental Divide. Jordan may also have been the first to use the common name, greenback cutthroat, for these trout. [1]

What is clear, from the similarity of characters, is that the greenback cutthroat did derive from the Colorado River cutthroat, and fairly recently, too, in terms of geological time. The triggering event was almost certainly headwater stream transfer across one or more of the low points along the Continental Divide. Such a transfer could not have occurred until these passes were free of the alpine glacial ice that was present off and on during the late-Pleistocene. [2] Headwater transfers could have carried the trout from the Colorado River basin to either the South Platte or Arkansas basins (I favor the latter as the initial step), followed in very short order by another transfer that took the trout from the first of these newly occupied basins to the next. Let's examine the possibilities.

It's apparent that the trout did *not* make it across any of the several fairly low passes (lowest Muddy Pass at 8,722 feet and highest Buffalo Pass at 9,915 feet) leading from the upper Colorado into North Park and the North Platte River drainage, because streams of the North Platte drainage never had native trout even though they could support trout and indeed now hold introduced species.

One possible crossover point does lie in this same area, however, in the northeastern-most corner of the Colorado River drainage, where waters of the Cache La Poudre River system, a major tributary of the South Platte, head up just across the Divide. Here lies La Poudre Pass, elevation 10,192 feet, where the contours are flat enough near the top that a ditch (the Grand River Ditch) rather than a tunnel, could be dug to transport water from the head of the Colorado River over into Long Draw Creek (aka Poudre Pass Creek), a headwater of the Cache La Poudre River. An ice field existed here during the Pleistocene, with ice filling La Poudre Pass and flowing into the valleys of the Colorado and Cache La Poudre rivers, feeding two of the largest glaciers in the Front Range. Scientists say it took 2,000 to 4,000 years for all this ice to recede when the Pleistocene ended, but the topography appears to be favorable for a headwater stream transfer to have occurred once it did. [3]

I do not believe trout could have crossed at Milner Pass (10,758 feet) in Rocky Mountain National Park

1. Jordan's report is in his "Report of Explorations in Colorado and Utah in the Summer of 1889" to the U.S. Fish Commission; see Jordan (1891b). Cope's report, titled "Report on the Reptiles and Fishes Obtained by the Naturalists of the Expedition," was included in the reports of the Hayden surveys; see Cope (1872). The army doctor who had sent the specimens Cope examined, William Alexander Hammond, later became famous in his own right. He served for a time as Surgeon General of the U.S. Army, and went on to become a pioneer in the field of neurology. His biography was written up by Blustein (1991), but other than noting that Hammond was quite active in organizing natural history collections during the period he was stationed at Fort Riley, she said nothing about the collection of the two *stomias* specimens. Dr. Behnke's remarks, cited in the opening paragraph of this Chapter, are in Behnke (1979) at page 120.

2. The location and extent of Pleistocene glacial ice in this region are discussed in detail in Westgate (1905), Capps (1909), Ives (1938), Ray (1940), Jones and Quam (1944), Richmond (1960, 1965), Madole (1976, 1980), and Meierding (1982).

3. In the first edition of his *Monograph of the Native Trouts of the Genus Salmo of Western North America*,

or any of the other high passes separating Middle Park in the Colorado River drainage from the east-flowing tributaries of the South Platte to well south of Denver. Valley glaciers occupied the upper reaches of most of these tributaries during the Pleistocene, and the majority of these originated in cirques with steep headwalls. I didn't probe back to the tops of all of these tributary systems when I visited greenback cutthroat country, but on those I did, the Continental Divide poses a high and formidable barrier indeed, with the swim for the trout up the headwater creek channels to potential crossover points looking to be impossibly steep.

The better possibilities lie to the south. A few miles south of Breckenridge, the Continental Divide jogs west as it shifts over from the Front Range to the Sawatch Range. Four passes occur here. Two of these, Boreas Pass (11,482 feet) and Hoosier Pass (11,541 feet), lead over from the Colorado River drainage to headwaters of the South Platte. Although both are splendid drives, just as author Marshall Sprague described them, my impression, especially on the Colorado River side of Hoosier Pass, was that the stretch nearest the top is pretty steep. But I crossed Hoosier Pass on the highway. Clergyman Bayard Taylor, who crossed it on horseback in 1866, described an easy climb with an ascent of no more than 90 feet per mile (1.7 percent) up a long, sloping meadow. [4]

A few miles farther west are two more, progressively lower, and maybe even easier gaps in the Continental

Dr. Behnke told of finding a pure or virtually pure population of brightly colored cutthroat trout in a small tributary of Long Draw Reservoir. These trout had meristic counts consistent with both greenback and Colorado River cutthroat trout, but spots that he thought were smaller than those of typical greenback specimens. Two barrier falls exist in the Cache La Poudre drainage downstream from Long Draw Reservoir, so it seemed improbable to him that this tributary could have been populated by upstream colonization of greenback cutthroats from the South Platte system. On the other hand, the Grand River Ditch, originally constructed in the 1890s to transport water across the Continental Divide to Long Draw Reservoir, could have also transported Colorado River cutthroat trout. He concluded that the tributary was populated with Colorado River cutthroats transported out-of-basin by the hand of man, albeit inadvertently. If Dr. Behnke's conclusion is correct, and the waters upstream from the Cache La Poudre barrier falls were indeed barren of trout prior to construction of the Grand River Ditch, that would rule out a headwater stream transfer across La Poudre Pass when Pleistocene ice receded. See Behnke (1979) at page 123 for this account.

4. Additional information about each of the mountain passes that are mentioned in this Chapter can be found in Sprague's book, *The Great Gates: the Story of the Rocky Mountain Passes* (1964). Bayard Taylor's description of his Hoosier Pass crossing is in a little book titled *Colorado: A Summer Trip* (1867). Taylor was on a lecture tour of Colorado, and sent dispatches back to *The New York Tribune*. His book was compiled from these dispatches.

Divide from the standpoint of trout making the swim, namely Fremont Pass at 11,318 feet and Tennessee Pass at 10,424 feet. Both of these passes lead from the Colorado River drainage into the headwaters of the Arkansas River, separated from the headwaters of the South Platte in this area by an intervening range of high mountains running north to south called the Mosquito Range.

It is by one or more of these four routes (five, if you include La Poudre Pass) that I believe cutthroat trout crossed over the Continental Divide in the late-Pleistocene. If I had to pick just one, I'd put my money on Tennessee Pass. We know from Chapter 12 that Colorado River cutthroat trout inhabit upper Clinton Creek, which is near the height of Fremont Pass and just a couple of miles away, so the trout could and did make the climb from the Colorado drainage side. But Tennessee Pass is lower still, and it also appears to have gentle-enough slopes, just under 3 percent on the climb up from the Colorado side and an almost ridiculously easy 0.5 percent on the descent to the Arkansas. [5]

But if the trout did indeed come via the upper Arkansas, how did they reach the South Platte? My candidate for that crossover lies farther down the Arkansas River valley. The highway route from either Fremont or Tennessee Pass leads you across Turquoise Creek and through historic Leadville, then south along the Arkansas River. Tight in against the east side of the valley roll the bare and weathered-looking peaks of the Mosquito Range. Over to the west are the rugged, craggy peaks of the Sawatch Range, including 14,431-foot Mt. Elbert, the highest peak in Colorado. You pass Lake Creek, which is the outlet of the Twin Lakes, then Clear Creek, Cottonwood Creek, Trout Creek (on the east side of the river just south of Buena Vista), Chalk Creek, and Brown Creek.

The valley drops in elevation as you continue south, becoming also gradually drier. You'll notice, too, that the lodgepole and ponderosa pines of the higher northern end have given way to the shrubby vegetation more typical of the arid west. On the left, the Mosquito Range has also lost some elevation, but on the right, the high peaks of the Sawatch Range march on and on: Mt. Yale, Mt. Columbia, Mt. Harvard, Mt. Princeton, Mt. Shavano.

Near the town of Salida, the South Arkansas River joins the mainstem. Here the Arkansas turns eastward and enters Royal Gorge. This marks the end of the upper valley. There's a way out to the south, over relatively low Poncha Pass, that takes you into the San Luis Valley and Rio Grande country (Chapter 14). But a few miles back up the Arkansas valley, just south of Buena Vista where Trout Creek comes down from the Mosquito Range, lies the passage that I believe is the

5. I estimated these slopes from elevation gains and losses over measured distances. On the climb up to Tennessee Pass from the Colorado drainage, I gained 1,530 feet in a 10-mile (52,800 feet) ascent, which is an overall gradient of 2.9 percent. On the Arkansas side, I lost just 272 feet in a 10-mile (52,800 feet) descent for an overall gradient of 0.5 percent.

key to the spread of the greenback cutthroat. There is no other place where the topography provides such an easy, natural passageway between the key drainages.

Trout Creek Pass, elevation 9,346 feet, offers low and easy access to South Park and the waters of the South Platte River. This is an ancient trail, and it's the way I think the fish traversed from the upper Arkansas into the South Platte drainage. Of course, if the initial invasion was from the Colorado basin to the South Platte, then Trout Creek Pass could just as well have served as a crossover point to the upper Arkansas drainage. Either way, if you've ever been to the valley of the Great Salt Lake, then crossing Trout Creek Pass into South Park should certainly give you a feeling of déjà vu. The South Platte side of the pass looks and feels so much like the verdant parts of the Bonneville Basin that you just *know* there had to be water here, that this just *had* to be the way the trout traveled.

Bayou Salado—that's the American fur trappers' rendition of the old Creole name Bayou Salade, meaning Salt Marshes, by which South Park was known to early French trappers. The Spanish called it Valle Salado, meaning Salt Valley. [6] If you had come in over Trout Creek Pass in the early years, that's about the first feature you would have encountered: the salt spring and associated marshes that provided those names. These salt springs and marshes, along with large tracts of native grasses in Bayou Salado, attracted deer, elk, antelope, and vast herds of buffalo, which in turn attracted the Indians. The Utes hunted and fished here, and contested for the park with the Arapaho, the Cheyenne, the Kiowa, the Comanches, and various other bands of plains Indians who also came in pursuit of game. The streams were full of trout, and were also rich with beaver.

It was the beaver that attracted the first whites. The area was claimed by France, and French trappers undoubtedly worked west up the Arkansas and into Bayou Salado in the 1700s. Spanish interest in the area seemed to be mostly in punishing marauding Indians. de Anza's military force passed through in 1779 in pursuit of Comanche raiders, and Zebulon Pike's small exploring party trailed another troop of Spanish cavalry up the Arkansas to the present site of Cañon City, then north into the park, in 1806. Spanish and Mexican gold seekers may have come as well; legends of lost mines and treasures persist to this day, but there are no records. Bayou Salado was sort of a no-man's-land with many sharing in its bounties, but not without conflict. Indian attacks were frequent, even by bands that would trade peacefully on one day, then try to "lift your ha'r," as the old trappers would say, the next. Daily life was precarious for any trapping, trading, or hunting party that entered Bayou Salado in the early 1800s.

6. A quick review of the history of this region: Although the Spanish had been established in Mexico and what is now the U.S. southwest since the 1500s, the French explorer LaSalle laid claim to all of the region from the Mississippi River east to the Rocky Mountains in 1682. France ceded its holdings to Spain in 1762 but wrested them away again in 1800, and Napoleon of France sold them to the United States (the Louisiana Purchase) in 1803. Just as President Thomas Jefferson had sent Lewis and Clark to explore the upper Missouri, so he also dispatched Army Lt. Zebulon Pike to explore the boundaries farther south in 1806. Pike passed through South Park in the winter of that year, but he entered Spanish territory and was imprisoned. Regarding Spanish New Mexico, that region came under Mexican rule in 1821 but was taken by the United States in the war with Mexico in 1845. My principal source for the history of South Park was Virginia McConnell Simmons' book, *Bayou Salado: The Story of South Park* (revised edition, 1982), with added material from Pike's report (see Jackson 1966) and later accounts by Farnham (1848 [1906]), Ruxton (1848), Sage (1859, reprinted in Hafen and Hafen 1954), Drannan (1910), and Bowles (1991).

The first American trapper to visit Bayou Salado may have been James Purcell, or Pursley, as Zebulon Pike recorded his name. The two met in Santa Fe during Pike's imprisonment there, and Purcell told Pike that he had been to the headwaters of the South Platte in 1805. He also said he had found gold there, a report that garnered no attention at the time, but would be remembered later. Other American trapping and hunting parties included Ezekial Williams of Manuel Lisa's Missouri Fur Trading Company in 1811–13 , James Pattie and Jim Beckwourth in 1826, and many others throughout the fur trapping era.

Parties and even single individuals intent on hunting, fishing, and camping came to Bayou Salado as well. Thomas Jefferson Farnham, guided by an old hunter and trapper named Kelly, led a small party through the park in 1839. Rufus Sage passed through in the winter of 1842, finding game, including buffalo, to still be abundant even at that season. Another who spent his time in Bayou Salado as a solitary hunter and camper, also in the winter, was the Britisher George Frederick Ruxton, who gained fame for the popular accounts he wrote about his travels. That was in 1847. Sir St. George Gore's hunting and fishing safari passed through in 1855.

Most of the visitors who left written accounts spoke of the abundance of game and their fabulous hunting experiences, but trout provided many a breakfast or dinner for the hunters and campers. For example, an issue of *The Rocky Mountain News* for 1873 reported that two Denver sportsmen on a vacation to South Park caught over 850 pounds of trout. [7] Trout abounded in the streams, even up in the headwaters—or at least they did until the mining boom engulfed the region. Remember James Purcell, who said he had found gold in the headwaters of the South Platte in 1805? Gold was discovered there again in 1859, and this time the rush was on. Mining camps and makeshift towns sprang up in the headwaters of Tarryall Creek below Boreas Pass and on headwater tributaries of the Middle Fork South Platte below Hoosier Pass. Tributaries of the upper Arkansas also saw their share of boom towns. Although a few of these towns did manage to endure, such as Leadville on the upper Arkansas and Alma and Fairplay in South Park, most faded into oblivion when boom turned to bust.

But not before the nearby streams and their trout populations were devastated. Water was diverted into sluices or flumes to carry ore to processing mills, or to provide water power for sawmills or steam power for stamp mills. Placer mining tore into stream beds and gravel bars, and hydraulic operations sluiced away the stream banks. In later years, large floating dredges operated in the Middle Fork South Platte near Fairplay. Timbered slopes were logged for fuelwood,

7. Historian John Monnett scoured old newspapers and other sources for stories of early-day trout fishing for his book, *Cutthroat and Campfire Tales: The Fly-Fishing Heritage of the West* (Monnett 1988). The examples I've used here are from reports he uncovered.

lumber, and supporting timbers for mines. When the railroads came to South Park in the 1870s and 1880s, slopes that still had trees were logged for railroad ties. Impacted streams were left flashy, sediment-laden, and largely sterile.

Farther downstream in South Park, away from the diggings, trout held out in the streams and big game herds continued to roam until livestock grazing took over as the principal land use. Cattle, sheep, and horses eventually displaced the buffalo, and hay fields sprang up in place of the native grasses. Even the salt spring that had contributed to the park's early name was commandeered for a time for commercial production of salt.

Along with the Trout Creek that leads up from the Arkansas to Trout Creek Pass, there is another Trout Creek in South Park itself. That one is a tributary of the Middle Fork South Platte, joining it at the old railroad stop at Garo. The narrow-gauge Denver, South Park and Pacific railroad built across South Park in 1879 and its tracks followed down that particular Trout Creek. Garo was located in what was still excellent fish and game country at the time, and Denver sportsmen could now arrive in a day. The train crews actually catered to them, stopping at all the favorite fishing spots and even stopping to load any game that was shot from coach windows as the train rambled along.

That's a far cry from what happened when the railroad reached the other Trout Creek, the one leading down to the Arkansas River from Trout Creek Pass. This Trout Creek had plenty of trout in it, too. But the Denver, South Park, and Pacific was now in a race to beat a rival line to Leadville, and it didn't spare the creek in its rush to lay track, or in stripping the slopes of the timber needed for ties. The result, again, was a flashy, sediment-laden, largely barren stream, much like the streams in the South Park mining districts.[8]

The Middle Fork South Platte River takes a southeasterly course as it flows through South Park. It picks up its Trout Creek tributary at Garo, the South Fork near Hartsel (where it becomes simply the South Platte), and other tributaries along the way. Antero Reservoir on the South Fork up near the old salt works, Spinney Mountain Reservoir, and Elevenmile Canyon Reservoir now hold back its waters before it exits South Park. But exit it does, finally, through Elevenmile Canyon, where the course turns north. It cuts up behind Colorado Springs and the southern portion of the Front Range, where it picks up the waters draining west and north from the Pike's Peak area (including a third Trout Creek that flows northward to join the South Platte), and also the waters of the Tarryall Creek drainage that heads near Boreas Pass on the Continental Divide. Cheesman Reservoir impounds the water again where

8. This account of the railroad's impact on the Trout Creek tributary to the Arkansas River is from Sprague (1964).

Arkansas River headwaters just below Fremont Pass. PHOTO BY AUTHOR

Lost Park Creek comes in, then the North Fork South Platte joins as the main stream makes its way through the mountains to Denver and beyond. Other tributaries join the mainstem along the way: Bear Creek, Clear Creek (where another of Colorado's big mining districts developed), the Boulder Creek system, Left Hand Creek, the St. Vrain system, the Big Thompson, and, at the town of Greeley, the Cache la Poudre system. Many of these waters offer fine fishing still, some even ranking in the blue-ribbon category. But the attractions now are introduced browns and rainbows, and, yes, cutthroat trout, too, in some reaches, but the non-native finespotted Snake River subspecies, not the native greenback.

HISTORICAL RANGE

Each of the drainage networks and all of the territory described in the section above was occupied by greenback cutthroat trout as the only native trout. I have mapped this historical distribution in Fig. 13-1. It all lies east of the Continental Divide, and almost all is within the bounds of the State of Colorado. A few of the historically occupied tributaries of the Cache la Poudre system do head up across the line in southeastern Wyoming, so a small portion of that state is greenback country as well. [9]

Most authorities map the historical distribution of greenback cutthroat trout in the mainstem South Platte River as extending downstream to the vicinity of Greeley, where the Cache La Poudre River comes in. The only real evidence for this that I know of consists of accounts in local newspapers from the era of the mining booms, but even these do not pinpoint the downstream distribution of the trout. According to historian John Monnett, who searched these old newspapers, one of the earliest fish tales was published in *The Rocky Mountain News* in February 1861, to the effect that a wagon load of trout was brought in from the South Platte drainage to be sold in an open air market in Denver. "Think of it," the article stated. "Fresh fish flowing out of the canyons of the Rocky Mountains." Another story, also unearthed by Monnett, involved a market fisherman who resorted to dynamite to fill his wagon with trout from the South Platte in the Denver area, even after this practice had been banned in Colorado. He and his partner were hauled into a Denver court for their offense, but got off with a warning. The next year they moved up the river from Denver a few miles, but were caught again. This time the punishment was stiff—a fine of $75. [10]

The historical distribution of greenback cutthroat trout in the mainstem Arkansas River is usually mapped as extending downstream to about the Cañon City area, where the river emerges from the Royal Gorge. My map pushes that historical distribution farther downriver to the Pueblo area, or even somewhat more. I base this on

9. Baxter and Stone (1995) stated that in the 1800s, greenback cutthroat trout occurred in Dale Creek, Albany County, Wyoming, and probably in Lonetree Creek in Albany and Laramie counties. They cite territorial governor J.W. Hoyt's report for 1878 to the U.S. Secretary of the Interior:

"The streams everywhere abound in fish of choice varieties, including the speckled trout. It is said that this most gamy and palatable of all the finny tribe has never been found in the North Platte or any of its affluents; but it abounds in Wyoming tributaries of the South Platte...."

10. See Monnett (1988).

passages in the writings of George Frederick Ruxton, who stayed at the old traders' pueblo at the mouth of Fountaine-qui-bouille (present-day Fountain Creek), where the modern city now stands, during the spring of 1847. He described the Arkansas at that location as "a clear, rapid river about a hundred yards in width," flowing through a quarter-mile wide river bottom timbered with a heavy growth of large cottonwoods. The previous winter had been unusually severe in the region, and the Arkansas had completely iced over. Ice-out occurred on March 24th, and shortly after, Ruxton wrote:

> "When the river was clear of ice I tried my luck with the fish, and in ten minutes pulled out as many trout, hickory shad, and suckers, but from that time never succeeded in getting a nibble. The hunters accounted for this by saying that the fish migrate up the stream as soon as the ice breaks, seeking the deep holes and bends of its upper waters, and that my first piscatory attempt was in the very nick of time, when a shoal was passing up for the first time after the thaw."[11]

The trout Ruxton caught would have been greenback cutthroat trout. His experience and the explanation offered by the local residents suggests that the trout made trophic migrations up and downstream past Pueblo during that period of history, hence the inclusion of the Pueblo area in Fig. 13-1.

Back in Pleistocene time, when the climate was wetter and cooler, the ancestor of the greenback cutthroat probably occupied the Arkansas and South Platte River systems farther downstream than they did historically,

11. See Ruxton (1848).

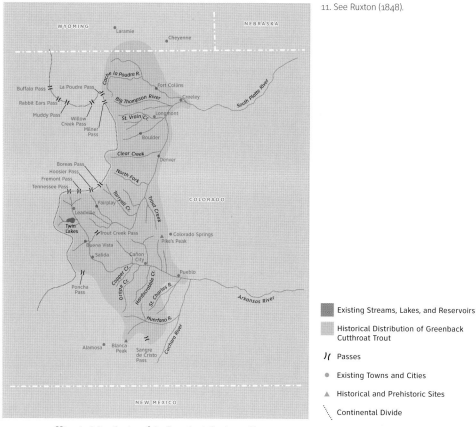

FIGURE 13-1. *Historical distribution of the Greenback Cutthroat Trout*

Legend:
- Existing Streams, Lakes, and Reservoirs
- Historical Distribution of Greenback Cutthroat Trout
-)(Passes
- Existing Towns and Cities
- ▲ Historical and Prehistoric Sites
- Continental Divide

then retreated upstream toward the mountains and foothills as the climate warmed. In the Arkansas River drainage, trout are known to have occupied the headwaters of the Huerfano River in historical times. The Huerfano heads up in a northwest-to-southeast trending valley (Wet Mountain Valley, aka Huerfano Park) lying between the Sangre de Cristo Range on the west and the Wet Mountains on the east. The Huerfano flows down the valley, then curves around the lower end of the Wet Mountains and heads northeast to enter the Arkansas River about 20–25 miles downstream from the City of Pueblo. In order to occupy Huerfano headwaters, either the trout had access to the mouth of that stream then colonized upstream, or they colonized via headwater transfer, perhaps from Grape Creek. Grape Creek has headwater forks that originate on slopes of both the Wet Mountains and the Sangre de Cristo Range. It then flows northward through Wet Mountain Valley and cuts across to the northeast to join the Arkansas River near Cañon City. A promontory divide lying across the valley separates the headwaters of Grape Creek from headwaters of the Huerfano River. [12]

The southern extent of historical greenback cutthroat distribution occurred somewhere near La Veta Pass. In 1853, the men of the Pacific railroad survey led by Capt. William Gunnison caught trout in Huerfano tributaries leading up to the pass, [13] but I have found

12. I have never had the opportunity to explore the Wet Mountain area myself, but Scott and Taylor (1975) have a good index map of Wet Mountain Valley and its surroundings that shows the principal drainages. Lewis H. Garrard, a young Cincinnatian who went west for his health in 1846–47 and traveled through the area with one of Céran St. Vrain's trading/trapping parties, wrote this glowing description of the Wet Mountains:

> "As the men...pointed out the different spurs, they expatiated more particularly on the Wet Mountain, with its lovely savannas, its cool springs and murmuring rills, its shady bowers of fragrant cedar and sheltered spots of grass, its rocky retreats and tumbling brooks, its grizzly bears and mountain sheep, its silky beaver and blacktailed deer with wide-spreading antlers, its monster elk and fleet-footed antelope, its luscious plums and refreshing grapes, its juicy cherries, delightful currants, and other attractions, making it the hunter's paradise."

Although Garrard never mentioned trout and fishing, this certainly sounds like trout country to me. For more of Garrard's adventures and descriptions of the country he passed through, see Garrard (1955).

13. Actually, Gunnison's party crossed Sangre de Cristo Pass, elevation 9,459 feet, which is the route of the old Taos Trail used by the Spanish and all subsequent trapping and trading parties to cross the Sangre de Cristo Range. La Veta Pass, elevation 9,383 feet, is located very close by and is 76 feet lower. La Veta Pass is the route of U.S. Highway 160. I will have more to say about the Gunnison party in Chapter 14, but for those who would like to read ahead, see Beckwith (1855) and Schiel (1859, [1959]).

nothing in the historical record to indicate that trout were present in the Cucharas River system, which is the next drainage south and is itself a tributary of the Huerfano. The Cucharas River joins the Huerfano after the latter stream has rounded the Wet Mountains and is flowing out across the plains on its way to the Arkansas.

LIFE HISTORY AND ECOLOGY

It has only been in recent years that much of anything on the life history and ecology of the greenback cutthroat trout has appeared in the literature. What little there is indicates few basic differences from other interior cutthroat subspecies.[14] They can (and historically did) display the same three basic life history strategies:

- A lacustrine, or lake-associated, life history, wherein the fish feed and grow in lakes or ponds, then migrate to tributary streams to spawn.
- A fluvial, or riverine (potamodromous), life history, wherein the fish feed and grow in main river reaches, then migrate to small tributaries of these rivers where spawning and early rearing occur.
- A stream-resident life history, wherein the trout complete all phases of their life cycle within the streams of their birth, often in a short stream reach and often in the headwaters of the drainage.

To date, all remnant and reintroduced populations exist in isolated headwater stream reaches or in mountain lakes, and so exhibit the stream-resident or lacustrine life history strategies. All populations that naturally exhibited fluvial life history behavior—for example,

14. Jordan (1891b) and Juday (1907b) made observations of size and food habits of greenback cutthroat trout in Twin Lakes. Bulkley (1959) gathered data on age and growth, food habits, and movements of a headwater stream population in Rocky Mountain National Park, and Nelson (1972) made similar observations of an unexploited lake population in the City of Boulder watershed. More recently, Scarneccia and Bergerson (1986), Harig et al. (2000), Harig and Fausch (2002), and Young and Guenther-Gloss (2004) have studied life history and habitat relationships, and Fausch and Cummings (1986), Cummings (1987), and Wang and White (1994) studied microhabitat use, agonistic behavior, and competitiveness of greenback cutthroat trout in sympatry with brook trout and brown trout. Trojnar and Behnke (1974) observed segregation of food and habitat between greenbacks, finespotted Snake River cutthroat trout, and brook trout cohabiting in a lake. Additional information on life history and ecology based on observations of broodstock fish and monitoring of reintroduced populations can be found in the various recovery plans written for this subspecies, in particular the 1983 and 1998 plans (see Greenback Cutthroat Trout Recovery Team 1983, 1998), and in Dwyer and Rosenlund (1988).

The Arkansas River again, near Salida, Colorado, where its course turns east.
PHOTO BY AUTHOR

those of the mainstem South Platte and upper Arkansas rivers—have been lost since the time of first settlement, and recovery plans have not addressed the resurrection of such populations. However, greenback cutthroat trout of one life history strategy can sometimes adapt to habitats and environments requiring a shift to another. This is an attribute that fishery managers, with patience and perseverance, have already been able to exploit in their recovery plans to increase the number of populations within the historical range. Most of these projects have involved the transfer of trout from stream-resident populations into lacustrine environments (ponds, small lakes, or reservoirs) to develop broodstocks, with the progeny being introduced into other lakes or into streams, where they revert back to a stream-resident life history.

The Stream-Resident Life History Strategy

The stream habitat requirements of greenback cutthroat trout appear to be little different from other interior cutthroat subspecies. Ideally, streams should be clear-running, cold, and well-oxygenated, with moderate gradients, rocky to gravelly substrates, balanced pool-to-riffle ratios, abundant riparian vegetation and large woody debris—in other words, the usual list. One relatively new bit is based on recent (circa 2000–04) studies of factors influencing success of cut-

throat reintroductions and the long-term viability of these populations. These studies indicate that for each population, there should be at least 5 stream miles of habitat of good enough quality to support 2,500 individuals greater than 3 inches in length.[15] The highest population density among currently known stream populations occurs in Hunters Creek in Rocky Mountain National Park, where a reach of only about a mile and a quarter supports a stable population which, in 1985, held an estimated 1,080 trout over 6 inches in length.

Spawning by these stream-resident trout can take place anytime from late spring to late summer, depending on elevation and water temperature in the spawning reaches. Water temperatures in the range of 41–46 degrees Fahrenheit (5 to 8 degrees Celsius) cue redd building and the deposition of eggs. In Hidden Valley Creek (elevation 8,825 feet) and Hunters Creek (elevation 9,500 feet), the trout spawn in mid June. In the uppermost headwaters of the North Fork Big Thompson River and in a headwater tributary of the South Fork Cache La Poudre River (both sites about 10,500 feet), the trout do not spawn until late July or August. Male trout in the North Fork Big Thompson population were observed to still be producing milt even in mid September.

Normally, one would not think that late spawning at these cold, high-elevation sites could allow enough

15. The studies referred to here include Harig et al. (2000), Hilderbrand and Kershner (2000a), Harig and Fausch (2002), and Young and Guenther-Gloss (2004).

time for the eggs to develop and fry to emerge and start feeding before winter conditions catch them. But when tested, eggs from that high-elevation South Fork Cache La Poudre population were found to require fewer degree-days to hatch than is typical for interior cutthroat trout, or for rainbow trout for that matter (256 degree-days for South Fork Cache La Poudre eggs vs. 312 degree-days for the eggs of lower-elevation greenbacks and for rainbow trout), which shortens incubation time by about a week in 46-degree Fahrenheit (8-degree Celsius) water. Although a week doesn't sound like much, it is evidently just enough of a cushion for the fry to successfully prepare for overwintering. When greenback cutthroats from lower-elevation populations have been stocked in these high-elevation sites, some have spawned successfully, but few, if any, of the eggs or alevins have survived.

Historically, greenback cutthroat trout were never noted for their large size. Even in the years when the market hunters and "sports" were taking such large quantities of trout, the maximum size of the fish typically reported was 1–2 pounds, or about a 15- to 18- or 19-inch trout. More commonly, though, the trout did not exceed about a foot in length, or less than half a pound in weight. As I mentioned in Chapter 12, if one wanted larger trout, one crossed over the Continental Divide to the upper Colorado River drainage.

This small average size may be due in large part to the diet of the greenbacks. Summer rearing and growth in stream-dwelling greenback cutthroat populations is fueled by a diet of terrestrial and aquatic insects. At the high-elevation North Fork Big Thompson site, the trout were found to feed almost entirely on terrestrials, mainly ants, wasps, bees, and two-winged insects. Trout in lower-elevation Hidden Valley Creek eat terrestrial insects as well, but at this site the nymphs and adults of aquatic insects provide variety in the fare. Yet individuals in some stream populations do grow to maximum lengths of 14 to 15 inches on these diets, which approaches those old-timers' reports. It is interesting that one population where trout of this size are fairly common is that same late-spawning population in the headwaters of the South Fork Cache La Poudre River. Water temperature in the headwater tributary where this population lives seldom exceeds 54 degrees Fahrenheit (12 degrees Celsius). It is probable that metabolism and growth processes are slowed in this environment, extending the life span of the trout and allowing them to attain these larger sizes. My guess is that the larger trout in this population may be 9 or 10 years old.

On the other hand, the size of the fish in lower-elevation sites may top out at about 9 to 10 inches and the maximum age may be no more than 6 or 7 years.

At Hunters Creek, elevation 9,500 feet, where about a mile and a quarter of stream supports a large population of greenbacks in the 6-inches-and-over category, a tagging study revealed that a group of trout ranging from 7 inches to 10 inches in length when first measured in June 1988 averaged only 6 grams heavier (that's just one-fifth of an ounce!) when measured again in June 1989, and the fish had no measurable change in length. This suggests to me that the trout had about topped out in terms of maximum length at the 7- to 10-inch size, and I'm guessing that the largest were pretty close to the end of their life span as well.

In a hatchery setting, female greenback cutthroats from stream-resident populations can attain sexual maturity and produce eggs at age 2, but in the wild, age 3 or even age 4 is more common. The typical first-maturing female in a wild population is 3 years old and approximately 7 inches in length. Such a trout would weigh about 38 grams if in good condition, and would carry anywhere from 68 to 84 eggs. [16] As in all other subspecies of cutthroat trout, larger females produce greater numbers of eggs. At a fish-culture facility, greenback cutthroats derived from Como Creek stock and reared to an average size of 254 grams (about 13 inches) produced 389 eggs per female. When reared for another year to 357 grams (about 14.5–15 inches), they produced 551 eggs per female.

16. How did I come up with that number for the weight of a 7-inch trout? And where did I get those values for the number of eggs it would carry if it were a female? I computed the weight of the trout with the aid of a *standard weight equation* that relates the weight of a trout in good condition to its total length. Actually, two such equations have been proposed by University of Wyoming researchers Carter Kruse and Wayne Hubert for interior subspecies of cutthroat trout, one equation for trout that live in streams and the other for trout living in lakes. For the equations themselves and for details of how they were developed, see their paper in Kruse and Hubert (1997). As for the number of eggs, the *relative fecundity* of trout typically falls within the range of 1.8 to 2.2 eggs per gram of female body weight. Once I had a value for the weight of the trout, I simply multiplied to get the range. I should say here that there are two ways to express the fecundity of trout. *Relative fecundity* is the number of eggs per gram of female body weight. As I indicated, this value typically falls within the range of 1.8 to 2.2 eggs per gram of body weight, but it tends to decrease in larger, older fish, in part because they produce larger eggs (although more of them) than smaller, younger fish. *Total fecundity* is the number of eggs produced per female in a population or group of fish. This value increases with fish size.

Trout Creek near the top of Trout Creek P
PHOTO BY AUTHOR

The Lacustrine Life History Strategy

Greenback cutthroat trout occurred historically in many lakes and ponds throughout the range of the subspecies, but in only one lake that would be considered large. Twin Lakes, in the upper Arkansas River drainage not far from Leadville, Colorado, actually consists of two connected lakes that total about 1,000 surface acres. They were formed by valley glaciers that extended down the valley of Lake Creek at least twice during the late-Pleistocene, almost but not quite reaching the Arkansas River. Two subspecies of cutthroat trout occurred in Twin Lakes: the now-extinct yellowfin cutthroat (Chapter 15) and the much smaller greenback cutthroat. Historical accounts testify to the abundance of trout in Twin Lakes and to the relative ease in catching them, but the greenbacks never exceeded 10–12 inches in length whereas the yellowfin could weigh in at 10 pounds or more. That may be because the two subspecies partitioned the feeding niches at Twin Lakes, thus avoiding competition (they evidently did not hybridize either, so they may have partitioned spawning niches as well). The yellowfin cutthroat was piscivorous in its feeding habits and could grow to a large size. The greenbacks, on the other hand, fed on insects and small benthic organisms and remained relatively small.[17]

Another example that reinforces the notion of the universally small size of greenback cutthroat trout is an adult male fish that was caught in a net in Bear Lake, Rocky Mountain National Park, in 1979. That fish was described as being so fat it was almost the shape of a sunfish, but its total length was only 12 inches.[18] That's about the maximum size attained by the rest of the Bear Lake population as well.

On the other hand, there are other examples to suggest that small size may not be an intrinsic trait of the subspecies—and that also refute the notion that greenbacks cannot or will not adopt a piscivorous or "meat-eater" lifestyle. These examples are all from Lytle Pond on the Fort Carson Military Reservation, where in September 1981, greenback cutthroats from Cascade Creek in the Arkansas River drainage were introduced to establish a wild broodstock. When the trout were first introduced, none of them exceeded 10 inches in length. However, by November 1983, one male trout had grown to a total length of 20 inches—that's a growth rate of at least 5 inches a year—and had attained a weight of 4.4 pounds. For other trout in the population, annual growth rates of 3 to 3.4 inches were recorded. As for eating habits and piscivory, in 1982, the year prior to the report of the 4.4-pound trout, another greenback cutthroat, taken illegally from Lytle Pond, was weighed in at 2.6 pounds. That trout had consumed

17. The information cited here is from the reports of Jordan (1891b) and Juday (1907b). Jordan made one further observation that speaks to the selective feeding niche of Twin Lakes greenbacks. He noted that when these fish were taken into the Leadville Hatchery, they would accept invertebrates, but not fish flesh, as food. John R. Trojnar, one of Dr. Behnke's graduate students, investigated another example of apparent feeding niche partitioning by greenback cutthroat trout in a lacustrine environment, this one in a reservoir in the North Platte River drainage where greenback cutthroats from a slightly introgressed hatchery stock, finespotted Snake River cutthroats, and brook trout had all been introduced (the North Platte drainage never had native trout). Trojnar found that the three populations partitioned the available food resources, with the brook trout utilizing bottom organisms, the greenback cutthroats feeding on benthic organisms, primarily *Daphnia* (exclusively *Daphnia* in August and September), and the finespotted Snake River cutthroats taking surface prey items, mainly terrestrial insects. That report is in Trojnar and Behnke (1974).

18. This description is in a fisheries management report for Rocky Mountain National Park prepared by Bruce D. Rosenlund of the U.S. Fish and Wildlife Service; see Rosenlund (1979).

a tiger salamander that was 4.5 inches long. Also, biologists began noticing variations in the Arkansas darter population of Lytle Pond after the greenbacks were introduced. They suspected that the greenbacks were eating these 2 to 2.5-inch native darters, but never confirmed their suspicions by performing stomach analyses on the greenbacks.

For spawning, greenback trout living in lakes respond to the same water temperature cues as stream-dwelling trout, moving into inlet or outlet streams to commence the spawning act when water temperatures reach the 41–46 degree Fahrenheit (5–8 degree Celsius) range. Timing is governed by the altitude of the site. Broodstock trout held at Lytle Pond, elevation 6,198 feet, spawn in early April. Trout in Upper Hutcheson Lake, elevation 11,161 feet, spawn in mid-July. This Upper Hutcheson Lake population is the highest known reproducing population of greenback cutthroat trout.

Greenback cutthroats rearing in lacustrine environments ought to exhibit about the same fecundity levels as stream-dwelling trout. In 1972, W.C. Nelson of the Colorado Division of Wildlife reported that a group of trout from Island Lake in the Boulder Creek watershed averaging 10.6 inches in total length produced an average of 299 eggs per female. If the standard weight of this group of trout was 203 grams (from the lacustrine

standard weight equation referenced in footnote 15, then these trout produced 1.5 eggs per gram of female body weight. This is a bit below the relative fecundity range normally stated for trout (1.8–2.2 eggs per gram of female body weight is the norm), but is consistent with relative fecundity values reported for other greenback cutthroats held in fish-culture facilities. As I noted above in the subsection on stream-resident trout, broodstock trout derived from Como Creek stock and reared to an average size of 254 grams produced 389 eggs per female, and those reared to 357 grams produced 551 eggs per female. These work out to relative fecundities of 1.5 eggs per gram of female body weight for both groups. [19]

STATUS AND FUTURE PROSPECTS

From the Brink of Extinction to the Brink of Recovery

When the first whites ventured into the Front Range region, the greenback cutthroat was the only trout (with the exception of the yellowfin cutthroat in Twin Lakes) to be found in the upper Arkansas and South Platte drainages. Early reports leave no doubt that trout were abundant in these waters. But, just as the Colorado River cutthroat is in the path of our current quest for energy sources, so the greenback was right in the path of Colorado's early economic development. The major

19. The report on fecundity of Island Lake greenbacks is in Nelson (1972). The data for Como Creek-derived broodstock trout were reported in Dwyer and Rosenlund (1988).

population centers and the lion's share of the agriculture sprang up in that zone along the east side of the mountains where the South Platte and the Arkansas flow.

Mining in the upper Arkansas basin and in tributaries of the South Platte tore up stream banks and stream beds, and introduced large amounts of sediment and toxic runoff that eliminated many greenback cutthroat populations. These impacts were evident early on. The prominent eastern journalist, Samuel Bowles, on a summer trip through the mountain regions of Colorado in 1868 in company with equally prominent state and national political figures, several times remarked on "waters troubled by the miners" giving their "mud color to the combined stream" when they discharged into larger bodies of water, or sluices of water that had been diverted for placer operations "bringing the pollutions of the world and of labor with it" when they flowed back into the streams. David Starr Jordan also commented on this during his survey of the streams of Colorado and Utah for the U.S. Commissioner of Fish and Fisheries in 1889. He noted that placer mining and the runoff from stamp mills had turned waters yellow and red with clay and rendered impacted streams uninhabitable for trout.[20] Early-day logging and livestock grazing contributed their shares to the destruction of stream habitat and loss of native trout populations as well.

The rapid growth along the Front Range also brought on stream diversion and removal of water for irrigation, which didn't escape notice in Jordan's report either. He wrote:

> "In the progress of settlement of the valleys of Colorado the streams have become more and more largely used for irrigation. Below the mouths of the cañons dam after dam and ditch after ditch turn off the water. In summer the beds of even large rivers…are left wholly dry, all the water being turned into these ditches…. Great numbers of trout, in many cases thousands of them, pass into these irrigating ditches and are left to perish in the fields. The destruction of trout by this agency is far greater than that due to all others combined, and is going on in almost every irrigating ditch in Colorado."

One could take issue with Jordan's last comment about losses in irrigation ditches outweighing all other agencies of trout destruction. During the gold rush years, market fisheries that harvested trout by the wagon load helped feed the hordes of hungry miners. Zealous sportsmen also took a heavy toll. Then, when it was apparent that the native trout had dwindled, came the introduction of non-native trouts. Remaining greenback cutthroats were quickly replaced in the larger streams by brown trout and rainbow trout. Brook trout just as quickly displaced native cutthroat populations in the smaller tributaries. Introduced rainbows and non-native species of cutthroat trout also contributed to introgression, virtually eliminating any pure populations of greenback cutthroat that remained. By 1937, some authorities pronounced the subspecies extinct.[21]

20. See Bowles (1991) and Jordan (1891b) for these remarks.

21. This conclusion was stated in a publication on Colorado trout authored by W.S. Greene of the Denver Museum of Natural History; see Greene (1937).

But not quite. A few small populations could still be found that were visually good representatives of the greenback cutthroat appearance, even if their genetic purity was questionable. One of these "virtually pure" populations occurred in the Forest Canyon section of the Big Thompson River in Rocky Mountain National Park. In 1959, the Big Thompson population was tapped for fish to stock in the Fay Lakes, also within the Park. The stocked fish did not establish themselves in the Fay Lakes, but they did take hold in Caddis Lake which they reached by migrating downstream. A self-reproducing population was discovered there in 1972. Another "virtually pure" population was discovered in Island Lake, a reservoir in the City of Boulder watershed in the headwaters of North Boulder Creek. In 1971, the Colorado Division of Wildlife obtained permission to use it as a broodstock lake for stocking other mountain lakes in northeastern Colorado as part a fishery management program. These lakes were and are open to anglers.

However, populations of unquestioned genetic purity were extremely rare. As of 1973, only two pure populations had been discovered, one in Como Creek, an isolated tributary of North Boulder Creek near Nederland, Colorado, and the other in the very headwaters of the South Fork Cache La Poudre River in Larimer County, Colorado. Thus, in 1973, when the U.S. Endangered Species Act became law, the greenback cutthroat

was included as an endangered species (the State of Colorado followed the federal lead in 1976 by placing the greenback cutthroat on its list, also as an endangered species).

This triggered a much larger recovery effort. A Greenback Cutthroat Trout Recovery Team was formed with membership including the Colorado Division of Wildlife, Colorado Cooperative Fishery Research Unit at Colorado State University, U.S. Fish and Wildlife Service, U.S. Forest Service, and National Park Service. A recovery plan was completed in 1977, and revisions were adopted in 1983 and 1998.[22] But early recovery work was hindered because the Endangered Species Act's prohibition against "take" of an endangered species placed stiff restrictions on most efforts to establish new populations by translocating trout from known pure populations, as well as transferring trout from these populations to establish broodstocks. To foster these kinds of actions, the status of the greenback cutthroat was changed to threatened in 1978, and much progress has been made since then.

Searching for additional pure populations in remote areas of the back country has continued under each of the recovery plans, and has contributed much to the success of these plans. For example, in 1977, state biologists turned up a small but pure population in Cascade Creek, a tiny headwater tributary of the South Huerfano River

22. The most recent version of the greenback cutthroat trout recovery plan, adopted in 1998, is cited in the bibliography under Greenback Cutthroat Trout Recovery Team (1998). Accounts of progress toward recovery under these plans have been written up by Stuber et al. (1988), Dwyer and Rosenlund (1988), Gardner and Herlinger (1996), Young and Harig (2001), and Young et al. (2002).

in the Arkansas River drainage. This population had evidently persisted through historical time because Cascade Creek is isolated from the South Huerfano River by a waterfall that prevented invasion by the brook trout that became dominant below the falls. And since then, several additional pure populations have been discovered in both the Arkansas and South Platte drainages.

But the major thrust was aimed at developing broodstocks of genetically pure greenbacks so that additional populations could be established. Trout from the pure-strain Como Creek population were captured in the fall of 1977 and taken to the federal Fish Cultural Development Center in Bozeman, Montana, to develop a South Platte broodstock. In 1981, trout were transferred from Cascade Creek to Lytle Pond on the Fort Carson Military Reservation for the purpose of establishing a broodstock for the Arkansas River drainage. Meanwhile, state biologists had also transplanted trout from Cascade Creek into McAlpine Lake in Huerfano County, also in the Arkansas River drainage, in order to have another broodstock source. I also just received word that the federal hatchery at Leadville, where greenbacks (and yellowfin cutthroats) were propagated in the early years, is undergoing renovation and will come full circle by once again housing an Arkansas River greenback broodstock once that renovation is completed, perhaps as early as the summer of 2007.

Over the years, at least three federal hatcheries and one state facility have reared greenback cutthroats from broodstock sources for the purpose of establishing new populations. Most of these trout have been stocked to create new self-sustaining populations to meet the objectives of recovery plans, but some stocking has also been done solely to provide sport fisheries. This approach was taken to help overcome initial resistance from anglers who objected to replacement of non-native species they could catch and keep with greenback cutthroats that perhaps they wouldn't be allowed to fish for at all, or even if they could, on a catch-and-release basis. Purveyors of sporting goods and services and local officials whose communities depended on outdoor recreation were also apprehensive initially about losing business if fishing opportunities went away.

These concerns appear to have faded with the opening of more lakes and streams on public lands to catch-and-release angling for greenbacks. The campaign for acceptance of the recovery program has also been aided by groups such as Trout Unlimited and the Federation of Fly Fishers, who have promoted the return of the greenbacks by such means as selling bumper stickers, window stickers, and even T-shirts, by distributing pamphlets, and by highlighting the recovery program and new angling opportunities in magazine and news articles. This effort received a major boost in 1994 when the Colorado legislature named the greenback cutthroat trout the Colorado State Fish.

Getting back to the recovery side of the equation, the most recent tally that I have seen was published in 2002, and set the number of genetically pure recovery populations either discovered or introduced at 68. [23] Of the 68 recovery waters listed in this tally, 41 waters were treated with antimycin or rotenone to remove non-native species (usually brook trout) before greenback cutthroats were introduced. This has done the job most of the time, but treatment with toxicants is not always sure-fire. In the greenback cutthroat recovery program, incomplete removal has resulted in brook trout rebounding in some instances to displace the introduced greenbacks. On other occasions, failure of artificial barriers or illegal reintroductions by anglers have permitted non-natives to reinvade. All told, 15 of the 41 treated waters have suffered setbacks owing to the reestablishment of non-native species.

Dr. Behnke provided one example of how quickly brook trout can reestablish and displace a greenback cutthroat population. In 1967, Black Hollow Creek, a small tributary of the Cache La Poudre River west of Fort Collins, was transformed into a greenback cutthroat sanctuary by constructing a barrier dam, then chemically treating the reach upstream to remove brook trout. Greenback cutthroats recovered from Albion Creek, along with other greenbacks from Como Creek, were released in this sanctuary in 1968. They flourished there until 1973, when brook trout were again discovered in the sanctuary reach. Four years later, in 1977, Dr. Behnke and his students electrofished the sanctuary reach for about a mile upstream from the barrier dam, but found only brook trout. [24]

Hidden Valley Creek, a greenback cutthroat sanctuary created in Rocky Mountain National Park back in 1973, provides another example. Hidden Valley Creek had also been chemically treated to remove brook trout, and things went fine for the introduced greenbacks for about three years. Then, in the fall of 1976, brook trout were discovered again in the treated reach, and each year thereafter their numbers increased. Biologists feared that, once again, the brook trout would completely displace the greenbacks. In 1982, in the hope that angling might be an effective way to control the brook trout while still protecting the greenback population, the Hidden Valley sanctuary was opened to angling under special and restricted regulations: catch and keep the brook trout, but catch-and-release the rare little greenbacks. This strategy seems to have worked. The brook trout have not disappeared, but neither has the greenback population been displaced. [25]

As these examples suggest, unless steps are taken to

23. See Young et al. (2002). In 2005, the U.S. Fish and Wildlife Service launched its latest 5-year review of the status of the greenback cutthroat, presumably to ascertain if listing is still warranted or if, in the Service's view, the subspecies is recovered sufficiently to come off the endangered species list. That review was completed just as this book was going to press, and most likely will have been released to the public by the time you read this. Check the Service's website at *http://mountain-prairie.fws.gov/ endspp/* for the latest information.

24. This account is at page 122 of Dr. Behnke's 1979 monograph.

25. Early on, many of the fly-fishers who sampled Hidden Valley's waters could not bring themselves to harvest the brookies, even after much pleading by park and state biologists. The catch-and-kill admonition prompted many strong and vocal protests including letters to national fishing magazines. This is understandable, given that fly-fishers have led the way in establishing a no-kill ethic. The catch-and-release philosophy is deeply ingrained, no matter what the species. But in cases like this, I think a little soul-searching is in order. Brook trout are not native to greenback cutthroat country. Their

prevent it, complete displacement seems to be the fate of greenback cutthroat populations in streams where brook trout have been introduced or can invade from other waters populated by brook trout. But maybe not always. In 1995 or 1996, I was told about a tiny population of greenbacks that has persisted for decades, apparently, in a stream segment that is dominated by brook trout. The thinking at the time was that this stream, on private land somewhere in the Tarryall Creek drainage, might serve as a research site where mechanisms of species interaction and persistence might be investigated. However, I have no information that any such studies were ever carried out there. [26]

By the mid-1990s, recovery activities had progressed so well that authorities were optimistic about soon delisting the greenback cutthroat. A new (and the most recent, to date) recovery plan was issued in 1998. The subspecies will be deemed recovered, this plan stated, when at least 20 *stable* genetically pure greenback cutthroat populations are established in at least 124 acres (50 hectares) of lakes and ponds and 31 miles (50 km) of stream within the historical range. At least five of these stable populations should occur in the Arkansas River drainage. As of 1998, the recovery program had established 18 stable populations, 15 in the South Platte drainage and 3 in the Arkansas River drainage. The recovery team was confident that

ready adaptability to the habitat and their propensity to spawn even in lakes and ponds (an old friend of mine once remarked, "Eastern brook trout will spawn in your toilet!") seriously threatens the existence of the rare and sensitive native cutthroat. I always liked the response the editor of *Fly Fisherman* magazine gave to one of those letters of protest. It was published in the June 1983 issue, and it read like this:

> "The greenback cutthroat was indigenous to the South Platte and Arkansas rivers in Colorado, but their numbers dwindled with the introduction of rainbow, brook, and brown trout. To eliminate infiltrating brook trout is to restore what was before man interfered."

26. In footnote 29 of Chapter 4, I told you about *interspecific competition* and the *competitive exclusion principle*, which states that two species cannot coexist unless their niches are sufficiently different that each species limits its own population growth more than it limits the growth of the other population. If this condition is not met, one of the species will inevitably be reduced to zero in the competition. What ecological scientists endeavor to do in studies of interspecific competition is elucidate the conditions under which one competing species either excludes the other, or the two achieve an accommodation that allows them both to coexist even in disproportionate numbers—as appears to be the case in this stream reach of the Tarryall drainage.

Boulder Creek, South Platte drainage, near the town of Hesse, Colorado. Como Creek, an upper tributary, is home to one of the few genetically pure, historical populations of greenback cutthroat to endure into recent decades. PHOTO BY AUTHOR

the two additional Arkansas drainage populations would be established by the year 2000, and delisting could then occur.

Some Bumps in the Road

I emphasized the word *stable* in the paragraph above for a reason. The recovery plan defined four quantitative thresholds for what would be counted as a stable self-reproducing population. Of the 68 historical or reintroduced populations tallied in 2002, most failed to meet one or more of these criteria:

- The population maintains a biomass of 22 kg of trout per hectare of habitat through natural reproduction.
- The population contains a minimum of 500 adults (individuals greater than 120 mm or about 5 inches in total length).
- The population should, through natural spawning, produce a minimum of two year classes within a five-year period.
- The habitat size occupied should be at least 2 hectares and be separated by a barrier to upstream fish migration.

Of the populations that did meet these thresholds of stability, only three had been established in the Arkansas River drainage, two short of the recovery plan target.

It was these two populations that the recovery team was working to verify when the entire recovery process hit a couple of bumps. The first of these was a report, issued in 2001 by R. Paul Evans and Dennis Shiozawa, geneticists at Brigham Young University, on the genetic purity of greenback specimens from 22 recovery waters.[27] They found that several populations previously thought to be genetically pure, including four populations that had been used to found broodstocks or other introduced populations, carried mitochondrial DNA markers from Yellowstone cutthroat trout or rainbow trout. Thus, they are not free of introgression as had been believed, which calls into question if they should be counted toward the recovery goal.

The second bump was also experienced in 2001, in the form of a negative critique of the plan's four criteria for defining a stable population. M.K. Young of the U.S. Forest Service and Amy Harig, then at Colorado State University and later with Trout Unlimited, published this critique in the journal *Conservation Biology*.[28] They believe that the targets specified in the four criteria either "ignore currently acknowledged fundamentals of population persistence or may be too low to secure populations from demographic instability or loss of genetic variation." The following four paragraphs summarize their major findings.

Biomass [29]

There is no certain relationship in conservation biology between biomass and population persistence, so the reason for including a biomass target of 22 kg of trout per hectare of habitat was unclear to Young and Harig

27. The genetics report referred to here was prepared for the Colorado Division of Wildlife. It is cited in the bibliography under Evans and Shiozawa (2001).

28. See Young and Harig (2001). This is an important paper and should be read by everyone interested in the recovery of at-risk salmonids.

29. The *density* of trout in a stream is a common measure used in fisheries management, and is expressed as number of trout per square yard or square mile of stream or in metric units, number per square meter or hectare (stream length times average width). The aggregate weight of this number of trout is the *biomass;* also called *standing crop* or *standing stock.*

and the recovery plan offered no explanation. Even so, they pointed out that this target value is less than half the median value for naturally reproducing trout populations in other Rocky Mountain streams. If anything, they suggest increasing this target to 50 kg of trout per hectare of habitat, which is closer to the median value for other trout populations in the region. [30]

Adult Abundance

The recovery plan's population size threshold of 500 adult fish (trout of at least 5 inches [120 mm] in length) was chosen to provide an effective population size (i.e., the number of adults that actually contribute genes to the next generation) of 240 fish. This criterion was criticized on three grounds. For openers, the effective population size for salmonids is typically a much smaller proportion of the total adult population than the 48 percent implied by this 240-fish target. For cutthroat trout, effective population size may be closer to about one-fifth the total number of adults. [31] That's because mating is not random, sex ratios may be unbalanced, and not all adults reproduce every year. Therefore, for an effective population size of 240, one would need an actual adult population closer to 1,200 trout. Second, an effective population size of 240 may itself be too low. As I discussed back in Chapter 11, an effective population size of 500 is recommended to

30. For an array of density and biomass data for trout in western streams, see Platts and McHenry (1988) and Scarnecchia and Bergersen (1987).

31. In footnote 33 of Chapter 11, I referred you to studies by Hilderbrand and Kershner (2000a), who examined the prognosis for long-term persistence of isolated cutthroat trout populations. They recommended a minimum population level of 2,500 adult trout to correspond to an effective population size of 500 individuals. Effective population size refers to the number of breeding individuals that actually contribute their genes to the next generation. Too low an effective population size leads to a loss of genetic variation in a population, which may in turn lead to lower overall fitness and a decrease in reproductive performance, all of which bode ill for long-term viability. An effective population size of 500 is generally accepted in conservation biology as the minimum level needed to offset these effects.

avoid long-term loss of genetic variation. That would push the target for total number of adults up to 2,500 trout. Third, Young and Harig modeled the variance in annual population growth using data available for 11 greenback cutthroat populations, and concluded that isolated populations with less than 2,000 individuals have less than a 50 percent probability of persistence for 100 years owing to nothing more than chance negative changes in population demographics (birth rates, death rates, and the like). They also pointed out that chance upsets to local environments can overwhelm isolated populations that are even larger. [32]

Year Class Success

Young and Harig's criticism of this threshold is based on their assertion that a population that produces only two successful year classes in a 5-year period is still at substantial risk. The chance of such a population suffering total reproductive failure over any given 5-year span is 8 percent. [33] While this may not sound like much, if the maximum age of adults in the population is, say, 7 or 8 years, and they may not spawn more than two or three times in their lifetimes, which are reasonable assumptions from our earlier discussion of greenback cutthroat life history traits, then even minor increases in mortality for any reason could extinguish the population. Young and Harig further

pointed out that, in trout, year class failures are rare in truly suitable habitats. They expressed the view that this recovery plan threshold reflects more the difficulty faced by the greenbacks trying to recruit in sites selected for reintroductions (often cold, high-elevation waters that may not allow sufficient time for successful early development of embryos or newly hatched fry) than a truly acceptable minimum for year class success.

Habitat Size

To orient you, the minimum of 2 hectares specified in the recovery plan criterion for occupied habitat area is 4.94 acres, or less than 0.008 square mile. This criterion was criticized as overly favoring recovery in lakes, because few of the habitat segments occupied by historical or introduced populations in streams meet this standard. Based on an average low-flow wetted width of 8.5 feet (2.6 m) for 10 streams with historical or introduced populations of greenback cutthroats on the Arapaho-Roosevelt National Forest, a typical stream segment would have to be 4.8 miles long to provide 2 hectares of habitat area. Harking back to Chapter 11 once again, where I discussed the work of Robert Hilderbrand and Jeff Kershner of Utah State University, who concluded that a stream length of 5.2 miles of suitable habitat is needed to support a population of 2,500 adult cutthroat trout at an abundance level of

32. Losses of rare and endangered trout populations to environmental upsets are well known. Young and Harig (2001) cited the example of the Main Diamond Creek, New Mexico population of Gila trout that was variously estimated at from 5,000 to 15,000 individuals strong. This population was totally eradicated in 1989 by a devastating flood that followed on the heels of a forest fire in the drainage.

33. Two years of successful reproduction in 5 years yields a probability of successful reproduction in any given year of $2 \div 5 = 0.4$, or 40 percent. Conversely, that's a probability of 0.6, or 60 percent, of *unsuccessful* reproduction in any given year. The chance of five consecutive failures is $0.6 \times 0.6 \times 0.6 \times 0.6 \times 0.6 = 0.08$, or 8 percent.

480 fish per mile (which, by the way, corresponds to an effective population size of 500 trout),[34] the 2-hectare habitat area minimum isn't that far off the mark. On the other hand, if the habitat or the population can support only 160 trout per mile, then 15.5 miles of stream length would be needed for 2,500 adult trout. That corresponds to a habitat area of about 6.5 hectares. The problem is, less than a quarter of the stream segments occupied by historical or introduced greenback populations presently meet even the 2-hectare criterion. As Young and Harig hinted, perhaps the recovery team should work toward seeing that they do.

Young and Harig also pointed out the lack of connectivity into metapopulation structure in any of the recovery populations. Although many of the recovery populations are clustered in the same geographical area, e.g., Rocky Mountain National Park and the Arapaho-Roosevelt National Forest, metapopulation connectivity does not exist among any of them, and, indeed, it does not appear to have been a focus in the recovery plan. They went on to point out that the small, isolated waters that still supported greenback cutthroats at the time the subspecies was originally listed as endangered may have been an historical artifact. These waters may have been simply the last remaining bastions of the subspecies. Confining all of the known historical populations and all of the introduced recovery populations to similar small, isolated segments prevents the subspecies from developing the mobile life histories that were historically common and may be necessary for long-term persistence.

They summed up their critique of the greenback cutthroat recovery criteria with these words: "We fear that the selected criteria may have been based on what contemporary populations might be capable of achieving, rather than on what would constitute long-term security for the subspecies." Several years have now gone by since their critique was published, without any word or published response from the recovery team that I am aware of. However, in December 2005, the U.S. Fish and Wildlife Service did announce that it was commencing a new formal status review to determine if listing of the greenback cutthroat is still warranted. As I write this one year later, that review has not yet been issued. So for now, the subspecies remains listed as threatened.

Lingering Threats

Restoring greenback cutthroat trout widely across the historical range may no longer be possible given the burgeoning pace of human development along the Front Range and the increasing demands for water and other natural resources that accompany human population growth.[35] Also, occupation of the larger stream systems by non-native species may now be so

34. See Hilderbrand and Kershner (2000a).

35. I don't have current figures on population growth or on the current pace and extent of human development along the Front Range, but you can find one reasonably good treatment of the subject in the November 1996 issue of *National Geographic* magazine. See Long (1996) for the citation.

entrenched as to be irreversible. But even in the more remote and protected areas where greenback cutthroat recovery efforts are presently focused, the subspecies faces continuing or impending threats.

One of these is associated with mining. I'm not talking here about the mining that may be occurring now or that may occur in the future, but rather the legacy of former operations that boosted Colorado's economy and spurred growth during the 19th century mining era. Stream ecologist J.D. Allan, University of Michigan School of Natural Resources and Environment, defines *legacy effects* as the consequences of disturbances that continue to influence environmental conditions long after the initial appearance of the disturbance. [36] That's an apt description of the acidic drainage that enters streams from abandoned and inactive mines and tailings, oftentimes for decades after the mining has stopped. There are few upper tributaries and gulches in the upper Arkansas watershed from the Continental Divide above Leadville southward along the Sawatch Range that do not have old abandoned mining camps and diggings—and attendant problems with acid mine drainage. The same is no doubt true of the South Platte generally south of Rocky Mountain National Park, because the upper tributaries along the Front Range were heavily worked as well.

Acid drainage can occur anywhere that sulfides of iron, copper, zinc, lead, or cadmium are exposed during mining. These are easily oxidized when exposed to air and water to produce acids that lower the pH of the water and can be directly toxic to most forms of aquatic life. The heavy metals themselves, when solubilized in any form and leached into streams, can also be toxic to fish. I mentioned back in Chapter 6 that the U.S. Geological Survey's Abandoned Mine Lands Initiative, a program that is supposed to map abandoned mine sites and tributaries affected by runoff from hard rock mining, had its teeth pulled by Congress a few years back and hasn't issued any new maps. But the Geological Survey, along with the U.S. Environmental Protection Agency, has been active in the upper Arkansas River basin on other aspects of the problem nevertheless. The Geological Survey has studied mechanisms of metal transport and transformation in streams affected by abandoned hard rock mines and is now focused on how to use this knowledge to develop effective remedial methods. The Environmental Protection Agency concentrates on the remediation efforts themselves. One remediation project that has been hailed as a success story took place in upper Chalk Creek, which enters the Arkansas River about 10 miles downstream from Buena Vista. Upstream on Chalk Creek is the old mining town of St. Elmo, and above that is the Mary Murphy Mine, said to have been one of the most productive in Colo-

36. See Allan, J.D. "Landscapes and riverscapes: the influence of land use on stream ecosystems" *Annual Reviews of Ecology, Evolution, and Systematics* 35 (2004): 257–284.

rado. Alas, it was also the single greatest contributor of heavy metals to the creek, creating a "dead zone" for young trout (brown trout in this case, not greenback cutthroats) that extended downstream for 12 miles. Remediation projects in 1991 and again in 1997–98 capped off mill tailings and diverted the Mary Murphy drainage source so that the "dead zone" has now retreated, although I don't think it has entirely disappeared even yet. [37]

As for the effects of acid mine drainage on greenback cutthroat trout, this certainly has to be a major factor to be guarded against in selecting recovery sites for new populations, and in any plans to extend the distribution of the subspecies. The 1998 recovery plan describes an experimental stocking of greenback cutthroats into Bard Creek, a fishless mountain stream that was known to have elevated levels of heavy metals due to past mining activity. Trout stocked at over 25 mm (about an inch) in length did survive to maturity and were able to spawn in Bard Creek despite the elevated heavy metal concentrations, but none of their progeny survived to swimup. For greenback cutthroat trout, eggs and alevins may be the life stages most sensitive to elevated levels of heavy metals. [38]

The other lingering and impending threat to greenback cutthroat trout that needs to be highlighted here is acid precipitation, more simply known as acid rain. In the late 1970s and early 1980s, when the first edition of this book was in preparation, this was the number-one environmental issue on everybody's minds. Concern abated after that, owing to the successes achieved by concerted efforts to reduce emissions from power plants, industrial stacks, and automobile tailpipes, as mandated by the Clean Air Act. But the problems never went completely away and, indeed, emissions that lead to acid precipitation have been edging back up recently. This is particular true along Colorado's Front Range, where existing greenback cutthroat populations are concentrated.

Acid precipitation is, simply, rain or snow that is contaminated with strong acids, particularly sulfuric acid and nitric acid. When fossil fuels that contain high amounts of organic sulfur and nitrogen (coal is a good example) are burned to release their energy, oxides of sulfur and nitrogen are given off as gases that exit through the stacks (automobile exhausts are another major source of the oxides of nitrogen). These are converted to aerosols of sulfate and nitrate in the atmosphere, and are scavenged out and converted to acids when it rains or snows, thus acid precipitation.

To quickly review some fundamentals, one way acidity and alkalinity can be expressed is on a pH scale running from 0 to 14, with the neutral point being pH 7. Decreasing numbers from 7 to 0 are more acid, each

37. For additional information about the Chalk Creek project and other work by the U.S. Geological Survey and Environmental Protection Agency in the upper Arkansas, check these websites: *http://toxics.usgs.gov/ toxics/* and *www.epa.gov/owow/ nps/Section319III/CO.htm*.

38. A summary of this experiment is at page 13 of the 1998 recovery plan; see Greenback Cutthroat Trout Recovery Team (1998). For a more thorough review of the impacts of mining and mineral extraction on trout and their habitats, see Nelson et al. (1991).

number representing a ten-fold increase in acidity. Thus, a pH of 5 is ten times more acid than a pH of 6, a pH of 4 is ten times more acid than a pH of 5, and so on. The same thing is true of alkalinity, except the numbers ascend the scale from 7 to 14. A limestone stream or lake would be slightly alkaline, somewhere between pH 7 and, say, 8.5. A softwater stream might range from pH 5.5 or 6 to pH 7. If there were such a thing as pure rainwater, its pH would be 5.6, or slightly acid. This is because water reacts with carbon dioxide in the air to form carbonic acid, a weak acid. Vinegar has a pH of about 3, lemon juice a pH of 2.3, and battery acid a pH of 1 (on the alkaline side, baking soda has a pH of about 8.4 and ammonia is pH 12). Salmonids generally start showing stress (arrested growth, reduced reproductive success, increased year-class failures) when water pH drops to about 5, and pH 4.5 water is lethal. But it's not just a matter of pH. Acidic waters also solubilize and leach out metals from lakebed and streambed sediments and surrounding soils, and these are toxic to fish, aluminum in particular. Thus, acid precipitation deals fish a double-whammy.

But wait. If even pristine rainwater is acidic, how come all our surface waters did not become acidic long ago? It's because lakes and waterways and the catchments that drain into them do have a buffering capacity, that is, the ability to neutralize acid, some much more so than others. Catchments underlain by limestone rock types have more buffering capacity than those underlain by granitic rock types that are often naturally acidic themselves. Viewed from an angler's perspective, the same geological settings that produce the most productive trout streams also produce the most buffering capacity against acid rain. [39] But when that buffering capacity is expended, as it can be after prolonged exposure to rains that are ever more acidic, or when the acidic snows of winter melt off in the spring, then lakes and waterways can become acidic very quickly and utterly devoid of aquatic life, as happened across northern Europe and in the northeastern United States and Canada in the 1970s and 1980s.

Most of the early national publicity generated by the acid rain menace focused on the northeastern U.S. and eastern Canada, because that's where the problem became the most severe and the effects most dramatically seen. But if it had remained just a regional problem, it would have no place in this book. Unfortunately, "the killer marched west," as one editorial writer put it several years ago. Colorado's Front Range geology is the granitic type that provides little in the way of buffering capacity for its mountain lakes and streams, although buffering capacity does increase as one moves down in elevation. And precipitation in the Front Range has indeed tipped to the more acidic side.

39. See footnote 40 of Chapter 9 for a review of what makes a productive trout stream.

One of a series of beaver ponds on Hidden Valley Creek, a refuge for greenback cutthroats that is open to angling to keep brook trout under control. It's catch-and-release only for the greenbacks. PHOTO BY AUTHOR

At Como Creek, home of one of the earliest rediscovered historical greenback cutthroat populations, a monitoring site operated by the University of Colorado reported a pH of 4.7 for the precipitation that fell on the catchment in 1984. Although these readings did back off some as emissions controls kicked in around the country, Front Range precipitation still measured pH 5.1 in 2001 (most recent data available). [40]

Most of the present recovery populations of greenback cutthroat trout are clustered in high-altitude habitats in Rocky Mountain National Park and the Arapaho-Roosevelt National Forest that have little buffering capacity against acid rain. As the Denver-to-Fort Collins metropolitan corridor continues to grow and develop, emissions are expected to increase even with controls in place, and that likely will increase acid deposition across this area as well. Concern about the susceptibility of greenback cutthroat trout prompted the U.S. Fish and Wildlife Service to test their sensitivity to acid pHs and to elevated aluminum concentrations. As with other salmonids, the threshold of sensitivity to pH occurred at pH 5.0. Reduced survival, impaired locomotion, and inhibition of feeding of

alevins and swim-up fry were detected at pH 5.0 in the absence of aluminum, but when as little as 50 micrograms per liter (0.05 part per million) of aluminum was present in the water these effects were observed at pH 6.0. The life stages most sensitive to acidic pH and elevated aluminum were the alevins and swim-up fry. [41]

All of these susceptibility tests were performed using standard continuous exposures of 7 days in a flow-through chamber. On the positive side, the investigators noted that, unlike other salmonids including other subspecies of cutthroat trout, the surviving greenbacks quickly resumed their normal feeding behavior and grew normally after exposure ceased. This suggests that the greenback subspecies may have at least some capacity to withstand pulses of acidic input, as would occur during spring snow runoff events.

Although the greenback cutthroat recovery team did make note of the results of these tests, no steps were included in the recovery plan to safeguard the integrity of existing populations or to offset the threat posed by increased acid precipitation to the long-term persistence of the subspecies.

40. Two overviews of the acid rain problem as it has affected the nation and the world, one easy and one more technical, appeared back in 1981. The easy read is a piece published in *National Geographic* magazine in November of that year. Its title: "Acid Rain: How Great a Menace?" See LaBastille (1981) for the complete citation. The more technical treatise, "Acidic Precipitation and its Consequences for Aquatic Ecosystems: A Review" by T.A. Haines, appeared in *Transactions of the American Fisheries Society*, also in November 1981. See Haines (1981) for that citation. Papers that document the westward spread of acid rain and the sensitivity of lakes and waterways in the Rocky Mountain west and the Front Range in particular include Baron (1983), Hinds (1983), Turk and Adams (1983), Kling and Grant (1984), and Nanus et al. (2003). The National Atmospheric Deposition Program maintains a website at *http://nadp.sws.uiuc.edu* where you can get additional information and reasonably up-to-date maps. There is also information on a U.S. Geological Survey site at *http://bqs.usgs.gov/acidrain/*.

41. See Woodward et al. (1991).

Rio Grande Cutthroat Trout, *Oncorhynchus clarkii virginalis*

CHAPTER 14

Rio Grande Cutthroat Trout

Oncorhynchus clarkii virginalis: Chromosomes, $2N=64$. Scales in lateral series 146–186 (Pecos River trout consistently have the highest counts, averaging 175 or more; trout associated with the Rio Grande usually average fewer). Scales above lateral line 39–47. Vertebrae 60–63. Pyloric caeca 33–59. Gill rakers 18–21. Basibranchial teeth typically feebly developed. Colors similar to *pleuriticus* (Chapter 12) and *stomias* (Chapter 13), in that red, orange, and golden-yellow hues are developed, but somewhat less intensely than in the latter two subspecies. Adult spotting pattern is distinctive in that the majority of the spots are concentrated on the caudal peduncle in a profuse, closely-set patch. The spots themselves are irregular or club-shaped rather than round, as in *pleuriticus* or *stomias*. Trout associated with the Pecos River drainage in New Mexico have larger spots than trout associated with the Rio Grande drainage in Colorado and New Mexico, and are quite like *stomias* in this regard. Contrary to what one might assume from the geographic proximity—or separation—of these drainages, trout thought to be native to the Canadian River drainage are closer in appearance and characters to Rio Grande trout than they are to specimens from the Pecos River drainage.

Rio Grande Cutthroat Trout,
Oncorhynchus clarkii virginalis, Pecos strain

1. See Fountain, P. *The Eleven Eaglets of the West* (London: John-Murray, Albemarle Street, 1905). The passage quoted is at pages 236–238.

THE UPPER RIO GRANDE COUNTRY

Paul Fountain was an Englishman who produced four books on his travels in America in the late 1800s. Although the Rocky Mountain areas had been fairly well settled by Fountain's time, relative to conditions of today, he saw the west when it was still "emphatically a wilderness," as he put it. Here is how he described the San Luis Valley:

"The San Luis Park is bigger than all the others put together, and is probably not less than 100 miles across, with, as far as I could perceive, a tolerably oval shape. Like the others, it is quite flat: the eye cannot distinguish the slightest rise or fall of the ground in any direction. In some parts it is not so well wooded as the more north-ern Parks; and there are spots which are almost desert in character. Quite a number of streams lose themselves in sinks within its bounds; but there is nevertheless pasturage for immense numbers of cattle. As in the other Parks, there was, even at this time, a number of ranchers feeding their herds; but the greater part of the Park was still overrun by large crowds of deer and prongbucks..."[1]

The last Chapter told how I believe that the Arkansas River valley, between the Mosquito and Wasatch Mountain ranges, may have been the artery by which ancestral cutthroat trout spread throughout the eastern and southern drainages of Colorado and New Mexico. I described how, in crossing Tennessee Pass, which leads into the upper Arkansas valley from the Colorado River system, and again in crossing Trout Creek Pass into South Park and the South Platte River, I got the strong, almost overpowering feeling that I had traveled ancient waterways, places where headwater stream captures or perhaps pluvial lake refugia had existed to aid in the movement of fishes from one drainage to the next.

So it is with the southern exit from the Arkansas valley at Poncha Pass, elevation 9,010 feet. The climb up to the pass is moderately steep, but not formidable. Then, when you top out, you come onto one of those broad, high, grass and sagebrush covered valleys that slopes off ever so gently and stretches away for miles to the south. This is the San Luis Valley, once part of Spanish New Mexico. It is no longer overrun by "prongbucks" as it was in Paul Fountain's day, but the lay of the land is pretty much the same. It is dry country, not what you'd call desert, but dry nevertheless, lying as it does in the rain shadow of the San Juan Mountains. The Indians called it "Land of the Blue Sky People."

But it wasn't always so. In Ice Age times there was plenty of water here. Karel L. Rogers of Adams State College, Alamosa, Colorado, and a group of co-workers from other institutions involved in studies of ancient climates of the San Luis Valley recovered fossil fishes identified as cutthroat trout from 740,000-year-old strata. If you match this date against Fig. 2-6 back in

Chapter 2, you'll see that it was near the end of a major glacial advance. At that time, according to Rogers and co-workers, the San Luis Valley supported a montane forest and deep, permanent aquatic habitats. [2]

One early fishing writer, in discussing the exodus of the cutthroat, even talked about an ancient Lake San Juan, with an outlet north to the Arkansas River. This ancient lake breached its southern rim, much as Lake Bonneville breached over Red Rock Pass (see Chapter 2), and drained away to form the Rio Grande. Trout that had migrated into the ancient lake were thus isolated, and evolved into the present Rio Grande cutthroat. [3] His theory sounds plausible, but try as I might—and I checked with the University of Colorado, the U.S. Geological Survey, and several other sources—I could not find any evidence to confirm the existence of an ancient Lake San Juan. However, I did turn up what I think might have been the source for the writer's theory. There is indeed evidence that in Ice Age times, not one but perhaps two major lakes occupied the San Luis valley. In 1877, F.M. Endlich, a geologist serving with the U.S. Geological and Geographical Survey led by Ferdinand Hayden, named them Coronado's Lakes. [4] There is no indication that either of these lakes was open to the Arkansas River, other than by headwater stream capture. Their outlets led instead to the Rio Grande, through which they

2. See Rogers et al. (1985, 1992) for details.

3. This theory was expressed by Arthur H. Carhart in his book, *Fishing in the West* (1950); see Carhart (1950).

4. The Endlich report is at pages 103–235 (plus pages 9–33) in the Annual Report of the U.S. Geological and Geographical Survey of the Territories (the Hayden surveys) 9, 1875. See Annual Reports (1868–83).

finally drained away as the outlets downcut.

Another possible route that trout might have exploited is Cochetopa Pass, elevation 10,032 feet, leading across the Continental Divide from the Gunnison River in the Colorado River drainage. Dr. Behnke has expressed the view that the Rio Grande cutthroat was derived from the Colorado River cutthroat via headwater stream transfer from the Gunnison River, probably at an earlier time than the transfer that gave rise to the greenback cutthroat, although he does acknowledge that Poncha Pass could have provided an arterial from the upper Arkansas drainage. Although Cochetopa Pass is more than 1,000 feet higher than Poncha Pass, it is another of those apparently "easier ways" that an overland traveler would naturally try to use to cross the Continental Divide. It was a well-traveled Indian trail, and in 1768, when Juan Batista de Anza first recorded its presence from the floor of the San Luis Valley, he described it as an immense sag in the range, so low as to give the illusion that the mountains vanished altogether. Captain John Gunnison's survey party, looking for a transcontinental railroad route through the Rockies, took this way out of the San Luis Valley in 1853 to avoid tackling the rugged San Juan Range. The trouble with Cochetopa Pass as an overland route, as Gunnison soon discovered, is that it leads into the spectacular but virtually impassible Black Canyon of the Gunnison River, now a National Monument. As an ancient waterway, however, it's another, more plausible story. [5] In any event, crossing into the San Luis Valley brings us into the native range of the Rio Grande cutthroat trout.

Much of the San Luis Valley was once part of a million-acre Spanish land grant that stretched from the crest of the Sangre de Cristos west to the Rio Grande. In the early 1850s, after the United States had acquired New Mexico, American settlers began to enter the territory. To protect these new arrivals as well as the Spanish- and Mexican-Americans already there, the army built two forts, Fort Union near Mora, New Mexico, and Fort Massachusetts on the lower slope of 14,363-foot Blanca Peak, on the eastern edge of the San Luis Valley where it could command the route over Sangre de Cristo Pass.

Fort Massachusetts was situated on the bank of Ute Creek, a small stream that flows down off the slopes of Blanca Peak to join Sangre de Cristo Creek, which in turn enters Trinchera Creek, a tributary of the Rio Grande River. In 1853, Captain John Gunnison's Pacific railroad survey party stopped at Fort Massachusetts and collected specimens of trout from Ute Creek. Three years later, scientist Charles Girard examined these specimens and named them as a new species, *Salar virginalis*. In 1872, E.D. Cope

5. Dr. Behnke's view on the spread of cutthroat trout into the Rio Grande basin is in his two monographs at page 131 in the 1979 monograph and page 152 of the 1992 publication, and in his 2002 book at page 210. Crossing Cochetopa Pass from the Colorado River drainage leads to tributaries of Saguache Creek, a stream that teemed with trout at the time of the Gunnison railroad survey, as journals kept by expedition members attest (see Beckwith 1855 and Schiel 1859, [1959]), although Saguache Creek now ends in a sump on the valley floor. Two higher passes in this same general area, Salt House or Carnero Pass (10,500 feet) and Spring Creek Pass (also known as Rio Del Norte or Williams Pass, 10,898 feet), also cross over from the Gunnison river drainage into tributaries of the upper Rio Grande. But the San Juan Range, through which these passes lead, was covered by one of the largest ice bodies in the Rocky Mountains during Pleistocene time, so the passes and upper tributary valleys, even the valley where the Rio Grande itself heads up, were filled with a succession of ice cap and valley glacier complexes. Headwater transfers would have had to occur, if they did, during periods when the gaps and valleys were ice-free. References that discuss Pleisto-

examined trout taken from nearby Sangre de Cristo Creek and described them as yet another new species, *Salmo spilurus*. For a time in those early years, the name *virginalis* was also used for the trout of the Bonneville basin, owing to a mistaken idea about where Ute Creek was actually located. But that was eventually rectified, and it was also realized that the trouts named *virginalis* and *spilurus* were one and the same. Girard's name *virginalis* had priority, so that was the name that prevailed for the Rio Grande cutthroat trout, with Ute Creek as the type locality for the subspecies. [6]

I was out on the floor of the San Luis Valley one bright September afternoon taking photos of Ute Creek when a dark, ominous thunderhead began to build over Blanca Peak. The storm descended over the mountain like a heavy gray cloak. Lightning split the sky and cannon-shots of thunder echoed across the valley. It was as if God was venting his wrath on some offending soul he had cornered up there on the peak, for all the while the sun shone brightly everywhere else around! Over dinner, I remarked about this phenomenon to a group of Colorado Division of Wildlife biologists with whom I would be climbing into that very same area the next day. "The Sangres make their own weather," one of them replied with a grin.

cene glaciation in this area include Atwood and Mathur (1932), Carrara et al. (1984), Leonard (1984), and Richmond (1986).

6. The survey for a southern route for a railroad to California, roughly along the 38th parallel, was undertaken by a party of the U.S. Army Corps of Topographical Engineers under Capt. John Gunnison with Lt. Edward Beckwith as second-in-command, Richard Kern as topographer and artist, and a German, Jacob Schiel, as surgeon/naturalist and geologist. The party came up the Huerfano River and approached Sangre de Cristo Pass via one of its tributaries. On August 7, 1853, with the party still on the east side of the pass, Jacob Schiel wrote that the men availed themselves of "the pretty mountain streams...filled with the finest trout." These east-side trout would have been greenback cutthroats. The party crossed Sangre de Cristo Pass six days later, and camped the night of August 13th on Sangre de Cristo Creek. The next night, they camped on Ute Creek just below Fort Massachusetts. There is no mention by either Schiel, who kept his own journal, or Lt. Beckwith, who completed the official report of the expedition, of trout caught from Ute Creek, although both men did write glowingly about the trout fishing the party experienced in Saguache Creek and its tributaries after they had crossed the San Luis Valley on the approach to Cochetopa Pass. But even so, during the party's week-long sojourn at Fort Massachusetts, trout were collected from Ute Creek and shipped east. It was these specimens that Charles Girard examined and named *Salar virginalis*. See Beckwith (1855) and Schiel (1859, [1959]) for accounts of the survey party's movements and trout fishing activities. Girard (1856) and Cope (1872) have the descriptions of *virginalis* and *spilurus* respectively, and Jordan (1920) and Snyder (1919, 1922) have the corrections regarding type locality and priority in naming. After passing through the San Luis Valley, Gunnison's party worked its way west into Utah and the Great Basin. But there, disaster struck. Gunnison, with Richard Kern and several others, was out ahead of the main party reconnoitering in the Sevier River area when Indians attacked, killing them all. Second-in-command Beckwith completed the survey's work and wrote up its final report. Five years after the survey party passed through, Fort Massachusetts got a failing grade from the army and was replaced by Fort Garland at a new location about six miles to the south. *The Colorado Magazine*, a quarterly publication of the State Historical Society of Colorado, had a report on excavation work that was done at the Fort Massachusetts site in 1964; see Baker (1965).

Figure 14-1 depicts my interpretation of the historical range of the Rio Grande cutthroat trout. Trout of this subspecies likely occupied all waters capable of supporting trout in the upper Rio Grande and upper Pecos drainages within the shaded area. They also occurred in upper tributaries of the Canadian River, such as the Vermejo River in New Mexico and Colorado and the Mora River in New Mexico.[7] There is much good trout water in the upper Canadian River drainage, and that's what I have tried to show inside the boundary depicted on the map. But how much of this was actually occupied by cutthroat trout historically, or how far downstream they occurred in the Canadian River, is not really known.

At Poncha Pass, the northernmost extent of the range, and down the crest of the Sangre de Cristo Range to Sangre de Cristo and La Veta passes, the Rio Grande cutthroat shares its boundary with the greenback cutthroat subspecies. West from Poncha Pass and then south along the Continental Divide through the San Juan Mountains, the boundary is shared with the Colorado River cutthroat subspecies. On the west side of the Continental Divide in the Black Range, at the southern tip of the contiguous distribution, are the headwaters of the Gila River, home of the Gila trout. Even without the outlier enclaves shown in the figure, the Rio Grande cutthroat trout has the southernmost distribution of all the cutthroat subspecies.

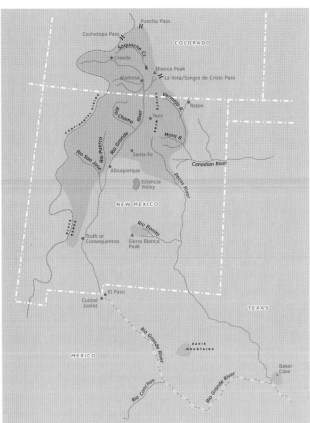

FIGURE 14-1. *Historical distribution of Rio Grande Cutthroat Trout*

Of the outlier distributions that are shown, the first is associated with Rio Bonito, a stream that heads up in the mountains north of Sierra Blanca Peak and flows east to the Pecos River in New Mexico. J.W. Daniel,

7. If you can find a copy, an entertaining book that highlights the abundance of trout in the upper Pecos and adjoining waters of New Mexico in the late 19th and early 20th centuries is Elliott Barker's *Beatty's Cabin*, published in 1953 (see Barker [1953] for the citation). Barker was raised on a ranch on the eastern flank of the Sangre de Cristo Range and packed into the upper Pecos country many times to hunt and to fish for the "black-spotted, red-bellied cutthroat trout." He also served as a federal forest ranger and Forest Supervisor in the area, and later as New Mexico State Game Warden. Another rare volume that also describes the country and speaks some of the trout and the fishing is *Campfires of a Naturalist*, by Clarence E. Edwords, published in 1893 (see Edwords [1893] for the citation). Still other early accounts of the fishing and the waters where the trout were found were published in the outdoor magazines of the period. These include *Apache* (1877) and *Carnifex* (1891).

who served as an assistant surgeon at Fort Stanton on the Rio Bonito during the Civil War period, reminisced about the trout fishing there in an 1878 issue of *Forest and Stream* magazine. The cutthroats are long gone from the Rio Bonito, but some years ago trout from that stream were planted in Indian Creek, a west-flowing stream on the slopes of Sierra Blanca Peak in the Mescalero Apache Indian Reservation. That population still exists and was examined by Dr. Behnke, who found it to be typical of the Pecos form of Rio Grande cutthroat trout in both appearance and meristic characters. [8]

The same J.W. Daniel who found trout in Rio Bonito, New Mexico, also caught "speckled trout" from Limpia Creek, a Pecos River tributary, while serving at Fort Davis in Texas during the Civil War period. A second account of trout caught from Limpia Creek also appeared in 1878, this one in a letter to the editor of *Forest and Stream* magazine from one N.A. Taylor, who stated that another surgeon, Dr. I.H. Hunter, had also fished Limpia Creek for "brook trout" during the Civil War period. [9] This stream heads up in the Davis Mountains, which comprise the second population enclave shown in Fig. 14-1.

The Davis Mountains are what ecologists call a *sky island*, jutting up as they do above the surrounding Chihuahuan Desert to an altitude of 8,378 feet at their highest point. Plants and animals living above about

8. Dr. Behnke gathered much information about the historical occurrence of cutthroat trout in Rio Bonito and the subsequent introduction of these trout into Indian Creek. This is discussed in detail in his two monographs; see Behnke (1979) at page 129 and Behnke (2002) at page 150. You can also find this information in a 1988 article Dr. Behnke prepared for *Trout*, the quarterly journal of coldwater fish conservation issued by Trout Unlimited; see Behnke (1988b). His examination of the Indian Creek population, derived from Rio Bonito stock, is in a report he prepared for the U.S. Fish and Wildlife Service in 1981; see Behnke (1981b). The recollections of Dr. J.W. Daniel appeared in *Forest and Stream* magazine, volume 10, number 48 for the year 1878 at page 339; see Daniel (1878).

9. N.A. Taylor's letter to the editor of *Forest and Stream* magazine actually appeared earlier than Daniel's account, which I cited in footnote 8 above. You can find it in *Forest and Stream* volume 10, number 13 at page 236. G.P. Garrett and G.C. Matlock of the Texas Parks and Wildlife Department also examined the evidence for the historical presence of Rio Grande cutthroat trout in Texas, and published their findings in *The Texas Journal of Science* in 1991. See Garrett and Matlock (1991) for the complete citation.

Ute Creek, Colorado type locality for Oncorhynchus clarkii virginalis, *the Rio Grande cutthroat trout. The reach shown here is about six miles downstream from the site of Fort Massachusetts where the original specimens were collected in 1853.* PHOTO BY AUTHOR

5,000 feet in these mountains are typical of Rocky Mountain montane forests, but are isolated from other similar assemblages by great distances, like an island in the sea. Up here, evergreen forests including stands of ponderosa pine and aspen support black bears, cougars, golden eagles, Mexican spotted owls, and several other rare bird species. Regarding the aspen and pine groves, I'm also reminded of a quote from M.R. Montgomery that "natural trout streams are never far from aspen, pine, or fir trees." There's a State Park in the Davis Mountains now and a land preserve of The Nature Conservancy that provide a measure of protection for this unique sky island enclave. But trout no longer swim in its streams. Livestock grazing and overuse of its water resources took their tolls on the stream habitat for trout long before these protections were ever put in place. [10]

J.W. Daniel also served at Fort Hudson, located on San Pedro Creek, a tributary of Devils River north of Del Rio, Texas, during the Civil War. While at this post, his 1878 reminiscence continued, he caught trout out of San Felipe Creek and also from Devils River, both Rio Grande tributaries. Devils River enters the Rio Grande about 45 miles downstream from the confluence of the Pecos River (nowadays, both the Pecos and the Devils River discharge into arms of Amistad Reservoir on the Rio Grande) and San Felipe Creek enters at Del Rio. Owing to these locations being so very far to the south, I would have dismissed this particular recollection as the site of an historical population enclave, but for two things. First, both San Felipe Creek and Devils River are spring-fed streams that produced good, reliable flows of cold water before settlers began to tap their water resources, so it's conceivable that both could have held trout as recently as the Civil War period. [11] Second, University of Texas archaeologists made a discovery there, in Baker Cave (approximate location marked with respect to the mouth of the Pecos River in Fig. 14-1), that indeed indicates trout were present not so very long before the historical period, and this lends credence to the Civil War report.

Baker Cave is a large rockshelter located in the canyon of San Pedro Creek. This whole area is a canyon-incised limestone plateau with hundreds of early-man sites dotting the canyon walls. Baker Cave is one of these, with evidence of occupation by several archaic and paleo-Indian cultures dating back more than 9,000 years. Among other things, a large stone hearth was discovered there that was filled with animal and fish bones—several species of sucker, grey redhorse, smallmouth buffalo (the fish, not the bison), bass (possibly the spotted bass), and sunfish. But what I want to tell you about did not come from the stone hearth, but, rather from a latrine area near the front of

10. The Montgomery quote is at page 98 of his book, *Many Rivers to Cross*; see Montgomery (1995). You can find out more about the Davis Mountains and The Nature Conservancy's land preserve on the Internet at *www.utexas.edu/ handbook/online/articles/DD/ rjd3_print.html, www.cdri.org/ Discovery/News.html,* and *www.nature.org/wherewework/ northamerica/states/texas/ preserves/art6647.html.*

11. This is a conclusion arrived at by Garrett and Matlock (1991), whom I cited in footnote 9 above. They, in turn, cited a reference work by G. Brune (1981) titled "Springs of Texas," which I was not able to locate.

the cave. There, archaeologists recovered 1,100-year-old *coprolites*, which is archaeology language for desiccated human feces. And among the food remains they found in these coprolites were—trout scales![12]

Now, I can't say I know that much about the mobility of these prehistoric hunter-gatherers, but the archaeologists say that they used what was at hand in their local environment. So the trout must have come from streams that were reasonably close by. And if trout were present 1,100 years ago, they just might have endured in cold, spring-fed streams into the historical period.

The possibility that one additional enclave may have existed where cutthroat trout were present historically even farther to the south was raised by reports that came out of the region of the Rio Conchos headwaters in Mexico, back in the middle and late 19th century. The Rio Conchos drainage heads up high on the eastern slopes of the Sierra Madre Range, between about 26 and 28 degrees north latitude. I didn't have the map room to illustrate the entire drainage for you, but Fig. 14-1 does show where the main river enters the Rio Grande south of the Davis Mountain enclave. On the Pacific side of the Sierra Madre crest opposite the Rio Conchos headwaters are located the streams where Mexican golden trout persist, and where other Mexican trouts of the rainbow lineage are native (I refer you back to Chapter 2 for a discussion of the origin and relationships of these Mexican trouts).

Three pieces of evidence point to the existence of native trout in streams of the Rio Conchos headwaters. The first is embodied in the language of the indigenous people. Joe Tomelleri, the renowned artist and fish illustrator, has made many trips into this part of Mexico. He tells me the indigenous people have a word, "aparique," with which they reply when proffered a photo or description of a trout.[13] That would suggest long familiarity with this type of fish.

The second and perhaps earliest piece of written evidence is an observation in the journal of John Woodhouse Audubon, the son of John James Audubon of *Birds of America* fame, which he recorded during a trip through the Rio Conchos headwaters in 1849. He wrote:

> "We looked in vain for fish in the most tempting of eddies and holes, but saw very few; little trout about five inches long were all that rewarded our search."

Audubon wrote the name of the tributary he was following when he made this observation as the Tomochic. Joe Tomelleri has scrutinized Audubon's map and travel itinerary in great detail. He believes Audubon was following up a creek called Agua Caliente on modern maps, and was "back in the belly of the Conchos" when he observed the "few little trout."[14]

The third piece of evidence was presented in 1886

12 Hester (1983) and Sobolik (1991) discuss the occupation of Baker Cave by archaic and paleo-Indian cultures. The Sobolik report, which is 141 pages long, details the discovery of the trout scales.

13. Joe Tomelleri, Leawood, Kansas, personal communication, May 20, 2005.

14. John Woodhouse Audubon was a painter and naturalist like his father, and participated in many natural history expeditions. He had also fished for trout in other places, so he would have known trout when he saw them. The journal of his trip through this area of Mexico with an accompanying map of his route was published in 1906 under the title *Audubon's Western Journal 1849–50* (see Audubon 1906 [1984]). Joe Tomelleri's website at *www.americanfishes.com* has much more information about the upper Rio Conchos drainage and on Audubon's movements in Mexico.

by distinguished paleontologist and naturalist E.D. Cope in a brief note in the prestigious journal *The American Naturalist*:

> "I owe to my friend, Professor Lupton, two specimens of a black-spotted trout from a locality far south of any which has hitherto yielded Salmonidae. They are from streams of the Sierra Madre, of Mexico, at an elevation of between 7,000 and 8,000 feet, in the southern part of the State of Chihuahua, near the boundaries of Durango and Sinaloa. The specimens are young, and have teeth on the basihyal bones, as in *Salmo purpuratus*, which they otherwise resemble." [15]

It was this reference to "black-spotted trout" resembling *Salmo purpuratus* (both terms were sometimes used for cutthroat trout in Cope's day) together with the observation of "teeth on the basihyal bones" (presumably meaning basibranchial teeth) that got everybody thinking cutthroat. But alas, the specimens were lost, so there is no way to verify that this was the case. Until recently, no other collections in the upper Rio Conchos drainage had included a native trout of any kind, although longnose dace and the Rio Grande sucker, two species that frequently occur with Rio Grande cutthroats in New Mexico, were collected there.

Cope's friend, Professor Lupton, was almost certainly Nathaniel Thomas Lupton, chairman of the chemistry department at Vanderbilt University in the 1870s and early 1880s. Lupton was also associated with men who invested in Mexican silver mines in the mountains of southern Chihuahua, and several times between 1879 and 1884, he journeyed into that region to assay silver ores and report on the various mines. The "black-spotted trout" were likely collected during his 1884 trip. Joe Tomelleri has been collaborating with Dean Hendrickson of the University of Texas and a small, binational group of scientists calling itself Truchas Mexicanas to search for native trout in Mexico. They have diligently traced Lupton's movements, and have ventured into the region themselves on several occasions since 1997 to make collections. The last two times in, in February 2005 and February-March 2006, they did find native trout in upper Rio Conchos stream reaches, but even though the trout did exhibit cutthroat-like markings under their jaws, these were not cutthroat trout. Rather, DNA testing indicated a close relationship to the trouts of the southern tributaries of the Rio Yaqui and Rio Maya basins. It will take more expeditions and additional collections to track down the elusive cutthroat trout, if indeed they do still exist. [16]

Showing as it does a core area of historical distribution in the north and outlier enclaves of historical distribution strung out over hundreds of miles to the south, Fig. 14-1 pretty clearly illustrates that the range of the cutthroat trout must have extended much further down the Rio Grande and Pecos River drainages, perhaps even beyond the confluence of these two streams, at some point back in prehistoric time. The

15. See Cope (1886).

16. The Truchas Mexicanas website at *www.utexas.edu/tmm/tnhc/fish/ research/truchas_mexicanas*, along with Joe Tomelleri's website at *www. americanfishes.com*, has illustrations and much more information about the history of the discovery and status of the Rio Conchos native trout, including a white paper on its conservation; and, of course, everything you might want to know about the work of the Truchas Mexicanas group.

cold, wet, pluvial climates that prevailed throughout this region during Pleistocene ice advances would have provided the proper conditions for a greatly extended range. Thus, in Pleistocene times when the climate was cooler and wetter, the Rio Grande may have carried its cutthroats southward clear into Mexico. [17]

I have already told you about the 740,000-year-old fossil trouts that Karel L. Rogers and collaborators found in the San Luis Valley. These were identified as cutthroat trout, but Dr. Behnke believes they were a primitive form that went extinct and was not the direct ancestor of the present Rio Grande cutthroat. If they had been, he reasons, after 740,000 years of separation, there should be a much greater degree of genetic differentiation from the Colorado River subspecies than actually exists. In fact, there is only a little. The close similarity of the Colorado River, Rio Grande, and greenback subspecies points to a late-Pleistocene time frame for the separation of these three forms. However, Dr. Behnke believes the Rio Grande cutthroat was isolated prior to the greenback because, as close as the three subspecies may be, the Rio Grande cutthroat shows slightly more divergence in characters from the parent Colorado River subspecies than does the greenback, indicating a somewhat longer time of separation. [18]

Other fossil trouts also identified as cutthroats have been found in Pleistocene-age sediments in the Estancia Valley, New Mexico. Pluvial lakes existed there off and on from about 130,000 years ago to about 10,000 years ago, but the dates for the trout fossils appear to be confined to the 25,000 to 11,000 year period. These are more likely to have been early-generation Rio Grande cutthroat trout. [19]

LIFE HISTORY AND ECOLOGY

No detailed studies of the life history and ecology of the Rio Grande cutthroat trout have been published. We do know that there are two distinct varieties of these trout, one associated with the upper Pecos River drainage in New Mexico, and the other with the upper Rio Grande drainage in Colorado and New Mexico. But the distinction lies in physical description (larger spots and more scales in the lateral series in Pecos River trout), not life history strategies. What little information is available indicates few basic differences from other interior cutthroat subspecies.

Historically, the subspecies displayed the same three basic life history strategies as the other interior forms:

- A stream-resident life history, wherein the trout complete all phases of their life cycle within the streams of their birth, often in a short stream reach and often in the headwaters of the drainage.

- A fluvial, or riverine (potamodromous), life history, wherein the fish feed and grow in main river reaches, then migrate to small tributaries of these rivers where spawning and early rearing occur.

17. References that provide additional details about the Pleistocene climate of at least the New Mexico portion of this range include Leopold (1951) and Antevs (1954).

18. Dr. Behnke explained his reasoning at page 152 of his 1992 monograph; see Behnke (1992).

19. You can learn more about the Estancia Valley Pleistocene lakes and their fossil cutthroat trout in Bachhuber (1989) and Behnke and Platts (1990). Pluvial lakes also occupied the Tularosa Basin and the Guzman Basin in southern New Mexico and northern Mexico during Pleistocene time, but I'm not aware of any trout fossils being found in any of these.

- A lacustrine, or lake-associated, life history, wherein the fish feed and grow in lakes or ponds, then migrate to tributary streams to spawn.

All populations that naturally exhibited fluvial life history behavior, e.g., those of the mainstem Rio Grande and Pecos rivers, have been lost since the time of first settlement, and restoration plans have not addressed the resurrection of such populations. There are no large natural lakes within the historical range, but small lakes and ponds do exist, some of which may have been populated with trout, so I am including the lake-associated life history strategy in the list above. Paul Fountain described one such lake that he found in the mountains overlooking the San Luis Valley:

> "The surface of this tarn, which is, so far as I know, nameless, is perfectly smooth, notwithstanding that a cascade falls into it at the south end. This, and all the other pools I examined, abound in fish, the most abundant kind being a species of trout, called salmon-trout throughout the West....Using a stick for a rod, a piece of string for a line, and a fly I made by wrapping threads from a red flannel shirt around the hook, the fish began to bite, and in two hours I landed a couple of dozen fish, in weight from a quarter of a pound to a pound each...."

Like the greenback cutthroat we discussed in the last Chapter, the few populations of Rio Grande cutthroat that managed to endure through the historical period did so in small, isolated headwater stream reaches, and so exhibit the stream-resident life history strategy. However, they do appear to retain at least some flexibility to adapt to other habitats and environments requiring a shift in life history strategy. Fishery managers have succeeded in establishing at least one broodstock population in Haypress Lake, Colorado, that originated from stream-resident trout, and trout from this and other broodstocks maintained in hatcheries have been stocked successfully in several wilderness lakes.

It was work performed in the field in connection with broodstock development that has given us the few specific facts that we do have about the life history and ecology of stream-resident populations.[20] Initially, state biologists from New Mexico field-spawned the trout at streamside to collect fertilized eggs for their broodstock program. They reported that the trout spawn just after the peak in snowmelt runoff when water temperatures warm into the 42–48 degree Fahrenheit (5.5–9 degree Celsius) range, which could occur anytime from April to July depending on latitude, elevation, and aspect of the stream (they also mentioned a day length effect, but I've not been able to garner any further details on this). They also reported that male trout spawned for the first time at age 2 and females at age 3, and the males were ready to spawn about two weeks earlier than the females. Sizes of mature trout handled during their work ranged from 4.2 inches to 10.3 inches (I'm presuming these are fork length values), and females within this size range produced an average of 175

20. The information cited here came from several agency reports including Cowley (1993) and Rinne (1995). The same material was also summarized in a formal status review of the Rio Grande cutthroat trout published by the U.S. Fish and Wildlife Service in 2002; see Williams (2002b). That review also cited a more recent agency report by the New Mexico Department of Game and Fish (2002), which I have never been able to obtain.

eggs per female. Dr. David Cowley, New Mexico State University, Las Cruces, worked with trout from Indian Creek on the Mescalero Apache Indian Reservation, and reported that 12 trout from that population (no sizes stated) produced an average of 311 eggs per female. Biologists observed that trout spawning in the streams chose sites in patches of gravel where the particle size ranged from a quarter-inch to about one and a half inches in diameter. At the Seven Springs Hatchery, New Mexico, Rio Grande cutthroat eggs took 32 days to hatch in 10-degree Celsius (50-degree Fahrenheit) water, which is 320 thermal units. No information was provided on time or thermal units required for the alevins to reach the free-swimming fry stage.

Trout living in small, high-elevation headwater streams must be opportunists when it comes to feeding. The streams in Rio Grande cutthroat country are mostly of the freestone variety, with the usual selection of aquatic life typical of such waters, but much of the food input comes from terrestrial sources: ants, beetles, grasshoppers, wasps, and the like. Based on 471 trout from three streams, Dr. Cowley developed a chart of age and size classes from which I extrapolated the following numbers to give you an idea of the rate of growth of Rio Grande cutthroats under reasonably good habitat and food conditions. By age 1 these trout will typically have attained a length of 2.5 inches (64

mm); 4.5 inches (114 mm) by age 2; 6.9 inches (174 mm) by age 3; and 8 inches (205 mm) by age 4. Based on these extrapolations, trout greater than 8 inches in length are at least 4 years old and may be age 5 or older. Maximum lengths reported for these trout are in the 10 to 11 inch range, and life spans are probably 6 to 8 years.

Rio Grande cutthroat trout evolved with several other fishes, including Rio Grande chub, creek chub, longnose dace, southern redbelly dace, Rio Grande sucker, and white sucker. These fishes represent taxa that other cutthroat subspecies, especially their fluvial and lacustrine life history forms, often use for food. It's not known if any of these were once important components of Rio Grande cutthroat diets, but they could have been when the fluvial life history strategy was prevalent. The U.S. Fish and Wildlife Service has pointed out that many of these fishes have themselves been extirpated or greatly reduced in numbers in the mainstem rivers and larger tributaries that were once populated with Rio Grande cutthroats.

Like all the other cutthroat subspecies, and indeed like all other species of trout, Rio Grande cutthroat trout require four types of habitat to carry out their life cycle: 1) suitable habitat for spawning and egg incubation, which I have already touched on briefly; 2) nursery habitat for newly emerged fry, which is generally found along stream margins where water velocities

Little Ute Creek, Colorado. The population living here was established by translocation. The peak in the background is Blanca Peak. PHOTO BY AUTHOR

are low; 3) habitat for summer rearing and growth; and 4.) overwintering habitat. The U.S. Forest Service, New Mexico Department Game and Fish, and Colorado Division of Wildlife each spent a lot of time in the 1970s and 1980s studying the physical characteristics of stream reaches occupied by Rio Grande cutthroats. Being a land management agency, the Forest Service's goal was to learn what the trout need so that habitat quality could be maintained. The state fish and wildlife agencies had that in mind as well for existing populations, but they also wanted to know what to look for in selecting candidate reaches for translocated populations. In 1981, David Langlois, the Colorado Division of Wildlife biologist who was overseeing that state's restoration program at the time, gave me this simple list of criteria for a good Rio Grande cutthroat stream:

- Perennial stream flow.
- The potential to maintain a biomass of 30 to 50 pounds of trout per acre of stream.
- Approximately 50:50 pool to riffle ratio.
- Gradient less than 10 percent, with the optimum between 3 and 5 percent.
- In-stream boulders, good riparian cover, and water over six inches deep in the pools, all to provide adequate cover for the trout.

Subsequent studies of stream habitat have refined these criteria. For example, considerably deeper pools than the minimum 6-inches specified above are almost certainly required in Rio Grande cutthroat streams to provide overwintering habitat. Lack of large, deep pools may be a limiting factor affecting Rio Grande cutthroat survival in headwater streams. [21]

STATUS AND FUTURE PROSPECTS

By the late 1960s, the Rio Grande cutthroat trout had vanished from so much of its historical range that it was placed on the U.S. Bureau of Sport Fisheries and Wildlife's list of endangered species. This view was reinforced in 1972 when two prestigious groups of scientists, the American Fisheries Society's Committee on Endangered Species and the Conservation Committee of the American Society of Ichthyologists and Herpetologists, included it on a list of threatened fishes of the United States. [22] Despite this, it was not included as either threatened or endangered when the U.S. Endangered Species Act became law in 1973. It was, however, listed as threatened by the State of Colorado, which had its own state-level endangered species legislation.

In the early to mid 1970s, the New Mexico Department of Game and Fish and the U.S. Forest Service conducted inventories of Rio Grande trout populations in New Mexico. In Colorado, that state's Division of Wildlife also commenced a program of Rio Grande cut-

21. A report prepared by Propst and McInnis (1975) provides an example of the type of stream analysis work that was carried out by the various management agencies in the 1970s and 1980s. Harig and Fausch (2002) have more recent information.

22. Dr. Behnke wrote a Rare and Endangered Species report that profiled the status of the Rio Grande cutthroat trout in 1967 (see Behnke 1967). The list of threatened species compiled jointly by the American Fisheries Society and American Society of Ichthyologists and Herpetologists is in Miller (1972).

throat inventory, restoration, and monitoring. In the years since, additional inventories and assessments have been made. The present consensus, if you can call it that, among the responsible agencies is that 161 populations now exist in Colorado and 106 populations in New Mexico. These include populations that existed naturally but had gone undiscovered until the inventories were conducted, as well as new populations that have been introduced since the programs started. Today, it is estimated that Rio Grande cutthroat trout occupy 480 miles of stream and 1,120 acres of lake in Colorado and 260 miles of stream but no lakes in New Mexico, which is somewhere around 5 to 7 percent of what the agencies are willing to accept as the stream miles and acreage of lakes that were occupied historically.[23]

Mind you, however, not all of these known populations are genetically pure, stable, or secure from introgression or extirpation. In fact, when you subtract the number of already introgressed populations, and the number that live under the threat of displacement or introgression, and the number with extremely low population sizes or that live in marginal habitats or limited stream reaches, you are left with 13 populations—yes, that's right, just 13 core populations, and that's the total for both states—that are pure, have over 2,500 adult fish, are secured above a barrier in good

23. Early inventories of Rio Grande cutthroat populations in New Mexico and Colorado were reported by McInnis and Stork (1974), Wheeless (1974), Propst and McInnis (1975), Hubbard (1976), Langlois (1979, 1980, 1981), and Josselyn (1982). The U.S. Forest Service published additional assessments in the mid-1990s (see Rinne 1995 and Stumpff and Cooper 1996), and the States of Colorado and New Mexico published updated reviews in 1998 (see Alves 1998 and Stumpff 1998). A coalition of conservation organizations presented its own independent assessment in 1998 as part of a petition to list the Rio Grande cutthroat trout under the U.S. Endangered Species Act (see Southwest Center for Biological Diversity et al. 1998), and the U.S. Fish and Wildlife Service followed with its own assessment in response to this petition as part of its formal status review (see Williams 2002b). The population numbers I cited in this paragraph came from that latter document. However, the most recent assessment of the subspecies was prepared for the U.S. Forest Service by researchers from New Mexico State University (see Pritchard and Cowley 2006). You can download your own copy from the Forest Service website at *www.fs.fed.us/r2/projects/scp/assessments/riograndecutthroattrout.pdf.*

habitat, and do not coexist with non-natives. I have listed these and the drainages in which they occur in Table 14-1.[24]

And even these 13 core populations live in complete isolation from one another. Not one has a semblance of metapopulation connectivity or gene flow that would increase its odds of long-term survivability. So even these populations could experience "runs of bad luck" (consecutive years of low birth rates and/or high death rates, extended periods of drought, post-wildfire effects, and the like) that could drop them into the extinction vortex.

So, how did the native Rio Grande cutthroat trout arrive at such a plight? The reasons cited are the same as for all the other interior cutthroat subspecies, with one or two unique wrinkles, perhaps. At or near the top of everybody's list is habitat degradation or outright loss caused by withdrawal of water from streams for irrigation. Livestock grazing also ranks high, and to these are usually added historical timber harvest practices and the impacts of early-day mining.

The advent of water withdrawal for irrigation in Rio Grande cutthroat country dates all the way back to 1350 A.D. or so, when the archaeological record shows that Native Americans occupying the middle Rio Grande region began to practice irrigated subsistence agriculture. But the pace accelerated considerably

TABLE 14-1 *Core Populations of Pure, Stable, and Secure Rio Grande Cutthroat Trout as Defined by the U.S. Fish and Wildlife Service**

STREAM	DRAINAGE	STATE	LAND OWNERSHIP
Cross Creek	Saguache Creek	Colorado	Rio Grande NF and Private
Medano Creek	San Luis Basin	Colorado	Rio Grande NF and Nat'l Park Service
San Francisco Creek	Trinchera Creek	Colorado	Private
El Rito Creek	Rio Chama	New Mexico	Carson NF
Bitter Creek	Red River	New Mexico	Carson NF
Columbine Creek	Red River	New Mexico	Carson NF
San Cristobal Creek	Rio Grande	New Mexico	Carson NF
Powderhouse Creek	Rio Chamito	New Mexico	Carson NF
Policarpio Creek	Rio Pueblo	New Mexico	Carson NF
Canones Creek	Rio Chama	New Mexico	Santa Fe NF
Rio Cebolla	Rio Jemez	New Mexico	Santa Fe NF
West Rio Puerco	Rio Grande	New Mexico	Santa Fe NF
Jacks Creek	Pecos River	New Mexico	Santa Fe NF

* Extracted from U.S. Fish and Wildlife Service Table 1 in "Endangered and Threatened Wildlife and Plants: Candidate Status Review for Rio Grande Cutthroat Trout." *Federal Register* 67, no. 112: 39936–39947 (June 11, 2002).

after Spanish colonization in the 1590s, and really kicked into high gear during the roughly 90 to 95 years between 1880 and 1973. According to one fairly recent analysis, more acres in the middle Rio Grande valley were under cultivation in 1880 than at any other time before or since, and livestock grazing was intense as well, with an estimated 2 million sheep and 200,000

24. I refer you again to the U.S. Fish and Wildlife Service status review issued in 2002 (see Williams 2002b). Credit is due to the biologists who prepared this particular review for specifying a minimum population size of 2,500 adult trout as one of their criteria for persistence and stability, based on the work of Hilderbrand and Kershner (2000a).

head of cattle, horses, and mules being run on the land. [25] The early 1900s saw the construction of numerous water supply and flood control dams in the Rio Grande headwaters to maintain and bolster these activities, and the livestock industry in particular continued to grow through the mid-1930s. It is now acknowledged that livestock numbers increased far beyond the carrying capacity of the range during this period with serious negative consequences for rangeland, riparian, and stream health. [26] Add to this the effects of the timber harvest and associated forest road-building practices of that period, and it's easy to see how widespread fragmentation and loss of Rio Grande cutthroat habitat occurred from the lowest stream reaches to the headwaters, across every trout-bearing watershed.

That same 1880 to 1973 period was also an era of wholesale stocking of non-native species. Rainbow, brook, and brown trout were introduced, beginning about the turn of the 20th century in New Mexico and perhaps a decade or two earlier in the Colorado portion of the range. They established themselves with relative ease. The rainbow trout could, and did, interbreed with the native cutthroats, resulting in introgressed populations pretty much everywhere the two came in contact. Brook and brown trout were able to displace the native cutthroat populations and, again, did so relatively quickly, especially in waters where human impacts were already making it difficult for the native

As I discussed back in Chapter 11 and again in Chapter 13, an effective population size of 500 is recommended to avoid long-term loss of genetic variation. Hilderbrand and Kershner's work indicates that an effective population size of 500 is achievable with, as a minimum, a total adult population of 2,500 trout.

25. The analysis referred to here was prepared by the Bosque Hydrology Group, a multi-agency and multi-university team formed in the early 1990s to tackle and resolve the problems involved in restoring riparian function and stream condition in the middle Rio Grande ecosystem, which was defined to include the Rio Grande drainage from Cochiti Reservoir to Elephant Butte Reservoir, New Mexico. The report containing this analysis is titled "Middle Rio Grande Bosque Ecosystem: Bosque Biological Management Plan," and is cited under Crawford et al. (1993) in the bibliography. You can download it off the Internet at *www.fws.gov/ bhg/*, but be warned, it's a *very* large pdf file. The report itself is 291 pages plus maps. A 2003 update was announced on the website as "coming soon," but was not posted as of December 1, 2006. The word *bosque*, by the way, is Spanish for riparian forest.

26. See Meehan and Platts (1978).

A Rio Grande cutthroat stream, population established by translocation, flows out of the San Juan Mountains on the west side of the San Luis Valley, Colorado.
PHOTO BY AUTHOR

cutthroats to endure. Displacement by brown trout was especially widespread in New Mexico waters.

Aside from human-caused habitat degradation, another human impact on native trout populations results from the high vulnerability of interior cutthroat subspecies to angling. Here's an example of how this can play out from one of Dr. Behnke's early reports. In New Mexico, the native Rio Grande cutthroat trout of the Rio Chiquito held their own and actually dominated introduced brown trout by more than 11 to 1, as determined by electrofishing surveys made in 1965 and 1966, when the stream was on private land and closed to all fishing. But in 1967, the U.S. Forest Service acquired the land and the stream was opened to harvest angling. Two years later, an electrofishing survey revealed that the ratio of cutthroat to brown trout had dramatically reversed. By 1969, owing to the differential vulnerability of the two species to angling, browns outnumbered the native cutthroats by 8 to 1. [27]

But it's the brown trout's very aloofness to the offerings of anglers that has made it a favorite, especially among fly-fishers, the world over. Nowhere is this so true as it is in New Mexico. Back in 1991, I picked up a copy of the book *Fly-Fishing in Northern New Mexico*, a project of Santa Fe's Sangre de Cristo Fly Fishers, edited by Craig Martin. [28] In the first Chapter, while noting that "releasing all native Rio Grande cutthroats

is a central theme," they quickly added, "we have a special concern for wild brown trout." There's the dilemma. Even though brook trout and rainbow trout have also been stocked in New Mexico waters, it seems to be the introduced brown trout, wild or not, that has been most pervasive in displacing that state's native cutthroat populations.

In 1998, the Southwest Center for Biological Diversity, an environmental and conservation advocacy group, along with four other such groups and two private individuals, petitioned the Secretary of the Interior to list the Rio Grande cutthroat trout as endangered under the U.S. Endangered Species Act. There was a court challenge when the Fish and Wildlife Service summarily dismissed the petition in a 90-Day Finding, and the Service finally did perform a detailed status review that was published in the Federal Register in 2002. But the decision, based on that review, was *not* to list this subspecies. [29]

Back in Chapter 3, where the U.S. Endangered Species Act and associated listing issues first came up in this book, I pointed out that the Fish and Wildlife Service can determine that a species is endangered or threatened due to one or more of five factors: 1) the present or threatened destruction, modification, or curtailment of its habitat or range; 2) overutilization for commercial, recreational, scientific, or educational

27. The results of the "before" surveys of Rio Chiquito trout were published in a New Mexico Department of Game and Fish agency report written by Little and McKirdy (1968). Behnke and Zarn (1976) had the "after" result at page 3 of their report. Native Rio Grande cutthroats can still be found in Rio Chiquito, but only in the uppermost reach of the stream at the end of the Forest Service road at an elevation of about 9,000 feet. The rest of the stream is now brown trout water.

28. See Martin (1991).

29. For the petition, see Southwest Center for Biological Diversity et al. (1998). For the U.S. Fish and Wildlife Service status review and decision not to list, see Williams (2002b). This may not be the end of the story, however. Early in 2003, the petitioners brought suit in federal court, claiming that the Fish and Wildlife Service ignored information indicating populations of Rio Grande cutthroat trout continue to be threatened by multiple factors. I've heard nothing yet about a decision, so stay tuned.

purposes; 3) disease or predation; 4) the inadequacy of existing regulatory mechanisms; and 5) other natural or manmade factors affecting its continued existence. With regard to the Rio Grande cutthroat trout, the Service did acknowledge that the distribution of this subspecies has been greatly reduced and many populations have been lost (owing, as history shows, to a combination of elements from factors 1, 2, 4, and 5). But the Service focused its analysis on just the 13 remaining core populations, which it said are enough to keep the subspecies off the list as either endangered or threatened. The Service concluded, essentially, that for these 13 core populations, the factors that led to the wholesale losses of the past have now been neutralized.

The Service did not regard any further loss or fragmentation of habitat to be a threat to the 13 core populations in the foreseeable future. It pointed instead to several watershed-scale projects on both private and National Forest lands that should, when completed, restore additional headwater stream reaches, establish new populations or extend existing ones into additional stream miles and acres of lake thus adding to the occupied range, and create at least some measure of metapopulation connectivity where none exists now. The Service highlighted three such projects that were under way in 2002, and two others that were in the planning stages:

- Costilla Creek watershed—joint project between Vermejo Park Ranch, a Ted Turner property, and the States of Colorado and New Mexico to restore habitat, remove non-natives and hybrids, and repopulate 13.6 miles of stream and 23.5 acres of lake with genetically pure Rio Grande cutthroat trout.

- Animas Creek—another project of the Turner Foundation on Ted Turner's Ladder Ranch to restore 29.8 miles of stream for pure Rio Grande cutthroat trout and connect this with Rio Grande cutthroats already present in Animas Creek on Forest Service land. [30]

- Comanche Creek watershed—a 6-partner project in the Valle Vidal Management Unit of Carson National Forest is restoring the watershed to full functional condition and improving 43.5 miles of stream habitat for Rio Grande cutthroat trout.

- Headwaters of East Fork Jemez and San Antonio rivers, Valle Caldera National Preserve—opportunity to restore headwater stream habitat and reconnect the two stream systems downstream from the preserve on National Forest land for a total of 69.6 miles of stream plus metapopulation connectivity for Rio Grande cutthroat trout.

- Rio Santa Barbara watershed, Carson National Forest—another opportunity to remove non-natives, repopulate with pure Rio Grande cutthroats, and reconnect several tributaries to create metapopulation structure.

The Costilla Creek and Animas Creek projects are taking place on private lands. It's ironic in a region where one would think that public lands would offer most in the way of protection, but in Rio Grande cutthroat country, large private land holdings have been instrumental in the protection and restoration of the native subspecies. I have already told you about the native cutthroat trout of the Rio Chiquito that held their own

30. The Rio Grande cutthroat population inhabiting the Forest Service reach of Animas Creek is the southernmost known native population of this subspecies.

and actually dominated introduced brown trout as long as the stream was on private land and closed to angling, but gave way quickly to displacement by brown trout when the land became public and the stream was made accessible to harvest anglers. In the formative years of Colorado's program to inventory and restore native Rio Grande trout populations, pure populations were found in creeks on the east side of the San Luis Valley on lands belonging to the Forbes Trinchera and Taylor ranches. The managers of these ranches cooperated generously with the Colorado Division of Wildlife. Placer Creek on the Forbes Trinchera Ranch and Indian Creek on the Taylor Ranch became the sources of many of the genetically pure translocated populations that now exist in other (public) waters in the Colorado portion of the Rio Grande cutthroat range. The State of Colorado also established a broodstock of Rio Grande cutthroat trout in Haypress Lake, also on private land on the Brown 4UR Ranch. Now it is the Turner properties that are keeping the precedent going in the New Mexico portion of the range.

With regard to the overutilization factor, there is no record that I am aware of to indicate that market fishing for Rio Grande cutthroat trout occurred during the settlement period, but fishing for subsistence and sport angling certainly did, and these took a heavy toll of the native trout well into the 1920s and 1930s. In his

book, *Beatty's Cabin*, published in 1953, former Forest Supervisor and later New Mexico State Game Warden Elliott Barker told how he and two teenage friends took 438 of the "black-spotted, red-bellied cutthroat trout" in 6 hours' fishing one summer day in 1903 in the Pecos River high country. All were kept and taken to their ranch homes where they were all consumed, none wasted. [31]

But that was then. Current fishing regulations in New Mexico and Colorado restrict many of the Rio Grande cutthroat streams to catch-and-release only. Although harvest angling is still allowed on other cutthroat waters in both states (a 2-trout limit in New Mexico and a 4-trout limit in Colorado), many anglers who do fish these waters practice catch-and-release regardless. And, because streams with pure Rio Grande cutthroats are mostly so remote, fishing pressure is light. Utilization for science and education is also light. Genetic testing of populations is now done non-lethally, and the take of trout for other tests where the fish must be sacrificed, e.g., for whirling disease, is carefully managed, as is the collection of trout from source populations for translocation. For these reasons, the U.S. Fish and Wildlife Service concluded that overutilization poses no threat to the 13 core populations.

Disease and predation was also ruled out as a threat to the 13 core populations, even though whirling

31. See Barker (1953). The story cited is in a Chapter titled "Boys, Bear, and Trout A'Plenty," starting at page 49 in my copy of the book.

disease was discovered in Rio Grande cutthroat country in the 1980s and by the late 1990s had spread into many Colorado and New Mexico waters and hatchery facilities. One infected facility was New Mexico's Seven Springs Hatchery, where that state's only Rio Grande cutthroat broodstock was maintained. That hatchery has now been renovated and is whirling disease-free. The status review did acknowledge that whirling disease can be a threat to Rio Grande cutthroat trout in waters where temperatures are relatively warm, habitat is degraded, and the parasite's secondary host, the aquatic worm *Tubifex tubifex*, can proliferate, e.g., in or downstream from beaver ponds or in open, trampled, and sedimented reaches that have not recovered from livestock grazing or past logging and road-building practices. But the Service concluded that whirling disease is no threat to the foreseeable existence of the 13 core Rio Grande cutthroat populations because they live in high-elevation headwater streams that typically have cold water and low sediment levels, conditions under which *Tubifex tubifex* does poorly. [32]

The Service also concluded that the States of New Mexico and Colorado and the U.S. Forest Service all have adequate regulatory mechanisms in place to protect and enhance Rio Grande cutthroat populations and their habitats. New Mexico has an approved management plan in place with a schedule of activities intended to improve the lot of the subspecies in that State through fiscal year 2005, which presumably will have been updated by the time you read this. In Colorado, the Rio Grande cutthroat is a "species of special concern" (downlisted from threatened in 1984) under that state's endangered species legislation, with a continuing program in place to inventory, restore, and monitor populations. As noted, angling for Rio Grande cutthroats is closely regulated in both states. The U.S. Forest Service lists the subspecies as a Management Indicator Species with habitat management objectives that are intended to maintain population viability and ensure that it does not become threatened or endangered. And finally, the status review stated that a range-wide conservation agreement was to be signed by all of these agencies in 2002 "to assure the long-term persistence of the subspecies, preserve its genetic integrity, and to provide adequate numbers and populations." [33]

Under the category of other natural and manmade factors, the Fish and Wildlife Service considered only four in its status review: wildfire, electrofishing, hatcheries, and public sentiment. Let me take up electrofishing first. This one could just as well have been dealt with by the Service in the category of overutilization for science and education, because electrofishing has long been an essential tool of fishery biologists to monitor population numbers, to collect fish for other

32. For a review of whirling disease and how it impacts trout, see footnote 21 of Chapter 4 and the references cited there.

33. I have cited the New Mexico State management plan in the bibliography under New Mexico Department of Game and Fish (2002), but I have to say I have not seen this document. My requests to the Department for a copy went unanswered, and I found no link to it anywhere on the Department's website. As for the range-wide conservation agreement, I had better luck with that document. I've cited it in the bibliography under Colorado Division of Wildlife et al. (2003), but I obtained my copy by writing to the U.S. Fish and Wildlife Service field office in Albuquerque, New Mexico.

scientific purposes, and to collect trout from source populations for translocations. It is also currently the primary tool for detecting and removing non-native trout from Rio Grande cutthroat streams. Indeed, though, electrofishing can have some detrimental effects on individual trout, mostly in the form of non-lethal injuries to the spine, although death can result if care is not taken. Accumulated effects of electrofishing might be significant for a small population, as when that population is electrofished again and again for several years running, but no research has yet demonstrated any negative effects at the population level. [34] For the 13 core Rio Grande cutthroat populations, therefore, the Service concluded that the use of electrofishing as a monitoring tool in their management poses no threat to their continued existence.

Hatchery management is another quick one to deal with. The Fish and Wildlife Service status review pointed out that hatchery-reared trout are not planted into the 13 pure, stable core populations. These are left to maintain themselves as wild stocks. Nor are rainbow, brook, brown, or non-native subspecies of cutthroat trout stocked anymore in any other water containing Rio Grande cutthroats. Therefore, hatchery management is not a threat to the 13 core Rio Grande cutthroat populations. However, the overall program of Rio Grande cutthroat management in both New Mexico and Colorado does depend on the use of hatchery-reared trout, and will continue to do so for the foreseeable future. New Mexico maintains a captive Rio Grande cutthroat broodstock at its Seven Springs Hatchery. Colorado maintains a wild broodstock at Haypress Lake, but also has captive broodstocks at its Poudre Rearing Unit and at its Fishery Research

34. The literature on electrofishing injury has grown to be quite extensive, especially in the decade of the 1990s, when much attention was focused on the subject. Many papers appeared in the fisheries journals during this period and almost every major angling magazine carryied at least one essay on the subject. To learn about the principles of electrofishing and how it is supposed to be done, one source I'd recommend is the Chapter titled "Electrofishing" by James B. Reynolds in the book *Fisheries Techniques*, 2nd edition; see Murphy and Willis (1996). For a sampling of the extensive literature available in the fisheries journals, see Hollender and Carline (1994), Snyder (1995), Kocovsky et al. (1997), Thompson et al. (1997a, 1997b), and Nielsen (1998).

Hatchery in Fort Collins. [35] Domestication and inbreeding in captive broodstocks, which ill-equip outplanted trout for life in the wild, are principal concerns for hatchery management programs. The transmission of diseases is another. Domestication and inbreeding are avoided by periodically introducing new trout into the broodstock from wild source populations that are rotated, so as not to deplete them. [36] The State of New Mexico has already had to disinfect and renovate its Seven Springs Hatchery to rid it of whirling disease.

Threats from wildfires are a more serious matter even for the 13 core Rio Grande cutthroat populations that the U.S. Fish and Wildlife Service regards as large, pure, stable, and secure. That's because the extirpation of trout populations as large as these or larger have already been documented in New Mexico. [37] In its Rio Grande cutthroat status review, the Service stated that it cannot determine if wildfire is a threat to the 13 core populations, but added that it is logical to assume a few isolated populations could be lost to the effects of fire in the foreseeable future.

Wildfire is a natural disturbance in forested watersheds. But the intensity of wildfires can be increased and made more catastrophic by human activities, such as livestock grazing and, although it may seem counterintuitive, by active fire suppression. And that is what has happened in the U.S. Southwest. I read somewhere that more lightning strikes occur in the southwest

35. In 1987, using Rio Grande cutthroat trout from the Indian Creek population on the Mescalero Apache Indian Reservation, the U.S. Fish and Wildlife Service tried to start a captive broodstock at its Mescalero National Fish Hatchery. But poor survival of eggs and young trout aborted this attempt, and in 2000, the hatchery was closed in a federal budget cutback. Later, the hatchery was renovated and reopened under tribal management. Late in 2005, I learned from Mike Montoya, manager of the tribal facility, that the Mescalero Apache Indian Tribe has begun a second effort to develop a Rio Grande cutthroat broodstock, again using trout from the Indian Creek population. Success there would go a long way toward advancing restoration of the subspecies.

36. At the U.S. Fish and Wildlife Service's Mora Fish Technology Center, where the Service maintains a broodstock of the endangered Gila trout, they fight domestication in another way. Conditions such as current flow, woody cover, substrate, and vegetation are set up to mimic as closely as possible what the trout would experience in a stream. That includes cohabiting with fish that occur naturally with the trout in the Gila River drainage, such as the desert and Sonora suckers.

37. See Propst et al. (1992), Rinne (1996), and Brown et al. (2001) for some of these accounts.

Beaver ponds, such as this one on a Rio Grande cutthroat stream in the San Juan Mountains, can supply good rearing habitat for the trout. PHOTO BY AUTHOR

than any other part of the U.S., and this region has the nation's highest incidence of lightning-caused fires. Historically, fires occurred every 4 to 5 years, but were relatively small, ground-level, low-intensity, understory fires that left open stands of larger trees. Prior to the 1950s, catastrophic crown fires that consumed the larger trees were extremely rare. Livestock grazing, practiced intensively from the 1880s through the 1930s, cropped off the grasses that supported the low-level fires, allowing woody brush, thick shrubs and seedlings to grow instead, and active fire suppression, a government policy since the early 1900s, prevented the natural low-level fires from burning this accumulation away. The result, across the southwestern landscape, has been much thicker stands that burn more intensely, creating large, hard to control, catastrophic crown fires that can spread and consume entire forest stands and burn across drainages. There is also an association with climate: larger fires occur in years of severe drought, which the southwest has also experienced recently. [38]

The fires themselves don't often kill the trout; it's what happens in the aftermath that brings about their demise. Fire season in the southwest is followed closely by a season of monsoon, bringing periods of often heavy precipitation and runoff into streams from the now denuded hill slopes. It's the heavy slurry of ash and sediment sluicing into the streams that can eliminate whole populations of trout. In addition, fire retardant chemicals dropped on the fire may also wash into streams and kill fish.

But catastrophic fire effects can cut both ways. They can also provide opportunities to reclaim streams that before the fires had been occupied by non-native trout. The Fish and Wildlife Service highlighted two such opportunities in its 2002 status review, both involving the Santa Fe National Forest in New Mexico. In 1996, the Dome Fire extirpated non-native trout residing in Capulin Creek on Forest Service and Bandelier National Monument lands. Plans were laid to repopulate the stream, once the instream habitat recovered sufficiently, with Rio Grande cutthroat trout. An e-mail from Sean Ferrell, Forest Fish Biologist for the Santa Fe National Forest, said that the first release of cutthroats could occur in the fall of 2005. So this project may have come to fruition by the time you read this. But the second opportunity highlighted in the status review was associated with the Viveash Fire of 2000 that burned approximately 29,000 acres in the drainage of Cow Creek, a Pecos River tributary, and extirpated the non-native trout in about 25 miles of that stream. Here, alas, the plan to reintroduce native Rio Grande cutthroats was derailed. Some of the affected reaches of Cow Creek are on private lands, and here the New Mexico Department of Game and Fish thought it prudent to restock the stream with non-native species, which it did in 2004.

Which brings me around to public sentiment,

38. You can get a better picture of the fire history of the region and of the interactions of vegetation type, forest structure, climate, livestock grazing, and fire suppression from these publications: Cooper (1960), Swetnam (1990), Covington and Moore (1994), Swetnam and Baisan (1996), and Touchan et al. (1996). Rinne (1996) has a good description of the effects of wildfire on fishes and aquatic invertebrates in the southwestern U.S.

the fourth of the other natural or manmade factors considered by the Fish and Wildlife Service in its 2002 status review. Recent years have seen a growing admiration for native cutthroat trout. Hardly an issue of any of the fly-fishing magazines goes by (and that's about the only kind of fishing or outdoor magazine I read anymore) without a piece of some kind on native cutthroat trout or their waters. The Rio Grande subspecies is the State Fish of New Mexico. The City of Albuquerque has constructed a Rio Grande cutthroat stream exhibit where city dwellers can view these trout in a simulated natural setting. Conservation organizations like New Mexico Trout have sold prints and posters to raise awareness and dollars for the conservation and preservation of native trout in New Mexico. The region's national forests have outreach and education programs, such as the Respect the Rio program of the Santa Fe National Forest, which highlights native fishes and respect for their habitat. The Federation of Fly Fishers has its Project Cuttcatch to raise awareness and appreciation of our native cutthroat trout. [39] In light of all this, the Fish and Wildlife Service concluded that public sentiment is not a threat to the 13 pure, stable, and secure core populations of Rio Grande cutthroat trout.

However, when it comes to *increasing* the number of stable, secure populations and expanding the subspecies range, a few obstacles still stand in the way.

One is resistance to the use of piscicides like antimycin and rotenone, which remains the most effective way to reclaim stream reaches from non-native species for the benefit of Rio Grande cutthroat trout. [40] In 2004, bowing to this resistance, the New Mexico Game Commission denied the use of antimycin in that state, which effectively stalled some large projects to create metapopulation structure in the Jemez and Sangre de Cristo mountains, according to Sean Ferrell, the Santa Fe Forest Fish Biologist (you'll recall that the Jemez project was one the Fish and Wildlife Service had highlighted in 2002 among its reasons for not listing the Rio Grande cutthroat). Unless support can be built for expanding the distribution of the Rio Grande cutthroat into more of its historical range, restoration efforts for this subspecies could be undermined.

One potential man-caused threat not addressed by the Fish and Wildlife Service in its 2002 status review is energy development; in particular, drilling for coalbed methane gas. What brought this to the forefront in Rio Grande cutthroat country was an application from Houston-based El Paso Corp. to drill for coalbed methane on the eastern 40,000 acres of the scenic and wildlife-rich 102,000-acre Valle Vidal Unit of Carson National Forest in northern New Mexico. That was followed by intensive efforts by the Bush White House to fast-track the approval process, and that in turn galvanized the opposition.

39. See Chapter 4 and footnote 36 of that Chapter for a discourse on the use of antimycin. I talked about rotenone and what I believe are spurious indictments against its use in Chapter 8 and footnotes 16 and 17 of that Chapter.

40. For more on Project Cuttcatch, see footnote 30 of Chapter 10. Rules and other necessary paperwork to qualify for a Project Cuttcatch award can be found on the Federation of Fly Fishers website at *www.fedflyfishers.org/Conserve/projects/Cuttcatch/Apply.htm.*

A. *Extent of development in 1986*   B. *Extent of development in 1999*   C. *Extent of development in 2001*   D. *Extent of development in 2004*

Ever since its acquisition in 1982 (ironically via donation from Pennzoil, an energy company), the Valle Vidal Unit has been managed by the Forest Service to emphasize protection and preservation of wildlife, scenic, and recreational values. Comanche Creek, a known Rio Grande cutthroat stream and site of one of the partnership projects cited by the Fish and Wildlife Service among its reasons why the subspecies needn't be listed, lies within the Valle Vidal but west of the acreage proposed for drilling. But McCrystal and North Ponil creeks flow through the heart of the proposed drilling area and both undoubtedly would have been affected. Neither of these drainages is known to have Rio Grande cutthroat populations now, but both are within the subspecies' historical range, and both are undisturbed enough at present to be eligible for the National Wild and Scenic Rivers list. Another concern

was for the area's premiere elk herd. Critics cited a study by the Wyoming Department of Game and Fish showing that elk abandon approximately 97 acres for every acre of drilling disturbance.

Just about everybody who is anybody rose up against drilling in the Valle Vidal, including Trout Unlimited, the Sierra Club, the National Wildlife Federation and other national environmental and conservation organizations, a host of New Mexico anglers, hunters, guides, outfitters, ranchers, environmental groups and outdoor enthusiasts, several local governments and chambers of commerce, the governor of New Mexico, and most (eventually all) of the state's congressional delegation. A principal concern of the opponents was that drilling would essentially industrialize the wild and scenic area with a spider web of well pads, roads, pipelines, compressor stations, power lines, and other infrastructure. An example of what such an impact actually looks like on the ground can be seen in Fig. 14-2. These are satellite images of Wyoming's Jonah Gas Field south of Pinedale, which lies within the native range of the Colorado River cutthroat trout. The images illustrate clearly how bringing a gas field into full production can wreak major changes on the landscape. Gas wells also pump great quantities of water out of the ground, raising additional concerns about depletion of local aquifers and the polluting effects of discharging this "produced water" into streams. [41]

But all ended well for the Valle Vidal, thanks to the fervor of the opposition and, I'd have to say, to the change in political climate that occurred in Washington, D.C. following the mid-term elections of 2006. A year earlier, New Mexico Representative Tom Udall introduced a bill into Congress that would withdraw the Valle Vidal from mineral and energy leasing. The Valle Vidal Protection Act of 2005 passed both the House and the Senate, and was signed into law by President Bush on December 12, 2006, thus ending the struggle.

Alas, however, for every victory there seems to be at least one setback. In this case, it has to do with the status of the core population in New Mexico's Rio Cebolla. You'll recall from Table 14-1 above that this is one of the thirteen populations cited by the Fish and Wildlife Ser-vice in its 2002 status review as being genetically pure, stable, and secure from invasion by non-native trouts. From 1989 through 1993, volunteers from New Mexico Trout, the conservation organization, had joined with the Forest Service in projects to repair habitat in Rio Cebolla. Brown trout were removed from the stream in 1994 and were replaced with Rio Grande cutthroat trout in 1995. For ten years, the cutthroats thrived in the stream free of brown trout, and it was indeed viewed as a major success story. But in 2005, I was advised [42] that brown trout had reappeared in an upper reach of Rio Cebolla, near where an old four-wheel drive track came close to the stream. Nobody has come right out and said so, but this might have been a deliberate act. The area is designated as roadless now, and is gated and fenced. But the gate had been forced open and two fences cut. Crews from the New Mexico Department of Game and Fish, New Mexico Trout, the U.S. Forest Service, and BLM mounted emergency electrofishing operations to remove as many of the brown trout as possible before they can spawn, but at this point the outcome is uncertain. Electroshocking may need to be continued for several years into the future to get all the encroaching browns. Until then, the Rio Cebolla core population must come off the Fish and Wildlife Service list because it is no longer secure from invasion by non-native trouts.

41. Refer back to Chapter 12 for more discussion of gas and oil drilling and production.

42. E-mail messages dated July 7, July 11, and July 21, 2005 from New Mexico Trout member Mike Maurer, Albuquerque, New Mexico.

Yellowfin Cutthroat Trout, *Oncorhynchus clarkii macdonaldi*, extinct

CHAPTER 15

Extinct Subspecies

"A curtain has fallen on [these] species, cutting [them] off forever from our view. Now [they are] gone for all time. [They] will never again be seen alive."

—Robert Silverberg,
The Auk, the Dodo, and the Oryx
New York: Thomas Crowell, 1967

ON EXTINCTION

Extinction, defined here as the act or process of dying out or vanishing completely from the planet, used to be thought of as a gradual, species-by-species occurrence, an inevitable consequence of the process of evolution and survival of the fittest. Natural changes in climate and habitat constantly take place, sometimes so gradually that in terms of a human lifespan we hardly notice them. Each species must adapt to these changes, selecting and enhancing the traits that favor continued existence and eliminating those that do not. Otherwise it dies out—it becomes extinct.

But that mechanism, in which many past species became extinct by evolving into new species, is now considered to be just the background process of life and evolutionary change on Earth. Paleontologists had also been aware that from time to time, large numbers of species representing great chunks of Earth's then-existing biodiversity disappeared from the fossil record all at once. At least twelve of these *mass extinctions*, five of them truly immense episodes, have occurred since life began on Earth. In the 1980s, evidence accumulated that these were caused by cataclysmic events, some of them extraterrestrial, such as the asteroid collision that ended the Age of Dinosaurs so abruptly 65 million years ago. Michael J. Benton, Department of Geology, The Queens University of Belfast, looks upon these mass-extinction events as re-set mechanisms. In his words:

> "Mass extinctions in effect 're-set the evolutionary clock:' they wiped out whole families or orders of plants and animals, irrespective of the Darwinian fitness of the individual organisms or of the species involved, and created opportunities for the survivors to radiate out into the 'empty' ecospace that was left."

Extinction by either of these mechanisms is a natural process. Extirpation, defined here as causing other species to die out or vanish completely from the planet through human actions, is not, even though the end result is just as final. Most natural scientists agree that we are in the midst of another immense mass extinction episode, with as many as half of Earth's plant and animal species already on the path to extinction within 100 years. Today, as in earlier mass extinctions, species are ceasing to exist before they can give rise to new species. But this time the mechanism is different. It's extirpation, not some calamity of nature such as triggered the mass extinctions of the past. [1]

We have already seen, in Chapters 3 through 14, how the collective actions of humans have impacted extant subspecies of cutthroat trout, reducing many to last-bastion habitats within their historical range, and indeed pushing some to the brink of extinction. In this, the last Chapter of the book, I will tell you what we know about two subspecies, the yellowfin

1. *National Geographic* magazine published two easy-to-read treatises on the subject of extinction in 1989 and 1999 (see Gore 1989 and Morell 1999 for the citations). Other more detailed and technical material can be found in Benton (1985), Elliott (1986), Jablonski (1986), and Raup (1986).

Twin Lakes, Colorado. These lakes once harbored the large yellowfin and smaller greenback cutthroat subspecies. They still serve anglers and recreationists, but their primary function now is as a pumped storage facility. PHOTO BY AUTHOR

cutthroat trout and the Alvord Basin cutthroat trout, which were extirpated almost as soon as we became aware of their existence.

YELLOWFIN CUTTHROAT TROUT

Oncorhynchus clarkii macdonaldi: Chromosome number unknown. Scales in lateral series 159–185 (mean 175), significantly fewer than greenback cutthroat trout (Chapter 13), with which the yellowfin cutthroat co-occurred. Scales above lateral line, average 42, also significantly fewer than greenback cutthroat trout. Gill rakers 20–22 (mean 21), more than typically found in the greenback subspecies. Colors silvery; silver-olive on the back with broad lemon-yellow tints on the sides, and golden-yellow fins. Spots small and pepper-like or star-shaped, smaller than the nostril, and profusely distributed posteriorly and on the dorsal and caudal fins. Spots did extend forward toward the head, but were sparse on the anterior half of the body. [2]

A few miles south of Leadville, Colorado, tucked up against the base of Mt. Elbert in the Sawatch Range, lies a pair of lakes known as the Twin Lakes. These lakes were formed in Pleistocene time, when glacier deposits blocked a creek tributary to the Arkansas River. Today, a manmade earthfill structure straddles the outlet and the Twin Lakes function as a Bureau of Reclamation pumped storage facility. Prospectors crawled over nearly every foot of this landscape during Colorado's early mining years. Camps and hardscrabble towns sprang up where strikes were made, each enjoying its brief time in the sun. A few of the earliest settlers catered to fishing parties and recreationists. "The Unsinkable Molly Brown" and other Colorado notables knew the Twin Lakes as a popular resort area at the turn of the 20th century. Campers and fishermen still enjoy the Twin Lakes, only nowadays fishermen work the lakes for big mackinaws and rainbow trout.

But the Twin Lakes lie in the heart of greenback cutthroat country, and, indeed, most of the earliest reports of fishing in the lakes alluded to the smallish native greenbacks. Charles Hallock's *Sportsman's Gazetteer and General Guide*, published in 1877, stated that Twin Lakes trout ran small, an obvious reference to the greenback cutthroat, which was never noted for its large size. If larger trout were desired, Hallock continued, anglers could always hike over Tennessee Pass to the Eagle River watershed, home of the Colorado River cutthroat, where trout often reached and exceeded 10 pounds. [3]

Yet stories of much larger trout in the Twin Lakes circulated almost from the beginning. Bayard Taylor was a clergyman from New Jersey who embarked on a lecture tour through the Rocky Mountain districts of Colorado in 1866. His party also fished its way along as it moved through the territory, including a stop at Twin Lakes, where it took up prearranged lodging at the Leonhardy household in the fledgling town of Dayton (now Twin Lakes). "Mr. Leonhardy had tempted us with descriptions of six and eight-pound trout," Taylor

2. This formal description combines David Starr Jordan's original description with that of Dr. Behnke, who examined Jordan's deposited specimens. Jordan originally named the yellowfin cutthroat trout *Salmo mykiss macdonaldi*, but that name, of course, has given way to *Oncorhynchus clarkii macdonaldi*. Jordan's original description is in Jordan and Evermann (1890); repeated in Jordan (1891b). Dr. Behnke's examination was first reported in his 1979 monograph and was repeated in his subsequent books in 1992 and 2002 (see Behnke 1979, 1992, 2002).

3. I referred to Charles Hallock's *Sportsman's Gazetteer and General Guide* back in Chapter 12, footnote 38. The original edition was published in 1877. I have never been able to find a copy of that one, but I have seen a later edition, published in 1883. You can find that citation in the bibliography under Hallock (1883).

wrote in a dispatch to *The New York Tribune*. Although fishing was good, catching one of the big ones was not to be in the cards. "The large specimens of trout did not bite," Taylor's dispatch continued, "they never do when there is a special reason for desiring it—but we had no right to complain." [4]

The first *official* mention of a large trout in the Twin Lakes that might be different from the smaller greenback cutthroat appeared in the 1885–86 report of the Colorado Fish Commissioner, John Pierce. The Colorado Fish Commission began propagating this trout along with the greenback cutthroat in 1885, using eggs taken from Twin Lakes fish. Pierce urged the U.S. Fish Commission to establish a propagation station there as well, which it finally did, outside of Leadville, in 1890. Meanwhile, in 1889, David Starr Jordan was leading a fisheries survey of the west for the U.S. Fish Commission, and one of his stops was at Twin Lakes. Jordan set the scene:

"These two lakes, formed by a moraine dam at the foot of Mt. Elbert and Mt. Grizzly, are the largest lakes on the east side of the divide in Colorado. They are separated from one another by another moraine across which they are connected by a short (⅛ mile) stream. The lower lake is the larger, 3 miles long by 2 miles wide. The upper lake is 1 ½ miles long by 2 wide. The lower lake averages 40 ft. depth; its lower part being extensively shallow, the middle and south side very deep. The bottom is gravelly and covered with water plants. In some places are piles of boulders. The shallow north side has heavy weeds, growing up to 5 feet high in water 10 feet deep. Among these weeds the trout chiefly feed. *Gammarus* is very abundant. The upper lake is colder and not quite so well stocked with fish."

At least part of Jordan's collection work at Twin Lakes was accomplished with the assistance of George R. Fisher, an angler and correspondent from Leadville, who had drawn Jordan's attention to the existence of two varieties of trout in the lakes. Jordan had already collected greenbacks, but Mr. Fisher told of a silvery, yellow-finned variety that grew to a much larger size. Fisher himself had taken trout ranging from 7 to 10 pounds, and told Jordan of others weighing up to 13 pounds. The two men arranged for a morning's fly-fishing together. Although they didn't get any of the large specimens (which Jordan later said the locals took mostly by spearing, probably when the trout were on their spawning runs), they did capture several smaller specimens of the silvery, yellow-finned "mystery fish." About these, Jordan wrote:

"These represent a very distinct form or variety of the mountain trout, which we recognize as a distinct sub-species under the name *Salmo mykiss macdonaldi*. We have taken pleasure in naming the yellow-fin for the United States Fish Commissioner, the Hon. Marshall MacDonald, in recognition of his services in connection with the propagation of the American Salmonids.... Color, silvery olive; a broad lemon yellow shade along the sides, lower fins bright golden yellow in life, no red anywhere except the deep red dash on each side of the throat. Body posteriorly and on dorsal and caudal fin profusely speckled with small pepper-like spots,

4. The Taylor party stopped to fish at Twin Lakes in 1866. His description of the Twin Lakes sojourn, as well as his other experiences in Colorado, were later compiled in book form; see Taylor (1867). Two other early parties that fished at Twin Lakes were the Bowles party in 1868; see Bowles (1869, [1991]) and the Schenck party in 1871; see Schenck (1965). These parties too caught plenty of trout, but none of the larger specimens.

smaller than the nostril and smaller than any of the other forms of *S. mykiss*. Occasionally these spots extend forward to the head, but they are usually sparse on the anterior half of the body."[5]

Unfortunately, this description of Jordan's was based on appearance only, and contained none of the meristic characters relied upon by scientists lacking modern genetic analysis tools to delineate subspecies. Jordan did, however, preserve specimens of both the yellowfin cutthroat and the greenback cutthroat that he and his companion had captured. And furthermore, the preserved specimens of each type of trout were in the same size range, 6 to 12 inches, so that people who examined them in later years would not be influenced by differences in size. Dr. Behnke performed such an examination of Jordan's museum specimens some 75 years after they had been collected, and noted that significant differences did exist not only in color and spotting, but also in meristic characters. He reported:

"When I examined the...specimens of yellowfin trout and...greenback trout from Twin Lakes and critically compared the data, I had no doubt that Jordan was correct; the yellowfin trout and the greenback trout from Twin Lakes were two distinct groups of cutthroat trout.... Although long preserved, the specimens are in good condition and the small, 'star' shaped spots and silvery coloration of the yellowfin trout specimens are clearly recognizable and readily differentiated from the dark coloration and pronounced, large, rounded spots of the greenback trout specimens.... In the meristic characters, I noted more gill rakers in the yellowfin trout than in the greenback specimens...(20–22 [mean 21] vs. 18–21 [mean 19]). The yellowfin specimens have 159–185 (mean 175)

scales in the lateral series and average 42 scales above the lateral line. The greenback specimens have 170–202 (mean 186) scales in the lateral series and average 48 scales above the lateral line."[6]

So, two distinct and reproductively isolated forms of cutthroat trout actually did exist in Twin Lakes, at least in 1889. Had the yellowfin always been there, as Bayard Taylor's account from 1866 might suggest, and just never been considered distinctive enough to report prior to 1885? Or were they introduced from some other location by anglers or miners—or by private fish culturists, who by that time had sprung up all over Colorado to supply the demand of hungry residents for trout? Other types of trout had been planted in the Twin Lakes even before Jordan's visit in 1889. He noted in his official report that eastern brook trout and rainbows had been introduced, the brook trout concentrating mostly in the colder waters of the upper lake, while the rainbows distributed themselves more evenly in both.

Jordan also ventured a suggestion as to where the yellowfins might have come from. Commenting on specimens he collected from the Eagle River near the town of Gypsum, Colorado, he wrote:

"Trout from Eagle River show more resemblance to the yellow-fin of Twin Lakes in the small size of the spots and the plain coloration. Their place seems, however to be in var. *pleuriticus* with the others from the Colorado River."

5. Jordan's collection of work at Twin Lakes is included in his official report to the U.S. Fish Commission; see Jordan (1891b). He provided additional recollections in a letter to *Forest and Stream* magazine (see Jordan 1889) and in his autobiography, published in 1922 (see Jordan 1922). In addition to the report to the U.S. Fish Commission, a short paper by Jordan and Evermann (1890) also has the formal description of the yellowfin cutthroat trout, plus a few observations of its life history characteristics. William J. Wiltzius, a wildlife researcher with the Colorado Division of Wildlife, did considerable digging into the historical record to piece together the story of the yellowfin cutthroat trout. His findings can be read in Behnke and Wiltzius (1982) and Wiltzius (1985). I could not locate Colorado Fish Commissioner Pierce's report for 1885–86, so I relied on Behnke and Wiltzius (1982) and Wiltzius (1985) for what it had to say.

6. This account is in Behnke (1979) at page 133.

Had Jordan had meristic character data to guide him, his placement of the Eagle River trout in the subspecies *pleuriticus* (Chapter 12) would undoubtedly have been more positive, since the characters of the yellowfin cutthroat differ significantly from *pleuriticus*, just as they do from the greenback subspecies.

Gordon Land, a fish culturist who held the position of Colorado Fish Commissioner in 1890, believed the yellowfins were native to the entire upper Arkansas River drainage, not just to the Twin Lakes. He wrote letters to the editors of three major outdoor magazines that year (actually, the same letter to all three publications) stating that he had found yellowfin trout in all tributaries of the upper Arkansas, and had even taken yellowfin cutthroat spawn from one of those tributaries, Chalk Creek, for use in his own fish-culture operations.[7]

I'd like to suggest another possible origin. The upper Arkansas River drainage has likely been the conduit for more than one radiation of ancestral cutthroat trout since mid-Pleistocene time. In Chapter 2 and again in Chapter 14, I told you about the 740,000-year-old fossil trout assigned to the cutthroat lineage that were found in the San Luis Valley south of the upper Arkansas drainage. I also cited Dr. Behnke's reasons for believing these mid-Pleistocene cutthroats could not have been the direct ancestors of the more recently evolved Rio Grande cutthroat that is the present "native" of the San Luis Val-

ley. By the same reasoning, they could not be the direct ancestors of the more recently evolved Colorado River cutthroat or greenback cutthroat subspecies. Supposing, though, that the yellowfin cutthroat was a relict of one of those earlier radiations that did not go completely extinct, but survived in the upper Arkansas drainage by partitioning habitat niches with the more recently arrived greenbacks. That would explain the two distinct forms of cutthroat trout in the Twin Lakes, as well as Gordon Land's observations of similar trout in other upper Arkansas tributaries. In fact, the upper Arkansas would have had to be the refugium for both subspecies, because the Twin Lakes were glaciated during the latest Pleistocene ice advance and could not have been occupied by either subspecies until the ice retreated.

But regardless of its origin, the yellowfin cutthroat did maintain reproductive isolation from the greenback cutthroats in Twin Lakes, and indeed, it exhibited a completely different life history. For one thing, yellowfin cutthroats spawned very shortly after ice-out, much earlier than the greenbacks, and much of this spawning evidently took place on gravel shoals in the larger lake. For another, the yellowfin cutthroats were fish eaters, which probably accounts for their large maximum size, whereas the greenback cutthroats in Twin Lakes fed mainly on the abundant *Gammarus* scuds. This difference in diet also reflected itself in the

7. These letters from Colorado Fish commissioner Gordon Land were sent to *Field and Farm* magazine, a Colorado publication, to *Forest and Stream* magazine, and to *Sports Afield* magazine. I've cited the latter two in the bibliography under Land (1890).

Do descendents of yellowfin cutthroat trout still swim in these waters? This is upper Lake Fork Creek, where a search for relict yellowfins centered in the early 1980s. PHOTO BY AUTHOR

color of the flesh of the two fishes. Yellowfins were noted for having pale flesh. Twin Lakes greenbacks, on the other hand, had the reddish flesh so characteristic of trout feeding on crustaceans.

Jordan also observed what might be thought of as a reciprocal predator-prey relationship between the yellowfin cutthroat trout and the suckers resident in the Twin Lakes. He wrote that "this fish [the yellowfin] feeds freely on young suckers and even on young trout." On the other hand, when the yellowfins spawned, "the suckers infest its spawning beds, devouring the eggs." [8]

Beginning in 1885 with State Commissioner Pierce's egg-taking activities and continuing with the federal hatchery near Leadville, both greenback and yellowfin cutthroat eggs were taken at Twin Lakes. But that activity ceased after 1897 due to dwindling numbers of native trout in the spawning runs and increasing local opposition. [9] Egg taking commenced again for the Leadville hatchery in 1899, but now the source was a set of lakes on Colorado's Grand Mesa, where private fish-culture operations had been ongoing since about 1890.

It is here that the story of the yellowfin cutthroat begins to get fuzzy. Yellowfin cutthroats disappeared from Twin Lakes sometime around the turn of the 20th century. In 1902 and 1903, Chauncey Juday of the U.S. Bureau of Fisheries sampled Twin Lakes and found no yellowfin cutthroats at all. [10] Rainbow trout had become dominant by then, and even the remaining greenbacks were hybridized. Lake trout too had been introduced, and, being the efficient piscivores that they are, had undoubtedly wrested that feeding niche away from any remaining yellowfins, effectively sealing their fate.

Yet U.S. Fish Commission reports continued to list yellowfin trout, *Salmo macdonaldi*, as one of the species propagated at Leadville through the year 1905. Reports from the Leadville Hatchery itself, however, listed only "blackspotted trout" derived from the Grand Mesa lakes as being propagated there after 1899. Apparently believing that some of the "blackspotted trout" from the Grand Mesa lakes were indeed yellowfins, cutthroat egg shipments that were labeled as yellowfin trout were made to Germany in 1899 and 1902. Eggs labeled "blackspotted trout" also went to Belgium in 1902 and to Wales in 1904, but the fates of these exports to Europe are unknown.

The native trout of the Grand Mesa lakes region is the Colorado River cutthroat subspecies, and it is most likely that the "blackspotted trout" handled at the egg-taking facilities there and propagated at the Leadville Hatchery belonged to that subspecies. But isn't it possible that yellowfin trout were stocked in the Grand Mesa lakes prior to their disappearance from the Twin Lakes? There are at least three hints in old outdoor

8. These observations were reported in Jordan and Evermann (1890).

9. Trapping the trout to take their eggs and milt for the hatchery prevented them from entering the outlet stream. As the spawning runs declined, this increasingly upset the locals who were accustomed to harvesting the spawning trout with their pitchforks and spears. Shots were fired to discourage the hatchery personnel, and on one occasion a trap was actually dynamited. These actions prevailed, and the hatchery looked elsewhere for its trout eggs. For a few years, the main source of supply of cutthroat eggs was the Grand Mesa lakes, where eggs from Colorado River cutthroat trout were available. But the feds lost their right to this resource in 1910, and turned to Yellowstone Lake for its egg supply. By 1923, no trout native to Colorado was being reared at the Leadville Hatchery at all, and only brook trout, rainbow trout, and Yellowstone cutthroat trout were being distributed from that facility. This information was derived in part from Wiltzius (1985) and also from a history of the Leadville National Fish Hatchery compiled by Rosenlund and Rosenlund (1989).

10. See Juday (1907b).

magazines and official reports suggesting that yellow-fin cutthroat trout from the Leadville Hatchery were indeed stocked in other waters prior to 1897, but there is no official record, or even an anecdotal one, that any of the Grand Mesa lakes ever received these trout. [11]

So the known, authenticated history of the yellowfin cutthroat is a short one. It was known to have occurred in the Twin Lakes and was described for science as a distinct subspecies based on specimens that remain well-preserved and can be viewed in museum collections. [12] Yet within fifteen years or so of its "official" discovery, the yellowfin cutthroat was extinct in the Twin Lakes, a victim of extirpation by enthusiastic harvesting of spawners from the spawning beds and the stocking of non-native species, principally rainbow trout and lake trout.

But the yellowfin cutthroat story has never been fully put to rest. The clouded, later history of this trout has been enough to keep rumors of its persistence alive. The connection with the Grand Mesa lakes thrived for a time. The U.S. Fish Commission report for 1931 included a statement that the U.S. Forest Service was propagating yellowfin cutthroat trout at the Grand Mesa lakes along with rainbow trout, brook trout, and other "blackspotted trout," but it didn't say where any of these trout were stocked. And continuing the thread, angler-author Arthur H. Carhart wrote, in 1950:

11. Three pieces of evidence that yellowfin cutthroats from the Leadville Hatchery were stocked in waters other than the Twin Lakes were cited in Wiltzius (1985). One was a note in the July 21, 1894 issue of *Field and Farm* magazine stating that "The State of Colorado has secured 10,000 eggs of yellow-fin trout, a rare fish.... Mr Callicotte [then the Colorado State Fish Commissioner] will endeavor to stock a number of the streams with yellow-fin species." Another is a photograph, reproduced in Wiltzius (1985) but originally published in *Sports Afield* magazine June 15, 1891 (vol. 6, no.6, page 161), showing trout of 6¾, 4¾, and 3½ pounds with silvery colors and small spots typical of the yellowfin cutthroat that were caught from a ranch pond near Buena Vista, still in the Arkansas River drainage but many miles south of the Twin Lakes. The third piece of evidence was found in reports of the U.S. Fish Commission for fiscal years 1896, 1897, and 1898, where stocking locations for yellowfin cutthroats were listed as the Twin Lakes and their tributaries, and also Mammoth Creek near Nederland, Colorado. The latter site is up near Boulder, in the drainage of the South Platte.

12. Seven specimens of yellowfin cutthroat trout were collected by Jordan from the Twin Lakes. Five of these specimens are in the National Museum of Natural History, Smithsonian Institution, Washington, D.C., and two are now housed at the California Academy of Sciences, San Francisco.

"The only place I ever caught yellow-fins has been from Island Lake on the Grand Mesa in western Colorado. They are a fighting fish and beautiful, reportedly growing to large size."[13]

Still other trout stalkers have sought out the subspecies in other remote and not-so-remote locations. I got caught up in one of those searches myself, albeit briefly, back in the late summer of 1981, when I was in Colorado for the first edition of this book. Robin Knox, one of my contacts at the Colorado Division of Wildlife, had told me that the year before, spurred by a persistent group of local sportsmen who believed yellowfin cutthroats might still be found in mountain streams west of Leadville, State biologists had teamed up with them to electrofish Lake Fork Creek, a tributary of Turquoise Reservoir. They brought no cutthroats to hand, only brook trout, but in one pool they did "turn over" a single yellowish trout that managed to elude capture. That ended the formal search, but the unidentified yellowish trout continued to intrigue the searchers.

The spokesman for the group of anglers who had instigated that search was Del Canty, a trophy fly-fisherman of some note who in fact still holds the State of Utah hook-and-line record for the 26-pound, 2-ounce rainbow trout he took from Flaming Gorge Reservoir in 1979. He had also had some training in fishery biology, so was considered a reasonably knowledgeable spokes-man. He filled me in on the long, long story, one that spanned nearly three decades, that had led, finally, to that yellowish trout in the waters of Lake Fork Creek.[14]

The story goes all the way back to the opening day of trout season, 1954. Canty and an older fishing partner, Frank Bozig, who was also "a real student of trout," in Canty's words, were fishing Twin Lakes near a spring area on the southwest shore of the upper lake. They had already captured several of the lake's rainbow trout and also a couple of lake trout when Bozig hooked a large fish that proved to be a cutthroat. This excited him greatly, not only for its size, which pulled their de-liar scale down to 5½ pounds, but also because, he proclaimed, it was one of the "real natives" of the lakes, the first he had seen in many years.

Bozig was readily familiar with the small "river cutthroats," his term for the greenback cutthroats that had once been more common in the waters around Leadville, but this was not one of them. The two men spent some time examining the large trout. Canty recalled that the feature that impressed him most was the large, rakish, "predatory head," as he described it, with strong, well-formed teeth. Next, were the small black spots clustered on the rear of the body, each surrounded by a light halo. Some spots were present on the belly near the anal fin as well, he remembered. Colors were silvery olive above; sides silver with just a faint hint of magenta,

13. See Carhart (1950) at pages 19–20. This, I believe, is the same Arthur H. Carhart that I introduced you to in Chapter 12, in connection with the Forest Service-commissioned landscape architecture study that preserved Trappers Lake, Colorado from development back in 1919.

14. Interview with Del Canty, Leadville, Colorado, September 1981. Also a letter from Mr. Canty to Dr. Behnke dated October 11, 1980, which Mr. Canty kindly shared with me. My contact at the Colorado Division of Wildlife was Robin Knox of the Denver office.

except that the area from the pelvic fins back to the caudal fin had a yellow hue, and the caudal fin was quite yellow with flecks of orange at the base of the rays. The anal fin was similarly colored, as were the pectoral and pelvic fins, except the latter lacked the orange flecks and were perhaps less brightly yellow. The belly had orange-red areas just forward of the pelvic and anal fins.

This description matches Jordan's original description of the yellowfin cutthroat in some respects, but there are notable deviations. The faint magenta hue on the anterior sides of the trout is one and the red-orange on the belly is another; Jordan remarked on the lack of red anywhere except the cutthroat slashes. Jordan also did not mention halos around the spots. Nevertheless, the size of the fish and the predatory head were compelling features in Canty's mind. They were also features he was going to see again in just a few weeks' time.

In June of that same year, Bozig took Canty to Timberline Lake in the mountains northwest of Leadville where they crossed snow banks to watch cutthroats spawning in the inlet stream. These fish had been carried in on horseback and stocked there many years before, Bozig had told him. Other lakes had received them as well, but Timberline was the only one Bozig knew of where they could reproduce. "These fish were decidedly the same as the one I had just seen at Twin Lakes although generally not as big," Canty recalled,

"even though the largest of these might have gone about 24 inches."

There is much more to this story, including several unsuccessful attempts to get the Leadville Hatchery to keep specimens of the Timberline trout separate from other cutthroats for use as a "yellowfin" broodstock. The sportsmen kept the issue alive, and finally, through continuous badgering, got the state to respond with that electrofishing survey.

I had some time, so I decided to hike into Timberline Lake and fly fish both the lake and its outlet, Lake Fork Creek. All I caught were small brook trout until I reached the pool on Lake Fork Creek that Canty had described for me, where the electrofishing crew had turned up that yellowish trout the year before. There was a crowd of small brook trout in this pool as well, but on one cast, as my fly floated along just inches from a deep cutbank, a yellow form, easily discernable among the dark brookies, flashed out and grabbed the fly!

Never was an angler so excited over hooking and playing an 8-inch trout! The little dickens was strong and put up quite a tussle, but I finally managed to get it up on the bank for inspection. Try to imagine the play of emotions as I realized what I was looking at was—a small, buttery-colored brown trout! Browns had been stocked in Turquoise Reservoir and they spawned in Lake Fork Creek, but not this far upstream, or so I was told. What

The Alvord Basin, view east across the ancient lake bed from atop Steens Mountain, Oregon.
PHOTO BY AUTHOR

this one was doing way up here in water so infested with brook trout was beyond me. I snapped its photograph to share with Del Canty and his group, and hiked out.

Could any descendents of the yellowfin cutthroat still swim in some remote stream or pond, waiting to be discovered? I'm reminded of the ivory-billed woodpecker, long thought to be extinct but recently rediscovered in remote swampy woodlands in Arkansas. Still, after my experience with the brown trout of Lake Fork Creek, I come down on the side of those who believe it is not likely. True, the origin and later history of the yellowfin cutthroat are clouded in uncertainty. But I believe it is a trout that has passed into the ages.

ALVORD CUTTHROAT TROUT

Oncorhynchus clarkii alvordensis: Chromosome number unknown. Scales in lateral series 125–150 (mean 137), significantly fewer than in Lahontan cutthroat trout (Chapter 5). Scales above lateral line 33–38 (mean 35). Pyloric caeca 35–50 (mean 42), significantly fewer than in Lahontan cutthroat trout, and exceedingly long. Gill rakers 20–26 (mean 23), about the same as in Lahontan cutthroat trout. Basibranchial teeth feebly developed (and lacking in about half of the type specimens). Colors, olive above with blue and gold reflections; sides brassy yellow with a deep rose or brick-red band, especially on larger specimens whose opercles are similarly colored; lower sides brassy on a gray background. Cutthroat marks intense red; lower fins bright to deep purplish rose. Spots few and of moderate size with a trace of a pale ring around them, almost all occurring above the lateral line except for a few below the lateral line on the caudal peduncle. [15]

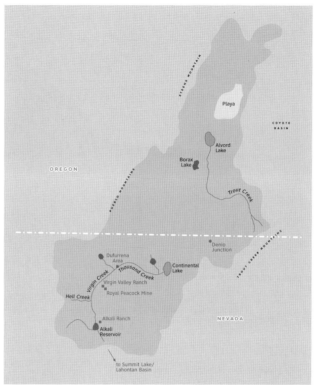

FIGURE 15-1. *Alvord Basin showing Virgin Creek and Trout Creek drainages*

15. The Alvord cutthroat trout was formally described as a subspecies by Dr. Behnke in his book, *Trout and Salmon of North America* (Behnke 2002). Dr. Behnke assigned the subspecies name *alvordensis*, first used by Carl L. Hubbs, University of Michigan, in his field notes of August 2, 1934 for the trout he and his family collected from Trout and Virgin creeks in the Alvord basin. The description I have written here combines Hubbs' original description with that of Dr. Behnke, who examined the deposited specimens and provided the meristic counts. The Hubbs field notes and specimens are on file at the University of Michigan Museum of Zoology, Ann Arbor, Michigan.

▮ Streams, Lakes, and Reservoirs

▯ Ephemeral Lakes

▮ Basin Boundary

▯ Playas

● Cities and Towns

■ Ranches and Other Locations

The Alvord Basin in northwestern Nevada and southeastern Oregon is desert country now, what with only about 6 inches of rainfall a year. When I inquired about the road to Denio on my first trip into that country many years ago, one of the old gents sitting on the

Alvord Cutthroat Trout, *Oncorhynchus clarkii alvordensis*, extinct

porch steps responded, "Whatta y' wanna go there fer? There's nothin' down there but rattlesnakes 'n' sagebrush." But back in Pleistocene time, when Lake Bonneville and Lake Lahontan existed, another lake, about 75 miles long on its north-south axis and maybe 5 to 10 miles wide, also filled the Alvord Basin, and its tributary streams ran robust and cold.

I mapped the location of the Alvord Basin in relation to the Lahontan, Coyote, and Humboldt basins back in Chapter 5, Fig. 5-1, and I show the basin again, with the additional points of reference that figure into this story, in Fig. 15-1. The Alvord Basin lies just north of the Lahontan Basin, with which it is contiguous along the high-elevation area north of Summit Lake and along the dividing ridges of the Trout Creek Mountains that separate it from the upper Quinn River drainage. The breathtaking thrust of Steens Mountain and the high and also imposing Pueblo Mountains form its western edge and separate it from Catlow Valley and the Malheur Basin of Oregon. The Coyote Basin, into which drain Willow and Whitehorse creeks (Chapter 7), lies across a low divide to the east.

An ancient gully at the northern end of the Alvord Basin angles off to the east in the direction of the Owyhee River, suggesting that at some point back in prehistory there might have been a waterway connecting the ancient Alvord Basin with the Snake/Columbia system. Maybe this was the outlet of ancient Lake Idaho and the Snake River system prior to its capture by the Columbia River at the beginning of the Pleistocene, as I discussed back in Chapter 2. There is also evidence of ancient overflow through Sand Gap, a low point in the divide separating the Alvord and Coyote basins, that would also have drained to the Owyhee River.

Near the headwaters of Virgin Creek, on the divide that separates the Alvord Basin from the Lahontan Basin and the Summit Lake drainage, is another ancient channel. This one once carried the outlet of pluvial Lake Parman into the Alvord Basin via Virgin Creek. Lake Parman formed after a landslide blocked the Mahogany Creek drainage, which had, up to that time, flowed southward through Soldier Meadow to discharge into Lake Lahontan. Summit Lake, a closed basin of its own at present, is the remnant of pluvial Lake Parman. [16]

This is the route by which cutthroat trout most likely gained access to the Alvord Basin. Although Dr. Behnke believes the Alvord Basin trout differentiated sufficiently enough while isolated in the Alvord Basin to merit separate subspecies designation, which he conferred in his 2002 book, he also points out that its characters are most closely aligned with the native cutthroat trout of the western Lahontan Basin (an average of 23 gill rakers in the Alvord cutthroat vs. an average of 24 in the Lahontan cutthroat, for example). Only two fishes besides the cut-

16. For further insights into the ancient drainage patterns of pluvial Lake Parman and the Summit Lake basin, see Layton (1979), Miflin and Wheat (1979), and Curry and Melhorn (1990).

Trout Creek, Oregon, where it emerges from the Trout Creek Mountains. This stream once held Alvord cutthroat trout. PHOTO BY AUTHOR

throat trout, the Alvord chub and the endangered Borax Lake chub, are native to the Alvord Basin. These fishes also are most likely derived from a western Lahontan Basin ancestor that gained access to the Alvord Basin via the same Lake Parman conduit.

Although there is no direct evidence, the earliest human inhabitants may have captured trout from Alvord Basin streams as part of their annual subsistence diet. [17] The early ranchers and their families certainly knew about the trout and caught them for food and pleasure. It wasn't until 1934, however, that the first collections for science were made in the basin by Dr. Carl Hubbs and his family. The Hubbs party collected trout from Trout Creek, Oregon, and from Virgin Creek and Thousand Creek, Nevada. [18]

But even then, extirpation of the subspecies was under way. In Trout Creek, according to information Hubbs received, rainbow trout had been stocked in 1929, either by local ranchers, the Oregon Fish and Game agency, or both. The trout he collected there in 1934 were clearly hybrids of rainbow trout and the native cutthroat. Over in the Virgin Creek drainage, a long-ago rockslide had occurred about midway up the stream course that forced the water to flow through the slide, thus separating the stream into two isolated segments. Rancher Tom Dufurrena (his ranch is now the sub-headquarters of the Charles Sheldon National Antelope Refuge) told Hubbs that no stocking had occurred anywhere in Virgin Creek until he, Dufurrena, had released 6,000 rainbow trout fingerlings into the lower segment in 1933. But when the Hubbs family collected a year later, they found rainbow trout (at least one to 11 inches) and hybridized cutthroats in the lower segment of Virgin Creek and also in Thousand Creek, which indicated that stocking had probably occurred earlier. They did find trout that Hubbs believed to be pure Alvord Basin cutthroats, but only in the unstocked upper segment of Virgin Creek above the rockslide.

Ah, but that bastion was eventually breached as well. Harry Wilson, the present owner of the Virgin Valley Ranch along with the Alkali Ranch that controls access to the Virgin Creek headwaters (you rock hounds may also know him as proprietor of the world famous Royal Peacock opal mine, also located in the Virgin Creek valley), told Nevada Division of Wildlife biologists that his father had stocked rainbow trout in Alkali Reservoir in 1942, and these had spilled over to spread throughout the upper segment of the creek. When scientists visited the area subsequently (Carl E. Bond of Oregon State University in 1953 and 1959; Behnke in 1971 and 1972; and J.E. Williams and Bond in 1978, 1979, and 1980), they found mostly rainbow trout in Alvord Basin streams. The few specimens that exhibited any cutthroat-like appearance were clearly introgressed. They

17. Up in a tributary drainage of Virgin Creek with the forbidding name of Hell Creek is located an early-man site known to archaeologists as Last Supper Cave. Artifacts and faunal remains dating back almost 10,000 years have been found there. No fish remains are listed among the faunal records, but freshwater mussels of the genus *Margaritifera* have been found, which were evidently harvested from Hell Creek. These mussels live in pure, clear, coldwater streams typically in association with trout. You can read more about Last Supper Cave in Layton (1979) and Grayson (1988).

18. Carl Hubbs, his colleagues, and his family made several extended trips into the American West to collect and catalog its fishes, especially those of the arid regions. He and his family visited both the Alvord Basin and the Coyote Basin in the summer of 1934. To read more about these expeditions and what the country was like in the years between World Wars I and II, see the Chapter, "Ichthyological Exploration of the American West: The Hubbs-Miller Era, 1915–50 ," written by Robert Rush Miller, Clark Hubbs, and Francis H. Miller, in the book *Battle Against Extinction: Native Fish Management in the American West*" (1991).

concluded that the native cutthroat trout of the Alvord Basin was extinct in pure form, a victim of introgression by the introduced rainbow trout. [19]

But the story doesn't end there. Rumors of large reddish cutthroat trout continued to circulate, seemingly pointing to a difficult-to-access gorge in that upper segment of Virgin Creek. In 1984, Nevada Division of Wildlife biologists finally got in there and, sure enough, discovered two cutthroat specimens among the rainbow trout and other, obviously hybridized individuals, that appeared to be identical in colors and spotting pattern to Hubbs' original description. Photos were sent to Dr. Behnke, who concurred, but felt he should examine specimens himself in order to be sure.

Greatly excited that a few possibly pure Alvord cutthroats might still exist, Dr. Behnke relayed the information to Robert H. Smith, a retired U.S. Fish and Wildlife Service biologist who thrived on the search for rare native trout. [20] Guided by Harry Wilson's son Walt, since the stream reach to be searched required access across Wilson's land, Smith and his fishing partner, Jack James, hiked into the upper Virgin Creek gorge in mid-July 1985. There, they captured three specimens, 15, 18, and 20 inches in length, along with a number of smaller "hybrids" (later determined to be typical rainbow trout) that they carefully preserved in formaldehyde and sent to Dr. Behnke. Dr. Behnke's

examination of the three large cutthroats confirmed that they were not only visually identical to the Hubbs specimens, but were meristically identical as well— in his opinion "pure or virtually pure" Alvord cutthroat trout. Dr. Behnke aged these specimens at 5, 6, and 7 years respectively, surprisingly long-lived and large given the meager waters in which they had been living.

The euphoria was to be short-lived, however. Evidently, only a few of the oldest cutthroats had remained pure or virtually pure in upper Virgin Creek. All authorities agreed that if the Alvord Basin cutthroat was to be saved from total introgression, pure specimens should be removed from Virgin Creek and transferred to a new stream, one that was barren of other fishes. And, owing to the advanced ages of the purest-looking trout, the sooner this was done the better. No transfer could be made in 1985, although another electrofishing survey conducted that year confirmed that few of the pure-looking cutthroats remained. But in October 1986, after an attempt in late summer had to be aborted, 60 trout were finally collected, from which 26 of the phenotypically best specimens were selected for transfer. After a small piece of muscle tissue was removed from each trout for genetic analysis, these 26 specimens were released in Jackson Creek, a fishless stream located in the Jackson Mountains of Nevada, in the Quinn River drainage.

19. The history of fish collections, and of trout stocking, in Alvord Basin streams is told in Williams and Bond (1983), Behnke (1979, 1981a, 1992, 2002), and Tol and French (1988).

20. Robert H. Smith, a graduate of Dartmouth, was a wildlife biologist for all of his career with the U.S. Fish and Wildlife Service, during which he made pioneering aerial surveys of breeding waterfowl in the Arctic region, and discovered the northern breeding grounds of the whooping crane. After retirement in the late 1960s, he turned his interest to native trout. He caught and photographed all of the native species and subspecies of North America, and published his accounts of seeking them out in the book *Native trout of North America* (Smith 1984, [1994]). This book became something of a best-seller when it appeared in 1984, and a second edition was published in 1994. It's the second edition that relates his experiences in the upper Virgin Creek gorge where he collected the trout specimens for Dr. Behnke's examination and later assisted the Nevada Division of Wildlife in collecting visually good specimens for a translocation attempt.

The first bad news came from the genetic analysis. You'll remember that the trout selected for transfer were the best of the lot appearance-wise and matched the Hubbs description as closely as could be discerned in the field. But the genetic analysis revealed that about half of their genes had come from rainbow trout parents. In other words, introgression had proceeded to the point where appearance phenotype was no longer an indicator of the trout's genotype. They may have looked like Alvord cutthroats, but they were really hybrids.

But even if the transferred trout were introgressed, preserving at least a part of the Alvord cutthroat evolutionary line would be a worthwhile endeavor. After all, in Dr. Behnke's words, "any trout that can produce twenty-inch specimens in a harsh desert environment must have something going for it." It was expected that the transferred trout would spawn in Jackson Creek in 1987. But the transfer failed. None of the transferred trout were ever seen again, nor were any juveniles ever observed. It was believed that the Alvord cutthroat had slipped at last into extinction. [21]

For this book too, in order to meet its layout and publishing schedule, that's where the story must end. But one tiny glimmer of hope may remain, in the form of a long-ago transfer into another drainage basin. In the fall of 2005, in his "About Trout" column in *Trout*, the magazine published by Trout Unlimited, Dr.

Behnke pointed to a statement in an old paper by Carl Hubbs and Robert Rush Miller to the effect that:

> "Local testimony definitely indicates that the trout of the tributaries of the Catlow Valley have been introduced. Native cutthroat were first brought in from Trout Creek of the Alvord system...." [22]

Catlow Valley lies in southeastern Oregon between Steens Mountain on the east and Hart Mountain on the west. Five streams drain into Catlow Valley. Four of these, Home, Skull, Threemile, and Rock creeks, are inhabited by redband rainbow trout, which Dr. Behnke is convinced are natives, thus casting doubt on at least part of the Hubbs and Miller testimony. The fifth, Guano Creek, a stream that flows into Catlow Valley from Hart Mountain, may have had no trout in the early days, and for this reason was Dr. Behnke's choice as the most likely stream for the Alvord Basin trout to have been released—provided, of course, that the rest of the testimony reported by Hubbs and Miller was accurate.

However, Guano Creek has also been stocked in years past—with Willow/Whitehorse cutthroat trout from the Coyote Basin in 1957, and in subsequent years with Lahontan cutthroats probably from the same hatchery-origin strain that inhabits Mann Lake (I told you about Mann Lake and its trout back in Chapter 5), and also with rainbow trout. On a visit to the headwaters of Guano Creek in the summer of 2006, Dr. Behnke found trout that, despite the stocking with

21. This installment of the Alvord cutthroat's story is told in detail in the second edition of Smith (1984, [1994]), in Tol and French (1988), and in Behnke (1986, 1992, 2002). Bartley and Gall (1991) have the results of the genetic analysis that showed these trout to be introgressed.

22. See Hubbs and Miller (1948) to review this testimony in context, and Behnke (2006) for his article in "About Trout."

rainbow trout, were visually good examples of Lahontan cutthroat, including some that had the appearance of the Alvord cutthroat illustrated in his book, *Trout and Salmon of North America*. He also obtained some local testimony as to how Alvord trout *might* have been transferred into the Guano Creek headwaters. Remember my story back in Chapter 5 about General George Crook, who always carried his hunting and fishing gear with him even on campaign, and was fishing the streams of the Tongue River headwaters when Custer met his fate on the Little Bighorn? In 1866 and 1867, that same George Crook, then a Lt. Colonel, was skirmishing against the Modoc and Paiute Indians out of Camp Warner, an army post located on the headwaters of Guano Creek. A military road connected the post with the Alvord Basin and Trout Creek. Could it be that Crook had trout transported from the Alvord Basin to Guano Creek in order to have fishing near the post? That's one of Dr. Behnke's speculations that would fit in with the Hubbs and Miller testimony. If so, then at least some part of the Alvord cutthroat may persist in the headwaters of Guano Creek. [23]

23. For more on Dr. Behnke's trip to the Catlow Basin and Guano Creek, see Behnke (2007).

Epilogue

"That the trout has been able to survive at the hands of man is a tribute to the toughness of its breed and to the efforts of those who have worked to preserve it. But it is late in The Year of the Trout and the future is far from assured."

—STEVE RAYMOND,
The Year of the Trout, 1985

In 1989, a couple of years after the first edition of this book and four years after Steve Raymond's *The Year of the Trout* appeared, I was invited to speak at the Wild Trout IV Symposium sponsored by the U.S. Fish and Wildlife Service, Trout Unlimited, and several other government and non-government organizations, at Yellowstone National Park. My paper would be scheduled in a session devoted to "Fish Fixes," the organizers explained, and they hoped I would tell the audience all about the successes that had been achieved in restoring and rebuilding wild cutthroat populations.

The trouble was, only two or three such "fixes" came readily to mind in 1989. There was the State of Idaho's experiment with special angling regulations in the early 1970s that had so impressively brought back the wild westslope cutthroat fisheries in Kelly Creek and the upper St. Joe River. But that "fix" had already been so thoroughly dissected and discussed in professional meetings, journal papers, and articles in angling magazines and newsletters that, frankly, I wasn't sure just what I could add. There was also the revival of the Lahontan cutthroat fishery in Pyramid Lake that began in the late 1970s. But again, that one had been heavily publicized already and it probably didn't count anyway, having been established and still largely maintained by hatchery stocking. And then there were the successes being claimed at the time in greenback

cutthroat country in establishing broodstocks and new populations in the wild—in Hidden Valley Creek, for example (open to anglers under special regulations), and in Bear Lake (closed to angling), both in Rocky Mountain National Park. Beyond these, there were only the promised "fixes" that had been written into the handful of federal and state recovery plans available in 1989. So those were the things I talked about. It wasn't much of a presentation, but then again, perhaps it did serve to highlight just how little progress had actually been made up to that time in reversing the decline of this once so abundant native.

The truth is, our westward expansion and the burgeoning growth and development that followed had not been kind to our native cutthroat trout. As I've explained elsewhere in these pages, when people of largely European extraction first ventured into the western regions of the continent, the cutthroat was the most broadly distributed, and probably also the most abundant, of North America's native trouts. It was the only trout to be found in what are now the States of Colorado, Utah, and Wyoming, and was the predominate species in Nevada, Idaho, Montana, and New Mexico. It was also ubiquitous, even if not the predominate species, in Oregon, Washington, British Columbia, Alberta, parts of California, and southeastern Alaska. The cutthroat was the first western trout to be recorded by European man in the New World, and was the first trout encountered by Lewis and Clark in their westward explorations. It provided food for the fur trappers, explorers and surveyors, and immigrant parties on occasion, and was exploited by the miners, the loggers, the railroad builders, and the early settlers both for food and for sport.

But all that exploitation, coupled with the wanton destruction of habitat, the extensive diversion of waters for human use, and the widespread introduction of non-native species of trout and charr—all of these by-products of unbridled growth and development took a heavy toll. Both the abundance and the distribution of native cutthroat trout, especially of the interior subspecies, shrank precipitously. Two cutthroat subspecies, the yellowfin and the Alvord Basin form, winked out completely during this period, and at least two others, the Bonneville and the greenback, were thought by many to have gone extinct as well until small, remnant populations that matched the old museum specimens were discovered in remote locations. In 1973, when the U.S. Endangered Species Act became law, only three cutthroat subspecies, the Lahontan, the Paiute, and the greenback, were actually placed on the list (initially as endangered but later changed to threatened), but others were carried as candidates for listing and six subspecies did receive listings as threatened or as spe-

cies of special concern because of low numbers under state statutes. By 1989, only the coastal, Yellowstone, and finespotted Snake River forms were still believed to be secure within their respective historical ranges, and subsequently, we learned that not even the coastal and Yellowstone subspecies were in as good a shape as advertised. Most of the cutthroat subspecies occupied only tiny fractions of their original historical ranges, and in terms of abundance, most populations still hung by the slenderest of threads. With this as a backdrop, it wasn't easy to come up with a positive spin on cutthroat "fixes" in 1989.

There is much more I could talk about now, of course, because progress toward recovery and restoration has indeed been made in the almost twenty years since Wild Trout IV. For one thing, those searches of remote backcountry areas had been continued, resulting in the discovery of additional small, genetically pure populations to add to the list of those we knew about in 1989. Some of the strongest of these have been tapped for "nearest neighbor" transplants into other streams that had lost their own cutthroat populations, thereby restoring populations to those streams. New broodstocks have also been established to produce trout for additional restoration projects and to provide recreational fishing opportunities in waters reclaimed for native cutthroat trout. These strategies appear to be

working especially well in Bonneville cutthroat country, where populations are being reintroduced in most
every part of the subspecies' historical range, and they
are also achieving results in many segments of the historical range of the Colorado River cutthroat, especially
in Utah and Wyoming. One project that is of special
interest to me is the Colorado River cutthroat restoration project in LaBarge Creek, Wyoming, where I had
the close encounter with the momma moose and calf
that I told you about in Chapter 12. The reintroduction
of Colorado River cutthroat trout to the upper 27 miles
of mainstem and 31 miles of upper tributaries in this
drainage, slated to take place in 2007 if the project is
still on schedule, will accomplish the largest and most
extensive metapopulation restoration yet to be undertaken, certainly in Wyoming, and maybe anywhere
in the cutthroat's broad range. Another momentous
project, if it succeeds, will be the restoration of upriver
spawning runs of Lahontan cutthroat trout past Derby
Dam in the Truckee River system of Nevada.

Restoration and recovery efforts got an additional
strong boost in 1995 when the Aquatic Conservation
Strategy for Pacific Northwest Fish (acronym PACFISH)
and the Inland Native Fish Strategy (acronym INFISH)
were published by the federal government. The full
force and effect of these strategies depends on who's
in power in Washington, D.C., of course, but even

so, they strengthened the focus of the federal land
management agencies on restoring, improving, and
safeguarding aquatic habitats for native fishes including cutthroat trout. And frankly, the raft of petitions
from concerned conservation organizations and
individuals to list cutthroat subspecies under the U.S.
Endangered Species Act didn't hurt either, even though
most have been rejected by the U.S. Fish and Wildlife
Service and they did bring on a backlash of efforts to
gut the Act that continues to this day in our nation's
capitol. The threat of listing, reinforced by these
petitions, has acted as a hammer over state, local, and
federal management and planning agencies, prompting them to get serious about preservation, restoration,
and recovery efforts, if only to keep the populations
under their respective purviews off the endangered
species list. Hey, whatever works!

So, the list of successes has indeed been growing.
But too often still, despite all the progress, native cutthroat populations are limited to small, isolated, and
fragmented headwater reaches. Sadly, too, for every
success that's recorded, there seems inevitably to be
a frustrating and discouraging setback. Perhaps the
most jarring of these was the discovery of illegally
introduced, predaceous lake trout in Yellowstone Lake
where they were found to be feeding heavily on Yellowstone cutthroat trout and swiftly reducing cutthroat

numbers in the lake. They could easily and quickly eradicate the lake's magnificent cutthroat populations if not held in check by the intensive but expensive gill-netting program that authorities have mounted. On top of that, whirling disease has recently struck the system and may have already adversely affected at least one of the lake's principal spawning tributaries.

Another disappointing setback occurred in New Mexico's Rio Cebolla, home of one of just 13 core populations of Rio Grande cutthroat trout that the Fish and Wildlife Service says are genetically pure, stable, and secure from invasion by non-native trouts. It was the integrity of these 13 core populations that the Service fell back on in rejecting a petition to list the Rio Grande cutthroat as threatened under the Endangered Species Act. But alas, in 2005, brown trout were discovered in the Rio Cebolla enclave, possibly the result of another illegal introduction, thus compromising the security of this core population. Authorities have been using electroshocking to remove as many of the brown trout as possible, but for now, there is one less core population within the Rio Grande cutthroat range.

Perhaps more disappointing is the stubborn opposition to the use of fish toxicants, even under the most carefully controlled conditions, that may have torpedoed a plan to remove introduced trout and reintroduce the threatened Paiute cutthroat trout into the one stream reach that comprised its original historical range. Since that reintroduction step is crucial for downlisting the Paiute cutthroat from its status as threatened under the U.S. Endangered Species Act, it may not now be possible to fulfill the recovery plan for this subspecies.

Two setbacks to the greenback cutthroat recovery plan surfaced in 2001. First of all, genetic analysis revealed that several populations previously thought to be genetically pure, including four populations that had been used to found broodstocks and other reintroduced populations, carry mitochondrial DNA markers from Yellowstone cutthroat trout or from rainbow trout. Thus, they are not free of introgression as had been believed, and ought not to be counted toward the recovery goal. But of even greater concern, scientists also published some pretty serious criticisms of the recovery criteria themselves that year. None of the targets specified in the plan for the size of recovery populations, population biomass, year class success, or habitat size are consistent with well-established fundamentals for population persistence, nor are those targets set high enough to secure populations from demographic instability or loss of genetic variation. Furthermore, there is no provision at all in the plan for reestablishing metapopulation connectivity. A new status review should be out for the greenback subspecies by the time you read this. It will be interesting to see if and how these issues are addressed.

These are all adversities that could be overcome, given time, but time itself may once again be running against the cutthroat trout and those working to preserve it. New threats loom over us that could quickly wipe out all the hard-earned gains and sweep away even the core populations that we have come to rely upon for recovery. Two of these threats are not really new to the interior west, namely 1) our seemingly ever-increasing need for energy, and 2) the continuing pressure and competition for water supplies for human use. Both of these are tied to our population growth, which continues ever upward, and also to the growth of our economy, meaning that the threats they pose to native trout populations and their habitats are only going to intensify. Status reviews, conservation agreements, and recovery plans for our cutthroat subspecies must proactively and explicitly address these increasing demands and the additional threats they pose to native cutthroat populations and habitats, and build in action plans to avoid or offset the adverse impacts.

But even more insidious are the impacts expected from global warming. We're already behind the curve on this one. Cold-adapted species in mountainous regions have already experienced dramatic decreases in their range, and species that have the mobility are already shifting their distribution northward. Springtime conditions are coming earlier in the northern latitudes, including our own Rocky Mountain region. Winter rains will intensify, replacing snow across the northern latitudes, while even deeper drought conditions will prevail in the Pacific southwest. These effects will completely alter stream hydrology and instream habitat conditions, meaning fewer and fewer stream and lake habitats that are suitable for coldwater fishes within the historical range of the cutthroat trout.

According to the latest reports from the authoritative Intergovernmental Panel on Climate Change, it's already too late for us to avoid some of these global warming impacts. And the longer we wait to arrest discharges of the greenhouse gases that bring about global warming, the more serious these impacts will become. But none of this is addressed in any of the status reviews, conservation agreements, or recovery plans that I am familiar with, and I have read them all. Like the other threats, the implications of recent global warming scenarios must be addressed, proactively and explicitly, and action plans built in to offset as many of the impacts as possible. If we can't catch up with this, if we can't at least draw even with the curve on global warming impacts, then the Year of the Trout will surely come to a close for our native cutthroat trout.

Getting back to that Wild Trout IV Symposium, the

highlight for me from that conference was the banquet speech given by the late Ernest Schwiebert, the famous angler and author. His subject, too was the native cutthroat trout, and one of his remarks has stayed with me ever since. It captured the appreciation, the admiration, and the sentiment that compelled me to assemble this book. This was his closing observation:

> "And when we still catch a cutthroat in these Shining Mountains, we too often catch a wild fish that has survived in spite of our sorry husbandry, and it is not merely a fish we are holding in our hands—it is both a poem and a living piece of history."

But I fear this sentiment has faded from the memories of too many who should care. Recent issues of two of my favorite fly-fishing magazines carried articles on the brown trout as the trout spirit most kindred to ourselves. Of all this country's fish, none is as American as the brown trout, one article asserted. America is a land of immigrants, the other article proclaimed, and so is the brown trout, an immigrant that has, like ourselves, succeeded in establishing itself in so many different settings all across the country—but, I have to add, again like ourselves, by often displacing the indigenous occupants of the new homelands they settled in. So even in the best-case scenario, even if we do succeed in retaining habitats in the American west that are suitable for trout, I don't expect our wild native cutthroats will ever again approach their historical levels of abundance and distribution. Too many populations of introduced trout have been established, with constituencies that would shield them from any attempts to remove them in favor of native cutthroats.

Don't get me wrong here, I am not against brown trout, or introduced brookies, or rainbow trout either. But I am *for* our native cutthroat trout. I do believe we should reserve a generous share of our western landscape for this once-so-abundant native, and take all necessary steps to ensure it remains a wild, living part of our western heritage.

Bibliography

ADAMS, C.G., R.H. BENSON, R.B. KIDD, W.B.F. RYAN AND R.C. WRIGHT. 1977. The Messinian salinity crisis and evidence of late Miocene eustatic changes in the world oceans. *Nature* 269: 383–386.

ADAMS, K.D., S.G. WESNOUSKY AND B.G. BILLS. 1999. Isostatic rebound, active faulting, and potential geomorphic effects in the Lake Lahontan basin, Nevada and California. *Geological Society of America Bulletin* 111: 1739–1756.

ADRIANO, D. 1956. *Fisheries investigations of Henry's Lake.* Boise: Idaho Department of Fish and Game, Annual Progress Report, Federal Aid in Fish and Wildlife Restoration Project F–13–R–2.

AHO, R.S. 1976. A population study of the cutthroat trout in an unshaded section of stream. Master's thesis, Oregon State University, Corvallis.

AIKENS, C.M. 1984. *Archaeology of Oregon.* Portland, Oregon: U.S. Department of the Interior, Bureau of Land Management.

AKCAKAYA, H.R. 1994. *RAMAS/GIS: linking landscape data with population viability analysis (version 1.0).* Setauket, New York: Applied Biomathematics.

ALDEN, W.C. 1932. *Physiography and glacial geology of eastern Montana and adjacent areas.* Washington, D.C.: U.S. Geological Survey Professional Paper 174.

ALEXANDER, C.B. AND R.L. SCARPELLA. 1999. Recovering Summit Lake's Lahontan cutthroat trout. *Endangered Species Bulletin* 24, no. 4: 14–15.

ALLAN, J.D. 1995. *Stream ecology: structure and function of running waters.* New York: Chapman & Hall.

———. 2004. Landscapes and riverscapes: the influence of land use on stream ecosystems. *Annual Reviews of Ecology, Evolution, and Systematics* 35: 257–284.

ALLEN, J.E. 1979. *The magnificent gateway.* Forest Grove, Oregon: Timber Press.

ALLEN, J.E. AND M. BURNS. 1986 [reprinted 2000]. *Cataclysms on the Columbia.* Portland, Oregon: Timber Press.

ALLENDORF, F., N. RYMAN, A. STENNEK AND G. STAHL. 1976. Genetic variation in Scandinavian brown trout (*Salmo trutta* L.): evidence of distinct sympatric populations. *Hereditas* 83: 73–82.

ALLENDORF, F.W. AND F.M. UTTER. 1979. Population genetics. Pages 407–454 in W.S. Hoar, D.S. Ranaal and J.R. Brett, editors. *Fish physiology, Vol. 8.* New York: Academic Press.

ALLENDORF, F.W. AND G.H. THORGAARD. 1984. Tetraploidy and the evolution of salmonid fishes. Pages 1–53 in B.J. Turner, editor. *The evolutionary genetics of fishes.* New York: Plenum Press.

ALLENDORF, F.W. AND M.M. FERGUSON. 1990. Genetics. Pages 35–63 in C.B. Schreck and P.B. Moyle, editors. *Methods for fish biology.* Bethesda, Maryland: American Fisheries Society.

ALLENDORF, F.W. AND R.F. LEARY. 1988. Conservation and distribution of genetic variation in a polytypic species, the cutthroat trout. *Conservation Biology* 2: 170–184.

ALLENDORF, F.W., R.F. LEARY, P. SPRUELL AND J.K. WENBURG. 2001. The problems with hybrids: setting conservation guidelines. *Trends in Ecology and Evolution* 16: 613–622.

ALT, D. 2001. *Glacial Lake Missoula and its humongous floods.* Missoula, Montana: Mountain Press Publishing Company.

ALVAREZ, L.W., W. ALVAREZ, F. ASARO AND H.V. MICHEL. 1980. Extraterrestrial cause for the Cretaceous-Tertiary extinction. *Science* 208: 1095–1108.

ALVAREZ, W. 1997. *T rex and the crater of doom.* Princeton, New Jersey: Princeton University Press.

ALVES, J. 1992. *Rio Grande cutthroat trout management plan.* Denver: Colorado Division of Wildlife.

———. 1998. *Status of Rio Grande cutthroat trout in Colorado.* Denver: Colorado Division of Wildlife.

AMERICAN WILDLANDS AND 6 co-petitioners. 1997 [amended 1998]. Petition for a rule to list the westslope cutthroat trout (*Oncorhynchus clarki lewisi*) as threatened throughout its range. Bozeman, Montana, American Wildlands and 6 co-petitioners.

ANDRUSAK, H. AND T.G. NORTHCOTE. 1971. Segregation between adult cutthroat trout (*Salmo clarki*) and Dolly Varden (*Salvelinus malma*) in small coastal British Columbia lakes. *Journal of the Fisheries Research Board of Canada 28*: 1259–1268.

ANNUAL REPORTS. 1868–1883. *U.S. Geological and Geographic Survey of the Territories (Hayden Surveys).* Washington, D.C.: U.S. Government Printing Office.

ANONYMOUS. 1904. *An Illustrated History of the Big Bend Country embracing Lincoln, Douglas, Adams, and Franklin Counties, State of Washington.* Spokane: Western Historical Publishing Company.

ANTEVS, E. 1948. The Great Basin with emphasis on glacial and postglacial times III, the Great Basin. *Bulletin of the University of Utah 38*, no. 20: 167–191.

———. 1954. Climate of New Mexico during the last glacio-pluvial. *Journal of Geology 62*: 182–191.

"Apache." 1877. Rio Grande trout. *Forest and Stream 9*, no. 4: 67.

ARBOGAST, B.S., S.V. EDWARDS, G. WAKELEY, P. BEERLI AND J.B. SLOWINSKI. 2002. Estimating divergence times from molecular data on phylogenetic and population genetic time scales. *Annual Review of Ecology and Systematics 33*: 707–740.

ARMANTROUT, N.B. AND M. CROUSE. 1981. *Whitehorse cutthroat trout.* U.S. Department of the Interior Bureau of Land Management.

ARMSTRONG, R.H. 1971. Age, food, and migration of sea-run cutthroat trout, *Salmo clarki*, at Eva Lake, southeastern Alaska. *Transactions of the American Fisheries Society 100*: 302–306.

ARNOLD, M.L. 1997. *Hybridization and evolution.* Oxford, U.K.: Oxford University Press.

ASHLEY, R.R. 1970. *Habitat management plan—Paiute trout.* Reno, Nevada: USDA Forest Service, Toiyabe National Forest.

ATWATER, B.F. 1986. Pleistocene glacial-lake deposits of the Sanpoil River valley, northeastern Washington. *U.S. Geological Survey Bulletin 1661.*

———. 1987. Status of Glacial Lake Columbia during the last floods from Glacial Lake Missoula. *Quaternary Research 27*: 182–201.

ATWOOD, W.W. AND K.F. MATHUR. 1932. *Physiography and Quaternary geology of the San Juan Mountains, Colorado.* Washington, D.C.: U.S. Geological Survey Professional Paper 166.

AUDUBON, J.W. 1906 [1984]. *Audubon's western journal 1840–1850.* Cleveland, Ohio, A.H. Clark, Company [reprinted 1984 by University of Arizona Press, Tucson].

AVERETT, R.C. AND C. MACPHEE. 1971. Distribution and growth of indigenous fluvial and adfluvial cutthroat trout, *Salmo clarki*, St. Joe River, Idaho. *Northwest Science 45*: 38–47.

AVERY, E.L. 1983. *A bibliography of beaver, trout, wildlife, and forest relationships with special reference to beaver and trout.* Madison, Wisconsin: Wisconsin Department of Natural Resources, Technical Bulletin 137.

AYALA, F.J. 1982. *Population and evolutionary genetics: a primer.* Menlo Park, California: The Benjamin Cummings Publishing Co.

BABCOCK, W.H. 1971. Effect of a size limit regulation on the trout fishery in Trappers Lake, Colorado. *Transactions of the American Fisheries Society 100*: 50–54.

BACHHUBER, F.W. 1989. The occurrence and paleolimnological significance of cutthroat trout (*Oncorhynchus clarki*) in pluvial lakes of the Estancia Valley, central New Mexico. *Geological Society of America Bulletin 101*: 1543–1551.

BACHMAN, R.A. 1984. Foraging behavior of free-ranging wild and hatchery brown trout in a stream. *Transactions of the American Fisheries Society 113*: 1-32.

BAILEY, R. 1978. Restoration of a wild Lahontan cutthroat trout fishery in the Truckee River, California–Nevada. Pages 53–55 in J.R. Moring, editor, *Proceedings of the Wild Trout–Catchable Trout Symposium.* Portland: Oregon Department of Fish and Wildlife.

BAILEY, R.G. 1935 [1947]. *River of No Return: a century of central Idaho and eastern Washington history and development.* Lewiston, Idaho: R.G. Bailey Printing Company [revised edition published 1947].

———. 1995. *Description of the ecoregions of the United States.* 2nd edition. Washington, D.C.: USDA Forest Service, Miscellaneous Publications 1391.

BAILLIE-GROHMAN, W.A. 1884. *Camps in the Rockies.* New York: Charles Scribner's Sons.

BAKER, B.M. 1998. *Genetic analysis of two naturally spawning populations of Crescent Lake cutthroat trout, Barnes Creek and Lyre River.* Olympia: Washington Department of Fish and Wildlife, Fish Program.

———. 2000. *Genetic analysis of Beardslee rainbow trout from Lake Crescent.* Olympia: Washington Department of Fish and Wildlife, Fish Program.

BAKER, G.R. 1965. Excavating Fort Massachusetts. *The Colorado Magazine 42*, no. 1: 1–15.

BAKER, J. 2001. The development of species-specific markers and their application to a study of temporal variation in hybridization within a coastal cutthroat trout (*Oncorhynchus clarki clarki*) population. Master's thesis, University of Washington, Seattle.

BAKER, T.T. AND 8 co-authors. 1996. Status of Pacific salmon and steelhead escapements in southeastern Alaska. *Fisheries* 21, no. 10: 6–18.

BAKUN, A. 1973. *Coastal upwelling indices, west coast of North America 1946–1971*. U.S. Department of Commerce, NOAA Technical Report NMFS-SSRF-671.

BAKUN, A. 1975. *Daily and weekly upwelling indices, west coast of North America 1967–1973*. U.S. Department of Commerce, NOAA Technical Report NMFS-SSRF-693.

———. 1990. Global climate change and intensification of coastal upwelling. *Science* 247: 198–201.

BALDWIN, C.M. 1998. Strawberry Reservoir food web interactions: a bioenergetic assessment of salmonid diet and predator-prey supply and demand. Master's thesis, Utah State University, Logan.

BALDWIN, C.M., D.A. BEAUCHAMP AND C.P. GUBALA. 2002. Seasonal and diel distribution and movement of cutthroat trout from ultrasonic telemetry. *Transactions of the American Fisheries Society* 131: 143–158.

BALDWIN, C.M., D.A. BEAUCHAMP AND J.J. VAN TASSELL. 2000. Bioenergetic assessment of temporal food supply and consumption by salmonids in the Strawberry Reservoir food web. *Transactions of the American Fisheries Society* 129: 429–450.

BALL, O.P. 1955. Some aspects of homing in cutthroat trout. *Proceedings of the Utah Academy of Sciences, arts, and Letters* 32: 75–80.

BALL, O.P. AND O.B. COPE. 1961. *Mortality studies on cutthroat trout in Yellowstone Lake*. Washington, D.C.: U.S. Fish and Wildlife Service, Bureau of Sport Fisheries and Wildlife, Research Report 55.

BALTZ, D.M. AND P.B. MOYLE. 1993. Invasion resistance to introduced species by a native assemblage of California stream fishes. *Ecological Applications* 3: 246–255.

BAMS, R.A. 1969. Adaptations in sockeye salmon associated with incubation in stream gravels. Pages 71–87 in T.G. Northcote, editor. *Salmon and trout in streams*. Vancouver, British Columbia: H.R. MacMillan Lectures in Fisheries, University of British Columbia.

BARKER, E.S. 1953. *Beatty's cabin*. Albuquerque: University of New Mexico Press.

BARNES, L.G., S.A. MCLEOD AND R.E. RASCHKE. 1985. A Late Miocene marine vertebrate assemblage from southern California. *National Geographic Research* 21: 13–20.

BARNOSKY, C.W., P.M. ANDERSON AND P.J. BARTLEIN. 1987. The northwestern U.S. during deglaciation: vegetational history and paleoclimatic implications. Pages 289–321 in W.F. Rudman and H.E. Wright, Jr., editors. *North American and adjacent oceans during the last deglaciation*. Volume K-3, The geology of North America. Boulder, Colorado: Geological Society of America.

BARON, J. 1983. Comparative water chemistry of four lakes in Rocky Mountain National Park. *Water Resources Bulletin* 19: 897–902.

BARRIE, J., K. CONWAY, R. MATHEWES, H. JOSENHANS AND M. JOHNS. 1993. Submerged Late Quaternary terrestrial deposits and paleoenvironment of northern Hecate Strait, British Columbia continental shelf, Canada. *Quaternary International* 20: 123–129.

BARTHOLOMEW, J.L. AND J.C. WILSON, EDITORS. 2002. *Whirling disease: reviews and current topics*. Bethesda, Maryland: American Fisheries Society Symposium 29.

BARTLEY, D.M. AND G.A.E. GALL. 1989. *Biochemical genetic analysis of native trout populations in Nevada, report on populations collected from 1976–1988*. University of California, Davis, Fisheries Biology Research Facility, Department of Animal Science, report on contract 86-98 to Nevada Division of Wildlife, Reno.

BARTLEY, D.M. AND G.A.E. GALL. 1991. Genetic identification of native cutthroat trout (*Oncorhynchus clarki*) and introgressive hybridization with introduced rainbow trout (*O. mykiss*) in streams associated with the Alvord Basin, Oregon and Nevada. *Copeia* 1991, no. 3: 854–859.

———. 1993. *Genetic analysis of threatened Nevada trout: report on populations collected from 1988–1992*. University of California, Davis, Fisheries Biology Research Facility, Department of Animal Science, report on contract 86-98 to Nevada Division of Wildlife, Reno.

BARTLEY, D.M., G.A.E. GALL AND A. MARSHALL-ROSS. 1987. *Biochemical genetic analysis of Nevada trout populations, October 1987*. University of California, Davis, Fisheries Biology Research Facility, Department of Animal Science, report on contract 86-98 to Nevada Division of Wildlife, Reno.

BAX, N.J., E.O. SALO, B.P. SNYDER, C.A. SIMENSTAD AND W.J. KINNEY. 1980. Salmon outmigration studies in Hood Canal: a summary—1977. Pages 171–201 in J. McNeil and D.C. Himsworth, editors. *Salmonid ecosystems of the North Pacific*. Corvallis: Oregon State University Press.

BAXTER, G.T. AND M.D. STONE. 1995. *Fishes of Wyoming*. Cheyenne: Wyoming Game and Fish Department.

BEACHAM, T.D. AND C.B. MURRAY. 1990. Temperature, egg size, and development of embryos and alevins of five species of Pacific salmon: a comparative analysis. *Transactions of the American Fisheries Society* 119: 927–945.

BEAL, F.R. 1959. For the cutthroat, a brighter outlook. *Wyoming Wildlife* 23, no. 8: 7–9.

BEAMISH, R.J., EDITOR. 1995. *Climate change and northern fish populations*. Ottawa: Canadian Special Publication of Fisheries and Aquatic Sciences 121.

BEAMISH, R.J. AND 5 co-authors. 1999. The regime concept and natural trends in the production of Pacific salmon. *Canadian Journal of Fisheries and Aquatic Sciences* 56: 516–526.

BEAUCHAMP, D.A., C.M. BALDWIN, J.L. VOGEL AND C.P. GUBALA. 1999. Estimating diel, depth-specific foraging opportunities with a visual encounter rate model for pelagic piscivores. *Canadian Journal of Fisheries and Aquatic Sciences* 56 (Supplement 1): 128–139.

BEAUCHAMP, D.A., S.A. VECHT AND G.L. THOMAS. 1992. Temporal, spatial, and size-related foraging of wild cutthroat trout in Lake Washington. *Northwest Science* 66: 149–159.

BECK, C. AND G.T. JONES. 1997. The terminal Pleistocene/early Holocene archaeology of the Great Basin. *Journal of World Prehistory* 11: 161–236.

BECKWITH, Lt. E.G. 1855. Report upon the route near the thirty-eighth and thirty-ninth parallels explored by Capt. J.W. Gunnison, Corps of Topographical Engineers, Part 1, and Upon the route near the forty-first parallel, Part 2. In *Reports of explorations and surveys to ascertain the most practical and economical route for a railroad from the Mississippi River to the Pacific Ocean, 1853–4*, Volume 2. Washington, D.C.: U.S. Secretary of War. A.D.P. Nicholson, Printer.

BEHNKE, R.J. 1960. Taxonomy of the cutthroat trout of the Great Basin with notes on the rainbow series. Master's thesis, University of California, Berkeley.

———. 1967. *Rare and endangered species report: the Rio Grande cutthroat trout (Salmo clarkii virginalis)*. Fort Collins: Colorado Cooperative Fishery Unit, Colorado State University.

———. 1970. *Rare and endangered species report: the Bonneville cutthroat trout, Salmo clarki utah*. Fort Collins: Colorado Cooperative Fishery Unit, Colorado State University.

———. 1972. The salmonid fishes of recently glaciated lakes. *Journal of the Fisheries Research Board of Canada* 29: 639–671.

———. 1973. *Rare and endangered species report: west-slope cutthroat trout*. Fort Collins: Colorado Cooperative Fishery Unit, Colorado State University.

———. 1979 [1981]. *Monograph of the native trouts of the genus Salmo of western North America*. Fort Collins, Colorado: U.S. Fish and Wildlife Service [revised and reprinted 1981].

———. 1980. *Endangered and threatened fishes of the upper Colorado River basin*. Fort Collins: Colorado State University, Cooperative Extension Service, Bulletin 503A.

———. 1981a. Systematic and zoogeographical interpretation of Great Basin trouts. Pages 95–124 in R.J. Naiman and D.L. Soltz, editors. *Fishes in North American deserts*. New York: John Wiley & Sons.

———. 1981b. *The cutthroat trout of Indian Creek, Mescalero Apache Indian Reservation, Tularosa Basin, Otero County, New Mexico*. Fort Collins: Colorado State University, Department of Fishery and Wildlife Biology, report prepared for U.S. Fish and Wildlife Service, Gallup, New Mexico.

———. 1986. About Trout. Alvord cutthroat. Trout (Trout Unlimited's Journal of Coldwater Fisheries Conservation) 27, no. 2: 50–54.

———. 1988a. Phylogeny and classification of cutthroat trout. Pages 1–7 in R.E. Gresswell, editor. *Status and management of interior stocks of cutthroat trout*. Bethesda, Maryland: American Fisheries Society Symposium 4.

———. 1988b. About Trout. Rio Grande cutthroat. Trout (Trout Unlimited's Journal of Coldwater Fisheries Conservation) 29, no. 3: 43–47.

———. 1990. About Trout. The family tree: origins of trout and salmon. Trout (Trout Unlimited's Journal of Coldwater Fisheries Conservation) 32, no. 2: 37–40.

———. 1992. *Native trout of western North America*. Bethesda, Maryland: American Fisheries Society Monograph 6.

———. 1993. About Trout. Lahontan cutthroat trout: a megafish for megatrends. *Trout* (Trout Unlimited's Journal of Coldwater Fisheries Conservation) 35, no. 2: 69–74.

———. 1997. About Trout. Movement, migration and habitat. *Trout* (Trout Unlimited's Journal of Coldwater Fisheries Conservation) 39, no. 1: 45–46, 53.

———. 2002. *Trout and salmon of North America*. New York: The Free Press.

———. 2006. About Trout: Ivory-billed trout? Trout (Trout Unlimited's Journal of Coldwater Fisheries Conservation) 47, no. 4: 56–58.

———. 2007. About Trout: Toward definitiveness. Trout (Trout Unlimited's Journal of Coldwater Fisheries Conservation) 48, no. 1: 54–55.

BEHNKE, R.J. AND D.E. BENSON. 1980. *Endangered and threatened fishes of the upper Colorado River basin*. Fort Collins, Colorado: Colorado State University Cooperative Extension Service, Bulletin 503A.

BEHNKE, R.J. AND M. ZARN. 1976. *Biology and management of threatened and endangered western trouts.* Fort Collins, Colorado: USDA Forest Service, General Technical Report RM-28.

BEHNKE, R.J. AND W.J. PLATTS 1990. The occurrence and paleolimnological significance of cutthroat trout (*Oncorhynchus clarki*) in pluvial lakes of the Estancia Valley, central New Mexico: discussion and reply. *Geological Society of America Bulletin* 102: 1731–1732.

BEHNKE, R.J. AND W.J. WILTZIUS. 1982. The enigma of the yellowfin trout. *Proceedings of the Annual Meeting, Colorado-Wyoming chapter American Fisheries Society* 17: 8–15.

BEISINGER, K.E. 1961. Studies on the relationship of the redside shiner (*Richardsonius balteatus*) and the longnose sucker (*Catostomus catostomus*) to the cutthroat trout (*Salmo clarki*) population in Yellowstone Lake. Master's thesis, Utah State University, Logan.

BEISSINGER, S.R. AND D.R. MCCOLLOUGH. 2002. *Population viability analysis.* Chicago, Illinois: University of Chicago Press.

BELSKY, A.J., A. METZKE AND S. USELMAN. 1999. Survey of livestock influences on streams and riparian ecosystems in the western United States. *Journal of Soil and Water Conservation* 54: 419–431.

BENDIRE, C.E. 1882. Notes on Salmonidae of the upper Columbia. *Proceedings of the U.S. National Museum* 4: 81–87.

BENEDICT, J.B. 1992. Along the Great Divide: paleoindian archaeology of the high Colorado Front Range. Pages 343–359 in D.J. Stanford and J.S. Day, editors. *Ice Age hunters of the Rockies.* Niwot, Colorado: Denver Museum of Natural History and University Press of Colorado.

BENSON, L.V. AND M.D. MIFFLIN. 1986. *Reconnaissance bathymetry of basins occupied by Pleistocene Lake Lahontan, Nevada and California.* Washington, D.C.: U.S. Geological Survey Water-Resources Investigations Report 85-4262.

BENSON, L.V., P.A. MEYERS AND R.J. SPENCER. 1991. Changes in the size of Walker Lake during the past 5000 years. *Paleogeography, Paleoclimatology, Paleoecology* 81: 189–214.

BENSON, L.V. AND Z. PETERMAN. 1995. Carbonate deposition, Pyramid Lake subbasin, Nevada: 3. The use of 87Sr values in carbonate deposits (tufas) to determine the hydrologic state of paleolake systems. *Paleogeography, Paleoclimatology, Paleoecology* 119: 201–213.

BENSON, N.G. 1960. Factors influencing production of immature cutthroat in Arnica Creek, Yellowstone Park. *Transactions of the American Fisheries Society* 89: 168–175.

———. 1961. *Limnology of Yellowstone Lake in relation to the cutthroat trout.* Washington, D.C.: U.S. Fish and Wildlife Service, Bureau of Sport Fisheries & Wildlife, Research Report 56.

BENSON, N.G. AND R.V. BULKLEY. 1963. *Equilibrium yield and management of cutthroat rout in Yellowstone Lake.* Washington, D.C.: U.S. Fish and Wildlife Service, Bureau of Sport Fisheries and Wildlife, Research Report 62.

BENSON, S.B. AND R.J. BEHNKE. 1960. *Salmo evermanni,* a synonym of *Salmo clarki henshawi. California Fish and Game* 47: 257–259.

BENTON, M.J. 1985. Interpretations of mass extinctions. *Nature* 314: 496–497.

BERGMAN, RAY. 1964. *Trout.* 2nd edition. New York: Alfred A. Knopf.

BERANEK, L.P., P.K. LINK AND C.M. FANNING. 2006. Miocene to Holocene landscape evolution of the western Snake River Plain region, Idaho: using the SHRIMP detrital zircon provenance record to track eastward migration of the Yellowstone hotspot. *Geological Society of America Bulletin* 118: 1027–1050.

BERMINGHAM, E., S.S. MCCAFFERTY AND A.P. MARTIN. 1997. Fish biogeography and molecular clocks. Pages 113–128 in T.D. Kocher and C.A. Stepien, editors. *Molecular systematics of fishes.* New York: Academic Press.

BERNARD, D.R. 1976. Reproduction by fluvial salmonids in Spawn Creek, Cache County, Utah. Master's thesis, Utah State University, Logan.

BERNARD, D.R. AND E.K. ISRAELSEN. 1982. Inter- and intrastream migration of cutthroat trout (*Salmo clarki*) in Spawn Creek, a tributary of the Logan River, Utah. *Northwest Science* 56: 148–158.

BETARBET, R. AND 5 co-authors. 2000. Chronic systemic pesticide exposure reproduces features of Parkinson's disease. *Nature Neuroscience* 3, no. 12: 1301–1306.

BEVERTON, R.H.J. AND S.J. HOLT. 1957 [1993]. *On the dynamics of exploited fish populations.* UK Ministry of Agriculture, Fisheries and Food, Fisheries Investigations (Series 2) 19 [reprinted 1993 by Chapman and Hall, London].

BILBY, R.E. 1984. Characteristics and frequency of cool-water areas in a western Washington stream. *Journal of Freshwater Ecology* 2: 593–602.

BINNS, N.A. 1967. *Effects of rotenone treatment on the fauna of the Green River, Wyoming.* Cheyenne: Wyoming Game and Fish Commission, Fisheries Research Bulletin 1: 1-224.

BINNS, N.A. 1977. *Present status of indigenous populations of cutthroat trout, Salmo clarki, in southwest Wyoming.* Cheyenne: Wyoming Game and Fish Department, Technical Bulletin No. 2.

———. 1981. *Bonneville cutthroat trout Salmo clarki utah in Wyoming.* Cheyenne: Wyoming Game and Fish Department, Fisheries Technical Bulletin 5.

BINNS, N.A. AND R. REMMICK. 1994. Response of Bonneville cutthroat trout and their habitat to drainage-wide habitat management at Huff Creek, Wyoming. *North American Journal of Fisheries Management* 14: 669–680.

BIODIVERSITY LEGAL FOUNDATION. 1998. Petition for a rule to list the Bonneville cutthroat trout (*Oncorhynchus clarki utah*) as threatened under the U.S. Endangered Species Act, 16 U.S.C. § 1531 et seq. (1973 as amended). Boulder, Colorado: Biodiversity Legal Foundation, petitioner.

BIODIVERSITY LEGAL FOUNDATION AND 3 co-petitioners. 1998. Petition for a rule to list the Yellowstone cutthroat trout (*Oncorhynchus clarki bouvieri*) as threatened under the U.S. Endangered Species Act, 16 U.S.C. § 1531 et seq. (1973 as amended). Boulder, Colorado: Biodiversity Legal Foundation, Alliance for the Wild Rockies, Montana Ecosystem Defense Council, and Mr. George Wuerthner, petitioners.

BIRT, T.P., J.M. GREEN AND W.S. DAVIDSON. 1990. Contrasts in development and smolting of genetically distinct sympatric anadromous and nonanadromous Atlantic salmon, *Salmo salar. Canadian Journal of Zoology* 69: 2075–2084.

BISSON, P.A. AND J.R. SEDELL. 1984. "Salmonid populations in streams in clearcut vs. old-growth forests of western Washington. Pages 121–129 in W.R. Meehan, T.R. Merrell and T.A. Hanley, editors. *Fish and wildlife relationships in old-growth forests.* Morehead City, North Carolina: Institute of Fisheries Research Biologists.

BISSON, P.A., K. SULLIVAN AND J.L. NIELSEN. 1988. Channel hydraulics, habitat use, and body form of juvenile coho salmon, steelhead, and cutthroat trout in streams. *Transactions of the American Fisheries Society* 117: 262–273.

BJORNN, T.C. 1957. *A survey of the fishery resources of Priest and Upper Priest lakes and their tributaries, Idaho.* Boise: Idaho Fish and Game Department, Completion Report.

———. 1961. Harvest, age structure and growth of game fish populations from Priest and Upper Priest lakes. *Transactions of the American Fisheries Society* 90: 27–31.

———. 1971. Trout and salmon movement in two Idaho streams as related to temperature, food, stream flow, cover, and population density. *Transactions of the American Fisheries Society* 100: 423–438.

———. 1975. The St. Joe River cutthroat fishery—a case history of angler preference. *Proceedings of the Western Association of State Game and Fish Commissioners* 55: 187–194.

BJORNN, T.C. AND J. MALLET. 1964. Movements of planted and wild trout in an Idaho river system. *Transactions of the American Fisheries Society* 93: 70–76.

BJORNN, T.C. AND D.W. REISER. 1991. Habitat requirements of salmonids in streams. Pages 83–138 in W.R. Meehan, editor. *Influences of forest and rangeland management on salmonid fishes and their habitats.* Bethesda, Maryland: American Fisheries Society, Special Publication 19.

BJORNN, T.C. AND R.F. THUROW. 1978. *Response of cutthroat trout populations to the cessation of fishing in St. Joe River tributaries.* Moscow, Idaho: University of Idaho College of Forestry, Wildlife and Range Sciences Bulletin 25.

BJORNN, T.C. AND T.H. JOHNSON. 1978. Wild trout management, an Idaho experience. Pages 31–39 in K. Hashagen, editor. *Wild trout management symposium proeedings.* San Jose, California: California Trout and American Fish Society.

BJORNN, T.C. AND 6 co-authors. 1977. *Transport of granitic sediments in streams and its effects on insects and fish.* Moscow, Idaho: University of Idaho College of Forestry, Wildlife and Range Sciences Bulletin 17.

BJORNSTAD, B.N., K.R. FECHT AND C.J. PLUHAR. 2001. Long history of pre-Wisconsin ice age cataclysmic floods: evidence from southeastern Washington State. *Journal of Geology* 109: 695–713.

BLACK, W.C. 1988. *Creekcraft: the art of flyfishing smaller streams.* Boulder, Colorado: Pruett Publishing Company.

BLACKETT, R.F. 1973. Fecundity of resident and anadromous Dolly Varden (*Salvelinus malma*) in southeastern Alaska. *Journal of the Fisheries Research Board of Canada* 30: 543–548.

BLACKWELDER, E. 1934. Supplemental notes on Pleistocene glaciation in the Great Basin. *Journal of the Washington Academy of Sciences* 24: 217–222.

———. 1948. The Great Basin with emphasis on glacial & postglacial times. The Great Basin, Part I. *Bulletin of the University of Utah* 38, no. 20: 1–16

BLAKLEY, A., B. LELAND AND J. AMES. 2000. *2000 Washington State salmonid stock inventory: coastal cutthroat trout.* Olympia: Washington Department of Fish and Wildlife.

BLOCK, D.G. 1955. Trout migration and spawning studies on the North Fork drainage of the Flathead River. Master's thesis, Montana State University, Bozeman.

BLUSTEIN, B.E. 1991. *Preserve your love for science: life of William A. Hammond, American neurologist.* New York: Cambridge University Press.

BOHLIN, T., C. DELLEFORS AND U. FAREMO. 1996. Date of smolt migration depends on body size but not on age in wild sea-run trout. *Journal of Fish Biology* 49: 157–164.

BOLING, D.M. 1994. *How to save a river: a handbook for citizen action.* Covello, California: Island Press.

BONNEY, O.H. AND L. BONNEY. 1970. *Battle drums and geysers: the life and journals of Lt. Gustavus Cheyney Doane, soldier and explorer of the Yellowstone and Snake River regions.* Chicago, Illinois: The Swallow Press, Inc.

BONNICHSEN, B. AND R.M. BRECKENRIDGE, EDITORS. 1982. *Cenozoic geology of Idaho.* Moscow: Idaho Bureau of Mines Geological Bulletin 26.

BOONE, L. 1988. *Idaho place names: a geographical dictionary.* Moscow: University of Idaho Press.

BORESON, K., U. MOODY AND K. MURPHEY. 1979. *Cultural resources overviews for the Bureau of Land Management Vail District, Oregon. Volume 1.* Sandpoint, Idaho: Cultural Resource Consultants, Inc., Cultural Resource Management Reports, no. 3.

BOSS, S.M. AND J.S. RICHARDSON. 2002. Effects of food and cover on the growth, survival, and movement of cutthroat trout (*Oncorhynchus clarki*) in coastal streams. *Canadian Journal of Fisheries And Aquatic Sciences* 59: 1044–1053.

BOTTOMLEY, R., R. GRIEVE, D. YORK AND V. MASAITIS. 1997. The age of the Popigai impact event and its relation to events at the Eocene/Oligocene boundary. *Nature* 388: 365–368.

BOUCHARD, D.P., D.S. KAUFMAN, A. HOCHBERG AND J. QUADE. 1998. Quaternary history of the Thatcher Basin, Idaho, reconstructed from the 87Sr/86Sr and amino acid composition of lacustrine fossils: implications for the diversion of the Bear River into the Bonneville Basin. *Paleogeography, Paleoclimatology, Paleoecology* 141: 95–114.

BOULTON, H.E. 1951. *Pageant in the wilderness: the story of the Escalante expedition to the interior basin, 1776.* Salt Lake City: Utah State Historical Society.

BOURGEOIS, J., T.A. HANSEN, P.L. WIBERG AND E.G. KAUFFMAN. 1988. A tsunami deposit at the Cretaceous-Tertiary boundary in Texas. *Science* 241: 567–570.

BOURKE, J.G. 1891 [1971]. *On the border with Crook.* New York: Charles Scribner's Sons [reprinted 1971 by University of Nebraska Press, Lincoln, Nebraska].

BOWEN, E. 1972. *The high Sierra.* New York: Time-Life Books.

BOWEN, G.J. AND 7 co-authors. 2002. Mammalian dispersal at the Paleocene/Eocene boundary. *Science* 295: 2062–2065.

BOWLER, B. 1975. Factors influencing genetic control in lakeward migration of cutthroat trout fry. *Transactions of the American Fisheries Society* 104: 474–482.

BOWLES, S. 1991 [1869]. *The parks and mountains of Colorado: a summer vacation in the Switzerland of America, 1868.* Norman: University of Oklahoma Press [originally published as a series of dispatches to Bowles' newspaper, *The Springfield (Massachusetts) Republican*, and then compiled and published in book form in 1869].

BOYD, R.T. AND Y.P. HAJDA. 1987. Seasonal population movements along the lower Columbia River: the social and ecological context. *American Ethnologist* 14: 309–326.

BOZEK, M.A. AND F.J. RAHEL. 1991. Assessing habitat requirements of young Colorado River cutthroat trout by use of macrohabitat and microhabitat analysis. *Transactions of the American Fisheries Society* 120: 571–581.

BOZEK, M.A. AND F.J. RAHEL. 1992. Generality of microhabitat suitability models for young Colorado River cutthroat trout (*Oncorhynchus clarki pleuriticus*) across sites and among years in Wyoming streams. *Canadian Journal of Fisheries and Aquatic Sciences* 49: 552–564.

BOZEK, M.A., L.D. DeBREY AND J.A. LOCKWOOD. 1994. Diet overlap among size classes of Colorado River cutthroat trout (*Oncorhynchus clarki pleuriticus*) in a high-elevation mountain stream. *Hydrobiologia* 273: 9–17.

BRADNER, E. 1969. *Northwest angling.* Portland, Oregon: Binfords & Mort, Publishers.

BRETZ, J.H. 1923. The channeled scablands of the Columbia Plateau. *Journal of Geology* 31: 617–649.

———. 1969. The Lake Missoula floods and the channeled scabland. *Journal of Geology* 77: 503–543.

BREWERTON, G.D. 1930 [1993]. *Overland with Kit Carson: a narrative of the Old Spanish Trail in '48.* New York: Coward-McCann, Inc. [reprinted 1993, University of Nebraska, Lincoln].

BRIAN, M.V. 1956. Segregation of species in the ant genus *Myrmica. Journal of Animal Ecology* 25: 319–337.

BRIGHT, R.C. 1967. Late Pleistocene stratigraphy of Thatcher Basin, southeastern Idaho. *Tebiwa* 10, no. 1: 1–7.

BROECKER, W.S. AND G.H. DENTON. 1990. What drives glacial cycles? *Scientific American* (January, 1990): 48–56.

BROOKS, C.E. 1986. *The Henry's Fork*. Piscataway, New Jersey: Winchester Press.

BROWN, D.K., A.A. ECHELLE, D.L. PROPST, J.E. BROOKS AND W.L. FISHER. 2001. Catastrophic wildfire and number of populations as factors influencing risk of extinction for Gila trout (*Oncorhynchus gilae*). *Western North American Naturalist* 61: 139–148.

BROWN, K.H. AND 5 co-authors. 2004. Genetic analysis of interior Pacific Northwest *Oncorhynchus mykiss* reveals apparent ancient hybridization with westslope cutthroat trout. *Transactions of the American Fisheries Society* 133: 1078–1088.

BROWN, L.G. 1984. *Lake Chelan fishery investigations*. Wenatchee, Washington: Chelan County Public Utility District no. 1 and Washington State Game Department.

BROWN, R.S. 1999. Fall and early winter movements of cutthroat trout, *Oncorhynchus clarki*, in relation to water temperature and ice conditions in Dutch Creek, Alberta. *Environmental Biology of Fishes* 55: 359–368.

BROWN, R.S. AND W.C. MACKAY. 1995a. Spawning ecology of cutthroat trout (*Oncorhynchus clarki*) in the Ram River, Alberta. *Canadian Journal of Fisheries and Aquatic Sciences* 52: 983–992.

———. 1995b. Fall and winter movements and habitat use by cutthroat trout in the Ram River, Alberta. *Transactions of the American Fisheries Society* 124: 873–885.

BRUHN, K. AND 6 co-authors. 1984. *A cutthroat collection: a guide to understanding and catching the mysterious trout*. Vancouver, British Columbia: Special Interest Publications.

BRUNE, G. 1981. *Springs of Texas*. Fort Worth, Texas: Branch-Smith, Inc. (Not seen; cited in Garrett and Matlock 1991).

BRUNS, E.M. (undated). Old Empire on the Carson River: my home town. *American Women's Diaries*, Segment 3, Western Women, Reel 24 (microfilm). New Canaan, Connecticut: Readex.

BRYAN, A.L. 1955. Archaeology of the Yale Reservoir, Lewis River, Washington. *American Antiquity* 20, no. 3: 281–283.

———. 1992. An appraisal of the archaeological resources of the Yale Reservoir on the Lewis River, Washington. *Archaeology in Washington* 4: 61–69.

BRYANT, E. 1848 [2000]. *What I saw in California*. New York: D. Appleton Company [reprinted as: *Rocky Mountain adventures*. Fairfield, Washington: Ye Galleon Press].

BRYANT, F.G. 1949. *A survey of the Columbia River and its tributaries with special reference to the management of its fishery resources*. Part 2, Washington streams from the mouth of the Columbia River to and including the Klickitat River. U.S. Fish and Wildlife Service, Special Scientific Report 62.

BRYANT, M.D., B.J. FRENETTE AND K.T. COGHILL. 1996. Use of the littoral zone by introduced anadromous salmonids and resident trout, Margaret Lake, southeast Alaska. *Alaska Fishery Research Bulletin* 3, no. 2: 112–122.

BUCKMAN, R.C. 1989. *Trout Creek Mountains wild trout investigations*. Hines, Oregon: Oregon Department of Fish and Wildlife Special Report.

BULKLEY, R.V. 1959. *Report on 1958 fishing studies by the Bureau of Sport Fisheries and Wildlife in Rocky Mountain National Park*. Washington, D.C.: U.S. Fish and Wildlife Service, Bureau of Sport Fisheries, Rocky Mountain Sport Fisheries Investigations, Administrative Report [not seen, cited in Behnke and Zarn 1976].

———. 1961. *Fluctuations in age composition and growth rate of cutthroat trout in Yellowstone Lake*. Washington, D.C.: U.S. Fish and Wildlife Service, Bureau of Sport Fisheries, Research Report 54.

———. 1963. *Natural variation in spotting, hyoid teeth counts, and coloration of Yellowstone cutthroat trout*. Washington, D.C.: U.S. Fish and Wildlife Service, Special Scientific Report, Fisheries 460.

———. 1966. *Coastal cutthroat ecology*. Corvallis: Oregon Game Commission, Fishery Research Report 4.

BULKLEY, R.V. AND N.G. BENSON. 1962 *Predicting year-class abundance of Yellowstone Lake cutthroat trout*. Washington, D.C.: U.S. Fish and Wildlife Service, Bureau of Sport Fisheries, Research Report 59.

BULMER, M. 1994. *Theoretical evolutionary ecology*. Sunderland, Massachusetts: Sinauer Associates, Inc.

BURTON, SIR RICHARD F. 1862. *The city of the saints, and across the Rocky Mountains to California*. New York: Harper and Brothers, Publishers.

BUSACK, C.A. AND G.A.E. GALL. 1981. Introgressive hybridization in populations of Paiute cutthroat trout (*Salmo clarki seleniris*). *Canadian Journal of Fisheries and Aquatic Sciences* 38: 939–951.

BUTLER, C. 1902. Notes on fishes from streams and lakes of northeastern California not tributary to the Sacramento Basin. *Bulletin of the U.S. Fish Commission* 22: 145–148.

BUTLER, V.L. AND N.J. BOWERS. 1998. Ancient DNA from salmon bone: a preliminary study. *Ancient Biomolecules* 2: 17–26.

BUYS, D.J. 2002. Competition between Bonneville cutthroat trout and brook trout in laboratory and field experiments. Master's thesis, Utah State University, Logan.

BUYS, D.J., J.L. KERSHNER AND T.A. CROWL. 1995. *Competitive interactions between Bonneville cutthroat trout and brook trout under various levels of habitat complexity, food availability, and fish densities.* Logan, Utah: Utah State University Department of Fisheries and Wildlife, Final Report to National Fish and Wildlife Foundation, Bring Back the Natives Program.

BYORTH, P.A. 1990. An evaluation of Yellowstone cutthroat trout production in three tributaries of the Yellowstone River, Montana. Master's thesis, Montana State University, Bozeman.

BYUN, S., B. KOOP AND T. REIMCHEN. 1997. North American black bear mtDNA phylogeography: implications for morphology and the Haida Gwai glacial refugium controversy. *Evolution* 51: 1647–1653.

CALHOUN, A.J. 1942. The biology of the black-spotted trout (*Salmo clarkii henshawi* Gill and Jordan) in two Sierra lakes. Ph.D. dissertation, Stanford University, Palo Alto, California.

———. 1944a. Black-spotted trout in Blue Lake, California. *California Fish and Game* 30: 22–42.

———. 1944b. The food of the black-spotted trout in two Sierra Nevada lakes. *California Fish and Game* 30: 80–85.

CAMPBELL, M.R., J. DILLON AND M.S. POWELL. 2002. Hybridization and introgression in a managed, native population of Yellowstone cutthroat trout: genetic detection and management implications. *Transactions of the American Fisheries Society* 131: 364–375.

CAMPTON, D.E. AND F.M. UTTER. 1985. Natural hybridization between steelhead trout (*Salmo gairdneri*) and coastal cutthroat trout (*Salmo clarki clarki*) in two Puget Sound streams. *Canadian Journal of Fisheries and Aquatic Sciences* 42: 110–119.

———. 1987. Genetic structure of anadromous cutthroat trout (*Salmo clarki clarki*) populations in the Puget Sound area: evidence for restricted gene flow. *Canadian Journal of Fisheries and Aquatic Sciences* 44: 573–582.

CAPPS, S.R., JR. 1909. *Pleistocene geology of the Leadville Quadrangle, Colorado.* Washington, D.C.: U.S. Geological Survey, Bulletin 386.

CARHART, ARTHUR H. 1950. *Fishing in the west.* New York: Macmillan.

CARL, L.M. AND J.D. STELFOX. 1989. A meristic, morphometric, and electrophoretic analysis of cutthroat trout, *Salmo clarki,* from two mountain lakes in Alberta. *Canadian Field-Naturalist* 103: 80–84.

CARNIFEX, J. 1891. Fishing along the Pecos, New Mexico. *Outing* 18, no. 4: 299–303.

CARRARA, P.E., S.K. SHORT AND R.E. WILCOX. 1986. Deglaciation of the mountainous regions of northwestern Montana, USA, as indicated by Late Pleistocene ashes. *Arctic and Alpine Research* 18: 317–325.

CARRARA, P.E., W.N. MODE, M. RUBIN AND S.W. ROBINSON. 1984. Deglaciation and postglacial timberline in the San Juan Mountains, Colorado. *Quaternary Research* 21: 42–55.

CAVENDER, T.M. 1986. Review of the fossil history of North American freshwater fishes. Pages 699–724 in C.H. Hocutt and E.O. Wiley, editors. *The zoogeography of North American freshwater fishes.* New York: John Wiley & Sons.

CAVENDER, T.M. AND R.R. MILLER. 1982. *Salmo australis,* a new species of fossil salmonid from southwestern Mexico. *Contributions from the Museum of Paleontology, University of Michigan* 26: 1–17.

CAWDERY, S.A.H. AND A. FERGUSON. 1988. Origins and differentiation of three sympatric species of trout (*Salmo trutta* L.) in Lough Melvin, Ireland. *Polish Journal of Hydrobiology* 35, Supplement A: 267–277.

CEGELSKI, C.C., M.R. CAMPBELL, K.A. MEYER AND M.S. POWELL. 2006. Multiscale genetic structure of Yellowstone cutthroat trout in the upper Snake River basin. *Transactions of the American Fisheries Society* 135: 711–726.

CENTER FOR BIOLOGICAL DIVERSITY AND 7 co-petitioners. 1999. Petition to list the Colorado River cutthroat trout (*Oncorhynchus clarki pleuriticus*) as a threatened or endangered species under the U.S. Endangered Species Act. City and State not specified, Center for Biological Diversity et al.

CHAMBERS, R.L. 1971. Sedimentation in Lake Missoula. Master's thesis, University of Montana, Missoula.

CHAMBERS, S.L. 1984. Endangered and threatened wildlife and plants: findings on pending petitions and descriptions of progress of listing actions. *Federal Register* 49, no. 14: 2485–2488.

CHANDLER, J.H., JR. AND L.L. MARKING. 1982. Toxicity of rotenone to selected aquatic invertebrates and frog larvae. *Progressive Fish-Culturist* 44: 78–80

CHAPMAN, D.W. 1988. Critical review of variables used to define effects of fines in redds of large salmonids. *Transactions of the American Fisheries Society* 117: 371–382.

CHAVEZ, A., TRANSLATOR, AND T.J. WARNER, EDITOR. 1995. *The Dominguez-Escalante journal: their expedition through Colorado, Utah, Arizona, and New Mexico in 1776.* Salt Lake City: University of Utah Press.

CHAVEZ, F.P., J. RYAN, S.E. LLUCH-COTA AND M. ÑIQUEN C. 2003. From anchovies to sardines and back: multidecadal change in the Pacific Ocean. *Science* 299: 217–221

"Chelano." 1904. Lake Chelan fishing. *Pacific Sportsman-The Outdoor Magazine of the Pacific Northwest* 2, no. 4: 204–205.

CHENEY, E., W. ELMORE AND W.S. PLATTS. 1991. *Livestock grazing on western riparian areas.* 2nd printing. Eagle, Idaho: Northwest Resource Information Center, produced for and issued by the U.S. Environmental Protection Agency.

CHRISTENSEN, S. 2002. *2002 cutthroat trout fish distribution survey report, Kendell Creek.* Idaho Falls, Idaho: USDA Forest Service, Caribou-Targhee National Forest.

CHRISTIANSEN, R.L. 1982. Late Cenozoic volcanism of the Island Park area, eastern Idaho. Pages 345–368 in B. Bonnichsen and R.M. Breckenridge, editors. *Cenozoic geology of Idaho.* Boise: Idaho Bureau of Mines and Geology, Bulletin 26.

CLAGUE, J.J., J. HARPER, R. HEBDA AND D. HOWES. 1982. Late Quaternary sea levels and crustal movements, coastal British Columbia. *Canadian Journal of Earth Sciences* 19: 597–618.

CLAGUE, J.J., R. BARENDREGT, R.J. ENKIN AND F.F. FOIT, JR. 2003. Paleomagnetic and tephra evidence for tens of Missoula floods in southern Washington. *Geology* 31: 247–250.

CLAGUE, J.J. AND T. JAMES. 2002. History and isostatic effects of the last ice sheet in southern British Columbia. *Quaternary Science Reviews* 21: 71–87.

CLANCY, C.G. 1988. Effects of dewatering on spawning by Yellowstone cutthroat trout in tributaries to the Yellowstone River, Montana. Pages 37–41 in R.E. Gresswell, editor. *Status and management of interior stocks of cutthroat trout.* Bethesda, Maryland: American Fisheries Society Symposium 4.

CLAPP, D.F., R.D. CLARK, JR. AND J.S. DIANA. 1990. Range, activity and habitat of large, free-ranging brown trout in a Michigan stream. *Transactions of the American Fisheries Society* 119: 1022–1034.

CLELAND, R.G. AND J. BROOKS, EDITORS. 1983. *A Mormon chronicle: the diary of John D. Lee, 1846–1876.* Volumes 1 and 2. Salt Lake City: University of Utah Press.

CLEMENS, W.A. AND G.V. WILBY. 1946. *Fishes of the Pacific coast of Canada. 2nd edition.* Ottawa: Fisheries Research Board of Canada Bulletin 68.

CLINE, G.G. 1963. *Exploring the Great Basin.* Norman, Oklahoma: University of Oklahoma Press.

CLINE, R., G. COLE, W. MEGAHAN, R. PATTEN AND J. POTYONDY. 1981. *Guide for predicting sediment yields from forested watersheds.* Missoula, Montana and Ogden, Utah: USDA Forest Service, Northern and Intermountain Regions.

COBOURN, J. 1999. Integrated watershed management on the Truckee River in Nevada. *Journal of the American Water Resources Association* 35: 623–632.

COFFIN, P.D. 1981. *Distribution and life history of the Lahontan/Humboldt cutthroat trout, Humboldt River drainage basin.* Reno: Nevada Department of Wildlife, Statewide Fisheries Program. Species Management Plan, Federal Aid Project F-20-17, Study IX, Job no. 1-P-1.

———. 1983. *Lahontan cutthroat trout fishery management plan for the Humboldt River drainage basin.* Reno, Nevada Department of Wildlife, Statewide Fisheries Program. Species Management Plan, Federal Aid Project F-20-17, Study IX, Job no. 1-P-1.

COFFIN, P.D. AND W.F. COWAN. 1995. *Lahontan cutthroat trout (Oncorhynchus clarki henshawi) recovery plan.* Portland, Oregon: U.S. Fish and Wildlife Service.

COLBORN, L.G. 1966. *The limnology and cutthroat trout fishery at Trappers Lake, Colorado.* Denver: Colorado Department of Game, Fish, and Parks, Special Report 9.

COLEMAN, M. E., AND V.K. JOHNSON. 1988. Summary of trout management at Pyramid Lake, Nevada, with emphasis on Lahontan cutthroat trout, 1954–1987. Pages 107–115 in R.E. Gresswell, editor. *Status and management of interior stocks of cutthroat trout.* Bethesda, Maryland: American Fisheries Society Symposium 4.

COLLEN, P. AND R.J. GIBSON. 2001. The general ecology of beavers (*Castor* spp.) as related to their influence on stream ecosystems and riparian habitats, and the subsequent effects on fish—a review. *Reviews in Fish Biology and Fisheries* 10: 439–461.

COLORADO DIVISION OF WILDLIFE AND 6 co-signers. 2003. *Conservation agreement for the range-wide preservation and management of the Rio Grande cutthroat trout (Oncorhynchus clarki virginalis).* Denver: Colorado Division of Wildlife, New Mexico Department of Game and Fish, U.S. Forest Service, U.S. Fish and Wildlife Service, Bureau of Land Management, National Park Service, and Jicarilla Apache Nation.

COLYER, W.C. 2002. Seasonal movements of fluvial Bonneville cutthroat trout in the Thomas Fork of the Bear River, Idaho-Wyoming. Master's thesis, Utah State University, Logan.

CONNELL, J.E. 2006. Notice of availability of the proposed Roan Plateau resource management plan amendment/final environmental impact statement, Colorado. *Federal Register* 71, no. 173: 52818–52820.

CONNOLLY, P.J. 1996. Resident cutthroat trout in the central Coast Range of Oregon: logging effects, habitat associations, and sampling protocols. Ph.D. dissertation, Oregon State University, Corvallis.

CONNOLLY, P.J., C.S. SHARPE AND S. SAUTER. 2002. *Evaluate status of coastal cutthroat trout in the Columbia River basin above Bonneville Dam: 2001 annual report.* Cook, Washington: U.S. Geological Survey Western Fisheries Research Center, Columbia River Research Laboratory. Available on the Internet at www.efw.bpa.gov. Click on Integrated Fish and Wildlife Program, then Technical Reports and Publications, then Published Reports, then Anadromous Fish. Enter the first author or report title in the search engine that opens.

CONNOLLY, P.J. AND J.D. HALL. 1999. Biomass of coastal cutthroat trout in unlogged and previously clear-cut basins in the central Coast Range of Oregon. *Transactions of the American Fisheries Society* 128: 890–899.

CONVERSE, Y. 2001. Endangered and threatened wildlife and plants: 12-month finding for a petition to list the Bonneville cutthroat trout as threatened throughout its range. *Federal Register* 66, no. 195: 51362–51366.

CONVERSE, Y. AND J. MIZZI. 1999. Conserving the Bonneville cutthroat trout. *Endangered Species Bulletin* 24, no. 5: 22–23.

COOK, E.F. AND E.J. LARRISON. 1954. Late Pleistocene age of the Snake River diversion. *Geological Society of America Bulletin* 65: 1241 (abstract).

COOK, S. AND R. MOORE. 1969. The effects of rotenone treatment on the insect fauna of a California stream. *Transactions of the American Fisheries Society* 98: 539–544.

COOPER, A. AND H.N. POINAR. 2000. Ancient DNA: do it right or not at all. *Science* 289: 1139.

COOPER, C.F. 1960. Changes in vegetation, structure, and growth of southwestern pine forest since white settlement. *Ecological Monographs* 30: 129–164.

COOPER, J.G. 1870. The fauna of Montana Territory. *American Naturalist* 3: 124–127.

COOPER, J.J. AND D.L. KOCH. 1984. Limnology of a desertic terminal lake, Walker Lake, Nevada, U.S.A. *Hydrobiologia* 118: 275–292.

COPE, E.D. 1872. Report on the recent reptiles and fishes of the survey, collected by Campbell Carrington and C.M. Dawes. Pages 467–476 in *Fifth annual report of the United States Geographical Survey of the Territories (Hayden's Survey), part 4.* Washington, D.C.: U.S. Government Printing Office.

COPE, E.D. AND H.C. YARROW. 1875. Report upon the collection of fishes made in portions of Nevada, Utah, California, New Mexico and Arizona during the years 1871, 1872, 1873, and 1874. Pages 635–703, 997–999, plates 26–32 in Wheeler, G.M. *Report upon geographical and geological exploration surveys west of the one-hundredth meridian.* Volume 5, Zoology. Washington, D.C.: U.S. Government Printing Office.

COPE, O.B. 1955. The Future of the Cutthroat trout in Utah. *Proceedings of the Utah Academy of Sciences, Arts, and Letters* 32: 89–93.

———. 1956. Some migration patterns in cutthroat trout. *Proceedings of the Utah Academy of Sciences, Arts, and Letters* 33: 113–118.

———. 1957a. *Races of cutthroat trout in Yellowstone Lake.* Washington, D.C.: U.S. Fish and Wildlife Service, Special Scientific Report-Fisheries 208: 74–84.

———. 1957b. Six years of catch statistics on Yellowstone Lake. *Transactions of the American Fisheries Society* 85: 160–179.

———. 1957c. The choice of spawning sites by cutthroat trout. *Proceedings of the Utah Academy of Sciences, Arts, and Letters* 34: 73–79.

CORDONE, A.J. AND T.C. FRANTZ. 1968. An evaluation of trout planting in Lake Tahoe. *California Fish and Game* 54: 68–69.

CORSI, C.E. 1988. The life history and status of the Yellowstone cutthroat trout (*Salmo clarki bouvieri*) in the Willow Creek drainage, Idaho. Master's thesis, Idaho State University, Pocatello.

COSTELLO, A.B. AND E. RUBIDGE. 2005. *COSEWIC status report on coastal cutthroat trout (Oncorhynchus clarkii clarkii).* Vancouver, British Columbia: University of British Columbia Native Fish Research Group, draft report for the Committee on the Status of Endangered Wildlife in Canada.

COURTILLOT, V. 1999. *Evolutionary catastrophes: the science of mass extinction.* Cambridge, U.K.: Cambridge University Press.

COVINGTON, W.W. AND M.M. MOORE. 1994. Southwestern ponderosa forest structure: changes since Euro-American settlement. *Journal of Forestry* 39, no. 1: 39–47.

COWAN, W. 1988. Biochemical genetics of Lahontan cutthroat trout (*Salmo clarki henshawi*) inhabiting the Summit Lake drainage basin, Humboldt County, Nevada. Senior project report, Humboldt State University, Arcata, California.

COWLEY, D.E. 1993. *Strategies for development and maintenance of a hatchery broodstock of Rio Grande cutthroat trout (Oncorhynchus clarki virginalis).* Las Cruces: New Mexico State University Department of Fishery and Wildlife Sciences, report prepared for the New Mexico Department of Game and Fish.

COWLEY, M.S. 1916 [1964]. *Wilford Woodruff (history of his life and labors as recorded in his daily journals).* Salt Lake City, Utah: Bookcraft [reprint].

CRANFORD, J.M. 1912. A new trout. *Forest and Stream* 78 (Feb. 24, 1912): 234–236. [The author's name, which was actually Crawford, was misspelled by the magazine.]

CRAWFORD, B.A. 1979. *The origin and history of the trout brood stocks of the Washington Department of Game.* Olympia: Washington State Game Department, Fishery Research Report.

———. 1998. *Status review of westslope cutthroat trout in Washington.* Olympia: Washington Department of Fish and Wildlife. Cover letter dated Dec. 31, 1998 and report to L. Kaeding, U.S. Fish and Wildlife Service, Bozeman Montana.

CRAWFORD, C.S. AND 5 co-authors. 1993. *Middle Rio Grande Bosque Ecosystem: bosque biological management plan.* Albuquerque, New Mexico: Biological Interagency Team. Available on the Internet at www.fws.gov/bhg/.

CRCT TASK FORCE. 1999 [2001]. *Conservation agreement and strategy for Colorado River cutthroat trout (Oncorhynchus clarki pleuriticus) in the States of Colorado, Utah, and Wyoming.* Fort Collins: Colorado Division of Wildlife.

———. 2004. *Colorado River cutthroat trout (Oncorhynchus clarki pleuriticus) management activities and accomplishment report: 1999–2003, Colorado, Wyoming, and Utah.* Fort Collins: Colorado Division of Wildlife.

CRCT COORDINATION TEAM. 2006. *Conservation strategy for Colorado River cutthroat trout (Oncorhynchus clarkii pleuriticus) in the States of Colorado, Utah, and Wyoming.* Fort Collins: Colorado Division of Wildlife.

CRESPI, B.J. AND M.J. FULTON. 2004. Molecular systematics of Salmonidae: combined nuclear data yields a robust phylogeny. *Molecular Phylogenetics and Evolution* 31: 658–679.

CRISP, D.T. 1981. A desk study of the relationship between temperature and hatching time for the eggs of five species of salmonid fishes. *Freshwater Biology* 11: 361–368.

———. 1988. Prediction, from temperature, of eyeing, hatching, and 'swim-up' times for salmonid embryos. *Freshwater Biology* 19: 41–48.

CROCKETT, S. 1999. The prehistoric peoples of Jackson Hole. Chapter 2 in Daugherty, J., editor. *A place called Jackson Hole: a historical resource study of Grand Teton National Park.* Moose, Wyoming: Grand Teton Natural History Association. Available on the Internet at www. cr.nps.gov/history/online_books/grte2/hrs2.htm.

CRUTZEN, P.J. 1987. Acid rain at the K/T boundary. *Nature* 230: 108–109.

———. 2000. Geology of mankind. *Nature* 415: 23.

CUMMINGS, T.R. 1987. Brook trout competition with greenback cutthroat trout in Hidden Valley Creek, Colorado. Master's thesis, Colorado State University, Fort Collins.

CUNJAK, R.A. 1996. Winter habitat of selected stream fishes and potential impacts from land-use activities. *Canadian Journal of Fisheries and Aquatic Sciences* 53, supplement 1: 267–282.

CUNJAK, R.A. AND G. POWER. 1986. Winter habitat utilization by stream resident brook trout (*Salvelinus fontinalis*) and brown trout (*Salmo trutta*). *Canadian Journal of Fisheries and Aquatic Sciences* 43: 1970–1981.

———. 1987. Cover use by stream-resident trout in winter: a field experiment. *North American Journal of Fisheries Management* 7: 539–544.

CURRAN, H. 1982. *Fearful crossing: the central overland trail through Nevada.* Las Vegas: Nevada Publications.

CURRENS, K.P., C.B. SCHRECK AND H.W. LI. 1990. Allozyme and morphological divergence of rainbow trout (*Oncorhynchus mykiss*) above and below waterfalls in the Deschutes River, Oregon. *Copeia* 1990: 730–746.

CURRY, B.B. AND W.N. MELHORN. 1990. Summit Lake landslide and geomorphic history of Summit Lake basin, northwestern Nevada. *Geomorphology* 4: 1-17.

CUSHING, C.E., EDITOR. 1997. *Freshwater ecosystems and climate change in North America: advances in hydrological processes.* New York: John Wiley & Sons.

CUSHMAN, D. 1966. *The Great North Trail: America's route of the Ages.* New York: McGraw-Hill Book Company.

DAHLEM, E.A. 1979. The Mahogany Creek watershed— with and without grazing. Pages 31–34 in O.B. Cope, editor. *Grazing and riparian/stream ecosystems: proceedings of the forum held at Denver, Colorado November 3–4, 1978.* Denver, Colorado: Trout Unlimited.

DANGBERG, G. 1975. *Conflict on the Carson.* Minden, Nevada: Carson Valley Historical Society.

DANIEL, J.W. 1878. Salmonidae in Texas. *Forest and Stream* 10, no. 48: 339.

DANSIE, A.J. 1987. Archaeofaunas: 26Pe450 and 26Pe366. Pages 137–147 in M.K. Rusco and J.O. Davis, editors. *Studies in archaeology, geology and paleontology at Rye Patch Reservoir, Pershing County, Nevada.* Carson City: Nevada State Museum, Anthropological Papers no. 20.

DAVIS, J.O. 1982. Bits and pieces: the last 35,000 years in the Lahontan area. Pages 53–75 in D.B. Madsen and J.F. O'Connell, editors. *Man and Environment in the Great Basin.* Washington D.C.: Society for American Archaeology, SAA Papers 2.

———. 1990. Giant meanders on Humboldt River near Rye Patch, Nevada, due to catastrophic flooding. *Geological Society of America Abstracts with Programs* 22, no. 6: A3097.

DEAGLE, B.E., T.E. REIMCHEN AND D.B. LEVIN. 1996. Origins of endemic stickleback from the Queen Charlotte Islands: mitochondrial and morphological evidence. *Canadian Journal of Zoology* 74: 1045–1056.

DE GROOT, J.D., S.G. HINCH AND J.S. RICHARDSON. 2007. Effects of logging second-growth forests on headwater populations of coastal cutthroat trout: a 6-year, multistream, before-and-after field experiment. *Transactions of the American Fisheries Society* 136: 211–226.

DE LA HOZ FRANCO, E.A. AND P. BUDY. 2005. Effects of biotic and abiotic factors on the distribution of trout and salmon along a longitudinal gradient. *Environmental Biology of Fishes* 72: 379–391.

DE QUEIROZ, K. AND J. GAUTHIER. 1992. Phylogenetic taxonomy. *Annual Review of Ecology and Systematics* 23: 449–480.

DESHAZO, J.J. 1980. *Sea-run cutthroat status report.* Olympia: Washington State Game Department, Fisheries Management Division Report 80-14.

DE STASO, J. III AND F.J. RAHEL. 1994. Influence of water temperature on interactions between juvenile Colorado River cutthroat trout and brook trout in a laboratory stream. *Transactions of the American Fisheries Society* 123: 289–297.

DEVLIN, R.H. 1993. Sequence of sockeye salmon type 1 and 2 growth hormone genes and the relationship of rainbow trout with Atlantic and Pacific salmon. *Canadian Journal of Fisheries and Aquatic Sciences* 50: 1738–1748.

D'HONDT, S., M.E.Q. Pilson, H. Sigurdsson, A.K. Hanson, Jr., and S. Carey. 1994. Surface-water acidification and extinction at the Cretaceous-Tertiary boundary. *Geology* 22: 983–986.

DIAMOND, J. 2005. *Collapse: how societies choose to fail or succeed.* New York: Viking Penguin.

DIANA, J.S. 1975. The movements and distribution of Paiute cutthroat trout, *Salmo clarki seleniris* Snyder, in the North Fork of Cottonwood Creek, Mono County, White Mountains, California. Master's thesis, California State University, Long Beach.

DIANA, J.S. AND E.D. LANE. 1978. The movements and distribution of Paiute cutthroat trout, *Salmo clarki seleniris*, in Cottonwood Creek, California. *Transactions of the American Fisheries Society* 107: 444–448.

DICKENS, G.R. 1999. The blast in the past. *Nature* 401: 752–755.

DICKERSON, B.R. AND G.L. VINYARD. 1999. Effects of high chronic temperatures and diel temperature cycles on the survival and growth of Lahontan cutthroat trout. *Transactions of the American Fisheries Society* 128: 516–521.

DICKERSON, R. 1997. *Nevada angler's guide: fish tails in the sagebrush.* Portland, Oregon: Frank Amato Publications.

DIMICK, R.E. AND F. MERRYFIELD. 1945. The fishes of the Willamette River system in relation to pollution. Oregon State College, Engineering Experiment Station, Bulletin 20, Corvallis.

DINGLE, H. 1996. *Migration: the biology of life on the move.* Oxford, U.K.: Oxford University Press.

DOBIE, J.F. 1947 [2001]. *The voice of the coyote.* New York: Curtis Publishing Company [Edson, New Jersey: Castle Books].

DODD, C.K., JR. 1985. Endangered and threatened wildlife and plants: review of vertebrate wildlife. *Federal Register* 50, no. 181: 37958–37967.

DOWLING, T.E., C.A. TIBBETS, W.L. MINCKLEY AND G.R. SMITH. 2002. Evolutionary relationships of the Plagopterans (Teleostei: Cyprinidae) from cytochrome *b* sequences. *Copeia* 2002: 665–678.

DOWNS, C.C. 1995. Age determination, growth, fecundity, age at sexual maturity, and longevity for isolated, headwater populations of westslope cutthroat trout. Master's thesis, Montana State University, Bozeman.

DOWNS, C.C., R.G. WHITE AND B.B. SHEPARD. 1997. Age at sexual maturity, sex ratio, fecundity, and longevity of isolated headwater populations of westslope cutthroat trout. *North American Journal of Fisheries Management* 17: 85–92.

DOWNS, J.F. 1966. *The two worlds of the Washoes.* New York: Holt, Rinehart and Winston.

DRANNAN, CAPT. W.F. 1910. *Thirty-one years on the plains and in the mountains, or, the last voice from the plains.* Chicago: Rhodes & McClure Publishing Company.

DREWERY, G. 1991. Endangered and threatened wildlife and plants: animal candidate review for listing as endangered or threatened species. *Federal Register* 56, no. 225: 53804–53836.

DREWERY, G. AND R. SAYERS. 1996. Endangered and threatened wildlife and plants: animal candidate review for listing as endangered or threatened species. *Federal Register* 61, no. 40: 7596–7613.

DRUMMOND, R.A. 1966. *Reproduction and harvest of cutthroat trout at Trappers Lake, Colorado.* Denver: Colorado Department of Game, Fish, and Parks, Special Report 10.

DRUMMOND, R.A. AND T.D. MCKINNEY. 1965. Predicting the recruitment of cutthroat trout fry in Trappers Lake, Colorado. *Transactions of the American Fisheries Society* 94: 389–393.

DRYER, T.J. 1980. First ascent of Mt. St. Helens [edited by H.M. Majors]. *Northwest Discovery* 1, no. 3: 164–180.

Dufek, D. and 6 co-authors. 1999. *Status and management of Yellowstone cutthroat trout (Oncorhynchus clarki bouvieri)*. Cheyenne: Wyoming Game and Fish Department.

Duff, D.A. 1988. Bonneville cutthroat trout: current status and management. Pages 121–127 in R.E. Gresswell, editor. *Status and management of interior stocks of cutthroat trout*. Bethesda, Maryland: American Fisheries Society Symposium 4.

———. 1996. Bonneville cutthroat trout *Oncorhynchus clarki utah*. Pages 35–73 in Duff, D.A., technical editor. *Conservation assessment for inland cutthroat trout: distribution, status, and habitat management implications*. Ogden, Utah: USDA Forest Service, Northern, Rocky Mountain, Intermountain, and Southwestern Regions.

Duffield, D.J. 1990. *Management plan and status report: native cutthroat trout of the Dixie National Forest*. Cedar City, Utah: USDA Forest Service, Dixie National Forest.

Dunbar, R.G. 1983. *Forging new rights in western waters*. Lincoln: University of Nebraska Press.

Dunbar, S., editor. 1927. *The journals and letters of major John Owen, pioneer of the northwest 1850–1871*. Volume 1. New York: Edward Eberstadt, Publisher.

Dunham, J., R. Schroeter and B. Rieman. 2003. Influence of maximum water temperature on occurrence of Lahontan cutthroat trout within streams. *North American Journal of Fisheries Management* 23: 1042–1049.

Dunham, J.B. 1996. The population ecology of stream-living Lahontan cutthroat trout (*Oncorhynchus clarki henshawi*). Ph.D. dissertation, University of Nevada, Reno.

Dunham, J.B., G.L. Vinyard and B. Rieman. 1997. Habitat fragmentation and extinction risk of Lahontan cutthroat trout. *North American Journal of Fisheries Management* 17: 126–133.

Dunham, J.B., M.E. Rahn, R.E. Schroeter and S.W. Breck. 2000. Diets of sympatric Lahontan cutthroat trout and nonnative brook trout: implications for species interactions. *Western North American Naturalist* 60: 304–310.

Dunham, J.B., M.M. Peacock, B.E. Rieman, R.E. Schroeter and G.L. Vinyard. 1999. Local and geographic variability in the distribution of stream-living Lahontan cutthroat trout. *Transactions of the American Fisheries Society* 128: 875–889.

Dunham, J.B., S.B. Adams, R. Schroeter and D. Novinger. 2002. Alien invasions in aquatic ecosystems: toward an understanding of brook trout invasions and their potential impact on inland cutthroat trout. *Reviews in Fish Biology and Fisheries* 12: 373–391.

Dunnigan, J.L. 1997. The spatial distribution of westslope cutthroat trout in the Coeur d'Alene River system. Master's thesis, University of Idaho, Moscow.

Dunning, J.B., B.J. Danielson and H.R. Pulliam. 1992. Ecological processes that affect populations in complex landscapes. *Oikos* 65: 169–175.

Dunraven, Earl of. 1876 [1917]. *The Great Divide*. London: Chatto & Windus [reprinted as Kephart, H., editor. 1917. *Hunting in the Yellowstone*. New York: Outing Publishing Company].

Dwyer, W.P. and B.D. Rosenlund. 1988. Role of fish culture in the reestablishment of greenback cutthroat trout. Pages 75–80 in R.E. Gresswell, editor. *Status and management of interior stocks of cutthroat trout*. Bethesda, Maryland: American Fisheries Society Symposium 4.

Dymond, J.R. 1931. Description of two new forms of British Columbia trout. *Contributions to Canadian Biology and Fisheries* 6: 393–395.

———. 1932. *The trout and other game fishes of British Columbia*. Ottawa: Canada Department of Fisheries.

Eaton, J.G. and R.M. Scheller. 1996. Effects of climate warming on fish thermal habitat in streams of the United States. *Limnology and Oceanography* 41: 1109–1115.

Ebbesmeyer, E.H. and R.M. Strickland. 1995. *Oyster condition and climate: evidence from Willapa Bay*. Fifth conference for shellfish growers. Seattle: University of Washington Sea Grant Program, Publication MR 95-02.

Editors. 1941. Queen of them all was Virginia City: history of lumbering in western Nevada. *The Timberman* 42, no. 8: 11–14, 50–62.

Edwords, C.E. 1893. *Campfires of a naturalist*. New York: D. Appleton Company.

Elias, S.A. 1996. *The Ice Age history of national parks in the Rocky Mountains*. Washington, D.C.: Smithsonian Institution Press.

Elliott, D.K., editor. 1986. *Dynamics of extinction*. New York: John Wiley & Sons.

Elliott, J. and R.W. Layton. 2004. *Lahontan cutthroat trout species management plan for the upper Humboldt River drainage basin*. Reno: Nevada Department of Wildlife. Available on the Internet at www.ndow.org/wild/conservation/fish/lct_smp. pdf. Eleven accompanying maps are also available for download as separate pdf files. Accessed March 2, 2005.

Elliott, J.M. 1978. Effect of temperature on the hatching time of eggs of *Ephemerella ignita* (Poda) (Ephemeroptera: Ephemerellidae). *Freshwater Biology* 8: 51–58.

————. 1989. The natural regulation of numbers and growth of brown trout, *Salmo trutta*, in two Lake District streams. *Freshwater Biology* 21: 7–19.

————. 1994. *Quantitative ecology and the brown trout.* New York: Oxford University Press.

ELLIOTT, J.M., U.H. HUMPESCH AND M.A. HURLEY. 1987. A comparative study of eight mathematical models for the relationship between water temperature and hatching time of eggs of freshwater fish. *Archiv fur Hydrobiologie* 109: 257–277.

ELLIS, M.M. 1940. *Pollution of the Coeur d'Alene River and adjacent waters by mine wastes.* U.S. Bureau of Fisheries, Special Scientific Report 1.

EMLEN, J.M., T.A. STREKAL AND C.C. BUCHANAN. 1993. Probabalistic projections for recovery of the endangered cui-ui. *North American Journal of Fisheries Management* 13: 467–474.

ENDLER, J.A. 1986. *Natural selection in the wild.* Princeton, New Jersey: Princeton University Press.

ENDLICH, F.M. 1877. Pages 103–235 (also pages 9–33) in *Annual Report of the U.S. Geological Survey of the Territories (the Hayden surveys) 9, 1875.*

ERVIN, K. 1989. *Fragile majesty: the battle for North America's last great forest.* Seattle, Washington: The Mountaineers.

EVANS, D.L. 1997. Designated critical habitat: Umpqua River cutthroat trout. *Federal Register* 62, no. 146: 40786–40791.

EVANS, R.P. AND D.K. SHIOZAWA. 2001. *The genetic status of greenback cutthroat trout (Oncorhynchus clarki stomias) populations in Colorado.* Provo, Utah: Brigham Young University, Department of Integrative Biology. Final Report prepared for Colorado Division of Wildlife, Denver.

EVERMANN, B.W. 1893. A reconnaissance of the streams and lakes of western Montana and northwestern Wyoming. *Bulletin of the U.S. Fish Commission* 11 (for 1891): 3–60 and accompanying plates.

————. 1896. A report upon salmon investigations in the headwaters of the Columbia River, in the State of Idaho, in 1895 together with notes upon the fishes observed in that State in 1894 and 1895. *Bulletin of the U.S. Fish Commission* 16: 149–202.

EVERMANN, B.W. AND H.C. BRYANT. 1919. California trout. *California Fish and Game* 5: 105–135.

EVERMANN, B.W. AND J.T. NICHOLS. 1909. Notes on the fishes of Crab Creek, Washington, with description of a new species of trout. *Proceedings of the Biological Society of Washington* 22: 1–2 and 91–94 (and accompanying Plate II).

EVERMANN, B.W. AND U.O. COX. 1896. A report upon the fishes of the Missouri River basin. Pages 325–429 in *U.S. Fish Commission of Fish and Fisheries, Part 20. Report of Commissioner for the Year Ending June 30, 1894.* Washington, D.C.: Government Printing Office

EWERS, J.C. 1958. *The Blackfeet: raiders of the northwestern plains.* Norman: University of Oklahoma Press.

FALCONER, D.S. 1989. *Introduction to quantitative genetics.* 3rd edition. New York: John Wiley and Sons.

FARNHAM, T.J. 1848 [1906]. *Travels in the great western prairies, the Anahuac and Rocky Mountains, and in the Oregon Country.* Volume 1. London: Richard Bentley [reprinted 1906 in R.G. Thwaites, editor. *Early western travels 1748–1846.* Cleveland, Ohio: The Arthur H. Clark Company].

FAUSCH, K.D. 1984. Profitable stream positions for salmonids: relating specific growth rate to net energy gain. *Canadian Journal of Zoology* 62: 441–451.

————. 1989. Do gradient and temperature affect distribution of, and interactions between brook charr and other salmonids in streams? *Physiology and Ecology Japan*, Special Volume 1: 303–322.

————. 1991. Trout as predators. Pages 65–80 in J. Stolz and J. Schnell, editors. *Trout: the wildlife series.* Harrisburg, Pennsylvania: Stackpole Books.

FAUSCH, K.D. AND T.R. CUMMINGS. 1986. *Effects of brook trout competition on threatened greenback cutthroat trout.* Fort Collins, Colorado: Final Report (Contract CX-1200-4-A043) to U.S. National Park Service Rocky Mountain Regional Office, Denver.

FAUSCH, K.D. AND T.G. NORTHCOTE. 1992. Large woody debris and salmonid habitat in a small coastal British Columbia stream. *Canadian Journal of Fisheries and Aquatic Sciences* 49: 682–693.

FEDERAL WRITERS' Project of the Works Progress Administration. 1938. *The Idaho encyclopedia.* Caldwell, Idaho: The Caxton Printers, Ltd.

FEMAT. 1993. *Forest ecosystem management: an ecological, economic, and social assessment.* Portland, Oregon: USDA Forest Service and 5 additional federal agencies. Report of the Forest Ecosystem Management Assessment Team.

FERGUSON, A. 1989. Genetic differences among brown trout, *Salmo trutta*, stocks and their importance for the conservation and management of the species. *Freshwater Biology* 21: 35–46.

FERGUSON, A. AND F.M. MASON. 1981. Allozyme evidence for reproductive isolated sympatric populations of brown trout *Salmo trutta* L. in Lough Melvin, Ireland. *Journal of Fish Biology* 18: 629–642.

FERGUSON, A. AND J.B. TAGGART. 1991. Genetic differentiation among the sympatric brown trout (*Salmo trutta*) populations of Lough Melvin, Ireland. *Biological Journal of the Linnean Society* 43: 221–237.

FERGUSON, G. 2003. *Hawk's Rest: a season in the remote heart of Yellowstone*. Washington, D.C.: National Geographic Adventure Press.

FERGUSON, M.M. AND F.W. ALLENDORF. 1992. Evolution of the fish genome. Pages 25–43 in P.W. Hochachka and T.P. Mommsen, editors. *Biochemistry and molecular biology of fishes*. Vol. 1. New York: Elsevier Science Publishers.

FERRIS, W.A. 1940 [1983]. *Life in the Rocky Mountains*. Denver, Colorado: Old West Publishing Company.

FIELD, M.C. 1957. *Prairie and mountain sketches*. Norman: University of Oklahoma Press.

FINNEY, B.P., I. GREGORY-EVANS, M.S.V. DOUGLAS AND J.P. SMOL. 2002. Fisheries production in the northeastern Pacific Ocean over the past 2,200 years. *Nature* 416: 729–733.

THE FITZPATRICKS. 1966. The tragic Wards. *True West* 13, no. 6 (July-August issue): 10–12, 56–60.

FLADMARK, K.R. 1979. Routes: alternative migration corridors for early man in North America. *American Antiquity* 44: 55–69.

FLANNERY, T. 2001. *The eternal frontier: an ecological history of North America and its people*. New York: Atlantic Monthly Press.

———. 2005. *The weather makers: how man is changing the climate and what it means for life on earth*. New York: Atlantic Monthly Press.

FLEENER, G.G. 1951. Life history of the cutthroat trout, *Salmo clarki* Richardson, in Logan River, Utah. *Transactions of the American Fisheries Society* 81: 235–248.

FLEISCHNER, T.L. 2003. Diversity deep and wild. *Conservation Biology* 17: 952–953.

FLETCHER, M.S., EDITOR. 1977. *The Wetherills of the Mesa Verde: autobiography of Benjamin Alfred Wetherill*. Lincoln: University of Nebraska Press.

FLINT, R.F. 1947. *Glacial geology and the Pleistocene Epoch*. New York: John Wiley and Sons.

FOLLETT, W.I. 1982. An analysis of fish remains from ten archaeological sites at Falcon Hill, Washoe County, Nevada, with notes on fishing practices of the ethnographic Kuyuidtkadl Northern Paiute. Pages 179–203 in E.M. Hattori, editor. *The archaeology of Falcon Hill, Winnemucca Lake, Washoe County, Nevada*. Carson City: Nevada State Museum, Anthropological Papers no. 18.

FOLMAR, L.C. AND W.W. DICKHOFF. 1980. The parr-smolt transformation (smoltification) and seawater adaptation in salmonids: a review of selected literature. *Aquaculture* 21: 1–37.

FOLSOM, M.M. 1970. Volcanic eruptions: the pioneer's attitude on the Pacific coast from 1800 to1875. *The Ore Bin (a publication of the Oregon Department of Geology and Mineral Industries)* 32, no. 4: 61–71.

FONTENOT, L.W., D.G. NOBLET AND S.G. PLATT. 1994. Rotenone hazards to amphibians and reptiles. *Herpetological Review* 25: 150–156.

FORSETH, T., O. UGEDAL AND B. JONSSON. 1994. The energy budget, niche shift, reproduction and growth in a population of Arctic charr, *Salmo alpinus. Journal of Animal Ecology* 63: 116–126.

FORSETH, T., T.F. NAESJE, B. JONSSON AND K HÅRSAKER. 1999. Juvenile migration in brown trout: a consequence of energetic state. *Journal of Animal Ecology* 68: 783–793.

FOSTER, L.E. 1978. Food habits of the cutthroat trout in the Snake River, Wyoming. Master's thesis, University of Wyoming, Laramie.

FOSTER, R.J. 1991. *Historical geology*. New York: Macmillan Publishing Company.

FOUNTAIN, P. 1905. *The eleven eaglets of the west*. London: John-Murray, Albemarle Street..

FOWLER, L.G. 1972. Growth and mortality of fingerling chinook salmon as affected by egg size. *Progressive Fish-Culturist* 34: 66–69.

FRALEY, J. AND B.B. SHEPARD. 2005. Age, growth, and movements of westslope cutthroat trout, *Oncorhynchus clarki lewisi*, inhabiting the headwaters of a wilderness river. *Northwest Science* 79: 12–21.

FRANCE, L.B. 1882 [1980]. A week on the Grand. *The American Angler*, August 19, 1882 [reprinted in *The American Fly Fisher* (the journal of the American Museum of Fly Fishing, Manchester, Vermont) 7, no. 4: 26–28].

———. 1884. *With rod and line in Colorado waters*. Denver, Colorado: Chain, Hardy & Company.

FRANKHAM, R. 1995. Effective population size/adult population size ratios in wildlife: a review. *Genetical Research* 66: 95–107.

FRANKLIN, I.R. 1980. Evolutionary change in small populations. Pages 135–149 in M. Soulé and B. Wilcox, editors. *Conservation biology: an evolutionary-ecological perspective*. Sunderland, Massachusetts: Sinauer Associates.

FREDGA, K. 1977. Chromosomal changes in vertebrate evolution. *Proceedings of the Royal Society of London B, Biological Sciences* 199: 377–397.

FREMONT, BREVET CAPTAIN JOHN C. 1845 [2002]. *Report of the Exploring Expedition to the Rocky Mountains in the Year 1842 and to Oregon and North California in the Years 1843–44*. U.S. Senate, 28th Congress, 2nd Session, Executive Document 174.

Washington, D.C.: Gales and Seton Printers [reprinted 2002 (without maps and some sketches) by The Narrative Press, Santa Barbara, California].

FRENCH, G. 1964. *Cattle country of Pete French*. Portland, Oregon: Binfords and Mort, Publishers.

FRENETTE, B.J. AND M.D. BRYANT. 1993. *Assessment of the resident cutthroat trout and Dolly Varden and introduced anadromous salmonids in Margaret Lake*. USDA Forest Service, Progress Report to the Ketchikan Ranger District, Ketchikan, Alaska.

FRESH, K.L., R.D. CARDWELL AND R.R. KOONS. 1981. *Food habits of Pacific salmon, baitfish, and their potential competitors and predators in the marine waters of Washington, August 1978 to September 1979*. Olympia: Washington Department of Fisheries, Progress Report 145.

FRITZ, W.J. AND J.W. SEARS. 1993. Tectonics of the Yellowstone hotspot wake in southwestern Montana. *Geology* 21: 427–430.

FUSS, H.J. 1982. Age, growth and instream movement of Olympic Peninsula coastal cutthroat trout (*Salmo clarki clarki*). Master's thesis, University of Washington, Seattle.

"Gage." 1906. Utah's trout waters. *Forest and Stream* 67 (July 14, 1906 issue): 58.

GALAT, L., E.I. LIDER, S. VIGG AND S.R. ROBERTSON. 1981. Limnology of a large, deep, North American terminal lake, Pyramid Lake, Nevada, U.S.A. *Hydrobiologia* 82: 281–317.

GALAT, D.L., G. POST, T.J. KEEFE AND G.R. BOUCH. 1985. Histological changes in the gill, kidney and liver of Lahontan cutthroat trout, *Salmo clarki henshawi*, living in lakes of different salinity-alkalinity. *Journal of Fish Biology* 27: 533–552.

GALL, G.A.E. AND E.J. LOUDENSLAGER. 1981. *Biochemical genetics and systematics of Nevada trout populations*. Fisheries Biology Research Facility, Department of Animal Sciences, University of California, Davis.

GARDNER, B. (undated). *Montana's fish species of special concern: westslope cutthroat trout*. Bigfork, Montana: Flathead National Forest, Swan Lake Ranger District. Available on the Internet at www.fisheries.org/AFSmontana/SSCpages/westslope_cutthroat_trout.htm. (accessed September 24, 2003).

GARDNER, H., PRODUCER/writer, and M. Herlinger, director/editor. 1996. *Incredible journey of the greenback cutthroats*. Denver, Colorado: Borderline Productions [video].

GARLICK, L. 1949. *Report of fishery investigations, Lake Crescent, Olympic National Park, with management recommendations*. Portland, Oregon: U.S. Fish and Wildlife Service.

GARRARD, L.H. 1955. *Wah-to-yah and the Taos Trail*. Norman, Oklahoma: University of Oklahoma Press [original manuscript published in 1850].

GARRETT, G.P. AND G.C. MATLOCK. 1991. Rio Grande cutthroat trout in Texas. *The Texas Journal of Science* 43: 405–410.

GARTE, S.J., EDITOR. 1994. *Molecular environmental biology*. Boca Raton, Florida: Lewis Publishers (CRC Press).

GEARY, M. 1999. *A quick history of Grand Lake*. Montrose, Colorado: Western Reflections Publishing Company.

GELATT, P.S. 2004. Endangered and threatened wildlife and plants: 90-Day Finding on a petition to list the Colorado River cutthroat trout. *Federal Register* 69, no. 76: 21151–21158.

GELWICKS, K.R., D.J. ZAFFT AND R.G. GIPSON. 2002. *Comprehensive study of the Salt River fishery between Afton and Palisades Reservoir from 1995 to 1999 with historical review: fur trade to 1998*. Cheyenne: Wyoming Game and Fish Department, Fish Division.

GERKING, S.D. 1959. The restricted movement of fish populations. *Biological Reviews of the Cambridge Philosophical Society* 34: 221–242.

GERSTUNG, E. 1979. *The Colorado River cutthroat trout (Salmo clarki pleuriticus), draft management plan*. Sacramento: California Department of Fish and Game.

———. 1982. Lahontan cutthroat trout: the big one that got away. *Outdoor California* 43, no. 4: 1–5.

GERSTUNG, E.R. 1986. *Fishery management plan for Lahontan cutthroat trout (Salmo clarki henshawi) in California and western Nevada waters*. Sacramento: California Department of Fish and Game in cooperation with Nevada Department of Wildlife, U.S. Forest Service, and U.S. Fish and Wildlife Service. Inland Fisheries Administrative Report, Federal Aid Project F33-R-11.

GERSTUNG, E.R. 1988. Status, life history, and management of the Lahontan cutthroat trout. Pages 93–106 in R.E. Gresswell, editor. *Status and management of interior stocks of cutthroat trout*. Bethesda, Maryland: American Fisheries Society Symposium 4.

———. 1997. Status of coastal cutthroat trout in California. Pages 43–56 in J.D. Hall, P.A. Bisson and R.E. Gresswell, editors. 1997. *Sea-run cutthroat trout: biology, management, and future conservation*. Corvallis: Oregon Chapter American Fisheries Society.

GIANNASI, D.E. 1990. The Clarkia fossil leaves: windows to the past. *Northwest Science* 64: 232–235.

GIERACH, J. 1989. *Flyfishing small streams*. Harrisburg, Pennsylvania: Stackpole Books.

GIGER, R.D. 1972. *Ecology and management of coastal cutthroat trout in Oregon.* Corvallis: Oregon State Game Commission, Fishery Research Report 6.

GILBERT, C.H. AND B.W. EVERMANN. 1894. A report upon investigations in the Columbia River basin with descriptions of four new species of fishes. *Bulletin of the U.S. Fish Commission* 14: 169–207.

GILDERHUS, P.A. 1982. Effects of an aquatic plant and suspended clay on the activity of fish toxicants. *North American Journal of Fisheries Management* 2: 301–306.

GILPIN, M.E. 1987. Spatial structure and population vulnerability. Pages 125–139 in M.E. Soulé, editor. *Viable populations for conservation.* New York: Cambridge University Press.

GILPIN, M.E. AND M.E. SOULÉ. 1986. Minimum viable populations: processes of species extinction. Pages 19–34 in M.E. Soulé, editor. *Conservation biology: the science of scarcity and diversity.* Sunderland, Massachusetts: Sinauer Associates, Inc.

GIRARD, C.F. 1856. Notice upon the species of the genus *Salmo* of authors observed chiefly in Oregon and California. *Proceedings of the Academy of Natural Sciences of Philadelphia* 8: 217–220.

GLICK, D., F. MONTAIGNE AND V. MORELL. 2004. The heat is on. *National Geographic* 206, no. 3: 2–75.

GLOVA, G.J. 1978. Patterns and mechanisms of resource partitioning between stream populations of juvenile coho salmon (*Oncorhynchus kisutch*) and coastal cutthroat trout (*Salmo clarki clarki*). Ph.D. dissertation, University of British Columbia, Vancouver, B.C.

———. 1986. Interaction for food and space between experimental populations of juvenile coho salmon (*Oncorhynchus kisutch*) and coastal cutthroat trout (*Salmo clarki*) in a laboratory stream. *Hydrobiologia* 132: 155–168.

———. 1987. Comparison of allopatric cutthroat trout stocks with those sympatric with coho salmon and sculpins in small streams. *Environmental Biology of Fishes* 20: 275–284.

GOBLE, D.D., J.M. SCOTT AND E.W. DAVIS, EDITORS. 2005. *The Endangered Species Act at thirty. Vol. 1: renewing the conservation promise.* Washington, D.C.: Island Press.

GOLD, J.R. 1977. Systematics of western North American trout (*Salmo*) with notes on the redband trout Sheepheaven Creek, California. *Canadian Journal of Zoology* 55: 1858–1873.

GOLD, J.R., G.A.E. GALL AND S.J. NICOLA. 1978. Taxonomy of the Colorado cutthroat trout (*Salmo clarki pleuriticus*) of the Williamson Lakes, California. *California Fish and Game* 64, no. 2: 98–103.

GOLD, J.R., J.C. AVISE AND G.A.E. GALL. 1977. Chromosome cytology in the cutthroat trout series *Salmo clarki* (Salmonidae). *Cytologia* 42: 377–382.

GOLD, J.R., W.J. KAREL AND M.R. STRAND. 1980. Chromosome formulae of North American fishes. *Progressive Fish-Culturist* 42, no. 1: 10–23.

GOLENBERG, E.M. 1991. Amplification and analysis of Miocene plant fossil DNA. *Philosophical Transactions of the Royal Society of London, Series B* 333: 419–427.

———. AND 6 co-authors. 1990. Chloroplast DNA sequence from a Miocene *Magnolia* species. *Nature* 344: 656–658.

GOOD, J.M. AND K.L. PIERCE. 1996. *Interpreting the landscape: recent and ongoing geology of Grand Teton and Yellowstone National Parks.* Moose, Wyoming: Grand Teton Natural History Association.

GOODHART, G.W. (as told to A.C. Anderson). 1940. *Trails of early Idaho.* Caldwell, Idaho: The Caxton Printers, Ltd.

GORDON, C.D., D.W. CHAPMAN AND T.C. BJORNN. 1970. The preferences, opinions, and behavior of Idaho anglers as related to quality in salmonid fisheries. *Proceedings of the Western Association of State Game and Fish Commissioners* 49: 98–114.

GORE, A. 2006. *An inconvenient truth.* New York: Rodale Books (also a documentary film by the same title).

GORE, R. (photos by Jonathan Blair) 1989. Extinctions. *National Geographic* 175, no. 6: 663–699.

GOWAN, C. AND K.D. FAUSCH. 1996. Mobile brook trout in two high-elevation Colorado streams: re-evaluating the concept of restricted movement. *Canadian Journal of Fisheries and Aquatic Sciences* 53: 1370–1381.

GOWAN, C., M.K. YOUNG, K.D. FAUSCH AND S.C. RILEY. 1994. Restricted movement in resident stream salmonids: a paradigm lost? *Canadian Journal of Fisheries and Aquatic Sciences* 51: 2626–2637.

GOWAN, F.R. AND E.E. CAMPBELL. 1975. *Fort Bridger: island in the wilderness.* Provo, Utah: Brigham Young University Press.

GRAHAME, J.D. AND T.D. SISK, EDITORS. 2002. *Canyons, cultures and environmental change: an introduction to the land-use history of the Colorado Plateau.* Flagstaff: Northern Arizona University, Land Use History of North America Program. Electronic document available on the Internet at www.cpluhna. nau.edu/ (accessed March 3, 2005).

GRAVES, S. 1982. *Merwin, Yale and Swift reservoir study, 1978–1979.* Olympia: Washington State Game Department.

GRAYSON, D.K. 1987. The biogeographic history of small mammals in the Great Basin: observations on the last 20,000 years. *Journal of Mammalogy* 68: 359–375

GRAYSON, D.K. 1988. *Danger Cave, Last Supper Cave, and Hanging Rock Shelter: the faunas.* New York: American Museum of Natural History Anthropological Papers, No. 66.

———. 1993. *The desert's past: a natural pre-history of the Great Basin.* Washington, D.C.: Smithsonian Institution Press.

GREEN, D.H. 1990. *History of Island Park.* Ashton, Idaho: Gateway Publishing.

GREEN, D.M. 2004. Designatable units for status assessment of endangered species. *Conservation Biology* 19: 1813–1820.

GREENBACK CUTTHROAT TROUT RECOVERY TEAM. 1983. *Greenback cutthroat trout recovery plan.* Denver, Colorado: U.S. Fish and Wildlife Service.

———. 1998. *Greenback cutthroat trout recovery plan.* Denver, Colorado: U.S. Fish and Wildlife Service.

GREEN RIVER WESTSIDE WORKING GROUP. 1993. *Interagency five year plan (1993–1997), Green River westside tributary enclave.* Pinedale: Wyoming Game and Fish Department, Bridger-Teton National Forest and Bureau of Land Management Pinedale Resource Area.

———. *Summary of accomplishments for the Colorado River cutthroat trout interagency five year plan (1993–1997), Green River westside tributary enclave.* Pinedale: Wyoming Game and Fish Department, Bridger-Teton National Forest and Bureau of Land Management Pinedale Resource Area.

GREENE, W.S. 1937. *Colorado trout.* Denver, Colorado: Denver Museum of Natural History, Popular Series 2 [document not seen; cited in Behnke and Zarn (1976)].

GRESSWELL, R.E. 1980. Yellowstone Lake: a lesson in fishery management. Pages 143–147 in W. King, editor. *Wild Trout II, proceedings of the symposium.* Yellowstone National Park, Wyoming: Trout Unlimited, Federation of Fly Fishers, and National Park Service.

———., EDITOR. 1988. *Status and management of interior stocks of cutthroat trout.* Bethesda, Maryland: American Fisheries Society Symposium 4.

———. 1991. Use of antimycin for removal of brook trout from a tributary of Yellowstone Lake. *North American Journal of Fisheries Management* 11: 83–90.

———. 1995. Yellowstone cutthroat trout. Pages 36–54 in M.K. Young, technical editor. 1995. *Conservation assessment for inland cutthroat trout.* Fort Collins, Colorado: USDA Forest Service, Rocky Mountain Forest and Range Experiment Station, General Technical Report RM-256.

GRESSWELL, R.E. AND J.D. VARLEY. 1988. Effects of a century of human influence on the cutthroat trout of Yellowstone Lake. Pages 45–52 in R.E. Gresswell, editor. *Status and management of interior stocks of cutthroat trout.* Bethesda, Maryland: American Fisheries Society Symposium 4.

GRESSWELL, R.E. AND S.R. HENDRICKS. 2007. Population-scale movement of coastal cutthroat trout in a naturally isolated stream network. *Transactions of the American Fisheries Society* 136: 238–253.

GRESSWELL, R.E., W.J. LISS AND G.L. LARSON. 1994. Life-history organization of Yellowstone cutthroat trout (*Oncorhynchus clarki bouvieri*) in Yellowstone Lake. *Canadian Journal of Fisheries and Aquatic Sciences* 51 (Supplement 1): 298–309.

GRESSWELL, R.E., W.J. LISS, G.L. LARSON AND P.J. BARTLEIN. 1997. Influence of basin-scale physical variables on life history characteristics of cutthroat trout in Yellowstone Lake. *North American Journal of Fisheries Management* 17: 1046–1054.

GRIFFITH, J.S. 1972. Comparative behavior and habitat utilization of brook trout (*Salvelinus fontinalis*) and cutthroat trout (*Salmo clarki*) in small streams in northern Idaho. *Journal of the Fisheries Research Board of Canada* 29: 265–273.

———. 1988. Review of competition between cutthroat trout and other salmonids. Pages 134–140 in R.E. Gresswell, editor. *Status and management of interior stocks of cutthroat trout.* Bethesda, Maryland: American Fisheries Society Symposium 4.

GRIFFITH, J.S. AND R.W. SMITH. 1993. Use of winter concealment cover by juvenile cutthroat and brown trout in the South Fork of the Snake River, Idaho. *North American Journal of Fisheries Management* 13: 823–830.

GRILES, J.S., SIGNATOR. 2003. Grazing administration—exclusive of Alaska. *Federal Register* 68, no. 235: 68451–68474.

GRISWOLD, K.E. 1996. Genetic and meristic relationships of coastal cutthroat trout (*Oncorhynchus clarki clarki*) residing above and below barriers in two coastal basins. Master's thesis, Oregon State University, Corvallis.

GROSS, M.R. 1987. Evolution of diadromy in fishes. Pages 14–25 in M.J. Dadswell and 5 co-editors. *Common strategies of anadromous and catadromous fishes.* Bethesda, Maryland: American Fisheries Society Symposium 1.

GROSS, M.R., R.M. COLEMAN AND R.M. MCDOWALL. 1988. Aquatic productivity and the evolution of diadromous fish migration. *Science* 239: 1291–1293.

GSRO (Governor's Salmon Recovery Office). 1999. *Extinction is not an option: the statewide strategy to recover salmon.* Olympia, Washington: Governor's Salmon Recovery Office.

GULLEY, D.D. AND W.A. HUBERT. 1985. *Evaluation of the cutthroat trout fishery of Berry Creek and Owl Creek in Grand Teton National Park.* Laramie: Wyoming Cooperative Fish and Wildlife Research Unit Completion Report.

GUNCKEL, S.L. AND S.E. JACOBS. 2006. *Population assessment of Lahontan cutthroat trout, 2006. Annual Progress Report, Fish Research Project Oregon.* Salem: Oregon Department of Fish and Wildlife.

GUTZWILLER, L.A., R.M. MCNATT AND R.D. PRICE. 1997. Watershed restoration and grazing practices in the Great Basin: Marys River of Nevada. Pages 360–380 in J.E. Williams, C.A. Wood and M.P. Dombeck, editors. *Watershed restoration: principles and practices.* Bethesda, Maryland: American Fisheries Society.

GUY, T.J. 2004. Landscape-scale evaluation of genetic structure among barrier-isolated populations of coastal cutthroat trout (*Oncorhynchus clarki clarki*). Master's thesis, Oregon State University, Corvallis.

GYLLENSTEN, U., R.F. LEARY, F.W. ALLENDORF AND A.C. WILSON. 1985. Introgression between two cutthroat trout subspecies with substantial karyotypic, nuclear and mitochondrial genomic divergence. *Genetics* 111: 905–915.

HADLEY, K. 1984. *Status report on the Yellowstone cutthroat trout (Salmo clarki bouvieri) in Montana.* Helena, Montana: report submitted to Montana Department of Fish, Wildlife and Parks.

HAFEN, L.R. AND A.W. HAFEN, EDITORS. 1954. *Rufus B. Sage, letters and scenes in the Rocky Mountains, 1836–1847.* Volume 5, Far West and Rockies Historical Series. Glendale, California: The Arthur H. Clark Company.

HAFEN, L.R. AND A.W. HAFEN, EDITORS. 1959. *The diaries of William Henry Jackson, frontier photographer. Diaries 1866–1874.* Volume 10, Far West and Rockies Historical Series. Glendale, California: The Arthur H. Clark Company.

HAGEN, G.O. 1951. *Lake Alice spawning operation—1951.* Pinedale: Wyoming Game and Fish Department.

HAGENBUCK, W.W. 1970. A study of the age and growth of the cutthroat trout from the Snake River, Teton County, Wyoming. Master's thesis, University of Wyoming, Laramie.

HAGGERTY, J.C. 1997. *Brooks Peninsula: an ice age refugium on Vancouver Island.* Victoria, British Columbia: B.C. Ministry of Environment, Lands and Parks.

HAIG-BROWN, RODERICK L. 1947. *The western angler.* New York: William Morrow & Company.

———. 1964 [1975]. *Fisherman's fall.* New York: Crown Publishers.

HAINES, T.A. 1981. Acidic precipitation and its consequences for aquatic ecosyststems: a review. *Transactions of the American Fisheries Society* 110: 669–707.

HALDANE, J.B.S. 1956. The relation between density regulation and natural selection. *Proceedings of the Royal Society (London), Series B* 145: 306–308.

HALEY, T. 1978. A review of the literature of rotenone. *Journal of Environmental Pathology and Toxicology* 1: 315–337.

HALL, H.D. 2006. Endangered and threatened wildlife and plants: 12-month finding for a petition to list the Yellowstone cutthroat trout as threatened. *Federal Register* 71, no. 34: 8818–8831.

HALL, J.D., P.A. BISSON AND R.E. GRESSWELL, EDITORS. 1997. *Sea-run cutthroat trout: biology, management, and future conservation.* Corvallis: Oregon Chapter American Fisheries Society.

HALL, J.D., G.W. BROWN AND R.L. LANTZ. 1987. The Alsea Watershed Study: a retrospective. Pages 399–416 in E.O. Salo and T.W. Cundy, editors. *Streamside management: forestry and fishery interactions.* Seattle: University of Washington, Institute of Forest Resources.

HALLERMAN, E., EDITOR. 2002. *Population genetics: principles and practices for fisheries.* Bethesda, Maryland: American Fisheries Society.

HALLERMAN, E. AND A. KAPUSCINSKI. 1993. Potential impacts of transgenic and genetically manipulated fish on natural populations: addressing the uncertainties through field testing. Pages 93–112 in J.G. Cloud and G.H. Thorgaard, editors. *Genetic conservation of salmonid fishes.* New York: Plenum Press.

HALLOCK, C. 1883. *The sportsman's gazetteer and general guide.* New York: Orange Judd Company.

HALLOCK, CHARLES 1884. *The American Angler.* October 4, 1884.

HAMILTON, H.L. 1941. The biological action of rotenone on freshwater animals. *Proceedings of the Iowa Academy of Science* 48: 467–479.

HAMILTON, S.J. 2004. Review of selenium toxicity in the aquatic food chain. *Science of the Total Environment* 326: 1–31.

HAMILTON, S.J., K.J. BUHL, N.L. FAERBER, R.H. WIEDMEYER AND F.A. BALLARD. 1990. Toxicity of organic selenium in the diet to chinook salmon. *Environmental Toxicology and Chemistry* 9: 347–358.

HAMILTON, S.J. AND V.P. PALACE. 2001. Assessment of selenium effects in lotic ecosystems. *Ecotoxicology and Environmental Safety* 50: 161–166.

HAMMOND, G.P. AND A. REY. 1940. *Narratives of the Coronado expedition*. Albuquerque: University of New Mexico Press.

HANSEN, H.P. 1947. Postglacial forest succession, climate, and chronology in the Pacific Northwest. *Transactions of the American Philosophical Society* 37, part 1: 1–ff.

HANSEN, W.R. 1985. Drainage development of the Green River Basin in southwestern Wyoming and its bearing on fish biogeography, neotectonics, and paleoclimates. *The Mountain Geologist* 22, no. 4: 192–204.

HANSKI, I. 1998. Metapopulation dynamics. *Nature* 396: 41–49.

HANSKI, I. AND M.E. GILPIN, EDITORS. 1996. *Metapopulation biology: ecology, genetics and evolution*. New York: Academic Press.

HANZEL, D.A. 1959. The distribution of the cutthroat trout in Montana. *Proceedings of the Montana Academy of Sciences* 19: 32–71.

HARDIN, G. 1960. The competitive exclusion principle. *Science* 131: 1292–1297.

HARIG, A.L. AND K.D. FAUSCH. 2002. Minimum habitat requirements for establishing translocated cutthroat trout populations. *Ecological Applications* 12: 535–551.

HARIG, A.L., K.D. FAUSCH AND M.K. YOUNG. 2000. Factors influencing success of greenback cutthroat trout translocations. *North American Journal of Fisheries Management* 20: 994–1004.

HARPER, D.D. AND A.M. FARAG. 2004. Winter habitat use by cutthroat trout in the Snake River near Jackson, Wyoming. *Transactions of the American Fisheries Society* 133: 15–25.

HARRISON, S. 1991. Local extinction in a metapopulation context: an empirical evaluation. *Biological Journal of the Linnean Society* 42: 73–88.

HARTLEY, S.E. 1987. The chromosomes of salmonid fishes. *Biological Review* 62: 197–214.

HARTMAN, G.F. AND J.C. SCRIVENER. 1990. Impacts of forest practices on a coastal stream ecosystem, Carnation Creek, British Columbia. *Canadian Bulletin of Fisheries and Aquatic Sciences* 223.

HARTMAN, G.F. AND T.G. BROWN. 1987. Use of small, temporary, floodplain tributaries by juvenile salmonids in a west coast rain-forest drainage basin. *Canadian Journal of Fisheries and Aquatic Sciences* 44: 262–270.

HARVEY, A.G. 1945. Meredith Gairdner, Doctor of Medicine. *British Columbia Historical Quarterly* 9, no. 2: 89–111.

HARVEY, B.C. 1998. Influence of large woody debris on retention, immigration, and growth of coastal cutthroat trout (*Oncorhynchus clarki clarki*) in stream pools. *Canadian Journal of Fisheries and Aquatic Sciences* 55: 1902–1908.

HARVEY, B.C., R.J. NAKAMOTO AND J.L. WHITE. 1999. Influence of large woody debris and a bankfull flood on movement of adult resident coastal cutthroat trout (*Oncorhynchus clarki clarki*) during fall and winter. *Canadian Journal of Fisheries and Aquatic Sciences* 56: 2161–2166.

HASKINS, R.L. 1993. *Current status of Bonneville cutthroat trout in Nevada*. Reno: Nevada Division of Wildlife.

HAUER, F.R. AND G.A. LAMBERTI, EDITORS. 1996 [2006]. *Methods in stream ecology*. [2nd edition, 2006]. New York: Academic Press.

HAWKINS, D.K. 1997. Hybridization between coastal cutthroat trout (*Oncorhynchus clarki clarki*) and steelhead (*O. mykiss*). Ph.D. dissertation, University of Washington, Seattle.

HAWKINS, D.K. AND T.P. QUINN. 1996. Critical swimming velocity and associated morphology of juvenile coastal cutthroat trout (*Oncorhynchus clarki*), steelhead trout (*Oncorhynchus mykiss*), and their reciprocal hybrids. *Canadian Journal of Fisheries and Aquatic Sciences* 53: 1487–1496.

HAWKINS, D.K. AND C.J. FOOTE. 1998. Early survival and development of coastal cutthroat trout (*Oncorhynchus clarki clarki*), steelhead (*Oncorhynchus mykiss*), and reciprocal hybrids. *Canadian Journal of Fisheries and Aquatic Sciences* 55: 2097–2104.

HAYDEN, F.V. 1872. *Preliminary report of the United States Geological Survey of Montana and portions of adjacent territories, 1871*. Washington, D.C.: U.S. Government Printing Office.

HAYDEN, P.S. 1968. The reproductive behavior of Snake River cutthroat trout in three tributary streams in Wyoming. Master's thesis, University of Wyoming, Laramie.

HAYS, J.D., J. IMBRIE AND N.J. SHACKLETON. 1976. Variations in the earth's orbit: pacemaker of the Ice Ages. *Science* 194: 1121–1132.

HAYWARD, T.L. 1997. Pacific Ocean climate change: atmospheric forcing, ocean circulation and ecosystem response. *Trends in Ecology and Evolution* 12, no. 4: 150–154.

HAZZARD, A.S. 1935. A preliminary study of an exceptionally productive trout water, Fish Lake, Utah. *Transactions of the American Fisheries Society* 65: 122–128.

HAZZARD, A.S. AND M.J. MADSEN. 1933. Studies of the food of the cutthroat trout. *Transactions of the American Fisheries Society* 63: 198–207.

HAZZARD, L.K. AND A. McDONALD. 1981. The Snake River cutthroat: a new trout for Colorado. *Colorado Outdoors* 30, no. 4: 1–3.

HEARN, W.E. 1987. Interspecific competition and habitat segregation among stream-dwelling trout and salmon: a review. *Fisheries* 12, no. 5: 24–31.

HEGGENES, J., T.G. NORTHCOTE AND A. PETER. 1991a. Spatial stability of cutthroat trout (*Oncorhynchus clarki*) in a small, coastal stream. *Canadian Journal of Fisheries and Aquatic Sciences* 48: 757–762.

———. 1991b. Seasonal habitat selection and preferences by cutthroat trout (*Oncorhynchus clarki*) in a small, coastal stream. *Canadian Journal of Fisheries and Aquatic Sciences* 48: 1364–1370.

HEIZER, R.F. 1951. Preliminary report on the Leonard Rockshelter site, Pershing County, Nevada. *American Antiquity* 17, no. 2: 89–98.

HENDERSON, R., J.L. KERSHNER AND C.A. TOLINE. 2000. Timing and location of spawning by nonnative wild rainbow trout and native cutthroat trout in the South Fork Snake River, Idaho, with implications for hybridization. *North American Journal of Fisheries Management* 20: 584–596.

HENDRICKS, S.R. AND R.E. GRESSWELL. 2001. Spatial and temporal variation in habitat use of coastal cutthroat trout in an isolated watershed. Pages 59–60 in Shepard, B., editor. *Practical approaches for conserving native inland fishes of the west.* Missoula: Montana Chapter American Fisheries Society.

HEPWORTH, D.K., C.B. CHAMBERLAIN AND M.J. OTTENBACHER. 1999. Comparative sport fish performance of Bonneville cutthroat trout in three small put-grow-and-take reservoirs. *North American Journal of Fisheries Management* 19: 774–785.

HEPWORTH, D.K., M.J. OTTENBACHER AND C.B. CHAMBERLAIN. 2002. A review of a quarter century of native trout conservation in southern Utah. *Intermountain Journal of Sciences* 8: 125–142.

———. 2004. *A 5-year review of the Colorado River cutthroat trout wild brood stock program at Dougherty Basin Lake, Utah, 1999–2003.* Salt Lake City: Utah Division of Wildlife Resources, Publication no. 04–05.

HEPWORTH, D.K., M.J. OTTENBACHER, C.B. CHAMBERLAIN AND J.E. WHELAN. 2003. *Abundance of Bonneville cutthroat trout in southern Utah, 2001–2002, compared to previous surveys.* Cedar City: Utah Division of Wildlife Resources, Publication no. 03–18.

HEPWORTH, D.K., M.J. OTTENBACHER AND L.N. BERG. 1997. Distribution and abundance of native Bonneville cutthroat trout (*Oncorhynchus clarki utah*) in southwestern Utah. *Great Basin Naturalist* 57: 11–20.

HERGER, L.G., W.A. HUBERT AND M.K. YOUNG. 1996. Comparison of habitat composition and cutthroat trout abundance at two flows in small mountain streams. *North American Journal of Fisheries Management* 16: 294–301.

HERSHLER, R. AND H. LIU. 2004. A molecular phylogeny of aquatic gastropods provides a new perspective on biogeographic history of the Snake River region. *Molecular Phylogenetics and Evolution* 32: 927–937.

HESTER, R.L. 1983. Late paleo-Indian occupations at Baker Cave, southwestern Texas. *Bulletin of the Texas Archeological Society* 53: 101–119.

HETHERINGTON, R., J. BARRIE, R. REID, R. MACLEOD AND D. SMITH. 2004. Paleogeography, glacially induced crustal displacement, and Late Quaternary coastlines on the continental shelf of British Columbia, Canada. *Quaternary Science Reviews* 23: 295–318.

HEWITT, EDWARD R. 1948 [1966]. *A trout and salmon fisherman for seventy-five years.* New York: Charles Scribner's Sons [reprinted 1966 by Abercrombie and Fitch].

HEWKIN, J. 1960. *John Day District annual report.* John Day, Oregon: Oregon State Game Commission.

HICKMAN, T. AND R.F. RALEIGH. 1982. *Habitat suitability index models: cutthroat trout.* Washington, D.C.: U.S. Fish and Wildlife Service, Western Energy and Land Use Team, Office of Biological Services, Report FWS/OBS-82/10.5.

HICKMAN, T.J. 1977. *Studies on relict populations of Snake Valley cutthroat trout in Utah.* Fort Collins: Colorado State University Department of Fishery and Wildlife Biology, prepared for U.S. Bureau of Land Management, Utah State Office, Contract no. UT-910-PH7-375.

———. 1978a. *Fisheries investigations in the Pilot Peak Mountain Range, Utah-Nevada 1977.* U.S. Department of the Interior, Bureau of Land Management, Contract UT-910-PH7-822.

———. 1978b. Systematic study of the native trout of the Bonneville Basin. Master's thesis, Colorado State University, Fort Collins.

———. 1984. *Status report for Salmo clarki utah (Bonneville cutthroat trout).* Sandy, Utah: report prepared for Office of Endangered Species, U.S. Fish and Wildlife Service, Denver, Colorado.

HICKMAN, T.J. AND D.A. DUFF. 1978. Current status of cutthroat trout subspecies in the western Bonneville Basin. *Great Basin Naturalist* 38:193–202.

HICKMAN, T.J. AND R.J. BEHNKE. 1979. Probable discovery of the original Pyramid Lake cutthroat trout. *Progressive Fish-Culturist* 41: 135–137.

HIGGINS, P., S. DOBUSH AND D. FULLER. 1992. *Factors in northern California threatening stocks with extinction.* Arcata, California: Humboldt Chapter American Fisheries Society.

HILBORN, R. AND C.J. WALTERS. 1992. *Quantitative fisheries stock assessment: choice, dynamics and uncertainty.* New York: Chapman and Hall.

HILDEBRAND, S.F. AND I.L. TOWERS. 1927. The food of trout in Fish Lake, Utah. *Ecology* 8: 389–397.

HILDERBRAND, R.H. 2003. The roles of carrying capacity, immigration, and population synchrony on persistence of stream-resident cutthroat trout. *Biological Conservation* 110: 257–266.

HILDERBRAND, R.H. AND J.L. KERSHNER. 2000a. Conserving inland cutthroat trout in small streams: how much stream is enough? *North American Journal of Fisheries Management* 20: 513–520.

HILDERBRAND, R.H. AND J.L. KERSHNER. 2000b. Movement patterns of stream-resident cutthroat trout in Beaver Creek, Idaho-Utah. *Transactions of the American Fisheries Society* 129: 1160–1170.

———. 2004a. Are there differences in growth and condition between mobile and resident cutthroat trout? *Transactions of the American Fisheries Society* 133: 1042–1046

———. 2004b. Influence of habitat type on food supply, selectivity, and diet overlap of Bonneville cutthroat trout and nonnative brook trout in Beaver Creek, Idaho. *North American Journal of Fisheries Management* 24: 33–40.

HILLIS, D.M. AND C. MORITZ, EDITORS. 1990. *Molecular systematics.* Sunderland, Massachusetts: Sinauer Associates, Inc.

HINDAR, K., B. JONSSON, N. RYMAN AND G. STAHL. 1991. Genetic relationships among landlocked, resident, and anadromous brown trout, *Salmo trutta* L. *Heredity* 66: 83–91.

HINDS, W.T. 1983. The incursion of acid deposition into western North America. *Environmental Conservation* 10: 53–58.

HINRICHS, K.-U., L.R. HMELO AND S.P. SYLVA. 2003. Molecular fossil record of elevated methane levels in late Pleistocene coastal waters. *Science* 299: 1214–1217.

HIRSCH, C.L., S.E. ALBEKE, AND T.P. NESLER. 2006. *Range-wide status of Colorado River cutthroat trout (Oncorhynchus clarkii pleuriticus): 2005.* Craig, Colorado: Colorado River Cutthroat Trout Conservation Coordinating Team Report.

HITT, N.P., C.A. FRISSELL, C.C. MUHLFELD AND F.W. ALLENDORF. 2003. Spread of hybridization between native westslope cutthroat trout, *Oncorhynchus clarki lewisi*, and nonnative rainbow trout, *Oncorhynchus mykiss. Canadian Journal of Fisheries and Aquatic Sciences* 60: 1440–1451.

HOAR, W.S. 1988. The physiology of smolting salmonids. Pages 275–343 in W.S. Hoar and D.J. Randall, editors. *Fish physiology.* Volume 11B. New York: Academic Press.

HOCUTT, C.H. AND E.O. WILEY, EDITORS. 1986. *The zoogeography of North American freshwater fishes.* New York: John Wiley & Sons.

HODSON, P.V., D.J. SPRY AND B.R. BLUNT. 1980. Effects on rainbow trout (*Salmo gairdneri*) of a chronic exposure to water borne selenium. *Canadian Journal of Fisheries and Aquatic Sciences* 37: 233–240.

HOELZEL, A.R. AND G.A. DOVER. 1991. *Molecular genetic ecology.* New York: Oxford University Press.

HOLBERTON, W. 1890 [1980]. Idaho trout. *Harper's Weekly* [reprinted in *The American Fly Fisher* (the journal of the American Museum of Fly Fishing) 7, no. 3: 18–19].

HOLDEN, P. AND 5 co-authors. 1974. Threatened fishes of Utah. *Proceedings of the Utah Academy of Sciences, Arts, and Letters* 51: 46–55.

HOLLENDER, B.A. AND R.F. CARLINE. 1994. Injury to wild brook trout by backpack electrofishing. *North American Journal of Fisheries Management* 14: 643–649.

HOLM, J. AND 7 co-authors. 2003. An assessment of the development and survival of wild rainbow trout (*Oncorhynchus mykiss*) and brook trout (*Salvelinus fontinalis*) exposed to elevated selenium in an area of active coal mining. Pages 257–273 in H.J. Browman and A.B. Skiftesvik, editors. *The Big Fish Band, Proceedings of the 28th Annual Larval Fish Conference, Bergen, Norway.*

HOLMES, K.L. 1955. Mt. St. Helens' recent eruptions. *Oregon Historical Quarterly* 56: 198–206.

HOLT, R.L. 1992. *Beneath these red cliffs: an ethnohistory of the Utah Paiutes.* Albuquerque: University of New Mexico Press.

HOOTON, B. 1997. Status of coastal cutthroat trout in Oregon. Pages 57–67 in J.D. Hall, P.A. Bisson and R.E. Gresswell, editors. 1997. *Sea-run cutthroat trout: biology, management, and future conservation.* Corvallis: Oregon Chapter American Fisheries Society.

HORAN, D.L., J.L. KERSHNER, C.P. HAWKINS AND T.A. CROWL. 2000. Effects of habitat area and complexity on Colorado River cutthroat trout density in Uinta Mountain streams. *Transactions of the American Fisheries Society* 129: 1250–1263.

HORBERG, L. 1954. Rocky Mountain and continental Pleistocene deposits in the Waterton region, Alberta, Canada. *Geological Society of America Bulletin* 65: 1093–1150.

HORNIG, C.E., D.A. TERPENING AND M.W. BOGUE. 1988. *Coeur d'Alene basin EPA water quality monitoring, 1972–1986*. Seattle, Washington: U.S. Environmental Protection Agency, EPA-910/9-88-216.

HOSEA, R.C. AND B. FINLAYSON. 2005. *Controlling the spread of New Zealand mud snails on wading gear*. Rancho Cordova: California Department of Fish and Game, Pesticides Investigations Unit, Office of Spill Prevention and Response, Administrative Report 2005-02.

HOUGHTON, S.G. 1976 [1994]. *A trace of desert water: the Great Basin story*. Glendale, California: Arthur H. Clark Company [reprinted 1994, Salt Lake City: Howe Brothers].

HOUSE, R.A. 1995. Temporal variation in abundance of an isolated population of cutthroat trout in western Oregon, 1981–1991. *North American Journal of Fisheries Management* 15: 33–41.

HOUSE, R.A. AND P.L. BOEHNE. 1986. Effects of instream structures on salmonid habitat and populations in Tobe Creek, Oregon. *North American Journal of Fisheries Management* 6: 38–46.

HOWETT, E.A. 1968. A historical geography of Alpine County and the Alpine Nevada Alliance. Master's thesis, University of California, Berkeley.

HSÜ, K.J., W.B.F. RYAN AND M.B. CITA. 1973. Late Miocene desiccation of the Mediterranean. *Nature* 242: 240–244.

HUBBARD, J.P. 1976. *Status and future of the Rio Grande cutthroat trout*. Santa Fe: New Mexico Department of Game and Fish, Endangered Species Program Report.

HUBBS, C.L. AND C. HUBBS. 1953. An improved graphical analysis and comparison of series of samples. *Systematic Zoology* 2: 49–57.

HUBBS, C.L. AND R.R. MILLER, 1948. Correlation between fish distribution and hydrographic history in the desert basins of western United States. Pages 17–166 in *The Great Basin, with emphasis on glacial and postglacial times*. Bulletin of the University of Utah, Biological Series 10.

HUBBS, C.L., R.R. MILLER AND L.C. HUBBS. 1974. Hydrographic history and relict fishes of the north-central Great Basin. *California Academy of Sciences Memoirs* 7: 1–259.

HUGHES, D. 1992. *The Yellowstone River and its angling*. Portland, Oregon: Frank Amato Publications.

———. 2002. *Trout from small streams*. Mechanicsburg, Pennsylvania: Stackpole Books.

HUME, M. (with H. Thommasen) 1998. *River of the angry moon: seasons on the Bella Coola*. Seattle: University of Washington Press.

HUNTINGTON, C., W. NEHLSEN AND J. BOWERS. 1996. A survey of healthy native stocks of anadromous salmonids in the Pacific Northwest and California. *Fisheries* 21, no. 3: 6–14.

HUSTON, J.E. 1969–1973. *Reservoir investigations*. Helena: Montana Fish and Game Department. Federal Aid to Fish Restoration, Project Reports F-34-R-2, R-3, R-4, R-5 and R-6.

HUSTON, J.E., P. HAMLIN AND B. MAY. 1984. *Lake Koosanusa fisheries investigations, final completion report*. Helena: Montana Department of Fish, Wildlife and Parks.

HUTCHINGS, J.M. 1857 [1962]. A jaunt to Honey Lake Valley and Noble's Pass. *Hutchings Illustrated California Magazine* 1, no. 12: 317–329 [reprint 1962 by Howell North, Berkeley, California].

HYNES, H.B.N. 1970. *The ecology of running waters*. Toronto, Ontario, Canada: University of Toronto Press.

IMBRIE, J. AND 8 co-authors. 1984. The orbital theory of Pleistocene climate: support from a revised chronology of the marine δ18O record. Pages 269–305 in A. Berger, J. Imbrie, J. Hays, G. Kukla and B. Saltzman, editors. *Milankovitch and climate. Part 1. NATO Advanced Science Institute Series*. Boston: D. Reidel Publishing Company.

INGLES, L.G. 1965. *Mammals of the Pacific states: California, Oregon, and Washington*. Stanford, California: Stanford University Press.

INTERNATIONAL COMMISSION ON ZOOLOGICAL NOMENCLATURE. 1999. *International code of zoological nomenclature, 4th edition*. London: International Trust for Zoological Nomenclature.

IRVINE, J.R. AND 5 co-authors. 2005. Canada's Species At Risk Act: an opportunity to protect "endangered" salmon. *Fisheries* 30, no. 12: 11–19.

IRVING, R.B. 1956. Ecology of the cutthroat trout of Henry's Lake, Idaho. *Transactions of the American Fisheries Society* 84: 275-196.

IRVING, W. 1836 [1950, 1987]. *Astoria: adventures in the Pacific Northwest*. Philadelphia: Carey, Lea and Blanchard. [Reprinted 1950 by Binfords & Mort, Portland Oregon, and 1987 by KPI, Limited, London and New York].

———. 1837 [1868, 1954, 2001]. *Adventures of Captain Bonneville*. Philadelphia: Carey, Lea and Blanchard. [Reprinted in 1868 by G.P. Putnam's Sons, New York, 1954 by Binfords & Mort, Portland, Oregon, and 2001 by The Narrative Press, Santa Barbara, California].

ISAAK, D.J. AND W.A. HUBERT. 2004. Nonlinear response of trout abundance to summer stream temperatures across a thermally diverse montane landscape. *Transactions of the American Fisheries Society* 133: 1254–1259.

ISRAEL, J.A., J.F. CORDES AND B. MAY. 2002. *Genetic divergence among Paiute cutthroat trout populations in the Silver King drainage and out-of-basin transplants.* Davis, California: University of California, Davis, Department of Animal Sciences, Genomic Variation Laboratory. Report not seen; cited in U.S. Fish and Wildlife Service (2003).

IUBNC (International Union of Biochemistry Nomenclature Committee). 1984. *Enzyme nomenclature 1984.* Orlando, Florida: Academic Press.

IVES, R.L. 1938. Glaciation of the headwaters of the Laramie, Cache La Poudre, and Colorado rivers, Colorado. *Geological Society of America Bulletin* 49: 1888–1889 (abstract only).

JABLONSKI, D. 1986. Background and mass extinctions: the alternation of macroevolutionary regimes. *Science* 231: 129–133.

———. 1997. Progress at the K-T boundary. *Nature* 387: 354–355.

JACKMAN, E.R. AND J. SCHARFF. 1967. *Steens Mountain in Oregon's high desert country.* Caldwell, Idaho: The Caxton Printers, Ltd.

JACKMAN, E.R. AND R.A. LONG. 1964. *Oregon Desert.* Caldwell, Idaho: The Caxton Printers, Ltd.

JACKSON, D., EDITOR. 1966. *The journals of Zebulon Montgomery Pike, with letters and related documents.* Two volumes. Norman: University of Oklahoma Press.

JACKSON, D.D. 1975. *Sagebrush country.* New York: Time-Life Books.

———. 1980. *Gold dust.* New York: Alfred A. Knopf.

JACKSON, W.H. (in collaboration with H.R. Driggs). 1929. *The pioneer photographer: Rocky Mountain adventures with a camera.* Yonkers-on-Hudson, New York: World Book Company.

JACOBS, S.E. 1981. Stream habitat utilization and behavior of sympatric and allopatric wild cutthroat trout (*Salmo clarki*) and hatchery coho salmon (*Oncorhynchus kisutch*). Master's thesis, Oregon State University, Corvallis.

JACOBSEN, R.D. 1982. New impacts by man in the upper Colorado River basin. Pages 71–80 in W.H. Miller, H.M. Tyus and C.A. Carlson, editors. *Fishes of the upper Colorado River system: present and future.* Proceedings of a symposium held at the Annual Meeting of the American Fisheries Society, September 18, 1981. Albuquerque, New Mexico: Western Division American Fisheries Society.

JAEGER, M.E., R.W. VAN KIRK AND T. KELLOG. 2000. Distribution and status of Yellowstone cutthroat trout in the Henry's Fork watershed. *Intermountain Journal of Science* 6: 197–216.

JAHN, L.A. 1969. Movements and homing of cutthroat trout (*Salmo clarki*) from open water areas of Yellowstone Lake. *Journal of the Fisheries Research Board of Canada* 26: 1243–1261.

JAKOBER, M.J., T.E. MCMAHON AND R.F. THUROW. 2000. Diel habitat partitioning by bull trout and cutthroat trout during fall and winter in Rocky Mountain streams. *Environmental Biology of Fishes* 59: 79–89.

JAKOBER, M.J., T.E. MCMAHON, R.F. THUROW AND C.G. CLANCY. 1998. Role of stream ice on fall and winter movements and habitat use by bull trout and cutthroat trout in Montana headwater streams. *Transactions of the American Fisheries Society* 127: 223–235.

JAMES, E.W. AND H.J. PEETERS. 1988. *California Mammals.* Berkeley: University of California Press.

JAMES, G.W. 1921 [1956]. *The lake of the sky, Lake Tahoe.* Pasadena, California: Radiant Life Press [reprinted Chicago: Charles T. Powner Co.].

JANIS, C.M. 1993. Tertiary mammal evolution in the context of changing climates, vegetation, and tectonic events. *Annual Review of Ecology and Systematics* 24: 467–500.

JARRETT, R.O. AND H.E. MALDE. 1987. Paleodischarge of the late Pleistocene Bonneville flood, Snake River, Idaho, computed from new evidence. *Geological Society of America Bulletin* 99: 127–134.

JAUQUET, J.M. 2002. Coastal cutthroat trout (*Oncorhynchus clarki clarki*) diet in south Puget Sound, Washington 1999–2002. Master's thesis, The Evergreen State College, Olympia, Washington.

JENKINS, T.M. 1969. Social structure, position choice and microdistribution of two trout species (*Salmo trutta* and *Salmo gairdneri*) resident in mountain streams. *Animal Behavior Monographs, volume 2.* London: Bailliere, Tindall and Cassell.

JESPERSEN, D.M. 1981. A study of the effects of water diversion on the Colorado River cutthroat trout (*Oncorhynchus clarki pleuriticus*) in the drainage of the North Fork of the Little Snake River in Wyoming. Master's thesis, University of Wyoming, Laramie.

JILLSON, W.R. 1917. The volcanic activity of Mount St. Helens and Mount Hood in historical time. *Geographic Reviews* 3: 482–483.

JOHNSON, G.L., D.H. BENNETT AND T.C. BJORNN. 1981. *Juvenile emigration of Lahontan cutthroat trout in the Truckee River-Pyramid Lake system.* Moscow: University of Idaho, College of Forest, Wildlife and Range Sciences.

JOHNSON, H.E. 1961. *Observations of the life history of cutthroat trout (Salmo clarki) in the Flathead River drainage: northwest Montana fish study.* Helena: Montana Montana Fish and Game Department. Federal Aid to Fish Restoration, Job Completion Report F-7-R-10.

————. 1963. Observations of the life history of cutthroat trout (*Salmo clarki*) in the Flathead River drainage, Montana. *Proceedings of the Montana Academy of Sciences* 23: 96–110.

JOHNSON, L. 2004. *Fly-fishing for coastal cutthroat trout.* Portland, Oregon: Frank Amato Publications.

JOHNSON, O. AND W.H. WINTER. 1846. *Route across the Rocky Mountains with a description of Oregon and California.* Lafayette, Indiana: John B. Semans, Printer.

JOHNSON, O.W. AND 5 co-authors. 1994. *Status review for Oregon's Umpqua River sea-run cutthroat trout.* Seattle: U.S. Department of Commerce, National Marine Fisheries Service, Northwest Fisheries Science Center. NOAA Technical Memorandum NMFS-NWFSC-15.

JOHNSON, O.W. AND 7 co-authors. 1999. *Status review of coastal cutthroat trout from Washington, Oregon, and California.* Seattle: U.S. Department of Commerce, National Marine Fisheries Service, Northwest Fisheries Science Center. NOAA Technical Memorandum NMFS-NWFSC-37.

JOHNSON, R., JR. 1999. *Arizona trout: a fly fishing guide.* Portland, Oregon: Frank Amato Publications.

JOHNSTON, J.M. 1979. *Sea-run cutthroat: Stillaguamish River creel census (1978) and harvest limit recommendations.* Olympia: Washington State Game Department.

————. 1982. Life history of anadromous cutthroat with emphasis on migratory behavior. Pages 123–127 in E.L. Brannon and E.O. Salo, editors. *Salmon and trout migratory behavior, proceedings of a symposium.* Seattle: University of Washington, School of Fisheries.

JOHNSTON, J.M. AND S.P. MERCER. 1976. *Sea-run cutthroat in saltwater pens: brood stock development and extended juvenile rearing (with life history compendium).* Olympia: Washington Game Department, Fisheries Research Report.

JOHNSTONE, H.C. AND F.J. RAHEL. 2003. Assessing temperature tolerance of Bonneville cutthroat trout based on constant and cycling thermal regimes. *Transactions of the American Fisheries Society* 132: 92–99.

JONES, A. 2000. Effects of cattle grazing on North American arid ecosystems: a quantitative review. *Western North American Naturalist* 60: 155–164.

JONES, D.E. 1972–1976. *Steelhead and sea-run cutthroat trout life history in southeast Alaska.* Juneau: Alaska Department. of Fish and Game, Progress Reports AFS-41.

JONES, J.D. AND C.L. SEIFERT. 1997. Distribution of mature sea-run cutthroat trout overwintering in Auke Lake and Lake Eva in southeastern Alaska. Pages 27–28 in J.D. Hall, P.A. Bisson and R.E. Gresswell, editors. *Sea-run cutthroat trout: biology, management, and future conservation.* Corvallis: Oregon Chapter American Fisheries Society.

JONES, K.K., J.M. DAMBACHER, B.G. LOVATT, A.G. TALABERE AND W. BOWERS. 1998. Status of Lahontan cutthroat trout in the Coyote Lake Basin, southeast Oregon. *North American Journal of Fisheries Management* 18: 308–317.

JONES, R.D., J.D. VARLEY, D.E. JENNINGS, S.M. RUBRECHT AND R.E. GRESSWELL. 1979. *Fishery and aquatic management program in Yellowstone National Park.* Yellowstone National Park, Wyoming: U.S. Fish and Wildlife Service Technical Report for 1979.

JONES, R.F. 1972. *Wapato Indians of the lower Columbia River valley: their history and prehistory.* Vancouver, Washington: privately printed.

JONSSON, B. 1982. Diadromous and resident trout, *Salmo trutta*: is their difference due to genetics? *Oikos* 38: 297–300.

JONSSON, B. AND N. JONSSON. 1993. Partial migration: niche shift versus sexual maturation in fishes. *Reviews in Fish Biology and Fisheries* 3: 348–365.

JONSSON, B. AND 7 co-authors. 1988. Life history variation of polymorphic arctic charr (*Slavelinus alpinus*) in Thingvallavatn, Iceland. *Canadian Journal of Fisheries and Aquatic Sciences* 45: 1537–1547.

JORDAN, D.S. 1889. Letter to the editor re: the yellow-finned trout of the Twin Lakes, Colorado, September 19, 1889. *Forest and Stream* 33, no 9: 167.

————. 1891a. A reconnaissance of the streams and lakes of the Yellowstone National Park, Wyoming, in the interest of the U.S. Fish Commission. *Bulletin of the U.S. Fish Commission* 9: 41–63.

————. 1891b. Report of explorations in Colorado and Utah during the summer of 1889, with an account of the fishes found in each of the river basins examined. *Bulletin of the U.S. Fish Commission* 9: 1–40.

————. 1894. Description of new varieties of trout. Pages 142–143 in *13th Biennial Report of the State Board of Fish Commissioners of California, 1893–1894*. Sacramento, California.

————. 1896. Notes on fishes little known or new to science. *Proceedings of the California Academy of Sciences, 2nd Series* 6: 201–244.

————. 1920. The trout of the Rio Grande. *Copeia* 85 (1920): 72–73.

————. 1922. *The days of a man: being memories of a naturalist, teacher, and minor prophet of democracy*. Two volumes. Yonkers-On-Hudson, New York: World Book.

JORDAN, D.S. AND B.W. EVERMANN. 1890. Description of the yellow-finned trout of Twin Lakes, Colorado. *Proceedings of the U.S. National Museum* 12: 543–454.

————. 1896. The fishes of North and Middle America. *Bulletin of the U.S. National Museum* 47, part 1.

————. 1898. The fishes of North and Middle America. *Bulletin of the U.S. National Museum* 47, part 2.

————. 1902. *American food and game fishes*. New York: Doubleday, Page and Company.

JORDAN, D.S., B.W. EVERMANN AND H.W. CLARK. 1930. Checklist of fishes and fishlike vertebrates of North and Middle America north of the northern boundary of Venezuela and Colombia. *U.S. Fish Commission Report 1928*, part 2.

JORDAN, D.S. AND C.H. GILBERT. 1883. Synopsis of the fishes of North America. *Bulletin of the U.S. National Museum* 16.

JORDAN, D.S. AND H.W. HENSHAW. 1878. Report upon fishes collected during the years 1875, 1876 and 1877 in California and Nevada. In *Annual report upon the geographical survey west of the one-hundredth meridian.*

Washington, D.C.: Annual Report of U.S. Army Chief of Engineers for 1878, Part 3, Appendix NN, pages 187–200.

JORDAN, D.S. AND J. GRINNELL. 1908. Description of a new species of trout (*Salmo evermanni*) from the upper Santa Ana River, Mount San Gorgonio, southern California. *Proceedings of the Biological Society of Washington* 21: 31–32.

JOSENHANS, H.W., D. FEDJE, K. CONWAY AND J. BARRIE. 1995. Postglacial sea levels on the western Canadian continental shelf: evidence for rapid change, extensive subaerial exposure, and early human habitation. *Marine Geology* 125: 73–94.

JOSENHANS, H.W. AND 5 co-authors. 1993. Surficial geology of the Queen Charlotte basin: evidence of submerged proglacial lakes at 170 m on the continental shelf of western Canada. *Geological Survey of Canada, Current Research, Part A*: 119–127 (Paper 93-1A).

JOSSELYN, S. 1982. *Rio Grande cutthroat trout management*. Monte Vista: Colorado Division of Wildlife, Nongame Investigations, Job Progress Report FW-23-R-1.

JOYCE, M.P. 2001. Reproduction of Snake River cutthroat trout in spring streams tributary to the Salt River, Wyoming. Master's thesis, University of Wyoming, Laramie.

JOYCE, M.P. AND W.A. HUBERT. 2004. Spawning ecology of finespotted Snake River cutthroat trout in spring streams of the Salt River Valley, Wyoming. *Western North American Naturalist* 64: 78–85.

JUDAY, C. 1907a. Notes on Lake Tahoe, its trout and trout fishing. *Bulletin of the U.S. Bureau of Fisheries* 26: 133–146.

————. 1907b. A study of Twin Lakes, Colorado, with special consideration to the foods of the trouts. *Bulletin of the U.S. Bureau of Fisheries* 26: 147–178.

JUNE, J.A. 1981. Life history and habitat utilization of cutthroat trout (*Salmo clarki*) in a headwater stream on the Olympic Peninsula, Washington. Master's thesis, University of Washington, Seattle.

KAEDING, L.R. 2001. Endangered and threatened wildlife and plants: 90-day finding for a petition to list the Yellowstone cutthroat trout as threatened. *Federal Register* 66, no. 371: 11244–11249.

————. 2006. Management implications. Page 10 in R.W. Van Kirk, J.M. Capurso and M.A. Novak, editors. 2006. *Exploring differences between fine-spotted and large-spotted Yellowstone cutthroat trout*. Symposium Proceedings, Idaho Chapter American Fisheries Society, Boise, Idaho. Available on the Internet at www.fisheries. org/idaho/.

KAEDING, L.R. AND G.D. BOLTZ. 2001. Spatial and temporal relations between fluvial and allacustrine Yellowstone cutthroat trout, *Oncorhynchus clarki bouvieri*, spawning in the Yellowstone River, outlet stream of Yellowstone Lake. *Environmental Biology of Fishes* 61: 395–406.

KAEDING, L.R., G.D. BOLTZ AND D.G. CARTY. 1994. Lake trout discovered in Yellowstone Lake threaten native cutthroat trout. *Fisheries* 21, no. 3: 16–20.

KALLEBERG, H. 1958. Observations in a stream tank of territoriality and competition in juvenile salmon and trout (*Salmo salar* and *S. trutta*). *Institute of Freshwater Research, Drottningholm* 39: 55–98.

KAPLINSKI, M.A. 1991. Geomorphology and geology of Yellowstone Lake, Yellowstone National Park, Wyoming. Master's thesis, Northern Arizona University, Flagstaff.

KAREIVA, P.M., J.G. KINGSOLVER AND R.B. HUEY. 1993. *Biotic interactions and global change*. Sunderland, Massachusetts: Sinauer Associates.

KARNELLA, C. 1996. Endangered and threatened species: endangered status for Umpqua River cutthroat trout in Oregon. *Federal Register* 61, no. 155: 41514–41522.

KATZ, M.E., D.K. PAK, G.R. DICKENS AND K.G. MILLER. 1999. The source and fate of massive carbon input during the latest Paleocene thermal maximum. *Science* 286: 1531–1533.

KAUFFMAN, J.B. AND W.C. KRUEGER. 1984. Livestock impacts on riparian ecosystems and streamside management implications: a review. *Journal of Range Management* 37: 430–438.

KEEFER, W.R. 1972. *The geologic story of Yellowstone National Park*. Washington, D.C.: U.S. Geological Survey, Bulletin 1347.

KELEHER, C.J. AND F.J. RAHEL. 1996. Thermal limits of salmonid distribution in the Rocky Mountain region and potential habitat loss due to global warming: a geographic information system (GIS) approach. *Transactions of the American Fisheries Society* 125: 1–13.

KELLER, G. AND 6 co-authors. 2004. Chicxulub impact predates the K-T boundary mass extinction. *Proceedings of the National Academy of Sciences (U.S.A.)* 101: 3753–3758.

KELLY, B.M. 1993. Ecology of Yellowstone cutthroat trout and evaluation of potential effects of angler wading in the Yellowstone River. Master's thesis, Montana State University, Bozeman.

KELLY, L.S. 1926. *"Yellowstone Kelly": the memoirs of Luther S. Kelly*. New Haven, Connecticut: Yale University Press.

KELSO, B.W., T.G. NORTHCOTE AND C.F. WEHRHAHN. 1981. Genetic and environmental aspects of the response to water current by rainbow trout (*Salmo gairdneri*) originating from inlet and outlet streams of two lakes. *Canadian Journal of Zoology* 59: 2177–2185.

KEMMERER, G., J.F. BOVARD AND W.R. BORMAN. 1924. Northwestern lakes of the United States: biological and chemical studies with reference to possibilities in production of fish. *Bulletin of the U.S. Bureau of Fisheries* 39 (for 1923): 51–140.

KENDRICK, G.D. 1984. *Beyond the Wasatch: the history of irrigation in the Uinta Basin and upper Provo River areas of Utah*. Interagency Agreement 3AA-40-00900 between Bureau of Reclamation, Upper Colorado Regional Office and National Park Service, Rocky Mountain Regional Office.

KENNEDY, C.J., L.E. McDONALD, R. LOVENRIDGE AND M.M. STROSKER. 2000. The effect of bioaccumulated selenium on mortalities and deformities in the eggs, larvae, and fry of a wild population of cutthroat trout (*Oncorhynchus clarki lewisi*). *Archives of Environmental Contamination and Toxicology* 39: 46–52.

KENNETT, J.P., K.G. CANNARIATO, I.L. HENDY AND R.J. BEHL. 2000. Carbon isotope evidence for methane hydrate instability during Quaternary interstadials. *Science* 288: 128–133.

KENT, R. 1984. *Fisheries management investigations in the upper Shoshone River drainage, 1978–1982*. Cheyenne: Wyoming Game and Fish Department, Fish Division Completion Report.

KERSHNER, J.L. 1995. Bonneville cutthroat trout. Pages 28–35 in M.K. Young, editor. *Conservation assessment for inland cutthroat trout*. Fort Collins, Colorado: USDA Forest Service, Rocky Mountain Forest and Range Experiment Station, General Technical Report RM-GTR-256.

KERSHNER, J.L., C.M. BISCHOFF AND D.L. HORAN. 1997. Population, habitat, and genetic characteristics of Colorado River cutthroat trout in wilderness and nonwilderness stream sections in the Uinta Mountains of Utah and Wyoming. *North American Journal of Fisheries Management* 17: 1134–1143.

KIEFLING, J.W. 1978. *Studies on the ecology of the Snake River cutthroat trout*. Cheyenne: Wyoming Game and Fish Department, Fisheries Technical Bulletin 3.

———. 1997. *A history of the Snake River spring creek spawning tributaries*. Cheyenne: Wyoming Game and Fish Department, Fish Division, Administrative Report.

KING, J.W. 1982. Investigation of the Lahontan cutthroat trout broodstock at Marlette Lake, Nevada. Master's thesis, University of Nevada, Reno.

KLAMT, R.R. 1976. The effects of coarse granitic sediments on the distribution and abundance of salmonids in the central Idaho batholith. Master's thesis, University of Idaho, Moscow.

KLAR, G.T. AND C.B. STALNAKER. 1979. Electrophoretic variation in muscle lactate dehydrogenase in Snake Valley cutthroat trout, *Salmo clarki* subsp. *Comparative Biochemistry and Physiology* 64B: 391–394.

KLEIN, W.D. 1974. *Special regulations and elimination of stocking: influence on fishermen and the trout population at the Cache La Poudre River, Colorado*. Denver: Colorado Division of Wildlife, Technical Publication 30.

KLING, G.W. AND M.C. GRANT. 1984. Acid precipitation in the Colorado Front Range: an overview with time predictions for significant effects. *Arctic and Alpine Research* 16: 321–329.

KNACK, M.C. AND O.C. STEWART. 1984. *As long as the river shall run*. Berkeley: University of California Press.

KNIGHT, C.A. 1997. Spawning attributes and early life history of adfluvial Bonneville cutthroat trout in the Strawberry basin. Master's thesis, Utah State University, Logan.

KNIGHT, C.A., R.W. ORME AND D.A. BEAUCHAMP. 1999. Growth, survival, and migration patterns of juvenile adfluvial Bonneville cutthroat trout in tributaries of Strawberry Reservoir, Utah. *Transactions of the American Fisheries Society* 128: 553–563.

KNOX, R., T. HICKMAN, D. LANGLOIS, T. LYTLE AND J. TORRES. 1980. *Colorado River cutthroat trout inventory*. Denver: Colorado Division of Wildlife, Endangered Wildlife Investigations Performance Report SE-3-2.

KOCH, D.L. 1973. Reproductive characteristics of the cui-ui lakesucker (*Chasmistes cujus* Cope 1883) and its spawning behavior in Pyramid Lake. *Transactions of the American Fisheries Society* 102: 145–149.

KOCOVSKY, P.M., C. GOWAN, K.D. FAUSCH AND S.C. RILEY. 1997. Spinal injury rates in three wild trout populations in Colorado after eight years of backpack electrofishing. *North American Journal of Fisheries Management* 17: 308–313.

KOEL, T.M., P.E. BIGELOW, P.D. DOEPKE, B.D. ERTEL AND D.L. MAHONY. 2005. Nonnative lake trout result in Yellowstone cutthroat trout decline and impacts to bears and anglers. *Fisheries* 30, no. 11: 10–19.

KOLAR, C.S. AND D.M. LODGE. 2001. Progress in invasion biology: predicting invaders. *Trends in Ecology and Evolution* 16: 2027–2042.

KOLBERT, E. 2006. *Field notes from a catastrophe: man, nature and climate change*. New York: Bloomsbury Publishing.

KONDOLF, G.M. AND M.G. WOLMAN. 1993. The sizes of salmonid spawning gravels. *Water Resources Research* 29: 2275–2285.

KOSTOW, K., EDITOR. 1995. *Biennial report on the status of wild fish in Oregon*. Portland: Oregon Department of Fish and Wildlife.

KOZFKAY, J.R., J.C. DILLON AND D.J. SCHILL. 2006. Routine use of sterile fish in salmonid sport fisheries: are we there yet? *Fisheries* 31, no. 8: 392–401.

KRAJICK, K. 2001. Thermal features bubble in Yellowstone Lake. *Science* 292: 1479–1480.

KRIJGSMAN, W., F.J. HILGEN, I. RAFFI, F.J. SIERRO AND D.S. WILSON. 1999. Chronology, causes and progression of the Messinian salinity crisis. *Nature* 400: 652–655.

KRUEGER, C.C. AND B. MAY. 1991. Ecological and genetic effects of salmonid introductions in North America. *Canadian Journal of Fisheries and Aquatic Sciences* 48, supplement 1: 66–77.

KRUSE, C.G. 1995. Genetic purity, habitat, and population characteristics of Yellowstone cutthroat trout in the Greybull River drainage, Wyoming. Master's thesis, University of Wyoming, Laramie.

KRUSE, C.G. AND W.A. HUBERT. 1997. Proposed standard weight (Ws) equations for interior cutthroat trout. *North American Journal of Fisheries Management* 17: 784–790.

KRUSE, C.G., W.A. HUBERT AND F.J. RAHEL. 1997a. Geomorphic influences on the distribution of Yellowstone cutthroat trout in the Absaroka Mountains, Wyoming. *Transactions of the American Fisheries Society* 126: 418–427.

KRUSE, C.G., W.A. HUBERT AND F.J. RAHEL. 1997b. Using otoliths and scales to describe age and growth of Yellowstone cutthroat trout in a high-elevation stream system, Wyoming. *Northwest Science* 71: 30–38.

———. 2000. Status of Yellowstone cutthroat trout in Wyoming waters. *North American Journal of Fisheries Management* 20: 693–705.

KUCERA, P.A., D.L. KOCH AND G.F. MARCO. 1985. Introductions of Lahontan cutthroat trout into Omak Lake, Washington. *North American Journal of Fisheries Management* 5: 296–301.

KYTE, F.T. 1998. A meteorite from the Cretaceous/ Tertiary boundary. *Nature* 396: 237–239.

LABAR, G.W. 1971. Movement and homing of cutthroat trout (*Salmo clarki*) in Clear and Bridge creeks, Yellowstone National Park. *Transactions of the American Fisheries Society* 100: 41–49.

LABASTILLE, A. 1981. Acid rain: how great a menace? *National Geographic* 160, no. 5: 652–680.

LAMARRA, V.C., C. LIFF AND J. CARTER. 1986. Hydrology of Bear Lake basin and its impact on the trophic state of Bear Lake, Utah-Idaho. *Great Basin Naturalist* 46: 690–705.

LAND, G. 1890. Letter to the editors re: the yellow fin trout of the Twin Lakes, Colorado. *Forest and Stream* 34, no. 1: 8 and *Sports Afield* 4, no. 4: 68–69.

LANDE, R. 1993. Risk of population extinction from demographic and environmental stochasticity and random catastrophes. *American Naturalist* 142: 911–927.

LANDER, F.W. 1861. *Maps and reports of the Fort Kearney, South Pass, and Honey Lake Wagon Road*. U.S. House of Representatives, 36th Congress, 2nd Session, Vol. 9, 1860–1861, House Executive Document 64. Washington, D.C.: Government Printing Office.

LANGFORD, N.P., EDITOR. 1905. *Discovery of Yellowstone Park: diary of the Washburn expedition to the Yellowstone and Firehole rivers in the year 1870*. Privately printed.

LANGLOIS, D. 1979, 1980, 1981. *Rio Grande cutthroat trout inventory, restoration, and monitoring*. Montrose: Colorado Division of Wildlife, Nongame Investigations Performance Reports FW-22-R, R-1, and R-2.

LANGLOIS, D. (AND FIVE CO-AUTHORS). 1994. *Colorado River cutthroat trout conservation strategy for southwestern Colorado.* Montrose: Colorado Division of Wildlife and U.S. Forest Service.

LARIVERS, I. 1962 [1994]. *Fishes and fisheries of Nevada.* Carson City: Nevada State Fish and Game Commission [revised edition published 1994 by University of Nevada Press, Reno].

———. 1966. Paleontological miscellanei. *Biological Society of Nevada Occasional Papers* 11: 1–6.

LARKIN, P.A. 1977. An epitaph for the concept of maximum sustained yield. *Transactions of the American Fisheries Society* 106: 1–11.

LA ROE, L.M. 2002. Salt Lake Valley's leap of faith. *National Geographic* 201, no. 2: 88–107.

LATTERELL, J.J. 2001. Distribution constraints and population genetics of native trout in unlogged and clear-cut headwater streams. Master's thesis, University of Washington, Seattle.

LATTERELL, J.J., R.J. NAIMAN, B.R. FRANSEN AND P.A. BISSON. 2003. Physical constraints on trout (*Oncorhynchus* spp.) distribution in the Cascade Mountains: a comparison of logged and unlogged streams. *Canadian Journal of Fisheries and Aquatic Sciences* 60: 1007–1017.

LAVENDER, DAVID. 1968 [1981]. *The Rockies.* New York: Harper & Row [reprinted 1981 by University of Nebraska Press Lincoln].

LAVIER, D. 1963. *The sea-run cutthroat.* Olympia: Game Bulletin of the Washington State Game Department 15, no. 3: 1–5.

LAYTON, T.N. 1979. Archaeology and paleo-ecology of pluvial Lake Parman, northwestern Great Basin. *Journal of New World Archaeology* 3, no. 3: 41–56.

LEA, T.N. 1968. Ecology of the Lahontan cutthroat trout, *Salmo clarki henshawi*, in Independence Lake, California. Master's thesis, University of California, Berkeley.

LEARY, R.F., F.W. ALLENDORF AND N. KANDA. 1998. *Lack of genetic divergence between westslope cutthroat trout from the Columbia and Missouri river drainages.* Missoula: University of Montana Division of Biological Sciences, Wild Trout and Salmon Genetics Laboratory Report 97/1.

LEARY, R.F., F.W. ALLENDORF, S.R. PHELPS AND K.L. KNUDSEN. 1984. Introgression between westslope cutthroat and rainbow trout in the Clark Fork River drainage, Montana. *Proceedings of the Montana Academy of Sciences* 43: 1–18.

———. 1985. Population genetic structure of westslope cutthroat trout: genetic variation within and among populations. *Proceedings of the Montana Academy of Sciences* 45: 37–45.

———. 1987. Genetic divergence and identification of seven cutthroat trout subspecies and rainbow trout. *Transactions of the American Fisheries Society* 116: 580–587.

———. 1988. Population genetic structure of westslope cutthroat trout: genetic variation within and among populations. *Proceedings of the Montana Academy of Sciences* 48: 57–70.

LEE, W.S. 1962. *The Sierra.* New York: G.P. Putnam's Sons, Inc.

LEIDER, S.A. 1997. Status of sea-run cutthroat trout in Washington. Pages 68–76 in J.D. Hall, P.A. Bisson and R.E. Gresswell, editors. 1997. *Sea-run cutthroat trout: biology, management, and future conservation.* Corvallis: Oregon Chapter American Fisheries Society.

LEMKE, R.W., W.M. LAIRD, M.J. TIPTON AND R.M. LENDVALL. 1965. Quaternary geology of northern Great Plains. Pages 15–27 in H.E. Wright, Jr. and D.G. Frey, editors. *The Quaternary of the United States.* Princeton: New Jersey, Princeton University Press.

LEMLY, A.D. 1997. A teratogenic deformity index for evaluating impacts of selenium on fish populations. *Ecotoxicology and Environmental Safety* 37: 259–266.

———. 2002. Symptoms and implications of selenium toxicity in fish: the Belews Lake case example. *Aquatic Toxicology* 57: 39–49.

LENTSCH, L.D. 1985. Evaluation of young-of-the-year production in a unique Colorado wild trout population. Master's thesis, Colorado State University, Fort Collins.

LENTSCH, L. AND Y. CONVERSE. 1997. *Conservation agreement and strategy for Colorado River cutthroat trout (Oncorhynchus clarki pleuriticus) in the State of Utah.* Salt Lake City: Utah Division of Wildlife Resources, Publication 97-20.

LENTSCH, L., Y. CONVERSE AND J. PERKINS. 1997. *Conservation agreement and strategy for Bonneville cutthroat trout (Oncorhynchus clarki utah) in the State of Utah.* Salt Lake City: Utah Division of Wildlife Resources, Publication 97-19.

LENTSCH, L.D., C.A. TOLINE, J. KERSCHNER, J.M. HUDSON AND J. MIZZI. 2000. *Range-wide conservation agreement and strategy for Bonneville cutthroat trout (Oncorhynchus clarki utah).* Salt Lake City: Utah Division of Wildlife Resources, Publication 00-19.

LEONARD, E. 1984. Climatic change in the Colorado Rocky Mountains: estimates based on modern climate at late Pleistocene equilibrium lines. *Arctic and Alpine Research* 21: 245–255.

LEONARD, J.E. 1950. *Flies.* New York: A.S. Barnes.

LEOPOLD, L.B. 1951. Pleistocene climate in New Mexico. *American Journal of Science* 249: 152–168.

LESTELLE, L.C. 1978. The effects of forest debris removal on a population of resident cutthroat trout in a small headwater stream. Master's thesis, University of Washington, Seattle.

LEVIN, D. 1980. The fish that wouldn't die. *Sports Illustrated* (March 17, 1980 issue): 50–52.

LI, W.H. AND D. GRAUR. 1991. *Fundamental of molecular evolution.* Sunderland, Massachusetts: Sinauer Associates, Inc.

LIKNESS, G.A. 1984. *The present status and distribution of the westslope cutthroat trout (Salmo clarki lewisi) east and west of the continental divide in Montana.* Bozeman: Montana Cooperative Fishery Research Unit, report to Montana Department of Fish, Wildlife and Parks.

LIKNESS, G.A. AND P.J. GRAHAM. 1988. Westslope cutthroat trout in Montana: life history, status, and management. Pages 53–60 in R.E. Gresswell, editor. *Status and management of interior stocks of cutthroat trout.* Bethesda, Maryland: American Fisheries Society Symposium 4.

LIM, S.T. AND G.S. BAILEY. 1977. Gene duplication in salmonid fish: evidence for duplicated but catalytically equivalent A(4) lactose dehydrogenase. *Biochemical Genetics* 15: 707–721.

LIM, S.T., R.M. KAY AND G.S. BAILEY. 1975. Lactose dehydrogenases of salmonid fish: evidence for unique and rapid functional divergence of duplicated H4 lactose dehydrogenase. *Journal of Biological Chemistry* 250: 1790–1800.

LINCOLN, R.F. AND A.P. SCOTT. 1984. Sexual maturation in triploid rainbow trout, *Salmo gairdneri* Richardson. *Journal of Fish Biology* 25: 385–392.

LINDAHL, T. 1993a. Instability and decay of the primary structure of DNA. *Nature* 362: 709–714.

———. 1993b. Recovery of antediluvian DNA. *Nature* 365: 700.

LINK, P.K. AND C.M. FANNING. 1999. Late Miocene Snake River flowed south into the Humboldt drainage: detrital zircon evidence. *Geological Society of America Abstracts with Programs* 31: A22.

LINK, P.K., H.G. McDONALD, C.M. FANNING AND A.E. GODFREY. 2002. Detrital zircon evidence for Pleistocene drainage reversal at Hagerman Fossil Beds National Monument, central Snake River Plain, Idaho. *Idaho Geological Survey Bulletin* 30: 106–119.

LINNAEUS, C. 1758. *Systema naturae per regna tria naturae, secundum classes, ordines, genera, species, cum characteribus, differentiis, synonymis, locis.* Laurentii Salvii, Holmiae, 12th edition.

LITTLE, R.G. AND H.J. McKIRDY. 1968. *Statewide fishing investigations, Rancho del Rio Grande Grant, Carson National Forest.* Santa Fe: New Mexico Department of Game and Fish, Federal Aid to Fish Restoration, Job Completion Report F-22-R-7 and F-22-R-8.

LITTLE SNAKE RIVER WORKING GROUP. 1994. *Conservation Plan for Colorado River cutthroat trout for the Little Snake River drainage, southeastern Wyoming.* Cheyenne: Wyoming Game and Fish Department, Medicine Bow National Forest, and Bureau of Land Management.

LOCH, J.J. AND D.R. MILLER. 1988. Distribution and diet of sea-run cutthroat trout captured in and adjacent to the Columbia River plume, May–July 1980. *Northwest Science* 62: 41–48.

LODGE, D.M. 1993. Biological invasions: lessons for ecology. *Trends in Ecology and Evolution* 8: 133–137.

LOGAN, G.A., J.J. BOON AND B. EGLINGTON. 1993. Structural biopolymer preservation in Miocene leaf fossils from the Clarkia site, northern Idaho. *Proceedings of the National Academy of Sciences* 90: 2246–2250.

LONG, M.E. 1996. Colorado's Front Range. *National Geographic* 190, no. 5: 80–103.

LOUDENSLAGER, E.J. AND G.A.E. GALL. 1980a. Geographic patterns of protein variations and subspeciation in the cutthroat trout, *Salmo clarki. Systematic Zoology* 29: 27–42.

———. 1980b. *Biochemical systematics of the Bonneville Basin and Colorado River cutthroat.* Fisheries Biology Research Facility, Department of Animal Science, University of California, Davis. Final Report to the Wyoming Game and Fish Department.

LOUDENSLAGER, E.J. AND R.M. KITCHIN. 1979. Genetic similarity of two forms of cutthroat trout, *Salmo clarki,* in Wyoming. *Copeia* 1979: 673–678.

LOUDENSLAGER, E.J. AND G.H. THORGAARD. 1979. Karyotypic and evolutionary relationships of the Yellowstone (*S. c. bouvieri*) and west-slope (*S. c. lewisi*) cutthroat trout. *Journal of the Fisheries Research Board of Canada* 36: 630–635.

LOVE, J.D. 1994. *Leidy Formation—new name for a Pleistocene glacio-fluviatile-lacustrine sequence in northwestern Wyoming.* Washington, D.C.: U.S. Geological Survey, Professional Paper 932–D.

LOVE, J.D. AND J.C. REED, JR. 1968 [1971, 1984]. *Creation of the Teton landscape: the geologic story of Grand Teton National Park.* Moose, Wyoming: Grand Teton Natural History Association [several updated editions of this book have been published].

LUKENS, J.R. 1978. Abundance, movements, and age structure of adfluvial westslope cutthroat trout in the Wolf Lodge Creek drainage, Idaho. Master's thesis, University of Idaho, Moscow.

LUPHER, R.L. AND W.C. WARREN. 1942. The Asotin stage of the Snake River canyon near Lewiston, Idaho. *Journal of Geology* 50: 866–881.

LUTERNAUER. J., K. CONWAY, J. CLAGUE, J. BARRIE AND B. BLAISE. 1989. Late Pleistocene terrestrial deposits on the continental shelf of western Canada: evidence for rapid sea level change at the end of the last glaciation. *Geology* 17: 357–360.

LYTLE, T.A., E.J. WAGNER, S.E. CHAPAL AND S. CULVER. 1982. *Colorado River cutthroat trout inventory.* Grand Junction: Colorado Division of Wildlife, Endangered Wildlife Investigations Performance Report SE-5-1.

MACIOLEK, J.A. AND P.R. NEEDHAM. 1952. Ecological effects of winter conditions on trout and trout foods in Convict Creek, California, 1951. *Transactions of the American Fisheries Society* 81: 202–217.

MACKLIN, J.H. AND A.S. CARY 1965. *Origin of Cascade landscapes.* Olympia: Washington Department of Conservation, Division of Mines and Geology, Information Circular 41.

MACPHEE, C. 1966. Influence of differential angling mortality and stream gradient on fish abundance in a trout-sculpin biotope. *Transactions of the American Fisheries Society* 95: 381–387.

MADOLE, R.F. 1976. Glacial geology of the Front Range, Colorado. Pages 297–318 in W.C. Mahoney, editor. *Quaternary stratigraphy of North America.* Stroudsburg, Pennsylvania: Dowden, Hutchinson and Ross, Inc.

———. 1980. Time of Pinedale deglaciation in north-central Colorado: further considerations. *Geology* 8: 118–122.

MADSEN, V.D. 1942. Investigations of the fishery of Fish Lake, Utah. Master's thesis, Utah State Agricultural College, Logan.

MAGEE, J.P., T.E. MCMAHON AND R.F. THUROW. 1996. Spatial variation in spawning habitat of cutthroat trout in a sediment-rich stream basin. *Transactions of the American Fisheries Society* 125: 768–779.

MALDE, H.E. 1965. The Snake River Plain. Pages 255–264 in H.E. Wright, Jr. and D.G. Frey, editors. *The Quaternary of the United States.* Princeton, New Jersey: Princeton University Press.

MALLEA-OLAETXE, J. 2000. *Speaking through the aspens: Basque tree carvers in California and Nevada.* Reno and Las Vegas: University of Nevada Press.

MALMQUIST, H.J. 1992. Phenotype-specific feeding behaviour of two arctic charr *Salvelinus alpinus* morphs. *Oecologia* 92: 354–361.

MANGUM, F.A. AND J.L. MADRIGAL. 1999. Rotenone effects on aquatic macroinvertebrates of the Strawberry River, Utah: a five-year summary. *Journal of Freshwater Ecology* 14: 125–135.

MANSFIELD, G.R. 1927. *Geography, geology, and mineral resources of part of southeastern Idaho.* Washington, D.C.: U.S. Geological Survey Professional Paper 152.

MANSON, C., signator. 2005. Endangered and threatened wildlife and plants: designation of critical habitat for the bull trout. *Federal Register* 70, no. 185: 56, 212-56, 311.

MANTUA, N.J., S.R. HARE, Y. ZHANG, J.M. WALLACE AND R.C. FRANCIS. 1997. A Pacific interdecadal climate oscillation with impacts on salmon production. *Bulletin of the American Meteorological Society* 78: 1069–1079.

MARCUS, J., EDITOR. 1997. *Mining environment handbook: effects of mining on the environment and American environmental controls on mining.* London, UK: Imperial College Press.

MARKING, L. 1988. *Oral toxicity of rotenone to mammals.* U.S. Fish and Wildlife Service, Investigations in Fish Control 94.

MARNELL, L.F. 1980. *Genetic reconnaisance of cutthroat trout, Salmo clarki (Richardson) in twenty-two westslope lakes in Glacier National Park, Montana.* Glacier National Park, Montana: National Park Service.

———. 1988. Status of the westslope cutthroat trout in Glacier National Park, Montana. Pages 61–70 in R.E. Gresswell, editor. *Status and management of interior stocks of cutthroat trout.* Bethesda, Maryland: American Fisheries Society Symposium 4.

MARNELL, L.F., R.J. BEHNKE AND F.W. ALLENDORF. 1987. Genetic identification of cutthroat trout, *Salmo clarki*, in Glacier National Park, Montana. *Canadian Journal of Fisheries and Aquatic Sciences* 44: 1830–1839.

MARSHACK, A. 1975. Exploring the minds of Ice Age man. *National Geographic* 147, no. 1: 65–89.

MARSHALL, G.T.H., A.R. BEAUMONT AND R. WYATT. 1992. Genetics of brown trout (*Salmo trutta* L.) stocks above and below impassable falls in the Conway River system, North Wales. *Aquatic Living Resources* 5: 9–13.

MARSHALL, T.L. 1973. Trout populations, angler harvest and value of stocked and unstocked fisheries of the Cache La Poudre River, Colorado. Ph.D. dissertation, Colorado State University, Fort Collins.

MARTIN, C., EDITOR. 1991. *Fly-fishing in northern New Mexico.* Albuquerque: University of New Mexico Press.

MARTIN, M.A., D.S. SHIOZAWA, E.J. LOUDENSLAGER AND N. JENSEN. 1985. Electrophoretic study of cutthroat trout populations in Utah. *Great Basin Naturalist* 45: 677–687.

MARTINEZ, A. 1988. Identification and status of Colorado River cutthroat trout in Colorado. Pages 81–89 in R.E. Gresswell, editor. *Status and management of interior stocks of cutthroat trout*. Bethesda, Maryland: American Fisheries Society Symposium 4.

MATHEWS, W.H. 1944. Glacial lakes and ice retreat in south-central British Columbia. *Transactions of the Royal Society of Canada* 38, section IV: 39–57.

MATTHEWS, W. 1994. *Navaho legends*. Salt Lake City: University of Utah Press.

MAULE, W.A. 1938. *A contribution to the geographic and economic history of the Carson, Walker, and Mono basins in Nevada and California*. USDA Forest Service, San Francisco.

MAY, B.E. 1996. Yellowstone cutthroat trout *Oncorhynchus clarki bouvieri*. Pages 11–35 in D.A. Duff, technical editor. *Conservation assessment for inland cutthroat trout: distribution, status and habitat management implications*. Ogden, Utah: USDA Forest Service, Northern, Rocky Mountain, Intermountain, and Southwestern Regions.

MAY, B.E., J.D. LEPPINK AND R.S. WYDOSKI. 1978. *Distribution, systematics and biology of the Bonneville cutthroat trout, Salmo clarki utah*. Salt Lake City: Utah Division of Wildlife Resources, Publication 78-15.

MAY, B.E., W. URIE AND B.B. SHEPARD. 2003. *Range-wide status review of Yellowstone cutthroat trout (Oncorhynchus clarki bouvieri): 2001*. Bozeman, Montana: Yellowstone Cutthroat Trout Interagency Coordinating Group (USDA Forest Service, Idaho Department of Fish and Game, Montana Department of Fish, Wildlife and Parks, and Wyoming Game and Fish Department).

MAYR, E. AND P.D. ASHLOCK. 1991. *Principles of systematic zoology, 2nd edition*. New York: McGraw-Hill.

MCAFEE, W.R. 1966a. Lahontan cutthroat trout. Pages 225–231 in A. Calhoun, editor. *Inland fisheries management*. Sacramento: California Department of Fish and Game.

———. 1966b. Paiute cutthroat trout. Pages 231–233 in A. Calhoun, editor. *Inland fisheries management*. Sacramento: California Department of Fish and Game.

MCCLEAVE, J.D. 1967. Homing and orientation of cutthroat trout (*Salmo clarki*) in Yellowstone Lake, with special reference to olfaction and vision. *Journal of the Fisheries Research Board of Canada* 24: 2011–2044.

MCCLELLAN, G.B. 1853. Diaries, May 20 to December 11, 1853, and July 15 to September 30, 1853. Washington, D.C.: Library of Congress, Manuscript Division, McClellan Papers.

MCCLINTOCK, W. 1910 [1999]. *The Old North Trail, or, life, legends and religions of the Blackfeet Indians*. London: Macmillan and Company [reprinted several times by the University of Nebraska Press, Lincoln]. Also accessible on the Internet at www.1st-hand-history.org/ONT/album1.html.

MCCULLOUGH, D.R., EDITOR. 1996. *Metapopulations and wildlife conservation*. Washington, D.C.: Island Press.

MCDONALD, E.V. AND A.J. BUSACCA. 1988. Record of pre-late Wisconsin giant floods in the channeled scabland interpreted from loess deposits. *Geology* 16: 728–731.

MCDOWALL, R.M. 1987. The occurrence and distribution of diadromy among fishes. Pages 1–13 in M.J. Dadswell and 5 co-editors. *Common strategies of anadromous and catadromous fishes*. Bethesda, Maryland: American Fisheries Society Symposium 1.

———. 1988. *Diadromy in fishes: migrations between freshwater and marine environments*. Portland, Oregon: Timber Press.

MCGINN, N.A., EDITOR. 2002. *Fisheries in a changing climate*. Bethesda, Maryland: American Fisheries Society Symposium 32.

MCHUGH, P. AND P. BUDY. 2005. An experimental evaluation of competitive and thermal effects on brown trout (*Salmo trutta*) and Bonneville cutthroat trout (*Oncorhynchus clarkii utah*) performance along an altitudinal gradient. *Canadian Journal of Fisheries and Aquatic Sciences* 62: 2784–2795.

———. 2006. Experimental effects of nonnative brown trout on the individual and population-level performance of native Bonneville cutthroat trout. *Transactions of the American Fisheries Society* 135: 1441–1455.

MCINNIS, M.A. AND E.F. STORK. 1974. *Distribution of native Rio Grande cutthroat trout in the Santa Fe National Forest*. Boulder, Colorado: Western Interstate Commission for Higher Education. Report prepared for the Santa Fe National Forest, Albuquerque, New Mexico.

MCINTYRE, J.D. AND B.E. RIEMAN. 1995. Westslope cutthroat trout. Pages 1–15 in M.K. Young, technical editor. *Conservation assessment for inland cutthroat trout*. Fort Collins, Colorado: USDA Forest Service, Rocky Mountain Forest and Range Experiment Station, General Technical Report RM-256.

MCKAY, S.J., R.H. DEVLIN AND M.J. SMITH. 1996. Phylogeny of Pacific salmon and trout based on growth hormone type-2 and mitochondrial NADH dehydrogenase subunit 3 DNA sequences. *Canadian Journal of Fisheries and Aquatic Sciences* 53: 1165–1176.

MCKEE, BATES. 1972. *Cascadia: the geological evolution of the Pacific Northwest*. New York: McGraw-Hill.

MCMULLIN, S.L. AND T. DOTSON. 1988. Use of McBride Lake strain Yellowstone cutthroat trout for lake and reservoir management in Montana. Pages 42–44 in R.E. Gresswell, editor. *Status and management of interior stocks of cutthroat trout.* Bethesda, Maryland: American Fisheries Society Symposium 4.

MCPHAIL, J.D. 1997. The origin and speciation of *Oncorhynchus* revisited. Pages 29–38 in D.J. Stouder, P.A. Bisson and R.J. Naiman, editors. *Pacific salmon and their ecosystems: status and future options.* New York: Chapman and Hall.

MCPHAIL, J.D. AND C.C. LINDSEY 1970. *Freshwater fishes of northwestern Canada and Alaska.* Ottawa, Canada: Fisheries Research Board of Canada, Bulletin 173.

MCVEIGH, H.P., R.A. HYNES AND A. FERGUSON. 1995. Mitochondrial DNA differentiation of sympatric populations of brown trout, *Salmo trutta* L., from Lough Melvin, Ireland. *Canadian Journal of Fisheries and Aquatic Sciences* 52: 1617–1622.

MEADOWS, B.S. 1973. Toxicity of rotenone to some species of coarse fish and invertebrates. *Journal of Fish Biology* 5: 155–163.

MECKLENBURG, C.W., T.A. MECKLENBURG AND L.K. THORSTEINSON. 2002. *Fishes of Alaska.* Bethesda, Maryland: American Fisheries Society.

MEDNIKOV, B.M., E.A. SHUBINA, M.H. MELNIKOVA AND K.A. SAVVAITOVA. 1999. The genus status problem in Pacific salmons and trouts: a genetic systematics investigation. *Journal of Ichthyology* 39: 10–17.

MEEHAN, W.R., EDITOR. 1991. *Influences of forest and rangeland management on salmonid fishes and their habitats.* Bethesda, Maryland: American Fisheries Society Special Publication 19.

MEEHAN, W.R. AND W.S. PLATTS. 1978. Livestock grazing and the aquatic environment. *Journal of Soil and Water Conservation* 33: 274–278.

MEEUWIG, M.H. 2000. Effects of constant and cyclical thermal regimes on growth and feeding of juvenile cutthroat trout of variable sizes. Master's thesis, University of Nevada, Reno.

MEFFE, G.K. AND C.R. CARROLL. 1994. *Principles of conservation biology.* Sunderland, Massachussetts: Sinauer Associates, Ltd.

MEGAHAN, W.F. AND G.L. KETCHESON. 1996. Predicting downslope travel of granitic sediments from forest roads in Idaho. *Water Resources Bulletin* 32: 371–382.

MEIERDING, T.C. 1982. Late Pleistocene glacial equilibrium line altitudes in the Colorado Front Range: a comparison of methods. *Quaternary Research* 18: 289–310.

MELOSH, H.J., N.M. SCHNEIDER, K.J. ZAHNLE AND D. LATHAM. 1990. Ignition of global wildfires at the Cretaceous/Tertiary boundary. *Nature* 343: 251–254.

MENDEL, G. 1865 [1901]. Versuche über Pflanzen-Hybriden. *Verhandlungen des naturforshenden Vereines in Brünn* 4: 3–47 [original in German; English translation published in 1901 in *Journal of the Royal Horticultural Society*, Vol. 26].

MERCER, S.P. 1980. *Sea-run cutthroat: development and evaluation of a new enhancement technique.* Olympia: Washington State Game Department, Fishery Research Report 81-17.

MERCER, S.P. AND J.M. JOHNSTON. 1979. *Sea-run cutthroat: development and evaluation of a new enhancement technique.* Olympia: Washington State Game Department, Fishery Research Report 80-9.

MERRIMAN, D. 1935. The effects of temperature on the development of eggs and larvae of the cutthroat trout (*Salmo clarkii clarkii* Richardson). *Journal of Experimental Biology* 12: 297–305.

MERRIT, J.I. 1985. *Baronets and buffalo: the British sportsmen in the American West 1832–1881.* Missoula, Montana: Mountain Press.

MESA, M.G. 1991. Variation in feeding, aggression, and position choice between hatchery and wild cutthroat trout in an artificial stream. *Transactions of the American Fisheries Society* 120: 723–727.

MEYER, J. AND S. FRADKIN. 2002. *Summary of fisheries and limnological data for Lake Crescent, Washington.* Port Angeles, Washington: Olympic National Park.

MEYER, K.A., D.J. SCHILL, F.S. ELLE AND J.A. LAMANSKY, JR. 2003b. Reproductive demographics and factors that influence length at sexual maturity of Yellowstone cutthroat trout in Idaho. *Transactions of the American Fisheries Society* 132: 183–195.

MEYER, K.A., D.J. SCHILL, F.S. ELLE AND W.C. SCHRADER. 2003a. A long-term comparison of Yellowstone cutthroat trout abundance and size structure in their historical range in Idaho. *North American Journal of Fisheries Management* 23: 149–162.

MEYER, K.A., D.J. SCHILL, J.A. LAMANSKY, M.R. CAMPBELL AND C.C. KOZPKAY. 2006. Status of Yellowstone cutthroat trout in Idaho. *Transactions of the American Fisheries Society* 135: 1329–1347.

MEYER, L.S., T.S. THUEMLER AND G.W. KORNELY. 1992. Seasonal movements of brown trout in northeast Wisconsin. *North American Journal of Fisheries Management* 12: 433–441.

MICHAEL, J.H., JR. 1989. Life history of anadromous coastal cutthroat trout in Snow and Salmon creeks, Jefferson County, Washington, with implications for management. *California Fish and Game* 75: 188–203.

MIFFLIN, M.D. AND M.W. WHEAT. 1979. Pluvial lakes and estimated pluvial climates of Nevada. *Bulletin of the Nevada Bureau of Mines and Geology* 94: 1–57.

MILLER, R.B. 1957. Permanence and size of home territory in stream-dwelling cutthroat trout. *Journal of the Fisheries Research Board of Canada* 14: 687–691.

MILLER, R.R. 1950. Notes on the cutthroat and rainbow trouts with the description of a new species for the Gila River, New Mexico. *Occasional Papers of the Museum of Zoology University of Michigan*, no. 529.

———. 1961. Man and the changing fish fauna of the American southwest. *Papers of the Michigan Academy of Science, Arts and Letters* 46: 365–404.

———. 1965. Quaternary freshwater fishes of North America. Pages 569–581 in H.E. Wright, Jr. and D.G. Frey, editors. *The Quaternary of the United States*. Princeton, New Jersey: Princeton University Press.

———. 1972. Threatened freshwater fishes of the United States. *Transactions of the American Fisheries Society* 101: :239–252.

MILLER, R.R., C. HUBBS AND F.H. MILLER. 1991. Ichthyolgical exploration of the American West: the Hubbs-Miller era, 1915–1950. Pages 19–40 in W.L. Minckley and J.E. Deacon, editors. *Battle against extinction: native fish management in the American West*. Tucson: University of Arizona Press.

MILLS, L.S. AND F.W. ALLENDORF. 1996. The one-migrant-per-generation rule in conservation and management. *Conservation Biology* 6: 1509–1518.

MILLS, L.S., M.E. SOULÉ AND D.F. DOAK. 1993. The keystone-species concept in ecology and conservation. *BioScience* 43: 219–224.

MILNE, W. 1987. A comparison of reconstructed lake level records since the mid-1800s of some Great Basin lakes. Master's thesis, Colorado School of Mines, Golden, Colorado.

MITCHELL, FINIS. 1975. *Wind River trails*. Salt Lake City, Utah: Wasatch Publishers, Inc.

MITCHELL, J.G. (photos by Joel Sartore). 2005. Tapping the Rockies. *National Geographic* 208, no. 1: 92–113.

MITCHELL, W.T. 1988. Microhabitat utilization and spatial segregation of juvenile coastal cutthroat and steelhead trout in the Smith River drainage, California. Master's thesis, Humboldt State University, Arcata, California.

MOCK, K.E., J.C. BRIM-BOX, M.P. MILLER, M.E. DOWNING AND W.R. HOEH. 2004. Genetic diversity and divergence among freshwater mussel (*Anodonta*) populations in the Bonneville Basin of Utah. *Molecular Ecology* 13: 1085–1098.

MONASTERSKY, R. 1997. Spying on El Niño: the struggle to predict the Pacific prankster. *Science News* 152, no. 17: 268–270.

MONNETT, J.H. 1988. *Cutthroat and campfire tales: the fly-fishing heritage of the West*. Boulder, Colorado: Pruett Publishing.

———. 1993. Mystery of the Bighorns: did a fishing trip seal Custer's fate? *The American Fly Fisher (journal of the American Museum of Fly Fishing, Manchester, Vermont)* 19, no. 4: 2–5.

MONTANA DEPARTMENT OF FISH, WILDLIFE AND PARKS. 2000. *Yellowstone cutthroat trout in Montana: distribution, status, conservation, and research efforts*. Helena: Montana Department of Fish, Wildlife and Parks, a compendium with 19 attachments submitted to the U.S. Fish and Wildlife Service August 2, 2000.

MONTGOMERY, D.R. 2000. Coevolution of the Pacific salmon and Pacific Rim topography. *Geology* 28: 1107–1110.

MONTGOMERY, M.R. 1995. *Many rivers to cross: of good running water, native trout, and the remains of wilderness*. New York: Simon and Schuster.

MOON, R. 1982. A frontier fly fisher, 1847. *The American Fly Fisher (journal of the American Museum of Fly Fishing, Manchester, Vermont)* 9, no. 3: 10–11.

MOORE, K.M.S. AND S.V. GREGORY. 1988. Summer habitat utilization and ecology of cutthroat trout fry (*Salmo clarki*) in Cascade Mountain streams. *Canadian Journal of Fisheries and Aquatic Sciences* 45: 1921–1930.

MOORE, V. AND D. SCHILL. 1984. *South Fork Snake River fisheries inventory*. Boise: Idaho Department of Fish and Game, River and Stream Investigations, Job Completion Report, Project F-73-R-5.

MORELL, V. (photos by Frans Lanting) 1999. The sixth extinction. *National Geographic* 195, no. 2: 42–56.

MORGAN, D.L. 1943. *The Humboldt: high road of the west*. New York: Rinehart & Company, Inc.

MORING, J.R. 1991. Life and death. Pages 105–111 in J. Stolz and J Schnell, editors. *Trout: the wildlife series*. Harrisburg, Pennsylvania: Stackpole Books.

MORING, J.R. AND R.L. LANTZ. 1975. *The Alsea Watershed Study: effects of logging on the aquatic resources of three headwater streams of the Alsea River, Oregon. Part 1—biological studies*. Corvallis: Oregon Department of Fish and Wildlife, Fisheries Research Report 9.

MORING, J.R., R.L. YOUKER AND R.M. HOOTON. 1986. Movements of potamodromous coastal cutthroat trout, *Salmo clarki clarki*, inferred from tagging and scale analysis. *Fisheries Research* 4: 343–354.

MORRISON, R.B. 1965. Quaternary geology of the Great Basin. Pages 265–285 in H.E. Wright, Jr. and D.G. Frey, editors. *The Quaternary of the United States*. Princeton, New Jersey: Princeton University Press.

MORROW, J.E. 1980. *The freshwater fishes of Alaska*. Anchorage: Alaska Northwest Publishing Company

MOTE, P.W., A.F. HAMLET, M.P. CLARK AND D.P. LETTENMAIER. 2005. Declining mountain snowpack in western North America. *American Meteorological Society Bulletin* 86: 39–49.

MOULTON, G.E., EDITOR. 1986–1997. *The journals of the Lewis and Clark expedition.* Volumes 2 through 11. Lincoln, Nebraska: University of Nebraska Press.

MOUNEKE, M.I. AND W.M. CHILDRESS. 1994. Hooking mortality: a review for recreational fisheries. *Reviews in Fisheries Science* 2, no. 2: 123–156.

MOYLE, P.B. 2002. *Inland fishes of California.* Revised and expanded. Berkeley: University of California Press.

MOYLE, P.B., R.M. YOSHIYAMA, J.E. WILLIAMS AND E.D. WIKRAMANAYAKE. 1995. *Fish species of special concern in California.* 2nd edition. University of California Davis, Department of Wildlife and Fisheries Biology.

MULLAN, J.W. 1975. Condition (K) as indicative of non-suitability of Snake River cutthroat trout in the management of high gradient, low diversity streams. *Proceedings of the Western Association of State Fish and Game Commissioners* 55: 267–274.

MULLAN, J.W., K.D. WILLIAMS, G. RHODUS, T.W. HILLMAN AND J.D. MCINTYRE. 1992. *Production and habitat of salmonids in mid-Columbia River tributary streams.* Leavenworth: Washington, U.S. Fish and Wildlife Service Monograph I.

MURATA, S., N. TAKASAKI, H. SAITOH AND N. OKADA. 1993. Determination of the phylogenetic relationships among Pacific salmonids using short interspersed elements (SINES) as temporal landmarks of evolution. *Proceedings of the National Academy of Science U.S.A.* 90: 6995–6999.

MURPHY, B.R. AND D.W. WILLIS, EDITORS. 1996. *Fisheries techniques.* 2nd edition. Bethesda, Maryland: American Fisheries Society.

MURPHY, M.L. AND J.D. HALL. 1981. Varied effects of clear-cut logging on predators and their habitat in small streams of the Cascade Mountains, Oregon. *Canadian Journal of Fisheries and Aquatic Sciences* 38: 137–145.

MURPHY, T.C. 1974. A study of Snake River cutthroat trout. Master's thesis, Colorado State University, Fort Collins.

MUTTKOWSKI, R.A. 1925. Food of the Yellowstone trout. *Roosevelt Wild Life Bulletin* 4, no. 2.

MYERS, G.S. 1949. Usage of anadromous, catadromous and allied terms for migratory fishes. *Copeia* 1949: 89–97.

MYERS, T.J. AND S. SWANSON. 1996a. Long-term aquatic habitat restoration: Mahogany Creek, Nevada, as a case study. *Water Resources Bulletin* 32: 241–252.

———. 1996b. Temporal and geomorphic variations of stream stability and morphology: Mahogany Creek, Nevada. *Water Resources Bulletin* 32: 253–265.

NAIMAN, R.J. AND R.E. BILBY, EDITORS. 1998. *River ecology and management: lessons from the Pacific coastal ecoregion.* New York: Springer-Verlag.

NAIMAN, R.J., R.E. BILBY, D.E. SCHINDLER AND J.M. HELFIELD. 2002. Pacific salmon, nutrients, and the dynamics of freshwater and riparian ecosystems. *Ecosystems* 5: 399–417.

NAKANO, S., K.D. FAUSCH, T. FURUKAWA-TANAKA, K. MAEKAWA AND H. KAWANABE. 1992. Resource utilization by bull char and cutthroat trout in a mountain stream in Montana, U.S.A. *Japanese Journal of Ichthyology* 39: 211–216. [In English].

NAKANO, S., T. KACHI AND M. NAGOSHI. 1990. Restricted movement of the fluvial form of red-spotted masu salmon, *Oncorhynchus masou rhodurus*, in a mountain stream, central Japan. *Japanese Journal of Ichthyology* 37: 158–163. [In English].

NANUS, L., D.H. CAMPBELL, G.P. INGERSOLL, D.W. CLOW, AND M.A. MAST. 2003. Atmospheric deposition maps for the Rocky Mountains. *Atmospheric Environment* 17: 4881–4892.

NARVER, D.W. 1975. *Notes on the ecology of cutthroat trout (Salmo clarki) in Great Central Lake, Vancouver Island, British Columbia.* Environment Canada, Fisheries and Marine Service, Technical Report 567.

NÄSLUND, I. 1993. Migratory behavior of brown trout, *Salmo trutta* L.: importance of genetic and environmental influences. *Ecology of Freshwater Fish* 2: 51–57.

NÄSLUND, I., G. MILBRINK, L.O. ERIKSSON AND S. HOLMGREN. 1993. Importance of habitat productivity differences, competition and predation for the migratory behavior of Arctic charr. *Oikos* 66: 538–546.

NATIONAL GEOGRAPHIC SOCIETY. 1993. *Water: the power, promise, and turmoil of North America's fresh water.* Special Edition. Washington, D.C.: National Geographic Society.

NEHLSEN, W., J.E. WILLIAMS AND J.A. LICHATOWICH. 1991. Pacific salmon at the crossroads: stocks at risk from California, Oregon, Idaho, and Washington. *Fisheries* 16, no. 2: 4–21.

NEI, M. 1987. *Molecular evolutionary genetics.* New York: Columbia University Press.

NELSON, J.S. AND M.J. PAETZ. 1992. *The fishes of Alberta.* 2nd edition. Edmonton and Calgary: University of Alberta Press and University of Calgary Press.

NELSON, J.S.. AND 6 co-authors. 2004. *Common and scientific names of fishes from the United States, Canada, and Mexico.* 6th edition. Bethesda, Maryland: American Fisheries Society Special Publication 29.

NELSON, K. 1993. *Status of the Bonneville cutthroat trout in the Bridger-Teton National Forest, Wyoming.* Jackson, Wyoming: USDA Forest Service, Bridger-Teton National Forest, Administrative Report.

NELSON, R.L., M.L. MCHENRY AND W.S. PLATTS. 1991. Mining. Pages 425–457 in W.R. Meehan, editor. *Influences of forest and rangeland management on salmonid fishes and their habitats.* Bethesda, Maryland: American Fisheries Society Special Publication 19.

NELSON, R.L., W.S. PLATTS, D.P. LARSEN AND S.E. JENSEN. 1992. Trout distribution and habitat in relation to geology and geomorphology in the North Fork Humboldt River drainage, northeastern Nevada. *Transactions of the American Fisheries Society* 121: 405–426.

NELSON, R.L., W.S. PLATTS AND O. CASEY. 1987. Evidence for variability in spawning behavior of interior cutthroat trout in response to environmental uncertainty. *Great Basin Naturalist* 47: 480–487.

NELSON, W.C. 1972. *An unexploited population of greenback cutthroat trout.* Fort Collins: Colorado Division of Wildlife.

NEWMAN, M.A. 1960. A comparative study of the residential behavior of juvenile salmonids. Ph.D. dissertation, University of British Columbia, Vancouver, B.C.

NEW MEXICO DEPARTMENT OF GAME AND FISH. 2002. *Long range plan for the management of the Rio Grande cutthroat trout in New Mexico.* Santa Fe: New Mexico Department of Game and Fish.

NICHOLAS, J.W. 1978a. *A review of literature and unpublished information on cutthroat trout (Salmo clarki clarki) of the Willamette watershed.* Portland: Oregon Department of Fish and Wildlife, Information Report Series, Fisheries 78-1.

———. 1978b. Life history differences between sympatric populations of rainbow and cutthroat trouts in relation to fisheries management strategies. Pages 181–188 in J.W. Moring, editor. *Proceedings of the wild trout-catchable trout symposium.* Portland: Oregon Department of Fish and Wildlife.

NICKELSON, T.E. (AND SIX CO-AUTHORS). 1992. *Status of anadromous salmonids in Oregon coastal basins.* Portland: Oregon Department of Fish and Wildlife.

NIELSEN, J.L., EDITOR. 1995. *Evolution and the aquatic ecosystem: defining unique units in population conservation.* Bethesda, Maryland: American Fisheries Society Symposium 17.

NIELSEN, J.L. 1998. Electrofishing California's endangered fish populations. *Fisheries* 23, no. 12: 6–12.

NIELSEN, J.L. AND G.K. SAGE. 2002. Population genetic structure in Lahontan cutthroat trout. *Transactions of the American Fisheries Society* 131: 376–388.

NIELSON, B.R. AND L. LENTSCH. 1988. Bonneville cutthroat trout in Bear Lake: status and management. Pages 128–133 in R.E. Gresswell, editor. *Status and management of interior stocks of cutthroat trout.* Bethesda, Maryland: American Fisheries Society Symposium 4.

NILSSON, N.A. 1967. Interactive segregation between fish species. Pages 295–313 in S.D. Gerking, editor. *The biological basis for freshwater fish production.* Oxford, U.K.: Blackwell Scientific Publications.

NILSSON, N.A. AND T.G. NORTHCOTE. 1981. Rainbow trout and cutthroat trout interactions in coastal British Columbia lakes. *Canadian Journal of Fisheries and Aquatic Sciences* 38: 1228–1246.

NORDING, H. 1983. Solution to the "char problem" based on Arctic char (*Salvelinus alpinus*) in Norway. *Canadian Journal of Fisheries and Aquatic Sciences* 40: 1372–1387.

NORRIS, R.D. AND U. ROHL. 1999. Carbon cycling and chronology of climate warming during the Paleocene/Eocene transition. *Nature* 401: 775–778.

NORTHCOTE, T.G. 1992. Migration and residency in stream salmonids—some ecological considerations and evolutionary consequences. *Nordic Journal of Freshwater Research* 67: 5–17.

———. 1995. Confessions from a four decade affair with Dolly Varden: a synthesis and critique of experimental tests for interactive segregation between Dolly Varden char (*Salvelinus malma*) and cutthroat trout (*Oncorhynchus clarki*) in British Columbia. *Nordic Journal of Freshwater Research* 71: 49–67.

———. 1997a. Why sea-run? An exploration into the migratory/residence spectrum of coastal cutthroat trout. Pages 20–26 in J.D. Hall, P.A. Bisson and R.E. Gresswell, editors. *Sea-run cutthroat trout: biology, management, and future conservation.* Corvallis: Oregon Chapter American Fisheries Society.

———. 1997b. Potamodromy in Salmonidae—living and moving in the fast lane. *North American Journal of Fisheries Management* 17: 1029–1045.

NORTHCOTE, T.G., A.E. PEDEN AND T.E. REINCHEN. 1989. Fishes of the coastal marine, riverine and lacustrine waters of the Queen Charlotte Islands. Pages 147–174 in G.G.E. Scudder and N.A. Gessler, editors. *The outer shore.* Skidgate, British Columbia: Queen Charlotte Islands Museum.

NORTHCOTE, T.G. AND G.F. HARTMAN. 1988. The biology and significance of stream trout populations (*Salmo* spp.) living above and below waterfalls. *Polish Archives of Hydrobiology* 35, no. 3–4: 409–442.

NORTHCOTE, T.G., S.N. WILLISCROFT AND H. TSUYUKI. 1970. Meristic and lactate dehydrogenase genotype differences in stream populations of rainbow trout below and above a waterfall. *Journal of the Fisheries Research Board of Canada* 27: 1987–1995.

NORTHCOTE, T.G. AND W.G. KELSO. 1981. Differential response to water current by two homozygous LDH phenotypes of young rainbow trout (*Salmo gairdneri*). *Canadian Journal of Fisheries and Aquatic Sciences* 38: 348–352.

NORTHWEST POWER AND CONSERVATION COUNCIL. 2004. *Upper Columbia River Fish and Wildlife Program: Upper Snake, Headwaters, Closed Basin Subbasins.* Portland, Oregon: Northwest Power and Conservation Council. Available on the Internet at www.subbasins. org or on CD from Northwest Power and Conservation Council, 851 SW Sixth Ave., Suite 1100, Portland, OR 97204.

NOVAK, M.A. 1989. *Investigations of cutthroat trout spawning in the Sanke River fishery, Grand Teton National Park.* Moose, Wyoming: U.S. Department of the Interior, National Park Service, Grand Teton national Park and Cheyenne, Wyoming, Wyoming Game and Fish Department.

NOVAK, M.A., J.L. KERSHNER AND K.E. MOCK. 2005. Molecular genetic investigation of Yellowstone cutthroat trout and finespotted Snake River cutthroat trout. Logan, Utah: U.S. Forest Service and Utah State University Aquatic, Watershed and Earth Resources Department. Report prepared for Wyoming Game and Fish Commission, Agreement 165/04. Available on the Internet at www.fs.fed.us/biology/resources/pubs/feu/ cutthroat_genetics_2005.pdf.

NOVICK, M. 2006. Spatial distribution patterns of coastal cutthroat trout in a Cascade Mountain stream. Master's thesis, Oregon State University, Corvallis.

NOWAK, G.M. 2000. Movement patterns and feeding ecology of cutthroat trout (*Oncorhynchus clarki clarki*) in Lake Washington. Master's thesis, University of Washington, Seattle.

NOWAK, G.M., R.A. TABOR, E.J. WARNER, K.L. FRESH AND T.P. QUINN. 2004. Ontogenetic shifts in habitat and diet of cutthroat trout in Lake Washington, Washington. *North American Journal of Fisheries Management* 24: 624–635.

NOWAK, G.M. AND T.P. QUINN. 2002. Diel and seasonal patterns of horizontal and vertical movements of telemetered cutthroat trout in Lake Washington, Washington. *Transactions of the American Fisheries Society* 131: 452–462.

NUNNEY, L. 1995. Measuring the ratio of effective population size to adult numbers using genetic and ecological data. *Evolution* 49: 389–392.

OGILVIE, R. 1989. Disjunct vascular flora of northwestern Vancouver Island in relation to Queen Charlotte Islands' endemism and Pacific coast refugia. Pages 127–130 in G.G.E. Scudder and N.A. Gessler, editors. *The outer shore.* Skidgate, British Columbia: Queen Charlotte Islands Museum.

OHNO, S. 1970a. The enormous diversity in genome sizes of fish as a reflection of Nature's extensive experiments with gene duplication. *Transactions of the American Fisheries Society* 99: 120–130.

———. 1970b. *Evolution by gene duplication.* New York: Springer-Verlag.

———. 1974. Protochordata, Cyclostomata, and Pisces. In B. John, editor. *Animal cytogenetics.* Vol. 4, Chordata. Berlin, Germany: Gebrüder-Bornträger.

OHNO, S., J. MURAMOTO, J. KLEIN AND N.B. ATKINS. 1969. Diploid tetraploid relationship in clupeoid and salmonid fish. Pages 139–147 in C.D. Darlington and K.R. Lewis, editors. *Chromosomes today.* Vol. 2. Edinburgh, UK: Oliver & Boyd.

OMERNIK, J.M. 1987. Ecoregions of the coterminous United States. *Annals of the Association of American Geographers* 77, no. 12: 118–125 (with map supplement).

O'NEAL, K. 2002. *Effects of global warming on trout and salmon in U.S. streams.* Washington, D.C.: Defenders of Wildlife and Natural Resources Defense Council. Available on the Internet at www.defenders.org/ fishreport.pdf.

ONO, R.D., J.D. WILLIAMS AND A. WAGNER. 1983. *Vanishing fishes of North America.* Washington, D.C.: Stonewall Press.

OREGON DEPARTMENT OF FISH AND WILDLIFE. 2005. *Oregon native fish status report 2005.* Public Draft Report. Salem: Oregon Department of Fish and Wildlife, Fish Division. Available on the Internet at www.dfw.state.or.us/fish/. Navigate to Native Fish, then to Oregon Native Fish Status Report.

OREGON NATURAL RESOURCES COUNCIL, THE WILDERNESS SOCIETY, AND UMPQUA VALLEY AUDUBON SOCIETY. 1993. Petition for a rule to list the North and South Umpqua river sea-run cutthroat trout as threatened or endangered under the Endangered Species Act and to designate critical habitat. Petition submitted to National Marine Fisheries Service, Seattle, Washington. Copy obtained from ONRC, 522 SW 5th Ave., Suite 1050, Portland, OR 97204.

OREGON NATURAL RESOURCES COUNCIL AND 14 co-petitioners. 1997. Petition to list sea-run cutthroat trout (*Oncorhynchus clarki clarki*) as threatened or endangered throughout its range in the states of California, Oregon and Washington, under the Endangered Species Act (ESA) and to designate critical habitat. Document submitted to National Marine Fisheries Service, Seattle, Washington. Copy obtained from ONRC, 522 SW 5th Ave., Suite 1050, Portland, OR 97204.

O'REILLY, P., T.E. REIMCHEN, R. BEECH AND C. STROMBECK. 1993. Mitochondrial DNA in *Gasterosteus* and Pleistocene glacial refugium on the Queen Charlotte Islands, British Columbia. *Evolution* 47: 678–684.

ORR, P.C. 1956. *Pleistocene man in Fishbone Cave, Pershing County, Nevada*. Carson City: Nevada State Museum, Department of Archaeology, Bulletin no. 2.

OSBORN, G. AND K. BEVIS. 2001. Glaciation in the Great Basin of the western United States. *Quaternary Science Reviews* 20: 1377–1410.

OSBORN, J.G. 1980. Effects of logging on cutthroat trout (*Salmo clarki*) in small headwater streams. Master's thesis, University of Washington, Seattle.

OSTBERG, C.O. AND R.J. RODRIGUEZ. 2002. Analysis of hybridization between native westslope cutthroat trout and introduced rainbow trout in North Cascades National Park, Washington. Paper presented at the Annual General Meeting, Western Division American Fisheries Society, Spokane, Washington, April 29–May 1, 2002.

——. 2006. Hybridization and cytonuclear associations among native westslope cutthroat trout, introduced rainbow trout, and their hybrids within Stehekin River drainage, North Cascades National Park. *Transactions of the American Fisheries Society* 135: 924–942.

OVERMEYER, P.H. 1941. George B. McClellan and the Pacific northwest. *Pacific Northwest Quarterly* 31: 3–60.

OVIATT, C.G., R.S. THOMPSON, D.S. KAUFMAN, J. BRIGHT AND R.M. FORESTER. 1999. Reinterpretation of the Burmester core, Bonneville basin, Utah. *Quaternary Research* 52: 180–184.

OVIATT, C.G. AND W.D. McCOY. 1992. Early Wisconsin lakes and glaciers in the Great Basin, U.S.A. Pages 279–287 in P.U. Clark and P.D. Lea, editors. *The last interglacial-glacial transition in North America*. Boulder, Colorado: Geological Society of America, Special Paper 270.

OVIATT, C.G., W.D. McCOY AND R.G. REIDER. 1987. Evidence for a shallow early or middle Wisconsin-age lake in the Bonneville Basin, Utah. *Quaternary Research* 27: 248–262.

PÄÄBO, S. 1989. Ancient DNA: extraction, characterization, molecular cloning, and enzymatic amplification. *Proceedings of the National Academy of Sciences* 86: 1939–1943.

PÄÄBO, S., R.G. HIGUCHI AND A.C. WILSON. 1989. Ancient DNA and the polymerase chain reaction: the emerging field of molecular archaeology. *Journal of Biological Chemistry* 264: 9709–9712.

PAINE, R.T. 1969. A note on trophic complexity and community stability. *American Naturalist* 103: 91–93.

PALMER, J. 1847. *Journal of travels over the Rocky Mountains to the mouth of the Columbia River*. Cincinnati, Ohio: J.A. and U.P. James.

PARDEE, J.T. 1910. The Glacial Lake Missoula, Montana. *Journal of Geology* 18: 376–386.

——. 1942. Unusual currents in Glacial Lake Missoula, Montana. *Geological Society of America Bulletin* 53: 1569–1600.

PARKINSON, E., R.J. BEHNKE AND W. POLLARD. 1984. *A morphological and electrophoretic comparison of rainbow trout (Salmo gairdneri) above and below barriers on five streams on Vancouver Island, B.C.* Vancouver: British Columbia Ministry of Environment, Fisheries Branch. Fisheries Management Report 83.

PARSONS, M., EDITOR. 1983. *Clarke County, Washington Territory*. Portland, Oregon: Washington Publishing Company.

PATTON, P.C., V.R. BAKER AND C. KOCHEL. 1978. New evidence for pre-Wisconsin flooding in the channeled scabland of eastern Washington. *Geology* 6: 567–571.

PAULEY, G.B., K. OSHIMA, K.L. BOWERS AND G.L. THOMAS. 1989. *Species profiles: life histories and environmental requirements of coastal fishes and invertebrates (Pacific Northwest)*. Sea-run cutthroat trout. U.S. Fish and Wildlife Service Biological Report 82 (11.86), U.S. Army Corps of Engineers TR EL-82-4.

PEACOCK, M.M. AND V. KIRCHOFF. 2004. Assessing the conservation value of hybridized cutthroat trout populations in the Quinn River drainage, Nevada. *Transactions of the American Fisheries Society* 133: 309–325.

PEACOCK, M., J.B. DUNHAM AND C. RAY. 2003. Recovery and implementation plan for Lahontan cutthroat trout in the Lahontan Basin: genetics management plan. Appendix D in *Walker River Basin Recovery Implementation Team*. Short-term action plan for Lahontan cutthroat trout (*Oncorhynchus clarki henshawi*) in the Walker River basin. Reno, Nevada: U.S. Fish and Wildlife Service.

PEALE, A.G. 1879. Report of A.G. Peale, MD, geologist of the Green River Division. Annual Report. U.S. Geological and Geographic Survey of the Territories (Hayden Survey) 11, 1877: 509–646 and plates 47–76. Washington, D.C.: U.S. Government Printing Office.

PEARCY, W.G., EDITOR. 1984. *The influence of ocean conditions on the production of salmonids in the North Pacific: a workshop, November 8–10, 1983, Newport, Oregon*. Corvallis: Oregon State University, Oregon Sea Grant Program, Publication ORESU-WO-83-001.

————. 1997. The sea-run and the sea. Pages 29–34 in J.D. Hall, P.A. Bisson and R.E. Gresswell, editors. *Sea-run cutthroat trout: biology, management, and future conservation.* Corvallis: Oregon Chapter American Fisheries Society.

PEARCY, W.G., R.D. BRODEUR AND J.P. FISHER. 1990. Distribution and biology of juvenile cutthroat trout *Oncorhynchus clarki clarki* and steelhead *O. mykiss* in coastal waters off Oregon and Washington. *Fisheries Bulletin* 88: 697–711.

PERKINS, E.D. 1849 [T.D. Clark, editor. 1967]. *Gold rush diary.* Lexington, Kentucky: University of Kentucky Press.

PERKINS, R.R., R.C. BUCKMAN, AND W.E. HOSFORD. 1991. *Trout Creek Mountains wild trout investigations.* Hines, Oregon: Oregon Department of Fish and Wildlife Special Report.

PETERSON, D.P., K.D. FAUSCH AND G.C. WHITE. 2004. Population ecology of an invasion: effects of brook trout on native cutthroat trout. *Ecological Applications* 14: 754–772.

PETIT, J.R. AND 18 co-authors. 1999. Climate and atmospheric history of the past 420,000 years from the Vostok ice core, Antarctica. *Nature* 399: 429–436.

PETTERSSON, J.C.E., M.M. HANSEN AND T. BOHLIN. 2001. Does dispersal from landlocked trout explain the coexistence of resident and migratory trout females in small streams? *Journal of Fish Biology* 58: 487–495.

PFEIFER, R. 1985. *Proposed management of the Snoqualmie River above Snoqualmie Falls.* Olympia: Washington Department of Wildlife, Fisheries Management Report 85-2.

PHELPS, S.R. AND F.W. ALLENDORF. 1982. Genetic comparison of upper Missouri cutthroat trout to other *Salmo clarki lewisi* populations. *Proceedings of the Montana Academy of Sciences* 41: 14–22.

PHILLIPPAY, M. 1970. *As I Remember.* Steamboat Springs, Colorado: The Steamboat Pilot.

PIELOU, E.C. 1969. *An introduction to mathematical ecology.* New York: Interscience Publishers.

PIERCE, B.E. 1984. The trouts of Lake Crescent, Washington. Master's thesis, Colorado State University, Fort Collins.

PIERCE, 1ST LT. H.H. 1883. *Expedition from Fort Colville to Puget Sound, Washington Territory, by way of Lake Chelan and Skagit River, during the months of August and September, 1882.* Washington, D.C.: U.S. Government Printing Office.

PIERCE, K.L. AND J.D. GOOD. 1992. *Field guide to the Quaternary geology of Jackson Hole, Wyoming.* Washington, D.C.: U.S. Geological Survey Open-File Report 92-504.

PIERCE, K.L. AND L.A. MORGAN. 1992. The track of the Yellowstone hotspot: volcanism, faulting, and uplift. Pages 1–53 in P.K. Link, M.A. Kuntz and L.B. Platt, editors. *Regional geology of eastern Idaho and western Wyoming.* Geological Society of America Memoir 179.

————. 1999. Drainage changes associated with the Yellowstone hotspot. *Geological Society of America Abstracts with Programs* 31: A443–A444.

PINKERTON, JOHN, EDITOR. 1808. *A general collection of voyages and travels.* Volume I. London: Longman, Hurst, Rees and Orme.

PISTER, P. 1990. Pure Colorado trout saved by California. *Outdoor California* 51, no. 1: 12–15.

PLATTS, W.S. 1957. The cutthroat trout. *Utah Fish and Game* 13, no. 1: 4, 7.

————. 1958. Age and growth of the cutthroat trout in Strawberry Reservoir, Utah. *Proceedings of the Utah Academy of Sciences, Arts, and Letters* 35: 101–103.

————. 1981a. Sheep and streams. *Rangelands* 3: 158–160.

————. 1981b. *Effects of sheep grazing on a riparian-stream environment.* Ogden, Utah: USDA Forest Service Intermountain Region, Research Note INT-307.

————. 1991. Livestock grazing. Pages 389–423 in W.R. Meehan, editor. *Influences of forest and rangeland management on salmonid fishes and their habitats.* Bethesda, Maryland: American Fisheries Society Special Publication 19.

PLATTS, W.S. AND M.L. MCHENRY. 1988. *Density and biomass of trout and char in western streams.* Ogden, Utah: USDA Forest Service Intermountain Region, General Technical Report INT-241.

PLATTS, W.S. AND R.L. NELSON. 1983. Population fluctuations and generic differentiation in the Humboldt cutthroat trout of Gance Creek, Nevada. *Transactions of the Cal-Neva Chapter American Fisheries Society* 1983: 15–20.

PLUHAR, C.J., B.N. BJORNSTAD, S.P. REIDEL, R.S. COE AND P.B. NELSON. 2006. Magnetostratigraphic evidence from the Cold Creek bar for onset of ice-age cataclysmic floods in eastern Washington during the early Pleistocene. *Quaternary Research* 65: 123–135.

PLUME, R.W. AND D.A. PONCE. 1999. *Hydrogeologic framework and ground-water levels, 1982 and 1996, middle Humboldt River Basin, north-central Nevada.* Carson City, Nevada: U.S. Geological Survey, Water-Resources Investigations Report 98-4029.

POAG, C.W. 1995. Upper Eocene impactites of the U.S. east coast: depositional origins, biostratigraphic framework, and correlation. *Palaios* 10: 16–43.

POAG, C.W., D.S. POWARS, L.J. POPPE AND R.B. MIXON. 1994. Meteoroid mayhem in Ole Virginny: source of the North American tektite strewn field. *Geology* 22: 691–694.

POFF, N.L. AND J.V. WARD. 1990. Physical habitat template of lotic systems: recovery in the context of historical pattern of spatiotemporal heterogeneity. *Environmental Management* 14: 629–645.

POFF, N.L., M.M. BRINSON AND J.W. DAY, JR. 2002. *Aquatic ecosystems and global climate change: potential impacts on inland freshwater and coastal wetland ecosystems in the United States.* Report prepared for Pew Center on Global Climate Change, Arlington, Virginia. Available on the Internet at www.pewclimate.org.

POINAR, H.N., M. HÖSS, J.L. BADA AND S. PÄÄBO. 1996. Amino acid racemization and the preservation of ancient DNA. *Science* 272: 864–866.

POJAR, J. 1980. Brooks Peninsula: possible Pleistocene glacial refugium on northwestern Vancouver Island. *Botanical Society of America Miscellaneous Series Publication* 158: 89.

POLLOCK, M.M., M. HEIM AND D. WEBSTER. 2003. Hydrologic and geomorphic effects of beaver dams and their influence on fishes. Pages 213–234 in S.V. Gregory, K. Boyer and A. Gurnell, editors. *The ecology and management of wood in world rivers.* Bethesda, Maryland: American Fisheries Society Symposium 37.

POPOV, B.H. AND J.B. LOW. 1950. *Game, fur animals and fish introductions into Utah.* Salt Lake City: Utah Department of Fish and Game, Publication No. 4.

PORTER, M.R. AND O. DAVENPORT. 1963. *Scotsman in buckskin: Sir William Drummond Stewart and the Rocky Mountain fur trade.* New York: Hastings House.

PORTER, S.C., K.L. PIERCE AND T.D. HAMILTON. 1983. Late Wisconsin mountain glaciation in the western United States. Pages 71–111 in S.C. Porter, editor. *Late Quaternary environments of the United States.* Volume 1, the Late Pleistocene. Minneapolis: University of Minnesota Press.

POTTER, C. 1982. Endangered and threatened wildlife and plants: review of vertebrate wildlife for listing as endangered or threatened species. *Federal Register* 47, no. 251: 58454–58460.

POWERS, D.A. 1991. Evolutionary genetics of fish. Pages 119–228 in J.G. Scandalios and T.F. Wright, editors. *Advances in genetics, Volume 29.* New York: Academic Press.

PRATT, P.P. 1970. *The autobiography of Parley Parker Pratt.* Salt Lake City: Utah, Deseret Book Company.

PRINGLE, C.S. 1905 [1989, 1993]. Across the plains in 1844. In S.A. Clarke, editor. *Pioneer days of Oregon history.* Volume 2. Portland: Oregon, J.K. Gill [reprinted 1989 and 1993 by Ye Galleon Press, Fairfield, Washington].

PRITCHARD, V.L. AND D.E. COWLEY. 2006. *Rio Grande cutthroat trout (Oncorhynchus clarkii virginalis): a technical conservation assessment.* Las Cruces: New Mexico State University Department of Fishery and Wildlife Sciences, report prepared for USDA Forest Service Rocky Mountain Region, Species Conservation Project. Available on the Internet at www.fs.fed.us/r2/projects/scp/assessments/index.shtml.

PROEBSTEL, D. 1998. Untitled letter report on genetic analysis of trout from Yakima and Wenatchee basin streams. Don Proebstel, World Salmonid Research Institute, Colorado State University, Fort Collins, Colorado to B.K. Ringel, U.S. Fish and Wildlife Service, Leavenworth, Washington.

PROEBSTEL, D., R.J. BEHNKE AND S.M. NOBLE. 1996. *Identification of salmonid fishes from tributary streams and lakes of the mid-Columbia basin.* Fort Collins: Colorado State University, Department of Fisheries and Wildlife Biology, draft report.

PROEBSTEL, D.S. AND S.M. NOBLE. 1994. Are "pure" native trout in the mid-Columbia River basin? Pages 177–184 in R. Barnhart, B. Shake and R.H. Hamre, editors. *Wild Trout V: wild trout in the 21st Century.* Mammoth Hot Springs: Yellowstone National Park. National Biological Survey, U.S. Fish and Wildlife Service, and 5 additional co-sponsors.

PROPST, D.L. AND M.A. MCINNIS. 1975. *An analysis of streams containing native Rio Grande cutthroat trout in the Santa Fe National Forest.* Boulder, Colorado: Western Interstate Commission for Higher Education. Report prepared for the Santa Fe National Forest, Albuquerque, New Mexico.

PROPST, D.L., J.A. STEFFERUD AND P.R. TURNER. 1992. Conservation and status of Gila trout, *Oncorhynchus gilae. The Southwestern Naturalist* 37: 117–125.

PROTHERO, D.R. 1998. The chronological, climatic, and paleogeographic background to North American mammalian evolution. Pages 9–36 in C.M. Janis, K.M. Scott and E.L. Jacobs, editors. *Evolution of Tertiary mammals of North America.* Vol. 1. New York: Cambridge University Press.

PUGET SOUND TASK FORCE. 1970. *Puget Sound and adjacent waters. Comprehensive study of water and related land resources.* Appendix 11, Fish and Wildlife. Vancouver, Washington: Pacific Northwest River Basins Commission.

PYNES, P. 2000. Erosion, extraction, reciprocation: an ethno/environmental history of the Navajo Nation's ponderosa pine forests. Ph.D. dissertation, University of New Mexico, Albuquerque.

———. 2002. Chuska Mountains and Defiance Plateau, Navajo Nation. In Grahame, J.D. and T.D. Sisk, editors. *Canyons, cultures and environmental change: an introduction to the land-use history of the Colorado Plateau.* Flagstaff: Northern Arizona University, Land

Use History of North America Program. Electronic document available on the Internet at www.cpluhna. nau.edu/ (accessed March 3, 2005).

PYRAMID LAKE FISHERIES. 1992. *Pyramid Lake fishery conservation plan*. Sutcliffe, Nevada: Pyramid Lake Paiute Tribe, Fisheries Department.

QUINLAN, R.E. 1980. A study of the biology of the Colorado River cutthroat trout (*Salmo clarki pleuriticus*) population in the North Fork of the Little Snake River drainage in Wyoming. Master's thesis, University of Wyoming, Laramie.

QUINN, T.P. AND N.P. PETERSON. 1996. The influence of habitat complexity and fish size on overwinter survival of individually marked juvenile coho salmon (*Oncorhynchus kisutch*) in Big Beef Creek, Washington. *Canadian Journal of Fisheries and Aquatic Sciences* 53: 1555–1564.

RADFORD, D.S. 1977. *An evaluation of Alberta's fishery management program for east slope streams*. Edmonton: Alberta Department of Recreation, Parks and Wildlife, Fishery Management Report 23.

RAE, W.E., EDITOR. 2001. *The best of Outdoor Life*. New York: The Lyons Press.

RALEIGH, R.F. 1971. Innate control of migrations of salmon and trout fry from natal gravels to rearing areas. *Ecology* 52: 291–297.

RALEIGH, R.F. AND D.W. CHAPMAN. 1971. Genetic control in lakeward migration of cutthroat trout fry. *Transactions of the American Fisheries Society* 100: 33–40.

RANKEL, G. 1977. *Fisheries management plan—Summit Lake Indian Reservation*. Reno: Nevada, U.S. Fish and Wildlife Service, Fisheries Assistance Office.

RAUP, D.M. 1986. Biological extinction in Earth history. *Science* 231: 1528–1533.

RAVILIOUS, K. 2002. Killer blow. *New Scientist* 174, no. 2341: 29–31.

RAWLEY, E.V. 1985. *Early records of wildlife in Utah*. Salt Lake City: Utah Division of Wildlife Resources, Publication 86-2.

RAY, L.L. 1940. Glacial chronology of the southern Rocky Mountains. *Geological Society of America Bulletin* 51: 1851–1918.

RAYMOND, S. 1996. *The estuary flyfisher*. Portland, Oregon: Frank Amato Publications.

REEVES, G.H., F.H. EVEREST AND J.R. SEDELL. 1993. Diversity of juvenile anadromous salmonid assemblages in coastal Oregon basins with different levels of timber harvest. *Transactions of the American Fisheries Society* 122: 309–317.

REHEIS, M. 1999. Highest pluvial-lake shorelines and Pleistocene climate of the western Great Basin. *Quaternary Research* 52: 196–205.

REHEIS, M.C., A.M. SARNA-WOJCICKI, R.L. REYNOLDS, C.A. REPENNING AND M.D. MIFFLIN. 2002. Pliocene to middle Pleistocene lakes in the western Great Basin: ages and connections. Pages 53–108 in R. Hershler, D. Madsen D. Currey, editors. *Great Basin aquatic systems history. Smithsonian Contributions to the Earth Sciences, number 33*. Washington, D.C.: Smithsonian Institution Press.

REHEIS, M.C. AND R.B. MORRISON. 1997. High, old pluvial lakes of western Nevada. Pages 459–492 in P.K. Link and B.J. Kowallis, editors. *Proterozoic to recent stratigraphy, tectonics, and volcanology, Utah, Nevada, southern Idaho and central Mexico*. Provo, Utah: Brigham Young University Geology Studies, vol. 42, part 1.

REICHARD, G.A. 1974 [1983]. *Navaho religion: a study in symbolism*. 2nd edition. Princeton, New Jersey: Princeton University Press [reprinted 1983 by University of Arizona Press, Tucson].

REIGER, G., EDITOR. 1972. *Zane Grey: outdoorsman*. New York: Prentice-Hall, Inc.

REINITZ, G.L. 1977. Electrophoretic distinction of rainbow trout (*Salmo gairdneri*), westslope cutthroat trout (*Salmo clarki*), and their hybrids. *Journal of the Fisheries Research Board of Canada* 34: 1236–1239.

REISNER, M. 1986 [reprinted 1993]. *Cadillac desert: the American west and its disappearing water*. New York: Viking Press.

REMMICK, R. 1981. *A survey of native cutthroat trout populations and associated stream habitats in the Bridger-Teton National Forest*. Cheyenne: Wyoming Game and Fish Department, Fish Division Administrative Report.

REPENNING, C.A., T.A. WEASMA AND G.R. SCOTT. 1995. *The early Pleistocene (latest Blancan-earliest Irvingtonian) Froman Ferry fauna and history of the Glenns Ferry Formation, southwestern Idaho*. Washington, D.C.: U.S. Geological Survey Bulletin 2105.

RESHETNIKOV, Y.S. AND 8 co-authors. 1997. An annotated checklist of the freshwater fishes of Russia. *Journal of Ichthyology* 37: 687–736.

RHYMER, J.M. AND D. SIMBERLOFF. 1996. Extinction by hybridization and introgression. *Annual Review of Ecology and Systematics* 27: 83–109.

RICHARDSON, F. AND R.H. HAMRE, TECHNICAL EDITORS. 1989. *Wild Trout IV: proceedings of the symposium*. Yellowstone National Park, Wyoming: U.S. Fish and Wildlife Service and 5 co-sponsors.

RICHARDSON, G.B. 1941. *Geology and mineral resources of the Randolph Quadrangle, Utah-Wyoming*. Washington, D.C.: U.S. Geological Survey Bulletin 923.

RICHARDSON, J. 1836 [reprinted 1978]. *Fauna Boreali–Americana; or the Zoology of the Northern Parts of British America: part third The Fish.* London: Richard Bently [reprinted by Arno Press, New York}.

RICHMOND, G.M. 1960. Glaciation of the east slopes of Rocky Mountain National Park. *Geological Society of America Bulletin* 71: 1371–1382.

———. 1965. Glaciation of the Rocky Mountains. Pages 217–230 in H.E. Wright, Jr. and D.G. Frey, editors. *The Quaternary of the United States.* Princeton, New Jersey: Princeton University Press.

———. 1986. Stratigraphy and correlation of glacial deposits of the Rocky Mountains, the Colorado Plateau and the ranges of the Great Basin. *Quaternary Science Reviews* 5: 99–127.

RICHMOND, G.M. AND D.S. FULLERTON. 1986. Introduction to Quaternary glaciations in the United States of America. *Quaternary Science Review* 5: 3–10 (and accompanying reports).

RICHMOND, G.M., R. FRYXELL, G.E. NEFF AND P.L. WEIS. 1965. The Cordilleran ice sheet of the northern Rocky Mountains and related Quaternary history of the Columbia Plateau. Pages 231–242 in H.E. Wright, Jr. and D.G. Frey, editors. *The Quaternary of the United States.* Princeton, New Jersey: Princeton University Press.

RICKER, W.E. 1954. Stock and recruitment. *Journal of the Fisheries Research Board of Canada* 11: 559–623.

———. 1972. Hereditary and environmental factors affecting certain salmonid populations. Pages 19–160 in R.C. Simon and P.A. Larkin, editors. *The stock concept in Pacific salmon.* Vancouver, B.C.: H.R. MacMillan Lectures in Fisheries, University of British Columbia.

RIEMAN, B.E. AND J.B. DUNHAM. 2000. Metapopulations and salmonids: a synthesis of life history patterns and empirical observations. *Ecology of Freshwater Fish* 9, no. 1–2: 51–64.

RIEMAN, B.E. AND K.A. APPERSON. 1989. *Status and analysis of salmonid fisheries: westslope cutthroat trout synopsis and analysis of fishery information.* Boise: Idaho Department of Fish and Game, Job Performance Report, Project F-73-R-11, Subproject II, Job 1.

RILEY, S.C., K.D. FAUSCH AND C. GOWAN. 1992. Movement of brook trout (*Salvelinus fontinalis*) in four small subalpine streams in northern Colorado. *Ecology of Freshwater Fishes* 1: 112–122.

RINGEL, B.K. 1997. *Analysis of fish populations in Icicle Creek, Trout Creek, Jack Creek, Peshastin Creek, Ingalls Creek, and Negro Creek, Washington 1994 and 1995.* Leavenworth: Washington, U.S. Fish and Wildlife Service.

RINNE, J.N. 1995. Rio Grande cutthroat trout. Pages 24–27 in M.K. Young, technical editor. *Conservation assessment for inland cutthroat trout.* Fort Collins, Colorado: USDA Forest Service, Rocky Mountain Forest and Range Experiment Station, General Technical Report RM-GTR-256.

———. 1996. Short-term effects of wildfire on fishes and aquatic macroinvertebrates in the southwestern United States. *North American Journal of Fisheries Management* 16: 653–658.

RINNE, J.N. AND P.R. TURNER. 1991. Reclamation and alteration as management techniques, and a review of methodology in stream renovation. Pages 219–244 in W.L. Minckley and J.E. Deacon, editors. *Battle against extinction: native fish management in the American west.* Tucson, University of Arizona Press.

RINNE, J.N., W.L. MINCKLEY AND J.N. HANSEN. 1981. Chemical treatment of Ord Creek, Apache County, Arizona to re-establish apache trout. *Journal of the Arizona-Nevada Academy of Sciences* 16: 74–78.

RIPLEY, E.A., R.E. REDMANN AND A.A. CROWDER. 1995. *Environmental effects of mining.* Delray Beach, Florida: St. Lucie Press.

RISELAND, J.L. 1909. *18th and 19th Annual Reports of the State Fish Commissioner for the Years 1907 and 1908.* Olympia: State of Washington Department of Fisheries and Game.

RIVERS, P.J. AND W.R. ARDREN. 1998. The value of archives. *Fisheries* 23, no. 5: 6–9.

ROBERTS, B.C. 1988. Potential influence of recreational use on Nelson Spring Creek, Montana. Master's thesis, Montana State University, Bozeman.

ROBERTS, B.C. AND R.G. WHITE. 1992. Effects of angler wading on survival of trout eggs and pre-emergent fry. *North American Journal of Fisheries Management* 12: 450–459.

ROBERTS, R.J. AND C.J. SHEPHERD. 1974. *Handbook of trout and salmon diseases.* Surrey, U.K.: Fishing News Books, Ltd.

ROBERTSON, G.C. 1978. Surficial deposits and geologic history, northern Bear Lake valley, Idaho. Master's thesis, Utah State University, Logan.

ROBERTSON, W.R.B. 1916. Chromosome studies I. Taxonomic relationships shown in the chromosomes of Telligidae and Acrididae. V-shaped chromosomes and their significance in Acrididae, Locustidae and Gryllidae: chromosomes and variation. *Journal of Morphology* 27: 179–391.

ROBINSON, R.S. 1950. The native cutthroat trout, *Salmo clarki*, of the Bonneville Basin and the Green River tributaries of Utah. Master's thesis, Oregon State University, Corvallis, Oregon.

RODGERS, D.W., W.R. HACKETT AND H.T. ORE. 1990. Extension of the Yellowstone Plateau, eastern Snake River Plain, and Owyhee plateau. *Geology* 18: 1138–1141.

RODRIGUEZ, M.A. 2002. Restricted movement in stream fish: the paradigm is incomplete, not lost. *Ecology* 83: 1–13.

ROGERS, K. AND J. WANGNILD. 2004 [revised 2005]. *Trapper Lake: resurrecting a native conservation population of Colorado River cutthroat trout.* Steamboat Springs and Meeker, Colorado: Colorado Division of Wildlife. Available on the Internet at http://wildlife. state.co.us/Research/CutthroatTrout/. Scroll down the document that opens to find a link to this paper.

ROGERS, K.L. AND 7 co-authors. 1985. Middle Pleistocene (Late Irvingtonian: Nebraskan) climate changes in south-central Colorado. *National Geographic Research* 1, no. 4: 535–563.

ROGERS, K.L. AND 11 co-authors. 1992. Pliocene and Pleistocene geologic and climate evolution in the San Luis Valley of south-central Colorado. *Paleogeography, Paleoclimatology, Paleoecology* 94: 55–86.

ROHLING, E.J. AND 5 co-authors. 1998. Magnitudes of sea-level lowstands of the past 500,000 years. *Nature* 394: 162–165.

ROHRER, R.L. 1983. *Evaluation of Henry's Lake management program.* Boise: Idaho Department of Fish and Game, Federal Aid in Sport Fish Restoration, Project F-73-R-5.

ROHRER, R.L. AND G.H. THORGAARD. 1986. Evaluation of two hybrid trout strains in Henry's Lake, Idaho, and comments on the potential use of sterile triploid hybrids. *North American Journal of Fisheries Management* 6: 367–371.

ROLLINS, P.H., EDITOR. 1935. *The discovery of the Oregon Trail: Robert Stuart's narratives.* New York: Charles Scribner's Sons.

ROMERO, N., R.E. GRESSWELL AND J.L. LI. 2005. Changing patterns in coastal cutthroat trout (*Oncorhynchus clarki clarki*) diet and prey in a gradient of deciduous canopies. *Canadian Journal of Fisheries and Aquatic Sciences* 62: 1797–1807.

ROSCOE, J.W. 1974. Systematics of the westslope cutthroat trout. Master's thesis, Colorado State University, Fort Collins.

ROSENBAUER, T. 1988. *Reading trout streams: an Orvis guide.* New York: Lyons & Burford, Publishers.

———. 1993. *Prospecting for trout: fly fishing secrets from a streamside observer.* New York: Dell Publishers.

ROSENFELD, J.S. AND 7 co-authors. 2002. Importance of small streams as rearing habitat for coastal cutthroat trout. *North American Journal of Fisheries Management* 22: 177–187.

ROSENLUND, B.D. 1979. *Rocky Mountain National Park Fisheries Management Report.* Lakewood, Colorado: U.S. Fish and Wildlife Service, Colorado Fisheries Assistance Office.

ROSENLUND, B.D. AND T.R. ROSENLUND. 1989. Leadville National Fish Hatchery 1889–1989. *Fisheries* 14, no. 3: 18–20.

ROY, J.E. 1974. *Memoires de L'Amerique Septentrionale, ou la suite des voyages de Mr. Le Baron de La Hontan.* Montreal: Editions Élysée.

ROYAL, L.A. 1972. *An examination of the anadromous trout program of the Washington State Game Department.* Olympia: Washington State Game Department.

RUBIDGE, E., P. CORBETT AND E.B. TAYLOR. 2001. A molecular analysis of hybridization between native westslope cutthroat trout and introduced rainbow trout in southeastern British Columbia, Canada. *Journal of Fish Biology* 59 (supplement A): 42–54.

RUBIDGE, E.M. AND E.B. TAYLOR. 2005. An analysis of spatial and environmental factors influencing hybridization between native westslope cutthroat trout (*Oncorhynchus clarkii lewisi*) and introduced rainbow trout (*O. mykiss*) in the upper Kootenay River drainage, British Columbia. *Conservation Genetics* 6: 369–384.

RUSCO, M.K. AND J.O. DAVIS. 1987. *Studies in archaeology, geology and paleontology at Rye Patch Reservoir, Pershing County, Nevada.* Carson City: Nevada State Museum, Anthropological Papers no. 20.

RUSSELL, I.C. 1903a. *Preliminary report on artesian basins in southwestern Idaho and southeastern Oregon.* Washington, D.C.: U.S. Geological Survey Water Supply and Irrigation Paper 78.

———. 1903b. *Notes on the geology of southwestern Idaho and southeastern Oregon.* Washington, D.C.: U.S. Geological Survey Bulletin 217.

RUSSELL, I.K. (in collaboration with H.R. Driggs). 1923. *Hidden heroes of the Rockies.* Yonkers-on-Hudson, New York: World Book Company.

RUSSELL, O. 1986. *Journal of a trapper.* Third printing. Lincoln, Nebraska: University of Nebraska Press.

RUXTON, GEORGE FREDERICK. 1848. *Adventures in Mexico and the Rocky Mountains.* London: J. Murray.

RUZYCKI, J.R., D.A. BEAUCHAMP AND D.L. YULE. 2003. Effects of introduced lake trout on native cutthroat trout in Yellowstone Lake. *Ecological Applications* 13: 23–37.

RUZYCKI, J.R., W.A. WURTSBAUGH AND C. LUECKE. 2001. Salmonine consumption and competition for endemic prey fishes in Bear Lake, Utah-Idaho. *Transactions of the American Fisheries Society* 130: 1175–1189.

RYAN, J.H. AND S.J. NICOLA. 1976. *Status of the Paiute cutthroat trout, Salmo clarki seleniris (Snyder), in California.* Rancho Cordova: California Department of Fish and Game, Inland Fisheries Administrative Report 76-3.

RYMAN, N. AND F. UTTER, EDITORS. 1987. *Population genetics and fishery management.* Seattle: Washington Sea Grant Program and University of Washington Press.

SABO, J.S. 1995. Competition between stream-dwelling cutthroat trout (*Oncorhynchus clarki clarki*) and coho salmon (*O. kisutch*): implications for community structure and evolutionary ecology. Master's thesis, University of Washington, Seattle.

SABO, J.S. AND G.B. PAULEY. 1997. Competition between stream-dwelling cutthroat trout (*Oncorhynchus clarki*) and coho salmon (*O. kisutch*): effects of relative size and population origin. *Canadian Journal of Fisheries and Aquatic Sciences* 54: 2609–2617.

SADLER, J.L. AND P.K. LINK. 1996. The Tuana Gravel: early Pleistocene response to longitudinal drainage of a late-stage rift basin, western Snake River Plain, Idaho. *Northwest Geology* 26: 46–62.

SAGE, RUFUS B. 1859. *Rocky Mountain life: or, startling scenes and perilous adventures in the far west during an expedition of three years.* Boston: Thayer and Eldridge.

SANDERSON, E.W. AND 5 co-authors. 2002. The human footprint and the last of the wild. *BioScience* 52: 891–904. Copies of this paper and its accompanying maps are available on the Internet at www.wcs.org/humanfootprint.

SANDLUND, O.T. AND 7 co-authors. 1987. Habitat use of arctic charr *Salvelinus alpinus* in Thingvallavatn, Iceland. *Environmental Biology of Fishes* 20: 263–274.

SANFORD, C.P.J. 1990. The phylogenetic relationships of salmonid fishes. *Bulletin British Museum of Natural History (Zoology)* 56: 145–153.

SCARNECCHIA, D.L. AND E.P. BERGERSEN. 1986. Production and habitat of threatened greenback and Colorado River cutthroat trouts in Rocky Mountain headwater streams. *Transactions of the American Fisheries Society* 115: 382–391.

———. 1987. Trout production and standing crop in Colorado's small streams, as related to environmental features. *North American Journal of Fisheries Management* 7: 315–330.

SCHEFFER, V.B. 1935. *Lake Crescent, Clallam County, Washington.* Portland, Oregon: U.S. Fish and Wildlife Service (12-page memo report on file at Olympic National Park, Port Angeles, Washington).

SCHENCK, A.B. 1965. Camping vacation, 1871. *Colorado Magazine* 42, no. 3: 185–215.

SCHIEL, JACOB H. 1859 [1959]. *Journey through the Rocky Mountains and the Humboldt Mountains to the Pacific Ocean.* Schaffhausen, Brodtmannschen Buchhandlens [translated from the German by T.H. Bornn and reprinted by University of Oklahoma Press, Norman, Oklahoma].

SCHILL, D.J., J.S. GRIFFITH AND R.E. GRESSWELL. 1986. Hooking mortality of cutthroat trout in a catch-and-release segment of the Yellowstone River, Yellowstone National Park. *North American Journal of Fisheries Management* 6: 226–232.

SCHLOSSER, I.J. 1995. Critical landscape attributes that influence fish population dynamics in headwater streams. *Hydrobiologia* 303: 71–81.

SCHLOSSER, I.J. AND P.L. ANGERMEIER. 1995. Spatial variation in demographic processes of lotic fishes: conceptual models, empirical evidence, and implications for conservation. Pages 392–401 in J.L. Nielsen, editor. *Evolution and the aquatic ecosystem: defining unique units in population conservation.* Bethesda, Maryland: American Fisheries Society Symposium 17.

SCHMETTERLING, D.A. 2000. Redd characteristics of fluvial westslope cutthroat trout in four tributaries to the Blackfoot River, Montana. *North American Journal of Fisheries Management* 20: 776–783.

———. 2001. Seasonal movements of fluvial westslope cutthroat trout in the Blackfoot River drainage, Montana. *North American Journal of Fisheries Management* 21: 507–520.

———. 2003. Reconnecting a fragmented river: movements of westslope cutthroat trout and bull trout after transport upstream of Milltown Dam, Montana. *North American Journal of Fisheries Management* 23: 721–731.

SCHMIDT, A.E. 1997. Status of sea-run cutthroat trout stocks in Alaska. Pages 80–83 in J.D. Hall, P.A. Bisson and R.E. Gresswell, editors. 1997. *Sea-run cutthroat trout: biology, management, and future conservation.* Corvallis: Oregon Chapter American Fisheries Society.

SCHMITTEN, R.A. AND J.R. CLARK. 1999. Endangered and threatened species; threatened status for southwestern Washington/Columbia River coastal cutthroat trout in Washington and Oregon, and delisting of Umpqua River cutthroat trout in Oregon. *Federal Register* 64, no. 64: 16397–16414.

SCHOENHERR, A.A. 1992. *A natural history of California.* Berkeley: University of California Press.

SCHRANK, A.J. AND F.J. RAHEL. 2004. Movement patterns of inland cutthroat trout (*Oncorhynchus clarki utah*): management and conservation implications. *Canadian Journal of Fisheries and Aquatic Sciences* 61: 1528–1537.

———. 2006. Factors influencing summer movement patterns of Bonneville cutthroat trout (*Oncorhynchus clarkii utah*). *Canadian Journal of Fisheries and Aquatic Sciences* 63: 660–669.

SCHRANK, A.J., F.J. RAHEL AND H.C. JOHNSTONE. 2003. Evaluating laboratory-derived thermal criteria in the field: an example involving Bonneville cutthroat trout. *Transactions of the American Fisheries Society* 132: 100–109.

SCHRECK, C.B. 1981. Parr-smolt transformation and behavior. Pages 164–172 in E.L. Brannon and E.O. Salo, editors. *Salmon and trout migratory behavior, proceedings of a symposium.* Seattle: University of Washington, School of Fisheries.

SCHRECK, C.B. AND R.J. BEHNKE. 1971. Trouts of the upper Kern River basin, California, with reference to systematics and evolution of western North American *Salmo. Journal of the Fisheries Research Board of Canada* 28: 987–998.

SCHRECK, C.B. AND P.B. MOYLE, EDITORS. 1990. *Methods for fish biology.* Bethesda, Maryland: American Fisheries Society.

SCHROETER, R.E. 1998. Segregation of stream-dwelling Lahontan cutthroat trout and brook trout: patterns of occurrence and mechanisms for displacement. Master's thesis, University of Nevada, Reno.

SCHUTZ, D.C. AND T.G. NORTHCOTE. 1972. An experimental study of feeding behavior and interactions of coastal cutthroat trout (*Salmo clarki clarki*) and Dolly Varden (*Salvelinus malma*). *Journal of the Fisheries Research Board of Canada* 29: 555–565.

SCHULLERY, P. 1980. "Their numbers are perfectly fabulous": Yellowstone angling excursions 1867–1925. *The American Fly Fisher* (the journal of the American Museum of Fly Fishing, Manchester, Vermont) 7, no. 2: 14–19.

———. 1982. Snake River cutthroats. *The American Fly Fisher* (the journal of the American Museum of Fly Fishing, Manchester, Vermont) 9, no. 3: 28–29.

———. 2006. *Cowboy trout: western fly fishing as if it matters.* Helena: Montana Historical Society Press.

SCHULTZ, L.P. 1935. Species of salmon and trout in the northwestern United States. *Pacific Science Congress Proceedings* 5: 3777–3782.

SCHULTZ, P.H. AND S. D'HONDT. 1996. Cretaceous-Tertiary (Chicxulub) impact angle and its consequences. *Geology* 24: 963–967.

SCHWIEBERT, E. 1978. *Trout* (two volumes). New York: E.P. Dutton.

SCIENTIFIC AMERICAN. 1982. *The fossil record and evolution.* San Francisco: W.H. Freeman and Company.

SCOPPETTONE, G.G., M.E. COLEMAN AND G.A. WEDEMEYER. 1986. *Life history and status of the endangered cui-ui of Pyramid Lake, Nevada.* Washington, D.C.: U.S. Fish and Wildlife Service, Fisheries and Wildlife Research Bulletin 1.

SCOTT, E.B. 1957. *The saga of Lake Tahoe: a complete documentation of Lake Tahoe's development over the last one hundred years.* Crystal Bay, Nevada: Sierra-Tahoe Publishing.

SCOTT, G.R. AND R.B. TAYLOR. 1975. *Post Paleocene tertiary rocks and Quaternary volcanic ash of the Wet Mountain Valley, Colorado.* Washington, D.C.: U.S. Geological Survey Professional Paper 868.

SCOTT, W.B. AND E.J. CROSSMAN. 1973. *Freshwater fishes of Canada.* Ottawa: Fisheries Research Board of Canada, Bulletin 184.

SCOTT, W.E., W.D. McCOY, R.R. SHROBA AND M. RUBIN. 1983. Reinterpretation of the exposed record of the last two cycles of Lake Bonneville, western United States. *Quaternary Research* 20: 261–285.

SCULLY, R.J. 1993. *Bonneville cutthroat trout in Idaho: 1993 status.* Boise: Idaho Department of Fish and Game.

SEALING, C. AND 5 co-authors. 1992. *Conservation plan for Colorado River cutthroat trout in northwest Colorado.* Grand Junction: Colorado Division of Wildlife, White River National Forest, Routt National Forest, Roosevelt-Arapahoe National Forest, and Bureau of Land Management Grand Junction and Craig Resource Areas.

SEKULICH, P.T. 1974. Role of the Snake River cutthroat trout (*Salmo clarki* subspp.) in fishery management. Master's thesis, Colorado State University, Fort Collins.

SETTLE, R.W., EDITOR. 1940 [reprinted 1989]. *The march of the mounted riflemen.* Glendale, California: Arthur H. Clark Company [reprint, University of Nebraska Press, Lincoln].

SEVON, M. 1988. *Walker Lake fisheries management plan.* Reno: Nevada Division of Wildlife.

SEVON, M., J. FRENCH, J. CURRAN AND R. PHENIX. 1999. *Lahontan cutthroat trout species management plan for the Quinn River/Black Rock basins and North Fork Little Humboldt River sub-basin.* Reno: Nevada Division of Wildlife.

SEWALL, J.O. 2005. Precipitation shifts over western North America as a result of declining Arctic sea ice cover: the coupled system response. *Earth Interactions* 9: 1–23.

SEWALL, J.O. AND L.C. SLOAN. 2004. Disappearing Arctic sea ice reduces available water in the American west. *Geophysical Research Letters* 31, L06209: 1–4.

SHAKLEE, J.B., F.W. ALLENDORF, D.C. MORIZOT AND G.S. WHITT. 1990. Gene nomenclature for protein-coding loci in fish. *Transactions of the American Fisheries Society* 119: 2–15.

SHARP, R.P. 1938. Pleistocene glaciation in the Ruby-East Humboldt Range, northeastern Nevada. *Journal of Geomorphology* 1: 296–323.

SHEBLEY, W.H. 1929. History of the fish and fishing conditions of Lake Tahoe. *California Fish and Game* 15: 194–203.

SHEDLOCK, A.M., J.D. PARKER, D.A. CRISPIN, T.W. PIETSCH AND G.C. BURMER. 1992. Evolution of the salmonid mitochondrial control region. *Molecular Phylogenetics and Evolution* 1: 179–192.

SHEDLOCK, A.M., M.G. HAYGOOD, T.W. PIETSCH AND P. BENTZEN. 1997. Enhanced DNA extraction and PCR amplification of mitochondrial genes from formalin-fixed museum specimens. *BioTechniques* 22: 394–400.

SHEEHAN, P.M., D.A. FASTOVSKY, R.G. HOFFMANN, C.B. BERGHAUS AND D.L. GABRIEL. 1991. Sudden extinction of the dinosaurs: latest Cretaceous, upper Great Plains, U.S.A. *Science* 254: 835–838.

SHEPARD, B., EDITOR. 2001. *Practical approaches for conserving native inland fishes of the west.* Missoula: Montana Chapter American Fisheries Society.

SHEPARD, B.B., B.E. MAY AND W. URIE. 2003. *Status of westslope cutthroat trout (Oncorhynchus clarki lewisi) in the United States: 2002.* Bozeman, Montana: Multi-State Assessment Team. Available on the Internet at www.fwp.state.mt.us/wildthings/westslope/content.asp.

———. 2005. Status and conservation of westslope cutthroat trout within the western United States. *North American Journal of Fisheries Management* 25: 1426–1440.

SHEPARD, B.B., B. SANBORN, L. ULMER AND D.C. LEE. 1997. Status and risk of extinction for westslope cutthroat trout in the upper Missouri River basin, Montana. *North American Journal of Fisheries Management* 17: 1158–1172.

SHEPARD, B.B., K.L. PRATT AND P.J. GRAHAM. 1984. *Life histories of westslope cutthroat trout and bull trout in the upper Flathead River basin, Montana.* Helena: Montana Department of Fish, Wildlife and Parks.

SHEPHERD, B.G. 1974. Activity localization in coastal cutthroat trout (*Salmo clarki clarki*) in a small bog lake. *Journal of the Fishery Research Board of Canada* 31: 1246–1249.

SHIELDS, G.O. 1889. *Cruising in the Cascades. A narrative of travel, exploration, amateur photography, hunting, and fishing.* Chicago and New York: Rand McNally & Company.

SHIOZAWA, D.K., J. KUDO, R.P. EVANS, S.R. WOOD AND R.N. WILLIAMS. 1992. DNA extraction from preserved trout tissues. *Great Basin Naturalist* 52: 29–34.

SHIOZAWA, D.K. AND R.P. EVANS. 1994a. *Relationships between cutthroat trout populations from thirteen Utah streams in the Colorado River and Bonneville drainages.* Ogden, Utah: Brigham Young University Department of Zoology, Final Report to Utah Division of Wildlife Resources, Contract 92-2377.

———. 1994b. The use of DNA to identify geographical isolation in trout stocks. Pages 125–131 in R. Barnhart, B. Shake and R.H. Hamre, editors. *Wild Trout V: Wild Trout in the 21st Century.* Mammoth Hot Springs, Wyoming: U.S. Fish and Wildlife Service, National Biological Service, and other sponsors.

———. 1995. *The genetic status of cutthroat trout from various drainages in the Wasatch-Cache National forest based on examination of mitochondrial DNA.* Ogden, Utah: Brigham Young University Department of Zoology, Interim Report to the U.S. forest Service, Contract 43-8490-4-0110.

SHIOZAWA, D.K., R.P. EVANS AND R.N. WILLIAMS. 1993. *Relationships between cutthroat trout populations from ten Utah streams in the Colorado River and Bonneville drainages.* Ogden, Utah: Brigham Young University Department of Zoology, Interim Report to Utah Division of Wildlife Resources, Contract 92-2377.

SHOUMATOFF, A. 1997 [1999]. *Legends of the American desert: sojourns in the greater southwest.* New York: Alfred A. Knopf, Inc. [reprinted by HarperCollins Publishers].

SIDOW, A., A.C. WILSON AND S. PÄÄBO. 1991. Bacterial DNA in Clarkia fossils. *Philosophical Transactions of the Royal Society of London, Series B* 333: 429–433.

SIGLER, W.F. 1953. *The rainbow trout in relation to the other fish in Fish Lake.* Logan, Utah: Utah State Agricultural College, Agricultural Experiment Station, Bulletin 358.

SIGLER, W.F. AND J.W. SIGLER. 1986. History of fish hatchery development in the Great Basin states of Utah and Nevada. *Great Basin Naturalist* 40: 583–594.

———. 1987. *Fishes of the Great Basin: a natural history.* Reno: University of Nevada Press.

———. 1990. *Recreational fisheries: management, theory, and applications.* Reno: University of Nevada Press.

SIGLER, W.F. AND R.R. MILLER. 1963. *Fishes of Utah.* Salt Lake City: Utah State Department of Fish and Game.

SIGLER, W.F., S.C. VIGG AND M. BRES. 1985. Life history of the cui-ui, *Chasmistes cujus* Cope, in Pyramid Lake, Nevada: a review. *Great Basin Naturalist* 45: 571–603.

SIGLER, W.F., W.T. HELM, P.A. KUCERA, S. VIGG AND G.W. WORKMAN. 1983. Life history of the Lahontan cutthroat trout, *Salmo clarki henshawi*, in Pyramid Lake, Nevada. *Great Basin Naturalist* 43: 1–29.

SIGURDSSON, H., S. D'Hondt and S. Carey. 1992. The impact of the Cretaceous/Tertiary bolide on evaporate terrane and generation of major sulfuric acid aerosol. *Earth and Planetary Science Letters* 109: 543–559.

SIGURJÓNSDÓTTIR, H. AND K. GUNNARSSON. 1989. Alternative mating tactics of arctic charr, *Salvelinus alpinus*, in Thingvallavatn, Iceland. *Environmental Biology of Fishes* 26: 159–176.

SILER, A.L. 1884. Depletion of fish in Panguitch and Bear Lakes, Utah. *Bulletin of the U.S. Fish Commission* 4: 51.

SILVERBERG, R. 1967. *The Auk, the Dodo and the Onyx*. New York: Thomas Crowell Company.

SILVERSTEIN, M. 1990. Chinookians of the Lower Columbia. Pages 533–546 in W. Suttles, editor. *Handbook of American Indians, vol. 7, Northwest Coast*. Washington, D.C.: Smithsonian Institution.

SIMMONS, V.M. 1966 [1982]. *Bayou Salado: the story of South Park*. [Revised edition published in 1982 by Century One Press, Colorado Springs, Colorado].

————. 2000. *The Ute Indians of Utah, Colorado, and New Mexico*. Boulder, Colorado: University Press of Colorado.

SIMON, J.R. 1946. *Wyoming fishes*. Cheyenne: Wyoming Game and Fish Department, Bulletin 4.

SIMON, R.C. AND A.M. DOLLAR. 1963. Cytological aspects of speciation in two North American teleosts, *Salmo gairdneri* and *Salmo clarki lewisi*. *Canadian Journal of Genetics and Cytology* 5: 43–49.

SIMPSON, C. AND C. SIMPSON. 1981. *North of the Narrows: men and women of the Upper Priest Lake country, Idaho*. Moscow: The University Press of Idaho.

SIMPSON, CAPT. J.H. 1876 [1983]. *Report of explorations across the Great Basin of the Territory of Utah for a direct wagon route from Camp Floyd to Genoa, in Carson Valley, in 1859*. Washington, D.C.: U.S. Army Corps of Topographical Engineers [reprinted 1983, University of Nevada Press, Reno].

SKAALA, Ø. AND K.E. JØRSTAD. 1987. Fine-spotted brown trout (*Salmo trutta*): its phenotypic description and biochemical genetic variation. *Canadian Journal of Fisheries and Aquatic Sciences* 44: 1775–1779.

————. 1988. Inheritance of the fine-spotted pigmentation pattern of brown trout. *Polish Archives of Hydrobiology* 35, no. 3–4: 295–304.

SKAALA, Ø. AND G. NAEVDAL. 1989. Genetic differentiation between freshwater resident and anadromous brown trout, *Salmo trutta*, within watercourses. *Journal of Fish Biology* 34: 597–605.

SKINNER, W.D. 1985. Size selection of food by cutthroat trout (*Salmo clarki*) in an Idaho stream. *Great Basin Naturalist* 45: 327–331.

SKÚLASON, S., S.S. SNORRASON, D.L.G. NOAKES, M.M. FERGUSON AND H.J. MALMQUIST. 1989. Segregation in spawning and early life history among polymorpjic arctic charr, *Salvelinus alpinus*, in Thingvallavatn, Iceland. *Journal of Fish Biology* 35, Supplement A: 225–232.

SLANEY, P. AND J. ROBERTS. 2005. *Coastal cutthroat trout as sentinels of lower mainland watershed health: strategies for coastal cutthroat trout conservation, restoration and recovery*. Surrey, B.C.: British Columbia Ministry of Environment, Lower Mainland Region 2. Available on the Internet at www.shim.bc.ca/cutthroat/ct.pdf.

SLANEY, T.L., K.D. HYATT, T.G. NORTHCOTE AND R.J. FIELDEN. 1996a. Status of anadromous salmon and trout in British Columbia and Yukon. *Fisheries* 21, no. 10: 20–36.

————. 1996b. Status of anadromous cutthroat trout in British Columbia. Pages 77–79 in J.D. Hall, P.A. Bisson and R.E. Gresswell, editors. 1997. *Sea-run cutthroat trout: biology, management, and future conservation*. Corvallis: Oregon Chapter American Fisheries Society.

SLOAT, M.R., R.G. WHITE, B.B. SHEPARD AND S. CARSON. 2001. Factors limiting the contemporary distribution of westslope cutthroat trout in the Madison River basin, Montana. Page 66 in B. Shepard, editor. *Practical approaches for conserving native inland fishes of the west*. Missoula, Montana Chapter AFS. Available on the Internet at www.fisheries.org/AFSmontana/Misc/Symposium%20Abstracts/Abstracts.pdf.

SMILEY, C.J. AND W.C. REMBER. 1985. Physical setting of the Miocene Clarkia fossil beds, northern Idaho. Pages 11–31 in C.J. Smiley, editor. *Late Cenozoic history of the Pacific northwest*. San Francisco, California: Pacific Division of the American Association for the Advancement of Science and California Academy of Science.

SMITH, D. 1993. *Mining America: the industry and the environment*. Boulder, Colorado: University Press of Colorado.

SMITH, G.R. 1981. Late Cenozoic freshwater fishes of North America. *Annual Review of Ecology and Systematics* 12: 163–193.

————. 1992. Introgression in fishes: significance for paleontology, cladistics, and evolutionary rates. *Systematic Biology* 41: 41–57.

SMITH, G.R. AND R.F. STEARLEY. 1989. The classification and scientific names of rainbow and cutthroat trouts. *Fisheries* 14, no. 1: 4–10.

SMITH, G.R. AND R.R. MILLER. 1985. Taxonomy of fishes from Miocene Clarkia lake beds, Idaho. Pages 75–83 in C.J. Smiley, editor. *Late Cenozoic history of the Pacific northwest: interdisciplinary studies of the Clarkia fossil beds of northern Idaho*. San Francisco, California: American Association for the Advancement of Science and California Academy of Science.

SMITH, G.R. AND 5 co-authors. 2002. Biogeography and timing of evolutionary events among Great Basin fishes. Pages 175–234 in R. Hershler, D.B. Madsen and D.R. Currey, editors. *Great Basin aquatic systems history. Smithsonian Contributions to the Earth Sciences, number 33*. Washington, D.C.: Smithsonian Institution Press.

SMITH, H.M. AND W.C. KENDALL. 1921. *Fishes of Yellowstone National Park with description of the park waters and notes on fishing*. Washington, D.C.: Department of Commerce, Bureau of Fisheries Document 904. Appendix III of the Report of the U.S. Commissioner of Fisheries for 1921.

SMITH, R.B. AND L.W. BRAILLE. 1994. The Yellowstone hotspot. *Journal of Volcanology and Geothermal Research* 61: 121–187.

SMITH, R.B. AND L.J. SIEGEL. 2000. *Windows into the Earth: the geological story of Yellowstone and Grand Teton National Parks*. New York: Oxford University Press.

SMITH, R.H. 1984 [1994]. *Native trout of North America*. Portland, Oregon: Frank Amato Publications [second edition by same publisher, 1994].

SNORRASON, S.S. AND 5 co-authors. 1989. Shape polymorphism in arctic charr, *Salvelinus alpinus*, in Thingvallavatn, Iceland. *Physiology and Ecology Japan* Special Volume 1: 393–404.

SNYDER, G.R. AND H.A. TANNER. 1960. *Cutthroat trout reproduction in the inlets to Trappers Lake*. Denver: Colorado Department of Game, Fish and Parks, Technical Bulletin 7.

SNYDER, J.O. 1914. A new species of trout from Lake Tahoe. *Bulletin of the U.S. Bureau of Fisheries* 32: 23–28.

———. 1917. Fishes of the Lahontan system of Nevada and northeastern California. *Bulletin of the U.S. Bureau of Fisheries* 35: 33–86.

———. 1919. Three new whitefishes from Bear Lake, Idaho and Utah. *Bulletin of the U.S. Bureau of Fisheries* 36: 3–9.

———. 1922. Notes on some western fluvial fishes described by Charles Girard in 1856. *Proceedings of the U.S. National Museum* 59: 23–28.

———. 1933. Description of *Salmo seleniris*, a new California trout. *Proceedings of the California Academy of Sciences* 209: 471–472.

———. 1934. A new California trout. *California Fish and Game* 20: 105–112.

———. 1940. The trouts of California. *California Fish and Game* 26: 96–138.

SNYDER, P.E. 1995. Impacts of electrofishing on fish. *Fisheries* 20, no. 1: 26–27.

SOBOLIK, K.D. 1991. *Prehistoric diet and subsistence in the lower Pecos as reflected in coprolites from Baker Cave, Val Verde County, Texas*. Austin: Texas Archeological Research Laboratory, University of Texas.

SOLA, L., S. CATANDELLA AND E. CAPANNA. 1981. New developments in vertebrate cytotaxonomy 3. Karyology of bony fishes—a review. *Genetics* 54: 285–328.

SOLOMON, D.J. 1985. Salmon stock and recruitment, and stock enhancement. *Journal of Fish Biology* 27 (Supplement A): 45–57.

SOLTIS, P.S., D.E. SOLTIS AND C.J. SMILEY. 1992. An *rbcL* sequence from a Miocene *Taxodium* (bald cypress). *Proceedings of the National Academy of Science* 89: 449–451.

SOULÉ, M.E., EDITOR. 1986. *Conservation biology: the science of scarcity and diversity*. Sunderland, Massachusetts: Sinauer Associates, Ltd.

———. 1987. *Viable populations for conservation*. Cambridge, U.K.: Cambridge University Press.

SOUTHWEST CENTER FOR BIOLOGICAL DIVERSITY AND 6 co-petitioners. 1998. *Petition to list the Rio Grande cutthroat trout (Oncorhynchus clarki virginalis) as an endangered species under the U.S. Endangered Species Act*. Tucson, Arizona: Southwest Center for Biological Diversity et al.

SPAIN, G. 1997. Oregon embarks on bold recovery plan for Pacific salmon: should it be used as an alternative to an ESA listing? *Endangered Species Update* 14, no. 5 & 6: 11–16.

SPAULDING, K.A., EDITOR. 1953. *On the Oregon Trail: Robert Stuart's journey of discovery*. Norman, Oklahoma: University of Oklahoma Press.

SPEAS, C. (AND FIVE CO-AUTHORS). 1994. *Conservation plan for Colorado River cutthroat trout (Oncorhynchus clarki pleuriticus) for the Little Snake River drainage, southeastern Wyoming*. Laramie, Wyoming: Little Snake River Interagency Working Group (Medicine Bow National Forest, Wyoming Game and Fish Department, and Great Divide Resource Area, Bureau of Land Management).

SPENCE, M.L. AND D. JACKSON, EDITORS. 1973. *The expeditions of John Charles Fremont*. Volume 2. Urbana, Illinois: University of Illinois Press.

SPINDEN, H.J. 1908. The Nez Perce Indians. *Memoirs of the American Anthropological Association* 2, part 3: 165–274.

SPLAWN, A.J. 1917 [1958]. *Ka-Mi-Akin: the last hero of the Yakimas*. Portland, Oregon: Kilham Stationery and Printing Company [2nd edition, 3rd printing, Caldwell, Idaho: The Caxton Printers Ltd.].

STANFORD ENVIRONMENTAL LAW SOCIETY. 2001. *The Endangered Species Act*. Stanford, California: Stanford University Press.

STANLEY, G.F., EDITOR. 1970. *Mapping the frontier: Charles Wilson's diary of the survey of the 49th parallel, 1858–1862, while secretary of the British Boundary Commission.* Seattle: University of Washington Press.

STANSBURY, H. 1852 [1988]. *Exploration and survey of the valley of the Great Salt Lake of Utah.* Philadelphia: Lippencott-Grambo and Company. [Reprinted 1988 under the title *Exploration of the valley of the Great Salt Lake.* Washington, D.C.: Smithsonian Institution Press].

STAPLES, B. 1991. *Snake River country: flies and waters.* Portland, Oregon: Frank Amato Publications.

STAPLES, B.A. 2003. West Yellowstone, Montana is "Trout Town, USA." *FFF Flyfisher* (Federation of Fly Fishers magazine) 36, no. 4: 28–31.

STARK, P. 1997. The Old North Trail. *Smithsonian* (Smithsonian Institution magazine), July 1997: 54–66.

STEARLEY, R.F. 1992. Historical ecology of Salmoninae, with special reference to *Oncorhynchus*. Pages 622–658 in R.L. Mayden, editor. *Systematics, historical ecology, and North American freshwater fishes.* Stanford, California: Stanford University Press.

STEARLEY, R.F. AND G.R. SMITH. 1993. Phylogeny of the Pacific trouts and salmons (*Oncorhynchus*) and genera of the family Salmonidae. *Transactions of the American Fisheries Society* 122: 1–33.

STEELQUIST, R. 1984. Admiral Beardslee goes fishing: the early exploitation of Lake Crescent trout. Pages 63–67 in J.M. Walton and D.B. Houston, editors. *Proceedings of the Olympic wild fish conference.* Port Angeles, Washington: Peninsula College and Olympic National Park.

STEFANICH, F.A. 1952. The population and movements of fish in Prickly Pear Creek, Montana. *Transactions of the American Fisheries Society* 81: 260–274.

STEFFERUD, J.A., D.L. PROPST AND G.L. BURTON. 1992. Use of antimycin to remove rainbow trout from White Creek, New Mexico. Pages 55–66 in D.A. Hendrickson, editor. *Proceedings of the Desert Fishes Council 22 and 23, 1990 and 1991 Annual Symposia, Bishop, California.*

STEWART, E.P. 1961. *Letters of a woman homesteader.* Lincoln: University of Nebraska Press.

STEWART, G.R. 1962. *The California trail.* New York: McGraw-Hill.

STEWART, I.T., D.R. CAYAN AND M.D. DETTINGER. 2004. Changes in snowmelt runoff timing in western North America under a "business as usual" climate change scenario. *Climate Change* 62: 217–232.

STOCKNER, J., EDITOR. 2003. *Nutrients in salmonid ecosystems: sustaining production and biodiversity.* Bethesda, Maryland: American Fisheries Society Symposium 34.

STOLLERY, D.J., JR. 1969. *Tales of Tahoe.* Sparks, Nevada: Western Printing and Publishing Company.

STOLZ, J. AND J. SCHNELL, EDITORS. TROUT: *the wildlife series.* Harrisburg, Pennsylvania: Stackpole Books.

STONE, L. 1874. Catalog of natural history specimens collected on the Pacific Slope in 1872 by Livingston Stone for the United States Fish Commission. Pages 200–215 in *U.S. Commission of Fish and Fisheries, Part II, Report of the Commissioner for 1872 and 1873.* Washington, D.C.: U.S. Government Printing Office.

STONECYPHER, R.W., JR., W.A. HUBERT AND W.A. GERN. 1994. Effect of reduced incubation temperatures on survival of trout embryos. *Progressive Fish-Culturist* 56: 180–184.

STOWELL, R. AND 5 co-authors. 1983. *Guide for predicting salmonid response to sediment yields in Idaho Batholith watersheds.* Missoula, Montana and Ogden, Utah: USDA Forest Service, Northern and Intermountain Regions.

STRAUSS, R.E. AND C.E. BOND. 1990. Taxonomic methods: morphology. Pages 109–140 in C.B. Schreck and P.B. Moyle, editors. *Methods for fish biology.* Bethesda, Maryland: American Fisheries Society.

STREKAL, T. AND 6 co-authors. 1992. *Cui-ui (Chasmistes cujus) second revision recovery plan.* Portland, Oregon: U.S. Fish and Wildlife Service.

STRONG, T.N. 1906 [1930]. *Cathlamet on the Columbia.* Portland, Oregon: Metropolitan Press [reprinted 1930 by Binfords and Mort, Portland].

STRONG, GEN. W.E. 1875 [1968]. *A trip to the Yellowstone National Park in July, August, and September, 1875.* Privately published [reprinted 1968, University of Oklahoma Press, Norman, Oklahoma].

STUBER, R.J., B.D. ROSENLUND AND J.R. BENNETT. 1988. Greenback cutthroat trout recovery program: management overview. Pages 71–74 in R.E. Gresswell, editor. *Status and management of interior stocks of cutthroat trout.* Bethesda, Maryland: American Fisheries Society Symposium 4.

STUMPFF, W.K. 1998. *Rio Grande cutthroat trout management.* Santa Fe: New Mexico Department of Game and Fish, Federal Aid to Fish Restoration Grant F-60-M, Project 11, Final Report.

STUMPFF, W.K. AND J. COOPER. 1996 Rio Grande cutthroat trout (*Oncorhynchus clarki virginalis*). Pages 74–86 in Duff, D.A., technical editor. *Conservation assessment for inland cutthroat trout: distribution, status, and habitat management implications.* Ogden, Utah: USDA Forest Service, Northern, Rocky Mountain, Intermountain, and Southwestern Regions.

SUBLETTE, J.E., M.D. HATCH AND M.E. SUBLETTE. 1990. *The fishes of New Mexico*. Albuquerque: University of New Mexico Press.

SUCKLEY, GEORGE. 1874. On the North American species of salmon and trout. Pages 91–160 in *Report of the Commissioner for 1872 and 1873, U.S. Commission of Fish and Fisheries, Part 2*. Washington, D.C.: Government Printing Office.

SUMNER, F.H. 1939. The decline of the Pyramid Lake fishery. *Transactions of the American Fisheries Society* 69: 216–224.

———. 1953. Migrations of salmonids in Sand Creek, Oregon. *Transactions of the American Fisheries Society* 82: 139–150.

———. 1962. Migration and growth of the coastal cutthroat trout in Tillamook County, Oregon. *Transactions of the American Fisheries Society* 91: 77–83.

SWALES, S., R.B. LAUZIER AND C.D. LEVINGS. 1985. Winter habitat preferences of juvenile salmonids in two interior rivers in British Columbia. *Canadian Journal of Zoology* 64: 1506–1514.

SWETNAM, T.W. 1990. Fire history and climate in the southwestern United States. Pages 6–17 in J.S. Krammes, technical coordinator. *Proceedings of a symposium on effects of fire in management of southwestern U.S. natural resources, November 15–17, 1988*. Tucson, Arizona: USDA Forest Service, General Technical Report RM-191.

SWETNAM, T.W. AND C.H. BAISAN. 1996. Historical fire regime patterns in the southwestern United States since A.D. 1700. Pages 11–32 in C.D. Allen, technical editor. *Fire effects in southwestern forests: proceedings of the second La Mesa Fire symposium*. Fort Collins, Colorado: USDA Forest Service, Rocky Mountain Forest and Range Experiment Station, General Technical Report RM-GTR-286.

SYTCHEVSKAYA, E.C. 1986. Paleogene freshwater fish fauna of the USSR and Mongolia. *Trudy Sovmestnaya Sovetsko-Mongol'skaya Paleontologicheskaya Ekspeditsiya* 29: 1–157.

TALABERE, A.G. 2002. Influence of water temperature and beaver ponds on Lahontan cutthroat trout in a high-desert stream, southeastern Oregon. Master's thesis, Oregon State University, Corvallis.

TANIGUCHI, Y., F.J. RAHEL, D.C. NOVINGER AND K.D. GEROW. 1998. Temperature mediation of competitive interactions among three fish species that replace each other along longitudinal stream gradients. *Canadian Journal of Fisheries and Aquatic Sciences* 55: 1894–1901.

TANNER, V.M. 1936. A study of the fishes of Utah. *Proceedings of the Utah Academy of Sciences, Arts, and Letters* 13: 155–183.

TAVE, D. 1986. *Genetics for fish hatchery managers*. Westport, Connecticut: AVI Publishing Company.

TAYLOR, B. 1867. *Colorado: a summer trip*. New York: G.P. Putnam and Son.

TAYLOR, D.W. 1960. Distribution of the freshwater clam *Pisidium ultramontanum*; a zoogeographic inquiry. *American Journal of Science* 258-A: 325–334.

———. 1985. Evolution of freshwater drainages and mollusks in western North America. Pages 265–321 in C.J. Smiley, editor. *Late Cenozoic history of the Pacific northwest*. San Francisco, California: Pacific Division Association for the Advancement of Science and California Academy of Sciences.

TAYLOR, D.W. AND G.R. SMITH. 1981. Pliocene mollusks and fishes from northeastern California and northwestern Nevada. *Contributions from the Museum of Paleontology, University of Michigan* 25, no. 18: 339–413.

TAYLOR, M.J. AND K.R. WHITE. 1992. A meta-analysis of hooking mortality of nonanadromous trout. *North American Journal of Fisheries Management* 12: 760–767.

TAYLOR, N.A. 1878. Letter to the editor. *Forest and Stream* 10, no. 13: 236.

THOMAS, D. 2002. Warming the fuel for the fire: evidence for the thermal dissociation of methane hydrate during the Paleocene-Eocene thermal maximum. *Geology* 30: 1067–1070.

THOMPSON, K.G., E.P. BERGERSEN AND R.B. NEHRING. 1997a. Injuries to brown trout and rainbow trout induced by capture with pulsed direct current. *North American Journal of Fisheries Management* 17: 141–153.

THOMPSON, K.G., E.P. BERGERSEN, R.B. NEHRING AND D.C. BOWDEN. 1997b. Long-term effects of electrofishing on growth and body condition of brown trout and rainbow trout. *North American Journal of Fisheries Management* 17: 154–159.

THOMPSON, R.S. 1992. Late Quaternary environments in Ruby Valley, Nevada. *Quaternary Research* 37: 1–15.

THORGAARD, G.H. AND S.K. ALLEN, JR. 1987. Chromosome manipulation and markers in fishery management. Pages 319–331 in N. Ryman and F. Utter, editors. *Population genetics and fishery management*. Seattle: University of Washington Press.

THUROW, R.F. 1982. *Blackfoot River fishery investigations*. Boise: Idaho Department of Fish and Game, Job Completion Report, Project F-73-R-3.

THUROW, R.F., C.E. CORSI AND V.K. MOORE. 1988. Status, ecology, and management of Yellowstone cutthroat trout in the upper Snake River drainage, Idaho. Pages 25–36 in R.E. Gresswell, editor. 1988. *Status and management of interior stocks of cutthroat trout*. Bethesda, Maryland: American Fisheries Society Symposium 4.

Thurow, R.F. and J.G. King. 1994. Attributes of Yellowstone cutthroat trout redds in a tributary of the Snake River, Idaho. *Transactions of the American Fisheries Society* 123: 37–50.

Tilman, D. and P. Kareiva, editors. 1997. *Spatial ecology: the role of space in population dynamics and interspecific interactions*. Princeton, New Jersey: Princeton Monographs in Population Biology 30.

Tipping, J. 1986. Effect of release size on return rates of hatchery sea-run cutthroat trout. *Progressive Fish-Culturist* 48: 195–197.

Tol, D. and J. French. 1988. Status of a hybridized population of Alvord cutthroat trout from Virgin Creek, Nevada. Pages 116–120 in R.E. Gresswell, editor. *Status and management of interior stocks of cutthroat trout*. Bethesda, Maryland: American Fisheries Society Symposium 4.

Toline, C.A., T. Seamons and J.M. Hudson. 1999. *Mitochondrial DNA analysis of selected populations of Bonneville, Colorado River, and Yellowstone cutthroat trout*. Logan: Utah State University Department of Fisheries and Wildlife, report to Utah Division of Wildlife Resources, Salt Lake City.

Tomasson, T. 1978. Age and growth of cutthroat trout, *Salmo clarki clarki* Richardson, in the Rogue River, Oregon. Master's thesis, Oregon Stste University, Corvallis.

Touchan, R., C.D. Allen and T.W. Swetnam. 1996. Fire history and climax patterns in ponderosa pine and mixed-conifer forests of the Jemez Mountains, northern New Mexico. Pages 33–46 in C.D. Allen, technical editor. *Fire effects in southwestern forests: proceedings of the second La Mesa Fire symposium*. Fort Collins, Colorado: USDA Forest Service, Rocky Mountain Forest and Range Experiment Station, General Technical Report RM-GTR-286.

Townley, J.M. 1980. *The Truckee basin fishery, 1844–1944*. Reno: Nevada Historical Society in cooperation with Desert Research Institute, Publication no. 43008.

Townsend, J.K. 1839 [1905, 1978]. *Narrative of a journey across the Rocky Mountains to the Columbia River*. Originally published in Philadelphia, 1839. [Reprinted (in edited form) 1905 in R.G. Thwaites, editor. *Early western travels, Volume 8*. Arthur H. Clark, Cleveland, Ohio; reprinted 1978, University of Nebraska Press, Lincoln, Nebraska].

Trelease, T.J. 1969a. The death of a lake. *Nevada Outdoors and Wildlife Review* 3, no. 1: 4–9. [Reprint of an article originally published in *Field & Stream* magazine, 1952].

———. 1969b. The rebirth of a lake. *Nevada Outdoors and Wildlife Review* 3, no. 1: 10–14.

Trimble, S. 1989. *The sagebrush ocean: a natural history of the Great Basin*. Reno: University of Nevada Press.

Tripp, D. and P. McCart. 1983. Effects of different coho stocking strategies on coho and cutthroat trout production in isolated headwater streams. *Canadian Technical Report of Fisheries and Aquatic Sciences* 1212.

Trojnar, J.R. and R.J. Behnke. 1974. Management implications of ecological segregation between two introduced populations in a small Colorado lake. *Transactions of the American Fisheries Society* 103: 423–430.

Trotter, P.C. 1989. Coastal cutthroat trout: a life history compendium. *Transactions of the American Fisheries Society* 118: 463–473.

———. 1991. Cutthroat trout *Oncorhynchus clarki*. Pages 236–265 in J. Stolz and J. Schnell, editors. *Trout. The wildlife series*. Harrisburg, Pennsylvania: Stackpole Books.

———. 1997. Sea-run cutthroat trout: life history profile. Pages 7–15 in J.D. Hall, P.A. Bisson and R.E. Gresswell, editors. *Sea-run cutthroat trout: biology, management, and future conservation*. Corvallis: Oregon Chapter, American Fisheries Society.

———. 2000. *Headwater fishes and their uppermost habitats: a review as background for stream typing*. Olympia: Washington Department of Natural Resources, Forest Practices Division, Report no. TFW-ISAG1-00-001. Available on the Internet at www.dnr.wa.gov/forestpractices/adaptivemanagement/cmer/publications/. Scroll down the list to find the report.

Trotter, P.C., B. McMillan, N. Gayeski, P. Spruell and M. Berkley. 1999. *Genetic and phenotypic catalog of native resident trout of the interior Columbia River basin. FY-98 report: populations of the upper Yakima basin*. Report prepared for Northwest Power Planning Council Upper Columbia Basin Fish and Wildlife Program, BPA contract 98-AP-07901. Available on the Internet at www.efw.bpa.gov. Click on Integrated Fish and Wildlife Program, then Technical Reports and Publications, then Published Reports, then Resident Fish. Enter the first author or report title in the search engine that opens.

Trotter, P.C., B. McMillan, N. Gayeski, P. Spruell and A. Whiteley. 2000. *Genetic and phenotypic catalog of native resident trout of the interior Columbia River basin. FY-99 report: populations of the Pend Oreille, Kettle, and Sanpoil river basins of Colville National Forest*. Report prepared for Northwest Power Planning Council Upper Columbia Basin Fish and Wildlife Program, BPA contract 98-AP-07901. Available on the Internet at www.efw.bpa.gov. Click on Integrated Fish and Wildlife Program, then Technical Reports and Publications, then Published Reports, then Resident Fish. Enter the first author or report title in the search engine that opens.

TROTTER, P.C., B. MCMILLAN, N. GAYESKI, P. SPRUELL AND M.K. COOK. 2002. *Genetic and phenotypic catalog of native resident trout of the interior Columbia River basin. FY-2001 report: populations of the Wenatchee, Entiat, Lake Chelan, and Methow river drainages.* Report prepared for Northwest Power Planning Council Upper Columbia Basin Fish and Wildlife Program, BPA contract 98-AP-07901. Available on the Internet at www.efw.bpa.gov. Click on Integrated Fish and Wildlife Program, then Technical Reports and Publications, then Published Reports, then Resident Fish. Enter the first author or report title in the search engine that opens.

TROTTER, P.C. AND P.A. BISSON. 1988. History of the discovery of the cutthroat trout. Pages 8–12 in R.E. Gresswell, editor. *Status and management of interior stocks of cutthroat trout.* Bethesda, Maryland: American Fisheries Society Symposium 4.

TROTTER, P.C., P.A. BISSON AND B. FRANSEN. 1993. Status and plight of the sea-run cutthroat trout. Pages 203–212 in J.G. Cloud and G.H. Thorgaard, editors. *Genetic conservation of salmonid fishes.* NATO Advanced Science Institute Series A, Vol. 248. New York and London: Plenum Press.

TRUCKEE RIVER BASIN RECOVERY IMPLEMENTATION TEAM. 2003. *Short-term action plan for Lahontan cutthroat trout (Oncorhynchus clarki henshawi) in the Truckee River basin.* Reno: Nevada, U.S. Fish and Wildlife Service.

TSCHUDY, R.H., C.L. PILLMORE, C.J. ORTH, J.S. GILMORE AND J.D. KNIGHT. 1984. Disruption of the terrestrial plant ecosystem at the Cretaceous-Tertiary boundary, western interior. *Science* 225: 1030–1032.

TSUYUKI, H. AND S.N. WILLISCROFT. 1973. The pH activity relations of two LDH homotetramers from trout liver and their physiological significance. *Journal of the Fisheries Research Board of Canada* 30: 1023–1026.

———. 1977. Swimming stamina differences between genotypically distinct forms of rainbow trout (*Salmo gairdneri*) and steelhead trout. *Journal of the Fisheries Research Board of Canada* 34: 996–1003.

TUOHY, D.R., EDITOR. 1978. *Honey Lake Paiute ethnography.* Part 1 (by F.A. Riddle) and Part 2 (by W.S. Evans, Jr.). Carson City: Nevada State Museum Occasional Papers no. 3.

———. 1988. Artifacts from the northwestern Pyramid Lake shoreline. Pages 201–216 in J.A. Willig, C.M. Aikens and J.L. Fagan, editors. *Early occupation in far western North America: the Clovis-Archaic interface.* Carson City: Nevada State Museum Anthropological Papers no. 21.

———. 1990. Pyramid Lake fishing: the archaeological record. Pages 121–158 in J.C. Janetski and D.B. Madsen, editors. *Wetland adaptations in the Great Basin.* Brigham Young University, Museum of Peoples and Cultures, Occasional Papers no. 1.

TURK, J.T. AND D.B. ADAMS. 1983. Sensitivity to acidification of lakes in the Flat Tops Wilderness area, Colorado. *Water Resources Research* 19: 346–350.

TURMAN, D.L. 1972. Studies on spawning boxes for coastal cutthroat trout. Master's thesis, Humboldt State University, Arcata, California.

TWAIN, MARK. 1872 [reprinted 1973]. *Roughing It.* American Publishing Company. [Reprinted in *The works of Mark Twain, Volume 2.* Berkeley, California: University of California Press].

URRUTIA, V. 1998. *They came to six rivers: the story of Cowlitz County.* Kelso, Washington: Cowlitz County Historical Society.

U.S. ARMY CORPS OF ENGINEERS. 2000. *Jackson Hole, Wyoming, environmental restoration and feasibility report, final report.* Walla Walla, Washington: U.S. Army Corps of Engineers. Available on the Internet at www.nww.usacoe.army.mil/reports/jackson/report.htm.

U.S. ENVIRONMENTAL PROTECTION AGENCY. 1976. *Quality criteria for water.* Washington, D.C.: U.S. Environmental Protection Agency.

———. 1998. *Level III ecoregions of the coterminous United States.* Corvallis, Oregon: U.S. Environmental Protection Agency National Health and Environmental Effects Laboratory.

U.S. FISH AND WILDLIFE SERVICE. 1985. *Paiute cutthroat trout recovery plan.* Portland, Oregon: U.S. Fish and Wildlife Service.

———. 1990. *Policy and guidelines for planning and coordinating recovery of endangered and threatened species.* Washington, D.C.: U.S. Fish and Wildlife Service.

———. 1999. *Status review for westslope cutthroat trout in the United States.* Portland, Oregon and Denver, Colorado: U.S. Fish and Wildlife Service, Regions 1 and 6.

———. 2001. *Status review for Bonneville cutthroat trout (Oncorhynchus clarki utah).* Portland, Oregon and Denver, Colorado: U.S. Fish and Wildlife Service, Regions 1 and 6.

———. 2004. *Revised recovery plan for the Paiute cutthroat trout (Oncorhynchus clarki seleniris).* Portland, Oregon: U.S. Fish and Wildlife Service, Region 1.

U.S. GENERAL ACCOUNTING OFFICE. 1988a. *Rangeland management: more emphasis needed on declining and overstocked grazing allotments.* Washington, D.C.: U.S. General Accounting Office, GAO/RCED-88-80.

———. 1988b. *Public rangelands: some riparian areas restored but widespread improvements will be slow.* Washington, D.C.: U.S. General Accounting Office, GAO/RCED-88-105.

————. 1991a. *Public land management: attention to wildlife is limited*. Washington, D.C.: U.S. General Accounting Office, GAO/RCED-91-64.

————. 1991b. *Rangeland management: comparison of rangeland condition reports*. Washington, D.C.: U.S. General Accounting Office, GAO/RCED-91-191.

UTAH BONNEVILLE CUTTHROAT TROUT CONSERVATION TEAM. 2004. *Bonneville cutthroat trout (Oncorhynchus clarki utah) conservation agreement and strategy in the State of Utah: post implementation assessment*. Salt Lake City: Utah Division of Wildlife Resources, Publication 04-31.

UTAH DIVISION OF WILDLIFE RESOURCES. 1991. *Dewatered stream miles*. Salt Lake City: Utah Department of Natural Resources, Division of Wildlife Resources, Instream Flow Task Force.

————. 1993. *Native cutthroat trout management plan*. Salt Lake City: Utah Department of Natural Resources, Division of Wildlife Resources.

UTLEY, R.M. 1997. *A life wild and perilous: mountain men and the paths to the Pacific*. New York: Henry Holt and Company.

UTTER, F.M. AND 5 co-authors. 1980. Population structure of indigenous salmonid species of the Pacific northwest. Pages 285–304 in W.J. McNeil and D.C. Himsworth, editors. *Salmonid ecosystems of the North Pacific*. Corvallis: Oregon State University Press.

VAJDA, V., J.I. RAINE AND C.J. HOLLIS. 2001. Indication of global deforestation at the Cretaceous-Tertiary boundary by New Zealand fern spike. *Science* 294: 1700–1702.

VANDER ZANDEN, M.J., S. CHANDRA, B.C. ALLEN, J.E. REUTER AND C.R. GOLDMAN. 2003. Historical food web structure and restoration of native aquatic communities in the Lake Tahoe (California-Nevada) basin. *Ecosystems* 6: 274–188.

VAN EIMEREN, P. 1996. Westslope cutthroat trout *Oncorhynchus clarki lewisi*. Pages 1–10 in D.A. Duff, technical editor. *Conservation assessment for inland cutthroat trout: distribution, status, and habitat management implications*. Ogden, Utah: USDA Forest Service, Northern, Rocky Mountain, Intermountain, and Southwestern Regions.

VAN KIRK, R. 2006. *Modeled population-level effects of chronic selenium exposure to Yellowstone cutthroat trout (Oncorhynchus clarkii bouvieri)*. Pocatello: Idaho State University Department of Mathematics. Report prepared for Greater Yellowstone Coalition, Idaho Falls, Idaho.

VAN KIRK, R.W. AND L. BENJAMIN. 2000. Physical and human geography of the Henry's Fork watershed. *Intermountain Journal of Sciences* 6: 106–118.

VAN KIRK, R.W. AND M. GAMBLIN. 2000. History of fisheries management in the upper Henry's Fork watershed. *Intermountain Journal of Sciences* 6: 263–284.

VAN KIRK, R.W., J.M. CAPURSO AND M.A. NOVAK, EDITORS. 2006. Exploring differences between fine-spotted and large-spotted Yellowstone cutthroat trout. *Symposium Proceedings, Idaho Chapter American Fisheries Society, Boise, Idaho*. Available on the Internet at www.fisheries.org/idaho/.

VAN KIRK, R.W. (AND 8 CO-AUTHORS). 1997. *Status of Yellowstone cutthroat trout in the upper Henry's Fork watershed*. Ashton, Idaho: Henry's Fork Foundation.

VAN VUREN, D.H. 2001. Spatial relations of American bison (*Bison bison*) and domestic cattle in a montane environment. *Animal Biodiversity and Conservation* 24: 117–124.

VARLEY, J.D. 1979. *Record of egg shipments from Yellowstone fishes, 1914–1955*. Yellowstone National Park, Wyoming: National Park Service, Information Paper 34.

VARLEY, J.D. AND R.E. GRESSWELL. 1988. Ecology, status, and management of the Yellowstone cutthroat trout. Pages 13–24 in R.E. Gresswell, editor. *Status and management of interior stocks of cutthroat trout*. Bethesda, Maryland: American Fisheries Society Symposium 4.

VARLEY, J.D. AND P. SCHULLERY. 1983. *Freshwater wilderness: Yellowstone fishes and their world*. Yellowstone National Park, Wyoming: The Yellowstone Library and Museum Association.

————., EDITORS. 1995. *The Yellowstone Lake crisis: confronting a lake trout invasion*. Yellowstone National Park, Wyoming: Yellowstone Center for Resources, Report to the Director of the National Park Service. Available on the Internet at www.nps.gov/Yell/publications/pdfs/laketrout2.pdf.

————. 1996. Yellowstone Lake and its cutthroat trout. Pages 49–73 in W.L. Halvorson and G.E. Davis, editors. *Science and ecosystem management in the national parks*. Tucson: University of Arizona Press.

————. 1998. *Yellowstone fishes: ecology, history, and angling in the park*. Mechanicsburg, Pennsylvania: Stackpole Books.

VARVIO, S., R. CHAKRABORTY AND M. NEI. 1986. Genetic variation in subdivided populations and conservation genetics. *Heredity* 57: 189–198.

VESTAL, E.H. 1947. A new transplant of the Piute cutthroat trout (*Salmo clarkii seleniris*) from Silver King Creek, Alpine County, California. *California Fish and Game* 33: 89–95.

VICK, S.C. 1913. *Classified guide to fish and their habitat in the Rocky Mountain parks*. Ottawa, Ontario: Dominion Parks Branch, Canada Department of the Interior.

VIGG, S. 1981. Species composition and relative abundance of adult fish in Pyramid Lake, Nevada. *Great Basin Naturalist* 41: 395–405.

VIGG, S. AND D.L. KOCH. 1980. Upper lethal temperature range of Lahontan cutthroat trout in waters of different ionic concentrations. *Transactions of the American Fisheries Society* 109: 336–339.

VINCENT, E.R. 1987. Effects of stocking catchable-size hatchery rainbow trout on two wild trout species in the Madison River and O'Dell Creek, Montana. *North American Journal of Fisheries Management* 7: 91–105.

VINCENT, R.E. AND W.H. MILLER. 1965. Altitudinal distribution of brown trout and other fishes in a headwater tributary of the South Platte River, Colorado. *Ecology* 50: 464–466.

VINYARD, G.L. AND A. WINZELER. 2000. Lahontan cutthroat trout (*Oncorhynchus clarki henshawi*) spawning and downstream migration of juveniles into Summit Lake, Nevada. *Western North American Naturalist* 60: 333–341.

VLADYKOV, V.D. 1963. A review of salmonid genera and their broad geographical distribution. *Transactions of the Royal Society of Canada* 1, series 4, section 3: 459–503.

VORE, D.W. 1993. Size, abundance and seasonal habitat utilization of an unfished trout population and their response to catch and release fishing. Master's thesis, Montana State University, Bozeman.

WACKERNAGEL, M. AND 10 co-authors. 2002. Tracking the ecological overshoot of the human economy. *Proceedings of the National Academy of Sciences (U.S.A.)* 99: 9266–9271.

WAGNER, E.J., R.E. ARNDT AND M. BROUGH. 2001. Comparative tolerance of four stocks of cutthroat trout to extremes in temperature, salinity, and hypoxia. *Western North American Naturalist* 61: 434–444.

WAITT, R.B., JR. 1980. About 40 last-glacial jokuhlaups through southern Washington. *Journal of Geology* 88: 653–679.

———. 1984. Periodic jokuhlaups from Pleistocene Lake Missoula: new evidence from varied sediments in northern Idaho and Washington. *Quaternary Research* 22: 46–58.

WAITT, R.B., JR. 1985. Case for periodic colossal jokuhlaups from Glacial Lake Missoula. *Geological Society of America Bulletin* 96: 1271–1286.

WAITT, R.B., JR. AND R.M. THORSON. 1983. The Cordilleran ice sheet in Washington, Idaho, and Montana. Pages 53–70 in S.C. Porter, editor. *Late Quaternary environments of the United States*. Volume 1, the Late Pleistocene. Minneapolis: University of Minnesota Press.

WALKER RIVER BASIN RECOVERY IMPLEMENTATION TEAM. 2003. *Short-term action plan for Lahontan cutthroat trout (Oncorhynchus clarki henshawi) in the Walker River basin*. Reno: Nevada, U.S. Fish and Wildlife Service. Available on the Internet at http://nevada.fws.gov/lctrit/Final/WRIT.pdf.

WALLACE, J.C. AND D. AASJORD. 1984. An investigation of the consequences of egg size for the culture of Arctic charr, *Salvelinus alpinus* (L.). *Journal of Fish Biology* 24: 427–435.

WANG, L. AND R.J. WHITE. 1994. Competition between wild brown trout and hatchery greenback cutthroat trout of largely wild parentage. *North American Journal of Fisheries Management* 14: 475–487.

WAPLES, R.S. 1991. *Definition of "species" under the Endangered Species Act: application to Pacific salmon*. Seattle: U.S. Department of Commerce, National Marine Fisheries Service, NOAA Technical Memorandum NMFS F/NWC-194.

WARING, R.H. AND J.F. FRANKLIN. 1979. Evergreen coniferous forests of the Pacific northwest. *Science* 204: 1380–1386.

WARNER, B.G. 1984. Late Quaternary paleoecology of eastern Graham Island, British Columbia. Ph.D. dissertation, Simon Fraser University, Burnaby, British Columbia.

WARNER, B.G., R.W. MATHEWES AND J.J. CLAQUE. 1982. Ice-free conditions on the Queen Charlotte Islands, British Columbia, at the height of late Wisconsin glaciation. *Science* 218: 675–677.

WARREN, J.W. 1981. *Diseases of hatchery fish*. Washington, D.C.: U.S. Fish and Wildlife Service.

WATERS, T.F. 2000. *Wildstream: a natural history of the free-flowing river*. St. Paul, Minnesota: Riparian Press.

WATSON, R. 1993. *The trout: a fisherman's natural history*. Shrewsbury, U.K.: Swan Hill Press.

WATSON, R.T., M.C. ZINYOWERA AND R.H. MOSS, EDITORS. 1998. *The regional impacts of climate change—an assessment of vulnerability*. A special report of IPCC Working Group II. Cambridge, U.K. and New York: Cambridge University Press.

WATSON, W.J. 1851 [1985]. *Journal of an overland journey to Oregon made in the year 1849*. Jacksonville, Oregon: E.R. Roe [reprinted 1985 by Ye Galleon Press, Fairfield, Washington].

———. 1984. Glacial chronology of the Ruby Mountains-East Humboldt Range, Nevada. *Quaternary Research* 21: 286–303.

WEBER, D.J. 1971. *The Taos trappers: the fur trade in the far southwest, 1540–1846*. Norman: University of Oklahoma Press.

WEDEMEYER, G.A., R.L. SAUNDERS AND W.C. CLARKE. 1980. Environmental factors affecting smoltification and early marine survival of anadromous salmonids. *Marine Fisheries Review* 42, no. 6: 1–14.

WEIER, T.E., C.R. STOCKING AND M.G. BARBOUR. 1974. *Botany: an introduction to plant biology*. 5th edition. New York: John Wiley & Sons.

WEIGEL, D.E., J.T. PETERSON AND P. SPRUELL. 2003. Introgressive hybridization between native cutthroat trout and introduced rainbow trout. *Ecological Applications* 13: 38–50.

WELCH, W.R. 1929. Trout fishing in California today and fifty years ago. *California Fish and Game* 15: 20–22.

WELSH, J.P. 1952. A population study of Yellowstone blackspotted trout (*Salmo clarki*). Ph.D. dissertation, Stanford University, Stanford, California.

WENBERG, J.K. 1998. The coastal cutthroat trout (*Oncorhynchus clarki clarki*): genetic population structure, migration patterns and life history traits. Ph.D. dissertation, University of Washington, Seattle.

WENBERG, J.K. AND P. BENTZEN. 2001. Genetic and behavioral evidence for restricted gene flow among coastal cutthroat trout populations. *Transactions of the American Fisheries Society* 130: 1049–1069.

WERNSMAN, G.R. 1973. The native trouts of Colorado. Master's thesis, Colorado State University, Fort Collins.

WESTGATE, L.G. 1905. The Twin Lakes glaciated area, Colorado. *Journal of Geology* 13: 285–312.

WHEELER, H.E. AND E.F. COOK. 1954. Structural and stratigraphic significance of the Snake River capture, Idaho-Oregon. *Journal of Geology* 62: 525–536.

WHEELER, S.S. 1974. *The desert lake: the story of Nevada's Pyramid Lake*. Caldwell, Idaho: The Caxton Printers, Ltd.

———. 1979. *The Black Rock Desert*. Caldwell, Idaho: The Caxton Printers, Ltd.

WHEELESS, C.H., JR. 1974. *Rio Grande cutthroat study (rare and endangered)*. Boulder, Colorado: Western Interstate Commission for Higher Education, report prepared for the Santa Fe National Forest, Albuquerque, New Mexico.

WHITE, D.A. AND C.J. SAKAMOTO. 1974. A probable relic population of the Bonneville cutthroat (*Salmo clarki utah*, Suckley: Salmonidae) in Salt Lake County, Utah. *Proceedings of the Utah Academy of Sciences, Arts, and Letters* 51, part 1: 66–68.

WHITE, S.M. 2003. A watershed perspective on the distribution and habitat requirements of young Bonneville cutthroat trout in the Thomas fork of the Bear River, Wyoming. Master's thesis, University of Wyoming, Laramie.

WHITEHOUSE, FRANCIS C. 1946. *Sport fishes of western Canada, and some others*. Published by the author, Vancouver, British Columbia.

WHITLOCK, C. 1993. Postglacial vegetation and climate of Grand Teton and southern Yellowstone National Parks. *Ecological Monographs* 63, no. 2: 173–198.

WHITLOCK, C. AND P.J. BARTLEIN. 1997. Vegetation and climate change in northwest America during the past 125kyr. *Nature* 388: 57–61.

WILEY, R.W. 1969. An ecological evaluation of the Snake River cutthroat fishery with an emphasis on harvest. Master's thesis, University of Wyoming, Laramie.

WILKIE, M.R., P.A. WRIGHT, G.K. IWAMA AND C.M. WOOD. 1993. The physiological response of the Lahontan cutthroat trout (*Oncorhynchus clarki henshawi*), a resident of highly alkaline Pyramid Lake (pH 9.4) to challenge at pH 10. *Journal of Experimental Biology* 175: 173–194.

WILLERS, W.B. 1991 [1981]. *Trout biology*, revised and augmented edition. New York: Lyons & Burford, Publishers [originally published 1981 as *Trout biology: an angler's guide*, University of Wisconsin Press, Madison].

WILLIAMS, J.E. AND C.E. BOND. 1983. Status and life history notes on the native fishes of the Alvord Basin, Oregon and Nevada. *Great Basin Naturalist* 43: 409–420.

WILLIAMS, J.E. AND W. NEHLSEN. 1997. Status and trends of anadromous salmonids in the coastal zone with special reference to sea-run cutthroat trout. Pages 37–42 in J.D. Hall, P.A. Bisson and R.E. Gresswell, editors. 1997. *Sea-run cutthroat trout: biology, management, and future conservation*. Corvallis: Oregon Chapter American Fisheries Society.

WILLIAMS, J.S. 1979. The geologic history of Bear Lake. *Bear Lake Magazine* Dec. 1, 1979 issue: 5–6.

WILLIAMS, R.N. 1991. *Genetic analysis and taxonomic status of cutthroat trout from Willow Creek and Whitehorse Creek in southeastern Oregon*. Boise, Idaho: Boise State University, Evolutionary Genetics Laboratory Report 91-3.

WILLIAMS, R.N. AND D.K. SHIOZAWA. 1989. *Taxonomic relationships among cutthroat trout of the western Great Basin: conservation and management implications*. Portland: Oregon Trout Technical Report no. 1.

WILLIAMS, R.N., D.K. SHIOZAWA AND R.P. EVANS. 1992. *Mitochondrial DNA analysis of Nevada cutthroat trout populations*. Boise, Idaho: Boise State University Evolutionary Genetics Laboratory Report 91-5.

WILLIAMS, R.N., R.P. EVANS AND D.K. SHIOZAWA. 1998. *Genetic analysis of indigenous cutthroat trout populations from northern Nevada*. Meridian, Idaho: Clear Creek Genetics Lab Report 98-1 to Nevada Division of Wildlife.

WILLIAMS, S., SIGNATORY. 2002a. Endangered and threatened wildlife and plants; withdrawal of proposed rule to list southwestern Washington/Columbia River distinct population segment of the coastal cutthroat trout as threatened. *Federal Register* 67, no. 129: 44933–44961.

———. 2002b. Endangered and threatened wildlife and plants; candidate status review for Rio Grande cutthroat trout. *Federal Register* 67, no. 112: 39936–39947.

———. 2003. Endangered and threatened wildlife and plants; reconsidered finding for an amended petition to list the westslope cutthroat trout as threatened throughout its range. *Federal Register* 68, no. 152: 46989–47009.

WILSON, D. 1992. *Sawdust trails in the Truckee Basin*. Nevada City, California: Nevada County Historical Society.

WILSON, E.N. (in collaboration with H.R. Driggs). 1919 [1991]. *The white Indian boy*. New York: World Book Company [reprinted 1991 by Paragon Press, Salt Lake City, Utah].

WILSON, E.O. AND W.H. BOSSERT. 1971. *A primer of population biology*. Sunderland, Massachusetts: Sinauer Associates, Inc.

WILSON, M.V.H. 1977. Middle Eocene freshwater fishes from British Columbia. *Life Sciences Contributions, Royal Ontario Museum* 113: 1–61.

———. 1980. Eocene lake environments: depth and distance-from-shore variation in fish, insect, and plant assemblages. *Paleogeography, Paleoclimatology, Paleoecology* 32: 21–44.

———. 1996. The Eocene fishes of Republic, Washington. *Washington Geology* 24, no. 2: 30–31.

WILSON, M.V.H. AND R.R.G. WILLIAMS. 1992. Phylogenetic, biogeographic, and ecological significance of early fossil records of North American freshwater teleostean fishes. Pages 224–244 in R.L. Mayden, editor. *Systematics, historical ecology, and North American freshwater fishes*. Stanford, California: Stanford University Press.

WILTZIUS, W.J. 1985. *Fish culture and stocking in Colorado, 1872–1978*. Fort Collins: Colorado Division of Wildlife, Division Report DOW-R-D-12-85.

WILZBACH, M.A. 1985. Relative roles of food abundance and cover in determining the habitat distribution of stream-dwelling cutthroat trout (*Salmo clarki*). *Canadian Journal of Fisheries and Aquatic Sciences* 42: 1668–1672.

WINGATE, G.W. 1886 [1999]. *Through the Yellowstone Park on horseback*. New York: O. Judd Company [reprinted 1999 by University of Idaho Press, Moscow].

WINOGRAD, I.J. AND 7 co-authors. 1992. Continuous 500,000-year climate record from vein calcite in Devils Hole, Nevada. *Science* 258: 255–260.

WINSHIP, G.P. 1964. *The Coronado expedition 1540–1542*. Chicago: The Rio Grande Press.

WINSTON, J. 1999. *Describing species: practical taxonomic procedure for biologists*. Irvington, New York: Columbia University Press.

WIPFLI, M.S., J.P. HUDSON, J.P. CAOUETTE AND D.T. CHALCONER. 2003. Marine subsidies in freshwater ecosystems: salmon carcasses increase the growth rates of stream-resident salmonids. *Transactions of the American Fisheries Society* 132: 371–381.

WIRGIN, I., L. MACEDA, J. STABILE AND C. MESING. 1997. An evaluation of introgression of Atlantic coast striped bass mitochondrial DNA in a Gulf of Mexico population using formalin-preserved museum collections. *Molecular Ecology* 6: 907–916.

WISHARD, L., W. CHRISTENSEN AND P. AEBERSOLD. 1980. *Biochemical genetic analysis of four cutthroat trout tributaries to the Idaho Blackfoot Reservoir system*. Olympia, Washington: Pacific Fisheries Research, final report to Idaho Department of Fish and Game, Soda Springs.

WISLIZENUS, F.A. 1912. *A journey to the Rocky Mountains in the year 1839*. St. Louis: Missouri Historical Society.

WOFFORD, J.E.B., R.E. GRESSWELL AND M.A. BANKS. 2005. Influence of barriers to movement on within-watershed genetic variation of coastal cutthroat trout. *Ecological Applications* 15: 628–637.

WOLBACH, W.S., R.S. LEWIS AND E. ANDERS. 1985. Cretaceous extinctions: evidence for wildfires and search for meteoritic material. *Science* 230: 167–170.

WOLFE, J.A. 1987. North American non-marine climates and vegetation during the late Cretaceous. *Paleogeography, Paleoclimatology, Paleoecology* 61: 33–77.

———. 1991. Paleobotanical evidence for a June "impact winter" at the Cretaceous/Tertiary boundary. *Nature* 352: 420–423.

WOLFF, S.W., T.A. WESCHE AND W.A. HUBERT. 1986. Assessment of a flow enhancement project as a riparian and fishery habitat mitigation effort. Pages 64–68 in *Proceedings of the twenty-first annual meeting of the Colorado-Wyoming Chapter American Fisheries Society, Fort Collins, Colorado*. Available on the Internet at http://library.wrds.uwyo.edu/wrp/86-20/86-20,html.

WONG, D.M. 1975. Aspects of the life history of the Paiute cutthroat trout, *Salmo clarki seleniris* Snyder, in North Fork Cottonwood Creek, Mono County, California, with notes on behavior in a stream aquarium. Master's thesis, California State University, Long Beach.

WOOD, J.W. 1974. *Diseases of Pacific salmon: their prevention and treatment*. 2nd edition. Olympia: Washington Department of Fisheries, Hatcheries Division.

WOOD, S.H. AND D.M. CLEMENS. 2002. Geologic and tectonic history of the western Snake River Plain, Idaho and Oregon. *Idaho Geological Survey Bulletin* 30: 69–103.

WOODRUFF, W. 1892. Utah fish and game notes. *Forest and Stream* 39, no. 12: 249.

WOODWARD, D.F., A.M. FARIG, E.E. LITTLE, B. STEADMAN AND R. YANCIK. 1991. Sensitivity of greenback cutthroat trout to acidic pH and elevated aluminum. *Transactions of the American Fisheries Society* 120: 34–42.

WOODWARD, D.F., P.M. MEHRLE, JR., AND W.L. MAUCK. 1981. Accumulation and sublethal effects of a Wyoming crude oil in cutthroat trout. *Transactions of the American Fisheries Society* 110: 437–445.

WRIGHT, H.E., JR. 1964. Origin of the lakes in the Chuska Mountains, northwestern New Mexico. *Geological Society of America Bulletin* 75: 589–598.

———., EDITOR. 1983. *Late Quaternary environments of the United States*. Vol. 1 (edited by S.C. Porter), The late Pleistocene. Minneapolis: University of Minnesota Press.

WRIGHT, H.E., JR. AND D.G. FREY, EDITORS. 1965. *The Quaternary of the United States*. Princeton, New Jersey: Princeton University Press.

WYATT, B. 1959. Observations on the movements and reproduction of the Cascade form of cutthroat trout. Master's thesis, Oregon State University, Corvallis.

WYDOSKI, R.S. 2003. Fecundity and recruitment potential of coastal cutthroat trout in Oregon and Washington. *California Fish and Game* 89: 107–127.

WYDOSKI, R.S. AND R.R. WHITNEY. 1979 [2003]. *Inland fishes of Washington*. Seattle: University of Washington Press [2nd edition 2003, American Fisheries Society, Bethesda, Maryland and University of Washington Press, Seattle].

YARROW, H.C. 1874. On the speckled trout of Utah Lake, *Salmo virginalis* Girard. Pages 363–368 in *U.S. Commission of Fish and Fisheries, Part II, Report of the Commissioner for 1872 and 1873*. Washington, D.C.: U.S. Government Printing Office.

YEKEL, S. 1980. *An evaluation of area fisheries in the Greybull River-Wood River drainages, Park County*. Cody: Wyoming Game and Fish Department, Fish Division Administrative Report.

YELLOWSTONE CUTTHROAT TROUT WORKING GROUP. 1994. *Yellowstone cutthroat trout (Oncorhynchus clarki bouvieri) management guide for the Yellowstone River drainage*. Helena: Montana and Cheyenne, Wyoming, Montana Department of Fish, Wildlife, and Parks and Wyoming Game and Fish Department.

YEOH, C.-G., T.H. KERSTETTER AND E.J. LOUDENSLAGER. 1991. Twenty-four-hour seawater challenge test for coastal cutthroat trout. *Progressive Fish-Culturist* 53: 173–176.

YOUNG, J.A. AND B.A. SPARKS. 1985. *Cattle in the cold Desert*. Logan, Utah: Utah State University Press.

YOUNG, K.A., S.G. HINCH AND T.G. NORTHCOTE. 1999. Status of resident coastal cutthroat trout and their habitat twenty-five years after riparian logging. *North American Journal of Fisheries Management* 19: 901–911.

YOUNG, M.K. 1994. Mobility of brown trout in south-central Wyoming streams. *Canadian Journal of Zoology* 72: 2078–2083.

———. 1995. Colorado River cutthroat trout. Pages 16–23 in M.K. Young, technical editor. *Conservation assessment for inland cutthroat trout*. Fort Collins, Colorado: USDA Forest Service, Rocky Mountain Forest and Range Experiment Station, General Technical Report RM-256.

———. 1996. Summer movements and habitat use by Colorado River cutthroat trout (*Oncorhynchus clarki pleuriticus*) in small, montane streams. *Canadian Journal of Fisheries and Aquatic Sciences* 53: 1403–1408.

———. 1998. Absence of autumnal changes in habitat use and location of adult Colorado River cutthroat trout in a small stream. *Transaction of the American Fisheries Society* 127: 147–151.

YOUNG, M.K. AND A.L. HARIG. 2001. A critique of the recovery of greenback cutthroat trout. *Conservation Biology* 15: 1575–1584.

YOUNG, M.K., A.L. HARIG, B. ROSENLUND AND C. KENNEDY. 2002. *Recovery history of greenback cutthroat trout: population characteristics, hatchery involvement, and bibliography*. Fort Collins, Colorado: USDA Forest Service, Rocky Mountain Research Station, General Technical Report RMRS-GTR-88WWW. Available on the Internet at www.fs.fed.us/rm/pubs/rmrs_gtr88.

YOUNG, M.K. AND P.M. GUENTHER-GLOSS. 2004. Population characteristics of greenback cutthroat trout in streams: their relation to model predictions and recovery criteria. *North American Journal of Fisheries Management* 24: 184–197.

YOUNG, M.K., R.N. SCHMAL, T.W. KOHLEY AND V.G. LEONARD. 1996. Colorado River cutthroat trout *Oncorhynchus clarki pleuriticus*. Pages 87–120 in Duff, D.A., technical editor. *Conservation assessment for inland cutthroat trout: distribution, status, and habitat management implications*. Ogden, Utah: USDA Forest Service, Northern, Rocky Mountain, Intermountain, and Southwestern Regions.

YOUNG, M.K., R.B. RADER AND T.A. BELISH. 1997. Influence of macroinvertebrate drift and light on the activity and movement of Colorado River cutthroat trout. *Transaction of the American Fisheries Society* 126: 428–437.

YOUNG, M.K., K.A. MEYER, D.J. ISAAK AND R.A. WILKISON. 1998. Habitat selection and movement by individual cutthroat trout in the absence of competitors. *Journal of Freshwater Ecology* 13: 371–381.

YOUNG, S.F., J.G. McLELLAN AND J.B. SHAKLEE. 2004. Genetic integrity and microgeographic population structure of westslope cutthroat trout (*Oncorhynchus clarki lewisi*) in the Pend Oreille basin in Washington. *Environmental Biology of Fishes* 69: 127–142.

YOUNG, W.P., C.O. OSTBERG, P. KEIM AND G.H. THORGAARD. 2001. Genetic characterization of hybridization and introgression between anadromous rainbow trout (*Oncorhynchus mykiss irideus*) and coastal cutthroat trout (*O. clarki clarki*). *Molecular Ecology* 10: 921–930.

ZACHOS, J.C., K.C. LOHMAN, J.C.G. WALKER AND S.W. WISE. 1993. Abrupt climate change and transient climates during the Paleogene: a marine perspective. *Journal of Geology* 101: 191–213.

ZACHOS, J., M. PAGANI, L. SLOAN, E. THOMAS AND K. BILLUPS. 2001. Trends, rhythms, and aberrations in global climate 65 Ma to present. *Science* 292: 686–693.

ZIMMERMAN, G.D. 1965. Meristic characters of the cutthroat trout. *Proceedings of the Montana Academy of Sciences* 25: 41–50.

ZINK, T. 2005. Restoration on the reservation: protecting native fish on native lands. *Trout* (Trout Unlimited's Journal of Coldwater Fisheries Conservation) 47, no. 2: 28–33, 56.

Conversion Table

**ENGLISH UNITS TO METRIC
(INTERNATIONAL SYSTEM) UNITS**

inches (in) $\times$ 25.4 = millimeters (mm)

inches (in) $\times$ 2.54 = centimeters (cm)

feet (ft) $\times$ 0.3048 = meters (m)

yards (yd) $\times$ 0.9144 = meters (m)

miles (mi) $\times$ 1.609 = kilometers (km)

square feet (ft^2) $\times$ 0.093 = square meters (m^2)

square yards (yd^2) $\times$ 0.836 = square meters (m^2)

square miles (mi^2) $\times$ 2.59 = square kilometers (km^2)

square miles (mi^2) $\times$ 259 = hectares (ha)

cubic feet ($ft3$) $\times$ 0.0283 = cubic meters (m3)

cubic yards $\times$ 0.765 = cubic meters (m3)

acres $\times$ 4047 = square meters (m2)

acres $\times$ 0.4047 = hectares (ha)

ounces (oz) $\times$ 28.35 = grams (g)

pounds (lb) $\times$ 454 = grams (g)

pounds (lb) $\times$ 0.454 = kilograms (kg)

tons (std U. S.) $\times$ 1.016 = metric tons (tonnes)

To convert from metric to English units, divide by the numbers shown (e.g., millimeters $\div$ 25.4 = inches

Common and Scientific Names of North American Fishes (Other Than Cutthroat Trout) Mentioned in the Text

FAMILY ANGUILLIDAE

American eel, *Anguilla rostrata*

FAMILY CLUPEIDAE

Pacific herring, *Clupea pallisii*

FAMILY CYPRINIDAE

Goldfish, *Carassius auratus,*
Lake chub, *Couesius plumbius*
Common carp, *Cyprinus carpio*
Alvord chub, *Gila alvordensis*
Utah chub, *Gila atraria*
Tui chub, *Gila bicolor*
Borax Lake chub, *Gila boraxobius*
Rio Grande chub, *Gila pandora*
Southern redbelly dace,
    Phoxinus erythrogaster
Northern pikeminnow,
    Ptychochailus oregonensis
Longnose dace, *Rhinichthys cataractae*

Speckled dace, *Rhinichthys osculus*
Redside shiner, *Richardsonius balteatus*
Creek chub, *Semotilus atromaculatus*

FAMILY CATOSTOMIDAE

River carpsucker, *Carpoides carpio*
Utah sucker, *Catostomus ardens*
Longnose sucker, *Catostomus catostomus*
Desert sucker, *Catostomus clarkii*
White sucker, *Catostomus commerconii*
Sonora sucker, *Catostomus insigna*
Mountain sucker, *Catostomus platyrhynchus*
Rio Grande sucker, *Catostomus plebeius*
Tahoe sucker, *Catostomus tahoensis*
Shortnose sucker, *Chasmistes brevirostris*
Cui-ui, *Chasmistes cujus*
June sucker, *Chasmistes liorus*
Blue sucker, *Cycleptus elongates*
Smallmoutn buffalo, *Ictiobus bubalus*
Gray redhorse, *Moxostoma congestum*

FAMILY ICTALURIDAE

White catfish, *Ameiurus catus*
Black bullhead, *Ameiurus melas*
Brown bullhead, *Ameiurus nebulosus*
Channel catfish, *Ictalurus punctatus*

FAMILY OSMERIDAE

Rainbow smelt, *Osmerus mordax*
Longfin smelt, *Spirinchus thaleichthys*

FAMILY SALMONIDAE

Mexican golden trout,
    Oncorhynchus chrysogaster
Gila trout, *Oncorhynchus gilae gilae*
Apache trout, *Oncorhynchus gilae apache*
Chum salmon, *Oncorhynchus keta*
Coho salmon, *Oncorhynchus kisutch*
Rainbow trout, *Oncorhynchus mykiss*
Golden trout, *O. mykiss aquabonita*
Sockeye salmon, kokanee, *O. nerka*

Chinook salmon, *Oncorhynchus tshawytscha*
Bear Lake whitefish, *Prosopium abyssicola*
Bonneville cisco, *Prosopium gemmifer*
Bonneville whitefish, *Prosopium spilonotus*
Mountain whitefish, *Prosopium williamsoni*
Landlocked (Atlantic) salmon,
　　Salmo salar sebago
Brown trout, *Salmo trutta*
Arctic charr, *Salvelinus alpinus*
Bull trout, *Salvelinus confluentus*
Brook trout, *Salvelinus fontinalis*
Dolly Varden, *Salvelinus malma*
Lake trout, *Salvelinus namaycush*

FAMILY ESOCIDAE

Tiger muskellunge (hybrid),
　　Esox lucius × *E. masquiningy*

FAMILY GASTEROSTEIDAE

Threespine stickleback,
　　Gasterosteus aculeatus

FAMILY COTTIDAE

Mottled sculpin, *Cottus bairdii*
Paiute sculpin, *Cottus beldingii*
Bear Lake sculpin, *Cottus extensus*

FAMILY MORONIDAE

White bass, *Morone chrysops*

FAMILY CENTRARCHIDAE

Sacramento perch, *Archoplites interruptus*
Bluegill, *Lepomis macrochirus*
Smallmouth bass, *Micropterus dolomieu*
Spotted bass, *Micropterus punctulatus*
Largemouth bass, *Micropterus salmoides*
White crappie, *Pomoxis annularis*

FAMILY PERCIDAE

Arkansas darter, *Etheostoma cragini*
Yellow perch, *Perca flavescens*
Walleye, *Sander vitreus*

FAMILY AMMODYTIDAE

Pacific sand lance, *Ammodytes hesapterus*

Index

A

Abandoned Mine Land Initiative, 209, 416

Acid rain, 417–419

Adfluvial behavior: Bonneville cutthroat, 343–346; coastal cutthroat, 64, 82–86; Humboldt cutthroat, 203–205; Westslope cutthroat, 122–124, 130; Willow/Whitehorse cutthroat, 223; Yellowstone cutthroat, 283–284

American Fisheries Society, 14, 88–89, 131, 347; Committee on Endangered Species, 434

American Society of Ichthyologists and Herpetologists, 434

Anasazi tribes, 368

Antimycin, 139–140, 410

Apache tribes, 27, 327, 367, 433

Arapaho-Roosevelt National Forest, 414, 415, 419

Arapaho tribes, 372, 395

Audubon, John James, 6

Audubon, John Woodhouse, 429

B

Babbitt, Bruce, 188

Bass: large-mouth, 78, 332; white, 332

Bear Lake Trout Enhancement Program, 334

Behavior: adfluvial, 283–284, 343–346; agonistic, 14, 15, 70; allacustrine, 76, 115, 262–274; amphidromous, 38, 64, 66–76; anadromous, 37; catadromous, 38; defensive, 119; diadromous, 37, 38; dominance, 119; fluvial, 81–82, 118–122, 130–131, 165, 274–277, 280, 309–312, 340–343, 377–379; lacustrine, 76–81, 115–118, 131, 161–165, 262–274, 280, 314–315, 331–340, 371–377, 405–406; lacustrine-adfluvial, 262–274; lake-associated, 331–340, 371–377; overwintering, 73, 123; piscivorous, 337; potamodromous, 37, 65, 81; predatory, 337; riverine, 309–312, 340–343, 2274–27774–277; spawning, 66–68, 73, 74, 81, 85; stream-resident, 82–86, 122–124, 130, 165–167, 278–279, 283–284, 312–314, 343–346, 379–380, 402–404

Behnke, Robert, 10, 12, 18, 21, 24, 26, 27, 39, 40, 43, 45, 46, 52, 53, 57, 63, 77, 83, 109, 113, 118, 127, 132, 152, 158, 165, 196, 197, 203, 220, 221, 222, 231, 255, 256, 258, 259, 261, 270, 282, 299, 315, 326, 329, 349, 361, 370, 377, 410, 424, 431, 438, 453, 461, 463, 464

Bendire, Captain Charles, 257, 258, 259

Biodiversity Legal Foundation, 348

Bluegills, 78

"Bluenoses," 333

Bonneville, Benjamin, 6

Bridger, Jim, 262, 323, 365, 373

Bring Back the Natives program, 139, 213, 293

Bullheads, 78; black, 332

Bush, George W., 138

C

California Department of Fish and Game, 231, 237, 238, 241, 243

California Wilderness Act (1984), 241

Callibaetis, 266

Canyonlands National Park, 368

Carp, 336; common, 332

Carson, Kit, 149, 373, 375

Catfish: channel, 332

Charr, 31, 174–176; Dolly Varden, 76, 77

Cheyenne tribes, 255, 256, 372, 395

Chubs, 311, 337; lake, 264, 267; tui, 163, 164, 166; Utah, 271, 332, 334, 336, 374

Cisco: Bonneville, 335, 336

Clark, William, 5, 11, 62, 104, 255

Clean Air Act, 417

Clean Water Act, 137, 138, 288

Climate change, 135–143, 285

Clinton, Bill, 138

Colorado: Conservation Strategy for Colorado River Cutthroat Trout, 383; Glenwood Springs Hatchery, 360; self-reproducing fisheries in, 360

Colorado Cooperative Fishery Research Unit, 408

Colorado Division of Wildlife, 360, 375, 377, 406, 408, 434

Coloration: Alvord Basin cutthroat, 459; Bonneville cutthroat, 321, 331, 352; Colorado River cutthroat, 359, 376; finespotted Snake River cutthroat, 295; function of, 14; Greenback cutthroat, 389; Humboldt cutthroat, 191, 201; Lahontan cutthroat, 145, 150; Rio Grande cutthroat, 421; role in agonistic behavior, 14, 15; variety of, 16; Willow/Whitehorse cutthroat, 215; yellowfin cutthroat, 451, 453; Yellowstone cutthroat, 245, 252

Comanche tribes, 395

Confederated Tribes of the Goshute Reservation, 350, 351

Conservation Agreement and Strategy, 137, 138, 187, 212, 381

Conservation Strategy for Colorado River Cutthroat Trout, 383

Crook, General George, 255, 256, 283

Custer, General George, 256, 257

D

Dace, 82, 202, 311; longnose, 264, 266, 430; speckled, 166, 271, 332

Dangers to survival, *220*; acid rain, 417–419; agricultural practices, 94, 130, 211; beaver dams, 239–240; climate change, 35, 38, 39, 135–143, 285; competition for water use, 212; disease, 286, 440, 441, 443; egg-taking operations, 168–170; energy exploration/development, 356, 373, 384, 385, 386; excessive harvesting, 126–130; fire, 444; fish-eating birds, 268; fish-mining, 169; flood control projects, 76; forest management practices, 67, *67*, 93, 94, 130, 170–172, 234, 352, 368, 407; fungal attacks, 310; gas drilling, 445–447; grizzly bears, 268; habitat degradation/loss: 130-131,93-95; human development, 415; hybridization, 133–134; hydroelectric power projects, 76; introduction of non-native species, 174–176; land-use, 209; livestock grazing, 176–177, 206, 207, 219, 225, 227, 228, 234, 284, 345, 368, 369, 384, 397, 407, 428, 436, 437; market/sport exploitation, 168–170; mining, 172, 208–211, 233–234, 285, 352, 353, 396, 397, 407, 416, 417; near-ocean conditions, 99–101; New Zealand mud snail, 286, 287; pollution, 285; replacement by introduced species, 131–133; road building, 67, 93, 130, 385; salinity crises, 38; sewage discharges, 209; silt, 223; water development, 98–99, 373; water diversions, 95, 155, 160, 173–176, 208, 284, 384, 386, 387, 407, 436; water quality issues, 177–185; water table damage, 210; whirling disease, 286, 440, 441, 443

Daphnia, 266, 271, 315, 373

Dark Canyon Wilderness, 368

Desert Fishes Council, 347, 348

Devine, John, 219

Diet: Bonneville cutthroat, 331, 332, 338, 342, 345; coastal cutthroat, 77, 78, 82, 85–86; Colorado River cutthroat, 373, 376, 377, 380; effect of wind on, 266; finespotted Snake River cutthroat, 311, 315; Greenback cutthroat, 403, 405; Humboldt cutthroat, 202; non-piscivorous, 266; opportunistic, 202, 239, 433; Paiute cutthroat, 239; piscivorous, 77, 161, 337, 377, 405; Westslope cutthroat, 117, 163; yellowfin cutthroat, 454; Yellowstone cutthroat, 264, 266–268, 271, 277, 279

DNA. *See* Genetic composition

E

Electroshocking, 241

Endangered Species Act (1973), 26, 89, 93, 100, 111, 124, 137, 168, 182, 183, 187, 212, 225, 241, 289, 346, 347, 380, 408, 434, 438

Environmental issues/groups. *See also* Dangers to survival: agricultural practices, 94, 130, 211; Bear Lake Trout Enhancement Program, 334; Biodiversity Legal Foundation, 348; Bring Back the Natives program, 139, 213, 293; Center for Biological Diversity, 242; Clean Air Act, 417; Clean Water Act, 137, 138, 288; climate change, 35, 38, 39, 135–143; for coastal cutthroat, 88; Conservation Strategy for Colorado River Cutthroat Trout, 383; Federal Land Management Protection Act, 288; flood-contro76l; forest management practices, 67, 93, 94, 130, 170–172; grazing allotments, 234; grazing management, 227; habitat degradation/loss, 93–95, 130–131; hydroelectric projects, 76; legacy effects, 416; livestock grazing, 176–177, 206, 207; mining, 172, 208–211; National Wildlife Federation, 446; Nature Conservancy, 187, 188, 428; Sensitive Species Act, 288; sewage discharges, 209; Sierra Club, 446; Trout Creek Mountain Working Group, 228, 229; Trout Unlimited, 234, 351, 352, 409, 446; Trust for Public Land, 187; water quality, 155, 160, 177–185, 208; Wild and Scenic Rivers Act, 288; Wilderness Act, 137, 288

Eosalmo driftwoodensis, 34, *34*, 36, 37

Evermann, Barton, 56, 109, 110, 124, 252, 301

Evolution, 27–57; divergences in, 30, 31, 42, 43, 44, 45, 46, 52, 53; geological time periods in, 28–29

Extinction, 450–451. *See also* Trout, Alvord Basin cutthroat; Trout, yellowfin cutthroat; mass, 450

F

Fecundity: Bonneville cutthroat, 346; coastal cutthroat, 85; Colorado River cutthroat, 376; finespotted Snake River cutthroat, 310, 314; Greenback cutthroat, 404, 406; Humboldt cutthroat, 201, 204; Paiute cutthroat, 238, 239; Westslope cutthroat, 117, 123, 163; Willow/Whitehorse cutthroat, 223; Yellowstone cutthroat, 279

Federal Land Management Protection Act, 137, 288

Federal Land Policy and Management Act (1976), 212

Federation of Fly Fishers, 139, 409

Fluvial behavior: Bonneville cutthroat, 340–343; coastal cutthroat, 81–82; Colorado River cutthroat, 377–379; finespotted Snake River cutthroat, 309–312; Humboldt cutthroat, 199–203; Rio Grande cutthroat, 431–434; Westslope cutthroat, 118–122, 130–131, 165; Yellowstone cutthroat, 274–277, 280

France, Lewis, 366

Fremont, John C., 2, 13, 112, 148, 149, 159, 193, 232, 371, 372

G

Gammarus, 266, 273, 373, 376

Genetic composition, 19–24, 30, 41–43; Alvord Basin cutthroat, 459; Bonneville cutthroat, 321, 329; coastal cutthroat, 43, 59; Colorado River cutthroat, 43, 359; divergences in, 42, 43, 44, 45, 46, 52, 53, 141, 197; finespotted Snake River cutthroat, 43, 295, 302–305; Greenback cutthroat, 43, 389; Humboldt cutthroat, 52, 191, 197, 204, 205; Lahontan cutthroat, 43, 52, 145; Paiute cutthroat, 52, 231; Rio Grande cutthroat, 421, 430; variation, 141, 142; Westslope cutthroat, 42, 43, 45; Willow/Whitehorse cutthroat, 215; yellowfin cutthroat, 451; Yellowstone cutthroat, 43, 245, 261–262

Glacier National Park, 125

Graylings, 31

Great Basin National Park, 353

Greenback Cutthroat Trout Recovery Team, 408

Green River (Wyoming), 6

Grey, Zane, 361

H

Habitat: complementation, 83; damage from beavers, 239–240; degradation, 130–131, 345; destruction, 284; fragmentation, 130–131; loss/alteration, 93–95, 130–131; management, 186–187, 212–213, 226, 227; partitioning, 110; rearing, 201; recovery, 186–190, 211–214, 241, 242; restoration, 88, 207, 318, 334; seasonal, 343; water velocity in, 120

Hallock, Charles, 284, 451

Hartman, G.F., 84

Harvest and Hatcheries program, 101

Hucho, 36, 37

Hyallela, 266

Hybridization, 125, 208, 283, 284, 308, 405; criterion for, 231; defining, 133–134; introgressive, 133; rainbow-cutthroat, 19, 43, 45, 118, 175, 280

I

Idaho: fishing regulations, 291; Madison River Fishery, 272

Idaho Department of Fish and Game, 275, 292; hybrid stocking by, 273

Inland Native Fish Strategy (INFISH), 137, 138, 187, 212, 381

J

Jackson National Fish Hatchery, 300–308, 316

Jordan, David Starr, 12, 13, 40, 43, 108, 110, 124, 157, 158, 178, 252, 258, 301, 361, 392, 407, 452, 454, 455

K

Kelly, Luther "Yellowstone," 360

Kiowa tribes, 395

Kokanee, 337, 372

L

Lacustrine behavior: Bonneville cutthroat, 331–340; coastal cutthroat, 76–81; Colorado River cutthroat, 371–377; finespotted Snake River cutthroat, 314–315; Greenback cutthroat, 405–406; Humboldt cutthroat, 205–206; Lahontan cutthroat, 161–165; rainbow trout, 76; Rio Grande cutthroat, 431–434; Westslope cutthroat, 115–118, 131; Yellowstone cutthroat, 76, 262–274, 280

Lake-associated behavior: Bonneville cutthroat, 331–340; coastal cutthroat, 76–81; Colorado River cutthroat, 371–377; Humboldt cutthroat, 205–206; rainbow trout, 76; Trout, Lahontan cutthroat, 161–165; Westslope cutthroat, 115–118; Yellowstone cutthroat, 76, 262–274

Land exchanges, 227, 241

Land management, 293; agencies, 89

Lewis, Meriwether, 1, 5, 6, 11, 13, 62, 104

M

Mackinaws, 175

Mary's River (Nevada), 5, *5*

Meristic characters, 17, *17*, 18, 19; Alvord Basin cutthroat, 459; Bonneville cutthroat, *19 fig*, 321, 327; coastal cutthroat, *19 fig*, 59; Colorado River cutthroat, *19 fig*; criticism of use of, 18; finespotted Snake River cutthroat, 295; Greenback cutthroat, *19 fig*, 389; Humboldt cutthroat, 191; Lahontan cutthroat, 145, 158; Paiute cutthroat, 231; Rio Grande cutthroat, *19 fig*, 421; Westslope cutthroat, *19 fig*, 103; Willow/Whitehorse cutthroat, 215, 221–222; yellowfin cutthroat, 451; Yellowstone cutthroat, 245

Merriam, C. Hart, 219

Migration: Bonneville cutthroat, 338; coastal cutthroat, 71, 73, 74, 80, 82; by finespotted Snake River cutthroat, 313; oceanic, 71; refuge, 71, 83, 167; spawning, 118, 160, 314, 342, 375–376; trophic, 71, 83; Westslope cutthroat, 121–122

Montana: Fish Cultural Development Center, 409; fishing regulations, 290, 291

Montana Department of Fish, Wildlife and Parks, 125, 292, 293

N

National Environmental Policy Act, 137, 187, 212, 244, 288

National Fish and Wildlife Foundation, 139, 213, 318

National Forest Management Act (1976), 187, 212

National Marine Fisheries Service, 89, 91

National Parks: Canyonlands, 368; Glacier National Park, 125; Great Basin, 353; Olympic, 77; restrictive regulations in, 254; Rocky Mountain, 402, 410, 415, 419; Yellowstone, 7, 141, 251, 252, 254, 274, 277, 280, 281, 284, 288, 301

National Park Service, 288; conservation agreements with, 353; recovery programs of, 408

National Wildlife Federation, 446

Native Americans, 219; Anasazi, 368; Apache, 27, 327, 367, 433; Arapaho, 372, 395; Cheyenne, 255, 256, 372, 395; Comanche, 395; Confederated Tribes of the Goshute Reservation, 350, 351; intertribal conflict, 369; Kiowa, 395; Navajo, 327, 367, 369; Paiute, 146, 147, 148, 155, 160, 162, 181, 189, 325, 327, 360, 373; Pueblo, 3; Shoshone, 254; Sioux, 255, 256, 372; Ute, 325, 327, 360, 367, 368, 372, 395; Washoe, 152

Nature Conservancy, 187, 188, 428

Navajo Forest Products Industries, 369

Navajo tribes, 327, 367, 369

Nevada Division of Wildlife, 160, 196, 206, 211; 151

Nevada Wildlife Commission, 214

New Mexico: fishing regulations, 440; Seven Springs Hatchery, 433, 441

New Mexico Department of Game and Fish, 434

New Zealand mud snail, 286, 287

Northcote, T.G., 84

O

Ogden, Peter Skene, 193, 195, 219, 323

Olympic National Park, 77

Oncorhynchus australis, 40

Oncorhynchus chrysogaster, 40

Oncorhynchus clarkii. See Trout, cutthroat

Oncorhynchus clarkii alvordensis. See Trout, Alvord Basin cutthroat

Oncorhynchus clarkii behnkei. See Trout, finespotted cutthroat

Oncorhynchus clarkii bouvieri. See Trout, Yellowstone cutthroat

Oncorhynchus clarkii clarkii. See Trout, coastal cutthroat

Oncorhynchus clarkii henshawi. See Trout, Lahontan cutthroat

Oncorhynchus clarkii lewisi. See Trout, Westslope cutthroat

Oncorhynchus clarkii macdonaldi. See Trout, yellowfin cutthroat

Oncorhynchus clarkii pleuriticus. See Trout, Colorado River cutthroat

Oncorhynchus clarkii seleneris. See Trout, Paiute cutthroat

Oncorhynchus clarkii stomias. See Trout, Greenback cutthroat

Oncorhynchus clarkii utah. See Trout, Bonneville cutthroat

Oncorhynchus clarkii virginalis. See Trout, Rio Grande cutthroat

Oncorhynchus keta, 37

Oncorhynchus salax, 37, 45

Oregon Department of Fish and Wildlife, 217, 223, 224, 225

P

Paiute tribes, 146, 147, 148, 155, 160, 162, 181, 189, 325, 327, 360, 373

Perch: yellow, 78, 332, 336, 374

Pike, Zebulon, 395, 396

Pikeminnows, 82, 133

Potamodromous behavior: coastal cutthroat, 81–82; Humboldt cutthroat, 199–203; Rio Grande cutthroat, 431–434; Westslope cutthroat, 118–122, 130–131

President's Forest Plan, 100

Pueblo tribes, 3

R

Range, native: Alvord Basin cutthroat, 25, 52; Bonneville cutthroat, 25, 52; coastal cutthroat, 23, 24, 60–62; Colorado River cutthroat, 25, 26, 52; finespotted Snake River cutthroat, 25, 52; Greenback cutthroat, 25, 52; Humboldt cutthroat, 25, 52; Lahontan cutthroat, 25, 52; Paiute cutthroat, 25, 52; Rio Grande cutthroat, 25, 52; Westslope cutthroat, 25, 26; Willow/Whitehorse cutthroat, 25; yellowfin cutthroat, 25; Yellowstone cutthroat, 25, 45, 52, 56

Redbands, 14

Redfish, 159

Restricted movement paradigm, 83

Rhabdofario lacustris, 37, 38

Riverine behavior: Bonneville cutthroat, 340–343; coastal cutthroat, 81–82; finespotted Snake River cutthroat, 309–312; Humboldt cutthroat, 199–203; Rio Grande cutthroat, 431–434; Westslope cutthroat, 118–122, 130–131; Yellowstone cutthroat, 274–277

Rocky Mountain National Park, 402, 410, 415, 419

Rotenone, 240, 241, 242, 243, 337, 410

Russell, Osborn, 249, 301

S

Salar virginalis, 424, 425

Salmo carmichaeli, 299

Salmo clarkii, 62. *See* Trout, cutthroat

Salmo clarkii alpestris, 51

Salmo eremogenes, 259, 259

Salmo evermanni, 178

Salmo gairdneri, 62

Salmo mykiss. *See* Trout, rainbow

Salmo mykiss bouvieri, 258

Salmon, 31, 37, 195; Atlantic, 333, 374; chinook, 99, 113; coho, 69, 70, 374; landlocked, 374; Pacific, 89, 100, 333

Salmonidae, 30, 31, 33, 34, 36, 37, 38, 39, 42

Salmonid Enhancement Program, 87–88

Salmon trout, 301

Salmo purpuratus, 258, 430

Salmo spilurus, 425

Salvelinus, 36, 37, 45

Schwiebert, Ernest, 298, 299, 313

Scuds, 251, 267; *Gammarus*, 266, 273; *Hyalella*, 266

Sculpins, 69, 77, 164, 202, 336; Bear Lake, 335; mottled, 271, 311; Paiute, 166

Shiners, 166, 311; redside, 202, 264, 267, 271, 332, 336, 374

Shoshone tribes, 254

Sierra Club, 446

Sioux tribes, 255, 256, 372

Siphlonurus, 273

Smelt, 77

Smilodonichthys rastrosus, 37

Smith, Gerald, 38, 40, 45, 52, 54

Smith, Jedediah, 232, 323

Snyder, J. Otterbein, 12, 196, 199, 206, 231, 235, 236, 237

Spawning behavior/locations: beaver dams and, 239; Bonneville cutthroat, 335, 338, 342–343, 345, 346; coastal cutthroat, 66–68, 73, 74, 76, 77, 85; Colorado River cutthroat, 375–376, 379, 380; finespotted Snake River cutthroat, 310–311, 313, 314; gender ratios in, 117, 163, 275, 376; Greenback cutthroat, 402, 403, 406; Humboldt cutthroat, 199, 201–202, 202; Lahontan cutthroat, 156; loss of habitats for, 317–318; mortality, 265; Paiute cutthroat, 238–240; Rio Grande cutthroat, 432–434; Westslope cutthroat, 116–117, 118, 119, 123, 162, 163, 166; Willow/Whitehorse cutthroat, 223; Yellowstone cutthroat, 251, 264–266, 274, 278–279, 290

Species Management Plan, 211

Splake, 374

Sport Fish Restoration program, 213

Spotting patterns: Alvord Basin cutthroat, 459; Bonneville cutthroat, 321, 331; coastal cutthroat, 59, 71; finespotted Snake River cutthroat, 17, 295, 302, 316; Greenback cutthroat, 389; Humboldt cutthroat, 191; Lahontan cutthroat, 145; Paiute cutthroat, 17; patterns, 17; Rio Grande cutthroat, 421; Westslope cutthroat, 103, 105, 113–114, 115 *fig*; Willow/Whitehorse cutthroat, 215; yellowfin cutthroat, 451, 453; Yellowstone cutthroat, 17, 245, 252

Stewart, Sir William Drummond, 371

Sticklebacks, 77

Stream-resident behavior: Bonneville cutthroat, 343–346; coastal cutthroat, 64, 82–86; Colorado River cutthroat, 379–380; finespotted Snake River cutthroat, 312–314; Greenback cutthroat, 402–404; Humboldt cutthroat, 203–205; Lahontan cutthroat, 165–167; Paiute cutthroat, 238–240; Rio Grande cutthroat, 431–434; Westslope cutthroat, 122–124, 130; Willow/Whitehorse cutthroat, 223; Yellowstone cutthroat, 278–279, 283–284

Suckers, 82, 250, 311; longnose, 264, 267; mountain, 166, 271; Rio Grande, 430; Tahoe, 166; Utah, 271, 335, 336, 374

Suckley, George, 12, 331, 340–341, 365, 366

Sunfish: green, 332

T

Trout: Apache, 16; black-spotted, 13, 430, 455; blueback, 80; brook, 13, 132, 166, 208, 260, 272, 283, 313, 316, 327, 333, 410; bull, 109, 117, 120; Clark's, 13; classification changes, 12, 13; Gila, 16; golden, 18; harvest, 80; lake, 264, 270, 272, 280, 314, 333, 372, 374; Mexican golden, 29, 40; Montana black-spotted, 105; mountain, 1, 112; Quinn River, 197, 198; redthroat, 13; Rocky Mountain, 13; silver, 157, 158; speckled, 1, 13, 27; spotted, 240; spring, 159; steelhead, 13, 37, 45, 64, 98; Tahoe, 157, 158; Waha Lake, 257, 258, 259, 261; winter, 159

Trout, Alvord Basin cutthroat, 459–465; coloration, 459; Endangered Species Act status of, *10 tab*; extinction of, 25, 26, 52; genetic composition, 459; meristic characters, 459; native range of, 25, 52, 459, 461–463; spotting patterns, 459

Trout, Bonneville cutthroat, *320, 321*, 321–357; adfluvial behavior, 343–346; coloration, 321, 331, 352; dangers to survival, 352, 353, 356; diet, 331, 332, 335, 336, 338, 342, 345; differentiated groups of, 321, 329, 330; ecology of, 330–346; Endangered Species Act status of, *10 tab*; fecundity, 346; fluvial behavior, 340–343; future prospects for, 346–357; genetic composition, 321, 329; historical distribution, 326–330; introgression and, 327, 350; lacustrine behavior, 331–340; lake-associated behavior, 331–340; life cycle, 330–346; life span, 343; meristic characters, *19 fig*, 321, 327; migrations, 338; native range of, 25, 52, 322–326, *325, 328, 329, 333, 337, 339, 345, 347, 355*; as official state fish of Utah, 356; population decline in, 331; relict population, 346; riverine behavior, 340–343; sexual maturity, 335, 342, 346; similarity to Yellowstone cutthroat, 321; size, 331, 333, 343, 346; spawning behavior/locations, 335, 338, 342–343, 345, 346; spotting patterns, 321, 331; status of, 346–357; stream-resident behavior, 343–346; thermal tolerance, 345; translocations of, 349, 350; water systems used, 324, 326–330, 340–343, 344

Trout, brown, 175, 298, 300–308, 314, 333, 372; dominance by, 133; propagation of, 132; replacement of pure populations by, 327

Trout, coastal cutthroat, *58*, 59–101; amphidromous behavior, 38, 64, 66–76; artificial propagation of, 96–98; coloration, *58, 59*; dangers to survival, 67, *67*, 93–101; decline in genetically pure populations, 64; diet, 77, 78, 82, 85–86; ecology of, 64–86; Endangered Species Act status of, *10 tab*, 89–93; environmental issues/groups and, 88; fecundity in, 85; first cutthroat described for science, 62; fluvial behavior in, 81–82; future prospects for, 86–101; genetic composition, 43, 59; growth patterns, 71, 77, 85; habitat loss/alteration and, 93–95; hatchery impacts on, 96–98; historical distribution, 62–64; hybridization of, 64; lacustrine behavior, 76–81; lake-associated behavior, 76–81; life cycle of, 64–86; life span, 85; meristic characters, *19 fig*, 59; migrations, 71, 73, 74, 80, 82; native range of, *23*, 24, 60–62; nomadic life style of, 72; overharvesting, 95–96; overwintering, 73; population declines in, 93–101; potamodromous behavior, 81–82; predators to, 69; preference for smaller streams, 64; rearing, 77; recovery programs for, 100–101; riverine behavior, 81–82; sexual maturity, 73; spawning behavior/locations, 66–68, 76, 77, 81, 85; spotting patterns, *58*, 59, 71; status of, 86–101; stocks of, 89; stream-resident behavior of, 82–86; survival of, 68

Trout, Colorado River cutthroat, *358*, 359–388; coastal cutthroat and, *58*, 59; coloration, 17, 376; conservation populations of, 382; dangers to survival, 368, 369, 373, 384, 385, 386; depletion of, 379, 380; diet, 373, 376, 377, 380; dominance hierarchies in, 380; ecology of, 370–380; Endangered Species Act status of, *10 tab*; fecundity, 376; fluvial behavior, 377–379; future prospects for, 380–388; genetic composition, 42, 43, 359; historical distribution, 363–370; lacustrine behavior, 371–377; lake-associated behavior, 371–377; life cycle, 370–380; meristic characters, *19 fig*, 359;

native range of, 25, 26, 52, 363–370, *365, 377, 378, 382, 387*; sexual maturity, 380; size, 373, 375, 376, 377, 378; spawning behavior/locations, 375–376; status of, 380–388; stream-resident behavior, 379–380; thermal tolerance, 375; water systems used, 26, 363–370, 371–373, 377–379, 379

Trout, cutthroat, 9; agonistic behavior in, 14, 15; basibranchial teeth in, 16, *16*, 19; called "salmon-trout," 13; characteristics of, 14–24; classifying, 9–14; coloration, 16; dangers to, 35; defining, 9–14; development of habitats for, 29–39; drift feeding in, 85–86; early studies, 1–14; emergence/divergences of, 39–57; evolution of, 27–57; genetic composition, 19–24, 30, 41–43; hierarchical social units, 14; historical distribution, 24, *24*, 25, 26; hit and run feeding in, 86; hyoid teeth in, 16; identification of, 14–24; inland radiation of, 42; interior, 83; large spotted, 55, 56; length, 1; lineage, 9, 10, *10 tab*; maxillary length in, 16; meristic characters, 17, *17*, 17–19, 18; mountain, 51, 109; naming of, 11; as "native" western fish, 1–9; non-diadromous populations, 91; origins of, 29–39; prehistoric ranges of, 27–57; Quinn River, 215; relatives of, 10; reproductively isolated populations of, 84; thermal tolerance of, 110, 111; Twin Lakes, 105, 106; unrestricted resident populations, 83–84; water systems used, 46–54

Trout, finespotted Snake River cutthroat, 295, *296*, 297–319; adaptability of, 313, 314, 315; coloration, 295; diet, 311, 315; displacement of, 313; divergences in, 308; ecology of, 308–315; Endangered Species Act status of, *10 tab*; fecundity, 310, 314; fluvial behavior, 309–312; as form of Yellowstone subspecies, 316; future prospects of, 315–319; genetic composition, 42, 43, 295, 302–305; geological time and, 306–308; growth rates, 311; hatchery orgin, 315; historical distribution, 300–308; introgression and, 308, 316; lacustrine behavior, 314–315; life cycle, 308–315; life span, 314; meristic characters, 295; migrations, 313;

native range of, 25, 52, *299*, 300–308, *307, 312, 318*; natural mortality, 310; partitioning of food sources by, 315; post-spawning mortality, 310; reduction of population, 301; reproductive isolation of, 295; riverine behavior, 309–312; sexual maturity, 310; size, 309, 311; spawning behavior/locations, 310–311, 313, 314; spotting patterns, 17, 295, 302, 316; status of, 315–319; stream-resident behavior, 312–314; thermal tolerance, 312; water systems used, 25, 300–308

Trout, Greenback cutthroat, 389, *390*, 391–419; acid rain testing, 419; as Colorado state fish, 410; coloration, 17, 389; diet, 403, 405; displacement by brook trout, 411; ecology of, 401–406; Endangered Species Act status of, *10 tab*, 26; fecundity, 404, 406; future prospects for, 406–419; genetic composition, 42, 43, 389; historical distribution, 398–401; lacustrine behavior, 405–406; life cycle, 401–406; meristic characters, *19 fig*, 389; native range of, 25, 52, 391–398, *397*, 398–401, *401, 404, 411, 418*; near-extinction, 406–412; partitioning of food sources by, 315; pure populations of, 408, 410; recovery programs for, 408–419; sexual maturity, 404; similarity to Colorado cutthroat, 391–395; size, 403, 404, 405; spawning behavior/locations, 402, 403, 406; spotting patterns, 389; status of, 406–419; stream-resident behavior, 402–404; thermal tolerance, 402, 406; water systems used, 25, 26, 391–398, 398–401

Trout, Humboldt cutthroat, 191–214, *192*; adfluvial behavior, 203–205; changeable environment of, 203, 204; coloration, 191, 201; dangers to survival, 206–211; diet, 202; displacement by non-native species, 203, 208; ecology of, 199–206; Endangered Species Act status of, *10 tab*; fecundity, 201, 204; fluvial behavior, 199–203; future prospects for, 206–214; genetic composition, 52, 191, 197, 204, 205; historical distribution, 195–199; hybridization with Yellowstone cutthroats, 208; intrinsic productivity of, 204; introgression in, 203; lack of lake populations of, 199; lacustrine behavior, 205–206; lake-associated behavior, 205–206; life cycle,

199–206; life span, 201; meristic characters, 191; native range of, 25, 52, 193–195, *195, 196, 201, 206, 208, 214*; overwintering, 200; population management and, 211–212; potamodromous behavior, 199–203; recovery projects, 211–214; reduction in range and population, 206; riverine behavior, 199–203; spawning behavior/locations, 199, 201–202; spotting patterns patterns, 191; status of, 206–214; stream-resident behavior, 203–205; survival in harsh environments, 203, 205; thermal tolerance, 202, 203, 206; water systems used, 191, 193–195, 198, 203, 204, 205, 206

Trout, Lahontan cutthroat, *144, 146*, 240; coloration, 145, 150; dangers to survival, 160, 167–190; diet, 161, 163, 166; ecology of, 161–167; egg-taking operations, 168–170; Endangered Species Act status of, *10 tab*; fish-mining and, 169; fluvial behavior, 165; forest management practices and, 170–172; future prospects, 167–190; genetic composition, 43, 52, 145; growth patterns, 163–164; historical distribution, 150–161; introduction of non-native species and decline of, 174–176; lacustrine behavior, 161–165; lake-associated behavior, 161–165; life cycle, 161–167; life span, 165; market fishery and, 155; meristic characters, 18, 145, 158; myths of, 146–147; naming, 13; native range of, 25, 52, 150–161, *157, 165, 173, 184, 189*; overharvesting and, 168–170; piscivorous feeding, 266–267; population decline, 167, 168–178; population management and, 185–186; restoration projects, 179–190; spawning behavior/locations, 156, 160, 162, 163, 166; spotting patterns patterns, 145; status of, 167–190; stream-resident behavior, 165–167; timetable for recovery of, 187–190; water diversions and, 173–176; water quality issues and, 177–185; water systems used, 147–161, 165, 166, 167, 168, 169

Trout, Paiute cutthroat, *230*, 231–244; co-evolution with amphibians, 242; coloration, 231; dangers to survival, 233–234, 239–240; defensive behavior, 239; diet, 239; dominance hierarchies and, 239; ecology of, 238–240;

Endangered Species Act and, *10 tab*, 241; fecundity, 238, 239; future prospects for, 240–244; genetic composition, 52; historical distribution, 234–238; introgression and, 237, 240; life cycle, 238–240; meristic characters, 231; native range of, 25, 52, 232–238, *235*, 240, *244*; poaching problems for, 240; pure stocks of, 241; rarity of, 240; recovery plans for, 241, 244; refugial populations, 238; self-reproducing populations, 238; sexual maturity, 238, 239; spawning behavior/locations, 238–240; spotting patterns, 17, 231; status of, 240–244; stream-resident behavior, 238–240; transplantation of, 236, 237; water systems used, 232–234

Trout, rainbow, 11, 43, 45, 64, 166, 240, 333, 372, 377, 378; agonistic behavior in, 14, 15; dominance by, 133; encroachment by, 291, 298; interior, 14; lacustrine behavior, 76; management of, 113; Mt. Whitney strain, 114; population increases, 283; propagation of, 132; replacement for Yellowstone cutthroat, 281–282; replacement of pure populations by, 327; sterilization of, 337; stocking of, 137, 138, 208, 290

Trout, Rio Grande cutthroat, *420, 421, 422*, 422–447; coloration, 421; core populations, *436 tab*; dangers to survival, 428; diet, 433; displacement of, 438; ecology of, 431–434; Endangered Species Act status of, *10 tab*; first cutthroat seen by Europeans, 62; fluvial behavior, 431–434; future prospects for, 434–447; genetic composition, 421, 430; habitat requirements, 433–434; historical distribution, 426–431; lacustrine behavior, 431–434; life cycle, 431–434; meristic characters, *19 fig*, 421; native range of, 25, 52, 422–431, *437, 443*; potamodromous behavior, 431–434; riverine behavior, 431–434; scales, *19 fig*; spawning behavior/locations, 432–434; spotting patterns, 421; status of, 434–447; stream-resident behavior, 431–434; thermal tolerance, 432; water systems used, 25, 426–431

Trout, Westslope cutthroat, *102*, 103–143; climate change and, 135–143; coloration, 17, 103, 105; dangers to survival, 125–143; decline in population of, 125–143; diet, 117; discontinuous range of, 109; early taxonomic confusion over, 109; ecology of, 115–124; Endangered Species Act status of, *10 tab*; excessive harvesting of, 126–130; fecundity, 117, 123, 163; fluvial behavior, 118–122, 130–131; future prospects, 124–143; genetic composition, 42, 43, 45; genetic variation in, 142; habitat loss and; historical distribution, 106–115; hybridization ansd, 133–134; lacustrine behavior, 115–118, 131; lake-associated behavior, 115–118, 131; life cycle, 115–124; life span, 117, 118, 124; meristic characters, *19 fig*, 103; migrations, 118, 121–122; native range of, 25, 26, 106–115, *122*, *128*; overwintering, 123; potamodromous behavior, 118–122, 130–131; replacement by introduced species, 131–133; restoration projects involving, 136–141; riverine behavior, 118–122, 130–131; sexual maturity of, 122; spawning behavior/locations, 116–117, 123; spotting patterns, 103, 105, 113–114, *115 fig*; status of, 124–143; stream-resident behavior, 122–124, 130; water systems used, 25, 106–115, 109, 114, 115, 116, 128, 129

Trout, Willow/Whitehorse cutthroat, 152, 215–229, *223*; ability to survive in extreme conditions, 223; adfluvial behavior, 223; coloration, 215; dangers to survival, 225; degraded habitat of, 228; ecology of, 223–225; Endangered Species Act status of, *10 tab*; fecundity, 223; future prospects, 225–229; genetic composition, 215; growth patterns, 223; habitat management plans for, 226, 227; historical distribution, 220–223; life cycle, 223–225; meristic characters, 215, 221–222; native range of, 25, 217–220, *219*, *221*; population declines, 225; population surveys of, 224; sexual maturity, 223; spawning behavior/locations, 223; spotting patterns, 215; status of, 225–229; stream-resident behavior, 223; thermal tolerance, 223, 224, 227; transplantation of, 220; water systems used, 217, 225

Trout, yellowfin cutthroat, *448*, 451–459; coloration, 451, 453; diet, 454; disappearance of, 455–459; Endangered Species Act status of, *10 tab*; extinction of, 25, 26; genetic composition, 451; meristic characters, 451; native range of, 25, *450*; size, 454; spotting patterns, 451, 453; water systems used, 25

Trout, Yellowstone cutthroat, 245–294, 333; adfluvial behavior, 283–284; allacustrine behavior, 262–274; angling mortality, 289, 291; angling mortality and, 268, 269, 270; annual mortality of, 268, 270; colonization by, 254, 261; coloration, 17, 245, 252; diet, 264, 266–268, 271, 277, 279; displacement by non-native species, 284; ecology of, 262–279; encroachment on by non-native species, 280; Endangered Species Act and, *10 tab*, 287; fecundity, 279; fishing restrictions on, 289, 290; fluvial behavior, 274–277, 280; future prospects for, 279–294; genetic composition, 43, 245, 261–262; historical distribution, 252–262; hybridization with, 208; inability to adapt to foreign environments, 270; introgression and, 281, 284; lacustrine-adfluvial behavior, 262–274; lacustrine behavior, 76, 262–274, 280; lake-associated behavior, 76, 262–274; life cycle, 262–279; life span, 267, 274; McBride Lake strain, 280; meristic characters, 18, 245; native range of, 25, 45, 52, 56, 246–262, 255, *273*, *279*; non-piscivorous feeding by, 266; overwintering, 266; population reductions, 279, 280, 284; predator fish and, 269; probability of extinction, 289; pure stocks of, 270–271, 280; riverine behavior, 2274–2777 274–277; role in ecosystem function, 267; self-reproducing populations, 270; sexual maturity, 274, 278; size, 267, 271–272, 274; spawning behavior/locations, 251, 264–266, 274–276, 278–279, 290; spotting patterns, 17, 245, 252; standing population, 268; status of, 279–294; stream-resident behavior, 278–279, 283–284; thermal tolerance, 264, 266, 271, 274, 277, 278; treated as Sensitive Species, 289; water systems used, 25

Trout Creek Mountain Working Group, 228, 229

Trout Unlimited, 139, 189, 234, 351, 352, 409, 446

Trust for Public Land, 187

Twain, Mark, 193

U

U.S. Bureau of Fisheries, 252, 255, 259, 261, 301

U.S. Bureau of Indian Affairs, 369

U.S. Bureau of Land Management, 137, 138, 139, 187, 204, 212, 222, 346, 348; cattle exclosures and, 226, 227; livestock exclosures and, 213; search for relict fish populations by, 349; Vale Project, 226

U.S. Bureau of Reclamation, 257, 348; Newlands Project, 153, 173

U.S. Bureau of Sport Fisheries and Wildlife, 434

U.S. Department of the Interior, 124, 168, 241

U.S. Environmental Protection Agency, 385, 416

U.S. Fish and Wildlife Service, 111, 136, 137, 152; acid rain testing, 419; Allotment Management Plan survey, 229; and coastal cutthroats, 92–93; environmental suits against, 294; identification of genetically pure strains by, 281; and Lahontan cutthroat, 180, 181, 182, 183; recovery plan for Paiute cutthroat, 241; recovery programs of, 211, 351, 408; Rio Grande cutthroat and, 439, 440; stabilization of stream banks by, 234; status reviews by, 294, 347, 348, 380, 381; stocking recommendations from, 314; Willow/Whitehorse cutthroat and, 223; Yellowstone cutthroat and, 287–288

U.S. Fish Commission, 252, 301

U.S. Food and Drug Administration, 337

U.S. Forest Service, 125, 136, 138, 139, 187, 212; cattle exclosures and, 234; recovery programs of, 408; Sensitive Species Act and, 288, 293; trout inventories by, 434

U.S. Geological Survey, 210; Abandoned Mine Land Initiative, 209, 416

Utah: Bear Lake Trout Enhancement Program, 334; Conservation Strategy for Colorado River Cutthroat Trout, 383; Cooperative Fish and Wildlife Research Unit, 337; fishing regulations, 334; fish-stocking policies in, 350; School and Institutional Trust Land Administration, 368

Utah Division of Wildlife Resources, 335, 337, 349, 352

Utah Reclamation Mitigation and Conservation Commission, 356, 357

Utah Wilderness Alliance, 348

Ute tribes, 325, 327, 360, 367, 368, 372, 395

V

Valle Vidal Protection Act (2005), 447

W

Walker, Joseph, 232

Walleyes, 332

Washington: Lake Chelan Hatchery, 261

Washington Department of Fish and Wildlife, 111, 114

Washoe Act (1956), 181

Washoe tribes, 152

Water systems used: by Bonneville cutthroat, 324, 326–330; by Colorado River cutthroat, 26, 363–370; by cutthroat trout, 46–54; by finespotted Snake River cutthroat, 25, 300–308; by Greenback cutthroat, 25, 26; by Humboldt cutthroat, 191, 193–195, 198, 203, 204, 205, 206; by Lahontan cutthroat, 147–161, 165, 166, 167, 168, 169; by Paiute cutthroat, 232–234; by Rio Grande cutthroat, 25; by Westslope cutthroat, 25, 106–116, 128, 129; by Willow/Whitehorse cutthroat, 217, 225; by Yellowstone cutthroat, 25, 252, 253fig, 254–262

Western Association of State Fish and Game Commissioners, 314

Wetherill, Benjamin, 366, 367

Whirling disease, 286, 440

Whitefish, 30; Bear Lake, 335; Bonneville, 335; mountain, 166, 271, 311

Wild and Scenic Rivers Act, 137, 288

Wilderness Act, 137, 288

Winthrop National Fish Hatchery, 184

Work, John, 219

Wyoming: Conservation Strategy for Colorado River Cutthroat Trout, 383; Cooperative Fish and Wildlife Research Unit, 278; Daniel Hatchery, 355; fishing regulations, 292; Jackson National Fish Hatchery, 300–308, 316; stocking programs, 293

Wyoming Game and Fish Department, 292, 309, 318, 319, 349, 355

Y

Yellowstone National Park, 7, 141, 251, 252, 254, 274, 277, 280, 281, 284, 288, 301

Yellowstone River, 4, 4, 7

Young, Brigham, 323, 365

Colophon

The text of this book is set in $^{11.5}/_{14}$ point
Monotype Walbaum, designed by Justus Erich Walbaum.

The heads are set in *Clearview Text Extra Thin*,
designed by James Montalbano.

The cutthroat trout illustrations are by Joseph R. Tomelleri.

All files for production were prepared on Macintosh computers.

It was printed by Golden Cup, Inc.

The paper is 115 gsm Gold East Matt Art.

It is bound in Saifu cloth.

It was designed by Charles Nix, with Takaaki Okada
and Gary Robbins.